High-Power Laser Handbook

About the Editors

Hagop Injeyan, Ph.D., recently retired from Northrop Grumman Aerospace Systems, where he was employed since 1982 and had been a Technical Fellow since 1999. He is currently a faculty member at California State University, Los Angeles. Dr. Injeyan holds 22 U.S. patents and has more than 20 publications in international scientific journals and proceedings in the areas of chemical lasers, high-power solid-state lasers, and nonlinear optics. He has served as Program and General Chair for the Advanced Solid-State Photonics Conference (ASSP) and Subcommittee Chair for the Conference on Lasers and Electro-Optics (CLEO).

Gregory D. Goodno, Ph.D., is a Senior Scientist at Northrop Grumman Aerospace Systems, where he has been employed since 1999. He has published over 50 technical contributions in the areas of nonlinear ultrafast spectroscopy, beam combination, and high-power slab and fiber lasers. Dr. Goodno currently serves as Program Chair for the Advanced Solid-State Photonics Conference.

High-Power Laser Handbook

Hagop Injeyan, Ph.D. Editor
Gregory D. Goodno, Ph.D. Editor

New York Chicago San Francisco
Lisbon London Madrid Mexico City
Milan New Delhi San Juan
Seoul Singapore Sydney Toronto

Cataloging-in-Publication Data is on file with the Library of Congress

Copyright © 2011 by The McGraw-Hill Companies, Inc. All rights reserved. Printed in China. Except as permitted under the United States Copyright Act of 1976, no part of this publication may be reproduced or distributed in any form or by any means, or stored in a data base or retrieval system, without the prior written permission of the publisher.

1 2 3 4 5 6 7 8 9 0 CTP/CTP 1 7 6 5 4 3 2 1

ISBN 978-0-07-160901-2
MHID 0-07-160901-6

Sponsoring Editor
Michael Penn

Editing Supervisor
Stephen M. Smith

Production Supervisor
Richard C. Ruzycka

Acquisitions Coordinator
Michael Mulcahy

Project Manager
Kritika Kaul,
Glyph International

Copy Editor
Tara Joffe

Proofreader
Mohinder Bhatnagar,
Glyph International

Indexer
Robert Swanson

Art Director, Cover
Jeff Weeks

Composition
Glyph International

McGraw-Hill books are available at special quantity discounts to use as premiums and sales promotions, or for use in corporate training programs. To contact a representative, please e-mail us at bulksales@mcgraw-hill.com.

This book is printed on acid-free paper.

Information contained in this work has been obtained by The McGraw-Hill Companies, Inc. ("McGraw-Hill") from sources believed to be reliable. However, neither McGraw-Hill nor its authors guarantee the accuracy or completeness of any information published herein, and neither McGraw-Hill nor its authors shall be responsible for any errors, omissions, or damages arising out of use of this information. This work is published with the understanding that McGraw-Hill and its authors are supplying information but are not attempting to render engineering or other professional services. If such services are required, the assistance of an appropriate professional should be sought.

Contents

Contributors xv
Foreword xix
Preface xxi
Introduction xxiii

Part 1 Gas, Chemical, and Free-Electron Lasers

1 Carbon Dioxide Lasers *Jochen Deile,*
Francisco J. Villarreal 3
   1.1 Introduction 3
   1.2 General Characteristics 4
   1.3 CO_2 Laser Basics 4
   1.4 CO_2 Laser Types 9
       1.4.1 Diffusion-Cooled CO_2 Lasers 9
       1.4.2 Fast-Flow CO_2 Lasers 12
   1.5 Applications 13
   References 15

2 Excimer Lasers *Rainer Paetzel* 17
   2.1 Introduction and Principle of Operation ... 17
   2.2 Technology and Performance of
       Excimer Lasers 19
       2.2.1 Principal Design and Technology 19
   2.3 Excimer Laser Designed to Application ... 28
       2.3.1 High-Power Excimer Laser 28
       2.3.2 Microlithography 31
       2.3.3 LASIK 32
   2.4 Application of High-Power Excimer Lasers ... 33
       2.4.1 High-Resolution Micromachining ... 35
       2.4.2 Brighter Displays 38
   References 40

3 Chemical Lasers *Charles Clendening,*
H. Wilhelm Behrens 43
   3.1 Introduction 43
   3.2 General Background 44
   3.3 Hydrogen Fluoride and Deuterium
       Fluoride Lasers 45
       3.3.1 Energy Levels 47

	3.3.2 Small Signal Gain	51
	3.3.3 Chemically Excited Species Generation	53
	3.3.4 Kinetic Processes, Deactivation, and Energy Transfer	55
	3.3.5 Fluid Mechanics and Nozzle Design	57
	3.3.6 Variations on Continuous Wave HF and DF Devices	62
	3.3.7 HF and DF Laser Performance	63
3.4	Chemical Oxygen Iodine Laser (COIL)	65
	3.4.1 Energy Levels	65
	3.4.2 Small Signal Gain	66
	3.4.3 Deactivation Processes	69
	3.4.4 Iodine Dissociation	69
	3.4.5 Singlet Oxygen Generator	69
	3.4.6 COIL Laser Performance Characterization	71
	3.4.7 COIL Laser Performance	72
3.5	Other Chemical Laser Concepts	73
	3.5.1 DF-CO_2 Transfer Devices	73
	3.5.2 Carbon Monoxide Lasers	74
References		74

4 High-Power Free-Electron Lasers *George R. Neil* ... **77**
   4.1 Introduction ... 77
   4.2 FEL Physics ... 77
      4.2.1 Physical Mechanism ... 77
      4.2.2 Wavelength ... 78
      4.2.3 Gain and Bandwidth ... 79
      4.2.4 Practical Considerations ... 82
   4.3 Hardware Implementation ... 83
      4.3.1 Overview ... 83
      4.3.2 Injectors ... 84
      4.3.3 Accelerators ... 87
      4.3.4 Wigglers ... 89
      4.3.5 The Optical Cavity ... 89
      4.3.6 Energy Recovery ... 91
   4.4 Status ... 93
   References ... 95

Part 2 Diode Lasers

5 Semiconductor Laser Diodes *Victor Rossin, Jay Skidmore, Erik Zucker* ... **101**
   5.1 Introduction ... 101
   5.2 Historical Growth of Power ... 102

	5.3	High-Power Laser Diode Attributes	103
	5.4	Device Geometry and Wafer Fabrication Processes	104
	5.5	Vertical and Lateral Confinement Laser Diode Structures	106
	5.6	Efficiency of Laser Diodes	108
	5.7	High-Power Broad-Area Laser Diodes	110
	5.8	High-Power Bars	112
	5.9	High-Power, Single-Mode Laser Diodes	114
	5.10	Burn-In and Reliability of Laser Diodes	116
	5.11	Submount Design and Assembly	120
	5.12	Fiber-Coupled Package Design and Processes	122
	5.13	Performance Attributes	125
	5.14	Spatially Multiplexed High-Brightness Pumps	126
	5.15	Qualification and Reliability	127
	References		129

6 High-Power Diode Laser Arrays
Hans-Georg Treusch, Rajiv Pandey — **133**

	6.1	Introduction	133
	6.2	Diode Laser Bar Assembly	134
	6.3	Heat Removal	136
		6.3.1 Water Guidelines for Minichannel Heat Sinks	137
		6.3.2 Expansion-Matched Microchannel Heat Sinks	138
	6.4	Product Platforms	139
	6.5	Device Performance	141
		6.5.1 Wavelength, Power, Efficiency, and Mode of Operation	141
		6.5.2 Beam Quality and Brightness	141
		6.5.3 Wavelength Locking	142
		6.5.4 Lifetime and Reliability	144
	6.6	Product Performance	145
		6.6.1 Fiber Coupling of Individual Diode Bars	146
		6.6.2 Power Scaling	150
		6.6.3 Fiber-Coupled High-Power Diode Laser Devices	151
	6.7	Direct High-Power Diode Array Applications	153
		6.7.1 Industrial Applications	153
		6.7.2 Medical Applications	158
		6.7.3 Defense Applications	158
	References		159

Contents

Part 3 Solid-State Lasers

7 Introduction to High-Power Solid-State Lasers
Gregory D. Goodno, Hagop Injeyan **163**
- 7.1 Introduction 163
- 7.2 Laser Gain Materials 164
  - 7.2.1 Cross Section and Lifetime 164
  - 7.2.2 Host Materials 166
  - 7.2.3 High-Average-Power SSL Materials ... 167
  - 7.2.4 High Pulse-Energy and Peak-Power SSL Materials 170
- 7.3 Pumping, Cooling, and Thermal Effects 171
  - 7.3.1 Pump Sources 172
  - 7.3.2 Laser Extraction and Heat Removal .. 174
- 7.4 Laser Beam Formation 176
  - 7.4.1 Stable Resonators 177
  - 7.4.2 Unstable Resonators 177
  - 7.4.3 Master Oscillator Power Amplifiers . 179
- 7.5 Wavefront Correction 180
  - 7.5.1 Spatial Phase Plates 180
  - 7.5.2 Phase Conjugation 181
  - 7.5.3 Adaptive Optics 182
- 7.6 Conclusion and Future Directions 184
- References 185

8 Zigzag Slab Lasers *Hagop Injeyan, Gregory D. Goodno* **187**
- 8.1 Introduction 187
- 8.2 Zigzag Slab Principle and Advantages 187
  - 8.2.1 Zigzag Geometry 187
  - 8.2.2 Scaling Laws 189
- 8.3 Traditional Side-Pumped Slabs 192
  - 8.3.1 Architecture and Technical Issues .. 192
  - 8.3.2 Performance 195
- 8.4 End-Pumped Slabs 198
  - 8.4.1 Architecture and Technical Issues .. 198
  - 8.4.2 Performance 200
  - 8.4.3 Power Scaling 201
- References 204

9 Nd:YAG Ceramic ThinZag® High-Power Laser Development *Daniel E. Klimek, Alexander Mandl* ... **207**
- 9.1 Introduction and ThinZag Concept Development 207
  - 9.1.1 TZ-1 Module Development 209

		9.1.2	TZ-2 Module Development	211
		9.1.3	TZ-3 Module Development	216
		9.1.4	Coupling Three TZ-3 Modules	217
	9.2	Summary		222
	Acknowledgments			222
	References			222

10 Thin-Disc Lasers *Adolf Giesen, Jochen Speiser* 225

- 10.1 Introduction 225
- 10.2 History 225
- 10.3 Principles of Thin-Disc Lasers 226
- 10.4 Possible Laser Materials 229
- 10.5 Numerical Modeling and Scaling 230
  - 10.5.1 Average Temperature 230
  - 10.5.2 Influence of Fluorescence 232
  - 10.5.3 Thermally Induced Stress 234
  - 10.5.4 Deformation, Stress, and Thermal Lensing 235
  - 10.5.5 Design Study for High-Power Thin-Disc Lasers 238
  - 10.5.6 Numerical Modeling of Gain and Excitation 239
  - 10.5.7 Equation of Motion 240
  - 10.5.8 Coupled Quasi-Static Numerical Model 241
  - 10.5.9 Influence of ASE 242
  - 10.5.10 Interaction of ASE and Excitation ... 243
  - 10.5.11 Time Resolved Numerical Model ... 245
  - 10.5.12 ASE-Limit 248
- 10.6 Thin-Disc Laser in Continuous-Wave Operation 250
  - 10.6.1 High Average Power 250
  - 10.6.2 Fundamental Mode, Single Frequency and Second Harmonic Generation (SHG) 251
- 10.7 Thin-Disc Laser in Pulsed Operation 254
  - 10.7.1 Q-Switched Operation of the Thin-Disc Laser 255
  - 10.7.2 Cavity-Dumped Operation of the Thin-Disc Laser 256
  - 10.7.3 Amplification of Nanosecond, Picosecond, and Femtosecond Pulses 257
  - 10.7.4 High Pulse Energy Thin-Disc Lasers ... 259
- 10.8 Industrial Realizations 261

Contents

	10.9 Summary	261
	Acknowledgments	262
	References	262
11	**Heat-Capacity Lasers** *Robert M. Yamamoto, Mark D. Rotter*	**267**
	11.1 Introduction	267
	11.2 System Architecture	267
	11.3 Laser Performance Modeling	273
	11.3.1 Pump Absorption, Gain, and Extraction	273
	11.3.2 The Effects of Amplified Spontaneous Emission	278
	11.3.3 Wavefront Distortion and Depolarization	282
	11.4 Current State of the Art	290
	11.4.1 Power Extraction	290
	11.4.2 Wavefront Control	290
	11.5 Scaling Approaches	295
	11.6 Applications and Related Experimental Results	296
	11.6.1 Rapid Material Removal (Boring/Ablation)	296
	11.6.2 Aerodynamic Imbalance Due to Airflow Interaction	296
	11.6.3 Laser Used for Humanitarian Mine Clearing	297
	11.6.4 Self-Contained 400-kW Heat-Capacity Laser on a Military Vehicle	299
	11.7 Summary	299
	References	300
12	**Ultrafast Solid-State Lasers** *Sterling Backus*	**301**
	12.1 Introduction	301
	12.2 Ultrafast Laser Sources and Oscillators	302
	12.2.1 Kerr Effect	302
	12.2.2 Ultrafast Oscillators	303
	12.3 Ultrafast Amplification Techniques	304
	12.3.1 Chirped Pulse Amplification	305
	12.3.2 Aberrations	308
	12.3.3 Amplifier Schemes	308
	12.3.4 Regenerative Amplification	309
	12.3.5 Multipass Amplification	310
	12.3.6 Downchirped Pulse Amplification	311
	12.4 Thermal Mitigation	314
	12.4.1 Optical Parametric Chirped Pulse Amplification	315

	12.5	Pulse Measurement	317
	12.6	Applications	319
		12.6.1 Filaments	319
		12.6.2 Precision Machining with Minimum Collateral Damage	319
		12.6.3 Laser-Based Photon and Particle Sources	321
		12.6.4 High Harmonic Generation	321
	References		324

13 Ultrafast Lasers in Thin-Disk Geometry
Christian Kränkel, Deran J. H. C. Maas, Thomas Südmeyer, Ursula Keller **327**

	13.1	Introduction	327
	13.2	Pump Geometry	329
	13.3	Thermal Management in Thin-Disk Geometry	331
	13.4	SESAM Mode Locking	336
		13.4.1 Pulse Formation Mechanisms	336
		13.4.2 Different Operation Regimes	338
	13.5	Conclusion and Outlook	348
	References		349

14 The National Ignition Facility Laser
Richard A. Sacks, Christopher A. Haynam **357**

	14.1	Introduction	357
	14.2	Historical Background	358
	14.3	NIF Facility and Laser Overview	362
	14.4	1ω Bundle Performance and 1ω/3ω NIF Operating Envelopes	366
		14.4.1 Energetics and the Laser Performance Operations Model Calibration Results	367
		14.4.2 Power versus Energy Operating Envelopes for NIF	368
	14.5	Performance Qualification Shots for Ignition Target Pulse Shapes	371
		14.5.1 Master Oscillator and Pulse Shaping System	371
		14.5.2 Preamplifier Module Description and Performance	372
		14.5.3 Main Laser 1ω Performance	376
		14.5.4 Frequency Conversion Performance	378
	14.6	Focal Spot Beam Conditioning and Precision Pulse Shaping for Ignition Experiments	385
		14.6.1 Spatial Beam Conditioning with Phase Plates	388

		14.6.2	Temporal Beam Conditioning with One-Dimensional SSD	391
		14.6.3	Frequency Conversion of Spatially and Temporally Conditioned Pulses	393
		14.6.4	Temporal Pulse Shaping	394
	14.7	2010 NIF Status and Experiments		395
	Conclusion			405
	Acknowledgments			406
	References			406

Part 4 Fiber Lasers

15 Introduction to Optical Fiber Lasers
Liang Dong, Martin E. Fermann **413**

- 15.1 Background 413
  - 15.1.1 History 413
  - 15.1.2 Advantages of Fiber Lasers 415
- 15.2 Rare-Earth-Doped Optical Fibers 416
  - 15.2.1 Basics of Optical Fibers 416
  - 15.2.2 Properties of Rare-Earth-Doped Optical Fibers 420
  - 15.2.3 Power Scaling of Fiber Lasers 426
  - 15.2.4 Fibers for High-Power Fiber Lasers . 435
- 15.3 Optical Fiber Lasers 445
  - 15.3.1 Continuous Wave Fiber Lasers 445
  - 15.3.2 Q-Switched Lasers 453
  - 15.3.3 Mode-Locked Fiber Lasers 454
- References 458

16 Pulsed Fiber Lasers *Fabio Di Teodoro* **463**

- 16.1 Introduction 463
- 16.2 Challenges to Pulse Power Scaling 464
  - 16.2.1 Nonlinear Optical Effects 464
  - 16.2.2 Amplified Spontaneous Emission . . . 471
  - 16.2.3 Optical Damage 472
- 16.3 Fiber Laser Trades for High-Pulse-Power Operation 474
  - 16.3.1 Type of Fiber 474
  - 16.3.2 Amplified Spontaneous Emission Management 477
  - 16.3.3 MOPA versus Power Oscillators . . . 478
- 16.4 High Pulse Energy and Peak Power Fiber Amplifiers: Results 480
  - 16.4.1 Single-Stage Fiber Amplifiers 480

	16.4.2	Gain-Staged MOPAs	483
	16.4.3	Polarization-Maintaining MOPAs and Wavelength Conversion	488
	16.4.4	Eye-Safe, Pulsed-Fiber Laser Sources	491
16.5	Conclusions and Outlook	493
Acknowledgments		495
References		495

17 High-Power Ultrafast Fiber Laser Systems
Jens Limpert, Andreas Tünnermann — 499

- 17.1 Introduction and Motivation — 499
- 17.2 Nonlinear Effects as Basic Limitations of Ultrashort Pulse Amplification in Rare-Earth-Doped Fibers — 501
- 17.3 High-Repetition-Rate Gigawatt Peak Power Fiber Laser System — 505
- 17.4 Peak Power and Pulse Energy Scaling Considerations — 508
- 17.5 Average Power Scaling of Ultrashort Pulse Fiber Systems — 512
- 17.6 Conclusion and Outlook — 513
- References — 514

18 High-Power Fiber Lasers for Industry and Defense
Michael O'Connor, Bill Shiner — 517

- 18.1 Introduction — 517
- 18.2 Fiber Laser Engineering — 518
- 18.3 Power Scaling of Broadband Multimode Fiber Lasers — 520
- 18.4 Power Scaling of Broadband Single-Mode Fiber Lasers — 523
- 18.5 High-Power Fiber Lasers in Industrial Applications — 526
- 18.6 Defense Applications of High-Power Fiber Lasers — 528
  - 18.6.1 Fiber Lasers for Strategic versus Tactical Directed-Energy Applications — 528
- References — 530

Part 5 Beam Combining

19 Beam Combining
Charles X. Yu, Tso Yee Fan — 533

- 19.1 Introduction — 533
  - 19.1.1 Motivation for Beam Combining — 533
  - 19.1.2 Beam-Combining Performance Metrics — 534

19.2 Beam-Combining Techniques and
Theories 536
  19.2.1 Incoherent Beam Combining 537
  19.2.2 Coherent Beam Combining 537
  19.2.3 CBC Performance Degradations ... 543
  19.2.4 Wavelength Beam Combining 547
  19.2.5 WBC Performance Degradations ... 550
  19.2.6 Hybrid Beam Combining 553
19.3 Beam Combining of Specific Laser
Systems 553
  19.3.1 Fiber Laser Beam Combining 554
  19.3.2 Semiconductor Laser Beam
  Combining 561
  19.3.3 Solid-State Laser Beam
  Combining 564
19.4 Summary 565
Acknowledgments 566
References 566

Index **573**

Contributors

Sterling Backus *Vice President, Research and Development, Kapteyn-Murnane Laboratories, Inc., Boulder, Colorado* (CHAP. 12)

H. Wilhelm Behrens *Fluid and Thermophysics Department Manager, Northrop Grumman Aerospace Systems, Redondo Beach, California* (CHAP. 3)

Robert L. Byer *The William R. Kenan, Jr., Professor, School of Humanities and Sciences, Department of Applied Physics, Edward L. Ginzton Laboratory, Stanford University, Stanford, California* (FOREWORD)

Charles Clendening *Technical Fellow, Northrop Grumman Aerospace Systems, Redondo Beach, California* (CHAP. 3)

Jochen Deile *Manager, Laser Development, TRUMPF Inc., Farmington, Connecticut* (CHAP. 1)

Fabio Di Teodoro *Senior Scientist, Northrop Grumman Aerospace Systems, Redondo Beach, California* (CHAP. 16)

Liang Dong *Department of Electrical and Computer Engineering and Center for Optical Materials Science and Engineering Technologies (COMSET), Clemson University, Clemson, South Carolina* (CHAP. 15)

Tso Yee Fan *Associate Group Leader, MIT Lincoln Laboratory, Lexington, Massachusetts* (CHAP. 19)

Martin E. Fermann *IMRA America, Ann Arbor, Michigan* (CHAP. 15)

Adolf Giesen *Head, Institute of Technical Physics, German Aerospace Center (DLR), Stuttgart, Germany* (CHAP. 10)

Gregory D. Goodno *Senior Scientist, Northrop Grumman Aerospace Systems, Redondo Beach, California* (CHAPS. 7, 8)

Christopher A. Haynam *Associate Program Leader, ICF and HED Science Program (NIF), Lawrence Livermore National Laboratory, Livermore, California* (CHAP. 14)

Hagop Injeyan *Technical Fellow, Northrop Grumman Aerospace Systems, Redondo Beach, California* (CHAPS. 7, 8)

Contributors

Ursula Keller Institute of Quantum Electronics, Physics Department, Swiss Federal Institute of Technology (ETH Zurich), Switzerland (CHAP. 13)

Daniel E. Klimek Principal Research Scientist, Textron Defense Systems, Wilmington, Massachusetts (CHAP. 9)

Christian Kränkel Institute of Quantum Electronics, Physics Department, Swiss Federal Institute of Technology (ETH Zurich), Switzerland (CHAP. 13)

Jens Limpert Institute of Applied Physics, Friedrich Schiller University Jena, and Fraunhofer Institute for Applied Optics and Precision Engineering, Jena, Germany (CHAP. 17)

Deran J. H. C. Maas Institute of Quantum Electronics, Physics Department, Swiss Federal Institute of Technology (ETH Zurich), Switzerland (CHAP. 13)

Alexander Mandl Principal Research Scientist, Textron Defense Systems, Wilmington, Massachusetts (CHAP. 9)

George R. Neil Associate Director, Thomas Jefferson National Accelerator Facility, Newport News, Virginia (CHAP. 4)

Michael O'Connor Director, Advanced Applications, IPG Photonics Corporation, Oxford, Massachusetts (CHAP. 18)

Rainer Paetzel Coherent GmbH, Dieburg, Germany (CHAP. 2)

Rajiv Pandey Senior Product Manager, DILAS Diode Laser Inc., Tucson, Arizona (CHAP. 6)

Victor Rossin Senior Engineering Development Manager, Communications and Commercial Optical Products, JDSU, Milpitas, California (CHAP. 5)

Mark D. Rotter Member of the Technical Staff, Lawrence Livermore National Laboratory, Livermore, California (CHAP. 11)

Richard A. Sacks Senior Scientist and Technical Lead, ICF and HED Science Program (NIF), Lawrence Livermore National Laboratory, Livermore, California (CHAP. 14)

Bill Shiner Vice President, Worldwide Sales, IPG Photonics Corporation, Oxford, Massachusetts (CHAP. 18)

Jay Skidmore Senior Engineering Development Manager, Communications and Commercial Optical Products, JDSU, Milpitas, California (CHAP. 5)

Jochen Speiser Head, Solid State Lasers & Nonlinear Optics, Institute of Technical Physics, German Aerospace Center (DLR), Stuttgart, Germany (CHAP. 10)

Thomas Südmeyer Institute of Quantum Electronics, Physics Department, Swiss Federal Institute of Technology (ETH Zurich), Switzerland (CHAP. 13)

Hans-Georg Treusch Director, Trumpf Photonics, Cranbury, New Jersey (CHAP. 6)

Andreas Tünnermann Institute of Applied Physics, Friedrich Schiller University Jena, and Fraunhofer Institute for Applied Optics and Precision Engineering, Jena, Germany (CHAP. 17)

Francisco J. Villarreal *Chief Laser Scientist, TRUMPF Inc., Farmington, Connecticut* (CHAP. 1)

Robert M. Yamamoto *Principal Investigator, Lawrence Livermore National Laboratory, Livermore, California* (CHAP. 11)

Charles X. Yu *Technical Staff, MIT Lincoln Laboratory, Lexington, Massachusetts* (CHAP. 19)

Erik Zucker *Senior Director of Product Development, Communications and Commercial Optical Products, JDSU, Milpitas, California* (CHAP. 5)

Foreword

The *High-Power Laser Handbook*, edited by Hagop Injeyan and Gregory D. Goodno, is both comprehensive and timely. It is comprehensive in that the laser technologies discussed include gas, chemical, and free-electron lasers, with a special emphasis on solid-state laser technologies, including semiconductor diode lasers, solid-state lasers, and fiber lasers, as well as power scaling and applications of high-power lasers.

The book is timely because 2010 marked the 50th anniversary of the demonstration of the ruby laser by Theodore Maiman at the Hughes Research Laboratory in Malibu, California. From the beginning, it was recognized that the laser would become useful for military applications, from radar to cutting metal at a distance. It was also recognized that the laser would provide unprecedented peak powers that could fuse hydrogen isotopes. After 50 years, laser technology has matured to the point that the supposed pipe dreams of the 1960s are now becoming reality.

The laser is now an essential tool in scientific research, from biology, chemistry, and physics to applied physics and engineering. Laser technology has enabled subwavelength resolution microscopy, which, in turn, is rapidly becoming an essential tool in biology and neurology. Lasers are the most precise form of electromagnetic radiation, enabling optical clocks of unprecedented accuracy (less than one second in the age of the universe). Lasers also allow the most precise measurements of length ever attempted. For example, they can measure to less than one billionth of an optical wavelength in the 4-km-length arms of the Laser Interferometer Gravitation Wave Observatory, which searches for the direct detection of gravitational waves that reach across the universe. In addition, lasers can control molecules and atoms in order to alter and control chemical reactions or to cool atoms to a single quantum state, known as the Bose-Einstein condensation.

Lasers have also affected our ability to optimize materials. By laser peening jet engine turbine blades, we have improved engine performance and reliability. Laser cutting of metals is now the preferred tool for manufacturing. Laser marking of surfaces is ubiquitous and allows the labeling and tracking of a multitude of parts. The laser transit is the tool of choice for measuring direction and distance, while the laser level provides a low-cost, elegant way to level a ceiling.

Lasers affect our lives in everything from medicine to entertainment to communications. More than one billion scans per day are made at checkout counters around the world on laser scanners. Lasers also connect us through fiber optic communications, which is the backbone of the modern world, far exceeding the optimistic projections for speed and bandwidth of a decade ago. We now demand video links and download movies online rather than through packaged delivery.

Today's lasers are efficient and do real work. More than 20 different lasers are used to manufacture an automobile. Every cell phone, laptop computer, and television is manufactured using precise laser light to drill, melt, or correct a link in a miniature circuit or television screen.

The dream that someday the laser would be a precise weapon for defense applications is now becoming reality. Video demonstrations of laser beams illuminating and destroying missiles, mortar rounds, and artillery rounds in flight are now available on YouTube for all to witness. The year 2009 saw a step toward efficient, compact lasers for weapons with the demonstration of a 20 percent efficient, diode-pumped solid-state laser at greater than 100 kW average output power. How did we progress from 2 mW diode-pumped laser power in 1984 to greater than 100 kW power today? The chapters in this book discuss this progress in advanced solid-state lasers and help lead to an understanding of the key breakthroughs in laser technology that have enabled a factor of one million increase in laser power in just a quarter century.

The year 2009 saw the commissioning of the world's largest laser—the National Ignition Facility's megajoule-class laser—for laser fusion studies. This laser was designed to study all aspects of laser fusion using the unique properties of lasers to deliver greater than 1 MJ of ultraviolet light to a target in less than 3 ns. The preliminary experiments have been published and are very promising. The goal is to achieve a fusion burn in the laboratory as a step toward a detailed understanding of matter compressed to a density and temperature, which, in turn, will allow an efficient fusion burn. The next step is to design and engineer a laser that can drive the fusion process at 10 Hz rate for application to fusion energy.

The past 50 years have seen remarkable progress in laser technology. This book captures elements of that progress from experts who have participated in and contributed to laser technology. An understanding of the first 50 years of laser technology and its applications may offer a glimpse into the next 50 years. Of course, it is difficult to make predictions about the future. My guess is that we will grossly underestimate the progress in laser technology and the breadth of applications laser technology will enable.

Robert L. Byer

Preface

Over the past decade, extraordinary progress has been made in all aspects of high-power laser development. Technological advances in gas, solid-state, fiber, free-electron, chemical, and semiconductor lasers have enabled unprecedented power, efficiency, beam quality, and reliability. Concurrent with—and often driving—this progress in laser source development has been an increased penetration of lasers into a diverse range of commercial, military, and scientific applications, from traditional laser machining to more esoteric applications, such as directed-energy weapons.

Understanding the state of the art in high-power laser technology is critical not just for laser engineers but also for systems engineers, optical designers, applications engineers, and technical managers. Too often an applications engineer or system designer may consider the laser source a "black box," without having any real understanding of the mechanisms involved in generating and emitting photons. In many cases, this lack of awareness of the inherent advantages or limitations of a given laser technology may result in a suboptimal design.

This book presents a series of newly written chapters that have been solicited from recognized leaders in each technical area. The intent of this compilation is to provide a wide-ranging snapshot survey of the current state of the art in high-power laser development. The approach is principally phenomenological, with the goal of providing readers with an intuitive understanding of the key features of various laser technologies, while leaving fundamental physical derivations to the referenced literature for the interested reader. The intent of this streamlined approach is to allow for a greater breadth of coverage of high-power laser technologies and applications than is typically available in a single volume.

Specifically, the goals of the book are as follows:

- To describe typical and state-of-the-art performance parameters for each major class of lasers

- To provide an appreciation both for how different types of lasers work and for the engineering or physics constraints that limit their performance or usefulness
- To provide practical analytic tools, as well as examples of real-world applications, so readers can identify an appropriate laser source for their needs

We hope this book will serve as a useful reference, both to those working directly in the field of high-power laser development and to engineers who may not be laser experts but who wish to identify appropriate laser capabilities and technologies. The level of the book is appropriate for professional engineers who have some background in optical physics but who are not necessarily experts in lasers. This book may also be suitable as supplementary course material for a university-level class on laser technology.

Hagop Injeyan
Gregory D. Goodno

Introduction

Each chapter in this book serves as a stand-alone review of a specific laser technology. We have grouped sets of chapters covering similar classes or related technologies into parts, with the intent of providing some structure and a suggested reading order for those who are not laser experts.

The first part (Chaps. 1–4) covers the basic functionality and discusses recent developments in specialized technologies of high-power gas, chemical, and free-electron lasers. These technologies are, in some regard, "mature," in that research and development (R&D) investment in these areas collectively peaked some years ago, with interest since being diverted into newer solid-phase technologies. Still, recent and significant R&D activity, continued relevance to industrial and military applications, and advances in source generation warrant their inclusion in any treatise on high-power lasers.

The next part (Chaps. 5 and 6) covers the state of the art in semiconductor diode lasers, along with the associated technologies of packaging, reliability, and beam shaping and delivery. Diode lasers are by far the most widespread and economically significant laser technology ever developed. The emergence of high-brightness, fiber-delivered diode laser systems has opened many new applications in materials processing. Moreover, as optical pump sources, diode lasers have revolutionized the field of solid-state lasers and have enabled the new and promising field of fiber lasers. Chapter 5 introduces the basic concepts underlying semiconductor diode laser emitters, including their manufacture, packaging, performance, and scaling. Chapter 6 extends this discussion to the packaging, power scaling, and fiber coupling of diode emitter arrays in bars and stacks, as well as covering some of the applications that are enabled by high-brightness diodes.

The largest part in this book (Chaps. 7–14) covers solid-state lasers (SSLs). The size of this part reflects both the relative amount of R&D that has been invested in SSL technology over the past decade and the diversity of scaling approaches depending on whether the goal is high continuous wave (CW) powers, high pulse energies, or high peak powers. The continued high level of interest on the part of the R&D community has elevated SSLs above other technologies as the highest-performing, electrically powered lasers in terms of peak and average powers, pulse energy, and shortest pulse widths. This interest, in turn, has enabled an enormous variety of applications in materials processing, inertial fusion, defense, spectroscopy, and high-field physics research.

The SSL part begins with a short introduction (Chap. 7), providing a broad overview of high-power SSLs and their unique features. This chapter is followed by a discussion of CW power-scalable, zigzag slab lasers in Chaps. 8 and 9. The other primary CW-scalable SSL geometry—the thin-disk laser—is described in Chap. 10. Chapter 11 discusses the concept of the heat-capacity laser, an out-of-the-box approach to power scaling that is relevant to military applications and that allows operation in multisecond "burst" modes. Chapter 12 introduces ultrafast SSLs, in which the design imperatives for generating short laser pulses must be balanced against those arising from average power (pulse repetition rate) scaling. Chapter 13 explores this balance in detail by describing the capabilities of the thin-disk geometry toward high-repetition-rate, ultrafast pulses. Finally, Chap. 14 reviews the recently completed National Ignition Facility laser for fusion energy research, which represents the most elaborate and highest pulse energy laser system built to date.

The next part (Chaps. 15–18) covers the fastest-evolving high-power technology of the past few years—fiber lasers. Fibers can be regarded as a specialized subset of SSLs; however, due to their remarkable geometric properties of light guidance and heat removal, they provide a unique technology platform for power scaling and packaging that warrants a full part of their own. Chapter 15 provides a thorough introduction to fiber lasers, from the fundamental nature of light guidance and modes in fiber to the various types of fibers most commonly used. It also introduces the nonlinear effects that limit further scaling of their output power. In Chap. 16, these nonlinear limits are explored in greater detail as they pertain to the generation of high peak power directly in fiber. Chapter 17 extends this discussion of peak power scaling to ultrafast chirped pulse amplifiers, in which the pulse spectral fidelity plays a critical role toward short pulse generation. Finally, Chap. 18 reviews the state of the art in high-average (CW) power fiber laser performance and engineering, as well as gives an introduction to common industrial and defense applications.

This book concludes with Chap. 19, which reviews various methods for beam combining. Combining many lasers in parallel allows beam-combined systems to achieve many times the performance of any single laser. A number of state-of-the-art demonstrations of spatial brightness or pulse energy has resulted from the implementation of beam-combining methods. Although beam combining is not a laser technology per se, it is assuming greater importance as underlying laser technologies reach maturity without satisfying the demand for high-spatial brightness, which is predominantly driven by defense applications.

High-Power Laser Handbook

PART 1
Gas, Chemical, and Free-Electron Lasers

CHAPTER 1
Carbon Dioxide Lasers

CHAPTER 2
Excimer Lasers

CHAPTER 3
Chemical Lasers

CHAPTER 4
High-Power Free-Electron Lasers

CHAPTER 1
Carbon Dioxide Lasers

Jochen Deile
*Manager, Laser Development, TRUMPF Inc.,
  Farmington, Connecticut*

Francisco J. Villarreal
Chief Laser Scientist, TRUMPF Inc., Farmington, Connecticut

1.1 Introduction

The carbon dioxide (CO_2) laser has been studied intensively over the past several decades. Although no longer being studied by academia, these lasers are still the most utilized in industrial applications, in terms of both units and dollars. Typical applications include metal cutting and welding; processing of nonmetals, such as plastics, fabric, and glass; and marking and coding applications—as well as many medical, dental, and scientific applications. Overall, laser cutting makes up approximately 25 percent of all industrial laser applications, which totaled about US$6 billion annually worldwide in 2008. CO_2 lasers have been successful because they are so versatile; a CO_2 laser can process almost any material of almost any thickness.

Historically, CO_2 lasers have been able to produce more power, with higher beam quality, and at lower costs than other lasers. Multi kilowatt CO_2 lasers have been available since the early 1980s. One of the major breakthroughs for the CO_2 laser came with the improved excitation of the CO_2 molecule by the addition of nitrogen (N_2) to the laser gas.[1] Technological advances reduced the size of the laser and made it absolutely reliable in industrial environments. Another major breakthrough in terms of reliability was the introduction of radio frequency (rf)–excited designs. Although fiber beam delivery systems are not available for CO_2 lasers and even though other technologies now offer better efficiencies, CO_2 lasers will be around for a

4 Gas, Chemical, and Free-Electron Lasers

Characteristic	Range	Typical Values
Quantum efficiency	—	40%
Electro-optical efficiency	10–30%	20%
Wall-plug efficiency	8–15%	12%
Wavelength	9–11 µm	10.6 µm
Power levels, continuous wave	1 mW–100 kW	10–300 W 2–10 kW
Power levels, pulsed	Up to 10^{13} W	
Small-signal gain (g_0) Saturation intensity (I_s)	0.5–1.5 m^{-1} 100–1000 W/cm^2	
Beam quality (M^2)	1–10	1.2
Beam diameters (86% diameter)	3–30 mm	20 mm
Focus diameters	15–600 µm	200 µm
Polarization	—	Linear

TABLE 1.1 General Characteristics of CO_2 Lasers

long time due to certain characteristics, such as their 10-micrometer (µm) wavelength and their investment costs.

To design a reliable, industrial product, many disciplines must be mastered, including optical resonators, gas chemistry, thermodynamics, surface chemistry, radio frequency (rf)-or-direct current (dc) excitation, discharge physics, and beam shaping. Some of these topics will be discussed in the next sections of this chapter. However, because many other laser books and relevant literature discuss all aspects of CO_2 laser physics in greater detail,[2-5] our focus is on providing a general overview of information relevant to typical industrial applications.

1.2 General Characteristics

Table 1.1 provides an overview of the most important characteristics of CO_2 lasers.

1.3 CO_2 Laser Basics

The CO_2 molecule is a linear symmetric molecule with an axis of symmetry along the nuclei and a plane symmetry perpendicular to this axis. The laser's emission wavelength is determined by the low-lying vibrational and rotational energy levels of the CO_2 molecule.

A major breakthrough for the CO_2 laser came with the improved excitation of the CO_2 molecule by the addition of nitrogen to the

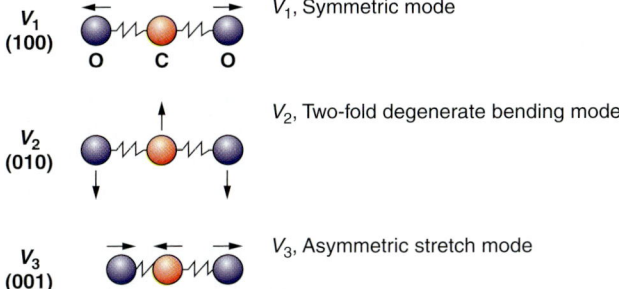

FIGURE 1.1 Normal modes of vibration of the CO_2 molecule.

laser gas.[1] Electric discharges excite the N_2 molecule very effectively. Because the N_2 molecule has two identical nuclei, its dipole radiation is forbidden. Thus, it can only decay by collision with the wall of the discharge vessel or with other molecules. The energy stored in the N_2 molecule can be easily transferred to the CO_2 molecule due to the close resonance of the N_2 vibration and the v_3 vibration levels of the CO_2 molecule (Fig. 1.2). The (00^01) level of CO_2 is only $\Delta E = 18$ cm^{-1} (where E is energy) higher than the v_1 vibrational level of nitrogen. Because this energy difference is much smaller than the average kinetic energy during collisions the CO_2 molecules can easily draw the vibrational energy of the N_2 to excite the v_3 vibration.[2]

A similar effect occurs between carbon monoxide (CO) and CO_2. CO is produced in the discharge by dissociation from CO_2; it is also often added to the laser gas mix of diffusion-cooled lasers. The cross section for excitation of the CO molecule in the electric discharge is rather large and the CO molecule can transfer energy to the v_3 vibration level because the energy difference between the CO vibrational level and the (00^01) level of CO_2 is $\Delta E = 170$ cm^{-1}, which is smaller than the average kinetic energy. The less-efficient energy transfer from CO to CO_2, as compared with the energy transfer from N_2 to

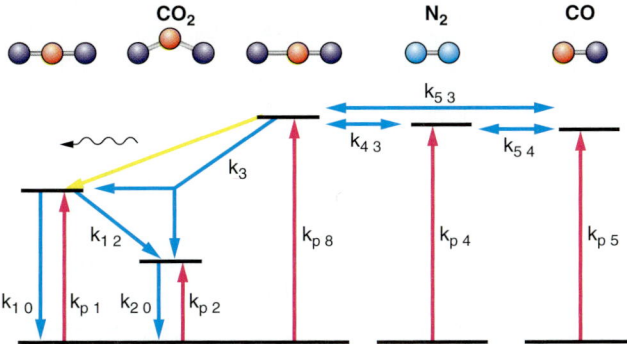

FIGURE 1.2 Vibrational energy levels of the CO_2, CO, and N_2 molecules.

CO_2, can be explained by the larger difference in energy levels and the fact that CO has a dipole moment and thus has spontaneous decay.

The energy transfer via N_2 and CO to CO_2 is much more efficient than the direct excitation of the CO_2 molecule; this is due to the much larger cross sections for vibrational excitation of N_2 and CO by electron impact. According to Hake and Phelps (1967), vibrational excitation of CO_2 molecules by electron impact is only efficient for a narrow range of electron energies.[6] The vibrational excitation of CO and N_2 by electron impact, however, is quite efficient for a wide range of electron energies. For optimum excitation of CO and N_2, the electron energies should range from 1 to 3 electron volts (eV). The range of electron energies can be adjusted by changing the pressure and composition of the laser gas mix.[2] Figure 1.3 shows the calculated small signal gain for various transitions in CO_2.

The gas discharge of CO_2 lasers is typically a Townsend discharge, which is a gas ionization process in which an initially very small amount of free electrons, accelerated by a sufficiently strong electric field, gives rise to electrical conduction through a gas by avalanche multiplication. When the number of free charges drops or the electric field weakens, the phenomenon ceases.

Rf discharges can be subdivided into inductive and capacitive discharges. For most lasers, only the capacitive discharges are relevant. The two important forms of the capacitively coupled RF discharge are named α- and γ-discharge, according to the Townsend coefficients α and γ, which describe where the electrons are generated.[4] The main difference between the two is the impedance of the sheaths, the power

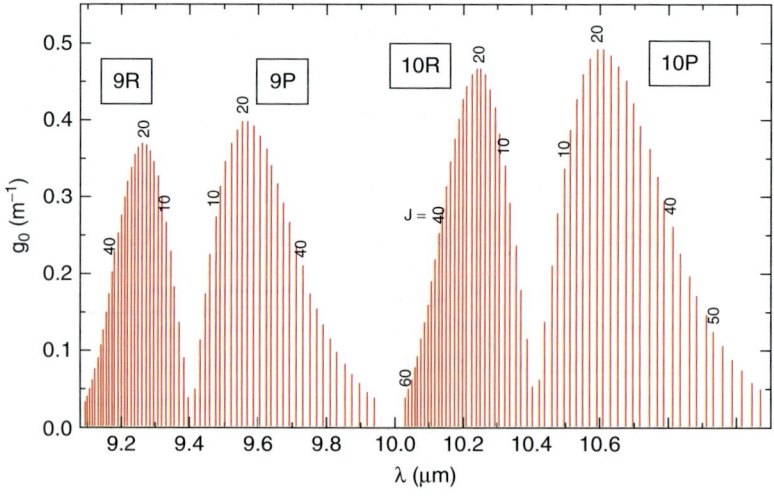

FIGURE 1.3 Calculated small-signal gain for the regular bands of the CO_2 laser at $T = 520$ K; RF density = 5 Wcm^{-3}; He = 73%; and $N_2 : CO_2 = 2.75$.[19]

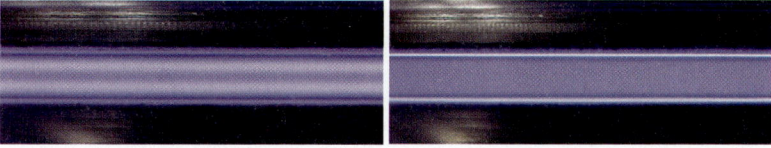

FIGURE 1.4 Left: A typical α-discharge in a 6-mm interelectrode gap. Right: A typical γ-discharge in a 6-mm interelectrode gap.

dissipation in the sheaths and their current densities. The two types can easily be distinguished by their intensity and luminosity distribution along the discharge length (see Fig. 1.4). The γ-discharge is also called *high-current discharge* because it has an order of magnitude higher current density than the α-discharge, which is called *low-current discharge*.

The gas used in CO_2 lasers is usually a mix of CO_2, N_2, and helium (He). To improve certain aspects of the laser's performance CO, xenon (Xe), and other gases are added as needed. The high thermal conductivity of helium, which is about six times higher than the thermal conductivity of N_2 and CO_2, reduces the gas temperature, because the temperature gradient between the laser gas and the cooled electrode surface is inversely proportional to the thermal conductivity. The thermal conductivity of helium is $\kappa_{He} = 0.17$ W/mK (watts per meter-kelvin) at 100°C. Helium's energy levels are all above 20 eV; thus, for a gas mix that is optimized for high laser power—namely, electron energy levels in the 1–3 eV range—helium does not significantly influence the discharge. Thermal conductivity is basically determined by the amount of helium in the gas mix, which results in improved heat removal and a reduction of the lower laser level's thermal population. In addition, the width of the gain profile is temperature dependent and increases with decreasing temperature. Helium also stabilizes the discharge since diffusion processes and thermal conductivity are important to stabilize the discharge by evening out local inhomogeneities.

Gas mixtures for diffusion-cooled lasers typically contain Xe. Three to five percent concentrations of Xe increase the laser's output power and efficiency. The increased efficiency results from the effect Xe has on the electron energy distribution in the discharge. Xenon has a relatively low ionization energy of 12.1 eV, which is about 2–3 eV less than that of the other gas components. Therefore, the number of electrons with energies above 4 eV decreases, and the number of electrons with energies below 4 eV increases.[7] The change in electron energy distribution has a favorable effect on the vibrational excitation of CO and N_2, as discussed earlier in this chapter.

Water (H_2O) has a strong influence on laser performance. In low-power, sealed-off lasers, H_2O is often added to the laser gas mix to suppress the dissociation of CO_2 molecules into CO and oxygen. In most high-power diffusion-cooled lasers with metal electrodes, the amount of water vapor in the system is so large that it has a negative effect on the laser's performance; thus, it must be removed. H_2O or hydrogen (H_2) contents above a critical level have a large impact on the relaxation of the upper laser level.

Water forms monolayers on the surfaces of the electrodes and the vacuum vessel. Removing these monolayers requires baking at high temperatures and low pressures. To prevent the moisture level in a laser system from increasing, water adsorbers, such as zeolites, can be added to the vacuum system of diffusion-cooled lasers. In fast-flow lasers, a small percentage of the laser gas is continuously replaced to prevent contamination and degradation.

If failures due to issues such as mirror damage are neglected, the lifespan of sealed-off CO_2 lasers will mainly be determined by the long-term stability of the CO_2 partial pressure and the contamination of the laser gas by leaks and outgassing of the materials used in the construction of the laser. Analogously, the gas exchange frequency of near sealed-off lasers depends on the same factors.[8,9]

The partial CO_2 pressure changes over time due to dissociation of the CO_2 molecules.[10,11] If no special preparations are taken, the initially reached equilibrium among CO_2, CO, and oxygen will be pushed further and further to the CO side, until efficient laser operation is no longer possible. The initial equilibrium is influenced by such factors as the gas mix, pressure, rf input power, and electrode materials. Typically 50 to 70 percent of the CO_2 is dissociated when the initial equilibrium is reached.

The dissociation of the CO_2 molecule is triggered by electron impact in the gas discharge:[12]

$$CO_2 + e^- \Leftrightarrow CO + O + e^- - 5{,}5 \text{ eV} \quad (1.1)$$

$$CO_2 + e^- \Leftrightarrow CO + O^- - 3{,}85 \text{ eV} \quad (1.2)$$

There are a few ways to stabilize the CO_2 partial pressure:

- Prevent the oxygen generated in the gas discharge from being consumed by either oxidation or adsorption processes.
- Use catalysts, such as gold, to accelerate the back reaction of CO and O_2 to CO_2.[11]
- Use gas additives, such as H_2O and H_2.[2]
- Use CO_2 donors.[13]
- Use a pre-dissociated gas mix.

To conserve the CO_2 partial pressure, it is critical to use the proper materials in constructing the laser. According to the law of mass action, the CO_2 partial pressure will decline if the oxygen partial pressure is reduced. Reduction of the oxygen partial pressure can be avoided by using nonoxidizing materials, such as quartz or ceramic, or by passivating the metallic materials used. Passivation of aluminum electrode surfaces, for example, can be achieved by chemical reactions between aluminum and a strong oxidizer, such as nitric acid.[8] Other methods include anodizing and applying conversion coatings. The CO_2 partial pressure can also be conserved by using a catalyst to accelerate the back reaction of CO and oxygen to CO_2.

The most commonly used method for stabilizing the CO_2 partial pressure is the use of pre-dissociated gas mixes, in which CO and sometimes oxygen are added to the gas mix. This approach avoids not only CO_2 dissociation but also the creation of oxygen. Avoiding the creation of oxygen is of interest because oxygen quenches both the upper laser level[14] and the exited N_2 molecule.[15]

1.4 CO_2 Laser Types

Like any other laser, the CO_2 laser has a limited efficiency. Efficiently removing the waste heat from the laser gas and keeping its temperature below 600 K is key to laser performance. Two types of CO_2 laser designs on the market efficiently remove the heat from the active medium: fast-flow and diffusion-cooled lasers. In fast-flow designs, the gas is circulated with speeds of up to half the speed of sound through the discharge area. The gas is then cooled in heat exchangers before returning to the discharge area. In diffusion-cooled designs, the laser gas is in contact with cooled surfaces, and the heat is removed by diffusion of the hot gas molecules to the water-cooled electrodes. These two categories of lasers are described in later sections of this chapter.

The different types of CO_2 lasers can be further categorized according to their excitation method, their design, and their operating parameters. The gas discharge for laser excitation can be direct current, medium frequency (0.3 to 3 MHz), radio frequency (3 to 300 MHz), or microwave (0.3 to 3 GHz) powered discharge. Further categories are sealed-off lasers, waveguide lasers, transversely excited atmospheric pressure (TEA) lasers, and gas dynamic lasers. The next sections focus on designs that are relevant for today's industrial applications.

1.4.1 Diffusion-Cooled CO_2 Lasers

The laser power (P_L) of diffusion-cooled lasers scales with the surface area A, which removes the heat from the gas, and the distance between the water-cooled electrode surfaces (the interelectrode gap) d:

$$P_L \propto A/d$$

The available power levels of diffusion-cooled lasers range from a few milliwatts up to 10 kW. In this section, we distinguish between high- (>1 kW) and low-power (<1 kW) lasers; we also describe the most common electrode geometries for diffusion-cooled high-power lasers.

FIGURE 1.5 Typical low-power diffusion-cooled laser: The fans on top provide the air cooling. The RF generator is integrated in the housing, and electrical control interfaces enable the laser power and pulse frequency to be controlled.[16]

Sealed-Off Low-Power Lasers

Diffusion-cooled lasers below the 1-kW power level are usually sealed-off lasers, meaning that the laser active gas is filled into the laser cavity once in the factory and is not replaced at regular intervals at the customer's location. The typical lifespan of the laser gas fill is about 30,000 hours. To achieve such a long laser gas lifespan, special care is required when preparing the materials that come in contact with the laser gas and when determining the leak rates of the vacuum system (as discussed in Sec. 1.3). Models with power levels less than 300 W are air cooled and do not require chilled water. The discharge is powered by an rf-generator that is integrated in the laser head (Fig. 1.5). The excitation frequency ranges from 40 to 80 MHz.

Due to their compact design, sealed-off lasers can easily be integrated into production lines for marking and coding systems, which are two of the major applications for these lasers. Other applications include cutting very thin sheet metal and metal foils or cutting nonmetals, such as plastics, fabrics, ceramics, and wood.

High-Power Diffusion-Cooled Lasers

Diffusion-cooled lasers above the 1-kW power level usually operate in a near sealed-off mode, meaning that the laser active gas is replaced in regular intervals at the customer's location. The typical gas exchange cycle is 72 hours. The gas is supplied in a small bottle attached to the laser. To maximize heat extraction from the gas volume, geometries that have a large ratio of surface area to gas volume are chosen. The most popular designs are slab and coaxial geometries (Fig. 1.6).

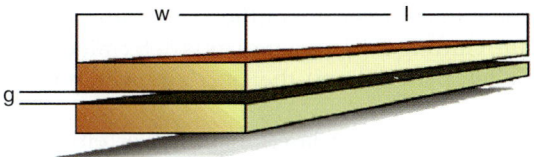

Figure 1.6 Typical geometries for high-power diffusion-cooled lasers: Slab-based design (left) and coaxial design (right). w: width; l: length; g: gap.

Typically, diffusion-cooled lasers with a slab geometry are also referred to as *waveguide lasers*. The interelectrode gap d is about 2 mm, which enables efficient cooling but also adds waveguide losses to the resonator losses. To minimize waveguide losses, the surface finish and positioning of the electrodes must be accurately controlled. Because coaxial designs have a π times larger surface area in the same footprint as a slab design, the interelectrode gap d can be increased by a factor of π without losing cooling capacity. The larger interelectrode gap size enables free space propagation and reduces resonator's internal losses. Because the electrodes are not part of the optical system of a resonator with free space propagation, the electrode surface finish and the positioning of the electrodes in relation to one another are less critical.

The stable-unstable hybrid resonators (Fig. 1.7a and b) generate beams that are not rotationally symmetric and that are therefore astigmatic and not usable for any application. A beam-shaping telescope is used to transform the astigmatic beam into a round, stigmatic beam. The beam quality post-telescope is about $M^2 = 1.1$

Excellent beam quality makes these lasers ideal for cutting sheet metal up to about a half inch thick. Cutting speeds in thin sheet metal

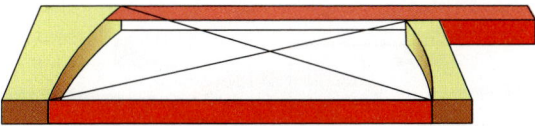

Figure 1.7a Stable-unstable hybrid resonator for a diffusion-cooled laser with a planar electrode structure.

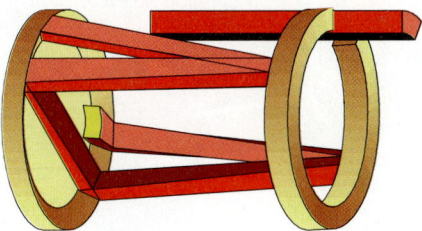

Figure 1.7b Stable-unstable hybrid resonator for a diffusion-cooled laser with a coaxial electrode structure.

are proportional to the intensity of the focus and not to the power of the laser beam. Cutting thicker materials is determined by the dynamics of the molten material and the laser's power, not by the laser's intensity.

Because diffusion-cooled lasers have lower efficiency than fast-flow lasers, the costs for cooling and pumping the gain medium per unit output power are higher. Since the costs of the other laser components are lower as compared with those for fast-flow lasers and because diffusion-cooled lasers have no moving parts (e.g., the turbo radial blower), the power level range at which fast-flow designs become more cost effective than diffusion-cooled lasers is about 3.5 to 4 kW.

1.4.2 Fast-Flow CO_2 Lasers

The most common design for industrial CO_2 lasers with power levels above 2 kW is the fast axial flow laser. This laser has been the workhorse of the industry and has revolutionized sheet metal processing. Power levels up to 20 kW are available as standard products, and lasers with up to 100-kW laser power have been built for special projects.

The rf power in fast-flow lasers is delivered to the gas discharge through electrodes that are attached to quartz tubes in which the discharge runs. To keep the laser gas cool enough to maintain efficient laser operation, the gas is removed from the discharge area, cooled in heat exchangers, and returned to the discharge area by a radial turbine blower (Fig. 1.8). Before the laser gas reenters the discharge area, the compression heat generated by the radial turbine blower is removed in a heat exchanger. The laser power scales with the amount of heat that can be removed from the laser gas (gas volume flow $\dot{V}$) and the

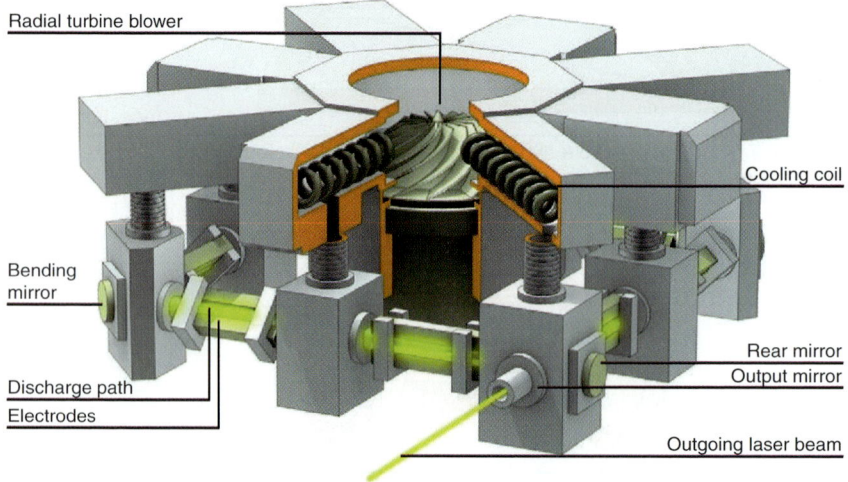

FIGURE 1.8 Multikilowatt fast-flow laser. (*Source: TRUMPF*[20])

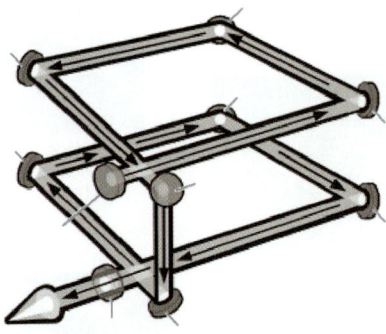 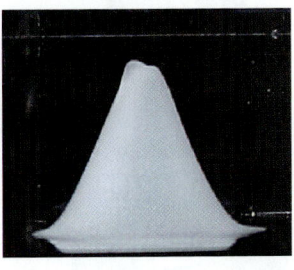

FIGURE 1.9 Left: Folded stable resonator of a multikilowatt fast-flow laser. Right: Intensity distribution of the outcoupled beam.

properties (f) of the gases, with f representing the degrees of freedom of the molecules, T the gas temperature and p the gas pressure:

$$P_L \propto pV(f-2)/T$$

The laser shown in Fig. 1.8 uses a stable resonator with a resonator length of about 6 m. To keep the laser compact, the optical resonator is folded (see Fig. 1.9). It generates a rotationally symmetric beam, and the modes are typically TEM_{00} or TEM_{01}. The output coupling degree is in the range of 40–60 percent. These lasers are available with different beam qualities to provide the best beam characteristics for different applications.

1.5 Applications

Typical applications of fast-flow and diffusion-cooled high-power lasers are shown in Fig. 1.10. The largest application by far is laser cutting; other applications include laser welding and surface modification. A detailed introduction to material processing with lasers is given in the Laser Institute of America's *Handbook of Laser Materials Processing*.[17]

For laser-cutting applications, most high-power lasers are integrated into two-dimensional flatbed cutting machines, while five-axis machines are used for three-dimensional applications. The versatility of the CO_2 laser enables four different cutting processes:

1. Sublimation cutting is used for nonmetals. Unlike the other processes, material is vaporized rather than melted. As a result, no assist gas is required to blow the molten material out of the kerf.

2. The most common process, used with mild steel, is oxidation cutting. As shown in Fig. 1.11, oxygen is the assist gas, and oxidation in the kerf provides additional heat that increases the cutting

14 Gas, Chemical, and Free-Electron Lasers

Deep penetration welding: 8–15 kW

Precision cutting and welding of automotive parts: 1–8 kW

Surface modification: 6–15 kW

Continuously welded tubes and profiles: 1–15 kW

Sheet metal cutting: 1–6 kW

FIGURE 1.10 Typical CO_2 laser applications in the multikilowatt range. (*Source: TRUMPF*)

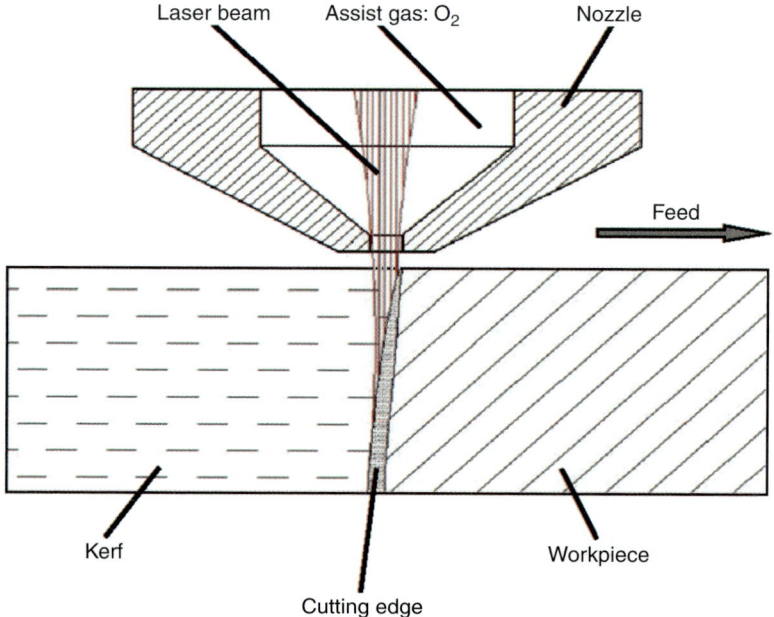

FIGURE 1.11 Oxygen is the assist gas in oxidation cutting. The oxidation reaction provides part of the heat making the cut. In fusion cutting, the assist gas is nonreactive and merely blows the molten material out of the kerf.

speed. If the finished kerf needs to be painted after cutting, the oxidation layer must be removed in a secondary process.

3. Fusion cutting uses a nonreactive assist gas; thus, all the energy comes from the laser itself. Mild steel, aluminum, stainless steel, and most alloys can be cut with this process. Because there is no oxidation layer, the workpiece can be painted or welded without further processing.

4. High-speed cutting, also known as *laser plasma cutting*, is a modified version of fusion cutting. A small pocket of vaporized material forms within the kerf, enhancing absorption. This process requires relatively high intensities, resulting in poor cut quality when compared with fusion cutting.

References

1. Patel, C. K. N. "Selective Excitation Through Vibrational Energy Transfer and Optical Maser Action in N_2-CO_2," *Phys. Rev. Lett.*, 13: 617–619, 1964.
2. Witteman, W. J. *The CO_2 Laser,* Springer Verlag, Berlin, 1987.
3. Cheo, P. K. *Handbook of Molecular Lasers,* Dekker, New York, 1987.
4. Raizer, Y. P. *Gas Discharge Physics,* Springer, Berlin, 1997.
5. Willett, C. S. *Gas Lasers: Population Inversion Mechanisms with Emphasis on Selective Excitation Processes,* Elsevier, 1974.
6. Hake, R. D., and Phelps, A. V. "Momentum-Transfer and Inelastic Collision Cross Sections for Electrons in O_2, CO, and CO_2," *Phys. Rev. Lett.*, 158: 70–84, 1967.
7. Novgorodov, M. Z., Sviridov, A. G., and Sobolev, N. N. "Electron energy distribution in CO_2 laser discharges," *IEEE Journal of Quantum Electronics*, QE-7(11): 508–512, 1971.
8. Laakmann, P., and Laakmann K. D. *Sealed-off RF-excited CO_2 lasers and method of manufacturing such lasers*, United States Patent 4, 393: 506, 1983.
9. Witteman, W. "High-Output Powers and Long Lifetimes of Sealed-Off CO_2 Lasers," *Appl. Phys. Lett.*, 11, 1971.
10. Macken, J. A., Yagnik, S. K. and Samis, M. A. "CO_2 Laser Performance with a Distributed Gold Catalyst," *IEEE J. Quantum Electron.*, 25: 1695-1703, 1989.
11. Heeman-Ilievva, M. B., Udalov, Y. B., Hoen, K., and Witteman, W. J. "Enhanced Gain and Output Power of a Sealed-Off RF-Excited CO_2 Waveguide Laser with Gold-Plated Electrodes," *Appl. Phys. Lett.*, 64: 673–675, 1994.
12. Smith, A. L. S., and Austin, J. M. "Dissociation Mechanism in Pulsed and Continuous CO_2 Lasers," *J. Phys. D: Appl. Phys.*, 7(2), 1974.
13. Malz, R., and Haubenreisser, U. "Use of Zeolites for the Stabilization of CO_2 Partial Pressure in Sealed-Off CO_2 Waveguide Lasers," *J. Phys. D: Appl. Phys.*, 24, 1991.
14. Center, R. E. "Vibrational Relaxation of CO_2 by O atoms," *J. Chem. Phys.*, 59, 1973.
15. McNeal, R. J., Whitson, M. E., and Cook, G. R. "Quenching of Vibrationally Excited N_2 by Atomic Oxygen," *Chem. Physics Lett.*, 16, 1972.
16. Universal Laser Systems. (Online) http://www.ulsinc.com/products/features/index.php, 2010.
17. Ready, J. F., and Farson, D. F. (eds.). *LIA Handbook of Laser Materials Processing,* Magnolia Publishing, 2001.
18. Vogel, H. *Gertson Physik,* Springer, Berlin, 1995.
19. Schulz, J. "Diffusionsgekuehlte, koaxiale CO_2-Laser mit hoher Strahlqualitaet," *Dissertation.* s.l. : RWTH Aachen, 2001. Bd. Dissertation.
20. TRUMPF: http://www.trumpf.com/en/press/media-services/press-pictures.html.

CHAPTER 2
Excimer Lasers

Rainer Paetzel

Coherent GmbH, Dieburg, Germany

2.1 Introduction and Principle of Operation

The excimer laser is today's most powerful, cost-effective, and dependable pulsed ultraviolet (UV) laser source. Since its first experimental realization in 1970 by Nikolai Basov et al.[1] at Moscow's Lebedew Institute of Physics, it has undergone rapid development. The unique output characteristics of excimer lasers enable innovations in growth industries as diverse as the medical, microelectronics, flat panel display, automotive, biomedical devices, and alternative energy markets.

Excimer lasers are gas lasers that, by nature, emit pulsed UV light. The term excimer is short for excited dimer, a class of molecules formed by the combination of two identical constituents in the excited state. The excimer laser's active medium is a combination of a rare gas, such as argon (Ar), krypton (Kr), or xenon (Xe), with a halide, such as fluorine (F_2) or chlorine (Cl_2). Under the appropriate conditions of electrical stimulation, an excited molecule is created that exists only in an energized state and that gives rise to laser light in the UV range. The exact wavelength of the laser light depends on the gas mixture used.[2]

Of the listed excimer wavelengths in Table 2.1, five are commercially relevant—xenon fluoride (XeF) at 351 nm, xenon chloride (XeCl) at 308 nm, krypton fluoride (KrF) at 248 nm, argon fluoride (ArF) at 193 nm, and fluorine (F_2) at 157 nm. Of these wavelengths, the 308 nm, 248 nm, and 193 nm cover the vast majority of products and applications.

The fundamental principle of operation of an excimer laser is illustrated by a simplified reaction scheme for KrF, as shown in Fig. 2.1.

The formation of the rare gas halogen molecule is dominated by two reaction channels. In the ion channel, a positive rare gas ion (Kr+) and a negative halide ion (F–) recombine in the presence of a buffer gas,

H$_2$	Ar$_2$	F$_2$	Xe$_2$	ArF	KrCl	KrF	XeBr	XeCl	XeF
116 nm	126 nm	157 nm	176 nm	193 nm	223 nm	248 nm	282 nm	308 nm	351 nm

TABLE 2.1 Excimer Gas Media and Emission Wavelengths

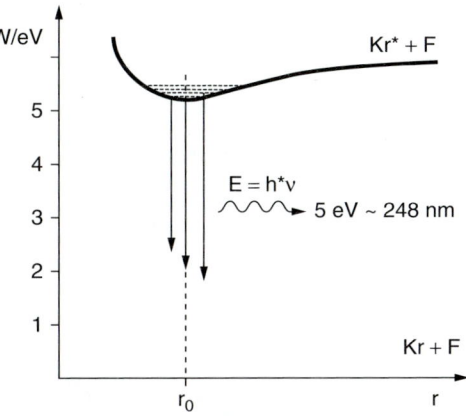

FIGURE 2.1
Schematic electron transition of a 248-nm excimer laser from the excited state to the ground state.

such as neon or helium. In the neutral channel, an excited-state rare gas atom (Kr*) reacts with a halogen molecule (F_2). These reactions take place on a nanosecond (ns) timescale, with upper-level production efficiencies of several tens of percent. The excited KrF* molecule remains unstable in the upper state and decays after several nanoseconds via emission of a photon into Kr and F. The Kr and F components that form the ground state are then available for another excitation cycle. Because the excitation rate must compete with fast quenching processes, collisions, and nonradiative decay, it requires high pump power densities, which can only be obtained in a pulsed system. Thus, excimer lasers intrinsically operate in the pulsed mode with high peak power.

2.2 Technology and Performance of Excimer Lasers

2.2.1 Principal Design and Technology

After more than 30 years of engineering, excimer lasers have reached high maturity. Several design aspects differentiate the excimer laser construction from that of other lasers. In addition, specific technologies have been developed to take advantage of the favorable operational conditions offered by the excimer laser. Because the laser's gas mixture contains one of the halogens fluorine or chlorine in small concentrations, it is of the utmost importance to select materials that will avoid consumptive reactions. Operation with relatively large gas volumes at gas pressures of up to 6×10^5 pascals (Pa) demands leak-tight, high-strength mechanical construction. Discharge voltages of more than 40 kilovolts (kV) are used for the excitation that determines the use of efficient high-dielectric-strength insulators. Performance thus relies on the excimer laser design, choice of materials, and production techniques.[3]

An important design factor of excimer lasers is the excitation method. The technology that is almost exclusively used in high-power industrial excimer laser systems is the high-pressure gas discharge

Gas, Chemical, and Free-Electron Lasers

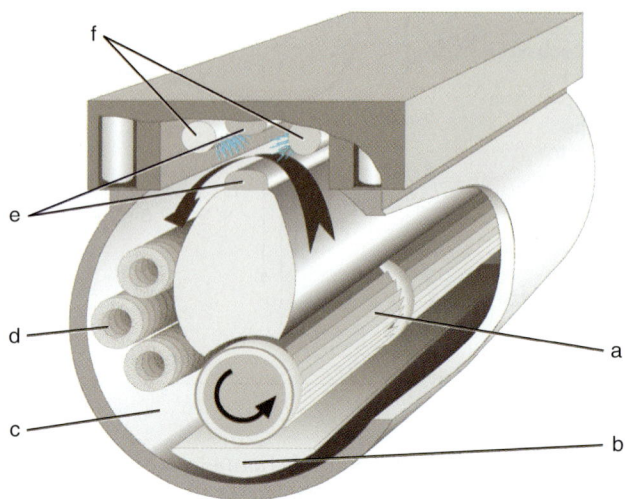

FIGURE 2.2 Example of an excimer laser tube with surface corona preionization. (*a*) Circulation fan, (*b*) electrostatic filter, (*c*) laser tube (pressure vessel), (*d*) heat exchangers, (*e*) electrodes, (*f*) surface corona preionization.

excitation. This method provides outputs of up to several joules and repetition rates in the kHz range. The discharge unit is integrated into the laser tube, which is designed as a high-pressure gas vessel (Fig. 2.2).

The laser gas mixture in the laser tube consists of a 0.05 to 0.50 percent halogen component, a 3 to 10 percent inert gas component, and the buffer gas (helium or neon) at a pressure of 3 to 6×10^5 Pa. Excimer lasers use short excitation pulses that terminate the discharge before the onset of instabilities, which leads to the typical short laser pulses of 10 to 30 ns.

Discharge Circuit

The technique to produce and control the homogeneous gas discharge is crucial for the performance of an excimer laser. The most important parts of this technique are the preionization of the laser gas, the discharge electrodes, the gas flow system, and the high-power discharge circuit.

In the context of an excimer laser, the term *preionization* means uniformly seeding the discharge volume with electrons and ions before initiating the main discharge. Sufficient electron density of 10^7 to 10^9 cm^{-3} is required to achieve a uniform glow discharge and to avoid instabilities. The electrode structures and preionization techniques used for excitation of the laser determine the cross section and quality of the discharge, as well as the laser's energy output and efficiency. Commercial high-power industrial excimer lasers usually employ either spark discharges or surface corona discharges for preionization of the laser gas. Specially designed excimers, as are used in fundamental laser studies, employ x-ray, or creeping discharge[4] for preionization or direct electron beam pumping for the main discharge to utilize a very large gain volume that achieves highest energies of up to hundreds of joules per pulse.

In a typical embodiment for preionization, a multitude of small preionization pins are arranged in a row with a dielectric surface adjacent to the discharge electrodes. Upon application of a fast voltage pulse, the pins act as small spark gaps, generating a surface-guided discharge about 10 ns before the main discharge. The UV radiation produced by the multiple discharges is sufficient to preionize large cross sections of the laser gas between the electrodes with a homogeneous initial seed density of at least 10^8 electrons/cm^3. In today's high-energy excimer laser designs, a dielectric material is placed between the preionization pins, and the resulting surface-guided discharge spreads over several millimeters instead of forming a very thin discharge channel. This design significantly reduces the consumption of the preionization pins, thereby enabling longer gas lifetimes and an electrode life of more than 10 billion pulses. Surface corona preionization (SCP) is typically used when high repetition rates using smaller discharge cross sections are required. SCP is the preferred design for low-energy lasers and lasers used in microlithography applications.

To remain within an optimum excitation energy density range, the active laser volume is scaled by means of the length of the electrodes, the gap of the electrodes, and the width of the discharge. In high-pressure gas-discharge lasers, the discharge electrodes are profiled to provide a highly uniform electric field distribution in the discharge region and to avoid field concentrations near the electrode edges, which would otherwise cause premature discharge instabilities and arcing. The electrode profile determines the discharge's maximum energy loading, as well as its width and profile, which in turn controls the profile of the laser beam. The electrodes must be able to withstand the adverse effects caused by the high-current discharge and must be made from a material that is chemically resistant to the fluorine or chlorine component used in the gas. Proprietary alloys have been developed that optimally meet the demands of fluorine or chlorine chemistry and thus minimize electrode erosion.

To provide high pump energy densities in a short time, as is required for population inversion, excimer lasers are usually pumped using a high-voltage capacitor circuit (pulser) that discharges the stored electrical energy directly into the active medium (Fig. 2.3). The pumping schemes involve efficient switching of this stored electrical energy into the discharge system in a very short time. Also required is a well-defined spatial and temporal profile, as this determines the discharge uniformity and, in turn, influences the extracted laser

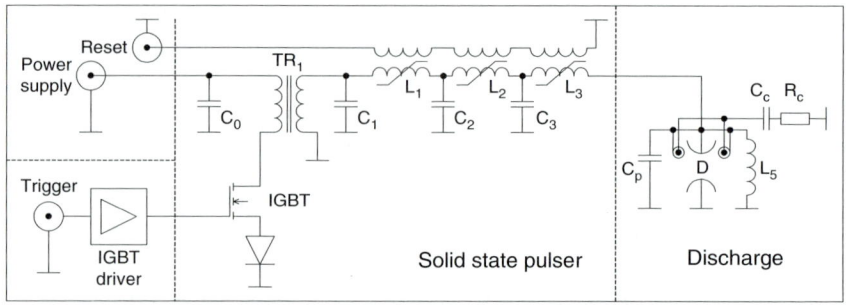

FIGURE 2.3 Example of a discharge circuit with a solid-state pulser and multistage compression. C_0: storage capacitor; IGBT (insulated gate bipolar transistor): solid-state switch; TR_1: transformer; C_1–C_3: magnetic compression circuit capacitors; L_1–L_3: magnetic compression inductors; Rc, Cc: circuit of corona preionizer; D: discharge electrodes (in laser tube); Cp: peaking capacitor, L_5: discharge coil.

beam's spatial uniformity. To obtain the required peak currents and voltage rise times, high-power excimer lasers use multistage pulse compression techniques and all-solid-state switching through modern semiconductor switches, such as thyristors, gate turn-offs (GTOs), or insulated gate bipolar transistors (IGBTs). Magnetic pulse compression circuits deliver the electrical energy to the laser cavity in multiple steps, as is done in a basic capacitor transfer (C-C transfer) circuit.

The primary solid-state switching circuit transfers the energy on a slow timescale from the primary energy store (C_0) to intermediate energy stores (C_1, C_2, C_3) in a multistage magnetic compression circuit. From there, the energy is rapidly transferred into the laser cavity for the discharge. The transformation of the pulse from a low-peak power to the fast high-peak power pulse required by the discharge is schematically shown in Fig. 2.4. The transfer from C_0 to C_1 uses a step-up transformer, which converts the primary charging voltage of C_0 to the required high voltage level of 20 to 40 kV in the secondary circuit. From C_0 to the final peaking capacitors (C_p), the pulse is typically compressed by a factor 50 to 100. The inductors L_1, L_2, and L_3 saturate after their hold-off time and then rapidly transfer the energy to the next stage. The saturable inductors are then reset to the original unsaturated state by the reset current that is actively supplied. An all-solid-state switching technology constitutes a major advancement toward highly reliable industrial excimer laser systems, because solid-state switches are maintenance free and have demonstrated a practically unlimited lifetime. Today all high-power industrial excimer lasers and high-repetition-rate lasers for microlithography applications use solid-state switching technology and routinely achieve maintenance-free operating times of several 10,000 hours.

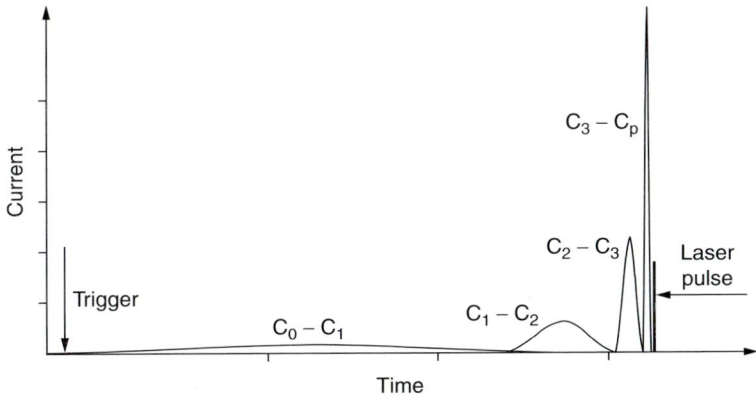

FIGURE 2.4 Compression of the discharge pulse from C_1 to discharge C_p.

For pulsed operation of gas-discharge lasers, the laser pulser's primary storage capacitor must be charged to the desired high-voltage level before energy transfer to the discharge. Resonant charging schemes are widely used in high-repetition-rate, high-power systems because of the high efficiency of direct current resonant charging (>90%). The advancement of switched-mode power supply (SMPS) technology makes this the method of choice for commercial excimer lasers.

Gas Circulation, Cooling, and Replenishment

Because the gas volume in the discharge area is no longer thermally homogenous after the discharge, it is completely exchanged between two successive laser pulses. A transverse circulation fan positioned within the laser tube completely replaces the gas volume between the main electrodes after each laser pulse, thereby providing a homogeneous gas flow over the entire electrode length. Between consecutive laser pulses, the gas exchange in the discharge must provide clearing of at least factor 2. Figure 2.5 shows the gas flow within the discharge region of a high-repetition-rate excimer laser.

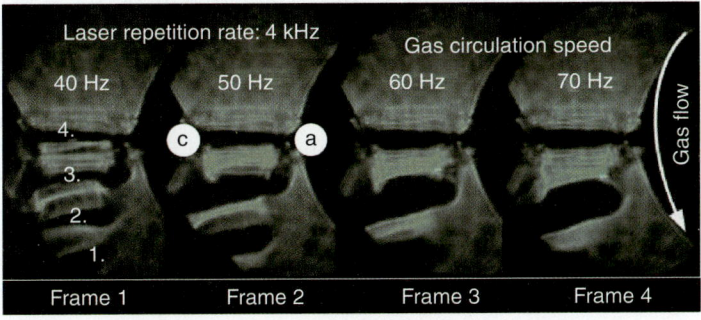

FIGURE 2.5 Principle of gas flow between electrodes.

The position of the electrodes—anode (a) and cathode (c)—is shown for reference in Frame 2 (Fig. 2.5). As shown in the figure, an expanded beam from a green helium-neon (HeNe) laser was directed longitudinally through the discharge; the picture frames were then recorded which synchronized with the discharge. The laser repetition rate was fixed to 4 kHz. The picture shows the variation of the gas speed with the driving motor set to frequencies of 40 Hz, 50 Hz, 60 Hz, and finally 70 Hz. The discharge region appears black because the change in its refractive index by the heated gas optically "blocks" the light. The 40-Hz setting (Frame 1) corresponds to a slow gas exchange speed, the gas volume of the discharge that is the 4th pulse and the leading pulses 3, 2, and 1 are seen. Because the spacing between the gas volumes is small, the laser action is affected by this disturbance of the gas and becomes unstable. With increasing gas flow speed, the clearing increases until, at 60 Hz, sufficient clearing and stable operation of the laser is observed. The typical gas circulation speed is about 25 m/s for high-energy lasers and up to 50 m/s for high-power and high-repetition-rate industrial excimer lasers. For high-power excimer lasers that use large-discharge cross sections and high repetition rates, clearing the gas volume between consecutive laser pulses becomes a demanding task. The built-in flow-loop system, which resembles a flow nozzle, optimizes the flow in the discharge region to avoid nonuniformity and flow separation on the electrode surfaces. Careful design of the gas flow-loop system using wind tunnel simulation allows gas speed and flow uniformity to be optimized to enable large cross sections and high repetition rates.

Excimer lasers typically operate with 2 to 4 percent conversion efficiency between the electrical input power and the UV output power. The surplus energy is removed efficiently as excess heat. The forced circulation in the laser tube brings the laser gas, heated by the laser discharge, to a heat exchanger, where it is recooled to the correct operating temperature. As with all gas laser cooling systems, efficient heat transfer between the laser gas and the heat exchanger represents a challenge. The heat exchanger, which in most designs uses water as a cooling medium in a closed- or open-loop system, needs sufficient contact area to provide high temperature stability, especially at high pulse repetition rates. On the other hand, the gas flow resistance across the heat exchanger must be small in order to be compatible with the cross flow fan characteristic. For optimum output, the laser tube windows must be protected against contamination from electrochemical erosion processes in the discharge. The laser tube window's outside surface is usually purged by dry pure nitrogen to remove all gaseous contaminations and impurities present in the environmental air. Consequently, purge systems have become a standard feature on all high-power and high-repetition-rate excimer lasers. In addition, active and passive contamination controls are necessary to keep the inside of the tube windows clean.

Window lifetimes of more than 10 billion pulses are achieved with optimized contamination control systems that enable 24-hour, 7-days-a-week excimer laser operation. Passive contamination control starts by selecting an enduring electrode material as well as suitable materials for the laser tube and its internal components. A thorough, controlled passivation procedure is applied to build up a halide layer on all internal parts of the laser tube and to avoid contamination build-up through reaction of the gas mixture's halogen components with the laser tube. Active contamination control is supported by electrostatic particle filtration and, in some cases, by cryogenic particle purification. In a typical embodiment, the pressure gradient generated by the circulation fan directs a fraction of the main gas flow toward the electrostatic or cryogenic precipitator. Driven by this pressure difference between intake and exit port of the precipitator, a steady gas flow through the device is achieved without any additional active fan or gas pump. Corona wires are used to charge the particles of the incoming laser gas. The gas flow speed within the precipitator is normally reduced to below 1 m/s to allow the charged particles to settle on the grounded purifier walls. The cleaned, particle-free gas is returned to the laser tube near the windows via baffle boxes, which are constructed like acoustic damping devices; these boxes are meant to create a turbulent-free gas volume in front of the laser windows and to prevent shock waves from transporting particles to the windows. Window lifetimes of 10 billion pulses and more are standard today in high-performance industrial excimer lasers with well-designed precipitation systems.

Ideally, halogen gas consumption due to electrode discharging is compensated for by halogen injections. Advanced self-learning replenishment algorithms add very small portions of halogen gas to the laser gas mixture without affecting the laser's energy stability during the injection phase. The replenishment rate depends on the laser's operating time, input energy, and performance parameters, such as the high voltage level or the temporal pulse width. The algorithms maintain the high voltage level and, therefore, keep all essential beam parameters stable throughout a period of up to one billion pulses with a single gas fill.

Laser Resonator

The typical resonator configuration for excimer lasers consists of planar optics. In this configuration, the rear mirror (RM) is a plane surface with dielectric coating that provides a high reflectivity of greater than 99 percent. The output coupler (OC) is also a plane mirror surface; the inner surface of the OC provides the reflectivity for the laser oscillator, whereas the outer surface is coated with a dielectric antireflection coating for optimum beam output (see Fig. 2.6). Depending on the excimer's wavelength and target energy, the OC's reflectivity can be as small as a few percent for a high-energy laser

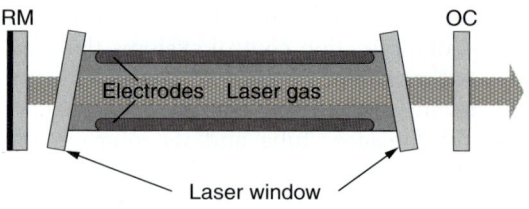

FIGURE 2.6 Planar resonator. RM: rear mirror; OC: output coupler.

using 248 nm or 308 nm or up to 50 percent for small lasers that operate at a lower energy regime.

The acceptance angle of the planar resonator is given by the geometry of the resonator; due to the short pulse length of typically 5 to 25 ns, there are only few roundtrips in the resonator. This leads to a multimode beam with a large beam cross section and a reasonably large beam divergence. Using the typical planar resonator, the excimer provides a beam divergence of 1 to 3 milliradians (mrad) and a large beam parameter product, which is calculated by beam size times beam divergence. For high-power excimer lasers, the beam parameter product is typically 50 mm·mrad. Although this is very different from other types of lasers, it has proven to be a fundamental advantage for many industrial large-area processing applications; the laser beam is considered to be a low-coherence source that avoids speckle and interference.

Figure 2.7 shows the beam profile of a high-energy excimer laser using the planar resonator for a typical excimer laser operating at 248 nm (KrF) and with a pulse energy of 1 joule (J). The measurement was taken with a standard beam profiler using beam attenuation and a charge-coupled device (CCD) camera. The beam cross section is 35 mm × 12 mm; typically the larger dimension is determined by the laser's electrode distance.

The energy distribution of the beam in this axis is a top-hat profile, which shows a plateau with high uniformity and symmetry. For many applications, this flat-top energy profile turns out to be very beneficial and yields uniformity in the working field without further beam homogenization. The profile in the orthogonal axis results from the discharge profile, which is mainly determined by the electrode gap, the electrode profile, and the operating parameters, such as gas composition and pressure. In particular, the electrode profile has evolved over the years to optimize the performance and lifetime of the different gases and the parameter range. This axis is approximated by a Gaussian beam shape.

For high-brightness applications of the excimer laser, the laser beam divergence may be reduced; for this, the acceptance angle within the resonator must be limited. For these high-brightness applications,

Excimer Lasers

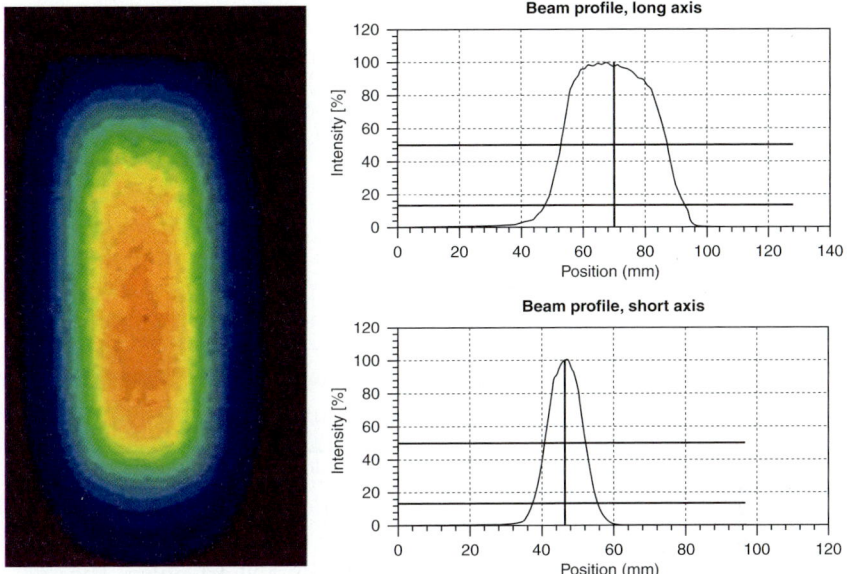

FIGURE 2.7 Beam profile of high-energy excimer laser at 248 nm and energy of 1000 mJ measured by CCD camera. Two-dimensional color display; one-dimensional intensity profile.

low-divergence resonators with curved optics have been developed and are used in various technical variants. In the basic concept, the resonator is made up of spherical curved mirrors that form a Cassegrain telescope with magnification M. The beam is expanded by the magnification factor within the resonator, and the output divergence is reduced (see Fig. 2.8). Practical values of M are in the range of 5 to 15, which leads to an increase in brightness of up to 2 orders of magnitude. Energy densities of more than 10 kJ/cm^2 are achieved in the focal spot. Variants of the low-divergence resonator use cylindrical optics that expand the beam only in one dimension and therefore reduce the laser beam divergence only for one desired beam axis. This is particularly useful for equalizing the beam parameter product of the laser for both axes or for achieving a highly focusable beam in one axis while leaving the other axis with high divergence.

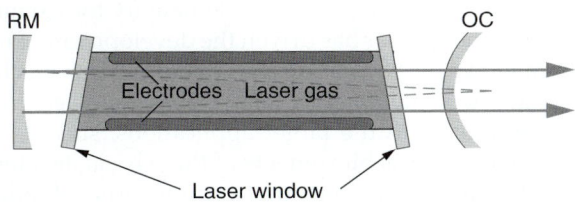

FIGURE 2.8 Low-divergence resonator. RM: rear mirror; OC: output coupler.

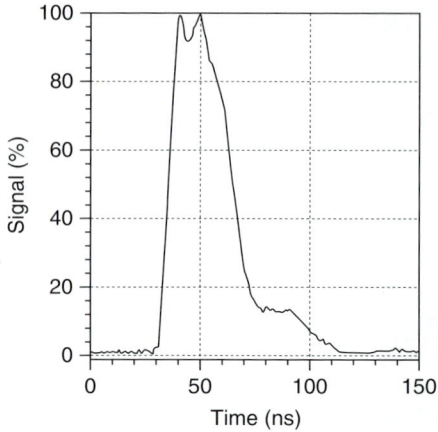

FIGURE 2.9   Pulse shape of KrF excimer laser operating at 248 nm and 650 mJ energy.

The excitation of the excimer is achieved by a short pulse. The resulting laser output is a pulse that starts after the laser threshold is exceeded and that then rapidly rises to its maximum intensity. A second and third maxima can be observed until finally all inversion is extracted within a few roundtrips in the resonator.

The typical output pulse of the 248-nm excimer laser is shown in Fig. 2.9 with a full-width, half-maximum (FWHM) pulse length of 22 ns. The pulse is modulated, and in this case, two peaks are seen. The separation between the peaks is 9 ns, which corresponds to the resonator length.

2.3   Excimer Laser Designed to Application

The development of excimer laser technology has been driven by several main applications. Each application poses different requirements on the laser to enable successful implementation in scientific, medical, and industrial fields.

2.3.1   High-Power Excimer Laser

High laser power in the UV region is the domain for the excimer, and the demand for higher power has driven the development of the excimer laser for many years. Several projects in the 1980s to reach multi-kilowatt output from the excimer laser were followed globally.[5] Although some interest in the target applications, such as isotope separation, has faded, the achievements of these basic developments are still utilized in the mature industrial excimer lasers of today. Typical output of 600 W is commercially available and proven in industrial operation. Development roadmaps show power levels of more

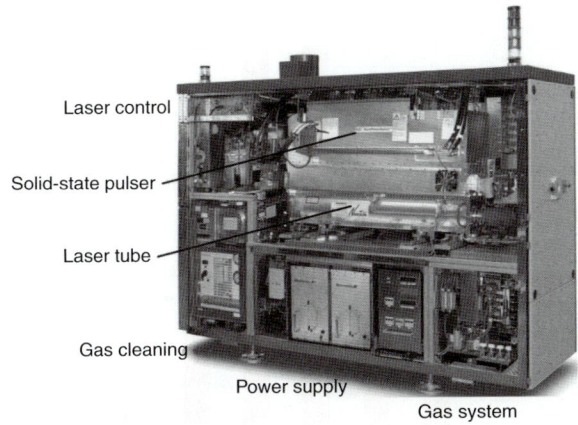

FIGURE 2.10   High-power industrial excimer laser.

than 1 kW are needed to achieve shorter takt times and high throughput for industrial applications.

As an example of a high-power industrial excimer laser, Fig. 2.10 shows an excimer laser from Coherent Inc. configured for 600-Hz operation. All laser modules are integrated into one laser chassis, which provides all utilities of gas, water, air flow, and electrical supplies and which serves as the laser tube's stable optical base. The center part of the excimer laser is the discharge unit, which comprises the laser tube, including the gas, and the discharge circuit. The laser uses solid-state switching in combination with magnetic pulse compression and voltage transformation, which eliminates routine maintenance of the excitation circuit. Maintenance costs are further reduced by an integral mechanical device that enables exchange of the laser tube without the pulser. High throughput (600-W power) and high stability (<0.5% root mean square), combined with long component lifetimes and a single discharge chamber configuration, significantly reduce operating costs.

For integration into the specific equipment and factory environment, laser control becomes more and more important. An integrated computer control with customized laser control boards and a real-time operating system manages all laser parameters on a shot-to-shot basis and actively stabilizes the energy, the time delay of the pulse, and other vital laser parameters. In addition, the controller provides fully automated data logging. Communication with the factory host via Ethernet protocol enables full integration into the manufacturing process.

The laser's peak-to-peak energy stability is shown in Fig. 2.11; the histogram is based on a test run of more than 62 million laser pulses. More than 99.999 percent of laser pulses are within the target energy window of ±1.5 percent. This stability is maintained under

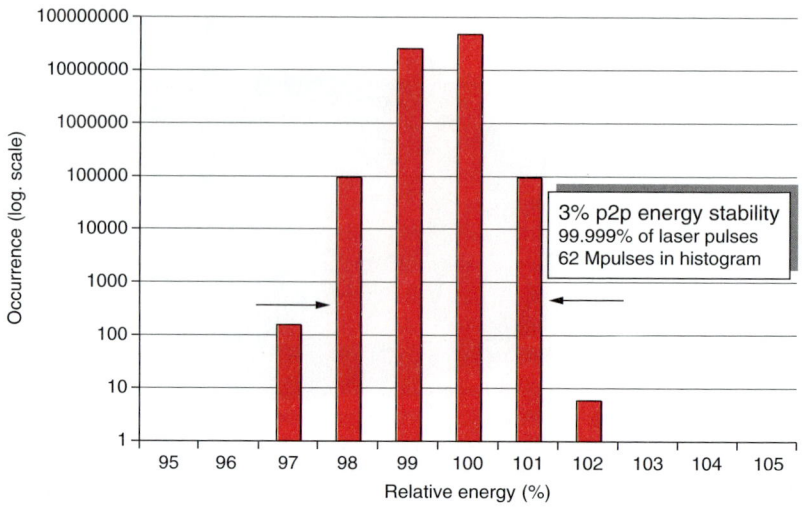

FIGURE 2.11 Histogram of peak-to-peak energy stability measured at 600-Hz, 540-W output power at 308-nm (XeCl) wavelength.

production conditions over a two-day working period to enable the uninterrupted production of the system. In the context of an annealing application, which is sensitive to the exact energy of a single laser pulse, the achievement of this very tight energy distribution is one of the key requirements to reach a high yield.

Since the first commercial introduction of the excimer laser in 1975 by Lambda Physik, its performance has been rapidly improved. Pulse energies have been increased, such that special lasers today deliver up to 10 J per pulse. In addition, repetition rates, as used in advanced microlithography, have increased up to 6 kHz. The tremendous improvement of excimer laser performance is best seen in the development of the available output power over time. In 1976, the first commercial products delivered up to 2 W of UV output using the most efficient excimer gas (KrF) with emission at 248 nm. Until 1985, these lasers were used solely for scientific applications, such as photo-ionization, UV chemistry, and spectrometry. In 1985, the power level of 100 W became commercially available; these lasers used the XeCl gas mixture with emission at 308 nm and started the industrial application of the excimer for polymer ablation and other materials processing.

Soon after power levels were further increased and stabilized. The 300-W power level, which became available in 2000, led to wide adoption of the excimer in the flat panel display industry. Only part of the widespread industrial applications of the excimer demand very high power in excess of 100 W. Figure 2.12 shows the roadmap of the output power of a commercial excimer laser. In 2010, the

Excimer Lasers

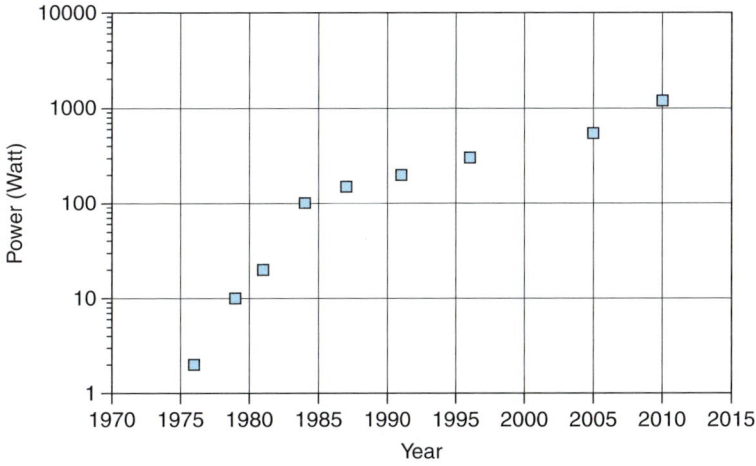

FIGURE 2.12 Power roadmap of commercial excimer laser.

power level reached 1200 W, and higher powers are on the roadmap to maximize the throughput and overall economics of industrial applications.

2.3.2 Microlithography

The excimer's deep UV (DUV) wavelength has a substantial advantage over the 365-nm I line of mercury lamps; this wavelength has allowed the excimer to achieve smaller features and as such has helped drive the evolution of large-scale integrated circuits, such as microprocessors and memory chips. The excimer laser, with its 248-nm wavelength, was chosen for mainstream microlithography in the early 1990s and, at that time, was considered for 250-nm features. These lasers are used for microlithography applications because they can deliver a very narrow spectral width that enables the high-contrast performance of the stepper lens as well as high power at a high repetition rate. Today after more than 20 years deployment of microlithography scanners based on 248 nm (KrF), the most advanced machines are using 193 nm (ArF) to reach the 65 nm, 45 nm, 32 nm, and 22 nm design nodes.

To achieve a high-resolution image, the laser's wavelength must have a very narrow spectrum in order to avoid color aberrations. Special line-narrowing schemes have been developed that reduce the output spectrum from the natural width of about 0.5 nm down to 0.1 picometer (pm). For the advanced 193-nm excimer laser used in microlithography, dual-chamber systems have become the standard. In these systems, the narrow-output spectrum is achieved by inserting dispersive elements into the resonator of the oscillator. The line-narrowing module typically consists of a prism beam expander

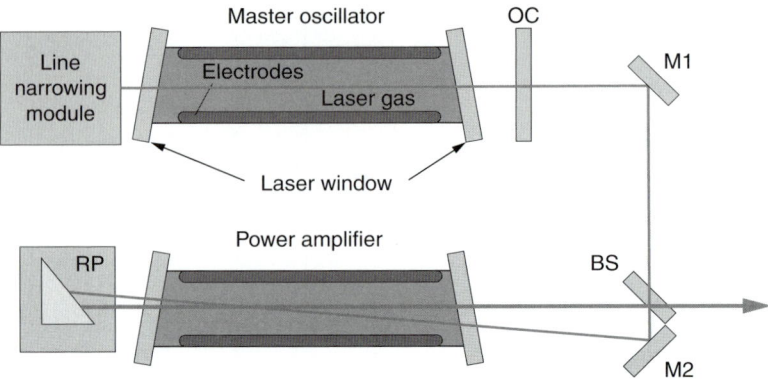

FIGURE 2.13 Typical dual-stage 193-nm laser for microlithography using a ring architecture. OC: output coupler; M1, M2: mirrors; BS: beam splitter; RP: retarder prism.

combined with a blazed grating in Littrow condition. The low output of this oscillator is then amplified in the second chamber to reach output powers of 90 W at a repetition rate of 6 kHz. The specific combination of the two chambers can be a straight master oscillator power amplifier (MOPA) configuration or it can use a ring geometry in the amplifier stage (see Fig. 2.13).

The master oscillator emits a low-power beam with a narrow line width, as determined by the line-narrowing module. The oscillator's output is then directed toward the amplifier by mirror M2 and gets amplified. The retarder prism (RP) deflects the beam, which then gets amplified to the full output. The beam splitter redirects a small portion of the output beam into the amplifier for a second and third loop of amplification.[6,7]

2.3.3 LASIK

LASIK (laser-assisted in situ keratomileusis) is the dominant photorefractive procedure employed worldwide to correct vision by direct ablation of corneal tissue. In 1983, Trokel and Srinivasan of IBM started their pioneering work on photorefractive surgery using the 193-nm excimer laser.[8] Ever since, this application has driven excimer laser development to provide extremely compact lasers that (1) are optimized for ArF operation at 193 nm, (2) meet all stringent requirements of the medical device regulations, and (3) provide long, maintenance-free operation and simplicity as demanded by medical applications. At the early stage of photorefractive eye surgery, larger lasers with higher energy were used. Today's trend, however, is to use small energies in the range of 3 to 5 mJ/pulse and repetition rates between 200 and 1000 Hz; these ranges, in combination with high-speed precision scanners, provide a short treatment time. A typical

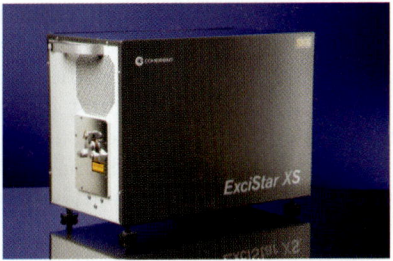

FIGURE 2.14   Compact excimer laser for refractive surgery (LASIK).

tabletop excimer laser, the ExciStar XS from Coherent Inc., is shown in Fig. 2.14. This compact model measures only 650 × 300 × 410 mm and easily integrates into medical systems. The laser delivers 5-mJ pulse energy that is stabilized by a built-in energy monitor and a feedback-loop with a typical repetition rate of 500 Hz. Long lifetimes of the gas and components lead to maintenance-free operation of more than one year.

2.4   Application of High-Power Excimer Lasers

The unique beam properties of today's excimer lasers allow these lasers to transform an unspecific material layer into a high-value functional surface. Representing today's most cost-effective and dependable pulsed UV laser technology, excimer lasers enable innovations in diverse growth industries, including the medical, microelectronic, flat panel display, automotive, biomedical device, and alternative energy markets. The combination of two fundamental aspects—wavelength and peak energy or peak power—determines the excimer laser's unique value, adding potential in high-tech industries that more than ever must balance product size, efficiency, and performance demands with process speed and production costs.

The unique interaction of pulsed short UV with materials such as polymer and poly(methyl methacrylate), or PMMA, was studied by Srinivasan as early as the 1980s.[9] The use of a high-intensity beam with photon energy that is higher than the bond energy of the substrate material (e.g., 5 eV for 248 nm) allows for the unique ablation mechanism of UV light and excimer lasers. From these studies, the term *cold ablation* was generated to describe the removal of polymer material by breaking chemical bonds rather than thermal decomposition. As the name indicates, there is minimal effect on the surrounding material due to the strong absorption of the UV laser radiation. The application of the 193 nm laser for photorefractive surgery is a direct result of these studies and has driven the development of the compact, low-powered excimer laser for the popular LASIK application.

The optical resolution that is achievable in laser material processing is proportional to the laser's wavelength. The short wavelengths of excimer lasers enable them to be among the most precise optical processing tools on the market. Depending on laser wavelength, material, and optical system, excimer laser–based material processing tools achieve feature sizes of 1 µm and smaller. The advantage of micromachining with a short UV wavelength is shown in the following equation for minimum feature size (MFS):

$$\text{MFS} \approx k_1 * \frac{\lambda}{NA}$$

where k_1 is processing factor, λ is wavelength, and NA is numerical aperture. With a practical numerical aperture of 0.12 and a process factor of 0.5, the 248-nm excimer laser yields a resolution of 1 µm.

Moreover, whereas the short wavelength translates into smaller lateral structures, it is the strong material absorption of the correspondingly high photon energy (e.g., 5.0 eV for 248 nm or 6.4 eV for 193 nm) that translates into very limited vertical material impact. In fact, the depth resolution of excimer laser thin-layer material processing is in the submicron range and can be as small as 50 nm/laser pulse depending on the material sample and the wavelength.

The short wavelength is directly absorbed by "transparent" materials, such as glass, quartz, Teflon, or transmissive conductor oxide (TCO) films, allowing UV laser radiation to directly interact with the small defined absorption volume and to minimize bulk heating effects. This advantage has led to a set of successful applications in which material processing with minimum heat effect is demanded. Figure 2.15 shows an example of the excimer ablation in glass. The crater's flatness indicates the high homogeneity of the excimer laser beam over the entire illumination area.

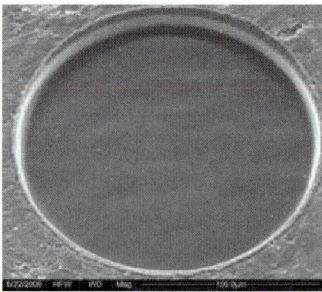

FIGURE 2.15 Excimer laser ablation pit in NIST (National Institute of Standards and Technology, Gaithersburg Maryland, USA); glass obtained at 193 nm after 50 laser pulses.

2.4.1 High-Resolution Micromachining

The unique ablation characteristics of pulsed UV lasers enable precise control of the spatial resolution—down to 1 μm in three dimensions. To take full advantage of the optical resolution, a typical setup employed for micromachining tasks using excimer lasers is used (Fig. 2.16). The output beam of the excimer is shaped by anamorphic telescopes and further homogenized by an optical beam homogenizer to illuminate the mask. The pattern that is to be "etched" onto the substrate is then determined by the mask pattern. With typical energy densities of 500 to 2000 mJ/cm^2, ablation rates of 0.1 to 0.4 μm/pulse are achieved in typical materials such as polyimide, positron emission tomography (PET), polyether ether ketone (PEEK), and Parylene. With this setup, the spatial resolution is mainly determined by the wavelength, the optical system, its numerical aperture, and aberrations. For a practical numerical aperture of 0.05 to 0.2, a resolution of 10 μm to less than 1 μm is typical for excimer applications. Thin films—such as indium tin oxide (ITO), which is typically used as TCOs in displays; silicon nitride; or Parylene buffer layer, which is typically used in microelectronics and display fabrication—are ablated with a single shot.

The ablation of thicker material is achieved by imposing multiple pulses onto one spot, which gives precise control of the ablation depth to a few 100 nm. The products manufactured by precision micromachining using the excimer laser cover various industries and products; the ongoing trend toward miniaturization drives new opportunities for microelectromechanical systems (MEMS), medical devices, and electronic components.[10–12]

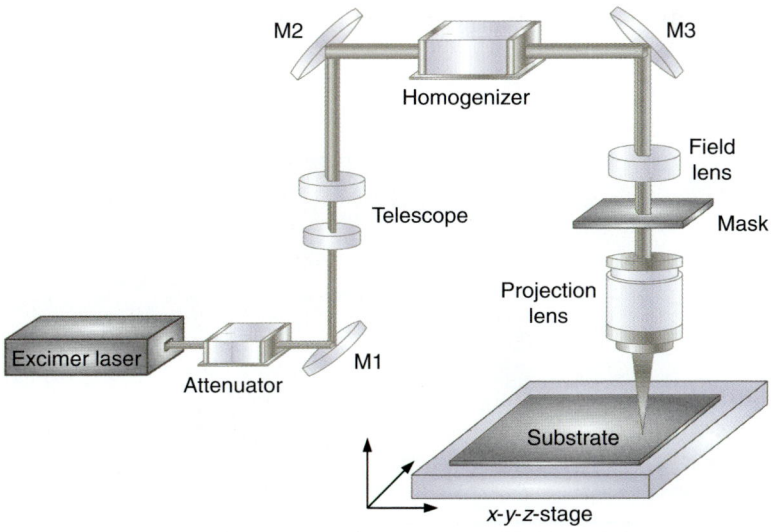

FIGURE 2.16 Typical layout for excimer micromachining system.

Ink-Jet Nozzle Drilling

One of the relevant industrial applications is the drilling of high-precision nozzles for ink-jet printers or dispensers of liquids and pharmaceuticals. Such nozzle arrays are typically fabricated on polyimide film with film thickness of about 10 µm. Depending on the specific application, the nozzles vary in size from 50 µm down to 2 µm and often demand a high precision on roundness, diameter on entrance and exit, and tight control of the taper angle. The tolerance of the hole size can be as small as 1 percent, and the variation on the taper angle can be less than 1 degree. The 248 nm laser with energy density of typically 1000 mJ/cm^2 has proved to be an excellent laser for achieving reproducible ablation within the tight tolerances demanded by 24-hour, 7-days-a-week industrial manufacturing conditions. Following the typical setup (Fig. 2.16), a complete nozzle plate, similar to that shown in Fig. 2.17, with as much as 300 nozzles is fabricated simultaneously. With an output power of 100 W, a high productivity is reached.

Three-Dimensional Patterning

To achieve dense, repeating, three-dimensional structures on a large scale, the synchronous image scan (SIS) should be used.[13] In this technique, the contour of the desired feature is sliced into multiple layers in the axis of the ablation z. The ablation per pulse is in the range of 0.1 to 0.2 µm. In this process, the substrate moves continuously during pulsed laser triggering, so that coincident with each projected laser pulse, the substrate moves by exactly one image repeat pitch. Changing the mask synchronously with the pulsing of the excimer laser yields the exact ablation pattern for each layer and results in the desired three-dimensional contour on the substrate.

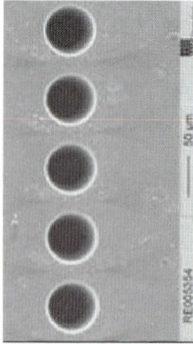 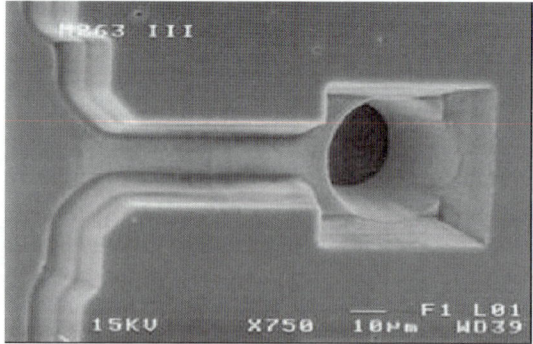

FIGURE 2.17 Nozzle array and single nozzle with flow channel. (*Source: LEXMARK*)

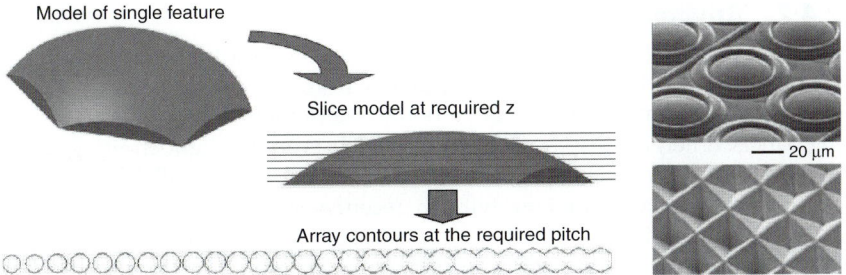

FIGURE 2.18 Principle of synchronous image scan (left) and examples of feature achieved by SIS (right).[13]

By using product-specific masks, large areas are machined with a repeating pattern to create master molds. The advantage of this method is that the high power of excimer lasers can be applied to produce large areas and a large quantity of products. Microlenses and trapezoidal three-dimensional structures in PMMA are shown in Fig. 2.18 as examples of applications produced with the SIS technique.

Direct Patterning of Sensor and Circuits

The high energy of excimer lasers enables large fields to be ablated with a single pulse. For a typical energy of 1 J, the ablation of about 1 cm^2 is achieved at a useful energy density of 800 to 1200 mJ/cm^2 (see Fig. 2.19). A complete sensor or circuit area is covered in a single laser shot, and with the proper choice of laser parameters, the complete patterning is achieved. In this single-shot ablation process, thin films of typically 50 to 100-nm thickness of various materials, such as gold, copper, ITO, and SiNx (silicon nitride) are removed. The single-shot ablation process is ideally suited for roll-to-roll processing by applying the pulse "on the fly" while the substrate continuously moves on. The high productivity of this model allows sensors, such as those used in medical, pharmaceutical, and electronic applications, to be produced cost effectively.

FIGURE 2.19 Sensor circuits structured by excimer laser, before singulation.

2.4.2 Brighter Displays

Over the past 10 years, the global flat panel display industry has shown tremendous growth in all display segments, from small mobile phones and car navigation displays to large home entertainment and advertising panels. Emerging display technologies, such as organic light-emitting diodes (OLEDs) or displays based on flexible substrates, will further drive the industry's rapid growth. In recent years, low-temperature polycrystalline silicon (LTPS) has demonstrated its advantages through successful implementation in applications such as highly integrated active-matrix liquid crystal displays (AMLCDs) and, most recently, active-matrix organic light-emitting diode displays (AMOLEDs).

Advances in manufacturing equipment have enabled the display industry to take advantage of larger glass sizes and improved economies of scale. For LTPS, this progress is manifested in the excimer laser source—in particular, output power, shot noise, and the optical beam delivery system that delivers the light to the substrate for the controlled crystallization process.[14] The annealing process demands tight control of every single laser pulse, making pulse-to-pulse (p2p) energy stability an extremely important laser parameter. Advances in this area over the past few years have substantially increased the available energy, thus satisfying the increasing demand for higher yields on larger glass plates and the demand for LTPS backplanes for AMOLED. The excimer laser process transforms the low-electron mobility silicon into an enabling thin film of 50-nm thickness, which supports fast voltage switching of the thin film transistor (TFT) for high-resolution AMLCDs, as well as the required current driving of emerging AMOLEDs.

The multihundred-watt output power allows for fast large-area processing. Dramatically increased electron mobility of greater than 100 cm^2/volt-second (Vs), which is 2 orders of magnitude higher than the electron mobility of amorphous silicon layers, has been achieved. The polysilicon layer (Fig. 2.20) permits electrons to move more easily through its highly ordered lattices.

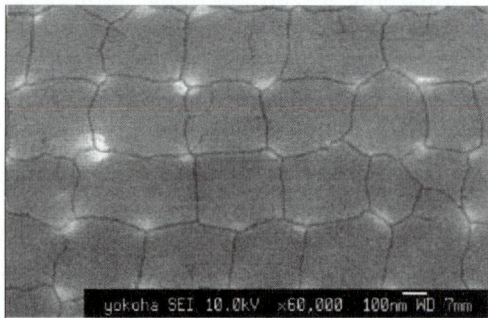

FIGURE 2.20 Highly ordered, polycrystalline silicon layer obtained after 308-nm excimer laser annealing and recrystallization.

With the increasing share of LTPS-based displays and AMOLED displays, manufacturing is adapting larger glass formats, such as Gen 5.5 (1300 × 1500 mm) for small notebook displays and Gen 8 (2200 × 2500 mm) for the cost-effective production of large-sized OLED television panels.[14] To meet the takt time requirements, the laser power is demanded to be greater than 1 kW and the line beam to be at least 750 mm in length.

Pulsed Laser Deposition

In the pulsed laser application, the excimer lasers high peak power UV output is used for the ablation of specific materials, which are then deposited on various other materials to produce unique thin films with tailored characteristics. In this way, a multitude of thin films can be deposited, with the advantage that the target material's stoichiometric properties are transferred to the substrate. The production of superhard coatings,[15] or diamond-like carbon (DLC), is one of the popular examples that has been applied to a wide range of industries. The emerging high-temperature superconductor (HTS) industry drives solutions ranging from magnetic energy storage to electrical energy transport grids operating at current densities 100 times higher than conventional copper-based systems. Technological advantages of using HTS-based systems, which are operable with liquid nitrogen cooling, include higher efficiencies, higher currents, higher power densities, and smaller weight and size as compared with conventional technologies. Figure 2.21 shows the amount of copper cable necessary to carry as much current as the small HTS tape containing a 1-µm-thick superconducting yttrium barium copper oxide (YBaCuO) layer. The future cost- and energy-saving potential of HTS are enormous, making these lasers a first-choice solution for breaking technological barriers. Pivotal to HTS commercialization are cost-efficient, high-performance thin-film deposition technologies.[14]

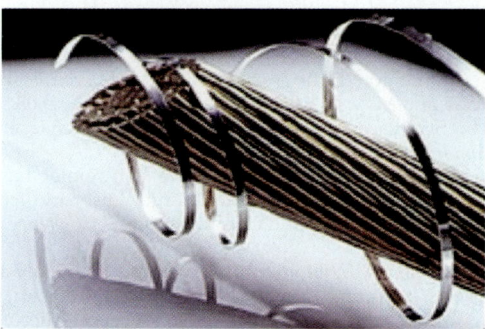

FIGURE 2.21 The thin high-temperature superconductor tape can carry as much power as the much larger copper wire.

Second-generation superconductor bands are built from a multilayer structure. The base is a stainless steel band that takes the mechanical forces onto which the multiple functional layers are deposited. The cerium oxide (CeO_2) buffer layer and the superconducting YBaCuO layer are typically produced by pulsed laser deposition (PLD) with 308-nm excimer lasers.[16] Within a vacuum chamber, the excimer laser ablates small portions from the target material (i.e., the YBaCuO) with each pulse. From the resulting plasma plume, a directed transfer of the target material is achieved, which is then deposited on the substrate. Tight control of the laser conditions leads to optimum ablation and a uniform deposition of the thin film.

The application range of high-power excimer lasers is steadily increasing, as driven by the industry trends toward miniaturization and cost-effective processing, including the elimination of wet chemistry. The excimer laser technology demanded by various applications and industries has been developed to a very mature stage. Future requirements of even higher power levels, in excess of 1 kW, and of further cost reductions in the laser process are the driving forces behind advancing excimer laser technology and paving the way for new opportunities.

References

1. Basov, N. G., et al.: "Laser Operating in the Vacuum Region of the Spectrum by Excitation of Liquid Xenon with an Electron Beam," *J. Exp. Theor. Phys. Lett.*, 12(S. 329), 1970.
2. Rhodes, C. D.: *Excimer Lasers—Topics in Applied Physics, Vol. 30*, 2nd ed., Berlin, Springer-Verlag, 1984.
3. Basting, D., and Marowsky, G.: *Excimer Laser Technology*, Berlin, Springer-Verlag, 2005.
4. Borisov V., et al.: "Conditions for the excitation of a wide-aperture XeCl laser with an average output radiation power of 1 kW," *Quantum Electronics*, 25: 408, 1995.
5. Godard, B., et al.: "First 1-kW XeCl Laser," *Proc. CLEO 93*, Baltimore, MD, 1993.
6. Yoshino, M., et al.: "High-Power and High-Energy Stability Injection Lock Laser Light Source for Double Exposure or Double Patterning ArF Immersion Lithography," *Optical Microlithography XXI*, eds. H. J. Levinson and M. V. Dusa, SPIE, Bellingham, WA, 2008.
7. Fleurov, V., et al.: "XLR 600i: Recirculating Ring ArF Light Source for Double Patterning Immersion Lithography," *Optical Microlithography XXI*, eds. H. J. Levinson and M. V. Dusa, SPIE, Bellingham, WA, 2008.
8. Trokel, S. L., Srinivasan, R., and Braren, B.: "Excimer laser surgery of the cornea," *Am. J. Ophthalmol.*, 96: 710–715, 1983.
9. Srinivasan, R., et al.: "Mechanism of the Ultraviolet Laser Ablation of Polymethyl Methacrylate at 193 and 248 nm: Laser-Induced Fluorescence Analysis, Chemical Analysis, and Doping Studies," *J. Opt. Soc. Am.*, 3: 785–791, 1986.
10. Herman, P. R., et al.: "VUV Holographic Gratings Etched by a Single F_2 Laser Pulse," *OSA Conference on Lasers and Electro-Optics*, Anaheim, CA, 1994.
11. Stamm, U., et al.: "Novel Results of Laser Precision Microfabrication with Excimer Lasers and Solid State Lasers," *1st International. Symposium on Laser Precision Microfabrication*, SOIE, Omiya, Saitama, Japan, 2000.
12. Paetzel, R.: "UV-Micromachining by Excimer Laser," *ICALEO*, 2005.

13. Abbott, C., et al.: "New Techniques for Laser Micromachining MEMS devices," *SPIE,* 4760: 281, 2002.
14. Herbst, L., Simon, R., Paetzel, R., Chung, S.-H., and Shida, J.: "Advances in Excimer Laser Annealing for LTPS Manufacturing," *IMID,* 2009.
15. Delmdahl, R., Weissmantel, S., and Reisse, G.: "Excimer Laser Deposition of Super Hard Coatings," *SPIE Photonics West,* 7581, 2010.
16. Usoskin, A., and Freyhardt, H. C.: "YBCO-Coated Conductors Manufactured by High-Rate Pulsed Laser Deposition," *MRS Bulletin*, 29(8): 583–589, 2004.

CHAPTER 3
Chemical Lasers

Charles Clendening

Technical Fellow, Northrop Grumman Aerospace Systems, Redondo Beach, California

H. Wilhelm Behrens

Fluid and Thermophysics Department Manager, Northrop Grumman Aerospace Systems, Redondo Beach, California

3.1   Introduction

Some chemical reactions are known to produce reactants whose nascent energy-level distribution is not in equilibrium in nature. Such chemical reactions could potentially provide a convenient energy source to produce the population inversions necessary to achieve lasing.

One important reason for the interest in chemical lasers is their use in mobile high-average-power systems. The most successful candidates have been the hydrogen halides, especially hydrogen fluoride (HF) and deuterium fluoride (DF), and the chemical oxygen iodine laser (COIL) devices. Both premixed and flowing mixing concepts have been developed. Megawatt-class continuous wave (CW) devices based on supersonic laser cavity gas flows have been demonstrated. There are also premixed devices that rely on electrically or photolytically driven initiation approaches; however, the initiators require power comparable to or greater than the laser output. Discussion of such primarily electrically initiated devices is very limited in this chapter.

There are several reasons for the interest in gas flow chemical lasers for use in high-energy laser (HEL) devices:

1. Chemical reactions supply the energy source.
2. Heat is removed continuously from the gain medium by the gas flow.

3. The supersonic gas flow's low density makes it relatively easy to minimize refractive index gradients and achieve acceptable beam quality characteristics.

The following sections provide a brief background to chemical lasers, followed by detailed discussions of the two most common chemical lasers—HF or DF and COIL. The chapter concludes with a short discussion of other chemical lasers that have been demonstrated to date.

3.2 General Background

In 1961, Penner and Polanyi[1] first suggested that chemically produced inversion could be used to create infrared lasing based on studies of low-pressure H + Cl_2 flames (low-pressure atomic hydrogen [H] and molecular chlorine [Cl2] flames). The first demonstration of such a chemical laser, which was conducted by Kasper and Pimental at the University of California at Berkeley in 1965,[2] consisted of a photolytically initiated hydrogen chlorine explosion.

Either exothermal chemical reactions directly produce the excited lasing species or reaction-produced excited species transfer energy to another laser species. In practice, however, very few such examples have proved to be scalable to high power. The set of such lasers could be significantly increased if one were willing to expand the definition of chemical lasers to include systems that rely on electrically or photolytically produced species to initiate the reaction chemistry or that actually supply a major reactant. An additional expansion would be possible if one were to include gas dynamic lasers (GDLs). GDLs use chemical combustion to produce hot gas mixtures in thermal equilibrium; they then expand those mixtures to supersonic conditions to exploit differences in molecular

Type	Example
Three-atom exchange	$F + H_2 \rightarrow HF^* + H$
	$O + CS \rightarrow CO^* + S$
Abstraction	$F + CH_4 \rightarrow HF^* + CH_3$
Photodissociation	$CF_3I + h\nu \rightarrow CF_3 + I(^2P_{1/2})$
Elimination	
Radical combination	$CH_3 + CF_3 \rightarrow HF^* + CH_2CF_2$
Insertion	$O(D) + CH_nF_{4-n} \rightarrow HF^* + OCH_{n-1}F_{3-n}$
Addition	$NF + H_2CCH_2 \rightarrow HF^* + CH_3C-N$
Photoelimination	$H_2C - CHCl + h\nu \rightarrow HCl^* + HCCH$

TABLE 3.1 Chemical Laser Classifications by Pimental[4]

relaxation rates in order to achieve population inversions. Pimental, who was a major early worker in the chemical laser field, identified several reaction types of chemical lasers (see Table 3.1).[3]

Noteworthy about the reactions in Table 3.1 is that most require that one of the reactants be a free radical or that an energetic photon (or electric discharge) be available to produce the reaction of interest. In practice, many of these reactions are "electric" lasers augmented chemically. For example, in the case of the photolytic iodine laser, the required ultraviolet (UV) photon is considerably more energetic than the resultant iodine atom–derived photon. In a few cases, it is possible to produce the radicals through purely chemical means, such as combustion-driven thermal dissociation of fluorine (F_2). It is also sometimes practical to minimize the number of required radicals by exploiting chemical chain reaction approaches.

More recently, one must also include net chemical reactions in liquid phase, such as

$$2H_2O_2 + 2KOH \rightarrow 2HO_2^- + 2H_2O + 2K^+ \tag{3.1}$$

Followed by

$$Cl_2 + 2HO_2^- \rightarrow O_2(^1\Delta) + H_2O_2 + 2\,Cl^- \tag{3.2}$$

This process produces the metastable excited oxygen electronic state, or the singlet delta; this state is the basis of COIL devices, which transfer energy from the singlet delta to iodine atoms. Other analogous excited metastable species of interest include nitrogen chloride (NCl) singlet delta and nitrogen fluoride (NF) singlet delta. Only the COIL devices manage to directly produce the excited species energy source, though in a sense even they involve a radical of sorts—the O_2H^- in solution.

This chapter's emphasis is on high-power CW mixing devices. HF and DF devices are taken as a general example and treated in some detail, and then COIL devices are considered. Finally, very brief discussions of other types of chemical lasers are presented. Readers interested in more details than the brief treatments in this chapter should consult Gross and Bott,[4] Stitch,[5] Cheo,[6] and Endo and Walter,[7] all of which contain more detailed discussions and comprehensive reference lists.

3.3 Hydrogen Fluoride and Deuterium Fluoride Lasers

The hydrogen halide laser (HF and DF) was one of the early chemical lasers to be demonstrated and one of the few lasers of any type that could be scaled to megawatt-class average powers. These molecular lasers lase based on vibrational and rotational transitions. Figure 3.1 shows a specific example of a DF* laser hardware layout, including where the particular reactions occur. The layout is shown to scale, with an adjacent drawing of the laser mixing nozzles magnified by a

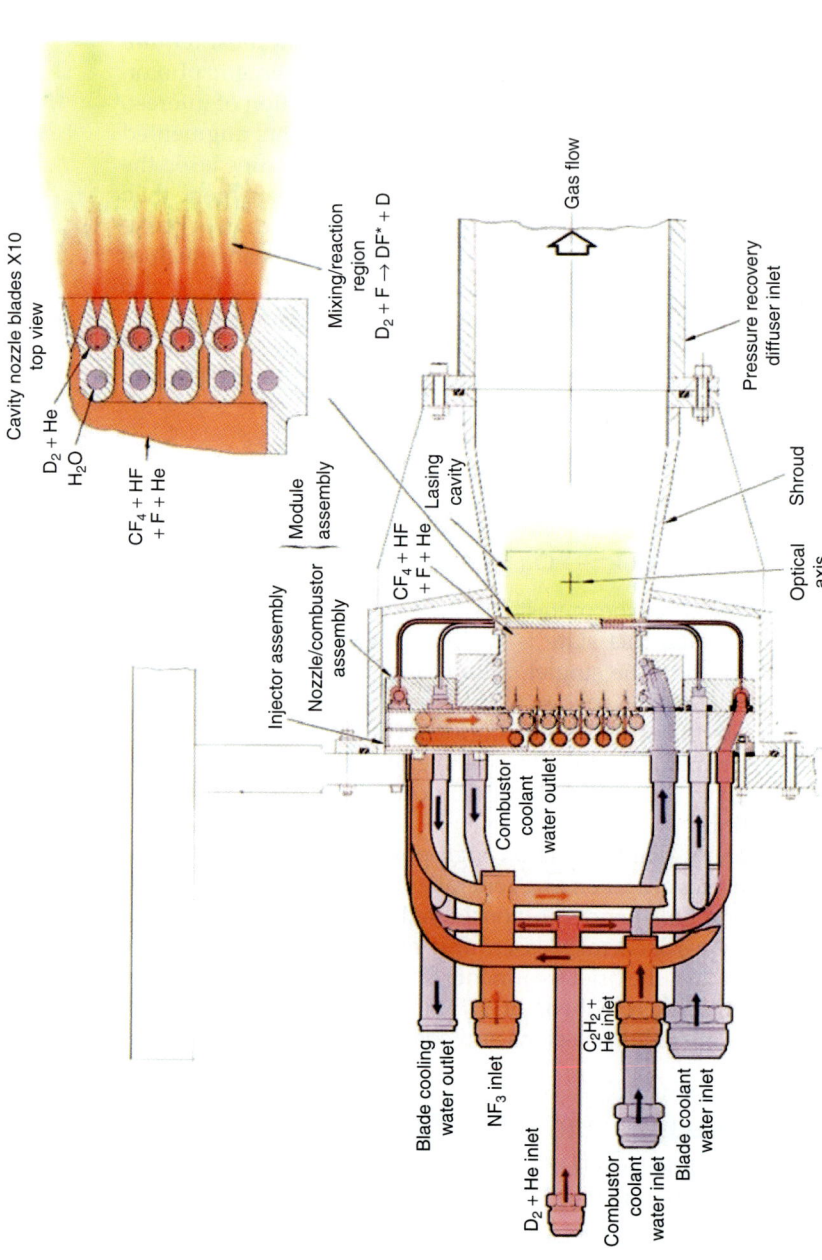

FIGURE 3.1 Deuterium fluoride (DF) laser module gas flow layout.

factor of 10. The reactions in the combustor and in the lasing cavity correlate with the reactions in Fig. 3.1.

An optical resonator then produces laser output based on the resultant gain. Figure 3.2 shows a typical high-power resonator system. The laser beam sizes are indicated in cross section at various locations. A key aspect of high-power chemical lasers is the use of an aerowindow to isolate the laser cavity (~10 torr) from the atmosphere. This isolation is typically achieved by creating a focus where the beam exits the cavity, as shown in Fig. 3.2.

The following discussion begins with a review of energy-level structure and small signal gain equations of HF and DF lasers; it then turns to population inversion generation and associated general fluid mechanic considerations. Finally, performance features associated with HF and DF laser operation are discussed.

3.3.1   Energy Levels

The lasing species in hydrogen fluoride and deuterium fluoride systems are the diatomic molecules HF and DF, respectively. The molecules are in their electronic ground state, and the energy levels of interest are the vibrational and rotational levels of the molecules. General diatomic molecule behavior is illustrated in Fig. 3.3. The possible motions of diatomic molecules can be divided into three primary components:

1. *Translational motion of the center of mass:* This motion, which can be treated as classical, is generally well characterized by a local static temperature. It can be ignored when considering laser transition energies, except for small secondary effects, such as Doppler broadening.

2. *Rotational motion about the various axes:* Quantization is required and is based on a rotational energy term of the type $(1/2)I\omega^2$, where I is the moment of inertia and ω is the angular frequency. Because the moment of inertia is small for the axis passing through both atoms, it can be ignored. Rotation about the other two axes produces significant contributions. Neglecting higher-order terms, the associated quantum levels are $B_J \times J \times (J + 1)$, where B_J is the molecule's rotational constant and J is the rotational quantum number. In the absence of lasing, the molecules are nearly in thermal equilibrium at the translational gas temperature statistical mechanics which implies that the fraction of the population in the Jth level F_J is given by Eqs. (3.3a) and (3.3b):

$$(2J + 1)\, e^{[-B_J \times J \times (J+1)/(kT)]} / Z \qquad (3.3a)$$

where $\qquad Z = \Sigma (2i + 1)\, e^{[-B_J \times i \times (i+1)/(kT)]} \qquad (3.3b)$

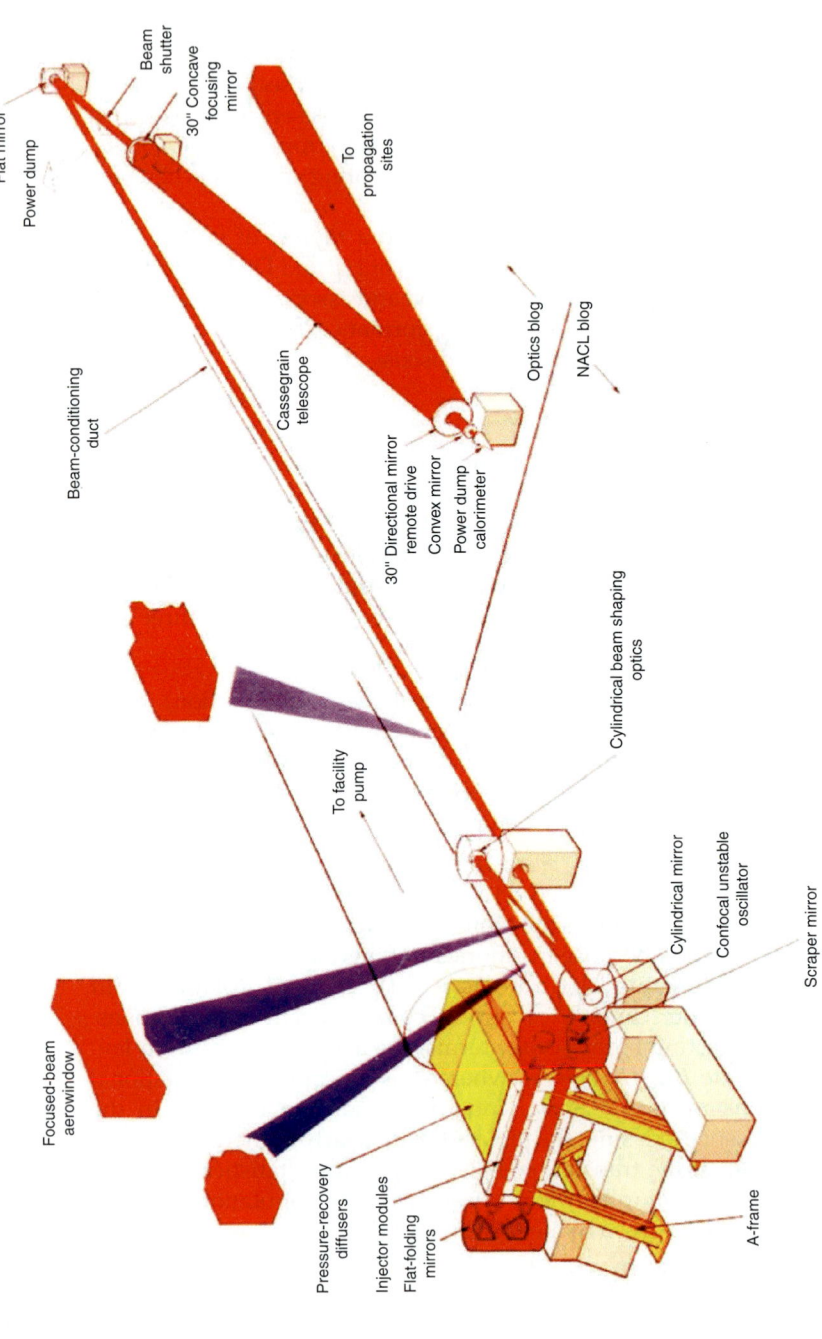

Figure 3.2 Example of a high-power DF laser resonator and beam path layout. The laser has two banks of nozzles stacked vertically.

FIGURE 3.3
Diatomic molecule motions that determine energy levels.

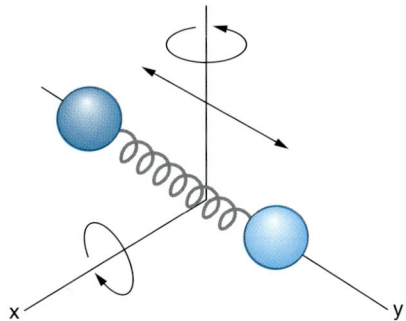

The partition function Z is approximately proportional to the temperature T and has a weak dependence on the vibrational level v, because B_J is also weakly dependent on v. When lasing, although deviations from this simple distribution can be taken into account in advanced models, the general trend remains.

3. *Vibrational motion:* Classically, vibrational motion can be viewed as a simple harmonic motion springlike mode. Quantum mechanically, it is characterized as a simple harmonic oscillator that sometimes includes anharmonic terms. The associated energy levels are simply:

$$\omega(v + 1/2) \tag{3.4}$$

where ω is vibrational quantum energy and v is vibrational quantum number.

In contrast to translational and rotational degrees of freedom, virtually all molecules are in the vibrational ground state (v = 0) in the absence of chemical pumping. When pumping or lasing is occurring, the assumption of a simple thermal behavior correlated with the translational temperature is invalid.

The rotational and vibrational energy level expressions were simplified and do not include the higher-order terms necessary to accurately determine energy levels. Equation (3.5) is a more accurate expression:

$$E(v, J) = \omega(v + 1/2) + X(v + 1/2)^2 + B_J \times J \times (J + 1)$$
$$+ B_{1J}(v + 1/2) \times J \times (J + 1) + \text{higher-order terms} \tag{3.5}$$

Typical values for HF and DF in the energy unit wave number are shown in Table 3.2.

		HF	DF
ω	vibrational level linear term	4138.73	3000.358
X	first anharmonicity correction	−90.05	−47.34
B_J	rotational constant	20.96	11.00
B_{1J}	rotational constant first correction	−0.7958	−0.2936

TABLE 3.2 HF and DF Energy Level Parameters. All Values are in Wavenumbers

The first term in Eq. (3.5) corresponds to the behavior of a simple harmonic oscillator with quantum number v. Allowable values for v are 0, 1, 2, ... The second term corresponds to a first-order anharmonic correction illustrating deviation from ideal harmonic behavior. The third term is associated with the rotational quantum number J and the associated rigid-rotor rotational energy. Allowable values for J are 0, 1, 2, ... The fourth term is the Coriolis effect correction to the rotational energy. More sophisticated expressions, including 16 terms or more, are also available. The interested reader can find a more detailed discussion of diatomic energy levels and spectroscopy in *Spectra of Diatomic Molecules*,[8] which also provides general information on diatomic molecule energy-level notation.

It should be noted that diatomic molecules containing hydrogen are atypical as compared with most single-bond diatomics. The former have rather large vibrational level separations, substantial amounts of anharmonicity, and very large rotational constants. This last feature is a major advantage from a laser standpoint.

The allowed molecular transitions for HF and DF are:

$$\Delta v = +/-1 \qquad (3.6a)$$

$$\Delta J = +/-1 \qquad (3.6b)$$

Moreover, the degeneracy of the $E(v, J)$ level is $2J + 1$.

In terms of energy levels, the primary allowed transitions are:

$E(v, J-1) \rightarrow E[(v-1), J]$, which is denoted as a P branch.

$E(v, J+1) \rightarrow E[(v-1), J]$, which is denoted as an R branch.

The P branch is the dominant lasing transition due to higher gain (see Sec. 3.3.2). Figure 3.4 shows energy-level diagrams appropriate for HF and DF lasers (note that vibrational levels are denoted by V as opposed to v).

Chemical Lasers

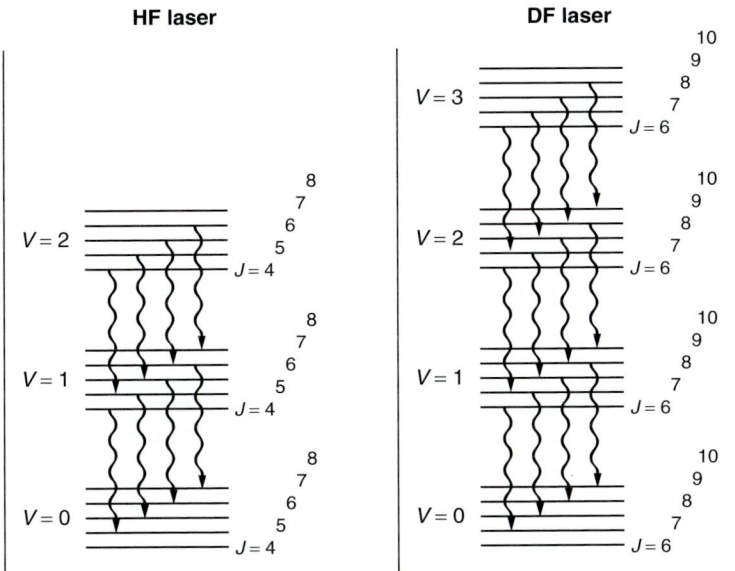

FIGURE 3.4 Energy-level diagram for representative laser transitions for hydrogen fluoride (HF) and DF lasers.

For the HF and DF P branch, the corresponding transition energies to the first order are as follows:

$$\omega_{photon} = 4138.73 - 180.1v + (2J - 1)B_e + \text{higher-order terms} \quad (3.7a)$$

$$\omega_{photon} = 3000.36 - 94.7v + (2J - 1)B_e + \text{higher-order terms} \quad (3.7b)$$

The corresponding wavelength range for HF CW lasers is 2.6 to 3.0 µm, whereas for DF lasers, it is 3.6 to 4.0 µm. Pulsed devices have a slightly wider range. The primary motivation for DF use as opposed to HF is to avoid atmospheric water band absorption. In the absence of deliberate line selection approaches, HF and DF devices usually lase simultaneously on multiple v-J transitions; Figs. 3.5 and 3.6 show typical spectra of HF and DF CW lasers, respectively. Atmospheric absorption information at these wavelengths can be found in Zissis and Wolf.[9]

3.3.2 Small Signal Gain

The general expression for small signal gain in a laser can be expressed as follows:

$$\gamma(v) = \{[N_U - N_L(g_U/g_L)] \times \lambda^2/(8\pi t_{spont})\}g(v) \quad (3.8)$$

where N_U = upper-level number density
N_L = lower-level number density

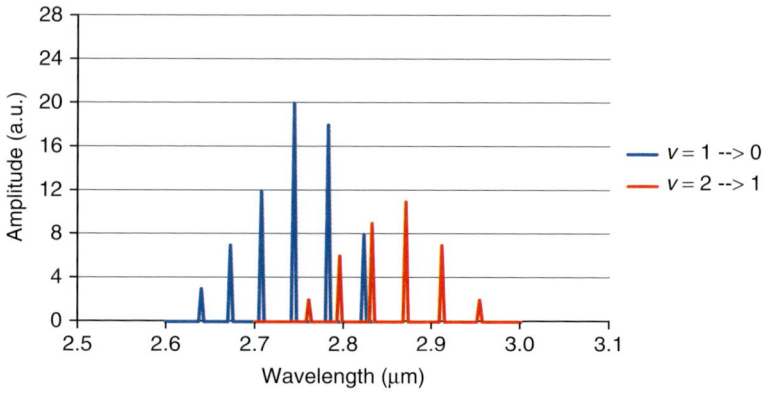

FIGURE 3.5 Typical lasing spectrum of HF continuous wave (CW) lasers.

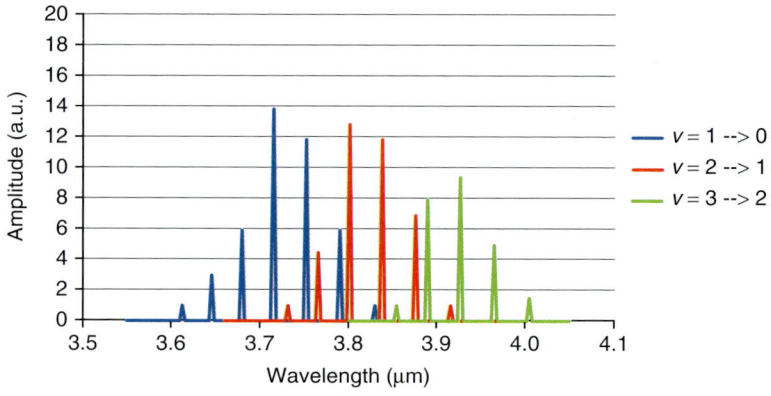

FIGURE 3.6 Typical lasing spectrum of DF CW lasers.

g_U = upper-level degeneracy factor
g_L = lower-level degeneracy factor
λ = laser wavelength
t_{spont} = upper-state spontaneous lifetime
$g(\nu)$ = normalized line shape

Limiting the discussion specifically to HF, we first consider the term in brackets and the transition $(v+1, J-1) \rightarrow (v, J)$. N_U can be expressed as the product of the total HF number density N and the fraction of these molecules in the $v+1$ and $J-1$ states.

One assumption that is frequently satisfied in the absence of lasing, and that is approximately satisfied during lasing, is that the rotational levels are in thermal equilibrium with an absolute temperature T

independent of vibrational level. Then, ignoring higher-order terms, thermodynamics tells us that the equilibrium rotational fraction for level J is given by Eq. (3.3a). Assuming the population of vibrational level v can be described as $F(v)$ and ignoring the v dependence of B_e and Z, the expression in brackets in Eq. (3.8) can be approximated as follows:

$$e^{(-J^2+J)B_e/(kT)}\{F(v+1) - F(v)[(2J-1)/(2J+1)]e^{-2JB_e/(kT)}\}/Z \quad (3.9)$$

For P-branch lasing, the exponential factor multiplying the second term allows one to achieve gain, even when the total population in vibrational level $v+1$—$F(v+1)$—is smaller than $F(v)$. This is called a *partial inversion* and the effect is substantial for HF and DF due to the large B_e values. Note that the analogous factor for R-branch lasing necessitates absolute inversion and increases the difficulty in achieving threshold.

The normalized line shape $g(v)$ includes line-width dependence. It is simple to calculate line width at very low pressures based on Doppler broadening. At line center, Eq. (3.10) applies.

$$g(0) = 2[\ln(2)/\pi]^{1/2}/\Delta v_D \quad (3.10)$$

$$\Delta v_D = 2v_0[2 \ln(2) \, kT/Mc^2]^{1/2} \quad (3.11)$$

where v_0 = centerline frequency
 T = absolute temperature
 M = molecular weight
 c = speed of light

At high pressures, which are not typical of CW devices but which are common to most pulsed devices, pressure broadening becomes dominant, and line width becomes inversely proportional to pressure. Taking into account pressure broadening for CW (as well as for pulsed systems) requires the use of Voigt functions, which combine pressure broadening and Doppler effects.[4] In practice, representative values of small signal gain are typically in the order of a few percent per centimeter in CW HF devices and are moderately higher in pulsed devices.

3.3.3 Chemically Excited Species Generation

The chemical reactions used to produce vibrationally excited HF are given in Eqs. (3.12) and (3.13), which are referred to as the cold and hot reactions, respectively.

$$F + H_2 \rightarrow HF^* + H + 31.5 \text{ kcal/mol (cold)} \quad (3.12)$$

$$H + F_2 \rightarrow HF^* + F + 98 \text{ kcal/mol (hot)} \quad (3.13)$$

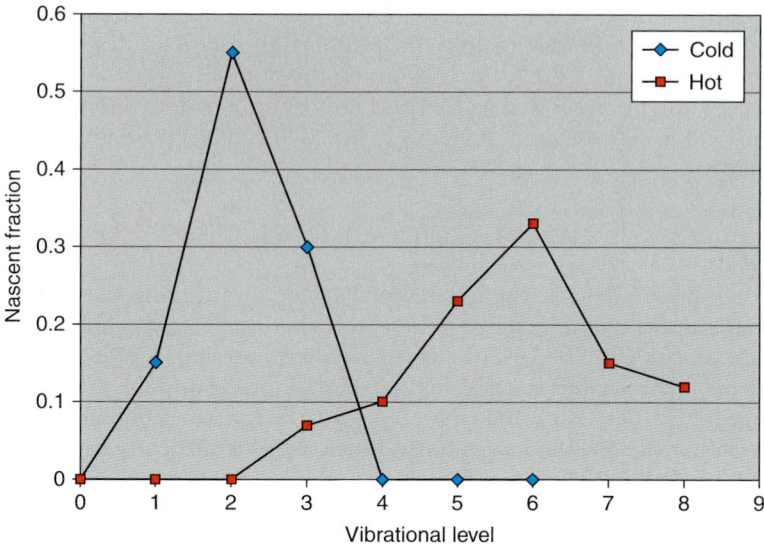

Figure 3.7 Estimated approximate nascent HF vibrational fractions at $T = 300$ K.

In these equations, the indicated exothermicity assumes complete relaxation of excited species. Two primary classes of lasers have been developed based on these reactions. The first are the cold reaction devices, in which molecular hydrogen is mixed with substantially dissociated atomic fluorine. Cold reaction measurements, originally performed by Polanyi et al.,[10] showed that the resultant HF is produced preferentially in excited molecular vibrational states. The estimated nascent population distributions for both reactions are shown in Fig. 3.7. Note that a substantial fraction of the total available reaction energy starts in vibrational levels and that the initial distributions indicate absolute inversions between certain vibrational levels. This suggests gain even without exploiting the possibility of lasing on partial inversions.

Although the hot reaction produces higher amounts of vibrational quanta per HF molecule and appears to be more advantageous, it is impractical to construct devices that are based predominantly on the hot reaction. Hydrogen's large bond energy (436 kJ/mol versus 157 kJ/mol for F_2) makes it very difficult to generate large amounts of hydrogen atoms. Furthermore, because the hot reaction has a tendency to produce a large fraction of its molecules in higher vibrational levels, those molecules have a tendency to deactivate much faster than at lower vibrational levels, as discussed below.

By contrast, cold reaction requires the production of large amounts of fluorine atoms, which is much more practical. In early devices, this production was accomplished electrically, using high-power electric

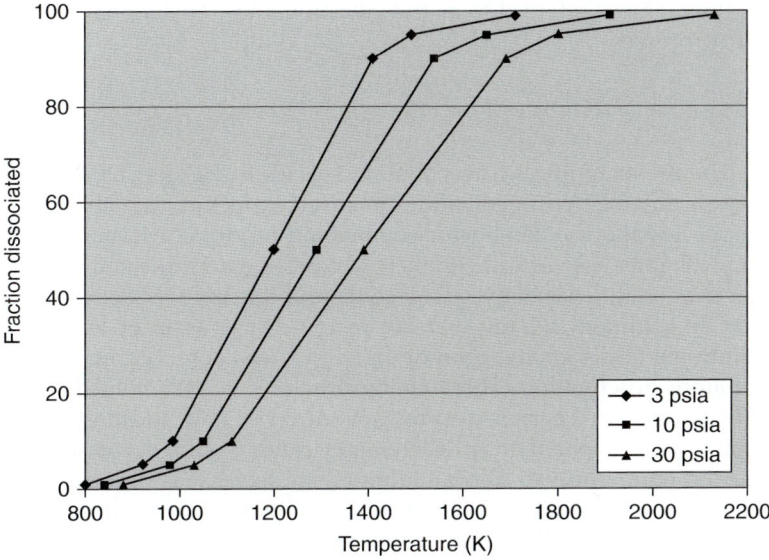

FIGURE 3.8 F_2 dissociation versus temperature and total fluorine pressure.

arcs to thermally dissociate the fluorine atom source. Later, chemical combustors were used for this purpose, because it is relatively easy to thermally dissociate F_2 molecules. Both F_2 and NF_3 have been used as fluorine atom sources, and various fuels have been used in the associated combustors, with part of the fluorine atoms being consumed in the combustion process and the excess delivered for subsequent reaction with the hydrogen (or deuterium) molecules. The equilibrium dissociation fraction depends on both temperature and fluorine partial pressure. Figure 3.8 shows scaling for typical operating parameters. Note that the indicated total pressure is only the partial pressure of the fluorine. Typically, as much as 1 order of magnitude or more of diluent gas is also present. The dissociation fraction α is defined as follows:

$$\alpha = [F]/(2([F_2] + [F]/2)) \quad (3.14)$$

where [F] and [F_2] are molecular concentrations or molar flow rates.

3.3.4 Kinetic Processes, Deactivation, and Energy Transfer

In addition to the pumping chemistry, other important kinetic processes must be considered when assessing chemical laser performance. Especially important are deactivation processes, in which a vibrationally excited molecule (vibrational level v) collides with another gas molecule (species M), which causes the excited molecule to transition to lower vibrational levels while also releasing heat into the flow. This is referred to as the vibrational to translational energy (VT) processes.

$$HF(v) + M \rightarrow HF(v-m) + M + \Delta Q \qquad (3.15)$$

The most significant deactivating species in HF and DF devices are typically the hydrogen halides themselves, including both the lasing species and the combustor combustion products. Measurements of kinetic rates associated with these processes have indicated that the deactivation rate has between a second- and third-power dependence on v. In addition, a large increase in the deactivation of HF(v) for vibrational levels greater than or equal to 3 was observed for hydrogen atom deactivation. These characteristics favor the cold reaction (F + H_2), discussed earlier, over the hot one (H + F_2). In addition, it was originally anticipated that deactivation rates would decrease with reduced temperatures. Although initially a decrease is observed, the deactivation rates have actually been found to reach a minimum and then increase with decreasing temperature (Fig. 3.9). This behavior illustrates the complex nature of deactivation processes.

Also important in understanding laser behavior are vibrational to vibrational energy transfer processes (V-V), in which two excited HF molecules collide and emerge with vibrational levels different from what they initially started with.

$$HF(v) + HF(v') \rightarrow HF(v+m) + HF(v'-m) \qquad (3.16)$$

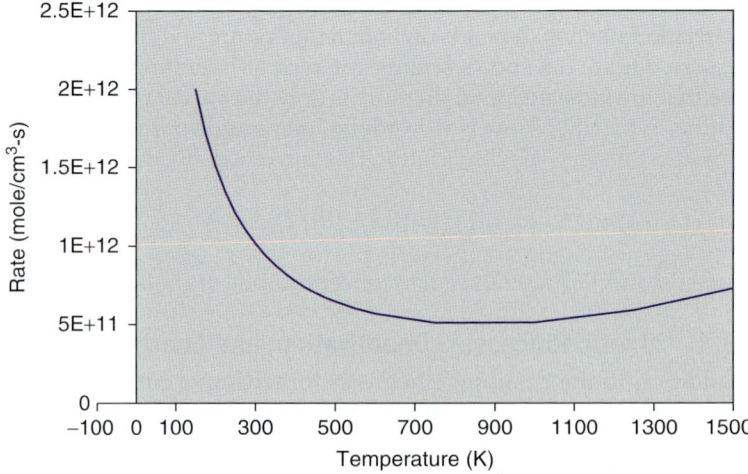

FIGURE 3.9 HF(v = 1) + HF → 2HF deactivation rate temperature dependence.

These processes are quite rapid. At pressures of interest in flowing devices, these processes substantially perturb the nascent fraction produced by the pumping reactions. More detailed discussions of HF and DF kinetic rates in general can be found in Cohen and Bott.[11]

It follows that for HF and DF devices, deactivation and energy transfer processes are quite important and substantially influence the design of these lasers. Specifically, they determine how high a partial pressure of HF or DF one can practically achieve in a laser device; they also dictate that if one wants to construct a high-power device, it is advantageous to have a high flow velocity to allow power extraction before deactivation depletes the excited species. To further illustrate this point, let us assume that the only process considered is a simple deactivation loss of HF($v = 1$) by HF at room temperature:

$$HF(v = 1) + HF(v = 0) \rightarrow HF(v = 0) + HF(v = 0) \qquad (3.17)$$

This process has a typical rate constant $k = 1 \times 10^{12}$ mole/s-cm^3. Even at room temperature and an HF partial pressure of 1 torr (molar density is 5.5×10^{-8} mole/cm^3), the corresponding $1/e$ decay time is only 18 μsec in the absence of other gases. At a velocity of 10^5 cm/s, the $1/e$ decay occurs in a flow distance of only 1.8 cm. This example illustrates the difficulty in pressure scaling and the motivation to flow at high velocity. It also illustrates the need to mix and extract power quickly in order to be competitive with deactivation losses.

3.3.5 Fluid Mechanics and Nozzle Design

The enhanced gain associated with Doppler broadening and favorable partial inversion at low temperatures makes it advantageous to operate HF and DF CW devices at temperatures far below those required to thermally dissociate fluorine. This is achieved by rapidly expanding the combustor flow in converging (subsonic) and then diverging (supersonic) nozzle geometries, which freezes the dissociation fraction while drastically dropping the pressure, static temperature, and density. In order to understand issues associated with such flowing laser devices, the following general review of concepts associated with one-dimensional fluid mechanics should be helpful.

At a given location, a gas is characterized by the fluid parameters and the relative mole fractions of the gas components. Variables include (1) static temperature T, (2) static pressure P, (3) density ρ, and (4) gas velocity U. Knowledge of the stoichiometry allows one to also calculate the average molecular weight W, the heat capacities at constant pressure C_P and temperature C_V, the specific heat ratio $\gamma = C_P/C_V$, and the speed of sound c. The gas equation of state, which is usually well approximated by the ideal gas law, allows calculation of the mass density and local molecular concentrations of the various gas constituents based on temperature and pressure.

When considering the evolution of kinetic processes, one simply uses the velocity U to relate position and time, using $dx = Udt$. At high velocities, where compressibility of the gas becomes significant, the flow behavior becomes complicated. This regime is usually defined to occur when the Mach number, $M = U/c$, becomes greater than ~0.3. For the case of a nonreacting flow with neither friction nor heat addition (isentropic), the flow is characterized by its stagnation properties, which correspond to flow conditions after the flow is isentropically brought to rest, given by:

$$T_0/T = 1 + 0.5\,(\gamma - 1)M^2$$
$$P_0/P = [1 + 0.5\,(\gamma - 1)M^2]^{\gamma/(\gamma-1)} \qquad (3.18)$$
$$\rho_0/\rho = [1 + 0.5\,(\gamma - 1)M^2]^{1/(\gamma-1)}$$

where P, T, and ρ are the static properties and P_0, T_0, and ρ_0 are the stagnation properties.

Gas flows that travel isentropically through a duct with variable cross section A satisfy Eq. (3.19):

$$dU/U = (dA/A)/(M^2 - 1) \qquad (3.19)$$

This expression illustrates the principle of operation behind the converging-diverging nozzle that is widely used in laser applications. In the converging section, the flow accelerates until it reaches the minimum area throat location, where the flow reaches $M = 1$. It then continues to accelerate beyond the throat in the expanding region, where M continues to increase to supersonic values, resulting in much lower pressure, static temperature, and density.

In parallel, one also flows the secondary flow of hydrogen that reacts with the fluorine atoms to produce the vibrationally excited HF and the associated heat of reaction. The addition of heat tends to drive the flow toward Mach 1 conditions, or the so-called *thermal choking case*. Avoiding this condition is a major concern in chemical laser designs. Thermal choking of supersonic flows leads to a variety of unfavorable behaviors, such as reduced velocity, increased density and pressure, higher temperatures, large optical path difference (OPD) effects associated with density variations, and feedback of flow behavior into upstream flow regions. To avoid thermal choking, an inert, diluent gas, such as helium or, more infrequently nitrogen, is used to increase the flow mixture's heat capacity, thus minimizing the effects of heat release. Alternatively, one can mitigate heat release through area expansion; however, this increases vacuum pumping demands. Figures 3.10 to 3.12 show the Mach number, temperature, and pressure dependence of the gas mixture as a function of position in a typical laser cavity with and without the addition of heat due to the secondary flow.

Chemical Lasers

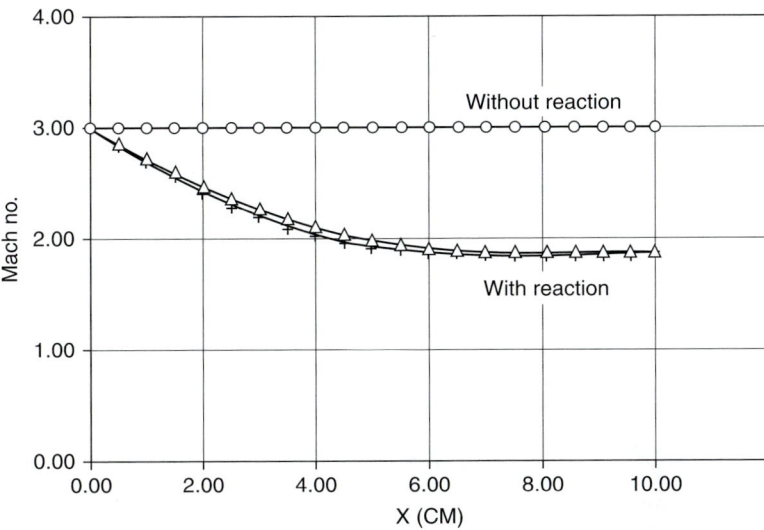

FIGURE 3.10 Mach number with and without reaction heat as a function of position in the laser cavity.

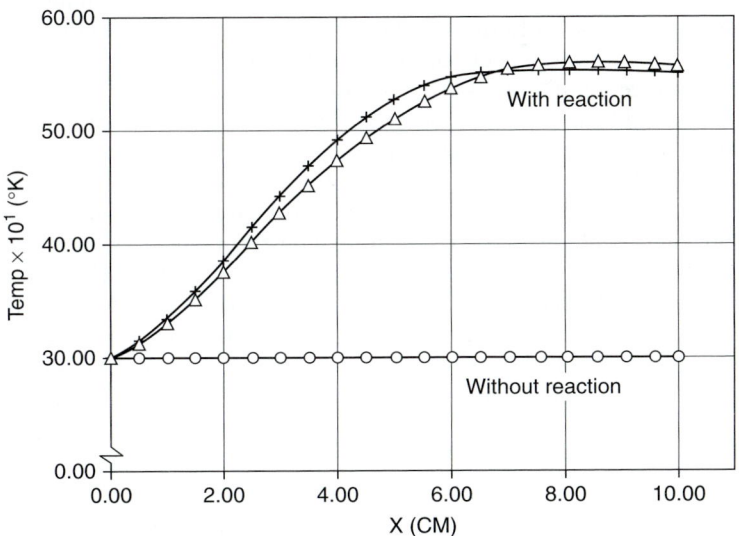

FIGURE 3.11 Gas temperature with and without reaction heat as a function of position in the laser cavity.

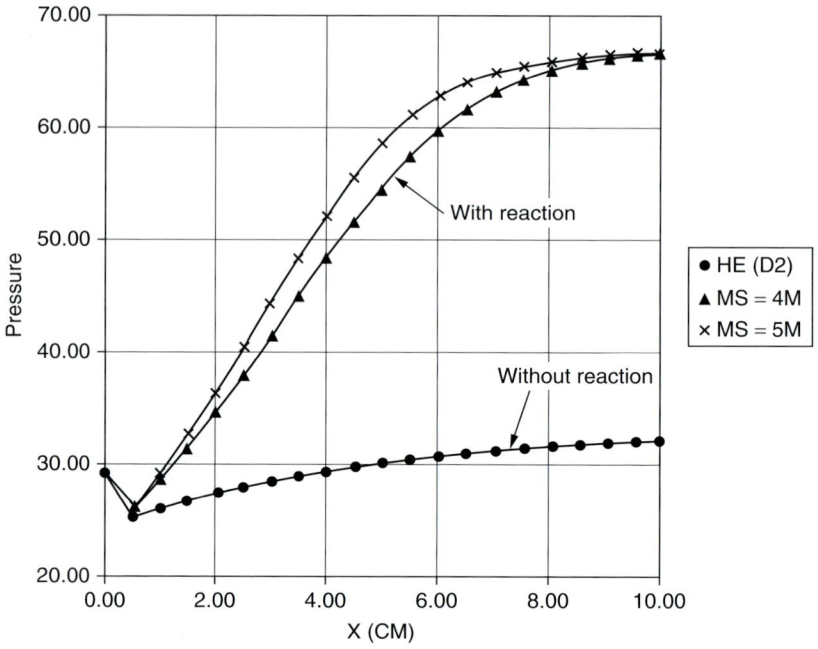

FIGURE 3.12 Cavity pressure with and without reaction heat as a function of position in the laser cavity.

The addition of the secondary flow also leads to the challenging problem of efficiently mixing the supersonic streams, allowing them to rapidly mix and react to produce the laser gain medium. Figure 3.13 schematically illustrates such a nozzle design.

Mixing is an essential factor in determining performance, and it must compete with deactivation. In general, there is a tradeoff between decreasing nozzle scale to minimize the mixing distance and increased viscous losses, cost, and complexity. Many nozzle variations have been developed to optimize performance in differing flow regimes. Figure 3.14 shows an exploded view of the Mid-Infrared Advanced Chemical Laser (MIRACL) DF nozzle module; Fig. 3.15 shows the entire laser nozzle assembly, which produces megawatt-class power levels. HF and DF mixing nozzles are generally thought to simultaneously achieve mixing by two parallel mechanisms:

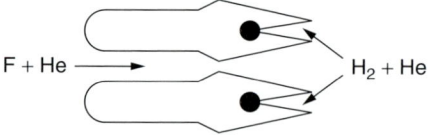

FIGURE 3.13 Schematic drawing of a typical nozzle design.

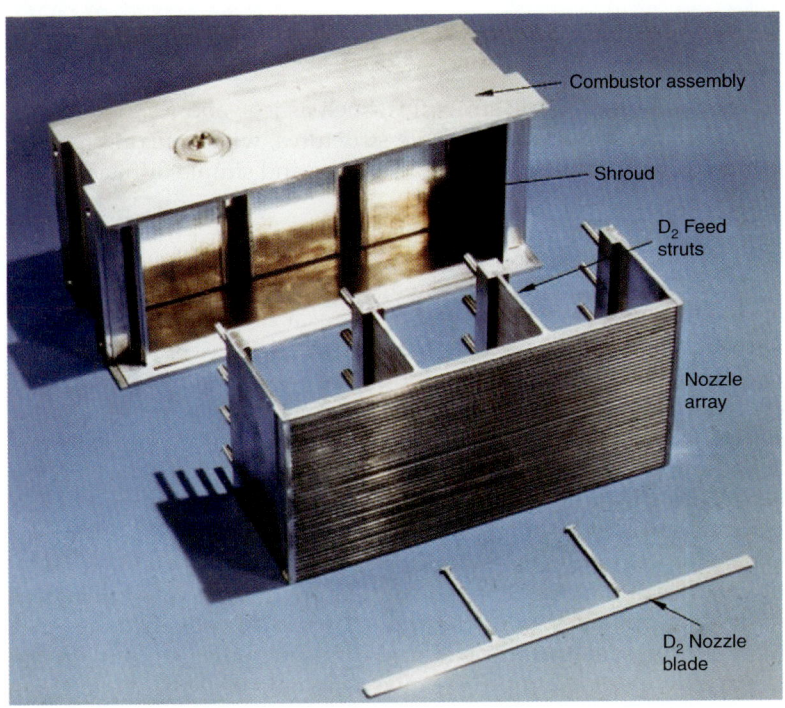

FIGURE 3.14 Exploded view of the Mid-Infrared Advanced Chemical Laser (MIRACL) nozzle module.

FIGURE 3.15 One of two MIRACL laser nozzle banks, consisting of 19 modules (not all shown).

(1) mixing of large-scale structures, such as jets; and (2) local, diffusional mixing between large-scale structures, whose mixing may be somewhat augmented by local turbulence and simple diffusion.

Pressure Recovery

CW laser devices typically operate at relatively low pressures, which means that in mobile systems, one must either use a chemical or microporous absorbent pump, such as a zeolite, or use an external pump, such as a single- or multiple-stage ejector system, to maintain the required low operating pressures. As shown in Fig. 3.16, an ejector consists of several sections, including a gas generator; supersonic mixing nozzles, which inject the gas into the subsonic laser flow; a constant-area supersonic diffuser region, in which the mixture is converted from supersonic to subsonic flow via two- or three-dimensional shock interactions; and a subsonic diffuser expansion region, which further slows the flow to yield an additional pressure increase. To gain added pressure recovery, the laser itself also contains supersonic and subsonic diffusers in series that work on the same principle. These diffusers, however, rely on the laser cavity itself to supply the mixed supersonic flow.

3.3.6 Variations on Continuous Wave HF and DF Devices

HF and DF Overtone Lasers

In addition to lasing on fundamental HF transitions ($\Delta v = 1$), it is also possible to lase on overtone ($\Delta v = 2$) transitions. However, the gain is reduced due to substantial reductions in the associated Einstein coefficients. Furthermore, it is necessary to use resonator concepts that accommodate the substantially reduced gain, while simultaneously suppressing the higher-gain fundamental laser transitions. Moderate-sized CW devices have been constructed using approaches somewhat similar to those used for more conventional low-pressure ($\Delta v = 1$) devices. Note that the above discussion has primarily used HF as an example; however, DF device approaches are very similar to HF ones.

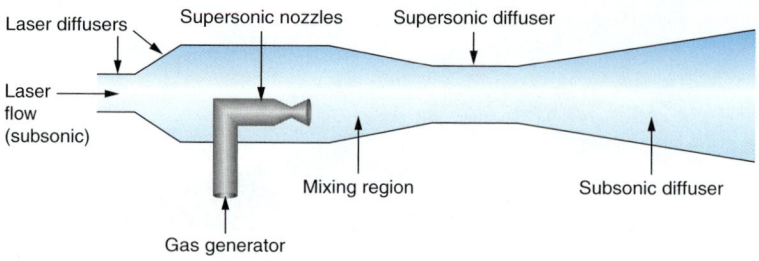

FIGURE 3.16 Schematic drawing of a typical laser pressure recovery system.

Pulsed HF and DF Lasers

Both HF and DF pulsed devices have been constructed. Typically, when H_2 (D_2) and F_2 are used, they rely on premixing with inhibitor gases, which are present to suppress premature reaction. A chain reaction is initiated via electrical production of either an electric discharge or a photolytic source. As Table 3.1 suggests, once one decides to use electrical initiation of the reaction, a wealth of alternatives are available for reactants. For laboratory applications, these alternatives (e.g., SF_6) may be much more attractive than the more conventional, efficient, but potentially more hazardous, reactants. When high average power is desired, an added difficulty arises. In this case, it is necessary to achieve a suitably high repetition rate. In practice, as the repetition rate and pressure increase, there is another challenging flow problem associated with quickly removing the previous pulse reaction products and heat. In the frequently used continuous-flow systems, one must either waste reactants or cope with some pressure feedback effects from the previous pulse.

3.3.7 HF and DF Laser Performance

The development of high-power HF and DF lasers began in the early 1970s and continued throughout the 1980s and 1990s, with the development of several multihundred kilowatt- to megawatt-class lasers. These included the Baseline Demonstration Laser (BDL) and the Navy-ARPA (Advanced Research Projects Agency) Chemical Laser (NACL), both built in the late 1970s; the MIRACL built in the early 1980s (Fig. 3.17); and the Alpha laser, developed by the U.S. Air Force in the

FIGURE 3.17 Mid-Infrared Advanced Chemical Laser.

FIGURE 3.18   Alpha laser facility (left) and Alpha laser nozzle (right).

1980s and early 1990s as part of the Strategic Defense Initiative (SDI) (Fig. 3.18). The primary motivation for these lasers has always been military applications. In the late 1990s, the Army Tactical High-Energy Laser (THEL) was introduced (Fig. 3.19), which consisted of the first complete laser weapon system that has successfully detected, tracked, and shot down numerous military projectiles, including rockets, artillery shells, and mortars. In spite of these successes, interest in HF and DF lasers has waned of late due to the logistic problems created by storing and transporting the reactive fuels (H_2/D_2 and F_2/NF_3) and disposing of the highly corrosive effluents (HF and DF) they produce.

FIGURE 3.19   Tactical High-Energy Laser system and beam director.

Chemical Lasers

Thus, current activity in HF and DF chemical lasers is limited to low-power laboratory devices and specialty applications, which typically use electrical discharge, rather than combustion, to dissociate SF_6 to produce fluorine atoms and which range in power from tens of watts to a few hundred watts.

3.4 Chemical Oxygen Iodine Laser (COIL)

The possibility of making a COIL device was first suggested by Derwent and Thrush[12] in 1971. The first lasing demonstration was made by McDermott and his team at the Air Force Weapons Lab (AFWL) in 1978.[13] Truesdell, Helms, and Hager[14] summarized development efforts within the United States through 1995. Technology improvements and scaling to large devices have continued at a variety of sites.

COIL devices are based on a very different chemistry from HF and DF lasers. Figure 3.20 shows a simplified block diagram of the operational approach. First, chlorine gas reacts with liquid basic hydrogen peroxide (BHP) in a gas-liquid reaction. The reaction then produces electronically excited singlet delta oxygen, which has a very long lifetime as compared with most electronically excited states. The reactor is denoted a singlet delta oxygen generator (SOG). Next, the excited oxygen, along with a suitable diluent, is transported to supersonic expansion nozzles, which mix the oxygen with molecular iodine and a suitable carrier gas. The singlet delta oxygen chemically dissociates the molecular iodine to produce atomic iodine, which has electronic states that favor the resonant transfer of energy from the singlet delta oxygen to the iodine. The resultant gain allows lasing on an atomic iodine transition. Typically, the ratio of oxygen to iodine atoms is relatively large, and many energy transfers occur per iodine atom until the singlet delta fraction is sufficiently depleted. As the block diagram in Fig. 3.20 shows, a practical device must also include provisions to circulate the BHP, remove the reaction heat, and recover pressure to ambient conditions.

3.4.1 Energy Levels

The lasing species in COIL devices is the iodine atom. In contrast to most other chemical lasers, the transition energy levels are electronic rather than molecular. The lasing transition is a magnetic dipole transition between two spin orbit terms of the ground state $5^2P_{1/2} \to 5^2P_{3/2}$. Because this transition is forbidden, it has a relatively long radiative lifetime of approximately 125 milliseconds (ms). The upper level is split into two hyperfine levels—total angular momentum quantum numbers, $F = 2$ and 3. The lower level is split into four hyperfine levels, $F = 1, 2, 3$, and 4. Associated degeneracies are $2F + 1$. The energy levels are shown schematically in Fig. 3.21, in which the

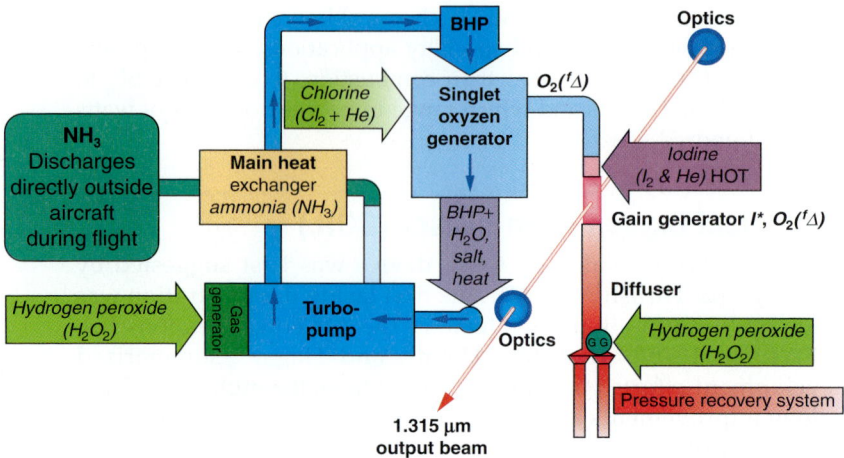

FIGURE 3.20 Chemical oxygen iodine laser (COIL) block diagram.

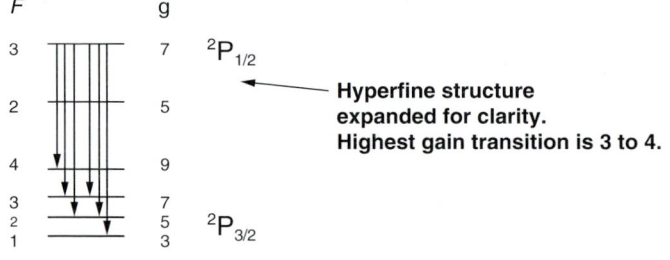

FIGURE 3.21 Iodine atom energy-level diagram.

separation of the hyperfine levels has been greatly exaggerated for clarity. The actual total separation in transition energies between the transitions $\Delta F = 0, +/-1$ is less than 1 part in 7000 for the levels. In the following sections, the upper and lower fine structure levels of iodine atoms will simply be abbreviated as I* and I.

3.4.2 Small Signal Gain

At the typical pressures that characterize chemically pumped devices, the gain profiles of the various lines are reasonably separated, and thermal equilibration of the hyperfine level populations can be assumed. Table 3.3 summarizes the A coefficients for the various transitions. These factors imply that chemical devices operate on a single transition of $F = 3$ to $F = 4$. The equilibration assumption implies that 7/12 of the excited iodine atoms will be in the $F = 3$ state but that only 9/24 of the lower state will be in the $F = 4$ state. Since

Transition	sec^{-1}
3-4	5.0
3-3	2.1
3-2	0.6
2-3	2.4
2-2	3.0
2-1	2.3

TABLE 3.3 Einstein A Coefficients

for upper and lower population densities N_U and N_L the gain is proportional to

$$N_U - (g_U/g_L)N_L \qquad (3.20)$$

Then, for the 3 to 4 transition, gain is proportional to

$$(7/12)[P_{1/2}] - (7/9)(9/24)[P_{3/2}] = (7/12)([P_{1/2}] - 1/2[P_{3/2}]) \qquad (3.21)$$

This implies that a partial inversion also produces gain in COIL devices.

The relatively narrow line width of the single-line COIL devices, as compared with other types of chemical lasers, has advantages in some applications. At very low pressures, for example, COIL devices are substantially Doppler broadened. At moderate pressures, pressure broadening also becomes important—and even potentially helpful from a hole-burning standpoint

Energy Pumping Reactions: Singlet Delta E-E Transfer

The source of energy for COIL devices is near-resonant energy transfer from electronically excited singlet delta oxygen [$O_2(^1\Delta)$] to ground-state iodine atoms, yielding $^2P_{1/2}$ iodine atoms and ground state $O_2(^3\Sigma)$ oxygen also denoted simply as O_2. The electronic energy level to electronic energy level (E-E) transfer energetics are illustrated in Fig. 3.22. Although the vibrational and rotational levels of oxygen are not shown in Fig. 3.22, vibrational levels may play some role in deactivation processes and iodine dissociation. The kinetic equation that describes the E-E transfer process is as follows:

$$I + O_2(^1\Delta) \rightarrow I^* + O_2 \qquad (3.22)$$

The reverse reaction is

$$I^* + O_2 \rightarrow I + O_2(^1\Delta) \qquad (3.23)$$

Gas, Chemical, and Free-Electron Lasers

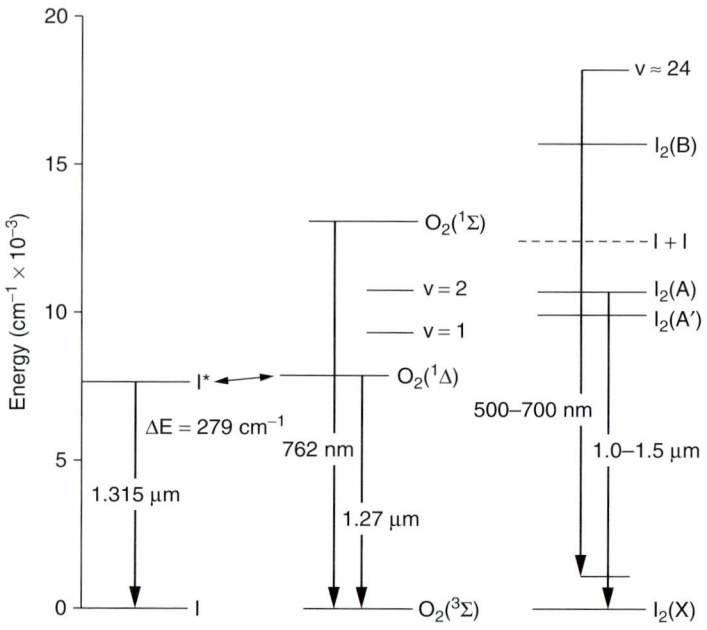

FIGURE 3.22 COIL energy levels and E-E transfer channel.

The forward reaction is reasonably rapid because of the small energy defect (279 cm^{-1}); it is faster than the reverse reaction because the transfer is downhill. In the absence of other processes, thermodynamic evaluation yields the equilibrium relationship for these equations:

$$[I^*]/[I] = 3/4 \; e^{(402/T)}[O_2(^1\Delta)]/[O_2(^3\Sigma)] \qquad (3.24)$$

where [I], [I*], [$O_2(^1\Delta)$], and [$O_2(^3\Sigma)$] are the species number densities and T is the absolute temperature in kelvins.

In practical devices, the amount of iodine present is a small percentage of the oxygen. In the absence of lasing deactivation, however, rates are usually slow enough that equilibrium can nearly be established, and Eq. (3.24) can be used to estimate the fraction of iodine atoms that are excited. In many supersonic devices, in the absence of lasing, most of the dissociated iodine atoms are excited. Because the gain threshold occurs when [I*]/[I] is 1/2, it is relatively easy to produce small signal gain. Even at room temperature, one only needs 15 percent of the total oxygen to be $O_2(^1\Delta)$, and less than 10 percent is sufficient at reduced temperatures. The ability of E-E transfer to repopulate $O_2(^3\Sigma)$ that has been depleted by stimulated emission as the flow passes through the resonator helps determine the characteristics of the resonator and the optimum iodine flow rate.

3.4.3 Deactivation Processes

Deactivation rates in a COIL device's laser cavity are considerably slower than HF and DF VT rates. However, because it is difficult to pressure scale singlet oxygen generators efficiently, it is advantageous to operate with relatively low-cavity Mach numbers. Furthermore, the total temperature of delivered SOG flows is low compared with HF and DF values; thus, even the reduced deactivation rates are an important concern, primarily because of the need to avoid thermal choking and to minimize temperature increases.

The most important deactivation processes in the laser cavity include the following:

$$I^* + H_2O \rightarrow I + H_2O \qquad (3.25)$$

$$I^* + O_2(^1\Delta) \rightarrow I + O_2(^1\Delta) \qquad (3.26)$$

In addition, considerable losses may be associated with the iodine dissociation process kinetics and possibly with deactivation by I_2.

3.4.4 Iodine Dissociation

$O_2(^1\Delta)$ serves the dual function of dissociating the I_2 molecules and exciting the I atoms. It is very fortuitous that when molecular iodine is mixed with $O_2(^1\Delta)$, it is chemically dissociated, especially because a single $O_2(^1\Delta)$ lacks the required energy (Fig. 3.22). This behavior was first reported by Ogryzlo and coworkers.[15] Although the dissociation process is not well understood, the original suggestion was that dissociation proceeded via $O_2(^1\Sigma)$, which was produced by the energy-pooling reactions shown in Eq. (3.27), plus the E-E transfer processes in which some energy loss in excess of the minimum of two $O_2(^1\Delta)$ molecules is required to dissociate I_2. However, it is currently believed that iodine dissociation is more complicated than a simple interaction with $O_2(^1\Sigma)$ and probably involves additional intermediate states that are most probably vibrational in nature.

$$O_2(^1\Delta) + O_2(^1\Delta) \rightarrow O_2(^3\Sigma) + O_2(^1\Sigma) \qquad (3.27)$$

3.4.5 Singlet Oxygen Generator

The mechanism for generation of $O_2(^1\Delta)$ consists of chlorine absorption in BHP and can be summarized by the net effective reactions that follow:

$$MOH + H_2O_2 \rightarrow HO_2^- + M^+ + H_2O, \text{ where } M = \text{Li, Na, or K} \qquad (3.28)$$

$$Cl_2 + HO_2^- \rightarrow O_2(^1\Delta) + 2\,Cl^- + H^+ \text{ (rate constant } k_1) \qquad (3.29)$$

$$HO_2^- + H^+ \leftrightarrow H_2O_2 \qquad (3.30)$$

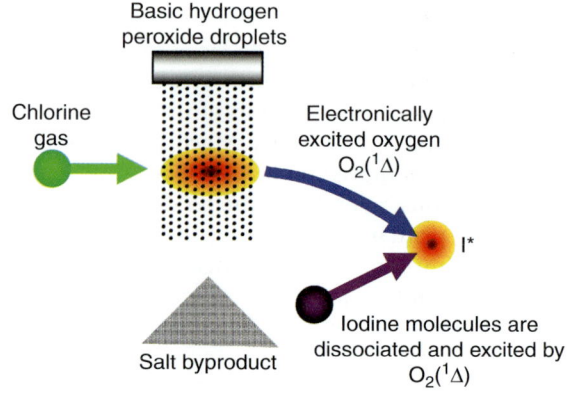

FIGURE 3.23   Schematic drawing of a singlet oxygen generator (SOG).

Reactions 29 and 30 occur near the liquid gas interface created by Cl_2 gas passing through a liquid-phase basic H_2O_2 prepared earlier using reaction 28, as shown schematically in Fig. 3.23.

The availability of efficient SOGs is what made COIL devices feasible. The primary quantities of interest when assessing SOG performance are (1) chlorine utilization, or the fraction of chlorine-reacted U; (2) singlet delta fraction F_Δ, or the fraction of oxygen in the $O_2(^1\Delta)$ state; (3) the amount of delivered impurities (e.g., H_2O); and (4) the transmitted gas pressure and temperature.

The rate of chlorine reaction is determined by the product of the chlorine and HO_2^- hydroperoxy ion concentrations. Assuming that absorption of chlorine into the liquid is the primary mechanism, the amount of chlorine available can be limited by several factors: (1) ability of the chlorine to penetrate the BHP surface layer, (2) solubility of the chlorine in the BHP, and (3) ability of the chlorine to diffuse from the gas phase to the liquid surface. The concentration of HO_2^- also determines the rate at which the reaction can occur. Diffusional modeling indicates that HO_2^- can be depleted so that it becomes the primary constraint on the reaction, unless surface stirring or replacement were to occur. It should also be noted that although the reaction in Eq. (3.29) occurs in the liquid phase, because $O_2(^1\Delta)$ can be deactivated rather rapidly by water in the liquid phase, it is essential that the reaction occur very near the surface so that the $O_2(^1\Delta)$ can escape back into the gas phase.

These requirements led to the development of a variety of reactor concepts that featured compact, large surface area liquid-gas interfaces. These interfaces maximize singlet delta fraction and effectively flow BHP surfaces to maximize chlorine utilization over a large molarity range. Examples of such interfaces include simple spargers (turbulent jet bubblers), various wetted-wall reactors, aerosols, and liquid-jet reactors.[16]

It is currently believed that the above reactions produce a near-unity $O_2(^1\Delta)$ fraction. The $O_2(^1\Delta)$ fraction can be reduced by several mechanisms: (1) deactivation within the liquid, leading to the so-called detachment yield; (2) deactivation by gas-phase surface film collisions; and (3) homogeneous deactivation in the gas phase. In most practical devices, the dominant singlet delta fraction loss mechanism is the homogeneous gas phase self-deactivation of $O_2(^1\Delta)$. Sophisticated SOG performance models can be used to accurately evaluate these processes. Many reported models concurrently model behavior in both the gas and liquid film streams and predict SOG performance characteristics. One such model[17] includes an effective resistance chlorine-oxygen mass transfer model; a local BHP HO_2^- diffusion model; and evaluation of $O_2(^1\Delta)$ detachment yield, surface deactivation, and gas phase deactivation.

SOG chlorine utilization is also a function of chlorine flow rate and BHP HO_2^- molarity. Because BHP is typically continuously replaced by a flowing process, the surface HO_2^- concentration is determined by the balance between reaction depletion and ion diffusion from within the liquid during the residence time that the BHP surface remains in the reaction zone. For a typical SOG, chlorine utilization is usually near unity levels at the very low chlorine flow limit but declines to values on the order of 0.8 to 0.9 at useful flow rates. At high initial surface $[HO_2^-]$ levels, utilization is typically a weak function of $[HO_2^-]$, and at reduced levels, it eventually decreases toward zero as $[HO_2^-]$ tends to zero. However, because most modern SOG concepts replace the BHP by flowing it in some manner, depletion is only important when ion diffusion is too slow to adequately maintain surface HO_2^-.

3.4.6 COIL Laser Performance Characterization

In addition to SOG parameters, net laser performance is frequently characterized by chemical efficiency, which is defined as the percentage of power output to the power output expected if 100 percent of the chlorine has reacted and each resultant oxygen molecule has produced one laser photon:

$$\text{Chemical efficiency} = P(\text{kW})/(91 \text{ kW} \times X_{Cl2}) \qquad (3.31)$$

The situation is often approximated in terms of a heuristic equation,[14] defined as follows:

$$\text{Chemical efficiency} = U \times (F_\Delta - N \times X_{I2} - F_{thres})\eta_{mix}\eta_{extract} \qquad (3.32)$$

where
- P = power
- X_{Cl2} = chlorine molar flow rate
- U = chlorine utilization
- F_Δ = SOG-delivered singlet delta fraction
- N = estimated number of $O_2(^1\Delta)$ consumed by dissociation costs and deactivation per initial I_2 molecule
- X_{I2} = iodine molar flow rate
- F_{thres} = lasing threshold singlet delta fraction
- η_{mix} = loss factor associated with imperfect mixing
- $\eta_{extract}$ = loss factor associated with imperfect optical extraction

In practice, the best reported small-scale device results have exceeded 0.3 (30%) chemical efficiency, based on this definition.

As is the case for HF and DF devices, very sophisticated laser cavity three-dimensional fluid mechanics computer models, including chemistry and physical optics, have been developed to predict performance. Their primary limitation appears to be uncertainties in kinetic processes and initial conditions, rather than in their ability to solve computational problems.

3.4.7 COIL Laser Performance

High-energy laser COIL technology has been developed primarily by the Air Force Research Labs (AFRL), which has led to the megawatt-class Airborne Laser (ABL). Practical engineered devices are fairly complicated. Figure 3.24 shows the Boeing 747 airplane, which houses the ABL, equipped with a beam director, in the nose of the airplane. The ABL fired in flight for the first time in August 2009 and was able to engage and destroy a ballistic missile in boost phase in February 2010, reemphasizing the potential of laser weapons.

FIGURE 3.24 Boeing 747 Airborne Laser (ABL) platform.

3.5 Other Chemical Laser Concepts

3.5.1 DF-CO_2 Transfer Devices

DF-CO_2 transfer devices are another chemical alternative associated with DF devices. A DF laser can relatively easily be converted to CO_2 lasing on the 10.6-μm transition by the addition of CO_2 to the conventional devices and appropriate changes to resonator optics. This conversion is possible because of a relatively efficient near-resonant vibrational energy transfer between the two molecules via the reaction:

$$DF(v) + CO_2(000) \rightarrow DF(v-1) + CO_2(001) \qquad (3.33)$$

Both CW flowing and pulsed devices have been demonstrated using this approach.

Other Hydrogen Halide Devices

It is also possible to construct other halogen halide chemical lasers. However, bond energies are such that there is not a simple analog to HF and DF cold reaction devices. Instead, one must also rely on the hot reaction; in practice, this forces one to rely primarily on chain reaction type devices. Table 3.4 summarizes the pertinent bond energies. Note that the H_2 bond is stronger than either the HBr or HCl bonds but weaker than the HF bond. However, it is still the case that cycling chain reactions are exothermal for both bromine and chlorine systems. In addition, HI in place of H_2 has been used to produce HCl lasers, though these have never been scaled as favorably as HF and DF devices.[5]

Molecule	kJ/mol
F_2	156.9
Cl_2	242.6
Br_2	193.9
H_2	436.0
HF	568.6
HBr	365.7
HCl	431.6
HI	298.7

TABLE 3.4 Bond Energies[18]

3.5.2 Carbon Monoxide Lasers

Chemically driven carbon monoxide devices have also been demonstrated. They typically rely on the highly exothermal pumping reaction:

$$CS + O \rightarrow CO^* + S, \Delta Q = 334 \text{ kJ} \quad (3.34)$$

where the CS radical is often produced by the reaction

$$CS_2 + O \rightarrow CS + SO \quad (3.35)$$

The pumping reaction produces highly vibrationally excited CO that is redistributed by V-V transfer processes. Furthermore, the VT deactivation rates for CO* are much more favorable than are those for HF. Unfortunately, due to the high O_2 bond energy, oxygen atoms are almost as difficult to produce as hydrogen atoms. Because the oxygen atoms are typically produced electrically, there is no real advantage to using an all–electrically driven CO laser. Furthermore, interest is limited by the relatively poor propagation characteristics of CO in the atmosphere.

References

1. Polanyi, J. C., "On iraser detectors for radiation emitted from diatomic gases and coherent infrared sources," *J. Chem. Phys.*, 34: 347, 1961; Penner, S. S., "Proposal for an infrared maser dependent on vibrational excitation," *J. Quant. Spectrosc. Radiative Transfer*, 1: 163, 1961.
2. Kasper, J. V. V., and Pimentel, G. C., "HCl Chemical Laser," *Phys. Rev. Lett.*, 14: 352, 1965.
3. Pimentel, G. C., "The significance of chemical lasers in chemistry," *IEE J. Quantum Electron*, 6: 174, 1970.
4. Gross, R. W. F., and Bott, J. F., *Handbook of Chemical Lasers*, John Wiley & Sons, New York, 1976.
5. Stitch, M. L., *Laser Handbook, Volume 3*, North-Holland Publishing Company, Amsterdam, 1979.
6. Cheo, P., *Handbook of Molecular Lasers*, Dekker, New York, 1987.
7. Endo, M., and Walter, R., *Gas Lasers*, CRC Press, New York, 2007.
8. Hertzberg, G., *Spectra of Diatomic Molecules*, Van Nostrand Reinhold Company, New York, 1950.
9. Zissis, G. J., and Wolf, W. L., *The Infrared Handbook*, Environmental Research Institute of Michigan, Ann Arbor, 1985.
10. Polanyi, J. C., and Woodall, K. B., "Energy distribution among reaction products VI F + H_2,D_2," *J. Chem. Phys.*, 57: 1574, 1972; Polanyi, J.C., and Sloan, J J., "Energy distribution amoung reaction products VII H + F_2," *J. Chem. Phys.*, 57: 4988, 1972.
11. Cohen, N., and Bott, J. F., "Review of Rate Data for Reactions of Interest in HF and DF Lasers," Aerospace Corporation TR SD-TR-82-86, Segundo, CA, 1982.
12. Derwent, R.G., and Thrush, B.A., "The radiative lifetime of the metastable iodine atom I($5^2P_{1/2}$)," *Chem. Phys. Lett.*, 9: 591, 1971.
13. McDermott, W., Pchelkin, W. E., Bernard, D. J., and Bousek, R. R., "An electronic transition chemical laser," *Appl. Phys. Lett.*, 32: 469, 1970.

14. Truesdell, K. A., Helms, C. A., and Hager, G. D., "History of chemical oxygen-iodine laser (COIL) development in USA," *Proceedings of the SPIE*, 2502(217), 1995.
15. Arnold, S.J., Finlayson, N., and Ogryzlo, E.A., "Some novel energy pooling processes involving $O_2(^1\Delta_g)$," *J. Chem. Phy.*, 44: 2529, 1966.
16. McDermott, W. E., "Generation of O_2(a1vg) a survey update," *Proceedings of the SPIE*, 2702(239), 1996.
17. Clendening, C. W., and Hartlove, J., "COIL performance model," *Proceedings of the SPIE,* 2702(226), 1996; Clendening, C. W., and Hartlove, J., "COIL performance modeling," *Proceedings of the SPIE*, 3268(137), 1998.
18. Dean, J. A., *Lange's Handbook of Chemistry*, McGraw-Hill, New York, 1992.

CHAPTER 4

High-Power Free-Electron Lasers

George R. Neil

Associate Director, Thomas Jefferson National Accelerator Facility, Newport News, Virginia

4.1 Introduction

The development of high-average-power free-electron lasers (FELs) has been underway for more than 30 years. And yet it has only been in the recent era that significant progress to high power has been achieved. This progress has been primarily due to the technical status of the available driver accelerator technology, especially the crucial electron injector, though other components have also played a limiting role. This chapter reviews the physics of FELs, as well as the technical approaches to high-power FELs, and discusses some of the applications of this technology.

4.2 FEL Physics

4.2.1 Physical Mechanism

Lasing of an FEL can be understood to result from the interaction of electromagnetic fields on a relativistic electron beam. In the simplest arrangement, a relativistic electron bunch is sent through a sinusoidal magnetic field produced by alternating magnets in a device called a *wiggler*. This causes the electrons to oscillate transversely. From the perspective of the electrons the wavelength of the wiggler (also called an *undulator*) is shortened by a Lorentz contraction of $(1 + \beta)\gamma$, where β is v/c, or the electrons' velocity along the axis divided by the speed of light, and γ is 1 plus the ratio of the electron's kinetic energy to its

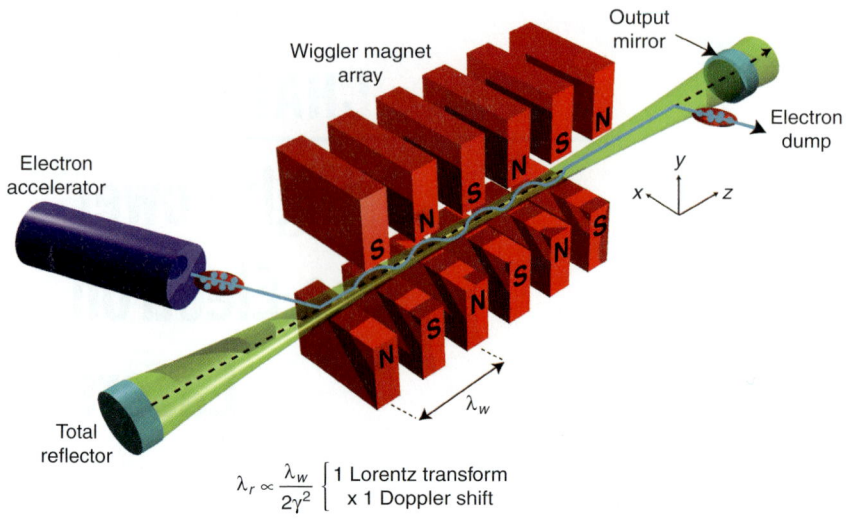

FIGURE 4.1 The free-electron laser interaction.

rest mass of 0.511 mega-electronvolts (MeV) (Fig. 4.1). In response to this transverse acceleration, the electrons radiate in a dipole pattern. Transformed back into the rest frame, this becomes Doppler shifted by another factor of γ and folded into a forward-directed $1/\gamma$ cone of radiation. This dipole radiation becomes the initial spontaneous emission from the laser.

Because the electrons are uniformly (at optical wavelength scales) distributed within the bunch, the initial light is relatively broadband and incoherent for wavelengths shorter than the bunch length. As the process continues, however, something remarkable happens. The electric field of the emitted photons when crossed with the wiggler field causes the electron density to be modulated at the optical wavelength. The once-smooth distribution of electrons becomes a set of microbunches radiating together in phase, thus establishing coherence in the emitted optical field. The bandwidth of the optical radiation narrows, the optical mode becomes well defined, and significant gain and energy extraction from the electrons can occur.[1,2]

4.2.2 Wavelength

The longitudinal bunching of electron motion can easily be derived from the equations of motion and the combined electromagnetic fields. However, it is more important to understand the physical principles at work; with that in mind, realize that the photon field constitutes a traveling wave of ponderomotive force. Electrons can fall down into this moving potential well and give up energy to the electromagnetic wave.

In fact, if the process is permitted to continue, the electrons can travel up the other side of the well and take energy back from the wave. A net transfer of energy from the electrons to the wave can only happen if the electrons are moving slightly faster than the ponderomotive wave. In this case, the electrons "surf" the wave but with roles reversed, as if the surfer were pushing the ocean wave rather than the other way around. The speed of this wave is wavelength dependent. The resonant wavelength (the one in which the electrons are traveling at the same velocity) is given by

$$\lambda_s = \frac{\lambda_w}{2\gamma^2}(1+K^2) \qquad (4.1)$$

where λ_s is the radiated wavelength; λ_w is the wiggler wavelength; and K is the strength parameter of the wiggler field, given by $K = 93.4 B_{rms}(T) \lambda_w(m)$, with B being the rms wiggler field (K is of order 1 and compensates for the fact that the electron trajectories in the wiggler are not exactly parallel to the axis). For example, if $B_{rms} = 0.2T$ and $\lambda_w = 0.05$ m, the resonant wavelength would be around 1.2 μm for an electron energy of 100 MeV. The resonant wavelength turns out to be the one in which the electrons slip backward exactly one optical wavelength for each wiggler period. When this occurs, a net transfer of energy between the electrons and the optical wave can occur, because the electrons' direction of transverse motion ends up always in the same direction as the transverse field of the optical wave ($qE \cdot dl$ is always positive; see Fig. 4.2).

In practical terms, K is a measure of the strength on the wiggler interaction and needs to be of order 1 to give reasonable gain.

It can be seen from a cursory examination of Eq. (4.1) that once constructed with a fixed wavelength, the wiggler has a limited range of control over the output wavelength either through the field strength in K by means of a power supply (if the wiggler is electromagnetic) or by changing the gap of a permanent magnet wiggler. The output wavelength can also be controlled through the input electron beam energy—hence, the statement that FELs can provide lasing at any wavelength. There are, however, practical and physics performance limitations to the operating range, which will become clearer in the discussions that follow.

4.2.3 Gain and Bandwidth

The small signal gain of an FEL is given by

$$g = 31.8 \, (I/I_A)(N^2/\gamma) B \eta_l \eta_f \eta_\mu \qquad (4.2)$$

where $I_A = 17$ kA, $B = 4\xi[J_0(\xi) - J_1(\xi)]^2$, and $\xi = K^2/[2(1+K^2)]$. The last three terms are degradations due to finite emittance, energy spread, and optical electron beam overlap. Here I is the peak current, N is the number of wiggler periods, and J is a Bessel function.

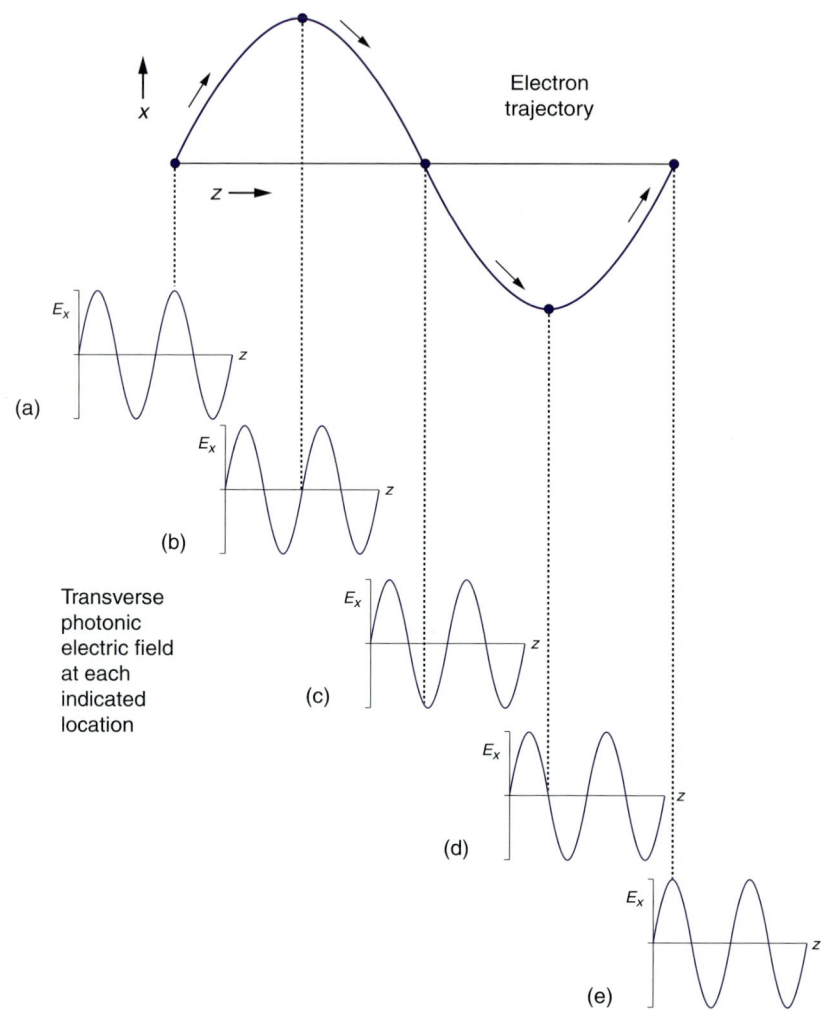

FIGURE 4.2 Resonant condition. The sinusoidal orbit of the electrons is illustrated over one wiggler period. We show an electron position against the resonant photon transverse electric field at each location. They both travel at the speed of light but, with a longer path to travel, the electron slips back one optical wavelength for each "wiggle." (*a*) The direction of the electron's transverse motion is in the same direction as the electric field, so the electron does work on the field—that is, it gives up some of its energy to the field. (*b*) The motion and electric field are at right angles, so no work is done. (*c*) Both the direction of motion and the sense of the field have reversed; so, again, there is transfer of energy from the electron to the field. (*d*) It is neutral again. (*e*) The electron is now back to the comparable position of (*a*) but has slipped back one optical period.

The gain is provided over a fractional bandwidth given by $1/2N$; as opposed to more conventional lasers, all the electron beam power can, in principle, be extracted from a narrow line within this bandwidth. If the input electron beam's energy spread is greater than or of the same order as this value, then the gain will be reduced; electrons whose energy falls outside this range will not significantly participate in the interaction.

Because the gain is limited to the $1/2N$ bandwidth, it is clear that as electrons give up energy, they eventually fall out of the resonance condition, as defined in Eq. (4.1). This is where the process stops unless something is done, such as changing the wiggler parameters as a function of distance along the wiggler, "tapering" the field strength to lower values to keep the (now-lower) energy electrons in resonance with the same wavelength. This was the approach initially investigated during the Strategic Defense Initiative (SDI) era for increasing the performance of high-power FELs.[3,4] The physics of this approach has been well demonstrated (although the product of the gain and efficiency for a given FEL system is constant, so that at some point, the gain is so low that optical losses prevent extraction of any more power). From a practical systems point of view, such an approach may not be advantageous, because the exhaust electron beam's energy spread also increases, which may render impractical the ability to recover the electron beam energy (see below).

Typically 50 or more wiggler periods are required to get sufficient gain such that extraction of 1 percent of the electron beam power is a reasonable expectation ($1/2N \sim 0.01$). The 99 percent of beam power that remains leaves the system at nearly the speed of light; because the lasing medium is in a vacuum, little distortion of the optical mode can occur. Optical mode distortion due to thermal effects in the lasing medium is a bane of conventional high-power, solid-state lasers, but it does not occur in FELs. However, uncompensated thermal distortion on FEL oscillator mirrors can lead to mode degradation, loss of gain, and so on. In very high-gain systems, the needle-thin electron beam only provides gain on axis, effectively providing a mode filter to keep the output at high-beam quality.

Not all of the electrons give up their energy equally; some are left out of the extraction process by virtue of having started at the wrong optical phase relative to the ponderomotive wave. As a consequence, these electrons remain at the initial energy or may even be slightly accelerated. As the process of energy extraction proceeds down the wiggler, the electron energy spread gradually increases. Experimentally it is observed that extrema electrons may have a total energy spread up to six times the average energy loss.[5] For this reason, once having lased, the electron bunch beam quality is usually unsuitable to permit reacceleration and reinsertion into the FEL a second time.

The electrons have one other potentially limiting physical parameter that is analogous to optical beam quality of photon beams and is called emittance. The *emittance* is defined as the product of the beam width and the divergence. The normalized emittance—that is, the emittance times γ—is a conserved quantity. In other words, after initial acceleration, the normalized emittance may only degrade or, at best, remain the same unless acted on by nonconservative forces. If the electron trajectories point outside the optical mode, then it is obvious that little gain could occur. The normalized emittance ε_n must lie within $\lambda_s/4\pi$ for gain to remain undegraded. Another way of looking at this is to realize that Liouville's theorem (and the second law of thermodynamics) says that it is not possible to make a brighter optical beam than the electron beam from which it is being made. The power out of the FEL is simply the electron beam power (voltage E times current $<I>$) times the FEL extraction efficiency.

4.2.4 Practical Considerations

Having described the FEL interaction, it should be recognized that the implementation of such a system can be either as an oscillator or as an amplifier, with each having attendant features and drawbacks. An oscillator has the following advantages: it does not require a seed laser, the required wigglers are short, and the peak currents required are modest. The oscillator's tunability is limited primarily by the mirrors and coatings used. High-power mirrors typically have 10 percent bandwidth due to their quarter-wave stack of dielectric reflection coatings. One mirror can be made partially transmissive to outcouple the light. Typically the gain at saturation is kept low (~20 percent), because for a given FEL, the product of the gain and efficiency is a constant. The small signal gain needs to be at least three times the gain at saturation for efficient energy extraction.

The price one pays for such advantages is dealing with the issue of thermal loading on the mirrors.[6] In addition, the alignment and figure tolerances of such an optical cavity are quite tight. Because the optical mode must match the electron beam in order to achieve high gain and energy extraction, the optical cavity, especially for high-power operation, tends to be long with a tight central core. The FEL interaction naturally produces harmonics at a power approximately 10^{-H} of the fundamental, where H is the harmonic number.[7] If low-order harmonics lie in the ultraviolet (UV) range, then care must be taken that the mirror coatings can live under the UV fluence.

Although an amplifier configuration does away with having to deal with high flux on mirrors (except the output mirror), it does necessitate a seed laser and significantly longer wigglers. In wavelength regimes where mirrors do not exist, operating in this manner is the only option. Typically such an FEL would provide a gain of 100 or more, and the electron beam–wiggler combination must provide for

High-Power Free-Electron Lasers

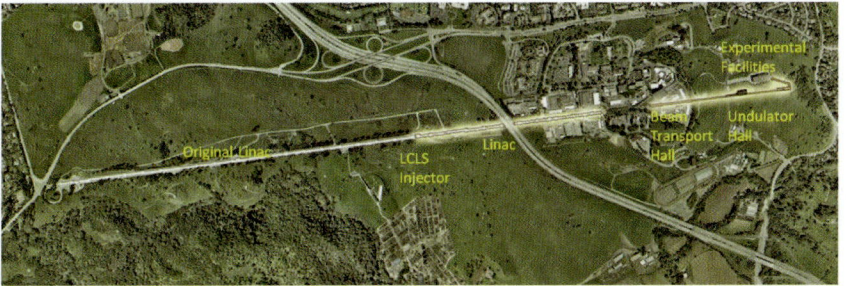

FIGURE 4.3 The Stanford Linear Accelerator Center with a layout of the Linac Coherent Light Source x-ray FEL. The LCLS utilized the final third of the SLAC linac with a new injector and undulator added. (*Courtesy John Galayda*)

this higher capability. It is feasible to operate the FEL as a self-amplified spontaneous emission mode in which the signal grows from noise, but this requires yet a longer wiggler; in addition, the output is likely to exhibit the characteristics of amplified noise. In some regimes, however, this may be the only option possible. For example, the Linac Coherent Light Source (LCLS) x-ray FEL at the Stanford Linear Accelerator Center (SLAC) uses a 120-m wiggler to produce a 10-keV x-ray photon pulse of extraordinary brightness and peak power grown strictly from noise (Fig. 4.3).[8] This is not likely to be the method of choice for a high-average-power system, however.

A hybrid design developed at Los Alamos National Laboratory (LANL) and called a Regenerative-Amplifier FEL (RAFEL) is another option.[9] The RAFEL is essentially a high-gain oscillator in which a small amount of feedback from the output is used to sustain the lasing. The system's high gain relieves, to a great extent, the tight tolerances on the optics and mitigates against thermal loading issues in the mirrors.

4.3 Hardware Implementation

4.3.1 Overview

To generate high-average-power FEL light, it is necessary to start with a very high-average-power–accelerated electron beam. Luckily, this technology has been extensively studied for decades because of its uses in nuclear physics; high-energy physics; and materials research on storage rings, neutron sources, and so on. There exist several gigawatt-level (10^9 watts of continuous beam power) average-power electron beams and hundreds of others at lesser high average powers. The goal of high-power FEL research has been to effectively harness such linear accelerator approaches to produce beams suitable for FELs. Even 1 percent energy extraction from such a beam would yield an incredible photon source.

An FEL system comprises an electron source (injector), an accelerator or linac with electron beam transport magnets, the wiggler, an optical system, perhaps an energy recovery system, and the dump. These are all supported by a number of auxiliary systems, such as power sources, cooling, alignment, controls, and so forth. A disadvantage of FELs is that all these systems are needed, even for low-power output. An advantage is that they do not get much bigger for high-average-power output. The discussion that follows covers technologies of these main subsystems as considered for high-average-power operation in the infrared to visible region. Other technologies may be more appropriate for other wavelength regions or for use at low average power.

4.3.2 Injectors

The injector is the most critical component in the entire FEL system, because the electron beam's quality can only degrade once the beam is formed. Because it is difficult to make high-quality continuous electron beams, the performance of most FELs is set by the injector's ability. In the search for suitable injectors, many approaches have been, and are being, adopted, but no clear winner for continuous operation has arisen from the group. Present candidates include high-voltage direct current (dc) guns with thermionic cathodes or photocathodes, copper radio frequency (RF) cavities with a photocathode, and superconducting RF guns with photocathodes.

The major issue that high-average current injector designers have is the continuous production of bunch charges that are so high that nonlinear space charge forces play a significant role in their control. Other, low-power FELs have dealt with this issue through several strategies: imposition of compensatory solenoid fields in a manner pioneered by Sheffield and Carlsten at LANL,[10] high initial cavity gradients, and tailoring of the density profiles longitudinally and transversely to linearize the forces. This design approach permits the production of electron bunches with 1 nanocoulomb (nC) of charge at a normalized emittance less than 1 mm·mrad. Such performance has yielded UV lasing using electron beam energies of only 45.2 MeV[11] and is presently driving the operation of the world's first hard x-ray laser, the LCLS at SLAC[8] (Fig. 4.3). High brightnesses at high bunch charge are significantly dependent on the high-cavity electric field gradients achievable in pulsed structures, because these gradients can accelerate the beam before space charge forces can work to degrade it. Typically a minimum of 20 to 40 MV/m is desired on the photocathode surface, although operating gradients of up to 125 MV/m at the cathode have been reported.[12] Unfortunately such high gradients cannot be maintained continuously—or even at high-duty factor—because of enormous associated ohmic losses in the RF cavities. Neither can such gradients be maintained in direct current fields, which are typically limited by field emission to 6 to 10 MV/m. Other strategies are employed in this case to help maintain brightness in continuous wave (CW) beam production.

A proponent of high-voltage dc guns with thermionic cathodes is the Budker Institute of Nuclear Physics in Russia. Researchers there have successfully produced up to 22 milliamperes (mA) average current with a normalized emittance of 30 mm·mrad.[13] Producing high average current in thermionic cathodes is straightforward. The main difficulty in applying this technology to short-wavelength FELs is that the emittance of such a system tends to be marginal for operating in the shorter infrared regions due to the degrading effect of the modulating grid. It is also technically difficult to produce the very short bunches needed for subsequent acceleration; therefore, RF buncher cavities are required in addition to accelerating cavities. The transport of the electrons through these cavities at low energies gives space charge forces an opportunity to degrade brightness.

To eliminate the need for grids and to accelerate the beam quickly so that space charge forces do not cause the electron beam quality to degrade while the electrons are at low energy, a group at Boeing produced a high-average-current RF photoinjector.[14] The copper cavity operated at 433 MHz. The injector used a mode-locked green laser on a CsKSn cathode to produce 25 percent duty factor pulses of 135 mA average current. The normalized emittance was 12 mm·mrad, which is suitable for short infrared (IR) lasing. The limitation of such a system was twofold—first, the cathode degraded due to the relatively poor vacuum environment in the RF cavity (roughly 3 hours in this case), and second, the RF power dissipation on the walls of the copper cavity was quite significant and led to difficulty in cooling the cavity, in addition to representing a significant overall power drain. Because of the power dissipation, average accelerating gradients are limited in such systems to around 6 MV/m. Nonetheless this effort, which was performed in 1986, remains a benchmark for this technology.

High-voltage direct current guns with photocathodes were used by the Thomas Jefferson National Accelerator Facility (Jefferson Lab) to produce a high-quality short-pulse beam of greater than 9 mA with long life. This long life was available because the geometry of dc guns is better for vacuum pumping.[15] The electron beam quality was suitable for lasing into the visible region, despite limitations in the voltage gradient to less than 4.5 MV/m due to high-voltage breakdown. This gradient limitation may be a factor in determining whether such a system can be scaled to yet higher currents, but efforts are underway at the Jefferson Lab to scale up the performance.

A technical challenge in the design of all photoinjectors is the need for an ultrahigh vacuum to avoid poisoning the cathode material. Vacuums of 10^{-9} to 10^{-10} torr are required for most cathode materials, with water vapor being a key poisoning element. Typically partial pressures of 10^{-11} torr of water are desired to maintain high quantum efficiency (up to 15 percent). Even when the cathode is not poisoned by an imperfect vacuum, however, back bombardment of ions created by the electron current onto the cathode surface can result in lifetime limitations. Lifetime is thus governed by total integrated charge delivered, rather than by time.

The laser source for photocathodes can be doubled, tripled, or quadrupled yttrium aluminum garnet (YAG) or yttrium lithium fluoride (YLF), depending on the cathode material. A number of different materials have found favor at different institutions: Cs_2Te has 13 percent quantum efficiency (QE) at 263 nm, with lifetimes of hundreds of hours; LaB_6, 0.1 percent QE at 355 nm, with lifetimes of 24 hours; K_2CsSb, 8 percent QE at 527 nm, with lifetimes of 4 hours; Cs_3Sb, 4 percent QE at 527 nm, with lifetimes of 4 hours; and GaAs(Cs), 5 percent QE at 527 nm, with lifetimes greater than 40 hours (see Refs. 16 and 17 for a review of many cathode materials). The lifetime data quoted here should be taken with some degree of skepticism, because little attempt has been made to unfold the effect of delivered charge and therefore back bombardment of the cathode life. Some cathode materials can be rejuvenated many times with oxygen cleaning and recesiation. Often, injector designs incorporate either a means to prepare and transfer new cathodes to the cavity or a cassette with multiple cathodes. For high-average current production, the use of UV laser sources is problematic because of the average power required, despite the relative robustness of the UV cathode materials. In the green (doubled YLF), it takes 22.4 W to produce 100 mA from a 1 percent QE cathode. Quadrupled YLF would require 44.8 W of short-pulse, mode-locked light to produce the same 100 mA at 1 percent QE. Such lasers are well beyond the commercial state of the art, and lifetime issues associated with the doubling crystals in the UV are an unsolved problem. Achieving the desired stability in phase and amplitude, as well as in reliability in the drive laser, is also not trivial. One would like amplitude stability of 0.5 percent or better and phase stability between the pulses of less than 1 picosecond (ps). Every doubling multiplies the amplitude noise by two.

To produce higher CW gradients while also delivering excellent vacuum around the cathode, groups are pursuing the development of a superconducting RF (SRF) injector cavity. To date, no SRF photogun has been demonstrated beyond some low-current demonstrations; however, such a development would have significant potential applications. A group at Forschungszentrum Dresden-Rossendorf who are pursuing such a development[18] believe it is possible to achieve nearly 20 MV/m on the cathode and 10 MV/m average in the cavity in a tesla-style 3½-cell 1300-MHz cavity. They have constructed a 1½-cell prototype. Although no fundamental physics issues have been identified, the engineering challenges are significant. First of all, it is difficult to hold the cathode accurately in the RF cavity surface and to prevent RF heating problems that would lead to the cavity going normal. It is also impossible to impose desired solenoid compensatory fields at the cathode because of the superconductor's shielding. Finally, the compatibility of the cathode itself with the superconducting environment is a potential issue. Ongoing research is aimed at answering these questions.

4.3.3 Accelerators

RF accelerators work by injecting short bunches of electrons in proper phase with an oscillating microwave field inside a cavity. The longitudinal electric field of the microwaves accelerates the electrons as energy is extracted from the microwaves. Electrons are such light particles that they travel at nearly the speed of light once they are greater than 1 MeV in energy; therefore, proper phasing of the microwave fields is straightforward. High-acceleration gradients are established by the fields: 60 MV/m or more in pulsed copper accelerators and 20 MV/m in modern CW SRF accelerators. High ohmic losses in copper cavities lead to severe heat loads in high-duty-factor copper accelerators, even with gradients reduced to 6 MV/m. As a consequence, most copper accelerators operate at duty factor of 10^{-3}, which is sufficient for scientific research applications but useless for high-average-power applications. An exception is the low-frequency 180-MHz recuperator system, developed at the Budker Institute; this system produces a continuous 30-mA 18-MeV electron beam for FEL lasing. Upgrades to higher energy are underway.

The difficulty with copper ohmic losses led to the development of superconducting accelerator cavities made of niobium (Fig. 4.4). The

FIGURE 4.4 Niobium cavities inside a cryomodule with the RF waveguide feeds in red. The electron beam enters from the pipe in the right foreground.

FIGURE 4.5 An aerial view of the Continuous Electron Beam Accelerator at Jefferson Lab. The FEL facility building is top center and CEBA is below ground in a 7/8 mile circumference oval. The nuclear physics end stations are in the three grass-covered domes at lower right. The cryogenic helium refrigerator is housed in the building group at the center of the oval.

low dissipation of niobium operated at 2 K allows CW operation at high gradients, though with the complication of a requirement for helium refrigeration.

The SRF linac structure, typified by the Continuous Electron Beam Accelerator (CEBA) at Jefferson Lab (Fig. 4.5),[19] produces 6-GeV electron beams for nuclear physics research using 1497-MHz cavities operated at 2 K. Ohmic losses are reduced to negligible levels by using SRF structures (6 W per cavity at typical gradients), while maintaining high-acceleration gradients (5 to 18 MV/m).[20] Among many additional factors, the gradient achievable depends on frequency, with higher frequencies producing higher gradients because of the reduced likelihood of a defect occurring over the cavity surface. As with copper accelerators, higher average current can be transported in lower-frequency cavities; for 100 mA and above, frequencies below 1500 MHz are desirable. It is worth noting that the first FEL,[21] the first tapered wiggler oscillator,[22] and the first visible lasing on a linac-based FEL[23] operated using the Stanford Superconducting Accelerator. Since its original demonstration, this linac has been a workhorse, serving several generations of FELs, because the CW beam yields high stability of power, wavelength, phase, and pulse length. In recent years, it has been extremely successful as a user facility, producing IR light for a number of two-photon experiments, as well as continuing to investigate the physics of the FEL interaction. It has since been removed from the Stanford campus and relocated at the Naval Postgraduate School in Monterey, California.

4.3.4 Wigglers

The wiggler represents a mature commercial technology. Wigglers have been constructed with both helical and planar symmetry, as well as with normal and superconducting electromagnets, permanent magnets, or hybrid combinations of the two. Ferrite elements are also used to concentrate the field. The commercial success of these devices has been due not so much to the market drive from the FEL community but rather to the second- and third-generation synchrotron light sources, which can have many insertion devices and for which the required quality of the magnetic field is very high.

The technology of choice is wiggler-period dependent, and for long-wavelength applications, electromagnetic wigglers prevail. For wiggler periods of 6 cm down to 2 cm or less, permanent magnets with hybrid wiggler technology take over. These systems use $SmCo_5$ or NdFeB permanent magnets with flux channeled by vanadium permendur, or similar materials, to produce $K \approx 1$ for approximately 1-cm gaps. Originally developed by Halbach,[24] these devices can produce significant gain in the infrared and visible spectra. The Jefferson Lab IR Demo wiggler, manufactured by STI Optronics, has $K = 1$ at a 12-mm gap with a 2.7-cm wavelength and 40.5 effective periods.

High-power applications demand that the wiggler gap be significant to avoid impingement of stray electrons into the radiation-sensitive material. Tunability is achieved by varying either the electron beam energy or the field strength. If the wiggler is adjustable, then it is much easier to tune the wavelength, because electron transport systems are chromatic and require retuning if the beam energy is adjusted outside a narrow range. Tuning hybrid wigglers is performed by adjusting the pole gap.

4.3.5 The Optical Cavity

An FEL's optical cavity is often more difficult to engineer than are those for conventional lasers. The FEL requires excellent overlap between the electrons and the optical mode in order to achieve high optical gain. The electron beam's dimensions are small, which implies that the mode must also remain small, with a relatively short Rayleigh range but modest mode size variations within the wiggler. A broad performance optimum occurs with a Rayleigh range of around $1/\pi$ of the wiggler length. Angular alignment tolerances can be very tight—on the order of microradians. If the electron beam is several hundred micrometers in diameter, one might expect that overlap must be held to a few tens of micrometers out of, say, a 10-m cavity length. In addition, the cavity length must match a subharmonic of the linac-operating frequency to a very high accuracy. It is not unusual to require a 10-m optical cavity length to be correct to within a micrometer. The range over which the optical cavity can be varied and still result in lasing is called the *detuning length*. In the infrared, the output's bandwidth may seem broad because it is Fourier transform–limited due to the subpicosecond pulse lengths (perhaps only 10 waves long). The bandwidth that is observed in the output is due to the interplay between the slippage of the electron pulse back one optical wavelength for each wiggler period and the optical cavity length, which may be shorter than the interpulse spacing by a small amount (see Fig. 4.6a and b).

The optical cavity must operate in a vacuum and usually must be remotely controlled because of the radiation environment. The low outcoupling and tight optical modes typically found yield high peak and average powers on the optics. Higher-energy machines produce significant fluxes of hard UV at the FEL harmonics, which can lead to mirror damage.[25–27] Outcoupling the power requires a transmissive optic (potential materials and heating issues), hole outcoupling (relatively inefficient), unstable ring resonator designs with a scraper (extra mirror bounces), or a grating (difficulty in survival at high fluence).

Early simulation studies determined that FEL oscillators can only tolerate around 0.2 waves of distortion.[6] This has been experimentally confirmed, with the FEL output showing saturation when thermal effects lead to greater distortion.[28] To control such distortion requires exceptional mirror coatings and advanced mirror designs or other techniques to minimize the impact of local heating.

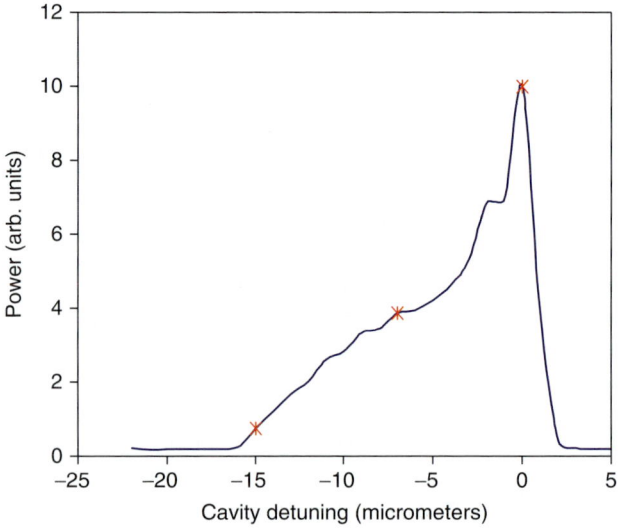

FIGURE 4.6a Power as a function of cavity length detuning in the IR Demo FEL.

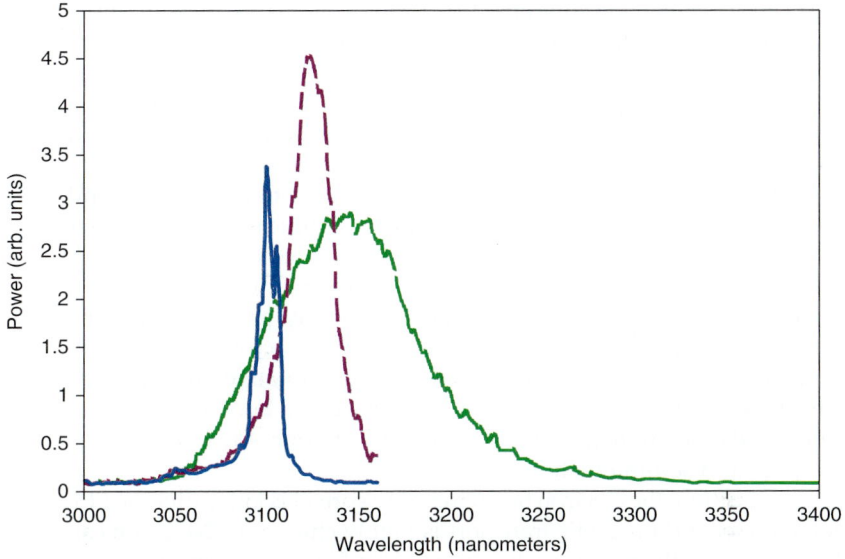

FIGURE 4.6b The corresponding spectra of operation at three points in the detuning curve. Narrowest (widest) spectrum is lowest (highest) power. At all times, the output is Fourier transform-limited—that is, the micropulse length changes as a function of detuning.

4.3.6 Energy Recovery

The low losses in superconducting cavities enhance the benefit of another technology—energy recovery. As was discussed earlier, the FEL interaction can remove approximately 1 percent of the electron beam power as light, while increasing the energy spread to 6 percent or more. That leaves 99 percent of the electron beam power available. When FELs are built for small, low-power facilities, the electron beam is typically dumped into a copper block after performing the lasing. In a CW high-power system, however, this dumping is extremely wasteful; therefore, the electron beam is reinserted into the linac 180 degrees out of phase with the RF fields, so that instead of being accelerated, the beam decelerates back down to the injection energy. The beam power is thus provided back to the RF fields. Because this process takes place in a nearly lossless superconducting cavity, the efficiency of such conversion is near perfect.

There are three benefits to such a procedure: (1) The electrical efficiency is substantially enhanced, because the only RF power that must be made up is the power lost to lasing and that beam power dumped at the end; (2) the power losses on the dump are substantially reduced, thereby simplifying the dump engineering design; and (3) because the electron beam energy is reduced below the photoneutron threshold of approximately 10 MeV, there is essentially no production of neutrons that could activate the environs. This third benefit is a very substantial factor in maintenance and radiation shielding. The price one pays for such benefits is the addition of a small amount of magnetic beam transport and the forced elimination of instabilities, which can result from feedback of the beam on itself. These issues have largely been resolved for optimized designs of low-frequency cavities.[29]

Optimization and control of the high-current transport to permit lasing and energy recovery are worthy of a significant paper on their own, and the scope is beyond what can be treated here. At high charge, there are issues associated with maintaining the electron beam quality as one accelerates, and especially as one bends, the beam. Some areas of this physics are still under active investigation and remain unresolved in terms of accurate quantitative predictability. The general strategy, though, is to allow the beam bunches to remain temporally long until just before the FEL interaction, so as to minimize external and self-interactions.

In addition, the electron beam's energy spread is large after the FEL interaction, and magnetic transport is highly chromatic. No beam can be lost during the transport, because even a few microamperes of current deposited locally in a vacuum pipe wall can burn a hole through it. One must also deal with the need to compress the beam's energy spread during the energy-recovery process. Otherwise the 6+ percent energy spread at 100 MeV would turn into 100 percent energy spread at 5 MeV. (See Fig. 4.7 for an overview of how this is accomplished.)

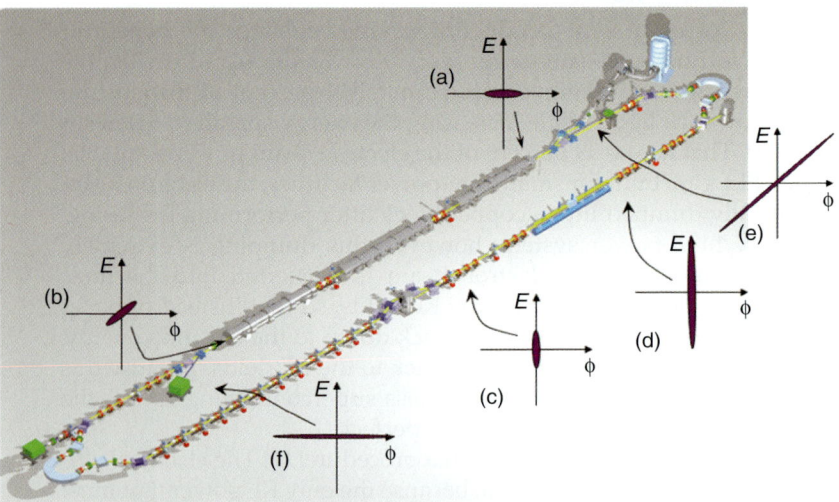

FIGURE 4.7 Requirements on phase space (energy vs. RF phase) shown at six points around the IR Demo energy recovering linac. (*a*) Long bunch in linac. (*b*) Chirped energy out of linac. (*c*) High-peak current (short bunch) at FEL. (*d*) Large energy spread out of FEL. (*e*) Energy compress using chirp while energy recovering. (*f*) "Small" energy spread at dump. (*Courtesy David Douglas*)

4.4 Status

Presently there exist only three FELs in the world operating at powers above 10 W—the Japan Atomic Energy Agency's (JAEA's) FEL, the recuperator FEL at the Budker Institute, and the Jefferson Lab's IR/UV Upgrade. These three FELs operate with energy recovery to improve overall efficiency, reduce RF power costs, and lower background radiation. The JAEA's system (Fig. 4.8) operates an FEL with millisecond pulses in the kilowatt range.[30] It is powered by an 8-mA, 17-MeV superconducting accelerator that produces 0.4-nC, 12-ps-long pulses. It first lased in August 2002 at 22 μm and extracted greater than 2.5 percent energy from the electron beam. CW operation was precluded by the capacity of the helium refrigerator.

The Budker system has energy recovered greater than 30 mA of average current. It also recently achieved two-pass acceleration and is on its way to a five-pass recirculation up to 80 MeV, followed by five passes down in energy (see Fig. 4.9) with multiple wiggler systems.[31] The Budker system has produced more than 400 W of average power at 60 μm. It uses 180-MHz copper-lined RF cavities. The system runs 1.5-nC, 70-ps-long pulses at 22.5 MHz.

The highest-power FEL in the world is Jefferson Lab's IR Upgrade FEL, which has produced 14 kW of average power at 1.6 μm (Fig. 4.10).[33] It is an upgrade of the IR Demo laser, which successfully demonstrated 2 kW of average power while energy recovering the electron beam energy.[32] The upgrade produces up to 9.3 mA of average current in 130 pC pulses at 75 MHz. The quality of the electron beam is sufficiently high

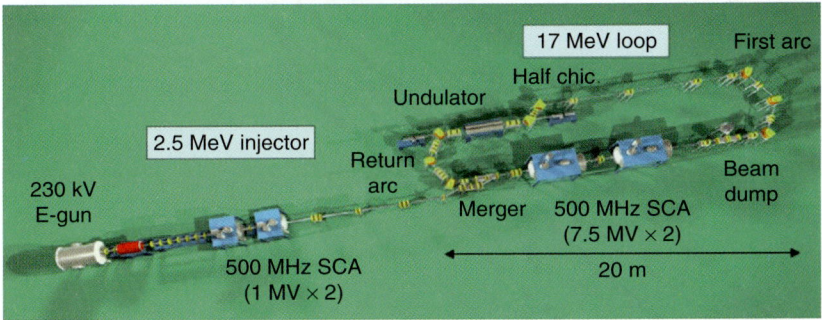

FEL wavelength is 22 μm and electron bunch charge is 0.5 nC.
The injector consists of 230 kV thermionic cathode DC gun, 83.3 MHz subharmonic buncher and two single-cell 500 MHz SCAs.
17 MeV loop consists of a merger chicane, two five-cell 500 MHz SCAs, a triple-bend achromat arc, half-chicane, undulator, return-arc, and beam dump.
First lasing in August 2002.

FIGURE 4.8 The JAEA FEL. The beam energy is 17 MeV with bunches of 0.4 nC at 20.8 MHz repetition rate. Light output is at 22 μm in 1 ms pulses at 10 Hz. (*Courtesy R. Hajima*)

94 Gas, Chemical, and Free-Electron Lasers

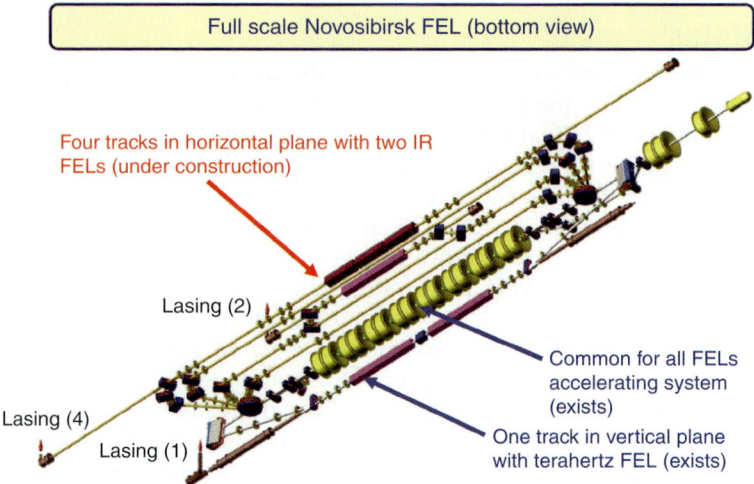

FIGURE 4.9 The Novosibirsk Recuperator. The beam energy is (40) 20 MeV with 1.5 nC pulses at (90) 22.5 MHz repetition rate. It produces 400 W at 60 μm. (Numbers in parentheses are under construction.) (*Courtesy V. Vinokurov*)

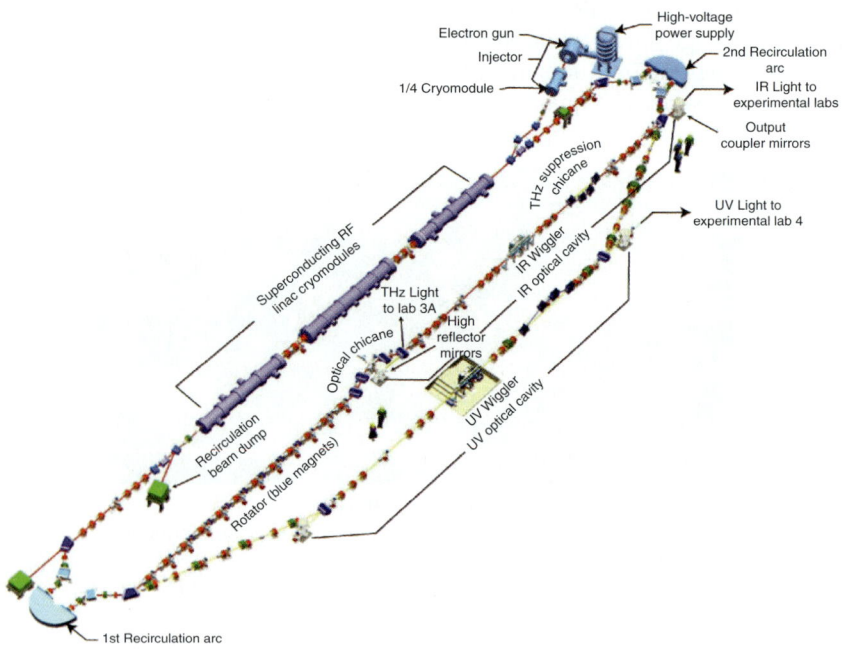

FIGURE 4.10 The Jefferson Lab IR/UV Upgrade FEL. The beam energy is 150 MeV with 135 pC pulses provided at up to 75 MHz, light output is 20/120/1 μJ/pulse in the UV/IR/THz bands at 250 nm, 1–14 microns, and 0.1–5 THz.

FIGURE 4.11 An image of the Jefferson Lab's IR Upgrade fifth through ninth harmonics in the visible while lasing at about 3 μm.

that it has also lased on the second,[34] third, and fifth[35] harmonics of the fundamental wavelength. A picture of the spontaneous emission from the harmonics is shown in Fig. 4.11. The linac also produced substantial power in the terahertz (THz) region from collective radiation in the magnetic bends.[36] A second FEL system (the UV Upgrade) on a parallel beamline has recently lased at 150 W average power in the ultraviolet.[37]

These systems have paved the way for future advances in high-average-power FELs and in the development of energy-recovered systems for high-brightness applications in photon research and development. Substantial development is still required before the FELs can reach their full potential. Perhaps in industrial applications, but certainly at the present level of performance, an exciting set of efforts is underway using free-electron lasers as the only real source of truly tunable high-average-power coherent radiation in the near- to mid-infrared region for scientific research.

References

1. Brau, C. A., *Free-Electron Lasers*, Academic Press, Boston, 1990.
2. Freund, H. P., and Neil, G. R., "Free Electron Generators of Microwave Radiation," *Electron Beam Generators of Microwave Radiation Proc. IEEE*, 87(5): 782–803, May 1999.
3. Feldman, D. W., Warren, R. W., Carlsten, B. E., Stein, W. E., Lumpkin, A. H., Bender, S. C., Spalek, G., et al., "Recent Results of the Los Alamos Free-Electron Laser," *IEEE J. Quantum Electron.*, QE-23: 1476–1488, 1987.
4. Christodoulou, A., Lampiris, D., Polykandriotis, K., Colson, W. B., Crooker, P. P., Benson, S., Gubeli, J., and Neil, G. R., "Study of an FEL Oscillator with a Linear Taper," *Phys. Rev. E.*, 66(056502), 2002.

5. Benson, S., Beard, K., Biallas, G., Boyce, J., Bullard, D., Coleman, J., Douglas, D., et al., "High Power Operation of the JLab IR FEL Driver Accelerator," Particle Accelerator Conference (PAC 07), Albuquerque, New Mexico, June 25–29, 2007.
6. McVey, B. D., "Three-dimensional simulations of free electron laser physics," *Nucl. Inst. and Meth.*, A250: 449–455, 1986.
7. Shinn, M., Behre, C., Benson, S., Douglas, D., Dylla, F., Gould, C., Gubeli, J., et al., "Xtreme Optics—The Behavior of Cavity Optics for the Jefferson Lab Free Electron Laser," Proceedings of the SPIE Boulder Damage Symposium XXXVIII, *SPIE* 6403: 64030Y-1, 2006.
8. Emma, P., "Commissioning Status of the LCLS X-Ray FEL," Working Paper TH3PBI01, Proceedings of the 2009 Particle Accelerator Conference, Vancouver, 2009.
9. Sheffield, R. L., Nguyen, D. C., Goldstein, J. C., Ebrahim, N. A., Fortgang, C. M., and Kinross-Wright, J. M., "Compact 1 kW Infrared Regenerative Amplifier FEL," *Free-Electron Laser Challenges*, P. G. O'Shea and H. E. Bennett, eds. *Proc. SPIE*, 2988: 28–37, 1997.
10. Carlsten, B. E., "New Photoelectric Injector Design for the Los Alamos National Laboratory XUV FEL Accelerator," *Nucl. Instrum. Meth.*, A285: 313–319, 1989.
11. O'Shea, P. G., Bender, S. C., Byrd, D. A., Early, J. W., Feldman, D. W., Fortgang, C. M., Goldstein, J. C., et al., "Demonstration of Ultraviolet Lasing with a Low Energy Electron Beam," *Nucl. Instrum. Meth.*, A341: 7–11, 1994.
12. Schmerge, J. F., Reis, D. A., Hernandez, M., Meyerhofer, D. D., Miller, R. H., Palmer, D. T., Weaver, J. N., et al., "SLAC RF Photocathode Gun Test Facility," *FEL Challenges*, Proceedings of SPIE Conference, SPIE, San Jose, California, February 13–14, 1997 (SPIE 2988: 90–96).
13. Gavrilov, N. G., Gorniker, E. I., Kayran, D. A., Kulipanov, G. N., Kuptsov, I. V., Kurkin, G. Y., Kolobanov, E. I., et al., "Status of the Novosibirsk High Power Free Electron Laser Project," *FEL Challenges*, Proceedings of SPIE Conference, San Jose, CA, February 13–14, 1997 (SPIE 2988: 185–187).
14. Dowell, D. H., Bethel, S. Z., and Friddell, K. D., "Results from the High Average Power Laser Experiment Photocathode Injector Test," *Nucl. Instrum. Meth.* A356: 167–176, 1995.
15. Hernandez-Garcia, C., O'Shea, P., and Sutzman, M., "Electron Sources for Accelerators," *Physics Today*, 61: 44, 2008.
16. Kong, S. H., Kinross-Wright, J., Nguyen, D. C., and Sheffield, R. L., "Photocathodes for Free Electron Lasers," *Nucl. Instrum. Meth.*, A358: 272–275, 1995.
17. Michelato, P., "Photocathodes for RF Photoinjectors," *Nucl. Instrum. Meth.*, A393: 455–459, 1997.
18. Michalke, A., Piel, H., Sinclair, C. K., Michelato, P., Pagani, C., Serafini, L., and Peiniger, M., "Photocathodes Inside Superconducting Cavities," Fifth Workshop on RF Superconductivity, Hamburg, Germany, 1991.
19. Krafft, G., and Bisognano, J., "On Using a Superconducting Linac to Drive a Short Wavelength FEL" *Proc. 1989 Particle Accelerator Conference*, 1256, 1989.
20. Neil, G. R., "Frontier Accelerator Technologies," Eighth International Topical Meeting on Nuclear Applications and Utilization of Accelerators (AccApp'07), Pocatello, Idaho, July 30–August 2, 2007.
21. Deacon, D. A. G., Elias, L. R., Madey, J. M. J., Ramian, G. J., Schwettman, H. A., and Smith, T. I., "First Operation of a Free-Electron Laser," *Phys. Rev. Lett.*, 38: 892–894, 1977.
22. Edighoffer, J. A., Neil, G. R., Hess, C. E., Smith, T. I., Fornaca, S. W., and Schwettman, H. A., "Variable-Wiggler Free-Electron-Laser Oscillation," *Phys. Rev. Lett.*, 52: 344–347, 1984.
23. Edighoffer, J. A., Neil, G. R., Fornaca, S., Thompson, H. R., Smith, T. I., Schwettman, H. A., et al., "Visible free-electron-laser oscillator (constant and tapered wiggler)," *Appl. Phys. Lett.*, 52: 1569–1570, 1988.
24. Halbach, K., "Design of Permanent Multipole Magnets with Oriented Rare Earth Cobalt Material," *Nucl. Instrum. Meth.*, 169: 1–10, 1980.
25. Couprie, M. E., Garzella, D., and Billardon, M., "Optical Cavities for UV Free Electron Lasers," *Nucl. Instrum. Meth.*, A358: 382–386, 1995.

26. Hama, H., Kimura, K., Hosaka, M., Yamazaki, J., and Kinoshita, T., "UV-FEL Oscillation Using a Helical Optical Klystron," *FEL Applications in Asia,* T. Tomimasu, E. Nishimika, T. Mitsuyu, eds., Ionics Publishing, Tokyo, 1997.
27. Yamada, K., Yamazaki, T., Sei, N., Suzuki, R., Ohdaira, T., Shimizu, T., Kawai, M., et al., "Saturation of Cavity-Mirror Degradation in the UV FEL," *Nucl. Instrum. Meth.,* A393: 44–49, 1997.
28. Neil, G. R., Benson, S. V., Shinn, M. D., Davidson, P. C., and Kloeppel, P. K., "Optical Modeling of the Jefferson Laboratory IR Demo FEL," *Modeling and Simulation of Higher-Power Laser Systems IV,* Proceedings of SPIE Conference, San Jose, CA, February 12–13, 1997, (SPIE 2989: 160–171).
29. Neil, G. R., and Merminga, L., "Technical Approaches for High Average Power FELs," *Rev. Modern Physics,* 74: 685, 2002.
30. Hajima, R., "Current Status and Future Perspectives of Energy Recovery Linacs," Working Paper MO4PBI01, Proceedings of the 2009 Particle Accelerator Conference, Vancouver, 2009.
31. Vinokurov, N., Dementyev, E. N., Dovzhenko, B. A., Gavrilov, N., Knyazev, B. A., Kolobanov, E. I., Kubarev, V. V., et al., "Commissioning Results with the Multipass ERL," Working Paper MO4PBI02, Proceedings of the 2009 Particle Accelerator Conference, Vancouver, 2009.
32. Neil, G. R., Bohn, C. L., Benson, S. V., Biallas, G., Douglas, D., Dylla, H. F., Evans, R., et al., "Sustained Kilowatt Lasing in a Free-Electron Laser with Same-Cell Energy Recovery," *Phys. Rev. Lett.,* 84: 662–665, 2000.
33. Benson, S., Beard, K., Biallas, G., Boyce, J., Bullard, D., Coleman, J., Douglas, D., et al., "High Power Operation of the JLab IR FEL Driver Accelerator," Particle Accelerator Conference (PAC 07), Albuquerque, New Mexico, June 25–29, 2007.
34. Neil, G. R., Benson, S. V., Biallas, G., Gubeli, J., Jordan, K., Myers, S., and Shinn, M. D., "Second Harmonic FEL Oscillation," *Phys. Rev. Lett.,* 87(084801): 2001.
35. Benson, S., Shinn, M., Neil, G., and Siggins, T., "First Demonstration of 5th Harmonic Lasing in a FEL," Presented at FEL 1999, Hamburg, Germany, August 23-26, 1999
36. Carr, G. L., Martin, M. C., McKinney, W. R., Jordan, K., Neil, G. R., and Williams, G. P., "High Power Terahertz Radiation from Relativistic Electrons," *Nature,* 420: 153–156, 2002.
37. Benson, S., Biallas, G., Blackburn, K., Boyce, J. Bullard, D., et al., "Demonstration of 3D Effects with High Gain and Efficiency in a UV FEL Oscillator," Proceedings of the 2011 Particle Accelerator Conference (PAC'11), New York, Mar. 28-Apr. 1, 2011.

PART 2
Diode Lasers

CHAPTER 5
Semiconductor Laser Diodes

CHAPTER 6
High-Power Diode Laser Arrays

CHAPTER 5
Semiconductor Laser Diodes
Laser Diode Basics and Single-Emitter Performance

Victor Rossin

Senior Engineering Development Manager, Communications and Commercial Optical Products, JDSU, Milpitas, California

Jay Skidmore

Senior Engineering Development Manager, Communications and Commercial Optical Products, JDSU, Milpitas, California

Erik Zucker

Senior Director of Product Development, Communications and Commercial Optical Products, JDSU, Milpitas, California

5.1 Introduction

Semiconductor laser diodes span a remarkably wide range of lasing wavelengths, materials systems, fabrication technologies, and applications. The primary use of *high-power* laser diodes has historically been as a pump source, or a laser that pumps or energizes another type of laser or optical amplifier. Diode-pumped, solid-state lasers and fiber lasers are the two main examples in the high-power or high-energy laser field. However, with advancements in robust, higher-brightness laser diode sources and more sophisticated fiber optic packaging techniques, semiconductor lasers are starting to see implementation as so-called direct-diode sources, where they are replacing traditional laser technologies, such as flash lamp–pumped or diode-pumped solid-state lasers and carbon dioxide (CO_2) gas lasers.

This chapter introduces the key attributes of the semiconductor laser diode to form the backdrop to its ubiquity. Although an overview of the physical mechanism of lasing in semiconductors is briefly presented, it is not the focus of this work. The wafer fabrication processes used to create the semiconductor laser chip are described, and the key processes that enable high-power laser performance are noted. State-of-the-art performance values for single-emitter lasers, both single spatial mode and multiple mode, are detailed. Understanding these values at the single-emitter level allows understanding of their scaling to one-dimensional and two-dimensional laser arrays, which are respectively known as laser "bars" and "stacks" and which are covered in Chap. 6. Basic single-emitter assembly concepts, fiber-coupled packaging, and reliability metrics and methods are also presented here.

5.2 Historical Growth of Power

The birth of the modern semiconductor laser took place in 1963 with two independent proposals for the double heterostructure laser design from Alferov and Kazarinov and from Kroemer.[1] Advances in two epitaxial growth techniques in the 1970s—molecular beam epitaxy (MBE) and metal organic chemical vapor deposition (MOCVD)—were important enabling technologies that allowed the creation of tightly controlled layer thickness and atomic composition, which are needed to grow quantum well (QW) active layers and which have the associated benefits of gain and reduction in threshold current. Significant commercialization of high-power laser diodes started in 1983, with the formation of Spectra Diode Labs.[2] The literature provides several excellent reviews of these early days.[1-3]

Continuous improvements in crystal growth technologies and the purity of materials sources drove improvements in the 1980s and 1990s. Advancements over the past 10 years have been driven by further refinements in laser design, which are focused on increased efficiency, improved facet passivation technology, robust die attach, and advanced heat sinking. As shown in Fig. 5.1, the past 17 years have seen steady growth in the reliable optical output power of commercial products. Both multimode and single-mode laser diode powers have increased by about 15 percent per annum. This growth rate is likely a function of increased investment in the required technologies. During the dot-com and telecom frenzy of the late 1990s, the advancement in 980-nm single-mode power increased to double the historic rate.

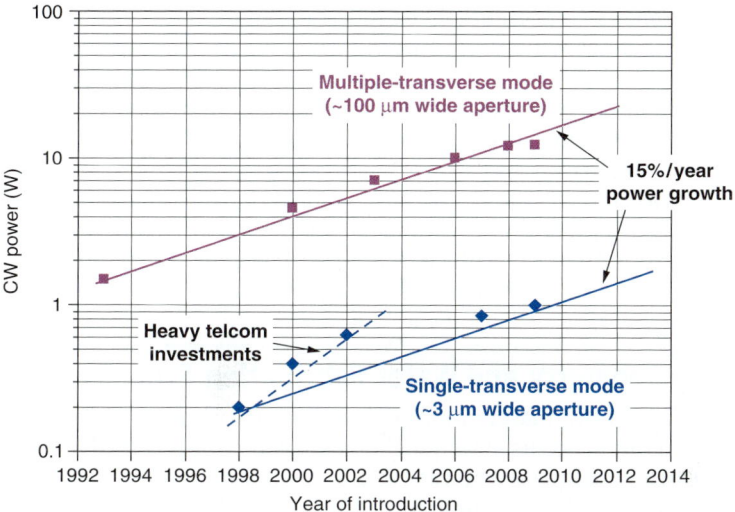

FIGURE 5.1 Growth in multimode and single-mode reliable continuous wave (CW) power for 9XX-nm.

5.3 High-Power Laser Diode Attributes

Several attributes distinguish the high-power laser class from other semiconductor lasers and are important to their utility. First is the rated optical power level at a specified reliability point. This is a key tradeoff in almost all high-power laser designs, because almost all high-power lasers can operate above the rated power, though at the expense of lower reliability. Power levels in the 10-W to 20-W continuous wave (CW) range are commercially available from a single-aperture source. The maximum power is typically limited by the linear power density (i.e., the power divided by aperture width) at the laser facet, where the light exits the confinement of the semiconductor waveguide and diffracts into air. More optical power may be obtained from a single chip either by increasing the width of the emission aperture or by forming monolithic arrays of these emitters on a laser "bar." Although these techniques increase the total power, they do so at the expense of brightness. Optical power is balanced by the etendue, or the two-dimensional spot size of the light in physical and numerical aperture space. The brightness of the laser, defined as the optical power divided by the etendue, is the physical parameter that dictates the extent to which various beam-combining methods may be used to form a single, higher-power beam from multiple single emitters. Recent advances in higher-brightness performance have come from both increases in reliable optical power from the chip and reduction of the etendue, especially in the far-field divergence of light emission.

The second attribute is the lasing wavelength. Here we focus on the commercially important 800 to 1000-nm band. The 808-nm laser has been the most widely used pump for Nd:YAG (yttrium aluminum garnet)-based solid-state lasers. More recently, 915-nm, 940-nm, and 976-nm lasers have been strongly growing due to their application in Er and Yb fiber laser and amplifier pumping and in Yb:YAG disk laser pumping.

A third attribute is the electrical-to-optical power conversion efficiency (PCE). Tremendous advances in PCE have occurred in the past several years, with hero results in the mid-70 percent range and commercial values in the mid-60 percent range.

5.4 Device Geometry and Wafer Fabrication Processes

A generic, high-power laser diode device geometry is shown in Fig. 5.2. Photon generation occurs at the junction between the *p*-type and *n*-type semiconductor materials when the diode is forward biased. Epitaxial growth of various layers simultaneously creates the *p*- and *n*-doped material and an optical waveguide in the "transverse" direction. Wafer-level processing creates the waveguide in the "lateral" direction. Cleaving of the wafer along mirror-smooth crystal planes creates parallel laser "facets," which form the laser resonator cavity.

Fabrication of the laser diode occurs at the wafer level, using semiconductor process steps similar to those used for silicon integrated circuit (IC) manufacturing. A typical process flow chart is shown in Fig. 5.3. Lasers are usually fabricated on 2-inch-, 3-inch-, or 4-inch-diameter *n*-type GaAs substrates. Various semiconductor layers are

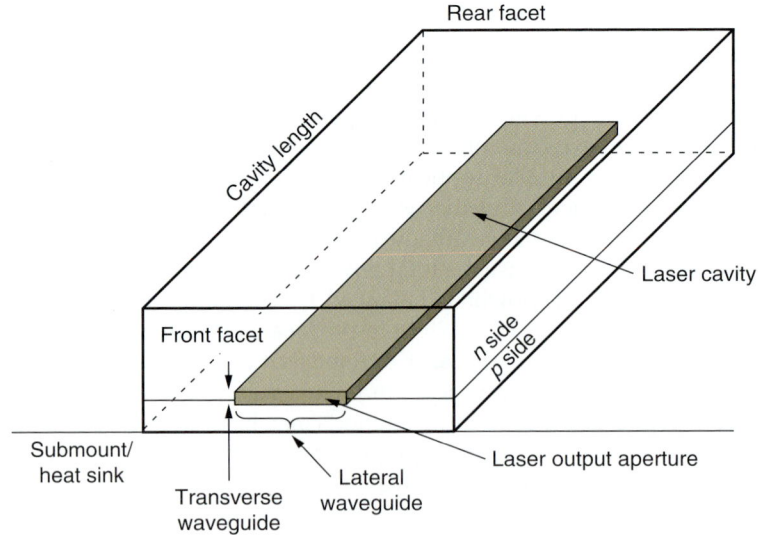

FIGURE 5.2 Illustration of a basic semiconductor laser and key terminology.

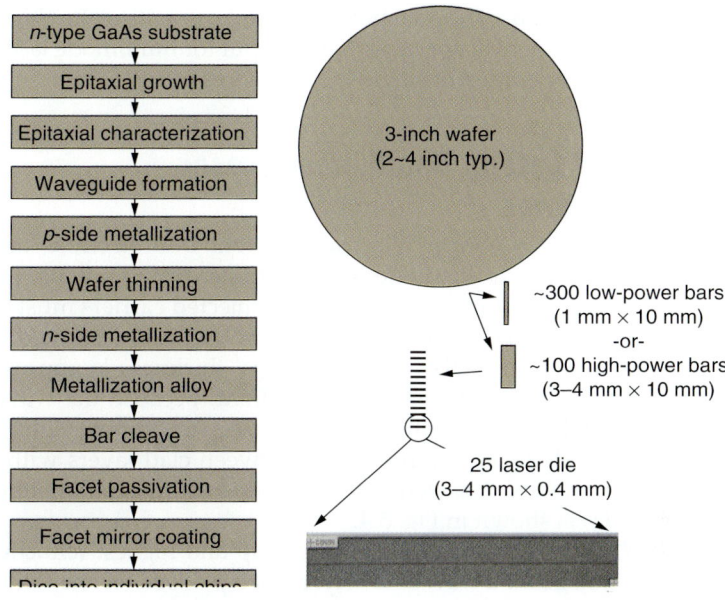

FIGURE 5.3 Typical wafer process flow for a semiconductor laser diode.

grown epitaxially by MOCVD or MBE and form the optical cladding and waveguide layers, as well as carrier confinement. This growth step is critical for both the laser's proper initial performance and its reliability. Sophisticated analytical tools are used to confirm material composition, layer thickness, doping levels, and defect density.

Structures for electrical and optical confinement in the lateral direction are then defined. Photolithography is used to pattern the desired laser geometry onto the wafer's surface. Various methods of dielectric deposition, etching, or ion implantation are used. On the p side of the wafer, a metallization stack is deposited to create an ohmic contact to the semiconductor, while also providing a stable surface for subsequent solder reflow or wire bonding. The wafer is then polished to a thickness of 100 to 150 μm for ease of subsequent cleaving and low electrical resistance through the substrate. Metallization is deposited on the n side of the thinned wafer and then briefly heated to alloy the contact to the semiconductor for low resistance. The wafer is cleaved into bars, which are then passivated and coated with a dielectric material to form the front (output) and rear mirror facets.

For high-power lasers, the foremost differentiating process is facet passivation and mirror coating. Because the facet power densities are so high—on the order of 100 MW/cm^2—care must be taken in the design and control of these processes. The details of these processes are tightly controlled trade secrets, and there are several competing methods. The purity and control of the epitaxial layers is the second key process required to ensure high reliability, high performance, and high yield.

Finally, the formation of the lateral waveguide is critical for high-brightness, low-numerical aperture (NA) output for multimode lasers and for kink-free operation of single-mode devices (see Sec. 5.9).

5.5 Vertical and Lateral Confinement Laser Diode Structures

Semiconductor lasers convert electrical current into electrons and holes that recombine at the diode junction to generate photons. For efficient operation, the optical mode and the injected carriers must be collocated and confined in space. Carriers are typically confined in one or more quantum wells (QWs). The QW thickness is approximately 10 nm or less and cannot confine light, because the wavelength is much larger than the QW thickness. To confine light, a vertical waveguide layer is sandwiched between clad layers with a lower refractive index. A sketch of this separate confinement heterostructure (SCH) is shown in Fig. 5.4.

The QW, which has the lowest energy gap and highest refractive index, is centered inside a waveguide layer that is p doped on one side of the QW and n doped on the other. Cladding layers have a higher energy gap and a lower refractive index than the waveguide layers. The thickness of the waveguide layers can be as thin as 50 nm or as thick as 1 μm or more. Because the optical mode is confined to the waveguide, its overlap with the gain-creating carriers confined to the QW is much less than 1. This overlap is called the transverse optical confinement factor (Γ) and can be as low as 1 percent. The waveguide index and energy gap can also be graded to increase the carrier capture in QW layer(s), referred to as a graded index separate confinement heterostructure (GRINSCH).[4]

For lateral optical and electrical confinement, additional postgrowth methods are used. The simplest lateral confinement can be achieved by blocking injection current outside the active stripe. One approach uses a dielectric layer on top of the semiconductor, with metal deposited through a window etched in the dielectric layer (Fig. 5.5a).[5] Alternatively, proton implantation may be used to create highly resistive regions in the cladding

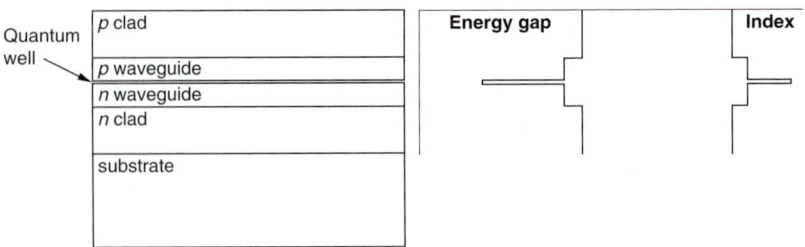

Figure 5.4 Layer structure of a separate confinement heterostructure (SCH) laser diode (left) and diagram of the energy gap and refractive index (right).

Semiconductor Laser Diodes

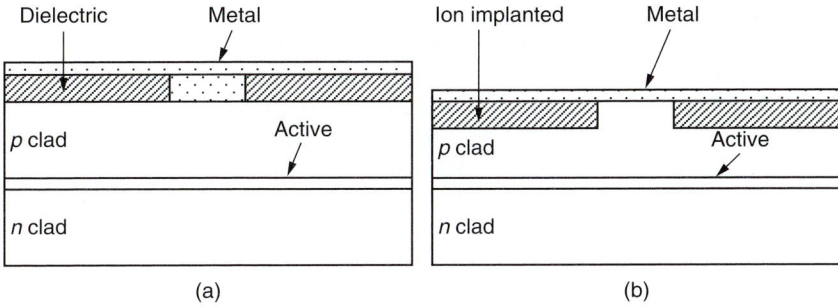

FIGURE 5.5 Laser structures with (a) dielectric current blocking and (b) ion implantation current blocking.

layer so that current flows only through the stripe that has no implantation (Fig. 5.5b).[6]

Current blocking results in a lateral distribution of injected carriers, which are defined by the stripe and broadened by current spreading under the blocking layers. This lateral distribution of injected carriers produces a distribution of gain and a modulation of complex effective index, which provides optical lateral confinement. Accordingly, these laser diode structures are called *gain guided* and are commonly used for broad-area lasers. These lasers have a wide aperture, allowing multiple lateral modes. For narrow-stripe, single-spatial-mode lasers, gain-guided structures are not suitable, because current spreading significantly widens the distribution of injected carriers. Moreover, carrier-induced reduction of the refractive index in the pumped region leads to antiguiding effects, further reducing optical confinement and the lateral contrast of the complex effective index.

To improve lateral confinement, a lateral refractive index step is introduced in index-guided structures. A simple, weakly index-guided ridge waveguide structure[7] can be formed by etching away portions of the cladding layers outside the stripe (Fig. 5.6a). Because semiconductor material is replaced with a lower-index dielectric, a lateral step in effective

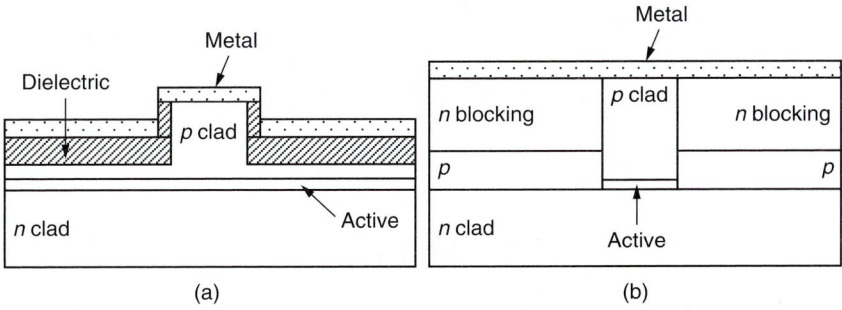

FIGURE 5.6 (a) Ridge waveguide laser and (b) buried-heterostructure laser.

index is achieved. The magnitude of the index step depends on how deeply the ridge is etched and is typically less than 10^{-2}.

A larger lateral index step and confinement can be achieved in buried-heterostructure lasers.[8] This structure is formed by a deep etch through the active layer and subsequent regrowth with wider-band-gap, lower-index layers to provide lateral mode confinement. Current blocking layers are also formed during regrowth (Fig. 5.6b). Although the regrowth process is well developed for InP/InGaAsP-based material systems, it is much more difficult in GaAs/AlGaAs material systems, due to oxidation of the aluminum-containing layers during the various process steps.

5.6 Efficiency of Laser Diodes

One of the major attributes of semiconductor diode lasers is the electrical-to-optical conversion efficiency, or wall-plug efficiency, which is the ratio of optical power P over electrical power (or the product of electrical current I and voltage V). High power conversion efficiency (PCE) is especially important for high-power lasers, because excessive heat can result in degradation of device performance. To dissipate excess heat, effective cooling is needed, which, in turn, requires more electric power and space, as well as additional packaging costs. Power loss in a semiconductor laser is divided among waste voltage, waste current, and optical loss. Waste voltage is a combination of the excess of the laser's turn-on voltage V_{to} compared with the voltage corresponding to the emitted photon energy V_λ and the voltage dropped over the series resistance R_s, which comprises semiconductor layers, metal layers, and wire bonds. Waste current is divided between the threshold current needed to reach the required gain and the leakage current that is commonly described by the internal quantum efficiency parameter η_i. Optical loss is due to distributed internal loss α_{int}—usually free carrier absorption—and external loss, such as light lost out the rear facet. A simple expression for PCE is

$$PCE = \frac{\eta_i\left(1 - \frac{I_{th}}{I}\right)}{\left(1 + \frac{\alpha_{int}}{\alpha_m}\right)\left(1 + \frac{V_{to} - V_\lambda}{V_\lambda} + \frac{IR_s}{V_\lambda}\right)} \tag{5.1}$$

where mirror loss α_m is

$$\alpha_m = \frac{1}{2L} \ln\left(\frac{1}{R_f R_r}\right) \tag{5.2}$$

$$V_\lambda = \frac{hc}{e\lambda} \tag{5.3}$$

Semiconductor Laser Diodes

Parameter	Value	Dimension
L	1.5	mm
R_f	0.01	
R_r	0.99	
λ	970	nm
I_{th}	280	mA
η_i	0.93	
α_{int}	2	1/cm
V_{to}	1.35	V
R_s	0.05	Ohm

TABLE 5.1 Typical Values for Laser Parameters

where L is laser cavity length; R_f and R_r are the reflectivities from front and rear facets, respectively; λ is the lasing wavelength; h is Plank's constant; c is the speed of light; and e is electron charge. For a typical 100-μm-wide, 970-nm laser (with parameters given in Table 5.1), the PCE is 63 percent at 2.5-W output power (2.65-A current). The remaining 37 percent of the power is waste heat and is distributed according to Fig. 5.7.

Equation (5.1) is only valid for the linear portion of the light versus current (L-I) characteristic. At higher power, self heating leads to rollover and additional losses. In general, lasers become less efficient at higher temperatures. Temperature dependences are usually described by two phenomenological parameters T_0 and T_1, which

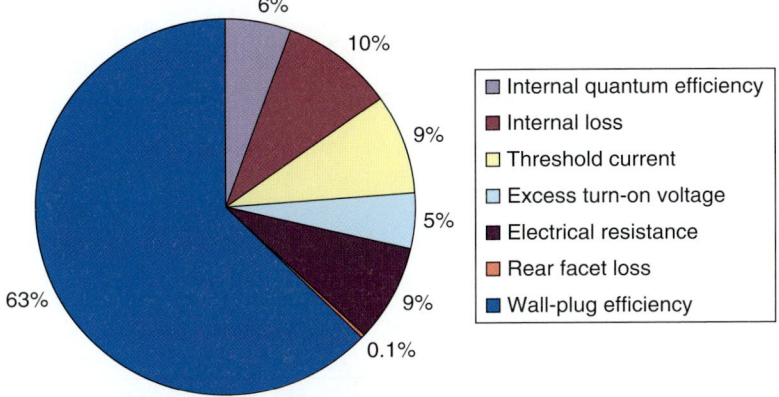

FIGURE 5.7 Distribution of total power from a laser diode. In this case, 63 percent of the input power generates useful light, while the remaining 37 percent results in waste heat from the indicated sources.

describe the increase of threshold current and the decrease of internal efficiency with temperature T:

$$I_{th} \sim \exp\left(\frac{T}{T_0}\right), \quad \eta_i \sim \exp\left(-\frac{T}{T_1}\right) \tag{5.4}$$

A significant effort has been made to increase the wall-plug efficiency of high-power lasers. Design optimization is focused on all contributors to the power waste: lowering of internal loss, increase of internal efficiency, and decrease of series resistance. A typical design tradeoff is in the clad-layer doping. Higher doping leads to lower series resistance and self heating, while also giving rise to higher internal losses due to free carrier absorption. Figure 5.8 shows the results of a design optimization for a 20-μm-wide, 915-nm AlGa(In)As/GaAs laser diode that peaks at a wall-plug efficiency of 73 percent at 25°C operating temperature.[9] This was achieved by decreasing internal loss to less than 1 cm^{-1} and achieving high values of $T_0 = 198$ K and $T_1 = 962$ K.

5.7 High-Power Broad-Area Laser Diodes

The maximum power from a diode laser P_{max} is proportional to the internal optical power density at catastrophic optical mirror damage P_{COMD}, according to the expression:[10]

$$P_{max} = \left(\frac{d}{\Gamma}\right) W \left(\frac{1-R}{1+R}\right) P_{COMD} \tag{5.5}$$

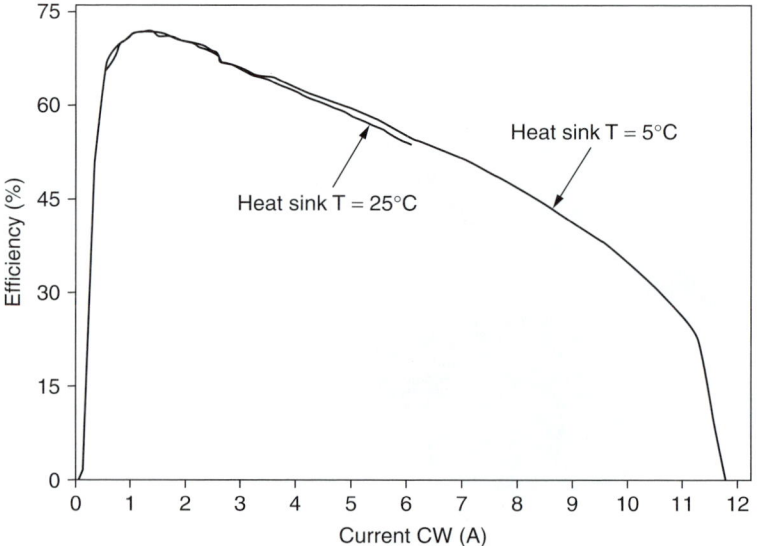

FIGURE 5.8 Dependence of CW power efficiency versus drive current recorded for an $L = 2$ mm, $W = 20$ μm laser diode at 5°C and 25°C.[9]

where W is the stripe width, R is the front facet reflectivity, d is the quantum well thickness, and Γ is the transverse optical confinement factor such that d/Γ is the equivalent spot size. Broad-area lasers are able to achieve high output powers due to a wide stripe width W. P_{COMD} is a function of the active region material and facet passivation techniques.[11–13] A P_{COMD} value as high as 24 MW/cm² was reported for CW operation of broad-area InGaAs/AlGaAs laser diodes lasing at 940 nm.[14] The transverse size of the optical mode can be increased using large optical cavity designs,[15] which provide a low optical confinement factor Γ. The optical confinement factor can be further reduced by designing an asymmetric waveguide structure[16] and using an optical trap layer.[17] Another design tool for achieving high optical power is to increase the laser cavity length. Larger cavity length leads to lower electrical and thermal resistances, which in turn result in reduced heating, higher efficiency, and higher thermal rollover powers. Efficient operation of long cavity length lasers requires low internal loss of less than 1 cm⁻¹. State-of-the-art broad-area lasers with stripe widths of approximately 100 µm have 4 to 5 mm cavity length and reach more than 20 W optical power in CW operation.[18–20] Reliable operating power for these lasers is rated in the 10 to 12 W range. Figure 5.9 shows 26.1 W of CW power achieved from a 5-mm-long, 90-µm-wide, 940-nm InGaAs/AlGaAs laser.[14]

Continuous wave power is limited not only by COMD but also by thermal rollover, as self heating results in decreasing efficiency. Pulse mode operation eliminates self heating, and much higher powers can be achieved. Figure 5.10 shows power as high as 32 W for 20-µs pulse duration and 1-kHz repetition rate for a 940-nm laser diode with a 100-µm emitting aperture.[21]

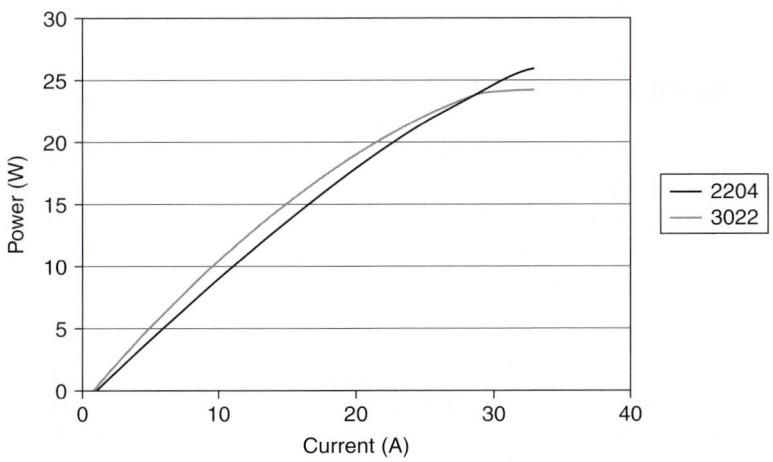

FIGURE 5.9 Experimental rollover light current characteristics for lasers with AlGaAs barrier series.[14]

Diode Lasers

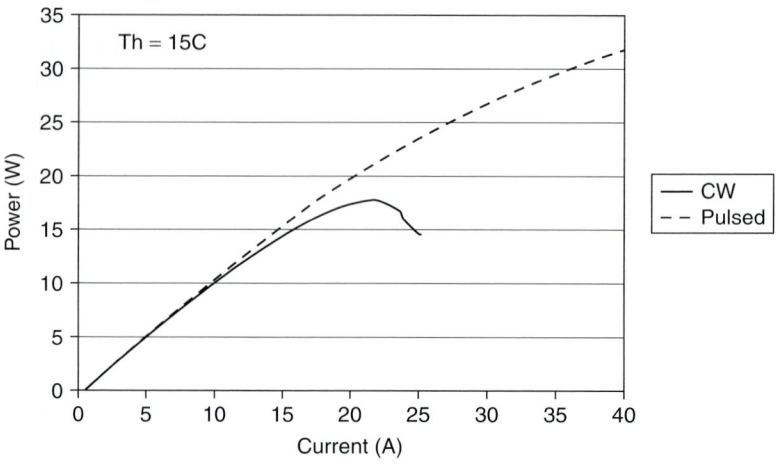

FIGURE 5.10 Light current characteristics of a 940-nm broad-area laser.[21]

The highest powers and efficiencies have been reported for diodes lasing in the 900 to 1000 nm wavelength range. These lasers have a strained InGaAs quantum well active region. For diodes lasing in the 800 to 870 nm wavelength range, the quantum well material is typically GaAs or AlGaAs, and the quantum well is lattice matched to the GaAs substrate. These short-wavelength lasers have higher threshold current density due to lower gain. They are more sensitive to temperature and less efficient than 9XX-nm lasers. Higher photon energy also causes lower COMD limits. Figure 5.11 shows power as high as 12.2 W from a 4.5-mm-long, 808-nm laser with 90-μm emitting aperture.[20] Reliable operating powers for 808-nm, 100-μm-wide broad-area laser diodes can reach the 5 to 6 W range, but more typically operate at 2 to 3 W.

5.8 High-Power Bars

Laser diodes can be arranged as an array on a single chip called a "bar." Important bar parameters are the fill factor, which is the ratio of the sum of emitter widths to the total bar width, and the pitch or spacing of the emitters. The standard bar width is 10 mm, although minibars with smaller widths are used for some applications. Low fill-factor bars allow high power per emitter, because there is less thermal cross talk between emitters. Linear power density as high as 85 mW/μm was reported for 9XX-nm bars with low fill factors in the 9 to 15 percent range.[22] To increase total output power, higher fill factors can be employed with a corresponding decrease in the linear power density per emitter, dropping to 45 mW/μm for a 50 percent fill-factor bar.[22] Power as high as 325 W was reported for a 1-cm-wide, 920-nm bar with 50 percent fill factor and proper cooling.[23] Figure 5.12 shows

Semiconductor Laser Diodes

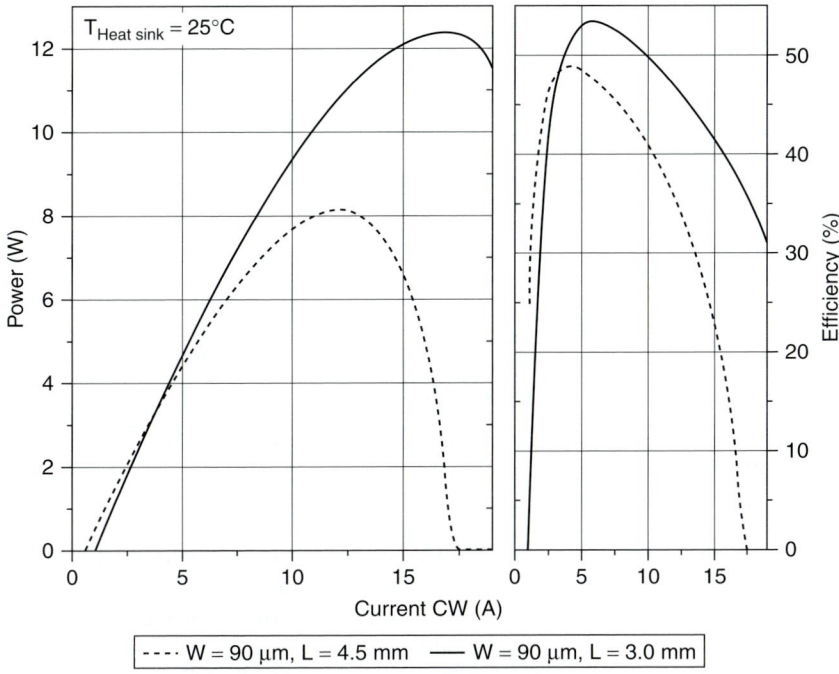

FIGURE 5.11 Room temperature power-current CW characteristics at 808 nm operating wavelength with different cavity lengths ($L = 4.5$ mm and $L = 3.0$ mm).[20]

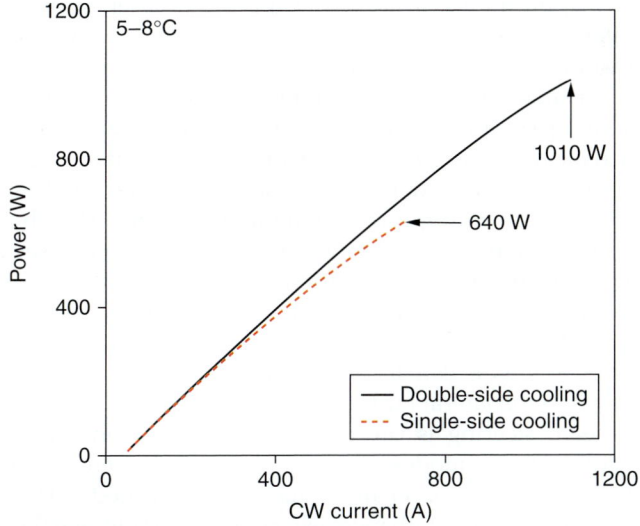

FIGURE 5.12 Light-current characteristics of a 940-nm bar (single-side, double-side cooling).[24]

Diode Lasers

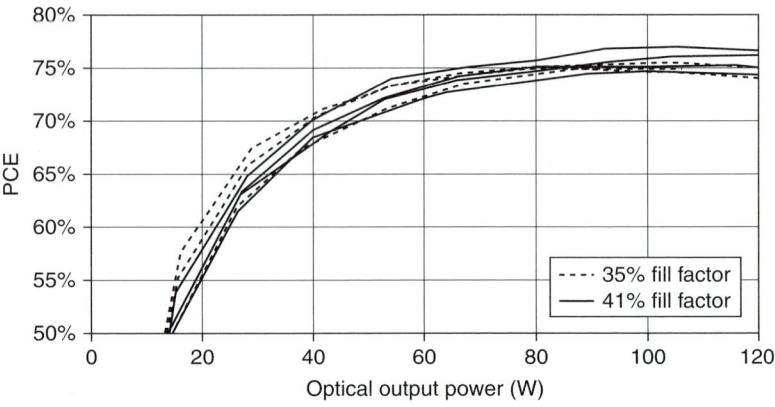

FIGURE 5.13 PCE characteristics of a 100-W, 940-nm bar with room-temperature cooling water.

demonstration of greater than 1000 W of power from a 940-nm bar with 83 percent fill factor and double-sided cooling.[24]

Significant advancements have been made on improving bar power conversion efficiencies.[25] Values of PCE greater than 70 percent for 940-nm, 80 to 120-W bars were reported.[26–28] Figure 5.13 shows PCE as high as 76 percent for a 100-W, 940-nm bar.

5.9 High-Power, Single-Mode Laser Diodes

For many applications, a narrow-stripe diode laser operating in a single lateral mode is necessary. The most widely used application is a 980-nm pump source in erbium (Er)-doped fiber amplifiers (EDFAs). Although a single-mode laser operates at lower absolute power than multimode broad-area lasers with wider emitting apertures, it can actually have higher brightness. The other important attribute of the single-mode laser is its stable diffraction-limited far field, which is essential for effective coupling into single-mode fiber. At high enough powers, diffraction-limited operation of a single-mode laser can be disrupted, which is usually observed as a nonlinearity, or "kink," in the light-versus-current characteristic. The kink power is an important parameter of single-mode lasers that limits usable power from the devices. A kink is usually accompanied by beam steering of several degrees in the lateral far field, as shown in Fig. 5.14.[29]

The beam steering is caused by coupling of fundamental and higher-order lateral modes.[29,30] A model for coherent coupling of fundamental and first-order mode has been proposed.[31] This model explains multiple kinks with increasing current by bringing the below-threshold first-order mode in and out of resonance, thus drawing power from the fundamental mode at each coherent kink. Increasing

Semiconductor Laser Diodes 115

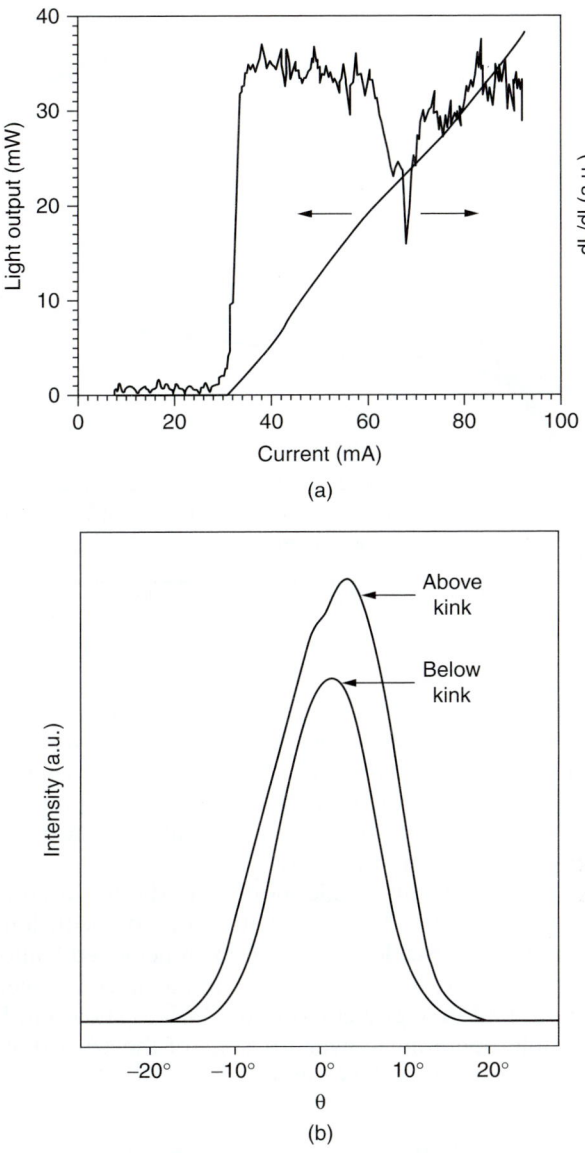

FIGURE 5.14 (a) Light output and dL/dI curves around kink, (b) far-field data below and above kink power level.[29]

kink power is an important design optimization for single-mode lasers.[32] Typically, to increase kink power, the stripe width must be narrowed to filter out higher-order modes. On the other hand, narrower stripe width leads to lower slope efficiency of the device due to higher loss of the fundamental mode. This trade-off leads to design

116 Diode Lasers

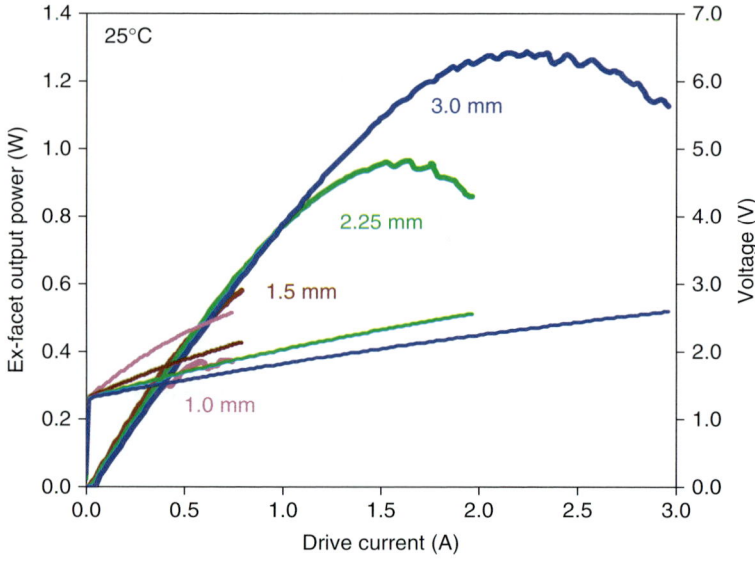

FIGURE 5.15 *L-I* and *I-V* characteristics for single-mode 980-nm lasers with different cavity lengths.

optimization between higher kink power and efficiency. A design that combines narrow and wide stripes in a three-section waveguide with a flared region achieves both high kink power and efficiency.[33] As with broad-area lasers, longer cavity length results in higher power of single-mode lasing due to lower series and thermal resistance and higher rollover power, as shown in Fig. 5.15.

State-of-the-art 980-nm single-mode laser diodes reach well over 1 W of output power.[34–36] For efficient operation with long cavity lengths, very low internal loss of 0.5 cm^{-1} was achieved,[37] allowing an extension of the cavity length to 7.5 mm and achieved power as high as 2.8 W, with kink-free power close to 2 W,[37] as shown in Fig. 5.16. Typical reliable operating power for state-of-the-art 980-nm pump laser diodes is in the 0.8 to 1 W range.

5.10 Burn-In and Reliability of Laser Diodes

Reliability of laser diodes is usually described by the so-called bath-tub curve, depicted in Fig. 5.17. As shown in the figure, there are three distinct regions. At early times, there is a region of decreasing failure rate, which is characteristic of infant failures. Then the failure rate stabilizes, and failures are distributed randomly with time. Finally, the failure rate increases due to the onset of wear-out failures. Usually diode lasers undergo a burn-in test that lasts long enough to screen out infant failures. For high-reliability applications, the burn-in

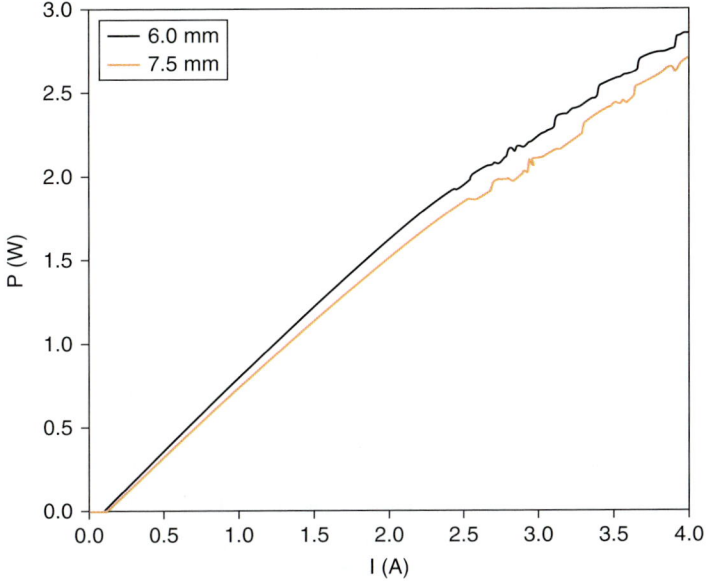

FIGURE 5.16 *L-I* curves up to a maximum current of 4 A for 6.0-mm and 7.5-mm chips.[37]

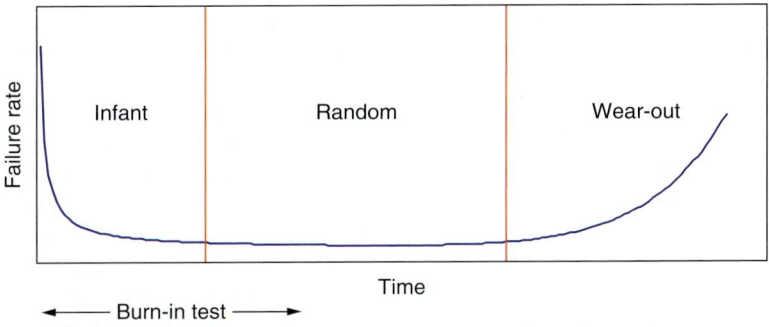

FIGURE 5.17 The bathtub curve of laser diode failure rate versus time.

test has a secondary objective of estimating the failure rate at the bottom of the bathtub curve to select the best wafers with the lowest failure rate. Figure 5.18 shows an example failure distribution of 76,000 lasers from the production burn-in test of 980-nm pump lasers at JDS Uniphase. This 1.5-mm-long chip, rated at 400-mW output power, was burned in at significantly higher temperature and current, as compared with operating conditions. As Fig. 5.18 shows, most of the failures occur in the first 20-hour time interval of burn-in. After 80 hours of burn-in test, the failure rate stabilizes as the bottom of the bathtub curve is reached.

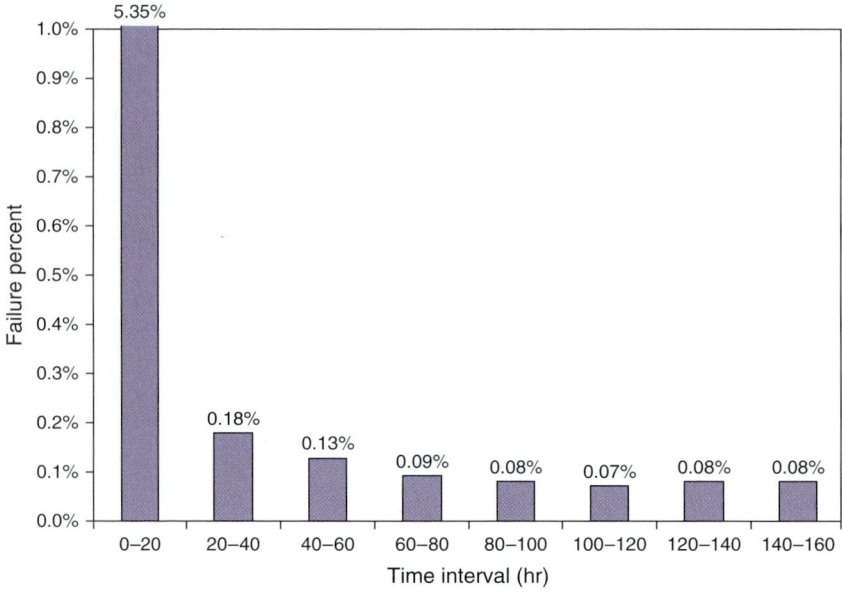

FIGURE 5.18 Failure distribution of a production burn-in test of 980-nm pump lasers.

Laser diode failure rate is an important parameter that must be estimated in order to predict performance in the field. An accelerated multi–cell life test is usually conducted, because running a lengthy test under actual operating conditions would be impractical.[38–40] Cell conditions are typically set at elevated temperature, current, and output power to accelerate the failure rate. Comparison (i.e., regression) of the failure rate across different cell conditions allows one to derive a failure acceleration model under various operating conditions. Table 5.2 summarizes the conditions and results of a 980-nm pump laser multi–cell life test conducted at JDS Uniphase. It is the same laser for which the burn-in failure distribution is shown in Fig. 5.18. All lasers were burned in to screen out infant failures.

This test was dominated by failures randomly distributed over time. No wear-out failures were observed at these conditions. The results in Table 5.2 clearly show a significant increase in failures at higher junction temperatures (estimated temperature of the *p-n* junction that takes into account thermal resistance between the junction and heat sink) and at higher currents. Dependence of failure rate λ on junction temperature T_j, current I, and power P is usually described as follows:

$$\lambda = \lambda_0 \exp\left(-\frac{E_a}{k_B T_j}\right) I^m P^n \tag{5.6}$$

Cell	Heat Sink Temperature (°C)	Current (mA)	Power (mW)	Junction Temperature (°C)	Number of Diodes	Time (hr)	Number of Failures
1	61	350	277	70	40	46,401	1
2	90	350	248	100	40	44,377	0
3	119	350	223	130	40	46,400	11
4	44	700	526	70	40	46,401	3
5	72	700	455	100	40	46,400	22
6	100	700	382	130	40	39,277	40
7	64	500	365	80	120	44,500	5
8	93	250	155	100	120	44,501	10
9	82	500	330	100	80	43,729	23
10	109	350	193	120	80	43,721	16
11	122	250	115	130	40	42,036	13
12	112	500	270	130	40	42,728	37
Total					720		181

TABLE 5.2 980-nm Pump Laser Multi–Cell Life Test Summary

where E_a is a thermal activation energy and m and n are exponents of current and power accelerations, respectively. To derive model parameters E_a, m, and n, the maximum likelihood method is used. This method finds the model parameters by maximizing the likelihood of observing actual number of failures on the test. For a random exponential distribution of failures, the time likelihood function L is

$$\ln(L) = \sum_i^{cells} n_i \ln(\lambda(T_{ji}, I_i, P_i)) - \lambda(T_{ji}, I_i, P_i) t_i \quad (5.7)$$

where n_i and t_i are the number of failures and total device hours in cell i. Application of this method to the results of Table 5.2 extracts the following model parameters:

$$E_a = 0.78 \ eV$$
$$m = 2.7 \quad (5.8)$$
$$n = 0$$

With Eq. (5.6), the estimated failure rate at operating conditions of 400 mW and 25°C is 53 failures-in-time (FIT) (53×10^{-9} hr^{-1}). This type of multicell methodology can be extended to packaged laser diodes, which is the subject of a subsequent section. But first we provide background on various package designs and processes.

5.11 Submount Design and Assembly

The chip-on-submount (COS) serves as a building block that may be used alone or integrated into a higher level of assembly. The semiconductor laser chip is soldered to the submount (or carrier) to provide a means for mechanical handling, burn-in testing, and thermal dissipation. The COS is typically clamped or bolted down (e.g., C-mount) or soldered to reduce thermal impedance (R_{th}). Hermetic designs (e.g., transistor outline [TO] cans) are sealed by a cap and window to protect against dust and moisture. Examples of these package types are shown in Fig. 5.19.

More than a decade ago, when high-reliability submarine components were first deployed, suppliers needed to eliminate field failures over a 25-year lifetime. Creep-resistant, high-shear-strength "hard solders," such as AuSn (80/20 percent weight) with a melting point (MP) of 280°C, became universally established for die bonding single-mode, 980-nm lasers. The advantage of flux- and whisker-free bonding is even more compelling when applied to high-power diodes with p-side-down bonds, where the laser facet is located only several micrometers from the solder interface. In this case, the facet must also be placed within plus or minus several micrometers from the submount edge

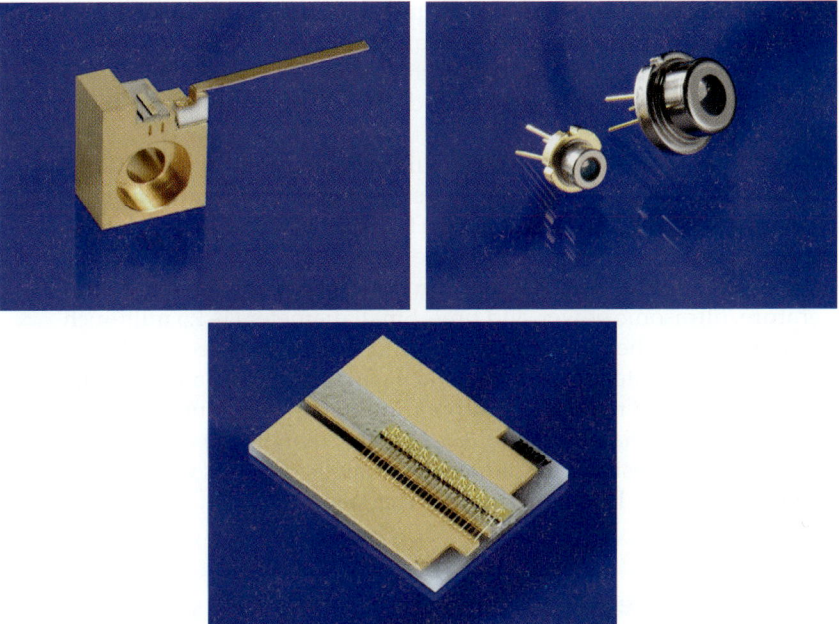

FIGURE 5.19   Photographs of chip on submount (COS) configurations: (*a*) C-mount for stand-alone operation, (*b*) sealed transistor outline (TO) can packages, and (*c*) a typical AlN submount that is subsequently soldered onto another package.

to avoid clipping the output emission while maintaining good heat conduction near the facet.

Die bonding must be optimized with temperature, force, and reduction gas to produce the uniform interfaces needed for high strength and thermal dissipation, as well as to avoid localized stresses that may degrade electro-optical parameters. To improve solder wetting, the top layer of the submount is gold-coated (Au) to avoid oxidation, followed by a diffusion barrier (e.g., Ni or Pt), followed by a thin adhesion layer (e.g., Ti) to resist peeling away from the underlying submount.

Due to the high AuSn MP, the submount bulk material must be nearly coefficient of thermal expansion (CTE) matched to GaAs (CTE = 5.7 ppm/°C), otherwise permanent mechanical stress will be induced during solder cool down that may degrade the diode lifetime. AlN, BeO, and 15 Cu/85 W (% wt) submounts are the most commonly deployed in the industry, though the future of BeO is questionable due to evolving environmental regulations. CuW or 20 Cu/80 Mo submounts have superior electrical performance, with the obvious drawback that insulation is needed beneath the submount. High-performance ceramics (e.g., chemical vapor deposition [CVD] diamond, boron nitride [BN] can be used as heat spreaders or combined with other materials to create an effective CTE matched closely to GaAs. However, interest in these ceramics is tempered by longer cavity diodes (i.e., lower thermal resistance) and challenged by high material and fabrication costs, which continue to delay serious reception of these advances by the industry.

Gold wire bonding to gold-metallized submounts (or lead frames inside the package) has proven to be an extremely mature and reliable process. Ball bonding is common for high-power diodes; whereas wedge bonding can reduce the wire height and length needed for high-speed applications. Standard process parameters are force, temperature, ultrasonic power, and time. Small diameter (1–1.5 milli-inch [mil]) wires may be dedicated for the chip exclusively because larger wires require greater bond force, which may introduce damage to the underlying active region. When bonding to the submount or leads, therefore, larger diameter wire (e.g., greater than 2 mil) or even ribbon bonding is desirable to reduce the number of bonds while maintaining acceptable PCE at higher drive currents.

5.12 Fiber-Coupled Package Design and Processes

A package (or housing) can add greater functionality to the COS (e.g., by adding monitor photodiode, thermistor, or thermal electric cooler [TEC]). However, these components are not typically employed for industrial applications, and TECs usually dissipate too much heat to be practical at ~10 W output power levels. The housing also enables fiber coupling, which is the focus of this section.

As mentioned previously, the industry has adopted designs and processes borrowed from the telecom and submarine industry, with strict assembly protocols and reliability standards. Best practices have been promulgated by customers who seek greater quality and reliability assurance to minimize the total cost of ownership.[41] Fortunately, there is particularly favorable synergy with 980-nm single-mode telecom pumps, and they share most of the same material components, assembly processes, and equipment. As shown by Fig. 5.20, a multimode fiber-coupled laser package resembles a single-mode pump, with the exception of the diode and corresponding larger fiber-core diameter.

The laser diode is bonded p-side down to a submount that is subsequently bonded to the base of the housing. A chisel lens or fast-axis collimating (FAC) lens, or fiber rod, is mounted near the laser facet for efficient coupling, and the fiber is bonded to the snout and strain relief that exits the wall. The outside package occupies ~15 × 13 × 8 mm^3 volume with a ~15-mm strain relief to meet fiber-integrity requirements. At this time, industrial laser diode manufacturers have not standardized form factors, as the telecom industry does.

The basic components and assembly methods used for fiber-coupling single-emitter diodes are summarized in Table 5.3. This list

Semiconductor Laser Diodes

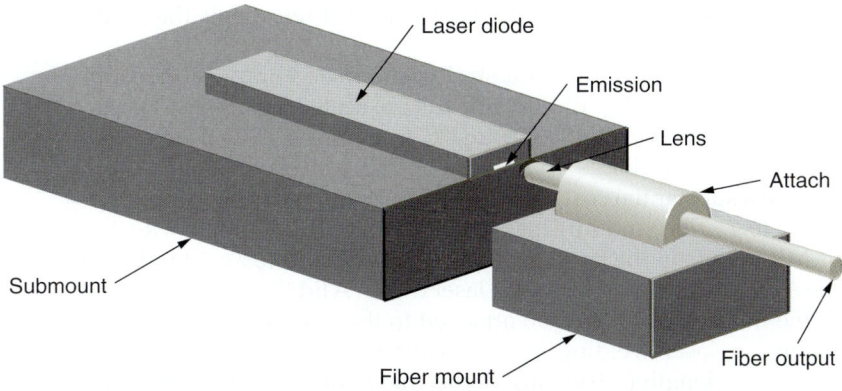

FIGURE 5.20 Illustration of fiber-coupled single-emitter laser diode. Alternatively, the chisel (or wedge) lens may be replaced by a separate fast-axis collimator (FAC) mounted to the submount with a separate fiber output. The white stripe is ~100 μm wide lasing aperture that nearly matches the fiber core diameter.

Submount	Fiber-Coupled Package			
Heat Sink AlN BeO CuW (15/85) CTE-matched composite	**Housing** Steel frame, Cu base Kovar frame, CuW base (20/80)	**Wire Bond** Au ball Ribbon	**Snout Seal** Metal or glass solder Epoxy	
Die-Bond Solder AuSn (80/20)	**Leads** Cu-core alloy Kovar	**Fiber Coupling** Chisel lens FAC + cleaved fiber	**Strain Relief** Epoxy (at strip region) Urethane boot	
Wire Bond Au ball	**Component Attach** Metal solder (< MP AuSn) (e.g., SAC, SnAg, BiSn)	**Lens Attach** Glass or metal solder Epoxy	**Lid** Steel or Kovar Getter (optional)	

TABLE 5.3 Summary of Key Components and Assembly Methods for Fabricating Fiber-Coupled Packages. The Chip-On-Submount (Left) May Stand Alone or Become a Component of the Fiber-Coupled Package

is not intended to be comprehensive; rather it represents the current state of the industry.

The housing consists of a steel frame that is brazed to a copper (Cu) base for low R_{th}. Electrical leads are sealed by low-MP glass. If low CTE is needed (e.g., ~7 ppm/°C to mount a TEC), Kovar housings with Cu/W bases and alumina electrical feedthroughs can be used at a substantial cost penalty. The COS and other internal components (e.g., fiber mount) inside the package are bonded with lower-MP solders in the range of 120°C < MP < 260°C to avoid reflowing the AuSn-solder joint below the laser diode. A lid that is CTE-matched to the housing frame is then attached to the seal ring, using resistive sealing or laser welding. Solder sealing is not a viable option for short-wavelength (< 980 nm) diodes due to oxygen added intentionally to prevent catastrophic optical damage (COD) failures.

A chisel lens or FAC collimates the light for efficient coupling into the fiber-tail assembly. Antireflection (AR) coatings are needed to increase coupling efficiency, as well as to prevent back reflections into the diode, which would degrade linearity and short-term power stability. Worse yet, laser diodes will fail catastrophically from transient high-peak-power pulses generated from the fiber laser; as such, manufacturers now offer optical isolation inside the pump (i.e., highly transmissive < 975 nm [laser diode] and highly reflective > 1050 nm [fiber laser]). Dichroic coatings are capable of creating greater than 30-dB isolation without any efficiency penalty.[42] State-of-the-art pumps routinely achieve 95 percent average coupling efficiency (AR-coated output) into 0.22 NA fiber and 92.5 percent into 0.15 NA fiber by employing FACs that improve coupling at low NA, as compared with chisel lens (due to reduced spherical aberrations of the former).

The fiber lens or FAC may be attached directly with low-MP lead-solder glass (~300°C) or ultraviolet epoxy, whereas AuSn soldering requires metallized fiber. With either lens design, the working distance remains less than 10 μm to avoid overfilling the fiber core in the lateral dimension. Each technique requires tailoring the cure or stress relief to stabilize the lens relative to the laser for eventual deployment, as well as to preserve low NA with varying case temperatures (0 to 75°C).

The fiber pigtail is secured to the package frame centered within the snout and generally maintains a hermetic seal under static or dynamic force. The most common bonding techniques include epoxy-, glass-, and solder-sealing of ferrules or directly to the fiber (metallized, to bond with metal solder). Additional strain relief allows the fiber to be coiled and assembled without weakening the fiber, as well as to protect against accidental tugging on the pigtail that might either break the fiber at greater than 5 newtons (N) or degrade coupling efficiency if the force is transmitted internally to the fiber attachment.

Moisture is well known to cause a variety of failure mechanisms in components and metallurgy.[43] For diodes, the most worrisome form of corrosion occurs at the laser facet that promotes COD, even when passivated by dielectric coatings. Moisture may be generated internally (e.g., from inadequate bakes prior to lid sealing) or from leaks created (e.g., via the fiber-snout or lid-housing interfaces, or electrical feed throughs). A getter can be sized to accommodate internal moisture accumulated over the device's lifetime, depending on environmental conditions and corresponding leak rates.[44–45] All the aforementioned sealing methods are capable of attaining internal moisture levels much less than 5000 ppm over the device lifetime via standard helium fine-leak screens with getters.

Catastrophic optical damage (COD) of the laser facet may also result from a photochemical phenomenon known as package-induced failure (PIF).[46] In the presence of organics, near-infrared photons produce carbon-rich hydrocarbons at the facet that absorb light until a thermal runaway melts the facet. Accordingly, organics (e.g., adhesives or epoxies) are frowned upon by the telecom industry. However, organics can be introduced safely in the presence of oxygen (since O_2 reacts with carbon-rich deposits to form harmless CO_2 and volatile hydrocarbons, thereby cleaning the facets and restoring their reliability). In fact, nearly-transparent epoxies reduce cladding-light absorption, which is an increasing benefit for higher-power and lower-NA fiber.

5.13 Performance Attributes

Customers seek high electro-optic performance, high coupling and thermal efficiency, and high power and linearity in a compact space and at low price. For a single-emitter fiber-coupled package with 100-μm core diameter output fiber (Fig. 5.21), output powers up to ~11 W and 50 percent PCE are commercially available.

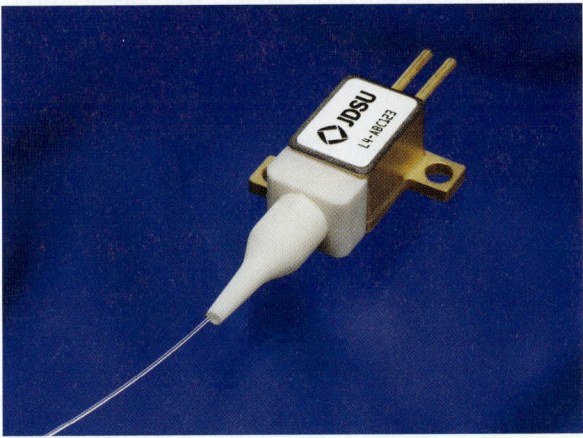

FIGURE 5.21 Photograph of a fiber-coupled, single-emitter laser diode package.[18]

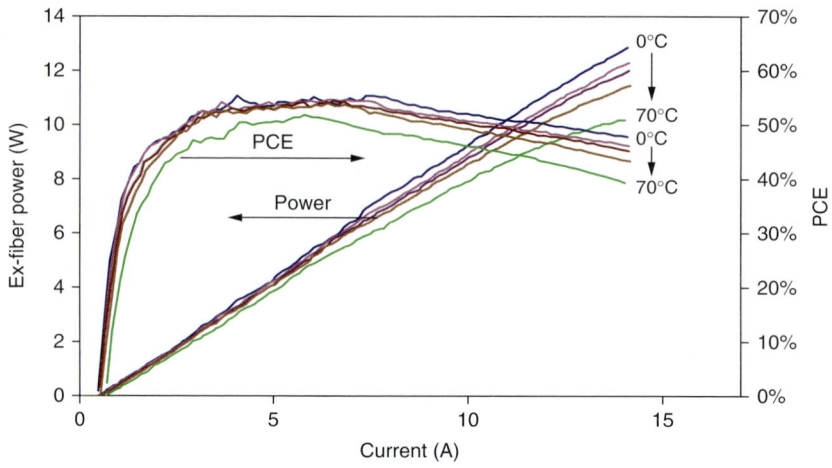

FIGURE 5.22 Output power and power conversion efficiency (PCE) for a fiber-coupled, singe-emitter package as a function of drive current at 0°C, 25°C, 35°C, 50°C, and 70°C case temperature (uncoated output).

Best-in-class average fiber-coupling efficiency is approximately 95 percent (AR-coated), and thermal impedance is approximately 7.9 K-mm/W (normalized to cavity length), with ~5.9 K-mm/W allocated to the junction-to-submount bottom interface and 2.0 K-mm/W to the package base.[18] As shown in Fig. 5.22, there is no appreciable rollover for case temperature less than 50°C.

For the microfiber laser market, single-emitter-based packages offer the lowest product price to optical output power ($/W) as compared with higher-brightness pumps. In high volumes, single-emitter pumps, such as the one shown in Fig. 5.21, presently cost about $100 per unit, or about $10/W. Over the past decade, cost per watt has steadily decreased (nearly 15 percent per year) via a combination of higher-power/diode, lower-cost packaging and offshore manufacturing.

5.14 Spatially Multiplexed High-Brightness Pumps

High-brightness (> 3 MW/cm^2-Sr) pumping of fiber lasers enables scaling to multi-kilowatt output power levels. Pumps with higher brightness offer several advantages for the fiber laser, such as smaller-diameter glass cladding, shorter fiber, and fewer combiners. As such, intensive research has been done in the pursuit of combining many single-emitter diodes into a larger single package with a single fiber output. Compared with laser bars, the advantages of this architecture include negligible thermal crosstalk between neighboring emitters, protection against cascading failed emitters, and, of course, the fact that their designs and processes are synergistic with fiber-coupled, single-emitter packages already reviewed.

All designs rely on spatial multiplexing the output emission from diodes stacked in a "staircase" formation. Each diode (usually) has a corresponding FAC and slow-axis collimating (SAC) lens to collimate and a mirror to point the emission beam. In some cases, the optical path length is held constant so that a common SAC can be shared among all emitters. The collimated output from all emitters is then focused onto the output fiber with matching NA. The literature is rife with various schemes to manipulate diodes attempting to strike a balance among the following trade-offs: cost per watt, PCE, and footprint at a target output power and brightness level.

The vertical spacing of the diodes mounted on the staircase is limited by mechanical stack-up tolerances (±50 μm). To minimize this penalty in aperture, the spacing may be increased tenfold to about 500 μm. With the output power chosen by the number of diodes N times coupling efficiency (CE), the total aperture height h is simply $h = N \times t$, where t is the step height. The fiber NA is defined by the application, so the coupling-lens focal length f_{cl} is $(h/2)/\text{NA}$. Fast-axis (FA) magnification M_y is chosen less than or equal to 40 to avoid overfilling the core while still remaining below the fiber NA. In the horizontal axis, the emitter aperture and far-field divergence approximately match the fiber-core diameter (105 μm) and NA (0.15), so the slow-axis (SA) magnification M_x is near unity. The key formulae are summarized simply as follows:

$$f_{cl} = (h/2)/\text{NA}, \text{ where } h = N \times t \tag{5.9}$$

$$M_y = f_{cl}/f_{fac} \leq 40 \text{ and } M_x = f_{cl}/f_{sac} \sim 1 \tag{5.10}$$

Based on the diode near-field and far-field characteristics and other constraints to meet good CE, a practical limit of 5 to 7 diodes can be combined. However, the brightness can be nearly doubled by polarization-beam combining (PBC) two arrays of emitters. Recently, IPG Photonics reported greater than 100 W with 50 percent PCE into 0.12 NA in a compact form factor (Fig. 5.23).[47]

5.15 Qualification and Reliability

Package housings and their internal components undergo first-article component qualifications, and the completed diode package must also be qualified prior to product release. Telcordia GR-468-CORE is regarded as a universal guideline that should be tailored to particular deployment conditions and lifetimes.[41] Among all endurance tests, temperature cycling (−40 + 85°C) and damp heat (85°C/85% relative humidity [RH]) are considered the most salient qualification tests, especially for uncontrolled (UNC) environments.

Customers realize that qualification testing imposes a necessary barrier to entry, but that qualification tests by themselves do not allow

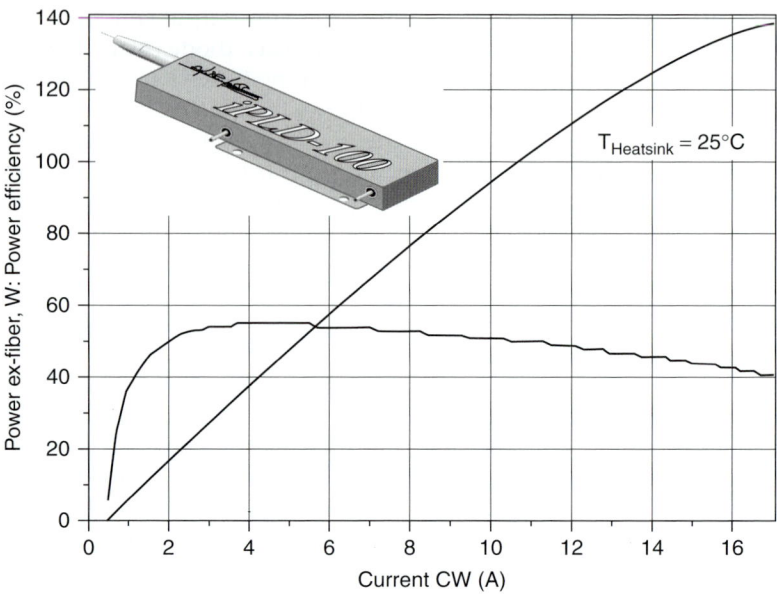

FIGURE 5.23 Spatial-multiplexed, high-brightness, multimode E-O characteristics. An output power of 100 W is achieved at 50 percent PCE at 25°C.

forecasting of the failure rate during field deployment. Package reliability is analogous to chip reliability in that a multicell life test should be used to derive a model with operating parameters, such as optical power and temperature. As compared with chip multicells, new failure mechanisms may appear inside the package at relatively lower temperatures (>85°C). High-temperature storage (HTS) may be extrapolated via activation energy to predict coupling stability versus time at the use conditions; however, it is insufficient for predicting the lifetime during operation if optical feedback or contamination from inside the package degrades the laser facet. Analogous to laser diode modeling, both random and wear-out failure mechanisms are presumed to occur that should be independently assessed during the multicell study. Package life testing is therefore needed to corroborate the results of the chip multicell model.

Typically, the greatest threat to long-term power stability is related to lens attachment. Low NA testing should be included to quantify the true stability, because standard 0.22 NA fiber obscures lens movement up to several micrometers and leads to optimistically forecasted alignment stability.

For most industrial applications, the failure rate from all package failure-mode contributions should be small compared with the laser diode failure modes. However, due to limited acceleration factors (i.e., maximum temperature) for package multicells, large sample sizes are needed to prove such low failure rates. The high expense of module life testing (e.g., capital equipment) necessitates judicious leveraging of common platform design and process reliability data. Unlike sudden, or "hard," chip failures, coupling stability is "soft," with a failure rate that gradually increases over time ($\beta > 1$); hence, field-failure statistics normally lead to an optimistic portrayal of the true coupling failure rate as specified (e.g., by end-of-life $\geq 10\%$ change in current or power), because field failures may go unnoticed or may not be reported in the early phase of deployment.

References

1. Alferov, Z. I., "Double Heterostructure Lasers: Early Days and Future Perspectives," *IEEE J. Sel. Top. Quant. Electron.*, 6: 832–840, 2000.
2. Jacobs, R. R., and Scifres, D. R., "Recollections on the Founding of Spectra Diode Labs, Inc. (SDL, Inc.)," *IEEE J. Sel. Top. Quant. Electron.*, 6: 1228–1230, 2000.
3. Welch, D. F., "A Brief History of High-Power Semiconductor Lasers," *IEEE J. Sel. Top. Quant. Electron.*, 6: 1470–1477, 2000.
4. Harder, C., Buchmann, P., and Meier, H., "High-Power Ridge-Waveguide AlGaAs GRIN-SCH Laser Diode," *Electron. Lett.*, 22: 1081–1082, 1986.
5. Dyment, J. C., "Hermite-Gaussian Mode Patterns in GaAs Junction Lasers," *Appl. Phys. Lett.*, 10: 84–86, 1967.
6. Dyment, J. D., D'Asaro, L. A., North, J. C., Miller, B. I., and Ripper, J. E., "Proton-Bombardment Formation of Stripe Geometry Heterostructure Lasers for 300 K CW Operation," *Proc. IEEE*, 60: 726–728, 1972.
7. Kaminow, I. P., Nahozy, R. E., Pollack, M. A., Stulz, L. W., and Dewinter, J. C., "Single-Mode CW Ridge-Waveguide Laser Emitting at 1.55 µm," *Electron. Lett.*, 15: 763–765, 1979.
8. Tsukuda, T., "GaAs-Ga$_{1-x}$Al$_x$As Buried Heterostructure Injection Lasers," *J. Appl. Phys.*, 45: 4899–4906, 1974.
9. Berishev, I., Komissarow, A., Mozhegov, N., Trubenko, P., Wright, L., Berezin, A., Todorov, S., and Ovtchinnikov, A., "AlGaInAs/GaAs Record High-Power Conversion Efficiency and Record High Brightness Coolerless 915-nm Multimode Pumps," *Proc. SPIE*, 5738: 25–32, 2005.
10. Botez, D., "Design Considerations and Analytical Approximations for High Continuous-Wave Power, Broad Waveguide Diode Lasers," *Appl. Phys. Lett.*, 74: 3102–3104, 1999.
11. Oosenburg, A., "Reliability Aspects of 980-nm Pump Lasers in EDFA Applications," *Proc. SPIE*, 3284: 20–27, 1998.
12. Ressel, P., Ebert, G., Zeimer, U., Hasler, K., Beister, G., Sumpf, B., Klehr, A., and Tränkle, G., "Novel Passivation Process for the Mirror Facets of Al-Free Active-Region High-Power Semiconductor Diode Lasers," *IEEE Photonics Technol. Lett.*, 17: 962–964, 2005.
13. Kawazu, Z., Tashiro, Y., Shima, A., Suzuki, D., Nishiguchi, H., Yagi, T., and Omura, E., "Over 200-mW Operation of Single-Lateral Mode 780-nm Laser Diodes with Window-Mirror Structure," *IEEE J. Sel. Top. Quant. Electron.*, 7: 184–187, 2001.
14. Petrescu-Prahova, I. D., Modak, P., Goutain, E., Silan, D., Bambrick, D., Riordan, J., Moritz, T., McDougall, S. D., Qiu, B., and Marsh, J. H., "High d/gamma Values in Diode Laser Structures for Very High Power," *Proc. SPIE*, 7198: 71981I-1–71981I-8, 2009.
15. Garbuzov, D. Z., Abeles, J. H., Morris, N. A., Gardner, P. D., Triano, A. R., Harvey, M. G., Gilbert, D. B., and Connoly, J. C., "High-Power Separate-Confinement Heterostructure AlGaAs/GaAs Laser Diodes with Broadened Waveguide," *Proc. SPIE*, 2682: 20–26, 1996.

16. Petrescu-Prahova, I. B., Moritz, T., and Riordan, J., "High-Brightness Diode Lasers with High d/Γ Ratio Obtained in Asymmetric Epitaxial Structures," *Proc. SPIE*, 4651: 73–79, 2002.
17. Petrescu-Prahova, I. B., Moritz, T., and Riordan, J., "High Brightness, Long, 940 nm Diode Lasers with Double Waveguide Structure," *Proc. SPIE*, 4995: 176–183, 2003.
18. Yalamanchili, P., Rossin, V., Skidmore, J., Tai, K., Qiu, X., Duesterberg, R., Wong, V., Bajwa, S., Duncan, K., Venables, D., Verbera, R., Dai, Y., Feve, J.-P., and Zucker, E., "High-Power, High-Efficiency Fiber-Coupled Multimode Laser-Diode Pump Module (9XX nm) with High-Reliability," *Proc. SPIE*, 6876: 687612-1–687612-9, 2008.
19. Pawlik, S., Guarino, A., Matuschek, N., Bätig, R., Arlt, S., Lu, D., Zayer, N., Greatrex, J., Sverdlov, B., Valk, B., and Lichtenstein, N., "Improved Brightness in Broad-Area Single Emitter (BASE) Modules," *Proc. SPIE*, 7198: 719817-1–719817-10, 2009.
20. Gapontsev, V., Mozhegov, N., Trubenko, P., Komissarov, A., Berishev, I., Raisky, O., Strouglov, N., Chuyanov, V., Kuang, G., Maksimov, O., and Ovtchinnikov, A., "High-Brightness Fiber Coupled Pumps," *Proc. SPIE*, 7198: 719800-1–719800-9, 2009.
21. Rossin, V., Peters, M., Zucker, E., and Acklin, B., "Highly Reliable High-Power Broad Area Laser Diodes," *Proc. SPIE*, 6104: 610407-1–610407-10, 2006.
22. Krejci, M., Gilbert, Y., Müller, J., Todt, R., Weiss, S., and Lichtenstein, N., "Power Scaling of Bars Towards 85mW per 1μm Stripe Width Reliable Output Power," *Proc. SPIE*, 7198: 719804-1–719804-12, 2009.
23. Lichtenstein, N., Manz, Y., Mauron, P., Fily, A., Schmidt, B., Müller, J., Arlt, S., Weiß, S., Thies, A., Troger, J., and Harder, C., "325 Watts from 1-cm Wide 9xx Laser Bars for DPSSL and FL Applications," *Proc. SPIE*, 5711: 1–11, 2005.
24. Li, H., Reinhardt, F., Chyr, I., Jin, X., Kuppuswamy, K., Towe, T., Brown, D., Romero, O., Liu, D., Miller, R., Nguyen, T., Crum, T., Truchan, T., Wolak, E., Mott, J., and Harrison, J., "High-Efficiency, High-Power Diode Laser Chips, Bars, and Stacks," *Proc. SPIE*, 6876: 68760G-1–68760G-6, 2008.
25. Stickley, C. M., Filipkowski, M. E., Parra, E., and Hach III, E. E., "Overview of Progress in Super High Efficiency Diodes for Pumping High Energy Lasers," *Proc. SPIE*, 6104: 610405-1–610405-10, 2006.
26. Kanskar, M., Earles, T., Goodnough, T., Stiers, E., Botez, D., and Mawst, L. J., "High-Power Conversion Efficiency Al-Free Diode Lasers for Pumping High-Power Solid-State Laser Systems," *Proc. SPIE*, 5738: 47–56, 2005.
27. Crump, P., Dong, W., Grimshaw, M., Wang, J., Patterson, S., Wise, D., DeFranza, M., Elim, S., Zhang, S., Bougher, M., Patterson, J., Das, S., Bell, J., Farmer, J., DeVito, M., and Martinsen, R., "100-W+ Diode Laser Bar Show > 71% Power Conversion from 790-nm to 1000-nm and Have Clear Route to > 85%," *Proc. SPIE*, 6456: 64560M-1–64560M-11, 2007.
28. Peters, M., Rossin, V., Everett, M., and Zucker, E., "High Power, High Efficiency Laser Diodes at JDSU," *Proc. SPIE*, 6456: 64560G-1–64560G-12, 2007.
29. Schemmann, M. F. C., van der Poel, C. J., van Bakel, B. A. H., Ambrosius, H. P. M. M. Valster, A., van den Heijkant, J. A. M., and Acket, G. A., "Kink Power in Weakly Index Guided Semiconductor Lasers," *Appl. Phys. Lett.*, 66: 920–922, 1995.
30. Guthrie, J., Tan, G. L., Ohkubo, M., Fukushima, T., Ikegami, Y., Ijichi, T., Irikawa, M., Mand, R. S., and Xu, J. M., "Beam Instability in 980-nm Power Lasers: Experiment and Analysis," *IEEE Photonics Technol. Lett.*, 6: 1409–1411, 1994.
31. Achtenhagen, M., Hardy, A. A., and Harder, C. S., "Coherent Kinks in High-Power Ridge Waveguide Laser Diode," *J. Lightw. Technol.*, 24: 2225–2232, 2006.
32. Achtenhagen, M., Hardy, A., and Harder, C. S., "Lateral Mode Discrimination and Self-Stabilization in Ridge Waveguide Laser Diodes," *IEEE Photon. Technol. Lett.*, 18: 526–528, 2006.
33. Balsamo, S., Ghislotti, G., Trezzi, F., Bravetti, P., Coli, G., and Morasca, S., "High-Power 980-nm Pump Lasers with Flared Waveguide Design," *J. Lightw. Technol.*, 20: 1512–1516, 2002.

34. Sverdlov, B., Schmidt, B., Pawlik, S., Mayer, B., and Harder, C., "1 W 980 nm Pump Modules with Very High Efficiency," *Proceedings of 28th European Conference on Optical Communications*, 5: 1–2, 2002.
35. Bettiati, M., Starck, C., Laruelle, F., Cargemel, V., Pagnod, P., Garabedian, P., Keller, D., Ughetto, G., Bertreux, J., Raymond, L., Gelly, G., and Capella, R., "Very High Power Operation of 980-nm Single-Mode InGaAs/AlGaAs Pump Lasers," *Proc. SPIE*, 6104: 61040F-1–61040F-10, 2006.
36. Yang, G., Wong, V., Rossin, V., Xu, L., Everett, M., Hser, J., Zou, D., Skidmore, J., and Zucker, E., "Grating Stabilized High Power 980 nm Pump Modules," *Proceedings of Conference on Optical Fiber Communications*, JWA30: 1–3, 2007.
37. Bettiati, M., Cargemel, V., Pagnod, P., Hervo, C., Garabedian, P., Issert, P., Raymond, L., Ragot, L., Bertreux, J.-C., Reygrobellet, J.-N., Crusson, C., and Laruelle, F., "Reaching 1W Reliable Output Power on Single-Mode 980 nm Pump Lasers," *Proc. SPIE*, 7198: 71981D-1–71981D-11, 2009.
38. Rossin, V. V., Parke, R., Major, J. S., Perinet, J., Chazan, P., Biet, M., Laffitte, D., Sauvage, D., Gulisano, A., Archer, N., and Kendrick, S., "Reliability of 980-nm Pump Laser Module for Submarine Erbium-Doped Fiber Amplifiers," *Optical Amplifiers and Their Applications*, S. Kinoshita, J. Livas, and G. van den Hoven, eds., Vol. 30 of Trends in Optics and Photonics, *Optical Society of America*, 216–219, 1999.
39. Pfeiffer, H.-U., Arlt, S., Jacob, M., Harder, C. S., Jung, I. D., Wilson, F., Oldroyd, T., and Hext, T., "Reliability of 980 nm Pump Lasers for Submarine, Long Haul Terrestrial, and Low Cost Metro Applications," *Proceedings of Conference on Optical Fiber Communications*, 483–484, 2002.
40. Van de Casteele, J., Bettiati, M., Laruelle, F., Cargemel, V., Pagnod-Rossiaux, P., Garabedian, P., Raymond, L., Laffitte, D., Fromy, S., Chambonnet, D., and Hirtz, J. P., "High Reliability Level on Single-Mode 980 nm–1060 nm Diode Lasers for Telecommunication and Industrial Application," *Proc. SPIE*, 6876: 68760P-1–68760P-8, 2008.
41. Telcordia, Generic Reliability Assurance Requirements for Optoelectronic Devices Used in Telecommunications Equipment GR-468-CORE, rev 1-2, December 1998, September 2004, respectively.
42. Wong, V., Rossin, V., Skidmore, J. A., Yalamanchili, P., Qiu, X., Duesterberg, R., Doussiere, P., Venables, D., Raju, R., Guo, J., Au, M., Zavala, L., Peters, M., Yang, G., Dai, Y., and Zucker, E. P., "Recent Progress in Fiber-Coupled Multi-Mode Pump Module and Broad-Area Laser-Diode Performance from 800 to 1500 nm," *Proc. SPIE*, 7198: 71980S-1–71980S-8, 2009.
43. Greenhouse, H., *Hermeticity of Electronic Packages*, Norwich, NY: William Andrew Publishing, 1999.
44. U. S. Department of Defense, *Test Method of Electronic and Electrical Component Parts*, MIL-STD-202G, September 12, 1963.
45. U. S. Department of Defense, *Test Method Standard for Microcircuits*, MIL-STD-883E, December 31, 1996.
46. Jakobson, P. A., Sharps, P. J. A., and Hall, D. W., "Requirements to Avert Packaged Induced Failures (PIF) of High Power 980nm Laser Diodes," *Proc. LEOS*, San Jose, CA, 1993.
47. Gapontsev, V., Moshegov, N., Trubenko, P., Komissarov, A., Berishev, I., Raisky, O., Strougov, N., Chuyanov, V., Maksimov, O., and Ovtchinnikov, A., "High-Brightness 9XX-nm Pumps with Wavelength Stabilization," *Proc. SPIE*, 7583: 75830A-1–75830A-9, 2010.

CHAPTER 6
High-Power Diode Laser Arrays

Hans-Georg Treusch

Director, Trumpf Photonics, Cranbury, New Jersey

Rajiv Pandey

Senior Product Manager, DILAS Diode Laser Inc., Tucson, Arizona

6.1 Introduction

During the past decade, significant increases in electro-optical efficiency of diode lasers—from values typically below 50 percent to record values of greater than 73 percent (see Chap. 5)—have enabled demonstrated maximum power levels of up to 1 kW from a 10-mm-wide laser bar in a lab environment. This increased efficiency, in turn, has resulted in reduced heat load and internal losses in the material. The latter has enabled laser resonator cavities that are longer than the typical 1 mm cavities of 10 years ago—up to 4–5 mm for the highest current power levels. Spreading out the heat by a factor of 4 with the larger footprint and cutting the heat load in half have resulted in the record value of 1 kW.[1]

In addition to the improved performance of high-power diode laser arrays with wavelength in the near-infrared (NIR), new materials have been developed to extend the range for the wavelength into the visible and midinfrared (MIR) regions. These new materials are aimed at new applications in the medical field, as well as at the pumping of eye-safe solid-state lasers in the MIR. The efficiency of these new materials is lower than traditional NIR diode lasers (Fig. 6.1), and high-yield assembly processes, as well as high-efficiency optical coupling methods, are required to establish usable products.

Diode Lasers

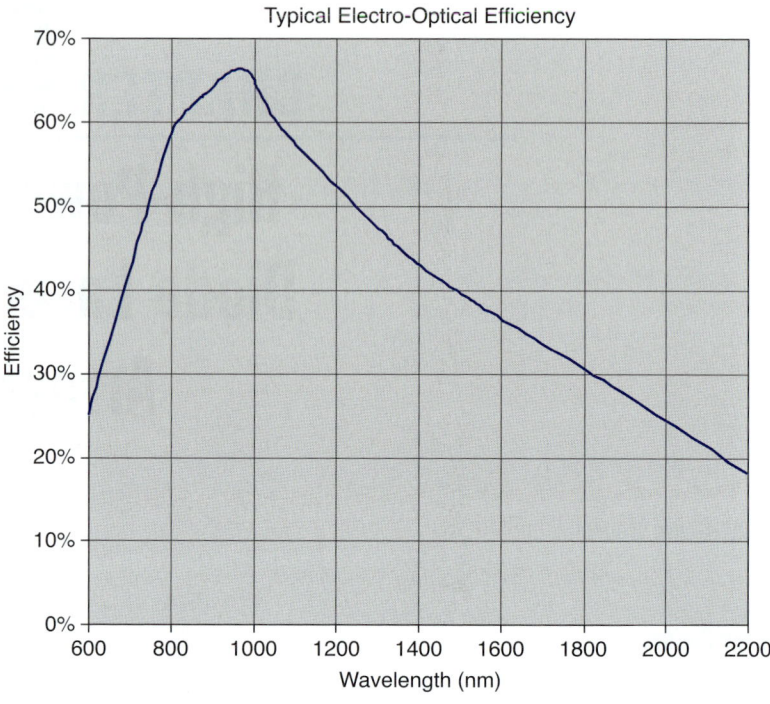

FIGURE 6.1 Typical electro-optical efficiencies of semiconductor material as a function of wavelength.

Early applications of diode laser arrays, such as pumping of solid-state laser rods and slabs, took advantage of the narrow wavelength and the reduced heat load in the laser crystals. With new applications in the area of materials processing, where diode lasers start to compete with lamp-pumped solid-state lasers, the brightness of the diode lasers has become the most important value to be conserved while scaling up the power to the multikilowatt level. The higher brightness level is also required for new pumping schemes needed for disc and fiber lasers.

The following sections will describe state-of-the-art high-power diode lasers and their manufacturing processes. Various forms of diode laser components, from a diode stack for short pulses (quasi-continuous wave, or QCW) to high-brightness fiber-coupled modules with continuous wave (CW) kilowatt output power levels, are introduced.

6.2 Diode Laser Bar Assembly

The performance (maximum power, wavelength, and reliability) of a diode laser device strongly depends on the temperature of the *p-n* junction, as described in Chap. 5. Therefore, all high-power diode laser bars, as well as single emitters, are assembled with the *p-n* junction very close (< 2 µm) to the heat sink or heat spreader (*p* side down). Solder and heat sink material must be chosen carefully to avoid any additional stress in the epitaxially grown layers, which would lead to wavelength distortions and localized changes in polarization.

In the past, two different approaches were widely used for bar packaging on a heat sink. The earlier process, developed in the late 1980s, was based on a soft solder (indium) and could use copper directly as the heat-sinking material (also called *direct bond*). Issues with reducing the indium surface and the interaction of indium with the necessary gold layers (brittle InAu intermetallic) required a very precise process control to achieve a highly reliable soft connection of the diode laser bar with the copper heat sink. The solder had to be soft, because the thermal expansion coefficient of GaAs and copper are different by a factor of 3. Although substantial progress was made addressing these packaging problems with indium, as the diode laser materials became more and more efficient and the diode bar drive currents reached beyond the 100-A mark, new reliability issues surfaced with the indium bonds. The high current density and the interest in the pulsed mode of operation, where the diode bar and the soft solder have to experience many full temperature cycles, caused the indium bond to fail within a couple of thousand hours of operation, due to solder migration and the well-known whisker formation.

The increased electro-optical efficiency of the diode laser materials favored a second approach in which an expansion-matched material is used to form a submount for the GaAs bar. With these submounts, a hard solder (AuSn) can be used to package the bar. Materials like CuW have been widely used for this approach, though with a disadvantage of reduced thermal conductivity. New submounts, including AlN and BeO materials, offer expansion matching in combination with electrical isolation to the subsequent metal heat sink (Fig. 6.2). These new ceramic devices work as simple submounts

FIGURE 6.2   Diode bar on CuW submount.

that reduce the stress on the bar and that can also carry other components, such as the N contact and optical components. The reliable AuSn solder joint approach offers an extended lifetime beyond 20,000 hours at higher operating currents, though with a slightly lower efficiency due to the increased thermal impedance. Indium solder still finds its application when highest efficiency and packaging density are required by the application, such as in continuous operation (see Sec. 6.4). Other submount materials, such as diamond and copper diamond compounds, offer even higher thermal conductivity than copper, but have poor electrical conductivity.

6.3 Heat Removal

The reliable output power of a high-power diode laser decreases with increasing temperature of operation. Two basic approaches are used to keep the temperature as low as possible. The first is to spread the heat in a block of material with high thermal conductivity (e.g., Cu) before removing the heat altogether (e.g., through transfer to air or water). Typical dimensions of such heat sinks generally range from several millimeters to a few centimeters; typical thermal impedance values are 0.5 to 0.7°C/W for a 10×2 mm^2 diode bar. Figure 6.3 illustrates various types of passively cooled heat sink—most common is the 1×1 inch footprint with different emission heights. The smaller footprint is typically used when multiple diodes are arranged in a horizontal array (see Sec. 6.4). Because the heat is generated on the front edge of the heat sink, where the diode bar is mounted, an extension to the front can reduce the thermal impedance by up to 20 percent.

For applications that require multiple diode bars, the challenge often is to arrange the bars in a small-volume array without compromising the effectiveness (i.e., the thermal impedance) of the heat sink. A standard approach is to employ modular, water-cooled (or active), minichannel heat sinks (Fig. 6.4). This stackable, modular technology

FIGURE 6.3 Passively cooled heat sinks with 1×1 inch footprint, plus one with 10×25 mm^2.

High-Power Diode Laser Arrays

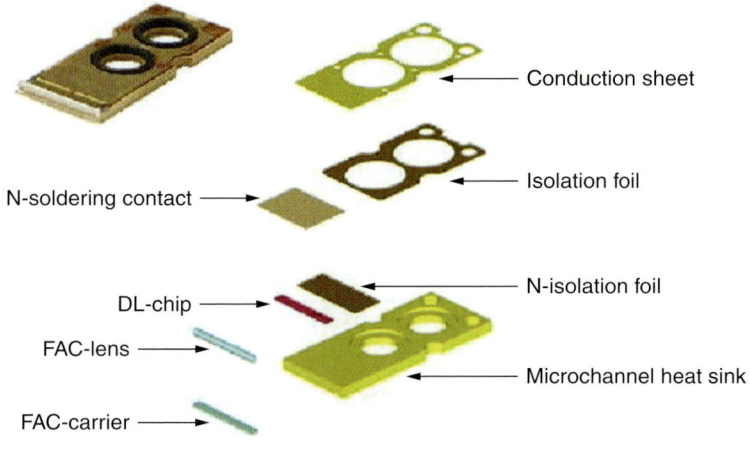

FIGURE 6.4 Components of a mini-/micro-channel heat sink, including a fast-axis collimation lens.

has been developed to the point that the thermal impedance is significantly lower than that of conventional single-bar platforms (typical values are 0.25 to 0.35°C/W, depending on the flow rate), which enables either increased power per diode laser bar or a longer lifetime of a diode laser at the same power level. To take full advantage of the improved cooling, the expected usage time of all water-cooled heat sinks must exceed the lifetime of the semiconductor material. The minichannel heat sinks most commonly employed by high-power laser diode suppliers are made out of copper because of its high thermal conductivity. The heat sink typically serves as the anode of the diode (without the thick Au plating required with alternative heat sink materials, such as Si, which are nonconductive). The dimension of the channel structure in the copper heat sink is typically in about 300 μm (which is 10 times or more than is employed in Si microchannel designs). As a result, the active copper heat sink can be operated at a pressure drop of approximately 15 psi with a 30-μm particle filter (compared with a greater than 45 psi with a 5-μm filter for Si microchannel heat sinks).

Although the copper minichannel technology offers clear benefits to users, early adopters have suffered from uneven reliability due to long-term corrosion effects. Detailed design optimization of the internal structures and advances in heat-sink fabrication and die-assembly processes have made today's devices capable of continuous operation well in excess of 10,000 hours, thus meeting the reliability requirements of most industrial applications.

6.3.1 Water Guidelines for Minichannel Heat Sinks[2]

As mentioned earlier, the water specification for vertical stacks, horizontal stacks, or any other assembly in a pump cavity depends on the

138   Diode Lasers

	Min. Resistivity	Max. Resistivity	pH Level	Expected Life
Vertical stacks	200 kΩ-cm (pitch < 2 mm)	500 kΩ-cm	6–7	> 10,000 hr
Horizontal or vertical stacks	50 kΩ-cm (pitch > 5 mm) 20 kΩ-cm (pitch > 10 mm)	150 kΩ-cm	6–7	> 20,000 hr

TABLE 6.1 Water Specifications for Actively Cooled Heat Sinks

pitch or the actual separation l_s between the heat sinks through the water. The distance is described as the length of the water along a dielectric passage (not including conducting spacers or manifolds).

In a standard vertical stack (with a pitch of 1.8 mm), the distance l_s equals 0.7 mm. The maximum recommended water resistivity is 500 kΩ-cm (see Table 6.1). The desired pH level of greater than 6 can be reached with a mixed-bed deionization cartridge. In a vertical stack with spacers (plastic inserts in the water passage), the distance l_s is increased to the distance l_{sp}, which includes the spacer thickness. Therefore, the water resistivity can be reduced by the ratio of l_s/l_{sp} until reaching a value of 100 kΩ-cm. No deionization cartridge is needed in this case. The same holds for all horizontal stacks with a typical distance l_s of more than 10 mm. The increased distance for horizontal stacks enables reduced water specifications and increased reliability in terms of the heat sink's expected lifetime. Horizontal stacks can be arranged with integrated optics to achieve vertically stacked beams with even higher brightness due to the increased fill factor and greater reliability than vertical stacks (see Chap. 7).

6.3.2 Expansion-Matched Microchannel Heat Sinks

The diode laser bar can be mounted directly with indium to the copper heat sink, or a CuW submount can be used to enable a hard solder (AuSn). Both solutions will have an electrical potential in the water and will need to follow the water specifications of Table 6.1. A new expansion-matched mini-channel heat sink avoids the electrical potential in the water and can therefore use any kind of coolant. The heat sink consists of a copper-AlN sandwich. The top and bottom copper layers are connected with an electrical feedthrough that is isolated from the center cooling structure made of copper. The center part is isolated by two AlN layers from the top and bottom layers. By adjusting the copper thickness on top to about 80 μm, the expansion is matched to the coefficient of GaAs (6.5 10-6/°C), as the coefficient of AlN is about 4.5 10-6/°C and copper is about 16 10-6/°C. The top copper layer is designed to accommodate the anode as well as the cathode for the laser

High-Power Diode Laser Arrays 139

FIGURE 6.5 Expansion-matched mini-channel heat sink. (*Courtesy of Curamik*)

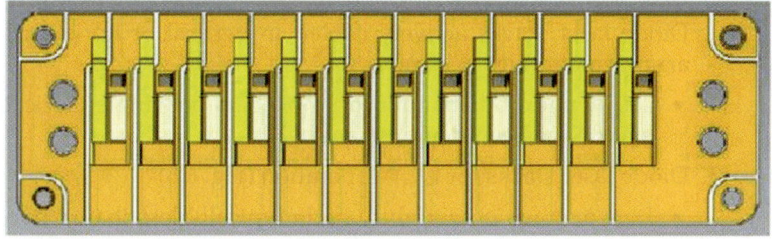

FIGURE 6.6 Schematic of a 12-bar horizontal stack based on a DCB (direct copper bond) substrate.

bar (see Fig. 6.5). These heat sinks can be stacked in the same way as the original copper mini-channel heat sinks (see Sec. 6.4).

The same technique can be used to generate a larger cooling platform for multiple laser diode bars; it also provides the interconnection of those bars in the top layer as well as efficient cooling for power levels beyond 1 kW. The schematic in Fig. 6.6 shows a horizontal stack of 12 diodes bonded to a DCB (direct copper bond) structure. No O-rings are needed, and compared with a vertical stack, the risk of leakage is reduced to a minimum.

6.4 Product Platforms

Based on the different cooling methods, the following general product platforms have been established in the market:

1. Diode bar on open heat sink, passively (Fig. 6.3) or actively cooled
   - 50–120-W CW power level for passively cooled and > 200 W for actively cooled platform

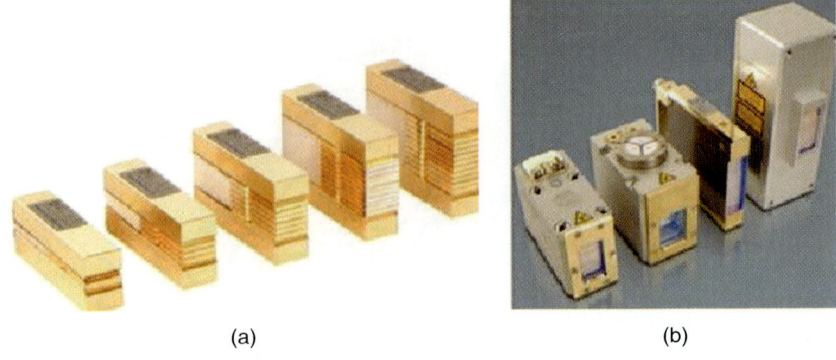

FIGURE 6.7 (a) Open frame stacks from 1–12 bars with and without fast-axis collimation lens; and (b) housed and sealed stacks with up to 70 bars, including both axis collimation.

2. Diode laser stacks actively cooled in vertical or horizontal arrangement (Figs. 6.7 and 6.8)
   - 200 W per diode bar and up to 70 bars per single vertical stack have been demonstrated
3. Diode laser stacks for QCW operation (Fig. 6.8)
   - Low average power, duty cycle typically less than 3 percent, with pulse duration less than 1 ms
   - Peak power > 250 W per bar for single waveguide design and > 600 W for a nano-stack design; multiple waveguides and p-n junction stacked in an epitaxially grown layer
   - Reduced cooling performance; highest packaging density

Typical bar pitch in an actively cooled high-power stack is greater than 1.5 mm and requires a flow rate of greater than 0.3 L/min per diode bar on a mini-channel heat sink. Depending on the inner structure of the mini-channel heat sink, the necessary pressure is in the

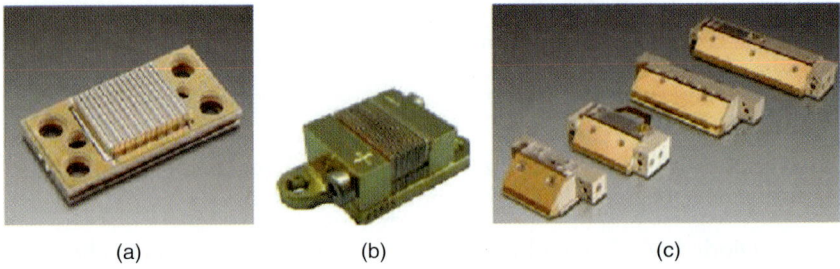

FIGURE 6.8 (a) and (b) Quasi-continuous wave (QCW) stacks with various pitch; (c) horizontal stacks with 3–8 diodes used for side pumping a laser rod, also includes part of the pump cavity.

range of 9 to 16 psi. With an increasing number of diodes in the stack, the water supply must switch from a single-sided supply to a double-sided supply, because the supply cross section for the water is limited by the mini-channel heat sink to an inlet diameter of about 5 mm. A pitch that is greater than or equal to 1.5 mm allows the attachment of the fast-axis collimation lens directly to the mini-channel heat sink via a glass submount. This method helps achieve the best beam pointing for the individual beam from the stack to less than 0.2 mrad.

6.5 Device Performance

6.5.1 Wavelength, Power, Efficiency, and Mode of Operation

Today, commercially available wavelength offerings range from 400 to 2200 nm. The highest power/bar is in the 880 to 980nm range, because this is the peak electro-optical efficiency range of high-power diode laser bars (as shown in Fig. 6.1). For example, in CW operating mode, at 980 nm laser diodes mounted on mini-channel-cooled heat sinks with AuSn bonding are now approaching 200 W/bar. However, in the 1800 to 2200 nm range, the maximum power of diode laser bars is usually less than 10 W. The practical limitations of waste heat removal from the diode bar limit its maximum performance. In this mode of operation, for maximum efficiency and lifetime, individual emitters on the 10-mm-wide laser diode bar are spaced so that thermal crosstalk and threshold current are minimized, while maximizing slope efficiency. For example, the most commonly used configuration for a 60-W, 808-nm wavelength bar is a 30 percent fill factor (19 emitters in which each emitter is 150 μm wide on a 500-μm pitch) and a 2-mm cavity length. This configuration allows for collimation of both fast and slow axes with commercially available microlenses.

However, in QCW mode, which is typically defined as duty cycles of less than 3 percent and pulse widths of less than 500 μs, the peak powers can reach in excess of 400 W/bar. This is because the average power is very low, and the thermal load on the laser bar is a tiny fraction of CW mode operation. Therefore, in QCW mode, the peak power is only limited by the optical intensity limits at the laser diode bar facet. Because facet optical intensity, and not thermal load, is the limiting factor, the laser diode bars operating in QCW mode typically have a much higher emitter count in a 10-mm bar (which is a much higher fill factor); fill factors of up to 80 percent are not uncommon. The higher emitter count (fill factor) spreads the peak power over more emitters, thus reducing peak power intensity on each emitter facet.

6.5.2 Beam Quality and Brightness

Despite the many advantages of high-power diode lasers, such as high electro-optical efficiency, compactness, and very high powers, they suffer from poor beam quality. Although the beam quality in the fast axis (assuming no bar smile) is diffraction limited ($M^2 < 1.2$), the beam quality in the slow axis is poor. For example, an industry standard of an 808-nm, 19-emitter bar with a 150-μm emitter width on a 500-μm pitch and a divergence angle of 6 degrees (90 percent power) has an M^2 of about 800. The degradation of beam quality is attributed to three factors: First is the large emitter width, which is needed to deliver the high power per emitter. Second is the emitter count in the diode laser bar. And third is the fill factor (30 percent, in this example). The emitter widths can be decreased but not by a large amount, because in high-power laser bars, the goal is to maximize power per emitter. As a result, the only two variables that can be optimized to improve beam quality are the emitter count and the fill factor. A lower emitter count and a lower fill factor laser bar improve the beam quality. The lower fill factor assumes that the nonemitting areas between emitters are filled after slow-axis collimation in order to recover beam quality. Another variable that is often used to improve beam quality by reducing the slow-axis divergence is the cavity length—a longer cavity length can reduce slow-axis divergence, while at the same time increasing power per emitter.

Low fill-factor bars, with emitter counts in the range of 5 to 10 and fill factors of about 10 percent with powers approaching 10 W per emitter in CW mode, are emerging as the preferred architecture for high-brightness applications. The low fill-factor bars aim to capture the beam quality of a single emitter, while delivering the power of a laser bar. For example, an 808-nm, 10 percent fill-factor bar with an emitter width of 100 μm and 10 emitters with a slow-axis divergence of 6 degrees (90 percent power) has a slow-axis M^2 of about 800. However, after slow-axis collimation (i.e., filling of the nonemitting areas), the M^2 value drops to 80, whereas the standard bar after slow-axis collimation has an M^2 equal to 240. For the same power output, the brightness of a low fill-factor bar is ~3 times higher than the standard bar.

Other techniques for improving beam quality and brightness are described further in Sec. 6.6.

6.5.3 Wavelength Locking

High-power diode lasers are multimode lasers; therefore, their spectral brightness is low. Although the centroid wavelength can be tuned fairly accurately at any given temperature, the FWHM (full width half maximum) is approximately 3 nm, and the FW $1/e^2$ (full width at $1/e^2$ of the maximum) is approximately 5 nm. Furthermore, the wavelength–temperature coefficient for these lasers is around 0.3 nm/°C. For some applications, this broad bandwidth and sensitivity to temperature create operational challenges. For example, pumping of standard ytterbium (Yb) fiber lasers in the 980-nm pump region requires a narrow bandwidth, due to the narrow absorption band. In some specific

High-Power Diode Laser Arrays 143

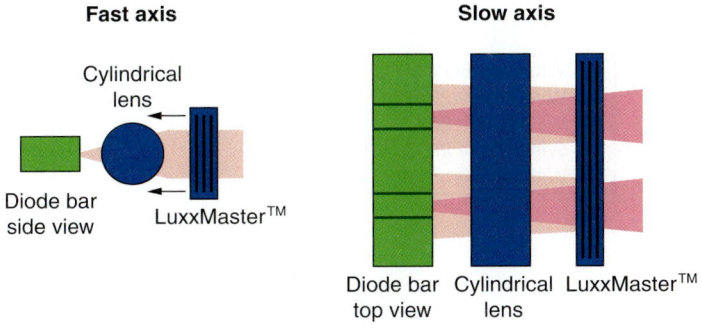

FIGURE 6.9 Schematic of a volume Bragg grating (VBG) attached in front of the fast-axis collimation lens.

applications, such as alkali-laser (rubium or cesium) pumping, which require 10 GHz bandwidth, these free-running lasers are completely unusable.[3] Wavelength locking is an effective method to overcome these challenges and target the high-power diode lasers for these applications. Wavelength locking is offered in two methods: either internal or external to the diode laser cavity.

- **Internal locking:** A grating for selective spectral feedback is etched in the structure of the semiconductor laser diode's active region.[4] Internal gratings reduce the wavelength temperature coefficient to 0.08 nm/K and can yield bandwidths of less than 1 nm.

- **External locking:** Optical components, such as volume Bragg gratings (VBGs) or volume holographic gratings (VHGs) can be attached to the array after fast-axis collimation of the diode laser bar, as shown in Fig. 6.9.

These commercially available wavelength locking components reduce the wavelength–temperature coefficient to ~0.01 nm/K. Figure 6.10 shows the wavelength locking performance of a high-power diode laser operating at 75 A. A slight bump on the right indicates that the laser is losing wavelength lock at higher operating temperature and that power is leaking to higher wavelengths. The wavelength-locked spectrum exhibits FWHM less than 0.5 nm and FW $1/e^2$ of less than 1 nm throughout the entire temperature range of 20 to 35°C.

The spectral stability of a wavelength-locked diode with respect to current is shown in Fig. 6.11. With wavelength locking, the diode laser shows a shift of 0.3 nm over a 20-A operating current range, which corresponds to a wavelength shift of about 0.015 nm/A. For a free-running laser bar, this value is typically 0.1 nm/A.

Diode Lasers

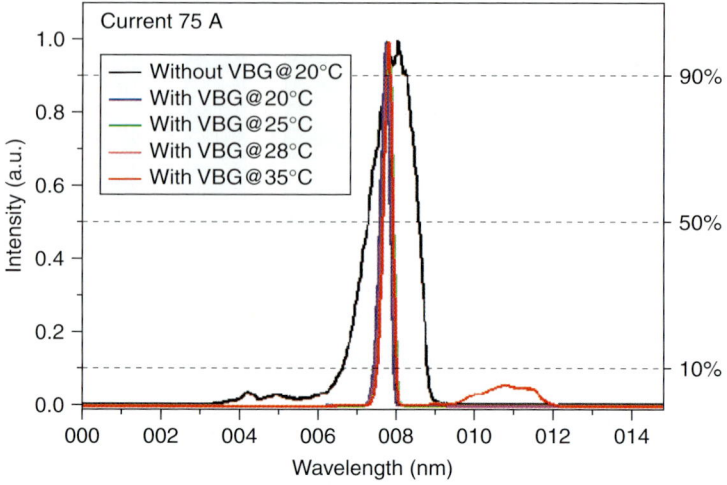

FIGURE 6.10 Laser diode bar wavelength-locked at 808 nm while operating at 75 A from 20–35°C.

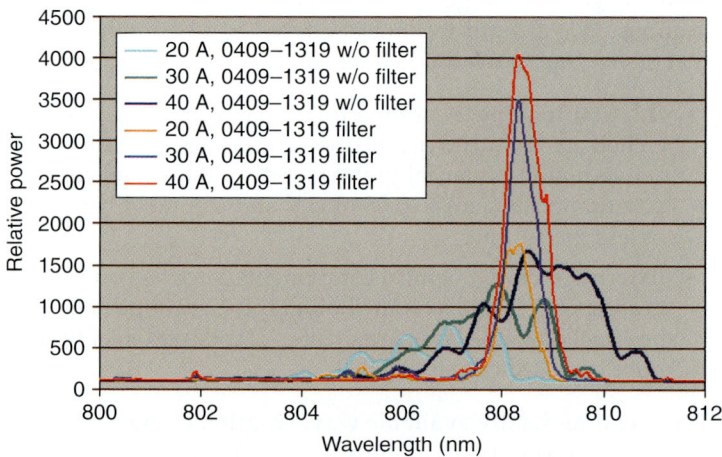

FIGURE 6.11 Spectral stability of diode laser with respect to operating current.

6.5.4 Lifetime and Reliability

The mean time to fail (MTTF) of a 50-W, CW-mode, 808-nm diode laser bar on a passively cooled heat sink is about 20,000 hours. The same bar 10 years ago would have barely lasted a few thousand hours. This tenfold increase in diode lifetime over the past decade is a result of continuous improvements in all process steps in the manufacturing of the packaged diode laser bar. Improvements in the epitaxial design, wafer processing, facet coating, facet passivation, diode bar metallization, bonding, and heat sink design have cumulatively contributed to long-term reliability.

Diode laser bars operating in QCW mode routinely have lifetimes of greater than 1 Gigapulses at peak powers of 200 W or higher. Long-term reliability of a laser diode bar is a function of three primary factors: (1) operating temperature, (2) operating power, and (3) operating current density. For example, a 50-W, 808-nm CW laser bar mounted on a passively cooled heat sink operating at 25°C will last twice as long as the same bar operating at 35°C. If the same bar is operated at 60 W instead of 50 W (i.e., the same heat sink operating temperature), then the junction temperature at the laser bar solder interface will rise by approximately 5 to 7°C above the 50 W operation junction temperature, which will reduce its lifetime. Furthermore, at 60-W operation, the current density is also higher, which accelerates aging of the bulk semiconductor material.

However, advances in the use of aluminum-free active regions and the increase of characteristic temperatures T_0 and T_1 have allowed the diode laser bar to operate at higher junction temperatures[5] without compromising efficiency. Advancements in antireflection (AR) coatings and facet passivation have increased the catastrophic optical mirror damage (COMD) threshold of emitters, which has allowed higher power per emitter in both CW and QCW modes of operation. The use of hard solder, such as AuSn, and of coefficient of thermal expansion (CTE)–matched heat sinks with lower thermal impedance has allowed the diode bar to operate reliably at higher powers.

6.6 Product Performance

Without first collimating the beam with a cylindrical lens, the large beam divergence (> 40°) perpendicular to the *p-n* junction (i.e., the fast-axis direction) allows only a limited number of applications. Side pumping of solid-state laser crystals, in which the diodes can be placed in very close proximity to the laser crystal, is one of those rare cases where the divergence is of benefit for uniform illumination of the crystal. The divergence in the lateral direction of a diode laser bar typically depends on the drive current or the current density, as the beam is first gain guided and to some extent index guided by the established temperature profile at higher output powers. The lateral divergence takes on values of between 4 and 10 degrees. These values for the divergence in both directions, as well as the dimensions of the emitting area, result in an astigmatic beam. The beam parameter product (full angle × diameter) is about 2 mm-mrad (M^2 about 1.3) in the fast-axis direction and up to 1700 mm-mrad (M^2 about 1000) in the slow-axis direction, which is too large for most applications. The beam quality in the slow-axis direction can be further improved by using an array of cylindrical lenses to collimate the individual emitters

146 Diode Lasers

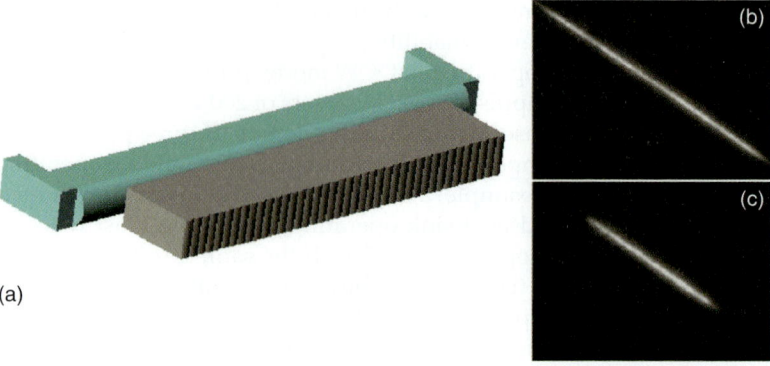

FIGURE 6.12 (a) Fast- and slow-axis collimation lens combination, (b) beam profile in the far field without slow-axis collimation, and (c) beam profile in the far field with slow-axis collimation.

(Fig. 6.12)—in other words, by increasing the optical fill factor of the beam from 20 or 30 percent to greater than 90 percent. The divergence is reduced to less than 3 degrees (50 mrad), and the beam parameter product is reduced to 500 mm-mrad.

The majority of diode bar applications require beam delivery through an optical fiber to conserve the initial brightness of the diode laser device. To achieve this task, the beam of an individual diode bar or the beams from a diode bar stack must be shaped to a uniform beam quality in both directions.

6.6.1 Fiber Coupling of Individual Diode Bars

During the 1990s, four slightly different methods were developed and used to homogenize the beam quality and preserve most of the brightness before coupling into the beam delivery fiber. In addition to these four methods which are explained in more detail below, an alternate low-cost approach was also used that does not maintain the brightness; this method coupled each emitter into a single fiber and used the fiber bundle as part of the beam delivery. Thus, for a typical diode laser bar, 19 individual fibers would be closely arranged in the area of a circle.

Southampton Beam Shaper[6]

The original beam shaper design (shown in Fig. 6.13a and 6.13b) is very simple: It consists of only two high-reflectivity (HR) flat mirrors that are aligned approximately parallel and separated by a small distance d. The mirrors are transversely offset from each other in both directions, so that small sections of each mirror are not obscured by the other. These unobscured sections form the input and output apertures of the beam shaper. An improved version of the two-mirror approach was designed later, using a plane parallel plate and adding

High-Power Diode Laser Arrays 147

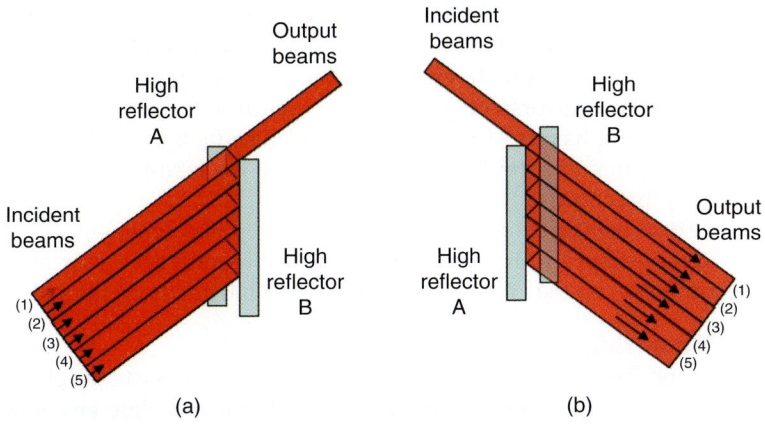

FIGURE 6.13 Two-mirror beam shaper: (a) plane view and (b) side view.

a HR pattern on both sides. The plate thickness was increased up to 5 mm in order to minimize the angle of incidence to a few degrees.

The action of the beam shaper is described with reference to Fig. 6.13a and 6.13b, which show, respectively, plane and side views of the beam shaper. In each case, the mirror surfaces are orthogonal to the plane of the figure. The incident beam can be considered to be composed of a number of adjacent beams. For the purpose of illustration, the incident beam has been arbitrarily chosen to consist of five parallel beams (1)–(5). Beam (1) is not incident on either mirror A or mirror B, because it passes above mirror A (see Fig. 6.13b) and by the side of mirror B (see Fig. 6.13a); thus, it emerges with no change to its original direction (assuming that any diffraction effects at the edge of mirror B are negligible). Beam (2), however, passes above mirror A but is incident on mirror B and is reflected so that it strikes mirror A immediately below Beam (1). Beam (2) is then reflected at mirror A and emerges from the beam shaper in the direction of Beam (1), though displaced beneath Beam (1). Beam (3) is reflected from mirror B so that it strikes mirror A underneath Beam (2); it is then reflected back to mirror B, where it is reflected onto mirror A, subsequently emerging parallel to Beams (1) and (2) but displaced underneath Beam (2). Beams (4) and (5) undergo similar multiple reflections at mirrors A and B and finally emerge, propagating beneath Beams (1), (2), and (3), as shown in Fig. 6.13b.

Thus, the action of the beam-shaping device is to effectively chop the incident laser beam into a specific number of beams and then to redirect and reposition these beams so that they emerge from the beam shaper stacked on top of one another. If the incident beam is initially many times diffraction-limited in one (x) direction (i.e., $M_x^2 \gg 1$), then the effect of the beam shaper is to decrease the width of the beam in the x direction, without significantly increasing its divergence. Thus, the overall result is that the composite beam that emerges from the beam shaper has a smaller value for M_x^2. In the y direction, the beam size is increased, but the divergence remains approximately constant (assuming that mirrors A and B are sufficiently parallel); hence, the emerging beam has its M_y^2 value increased. The factor by which M_y^2 is increased depends on the number of times the beam is cut in the lateral (x) direction.

The disadvantages of this design are the different path lengths of the individual beams and the losses due to multiple reflections (28x for a uniform beam) on the HR coating, which is not 100 percent reflective.

Step Mirror FhG-ILT[7]

The second design for a beam-shaping device also consists of reflecting surfaces. A first-step mirror (Fig. 6.14) divides the line emission from a diode laser bar into individual line segments, while a second-step mirror stacks the line segments in the direction of the better beam quality (similar to the double-mirror Southampton beam shaper). Each beam has the same path length and hits the mirror surfaces only twice. A typical step size is 1 mm to match the beam size after fast-axis collimation; thus, a 10-mm bar can be cut into 10 segments. This reduces the beam parameter product from 500 mm-mrad for a 30 percent fill factor bar to 50 mm-mrad in the lateral direction and increases the value to 20 mm-mrad in the vertical direction. To couple into a fiber with minimized losses, the sum of the beam parameter products must be the same or less than the product of the diameter times the fiber's NA. With a value of 70 mm-mrad for the diode beam and a typical NA of 0.2, the smallest fiber diameter that can be used with this approach is about 200 µm. Using a diode bar with fewer emitters and increased

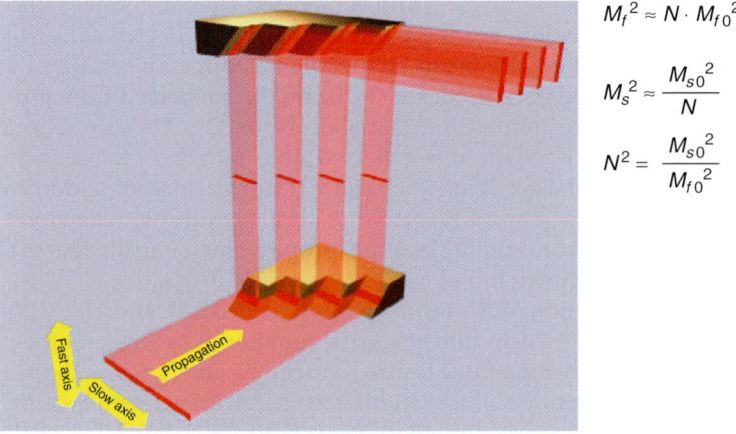

$$M_f^2 \approx N \cdot M_{f0}^2$$

$$M_s^2 \approx \frac{M_{s0}^2}{N}$$

$$N^2 = \frac{M_{s0}^2}{M_{f0}^2}$$

Figure 6.14 A step-mirror beam shaper rotating single line elements and formulas to calculate the number of steps.

emitter spacing to avoid thermal crosstalk, the step mirror is the best-adapted solution to couple into a 100-μm-diameter fiber, even with an NA of 0.12. The beam quality in the lateral direction is then given only by the single emitter to 10 mm-mrad, and 8 to 10 emitters can be stacked in the vertical direction. Demonstrations of 50 W from a single bar and 100 W from two polarization-coupled bars have been done from a 100-μm, 0.12-NA fiber for a single wavelength.

Beam Shaping with Refractive Optics

A beam-shaping solution for a higher fill-factor bar is shown in Fig. 6.15. After fast-axis collimation, the beams from individual emitters are deflected in different planes by a set of microprisms. The gained space between the emitters is used to collimate the individual beam in the slow axis with a two-dimensional array of lenses. The result is shown in Fig. 6.15 as an array of collimated beams that can be focused into a 200-μm, 0.2-NA fiber by a spherical lens.

The advantage of this and the next approach is that the lenses can be arranged in a straight beam path, which makes the alignment and the packaging easier.

One of the most common approaches in beam shaping a single-diode laser bar uses a tilted cylindrical lens array designed as an $M = 1$ telescope. The cylindrical lens array changes the divergence angles of the slow and fast axes and allows slow-axis collimation with a single cylindrical lens (Fig. 6.16). This optical setup is typically used with a 19-emitter bar and allows coupling into a 200-μm fiber. Even coupling into a 100-μm-core, 0.2-NA fiber is possible, because 9 of the individual beams can be overlapped with the other 10 beams by polarization coupling.

Polarization coupling (Fig. 6.17) is one method for increasing the brightness of diode laser bar devices. The polarization ratio of diode lasers is in the range of 92 to 98 percent and is increasing with shorter wavelengths in the range of 980 to 800 nm. Therefore losses in the range of 5 to 10 percent need to be considered when using this technique.

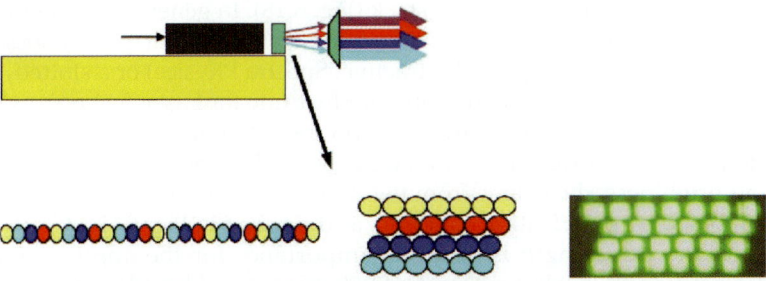

FIGURE 6.15 Beam shaping with refractive optics (prisms and slow-axis collimation).

FIGURE 6.16 Beam shaping with tilted cylindrical lens array.

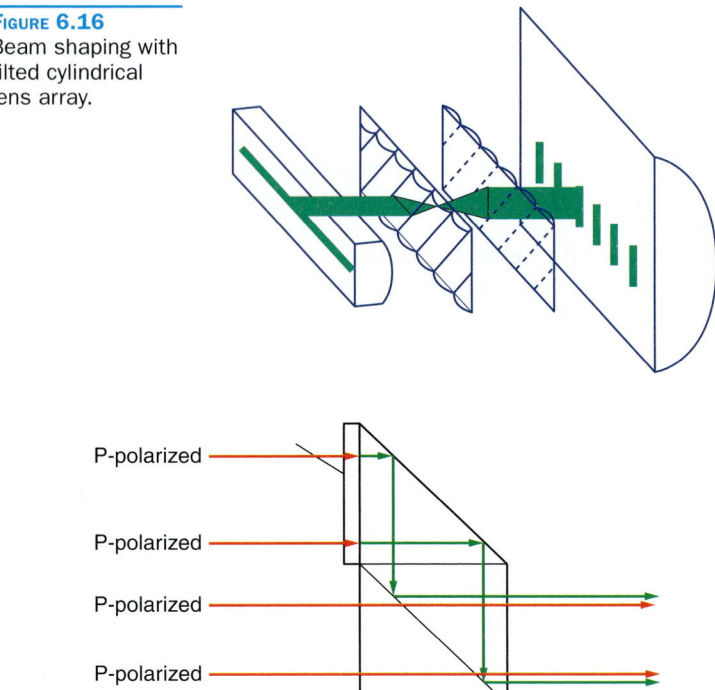

FIGURE 6.17 Polarization coupling scheme with half-wave plate and polarization cube.

6.6.2 Power Scaling

Multikilowatt power levels can be reached by using multiple diode laser bars. As shown in Sec. 6.4, the diodes can be arranged on minichannel heat sinks in a stacked format. The fill factor in such stacks only reaches values of up to 50 percent due to the pitch of the heat sinks and the beam size created by the fast-axis collimation lens. This fill factor can be increased by interleaving the beams of a second stack between the beams of the first stack (Fig. 6.18). In general, two methods are available for interleaving two stacks without power loss: using a stack of glass plates (refractive; Spectra Physics) or a slotted/striped mirror (reflective) (Fraunhofer Institute for Laser Technology) to interleave two stacks without power loss. Both techniques double the power and brightness of a stack. To further increase the power and brightness, the beams from two interleaved sets of stacks can be combined by polarization coupling, as described in Fig. 6.17.

If the wavelength is of minor importance for the application, the power and brightness can be further increased by adding multiple wavelengths to the beam. More than seven narrow-band diode wavelengths have been developed between 800 and 1030 nm, where

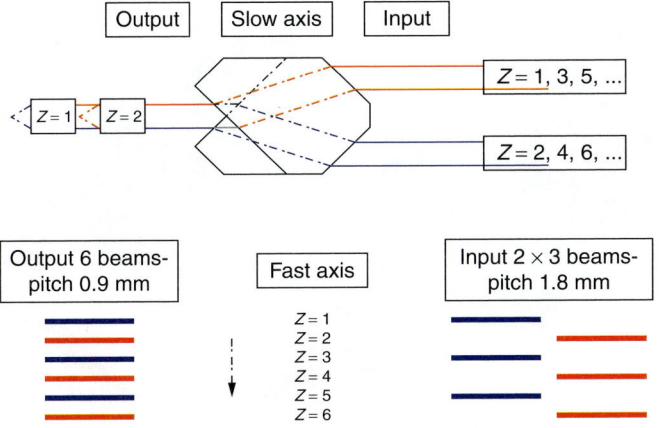

FIGURE 6.18 Refractive method of interleaving beams from a diode laser stack.

the electro-optical efficiency is greater than 55 percent and where power levels greater than 100 W per diode bar are available. The different wavelengths can be multiplexed with dielectric-edged mirror into a common beam. Assuming a 12-bar stack with an internal pitch of 1.8 mm and an average power of 120 W per diode bar, the total power in a 21 × 10 mm beam size will be 2880 W after interleaving, and 5742 W after polarization coupling, assuming a 5 percent coupling loss. Multiplexing all seven available wavelengths, the power will reach more than 35 kW CW from a 2-cm^2 aperture. The full beam parameter product (BPP) can be as low as 80 × 250 mm-mrad for the two axes, or a combined BPP of 330 mm-mrad. This enables fiber coupling into an 800-μm core diameter fiber with an NA of 0.22, which has an intrinsic BPP of 352 mm-mrad.

6.6.3 Fiber-Coupled High-Power Diode Laser Devices

Following the above described example of extreme power, fiber-coupled diode laser stacks have been developed and placed in the market as a replacement for lamp-pumped solid-state lasers. Because the beam quality of lamp-pumped solid-state lasers enabled fiber coupling into a 400- or 600-μm core fiber with an NA of 0.12, the diode laser stacks had to be designed for the same beam parameter product of less than 120 mm-mrad (full angle). An example of such a design is shown in Fig. 6.19. The slow-axis beam quality is improved by rearranging the emitter from the horizontal line in the diode bar to a vertical stack of emitters, using two stacks of parallel glass plates similar to the interleaving concept. By choosing the correct number of emitters for the diode bar, the stack's beam quality can be made uniform in both directions and, therefore, most effectively coupled into a fiber.

Diode Lasers

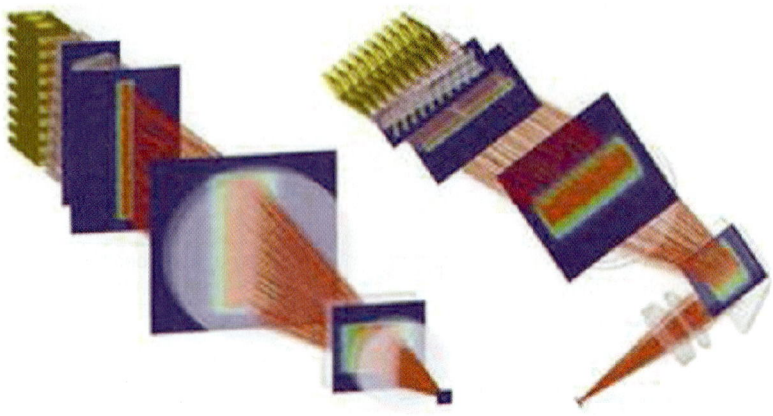

FIGURE 6.19 Fiber coupling scheme of stacks. (*Courtesy of LaserLine*)

Maximum output power	1000 W	2000 W	3000 W	4000 W
Beam quality	20 mm-mrad		30 mm-mrad	
Laser light cable	400 µm, NA = 0.1		300 µm, NA = 0.2 or 600 µm, NA = 0.1	
Spot at f = 100 mm	0.2 mm		0.3 mm	

TABLE 6.2 Typical Performance Parameters of a Fiber-Coupled Diode Laser System

The power scaling to multikilowatts can be done thereafter by polarization coupling and wavelength multiplexing. Power levels above 4 kW from a 600-µm fiber have been demonstrated with this fiber-coupled approach (see Table 6.2).

Because the power from a single 100-µm fiber of 0.12 NA was increased from a typical 10 W to more than 100 W for a single wavelength, an alternate design to actively cooled stacks became available. By using 19 individual modules, the fibers can be arranged in a close-packed circle of less than 600 µm and can therefore be coupled into the same size fiber, thus preserving the low 0.12 NA value. Thus, adding other wavelengths allows the power to increase by roughly 1.5 kW per wavelength, providing systems in the multikilowatt range. Figure 6.20 compares the size of a 3-kW lamp-pumped solid-state laser and a diode laser system of the same output power.

With the initial electro-optical efficiency of diode lasers currently in the range of 60 to 65 percent, the wall-plug efficiency of diode laser systems can be above the 40-percent mark, which is well above that for all diode-pumped solid-state lasers.

High-Power Diode Laser Arrays 153

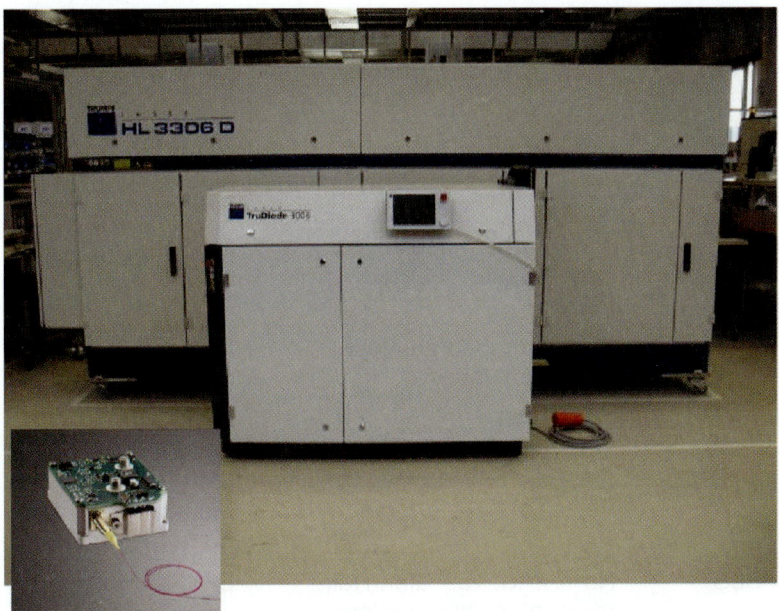

FIGURE 6.20 3-kW lamp-pump solid-state laser compared (back) with a 3-kW TruDiode system (front). The 100-W base module is shown in the inset.

6.7 Direct High-Power Diode Array Applications

High-power diode lasers were developed primarily for the pumping applications of solid-state lasers to replace less-efficient arc lamps. The narrow spectrum of diode lasers, as well as their high electro-optical conversion efficiency, enabled a significant improvement of solid-state laser technology. Without diode laser pumping, the current beam quality of solid-state lasers, as well as the technology of fiber lasers, would not be achievable. Table 6.3 summarizes the main applications of high-power diode lasers and includes direct applications in medical and industrial areas.

6.7.1 Industrial Applications

Pumping of multikilowatt solid-state lasers for industrial applications like welding, cutting, and so on is still the most important and growing market for high-power diode lasers. Early approaches of side pumping of Nd:YAG laser rods did not succeed, because the beam quality of the competing CO_2 lasers could not be reached, and the overall efficiency was typically below 20 percent. However, the development of high-brightness diode pump sources has enabled new and more efficient technologies, such as the thin-disk laser shown in Fig. 6.21 (see also Chap. 10) and fiber lasers (see Chaps. 15–18).

Wavelength	Application	Market
630–635, 652, 668	Photodynamic therapy	Medical
670	Cr^{3+}: LiSAF – fs-Laser	Diode-pumped solid-state laser (DPSSL)
689, 730	Age-related macular degeneration, Photodynamic therapy	Medical
780, $\Delta\lambda < 1$	Diode-pumped gas laser (rubidium vapor)	Defense (high-energy laser)
785, 792, 797	TM^{3+}: YAG ≥ 2 µm	DPSSL
795 $\Delta\lambda < 1$	Nd^{3+}: YLF	DPSSL
	$Rb^{3+}/Xe^{139}/$—Pumping	Medical; Instrumentation
805, 808	Nd^{3+}: YAG, Vascular, Hair removal, Ophthalmology	DPSSL; Medical
810 ± 10	Cosmetic, Hair removal, Dental, Biostimulation, Surgical	Materials processing; Medical
830	Prepress, Computer-to-plate (CTP), Direct-on-press (DOP)	Graphic arts
852, 868–885	Diode-pumped gas laser (cesium vapor)	DPSSL
	Nd^{3+}: XXX (various host crystals)	Defense
901	Yb^{3+}: SFAB	DPSSL
905	Laser range finding	Instrumentation
915	Yb: Glass, Fiber laser	DPSSL; Medical
940	Yb^{3+}: YAG, Disk, Varicose vein removal, Surgical applications	DPSSL; Medical
		Materials processing
968, 973–976	Yb^{3+}: YAG, Disk	DPSSL
	Yb^{3+}: Glass, Fiber laser, Dental, Surgical, Ophthalmology	Medical

TABLE 6.3 Summary of Diode Laser Applications, Sorted by Wavelength (*Continued*)

Wavelength	Application	Market
980 ± 10	Dental, Prostate treatment	Medical
		Materials processing
1064	Hair removal, Tattoo removal	Medical
1330–1380	Medical	Medical
1450–1470	Acne treatment, Turbulence detection, Er^{3+} pumping	Medical
1530, 1700	Medical	Medical
	Rangefinder, Missile defense	Defense
1850–2200	Surgical, Ho^{3+}: Pumping, Turbulence detection, Plastic welding/marking	Avionics
		DPSSL; Medical
		Materials processing

TABLE 6.3 (*Continued*)

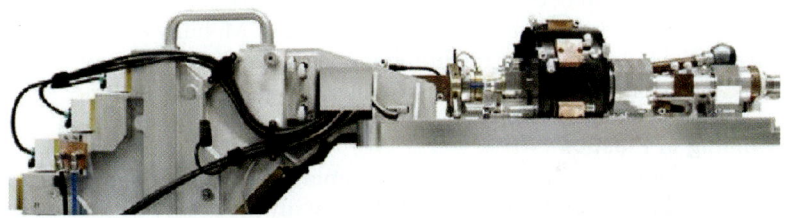

FIGURE 6.21 Multikilowatt diode-pumped disk laser: homogenized pump source (left) and laser cavity (right).

The development of high-brightness diode laser devices has also opened new markets for direct-diode industrial applications, as the power and beam quality of diode laser systems are matching the values of earlier-generation lamp-pumped solid-state lasers with a tenfold improvement in wall-plug efficiency.

Laser Welding

Joints that are produced through laser welding are characterized by high welding speed, high levels of stability, and very low distortion (Fig. 6.22). At the same time, excellent weld seam surfaces can be obtained. An almost maintenance-free operation, a lifetime of more than 30,000 operating hours, and the best efficiency of all lasers make the diode laser superior in the welding of thin sheet metals. As a comparison, lamp-pumped Nd:YAG lasers require a lamp change

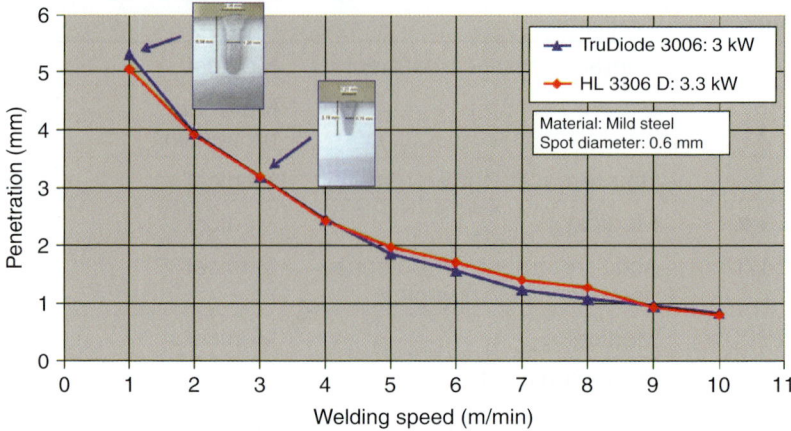

FIGURE 6.22 Comparison of weld depth in mild steel as a function of welding speed for a 3-kW diode laser and a 3.3-kW lamp-pumped solid-state laser.

approximately every 1,000 operating hours, resulting in operating costs an order of magnitude higher than those of diode lasers.

At power levels up to 2.3 kW, the size of a diode laser–based system is comparable to that of conventional welding systems, such as tungsten inert gas (TIG) or metal inert gas (MIG) welding. Mobility and compactness make the diode laser the number one choice for a variety of metal-welding applications as a particularly flexible tool in production.

Welding of Plastics

Welding of plastics combines the advantages of noncontact welding without forming fluff or excess melting with the ability of a measurable setting path. Laser welding is also unique in that it allows for noncontact welding with low thermal and mechanical load; this is especially advantageous to plastic housings with built-in electronic components, which may be damaged in conventional procedures, such as vibration welding or ultrasonic welding. Figure 6.23 shows

FIGURE 6.23
Remote car key laser welded. (*Courtesy of LaserLine*)

High-Power Diode Laser Arrays

an example of plastic welding a remote car key, which was one of the first diode laser welding applications in industry.

The advantage of the diode laser, in comparison with conventional solid-state lasers, is its shorter wavelength and "top hat" beam profile without intensity peaks. This avoids local overheating that might damage the welded components.

Local and Selective Heat Treating

A unique advantage of the diode laser hardening process over conventional heat treating processes is that it is possible to adjust its spot to the contour requiring hardening and, therefore, to achieve extremely high throughput. Its easy mode of operation allows the diode laser to be integrated easily into production processes and, if desired, to be used with an industrial robot (Fig. 6.24).

Compared with other lasers used for hardening, diode lasers have the added advantage of a shorter emission wavelength that is better absorbed by metals, as well as superior process stability. In addition, diode lasers do not require special absorption layers that can prevent temperature control by a pyrometer and that also may result in surface contamination.

Laser Brazing

In addition to requiring high strength and a small heat-affected zone, particularly high demands are made on the appearance of the weld seam in the case of visible seams. Laser brazing is an ideal approach for such situations. As an example, in the automotive industry, laser brazing is used to join the external visible parts of vehicles, such as the trunk lid, roof seams, doors, or C pillars (Fig. 6.25). Diode lasers are now considered proven technology for providing high levels of reliability and process stability for many applications that require three-shift production, such as the automotive industry.

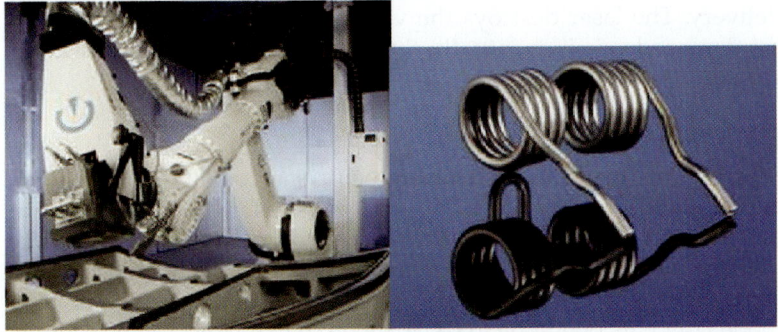

FIGURE 6.24 Laser hardening of tools and springs. (*Courtesy of LaserLine*)

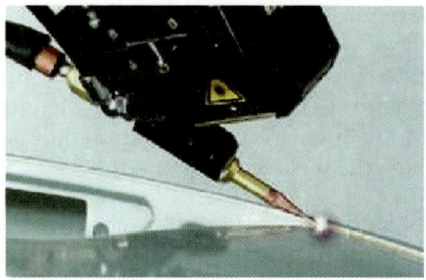

FIGURE 6.25 Laser brazing of an automotive part. (*Courtesy of LaserLine*)

6.7.2 Medical Applications

Diode lasers are used in a variety of medical applications, such as hair removal, tattoo removal, endovenous laser treatment (EVLT), photodynamic therapy, dental surgery, and cosmetic surgery. In hair removal, an 810-nm laser operating in pulse mode delivers light through a handheld device to the skin surface. The laser light is readily absorbed in the dark matter (melanin) of the hair follicle, removing the hair while sparing the rest of the skin. Tattoo removal is very similar to hair removal. The selective absorption of the laser light in the color-embedded skin tissue leads to fragmentation of the tissue; these fragments are then absorbed by the body and eliminated. The main difference between the two is the use of multiple wavelengths to remove the various colors of ink used in tattoos. A wavelength range between 670 and 890 nm is used to remove green and blue inks, while a range of 500 to 700 nm is used to remove red, orange, and purple inks. Black ink absorbs all wavelengths.

In dental surgery, such as periodontal (gum) surgery, a 980-nm fiber-coupled diode laser is used for precision cutting of soft gum tissue. This allows faster healing and relatively less scarring as compared with other techniques.

The treatment of varicose veins is another procedure that is now using diode lasers. In EVLT, an 808-nm laser beam operating in the 15 to 30 W range is delivered inside the varicose vein via a microfiber delivery. The laser destroys the varicose vein from the inside, and the damaged vein is eventually absorbed and eliminated as waste by the body.

6.7.3 Defense Applications

Diode lasers, by virtue of their high efficiency, small footprint, compactness, robustness, and low operating costs, are widely deployed in defense applications. Diode lasers mounted on ground and airborne military vehicles are used as illuminators. An illuminator typically consists of several stacks, with each stack consisting of both axis-collimated diode laser bars operating in QCW mode so that lasers can be conductively cooled. These stacks can deliver multikilowatt peak power at a chosen target.[8] Direct diode lasers with increased spatial brightness may also find applications in long-range target designation. For example, airborne target designation will require higher powers (> 5 W) because of the long distances between the aircraft and the target. Laser ignition of explosives is another application that removes the need to use electrical wiring for explosive chemicals and thus reduces the risk of accidental detonation. The laser beam delivers the required thermal intensity at the explosives for direct detonation and eliminates the need for other chemicals that were once used to trigger the explosion. This technique also eliminates the toxic waste.

References

1. Li, H., Chyr, I., Brown, D., Reinhardt, F., Romero, O., Chen, C.-H., Miller, R., Kuppuswamy, K., Jin, X., Ngugen, T., Towe, T., Crum, T., Mitchell, C., Truchan, T., Bullock, R., Wolak, E., Mott, J., and Harrison, J., "Next-Generation High-Power, High-Efficiency Diode Lasers at Spectra-Physics," *SPIE Proceedings*, 6824: 2008.
2. Treusch, G., Srinivasan, R., Brown, D., Miller. R., and Harrison, J., "Reliability of Water-Cooled High-Power Diode Laser Modules," *SPIE Proceedings*, 5711: 132–141, 2005.
3. Kohler, B., Brand, T., Haag, M., and Biesenbach, J., "Wavelength Stabilized High-Power Diode Laser Modules," *SPIE Photonics West*, San Jose, California, 2009.
4. Osowski, M. L., Hu, W., Lambert, R. M., Liu, T., Ma, Y., Oh, S. W., Panja, C., Rudy, P. T., Stakelon, T., and Ungar, J., "High Brightness Semiconductor Lasers," *SPIE Photonics West*, San Jose, California, 2007.
5. Crump, P. A., Crum, T. R., DeVito, M., Farmer, J., Grimshaw, M., Huang, Z., Igl, S. A., Macomber, S., Thiagarajan, P., and Wise, D., "High Efficiency, High Power, 808nm Laser Array and Stacked Arrays Optimized for Elevated Temperature Operation," *SPIE Photonics West*, San Jose, California, 2005.
6. Clarkson, W. A., and Hanna, D. C., "Two-Mirror Beam-Shaping Technique for High-Power Diode Bars," *Optics Lett.*, 21(6): 375–377, 1996. http://www.orc.soton.ac.uk/viewpublication.html?pid=518P.
7. Treusch, H.-G., Du, K., Baumann, M., Sturm, V., Ehlers, B., and Loosen, P., "Fiber-Coupling Technique for High-Power Diode Laser Arrays," *SPIE Proceedings* 3267: 98–106, 1998.
8. Rudy, P., "The Best Defense Is a Bright Diode Laser," *Photonics Spectra*, December 2005.

PART 3
Solid-State Lasers

CHAPTER 7
Introduction to High-Power Solid-State Lasers

CHAPTER 8
Zigzag Slab Lasers

CHAPTER 9
Nd:YAG Ceramic ThinZag® High-Power Laser Development

CHAPTER 10
Thin-Disc Lasers

CHAPTER 11
Heat-Capacity Lasers

CHAPTER 12
Ultrafast Solid-State Lasers

CHAPTER 13
Ultrafast Lasers in Thin-Disk Geometry

CHAPTER 14
The National Ignition Facility Laser

CHAPTER 7

Introduction to High-Power Solid-State Lasers

Gregory D. Goodno

Senior Scientist, Northrop Grumman Aerospace Systems, Redondo Beach, California

Hagop Injeyan

Technical Fellow, Northrop Grumman Aerospace Systems, Redondo Beach, California

7.1 Introduction

Recent years have witnessed rapid growth in both average and peak powers attainable from solid-state lasers (SSLs). Continuous SSL output powers with good beam quality have reached the 100-kW level,[1] and SSL pulse energies and peak powers have exceeded 1 MJ and 1 PW, respectively.[2,3] This progress has been the result of many years of iterative advances in materials and processing methods, coupled with revolutionary developments, such as diode-pumping, thermally scalable laser architectures, and wavefront correction techniques.

Solid-state lasers differ from gas or chemical lasers in several important respects. First, as the name suggests, the lasing material is solid phase and thus cannot be flowed during operation. Volumetrically deposited waste heat must be removed from the surfaces, typically leading to large thermal gradients during high-average-power (HAP) operation. Second, all SSLs are optically pumped. Hence, a key engineering consideration is selection of the optical pump source and optical conditioning to couple the pump photons to the gain material. Due to their optically pumped nature, HAP SSLs essentially function as brightness enhancers—that is, they convert low-spatial-brightness pump photons into an output beam with improved beam quality (BQ), but with lower total power due to imperfect efficiency. The overriding consideration that drives HAP SSL designs is minimization of the output beam's thermo-optic distortion so as to maximize the brightness increase (where *brightness* is loosely defined as the ratio of power to BQ^2). Finally, many SSL materials exhibit relatively long upper-state lifetimes or broad-gain bandwidths compared with other types of lasers. This allows SSLs to act as energy-storage devices, in that the energy accumulated during a long optical pumping cycle can be released very quickly in the form of a short, high peak-power pulse.

This chapter discusses considerations that typically drive the selection of the laser gain material, pump source, pump delivery optics, and the geometries for both heat removal and optical extraction. The chapter is intended to serve as a brief prelude and introduction to Chaps. 8–14, which describe some of the most successful state-of-the-art high-power SSL architectures. More general background for the design and engineering of solid-state lasers can be found in the classic textbook by Koechner.[4]

7.2 Laser Gain Materials

All SSL materials consist of an optically transparent host doped with active ions that absorb pump light and emit laser light. Since the invention of the laser, an enormous body of research has accumulated on various combinations of lasant:host materials optimized for particular features or applications.[5] We confine this chapter to a discussion of specific laser materials and properties that are most relevant for peak and average power scaling, along with the basic concepts underlying SSL laser emission.

7.2.1 Cross Section and Lifetime

The probability of an active ion absorbing or emitting a photon is proportional to its transition cross section σ. The cross section represents the gain per unit length per inversion density ΔN, so that the laser small-signal gain is $g_0 = \sigma \Delta N$. A high cross section is usually advantageous for an SSL, as fewer incident photons are needed to saturate any given transition, whether during pumping or stimulated emission. This relaxes the need for high laser intensities and reduces the propensity for optical damage of the material. Moreover, a large laser gain enables an SSL architecture to be more tolerant to optical losses without substantial sacrifice in efficiency, thus providing design flexibility for the optical configuration of the extracting beam.

Another key spectroscopic parameter is the fluorescence lifetime τ for spontaneous decay of the upper laser level via emission of a photon.

Introduction to High-Power Solid-State Lasers

Many SSL materials have long upper-state lifetimes, typically on the order of $\tau \sim 1$ ms. This allows them to act as "optical capacitors," storing pump energy during a long pump cycle that can be released quickly in the form of a short pulse. Even for continuous wave (CW) pumping, a long upper-state lifetime is advantageous because it reduces the amount of pump power needed to reach inversion. Heuristically, the inversion density that can be accumulated by a pump power density R (where R is the number of photons per unit time per unit volume) is $\Delta N = \tau R$.

One of the most useful figures of merit (FOMs) for SSL materials is the product $\sigma\tau$. Because $\sigma\tau = g_0/R$, this FOM indicates how much laser gain is obtained for a given pump rate. A material with high $\sigma\tau$ will lase very easily—that is, it requires less pump power density R to reach a certain gain g_0. Figure 7.1 shows values of σ and τ for some common SSL families of materials.

For high-pulse energy lasers, the energy storage capability of a material is of paramount interest. Obviously, high τ allows a material to store more energy for a given pump rate, which is unambiguously helpful for pulsed lasers. However, a high σ can be a disadvantage for energy storage. Depending on the geometry of the gain material, amplified spontaneous emission (ASE) can prematurely depopulate the upper laser level, clamping the obtainable inversion density and small-signal gain. Hence, ASE can severely limit the ability of a material to store energy for pulsed operation. This is also an issue for large-aperture CW lasers, in which high-transverse laser gain can lead to parasitic lasing or loss of efficiency. Still, even for high-energy

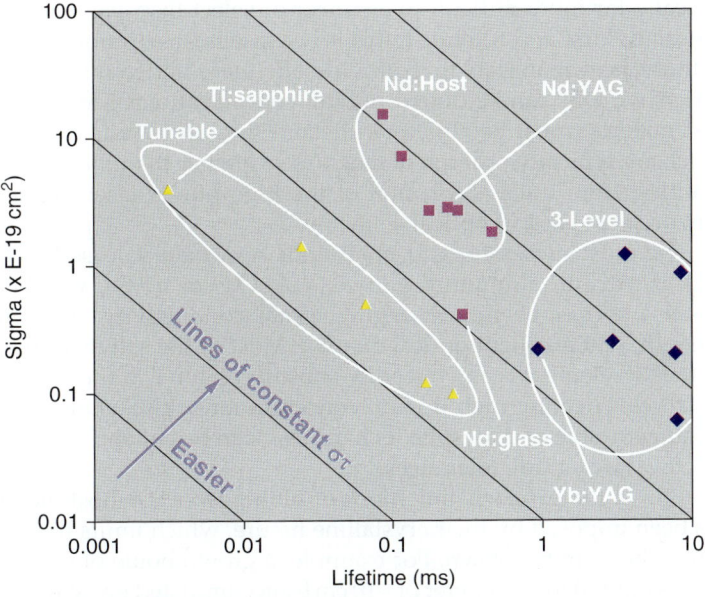

FIGURE 7.1 $\sigma\tau$ figure of merit for major solid-state laser (SSL) materials.

lasers, σ must be sufficiently high to provide reasonable levels of gain and to prevent the saturation fluence ($F_{sat} = h\nu/\sigma$) from exceeding the material's damage threshold.

7.2.2 Host Materials

The choice of host material is of particular importance for high-power SSLs. The highest grade of optical material purity is critical, both for purposes of laser damage resistance and to minimize transmission losses of the high-power extracting laser beam—in particular, absorption losses, which deposit excess heat in the material. The material must be able to be cut and polished to laser-grade specifications (typically better than 1/10 wave surface figure and 10/5 scratch-dig) with reasonable effort and yield. The mechanical properties of the host material are also of key importance for high power. High thermal conductivity will minimize the temperature increase associated with a given volumetric heat load that arises from lasing. The fracture toughness—that is, the peak surface tensile stress that the material can withstand—will determine the ultimate power density allowed for a particular geometry. Finally, the host material's thermo-optic properties (i.e., the change in index with temperature dn/dT and the coefficient of thermal expansion, CTE or α) drive the magnitude of laser wavefront distortion and depolarization for any given temperature increase. All these properties work together to determine the performance of a particular architecture.

Many of these considerations apply not only to the laser gain materials but also to any optical materials or coatings upon which the high-power laser beam is incident. However, laser gain materials are typically far more difficult to engineer or select than passive optical materials. First and foremost, this is because host selection is limited to those host materials that provide adequate lattice matches such that active ions can be doped in high concentrations. Moreover, the host material must be able to withstand the laser waste heat loads, which are typically order(s) of magnitude greater than the heat loads resulting from trace absorption of the high-power laser beam that may occur in passive optical elements.

The most successful and ubiquitous host material used in HAP SSLs is yttrium aluminum garnet ($Y_3Al_5O_{12}$, or YAG), which possesses a fortuitous mix of high thermal conductivity, mechanical strength, and excellent optical quality.[5] Most of the active lasing rare earth (RE) elements can be readily substituted for Y in the YAG crystal lattice, enabling high dopant concentrations. YAG is also readily manufacturable and, despite its hardness, can be cut and polished to exacting laser-grade tolerances.

One of the primary limitations of high-power SSL host materials has been imposed by their crystalline nature, which limits the size to which they can be grown. For example, a grown boule of crystalline YAG is limited to a diameter of ~10 cm by accumulated growth stresses

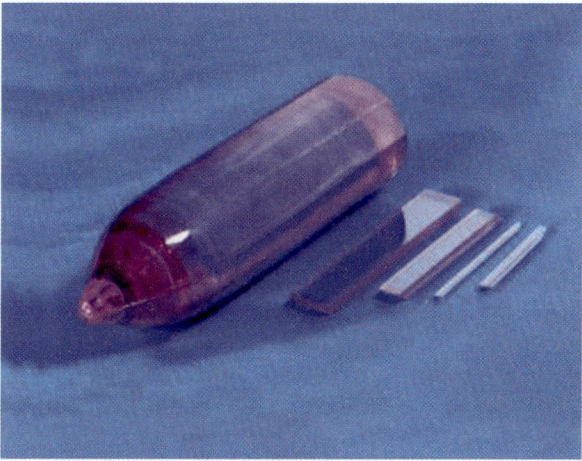

FIGURE 7.2 SSL slabs cut from Nd:YAG boules.

(Fig. 7.2). Boule length is also limited by doping gradients that arise from increasing concentrations of the dopant in the melt as growth proceeds.[6] The maximum clear aperture that can be cut from such a boule is typically about one-third its diameter due to the need to avoid low-optical quality areas in the boule that exhibit growth striations.[7]

A significant development in the past decade has been the emergence of high-optical-quality microcrystalline ceramics, which have largely displaced bulk crystalline hosts in high-power SSLs.[8] These ceramic materials are fabricated from high-purity crystalline nanopowders that are pressed and sintered into the desired final shape. Because interstitial regions between the individual microcrystal domains are much smaller than an optical wavelength, the sintered material can exhibit excellent transparency and homogeneity. The sintering fabrication process eliminates the size constraints imposed by crystal growth and has enabled the production of finished YAG pieces with greater than 10 x 10 cm^2 clear apertures, including co-sintered structures comprised of different doping concentrations or entirely different materials (see Chap. 11). The spectroscopic, thermal, and mechanical properties of finished ceramics tend to be nearly identical or superior to those of crystalline YAG. However, ceramic YAG has been shown to be somewhat more resistant to thermal stress fracture than crystalline YAG,[9] because there are no contiguous cleave boundaries and more energy is required to propagate a crack between crystal domains than in a single-crystal lattice.

7.2.3 High-Average-Power SSL Materials

Virtually all HAP SSLs are based around YAG that is doped either with Nd^{3+} or Yb^{3+} and that emits near 1064 nm or 1030 nm, respectively. Several factors are responsible for these two materials' dominance of

HAP SSLs. First, as described above, is YAG's favorable material properties, along with its ability to readily accept dopant concentrations exceeding 1 percent Nd and up to 100 percent (stoichiometric) Yb.[10] Second, both Nd and Yb exhibit favorable spectroscopic characteristics that are amenable to diode pumping and that result in highly efficient conversion of pump light to laser light.

Nd:YAG

Nd:YAG is historically the most common SSL gain material, having found widespread application in lamp-pumped rod lasers. This four-level laser can either be lamp pumped or diode pumped, most typically at the broad 808-nm band transition (Fig. 7.3). The upper laser level has a lifetime of 230 μs, providing reasonable energy storage capability for pulsed operation. Operated on the highest gain lasing transition at 1064 nm, the fraction of pump power that is converted to waste heat in the material (i.e., the quantum defect) is $1 - 808/1064 = 24\%$. Recent work has explored pumping directly into the upper laser level at 885 nm to reduce the quantum defect.[11]

In crystalline hosts, Nd^{3+} exhibits an extraordinarily large cross section for stimulated emission compared with other RE ions. For Nd:YAG at 1064 nm, $\sigma = 2.8 \times 10^{-19}$ cm^2, and it can even be several times larger for other host materials, such as YVO_4 (Fig. 7.1). For CW

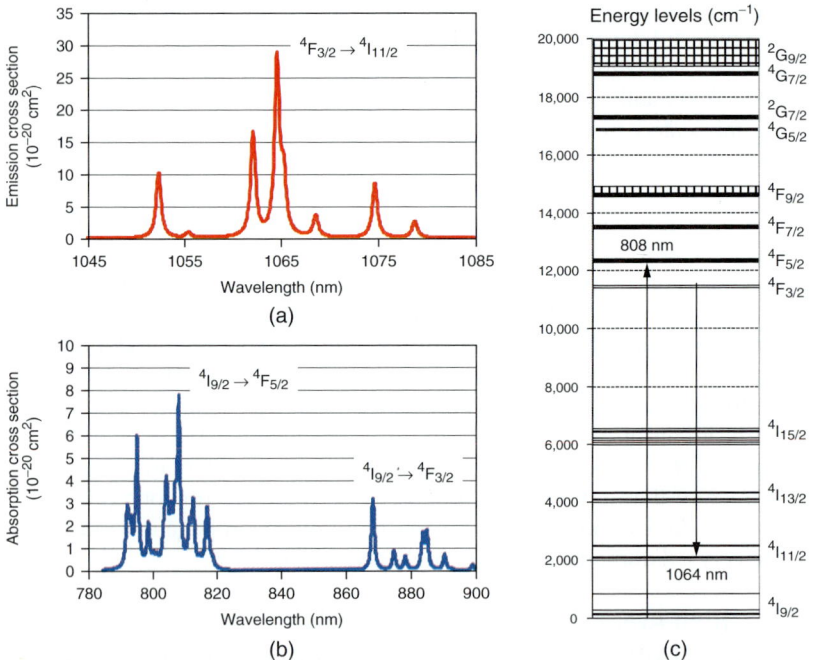

FIGURE 7.3 Nd:YAG spectroscopic parameters. (*a*) Emission cross section, (*b*) absorption cross section, (*c*) energy levels.

lasers, the large emission cross section enables high-gain extraction geometries with reduced sensitivity to optical losses and relatively low saturation intensity $I_{sat} = h\nu/\sigma\tau = 2.8$ kW/cm² for efficient extraction. The corresponding low saturation fluence makes Nd:YAG attractive for moderate energy pulse lasers, where efficiency and damage resistance are of paramount importance. However, the high cross section makes Nd:YAG generally ill suited for high pulse energies (> ~10 J) due to the onset of parasitics and ASE.

Yb:YAG

With the recent advent of diode pumping, Yb:YAG has emerged as an attractive alternative to Nd:YAG in numerous HAP SSL architectures.[12] Yb:YAG's predominant spectroscopic feature is its simple energy level structure, with essentially only two energy levels (Fig. 7.4). These levels are Stark-split into thermally populated manifolds, allowing energetically close pump and lasing transitions at 940 nm and 1030 nm, respectively. The corresponding ~9 percent quantum defect is two to three times smaller than for Nd:YAG, so that Yb:YAG is intrinsically high efficiency, generating relatively little waste heat per emitted photon. Yb:YAG is a quasi-three-level laser, with about 5 percent Boltzmann population in the terminal laser level at room temperature. Hence, bulk Yb:YAG SSLs typically exhibit rather high lasing thresholds, because the material must first be pumped to transparency before exhibiting net gain. Nevertheless, when operated high above threshold, Yb:YAG lasers can be extremely efficient (c.f., Chap. 10).

Yb:YAG's low emission cross section $\sigma = 2.2 \times 10^{-20}$ cm² leads to a low gain for most CW devices, requiring careful management of optical losses and typically multiple lasing passes to fully extract the material. Whereas Yb:YAG's long ~1-ms upper-state lifetime would

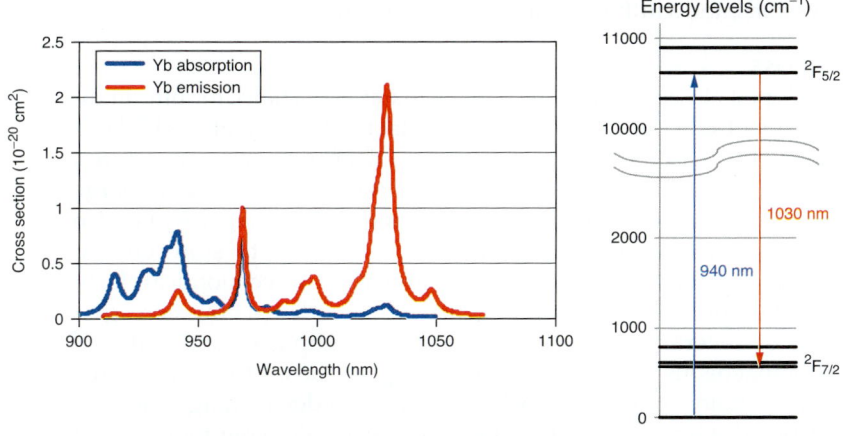

FIGURE 7.4 Yb:YAG energy levels, absorption, and emission cross sections.

seem attractive for pulse energy storage, the low cross section leads to saturation fluence $F_{sat} = h\nu/\sigma = 8.8$ J/cm², making efficient extraction without damage difficult outside clean room environments.

Cryogenic cooling of Yb:YAG is also commonly employed to reach higher powers than are easily attainable at room temperature.[13] Cooling the material to near-liquid nitrogen temperature (77 K) provides a number of benefits. First, the thermo-optic and thermomechanical properties of the YAG host material improve dramatically, with a factor of ~4 increase in thermal conductivity and a lowering of both dn/dT and CTE by factors of ~7 and 3, respectively. Second, the terminal laser level population is frozen to near zero, and the material becomes a true four-level laser with low lasing threshold. Finally, thermal broadening of the Yb gain spectrum that occurs at room temperature is eliminated, sharply narrowing the line width and increasing the peak gain by a factor of ~7. These benefits must be weighed against the cumbersome engineering required for implementation of cryo-cooling, which either requires a reservoir of liquid nitrogen with associated limited runtime, or integration of closed-loop chillers that are noisy and bulky and that severely affect overall system efficiency.

7.2.4 High Pulse-Energy and Peak-Power SSL Materials

Whereas Nd:YAG and Yb:YAG are the bases for most HAP SSLs emitting near 1 µm, other materials can provide improved performance for pulsed operation. For lasers intended to scale to high pulse energies or peak powers, the average power (i.e., the pulse repetition rate) is often of secondary importance. This opens the door to the use of host materials that are less thermally advantageous than YAG. The key material considerations for pulsed lasers are the energy storage and extraction capability; the ability to obtain large clear apertures free from any defects that might provide seeds for damage; and an emission bandwidth that can support short pulses.

Nd:glass

Nd-doped glasses have long been the material of choice for ultrahigh-energy pulsed lasers, such as the National Ignition Facility (NIF) laser (Chap. 14). Laser glass can be fabricated in meter-class apertures, which are beyond even the capability of ceramics. This enables large apertures to spread the laser energy to avoid damage, while also providing a large gain volume of Nd in which to store energy. Nd:glass's broad absorption spectrum allows for economical flashlamp pumping, and its inhomogeneously broadened emission spectrum of several nanometers can support subpicosecond pulses.[14] Inhomogeneous broadening from the glass host also reduces the peak emission cross section by nearly an order of magnitude, as compared with Nd:YAG (c.f., Fig. 7.1), hence allowing more stored energy without ASE depletion. However, the glass host material has thermal conductivity an order of magnitude smaller than YAG; though this does not affect the pulse energy, it typically limits pulse repetition frequencies to millihertz or lower.

Ti:Sapphire

If the SSL's purpose is to generate the highest peak power pulses, an alternative to increasing the pulse energy is to decrease the pulse duration. This is often much less expensive, because it reduces the required pumping power; moreover, ultrashort pulses enable many unique high-power applications (see Chap. 12). The ability of an SSL material to generate short pulses directly is limited by its gain bandwidth. Ti:sapphire is the most commonly used ultrashort-pulse material, owing to its nearly 1 octave of spectral coverage from 650 to 1100 nm.[15] Although the thermal properties of sapphire are even better than those of YAG, Ti:sapphire is not particularly well suited to energy storage, due to its short (~3 μs) upper-state lifetime, which typically requires pulse amplifiers to be pumped with short-pulse, Q-switched, and frequency-doubled (515 or 532 nm) YAG lasers. The large quantum defect between pump and emission wavelengths, along with the lack of economical high-power pump sources in the blue-green, typically limits average powers from Ti:sapphire to less than 100 W.

Other Pulsed Materials

Although the previously discussed materials dominate most high energy and peak power lasers, some less common materials warrant mention for their pulsed laser characteristics. Yb:SFAP [Yb^{3+}:$Sr_5(PO_4)_3F$] has been investigated as a diode-pumped material that is suitable for high-energy storage, due to its combination of large size, long upper-state lifetime, and intermediate saturation fluence, which balances energy storage against damage limits.[16,17] Yb-doped tungstates and sesquioxides have recently been developed for directly diode-pumped, short-pulse lasers. These materials exhibit microscopically disordered structures that result in the broad gain bandwidths needed to support ultrafast pulses. In particular, the Yb-doped sesquioxides Yb:Sc_2O_3 and Yb:Lu_2O_3 exhibit thermal conductivity slightly greater than YAG and have demonstrated potential for average power scaling in thin-disk geometries.[18]

7.3 Pumping, Cooling, and Thermal Effects

Careful management of thermal effects in the gain medium is the overriding engineering imperative for any successful high-power SSL design. There are two primary reasons temperature is so important for SSLs. First, thermal gradients must not be allowed to become large enough so as to pose a fracture risk to the laser material. Thermal stress fracture ultimately limits the power density attainable for any pumping and cooling geometry. However, often well before powers reach the level at which thermal fracture is a risk, thermo-optic wavefront distortions can severely limit the beam quality (BQ) of the extracting laser beam. This section describes how the need to minimize thermal aberrations drives the selection of geometries for pumping, cooling, and laser extraction.

7.3.1 Pump Sources

The lasing process unavoidably generates waste heat because of the energy difference between the pump and emission photons. This heat is deposited throughout the volume of the lasing material in proportion to the amount of pump light absorbed locally. Any nonuniformity in the profile of absorbed pump light across the laser clear aperture will translate into nonuniformities in heat deposition and development of thermal gradients that can aberrate the laser beam. Hence, a primary design consideration for high-power SSLs is to ensure that the material volume is pumped as close to uniform as possible across the extracting beam aperture. A second key consideration is to minimize the heat generation per emitted photon—that is, to pump the material with a photon as close in wavelength as possible to the emission wavelength. In this section, we discuss how these considerations affect selection of an appropriate pump source and conditioning optics.

Lamp Pumping

In 1960, Ted Maiman at Hughes Research Lab demonstrated the first laser, using a cheap and simple photographic flash lamp to pump a solid-state ruby crystal.[19] Although ruby was soon supplanted with more efficient and higher power Nd-doped materials, CW arc lamps and pulsed flash lamps filled with noble gases remained the predominant pump sources for SSLs until the development of high-power diodes in the 1990s. Nevertheless, lamp pumping severely limits the performance of high-power SSLs, and its use today is confined to either low-end, multimode lasers in the less than ~100 W range or to low-repetition-rate, high-pulse-energy lasers, in which the cost of sufficient diode pumps is prohibitive (including, interestingly enough, the multibillion-dollar NIF laser [Chap. 14]).

The primary disadvantage of lamp pump sources is their broadband emission spectrum, which spans the entire visible range from the ultraviolet (UV) to the near infrared (IR) (Fig. 7.5). For comparison, the absorption spectrum of Nd:YAG is also shown in Fig. 7.5. Only the small fraction of lamp power that coincides with an Nd absorption feature can be absorbed and be converted to laser light; the remaining power is simply wasted (Fig. 7.5, shaded regions). Regardless of the SSL gain material, this waste severely limits the

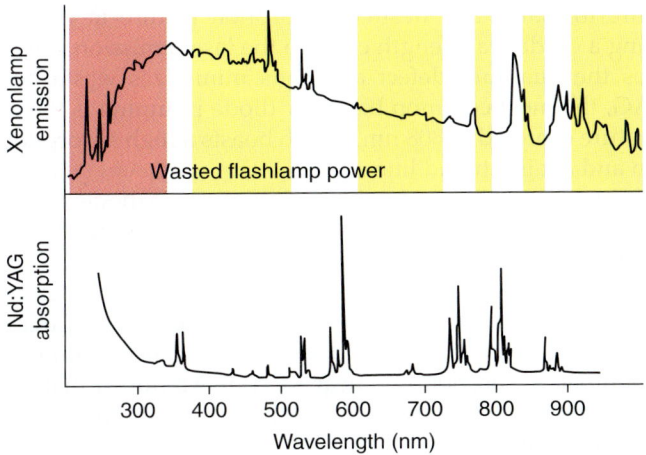

FIGURE 7.5 (top) Emission spectrum of a xenon-filled flash lamp. The shaded regions of the spectrum represent wasted energy that is not absorbed by Nd:YAG (bottom).

laser's efficiency. Even if a lamp photon happens to be at a favorable wavelength for absorption, it will most likely be at a transition to an energy level high above the upper laser level, leading to a large quantum defect and thus a large amount of heat deposited in the gain material for every emitted laser photon. From the standpoint of performance, this excess heat is the primary disadvantage of lamp pumping in comparison to the modern standard of diode pumping.

One final difficulty with lamp pumping is that lamps emit in all directions, with low spatial brightness, which severely constrains the geometric choices for optical coupling of the pump source into the gain medium. The most common choices are either to simply close-couple the lamp(s) against the gain medium, typically with a reflector to capture light emitted away from the desired direction, or to embed both the lamp and the gain medium (typically in the form of a rod) at the foci of an elliptical reflecting cavity, so that the lamp light is reimaged onto the rod. Neither of these geometries is advantageous for scaling to higher power, because they both constrain the geometries for laser beam extraction and heat removal.

Diode Pumping

The development of efficient, high-power laser diodes for pumping SSLs has revolutionized the development of HAP SSLs over the past 15 years. Owing to the importance of diode lasers both as pump sources and as high-power lasers in their own right, they are discussed in detail in Chaps. 5 and 6.

Diodes make ideal excitation sources for SSLs. Their emission spectrum can be engineered through choice of material and epitaxial structure to match any desired absorption feature in the near IR. Selecting a diode wavelength close to the laser emission wavelength reduces the quantum defect and thus minimizes waste heat. For Nd:YAG, the most common band for diode pumping is the absorption feature centered at 808 nm, which boasts a high absorption cross section and a fairly broad line width.

The narrow, engineerable emission spectrum of diodes has enabled use of SSL gain materials that simply could not be pumped effectively with lamps. The most prominent of these is Yb:YAG. Owing to Yb:YAG's simple energy-level structure, it is almost completely transparent except in the 900 to 1100 nm range (Fig. 7.4). Yet this makes Yb:YAG an ideal candidate for diode pumping, most commonly at the ~10-nm broad absorption line near 940 nm. Yb-doped materials can also be pumped at the narrower ~980-nm, zero-phonon line to minimize the quantum defect.

Another advantage of diode pumping is its highly directional (i.e., bright) emission. Although high-power diode emitters are multimode lasers, their beam quality is nevertheless sufficient to enable beam shaping using conventional optics or lens ducts. Hence, the pump intensity distribution within the SSL gain medium can be tailored to generate smooth excitation profiles to minimize thermal nonuniformities across the lasing aperture. Another advantage of high diode brightness is that the pump light can be focused to very high intensities, easily surpassing the tens of kW/cm^2 needed for efficient pumping of the quasi-three-level Yb transition. Diode focusability also eliminates the need for the SSL gain material to have a large surface area devoted to receiving pump light. With the use of focused diodes, pump light can be coupled into the gain medium through a relatively thin edge or tip, allowing large-area faces to be devoted to heat removal or laser extraction, with the corresponding thermal advantages.

Diodes have numerous other practical advantages as SSL pump sources. They may be scaled nearly arbitrarily in power (although not in brightness) by incoherently stacking multiple emitters or bars to form large arrays, with multikilowatt modules commonplace. They have benefited from years of investment in reliability engineering to achieve lifetimes typically measured in tens of thousands of hours. Finally, diodes are relatively compact and enable packaging of SSLs for platforms and environments (such as space), where lamp-pumped systems simply would not be feasible. In particular, fiber delivery of diode light allows unprecedented design flexibility and packaging convenience.

7.3.2 Laser Extraction and Heat Removal

Removal of heat through the laser material's surface creates thermal gradients that can aberrate the extracting laser beam and thus limit the output BQ, or even lead to catastrophic failure due to stress fracture.

All HAP SSL designs require some means of managing the impact of thermal gradients on the extracting laser beam's wavefront. There are two geometric considerations here.

The first consideration is to select a cooling geometry that minimizes the magnitude of the thermal gradients themselves. This leads to a gain material shape with a large surface area for heat removal, so that the surface heat flux is minimized. Furthermore, reducing the thickness of the gain material along the direction normal to the cooling surface will reduce the temperature rise. Hence, the desire to minimize thermal gradients in SSLs invariably leads to high-aspect ratio structures.

The second consideration is to select a laser extraction geometry that has little or no sensitivity to thermal gradients—in particular, one in which the extracting laser beam propagates with a vector component aligned with the primary thermal gradient. As an example, consider the slab geometry shown in Fig. 7.6, in which the slab is cooled from both top and bottom, thus creating a temperature gradient in the vertical direction. The extracting laser beam propagates from left to right. If the extracting beam simply propagates straight through the slab (Fig. 7.6a), then its center will sample hotter material than the edge. The optical path difference (OPD) across the beam due to slab thermal expansion (α) and index changes (dn/dT) is

$$\text{OPD} = [dn/dT + (n-1)\alpha]L\Delta T \tag{7.1}$$

With a slab length L of 10 cm and center-to-edge gradient ΔT of 40°C (which are typical numbers for a 4-kW slab), the OPD is on the order of ~50 µm, or 50 waves.[20] This much thermal focusing would prevent the beam from even propagating through the slab, much less with good beam quality.

Compare this with the zigzag geometry of Fig. 7.6b, in which the extracting beam reflects from top and bottom surfaces as it propagates. After one trip from top to bottom, each part of the beam has passed through the hot center and cold edges, hence experiencing

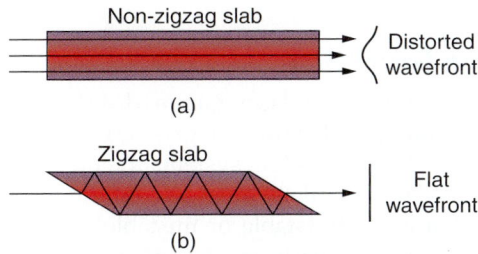

FIGURE 7.6 Comparison of (a) straight-through and (b) zigzag slab cooling and extraction geometries.

176 Solid-State Lasers

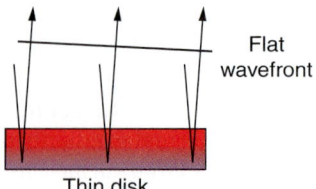

FIGURE 7.7 Thin-disk cooling and extraction geometry.

identical optical path length. With such a geometry, the extracting beam's wavefront is, to the first order, unaffected by the magnitude of the thermal gradient; therefore, the architecture can be scaled to high power (c.f., Chaps. 8 and 9). The same principle underlies the scalability of the thin-disk architecture (Fig. 7.7; also see Chap. 10).

In practice, for geometries such as those shown in Figs. 7.6 and 7.7, the quality of the extracted wavefront is driven by edge effects, mounting stresses, and uncontrolled nonuniformities in the thermal gradients. Just as it is critical to minimize nonuniformities in heat deposition during pumping, it is also important to ensure uniform heat removal through a spatially uniform, low-thermal impedance path from the cooled surface to the heat sink. This is relatively straightforward when the surface is cooled with direct liquid or gas flow. However, when the surface is conduction cooled, a host of engineering issues must be solved, including avoiding mechanical mounting stresses, CTE matching of the gain material to the substrate, uniform wetting of solder or other thermal interface materials, and preventing the extracting laser beam from coupling to the cooling substrate. Solutions to some of these issues are discussed in the context of specific architectures in Chaps. 8–10.

Finally, one noteworthy exception to these heat-removal considerations are heat capacity lasers, discussed in Chap. 11. These devices are uncooled and store heat during an operation time that is limited by the gain material's heat capacity. In the absence of surface cooling, the gain material is free from thermal gradients and expands uniformly without wavefront distortion. Hence, wavefront aberrations are driven primarily by nonuniformities in heat generation from pumping and laser extraction.

7.4 Laser Beam Formation

A low-aberration, high-power laser gain module incorporating favorable pumping, cooling, and extraction geometries forms the building block for any high-brightness SSL system. To generate a high-power output beam, the gain module(s) must be configured as part of either a resonant oscillator cavity (stable or unstable) or an amplifier. The optimum configuration choice is one that efficiently extracts the stored energy while minimizing losses and accumulated OPD, so as to generate the highest brightness output beam. This section discusses the considerations underlying the trade between oscillators and amplifiers for high-power lasers.

7.4.1 Stable Resonators

Stable resonators are geometrically stable in the sense that they confine a cone of rays upon reflection between two curved mirrors. This allows a near-planar wavefront to build up during laser oscillation, providing a simple, robust means of generating good beam quality. Stable resonators are typically configured to support only a single, TEM_{00} (Gaussian) mode via selective gain competition against higher-order modes. The TEM_{00} mode experiences higher net round-trip gain through improved geometric overlap with the pumped gain volume or lower clipping losses from intracavity apertures.[21]

Stable resonator ray confinement naturally leads to tightly focused spots within the cavity, with spot dimensions determined by diffraction and typically on the order of $\sim(\lambda L)^{1/2} = 1$ mm for 1-μm wavelengths and cavity lengths $L \sim 1$ m. With such small beam sizes, the resulting high intensity allows easy saturation and efficient extraction of the gain material. Their simplicity and robustness allow stable resonators to form the cornerstone of most low- to moderate-power SSLs. However, they are poorly suited for generating good beam quality from high-power SSLs with large gain apertures, because the fundamental stable mode cannot be easily scaled to diameters beyond the order of a few millimeters without impractically long cavity lengths or alignment sensitivities. Nevertheless, for applications where multimode output is acceptable, the high circulating power achievable in a high-Q stable resonator enables efficient extraction of low-gain materials, such as Yb:YAG, or low-gain extraction geometries, such as thin disks.

7.4.2 Unstable Resonators

When the output power from SSLs grows to the point at which thermal or damage limits become prohibitive for millimeter-class spots, another extraction geometry must be adopted. Unstable resonators are often employed for high-power SSLs, because they allow very large mode areas with excellent BQ.[21] Instead of supporting cavity modes whose size is determined by diffraction, unstable resonator modes are not geometrically confined. Laser oscillation initially builds up within a Fresnel core of diameter $\sim(\lambda L)^{1/2}$, in which diffractive beam spreading dominates the cavity mirror curvatures (Fig. 7.8). The mirror curvatures are chosen to magnify the beam by a factor of M upon each round trip, so that beam sizes are constrained only by the limiting aperture of the primary mirror or the intracavity gain element. The final beam is outcoupled either by spreading past the clear aperture of the secondary mirror or by using a larger secondary mirror with spatially varying reflectivity that tapers to zero

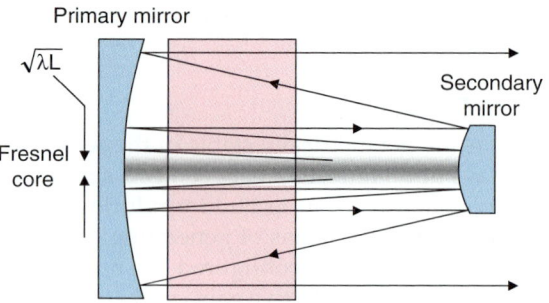

FIGURE 7.8 Unstable resonator cavity.

at the edges. This latter configuration is widely used in present-day unstable resonators to eliminate diffraction from the hard-edged aperture. For nonspatially varying reflectivities, the outcoupling fraction is approximately $1 - 1/M^2$.

Whereas unstable resonators provide large mode volumes, they also impose some challenges when used to extract high-power SSL gain materials. To maintain good wavefront control and single-mode output, any OPD imposed by the gain material must be small enough that it is overwhelmed by the mirror curvatures. Otherwise, OPD from the gain module can effectively form a lens over a small aperture that can drive the resonator over the stability boundary, hence forming a locally stable resonator. This, in turn, causes "filamentation" of the unstable resonator mode, in which multiple independent output beams with uncorrelated wavefronts colase over different subapertures of the gain medium.

To avoid such an event and to provide some robustness against thermal OPD, unstable resonators for SSLs are typically designed with high-curvature mirrors, leading to short resonator lengths and large magnifications. However, high M leads to large outcoupling fractions and, thus, to a requirement for correspondingly high laser gain to make up for the loss on each round trip to maintain laser oscillation. Hence, unstable resonators tend to achieve the most success with high gain materials, such as Nd, or with geometries that provide a long gain path for the extracting beam (e.g., zigzag slabs). To further increase laser gain, the modules are often operated in a pulsed or quasi-CW format, even when the goal is high average, rather than peak, power.[22]

This is not to say that unstable resonators cannot be made to work with low gain materials and module architectures such as Yb:YAG thin disks. Even with a low-gain SSL medium, unstable resonator extraction has been demonstrated successfully by combining multiple gain modules, or gain module passes, per resonator round trip, often with the use of image-relay optics to accommodate long physical beam paths without changing the resonator's optical length.[23] However, multiple gain passes per round trip leads to more traversals of the beam through the aberrated gain material before ejection from the resonator, which exacerbates the OPD that would have been picked up by the beam upon a single pass and which can limit the output beam's quality.

7.4.3 Master Oscillator Power Amplifiers

Master oscillator power amplifiers (MOPAs) provide versatile extraction configurations at the cost of some complexity (Fig. 7.9). A low-power beam with well-controlled spatial and temporal characteristics is formed using a master oscillator (MO). This beam is then amplified separately in one or more stages of power amplifiers (PAs). Separation of the beam formation in the MO and its amplification in the PA provides flexibility to independently optimize different output parameters that would be impossible to generate simultaneously from a single resonator. For example, fast pulses can be generated from small, low-power Q-switched or mode-locked MOs without concern for damage. Beam footprints can be optimally sized in the PA to achieve good saturation without the need to consider resonator mode effects. Due to the lack of feedback dynamics, it is straightforward to implement advanced methods for wavefront or polarization correction in the PA.

Although the MOPA concept is simple, its implementation can be cumbersome, due to the high gain often needed to bridge the orders-of-magnitude difference in power from the MO to PA. High gain can typically be obtained only by multiple amplifier stages or by multiple passes per amplifier, leading to complex optical beam paths. Faraday isolators are typically required to prevent feedback between the MO and successive gain stages and can themselves severely limit extracted power due to thermal lensing and depolarization.[24] Finally, many MOPAs employ near-counterpropagating beam passes to reach full saturation in the PA, imposing a requirement for some means of outcoupling the high-power laser light through either spatial or polarization multiplexing.

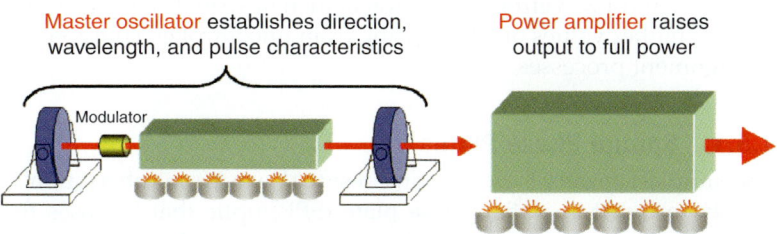

FIGURE 7.9   Master oscillator power amplifier (MOPA).

7.5 Wavefront Correction

Even with advantageous pumping, cooling, and beam extraction geometries, the magnitude of typical thermal excursions of the gain material during operation makes it very difficult to generate near-diffraction-limited beam quality directly from a large-aperture, HAP SSL device. Even if thermal gradients were reduced to a small fraction of the overall thermal change in optical path length, this would typically still be enough wavefront distortion to substantially degrade the beam quality. As a rule of thumb, a laser application whose efficacy is driven by the peak focused intensity can typically tolerate root-mean-square (RMS). OPD on the order of $\Delta\phi = 1/10$ wave. Using the Marechal approximation,[25] this OPD reduces the far-field peak intensity (or Strehl ratio) by $\sim 1 - \exp[-(2\pi\Delta\phi)^2] = 33$ percent compared with a planar wavefront beam.

In principle, OPD can be entirely eliminated by a combination of uniform pumping, purely one-dimensional heat removal, and an extraction path through the gain medium that has a vector component along the primary thermal gradient. Yet, in practice, it is nearly impossible to completely eliminate OPD. Edge effects that break the symmetry of one-dimensional heat removal will impose some OPD. Any nonuniformity in pumping or cooling along dimensions transverse to beam propagation will not be averaged out. Given that typical multikilowatt gain modules exhibit multiple tens of waves' increase in optical path due to temperature rises during operation, achieving residual OPD less than $\sim\lambda/10$ requires heat generation and removal to be uniform to within less than $\sim 1\%$ across the clear aperture. Due to uncontrolled variations in pump-diode emission, nonuniform aging, optical surface tolerances, surface wetting, and thermal contact, these tolerances are difficult, if not impossible, to achieve.

In the worst case, the difficulty of obtaining near-planar wavefronts increases linearly with the number of gain modules or gain passes in the beam path when assuming highly correlated aberrations (e.g., with multiple passes through the same gain module volume). In the best case, with uncorrelated aberrations, the difficulty increases as the square root of the number of gain module passes. Many high-BQ and high-power CW SSLs incorporate some additional means of wavefront correction in their system design to accommodate higher values of OPD arising from uncontrolled components or alignment processes.

7.5.1 Spatial Phase Plates

The simplest means for correcting residual wavefront aberration is simply to insert a spatial phase plate (SPP) optic that imposes the conjugate wavefront profile, so that downstream of this optic, the net laser wavefront is near-planar. In the simplest case, the SPP is simply a lens to correct for thermal focusing. Computer-controlled fabrication methods, such as magnetorheological finishing (MRF), provide the capability to manufacture custom surface-relief profiles in silica and other substrates, with spatial frequencies ~1 per mm and strokes (wavefront amplitude) of multiple waves.[26,27] SPPs have been demonstrated to increase brightness from both stable and unstable resonators.[23,28]

Although SPPs do provide simple methods of correction, they can be cumbersome to implement in a high-fidelity system. Gain module OPD can be rigorously calculated using numeric models, but in an HP SSL, residual OPD is often driven by uncontrolled component variations rather than by deterministic design; therefore, an SPP must be custom fabricated for each laser. This requires that the laser first be built and its wavefront measured at full power before the SPP can be made. Moreover, any change in the laser's thermal profile due to changes in operating power, component degradation, or the influence of the SPP itself on the extracting beam can invalidate the old wavefront map and require installation of a new SPP.[23] Finally, it is difficult to achieve $\sim\lambda/10$ fidelity given the accumulated tolerances in wavefront measurement, SPP manufacturing, and final installation and alignment; thus, even with an SPP, it is difficult to directly obtain near-diffraction-limited beams from large apertures.

To further correct laser wavefronts, dynamic methods are often employed that can respond in real time to changes in the laser's aberrations.

7.5.2 Phase Conjugation

Phase conjugate mirrors (PCMs) represent attractive dynamic methods for wavefront correction of high-power lasers. A PCM differs from a regular mirror in that it reflects the conjugate of an incident wavefront. For example, whereas an incident diverging beam would still be diverging after reflection from a regular mirror, it would be converging after reflection from a PCM. This phase conjugation provides automatic correction of laser and optic wavefront aberrations and beam jitters without active electronic controls.

One particularly successful implementation of PCMs in HP SSLs has used stimulated Brillouin scattering (SBS) in liquid Freon.[29] The basic concept, implemented in a MOPA configuration, is shown in Fig. 7.10. The low-power beam with a planar wavefront is incident on the PA from the left. Upon the first pass through the PA, the beam is amplified and aberrated. The aberrated beam is then focused into the cell containing a Brillouin-active material. Electrostriction of the material near the beam focus creates a longitudinal acoustic grating whose transverse phase profile is identical to the optical wavefront of the focused beam. After Bragg reflection from this moving grating, the return beam has the conjugate wavefront of the forward beam, so

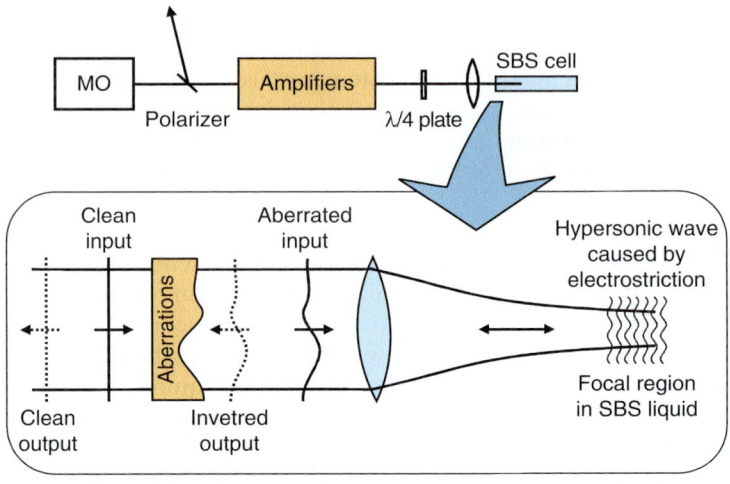

FIGURE 7.10 SBS-based phase conjugate mirror configured with a MOPA.

that upon the return transmission through the aberrated PA, the original planar wavefront is recovered at the MOPA output port.

PCMs can be formed using a variety of linear and nonlinear physical mechanisms and materials. Although SBS works particularly well with pulsed, high-peak-intensity SSLs, lower-threshold PCMs have been demonstrated using SBS in multimode fibers[30] and in free space using photorefractive or thermal gratings.[31]

Despite the attractive simplicity of a PCM, they are not always feasible to implement on an HP SSL. Each PCM mechanism, whether SBS, thermal, or photorefractive, constrains the incident laser's operating regime. For example, SBS in Freon has a high threshold and requires a long interaction length to build up sufficient reflectivity from the acoustic grating; thus, it does not work well with anything other than pulsed, single-frequency lasers with long coherence lengths. The dynamic range in power of most thermal PCM configurations is also limited. Typically the reflection from a PCM is significantly less than unity, requiring a high-gain geometry to avoid substantial loss of efficiency. Finally, the conjugation range of any PCM will be limited—essentially, the input wavefront aberrations must be of sufficiently low amplitude and spatial frequency such that the beam does not break into separate spots near the focus. This yields a set of conjugate returns from each spot with uncorrelated phases that will not yield planar output after the second PA pass.

7.5.3 Adaptive Optics

Adaptive optics (AO) provides a more flexible and engineerable means of wavefront control than phase conjugation.[32] This capability comes at the cost of added complexity in the form of active control

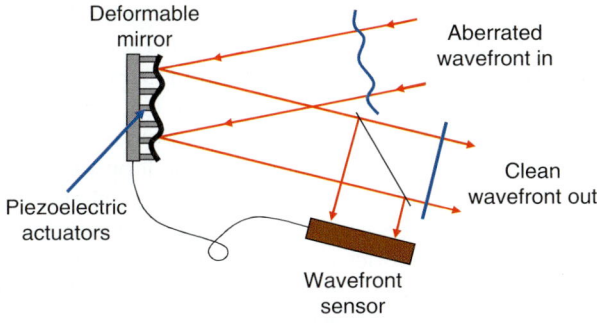

FIGURE 7.11 Closed-loop adaptive optics (AO) system for laser wavefront correction.

loops and actuators. An AO system involves actively, or adaptively, controlling the shape or orientation of optics in the beam path so as to reduce or eliminate OPD on the high-power output beam. Most often, the actuated optics are integrated with active sensing of the output beam wavefront as part of a continuous feedback loop. However, AO systems can also be configured as feed-forward devices based on such laser operating parameters as pump power levels.

A simple example of a closed-loop AO system is shown in Fig. 7.11.[33] An aberrated, high-power beam is incident on a deformable mirror (DM). The DM's optical surface consists of a thin, polished face sheet with a low-absorption high-reflectivity (HR) coating. The face sheet changes its shape in response to stress imposed by individually addressable piezoelectric actuators attached to its rear surface (Fig. 7.12). The beam reflected from the DM is sampled and its wavefront measured using a Shack-Hartmann sensor.[32] This wavefront information

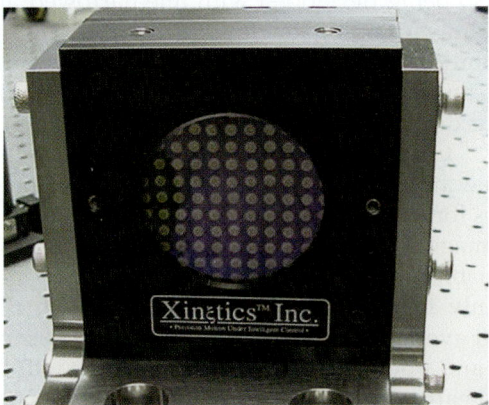

FIGURE 7.12 Deformable mirror for high-power SSL beam correction (Xinetics, Inc.). Individual actuators can be seen through the 1064-nm high-reflection coated face sheet, which is transparent at visible wavelengths.

is used to derive a set of control signals to change the DM's shape to impose the conjugate wavefront aberration on the beam. The new and (hopefully) reduced aberration wavefront is then sensed to close the feedback loop.

Integration of an AO system with a high-power SSL can be complex. A key consideration is to ensure that the loop rate and control bandwidth are sufficient to keep up with dynamic changes imposed by the laser. These changes can be due either to warm-up transients of the gain modules or optics during cycled operation or to turbulence driven by hot optics or mechanical parts near the beam path. Another consideration is to ensure that the number and stroke of actuators can appropriately compensate the spatial frequencies and amplitudes of the incident OPD. Finally, some high-power SSL designs integrate AO inside a resonant cavity—typically, an unstable resonator.[34] This can couple the AO system to resonator modes and extraction dynamics, which often require a complex control algorithm to generate stable output.

7.6 Conclusion and Future Directions

In this chapter, we introduced the underlying concepts and most widely used methods for achieving high power in SSLs. The selection of SSL material, pump source, heat removal and laser extraction geometries, and overall system architecture plays a critical role in the scalability of a design to high power. The following chapters provide design details for some of the most successful SSLs to date.

Much work is underway to continue developing SSLs to even higher power levels. Laser-pump diodes are rapidly becoming cheaper, brighter, and more reliable, which enables more controlled beam shaping and deterministic heat deposition profiles. High-power diodes are also being developed with line-narrowed and stabilized spectra, enabling pumping on low-quantum-defect spectral lines such as 885-nm for Nd:YAG, which reduces waste heat. Improved ceramic fabrication methods are yielding structures with gradient or heterogeneous doping profiles for improved pumping uniformity or reduced ASE.[35] Ceramic fabrication methods are also enabling production of new host materials with improved spectral and thermal characteristics for high-average-power ultrafast-pulse lasers.[36]

Laser damage resistance of optical elements is a key issue for HAP SSL reliability and usability. Historically, damage has been a concern mainly for pulsed lasers. However, with the advent of multikilowatt average powers, CW damage is emerging as a major engineering and operational issue.[37] Finally, as will be discussed in Chap. 19, beam combining of multiple HAP SSLs (or other HAP lasers, such as fibers or diodes) is a very active field, due to its promise of ultimate scalability by bypassing the limits of any specific laser architecture.

References

1. McNaught, S. J., Asman, D. P., Injeyan, H., et al., "100-kW Coherently Combined Nd:YAG MOPA Laser Array," *Frontiers in Optics*, paper FThD2, 2009.
2. Moses, E. I., Boyd, R. N., Remington, B. A., Keane, C. J., and Al-Ayat, R., "The National Ignition Facility: Ushering in a New Age for High Energy Density Science," *Phys. Plasmas*, 16: 041006, 2009.
3. Perry, M. D., Pennington, D., Stuart, B. C., et al., "Petawatt Laser Pulses," *Opt. Lett.*, 24: 160–162, 1999.
4. Koechner, W., *Solid State Laser Engineering*, 6th ed. Springer, Berlin, 2006.
5. Kaminskii, K., *Laser Crystals: Their Physics and Properties*, 2nd ed. Springer, Berlin, 1990.
6. "Nd:YAG," http://www.as.northropgrumman.com/products/synoptics_nd_yag/index.html, accessed November 18, 2010.
7. Ostermeyer, M., Mudge, D., Veitch, P. J., and Munch, J., "Thermally Induced Birefringence in Nd:YAG Slab Lasers," *Appl. Opt.*, 45: 5368, 2006.
8. Ikesue, A., and Aung, Y. L., "Ceramic Laser Materials," *Nature Photonics*, 2: 721, 2008.
9. Ueda, K., Bisson, J. F., Yagi, H., Takaichi, K., Shirakawa, A., Yanagitani, T., and Kaminskii, A. A., "Scalable Ceramic Lasers," *Laser Physics*, 15, 927: 2005.
10. Patel, F. D., Honea, E. C., Speth, J., Payne, S. A., Hutcheson, R., and Equall, R., "Laser Demonstration of $Yb_3Al_5O_{12}$ (YbAG) and Materials Properties of Highly Doped Yb:YAG," *IEEE J. Quant. Electron.*, 37: 135, 2001.
11. Lavi, R., and Jackel, S., "Thermally Boosted Pumping of Neodymium Lasers," *Appl. Opt.*, 39: 3093, 2000.
12. Lacovara, P., Choi, H. K., Wang, C. .A., Aggarwal, R. L., and Fan, T. Y., "Room-Temperature Diode-Pumped Yb:YAG Laser," *Opt. Lett.*, 16: 1089, 1991.
13. Fan, T. Y., Ripin, D. J., Aggarwal, R. L., Ochoa, J. R., Chann, B., Tilleman, M., and Spitzberg, J., "Cryogenic Yb^{3+}-Doped Solid-State Lasers," *IEEE J. Sel. Topics in Quant. Electron.*, 13: 448, 2007.
14. Weber, M. J., "Science and Technology of Laser Glass," *J. Non-Crystalline Solids*, 123: 208, 1990.
15. Moulton, P. F., "Spectroscopic and Laser Characteristics of $Ti:Al_2O_3$," *J. Opt. Soc. Am.*, B3: 125, 1986.
16. Bibeau, C., Bayramian, A., Armstrong, P., et al., "The Mercury Laser System—An Average Power, Gas-Cooled, Yb:S-FAP Based System with Frequency Conversion and Wavefront Correction," *J. Phys. IV France*, 133: 797, 2006.
17. Schaffers, K. I., Tassano, J. B., Bayramian, A. J., and Morris, R. C., "Growth of Yb:S-FAP [Yb^{3+}:$Sr_5(PO_4)_3F$] Crystals for the Mercury Laser," *J. Crys. Growth*, 253: 297, 2003.
18. Südmeyer, T., Kränkel, C., Baer, C. R. E., et al., "High-Power Ultrafast Thin Disk Laser Oscillators and Their Potential for Sub-100-femtosecond Pulse Generation," *Appl. Phys.*, B97: 281, 2009.
19. Maiman, T. H., "Stimulated Optical Radiation in Ruby," *Nature*, 187: 493, 1960.
20. McNaught, S. J., Komine, H., Weiss, S. B., et al., "Joint High Power Solid State Laser Demonstration at Northrop Grumman," 12th Annual Directed Energy Professional Society Conference, November 2009.
21. Siegman, A. E., *Lasers*, University Science Books, Sausalito, CA., 1986.
22. Machan, J., Zamel, J., and Marabella, L., "New Materials Processing Capabilities Using a High Brightness, 3 kW Diode-Pumped, YAG Laser," *IEEE Aerospace Conf.*, 3: 107, 2000.
23. Avizonis, P. V., Bossert, D. J., Curtin, M. S., and Killi, A., "Physics of High Performance Yb:YAG Thin Disk Lasers," *Conference on Lasers and Electro-optics*, paper CThA2, 2009.
24. Khazanov, E. A., Kulagin, O. V., Yoshida, S., Tanner, D. B., and Reitze, D. H., "Investigation of Self-Induced Depolarization of Laser Radiation in Terbium Gallium Garnet," *IEEE J. Quantum Electron.*, 35: 1116, 1999.
25. Born, M., and Wolf, E., *Principles of Optics*, 6th ed., 464, Pergamon Press, London, 1980.

26. Golini, D., Jacobs, S., Kordonski, W., and Dumas, P., "Precision Optics Fabrication Using Magnetorheological Finishing," Advanced Materials for Optics and Precision Structures, *SPIE Proc.*, CR67: 251, 1997.
27. Bayramian, A., Armstrong, J., Beer, G., et al., "High-Average-Power Femtopetawatt Laser Pumped by the Mercury Laser Facility," *J. Opt. Soc. Am.*, B25: B57, 2008.
28. Bagnoud, V., Guardalben, M. J., Puth, J., Zuegel, J. D., Mooney, T., and Dumas, P., "High-Energy, High-Average-Power Laser with Nd:YLF Rods Corrected by Magnetorheological Finishing," *Appl. Opt.*, 44: 282, 2005.
29. St. Pierre, R., Mordaunt, D., Injeyan, H., et al., "Diode Array Pumped Kilowatt Laser," *IEEE J. Sel. Topics Quantum Electron.*, 3: 53, 1997.
30. Riesbeck, T., Risse, E., and Eichler, H. J., "Pulsed Solid-State Laser System with Fiber Phase Conjugation and 315 W Average Output Power," *Appl. Phys.*, B73: 847, 2001.
31. Zakharenkov, Y. A., Clatterbuck, T. O., Shkunov, V. V., et al., "2-kW Average Power CW Phase-Conjugate Solid-State Laser," *IEEE J. Sel. Top. Quant. Electron.*, 13: 473, 2007.
32. Tyson, R. K., *Principles of Adaptive Optics, 2nd ed.*, Academic Press, San Diego, 1997.
33. Goodno, G. D., Komine, H., McNaught, S. J., et al., "Coherent Combination of High-Power, Zigzag Slab Lasers," *Opt. Lett.*, 31: 1247, 2006.
34. LaFortune, K. N., Hurd, R. L., Fochs, S. N., Rotter, M. D., Pax, P. H., Combs, R. L., Olivier, S. S., Brase, J. M., and Yamamoto, R. M., "Technical Challenges for the Future of High Energy Lasers," *SPIE Proc.*, 6454: 645400, 2007.
35. Soules, T., "Ceramic Laser Materials for the Solid-State Heat Capacity Laser," in *Frontiers in Optics*, OSA Annual Meeting, paper FWW2, 2006.
36. Schmidt, A., Petrov, V., Griebner, U., et al., "Diode-Pumped Mode-Locked Yb:LuScO$_3$ Single Crystal Laser with 74 fs Pulse Duration," *Opt. Lett.*, 35: 511, 2010.
37. Shah, R. S., Rey, J. J., and Stewart, A. F., "Limits of Performance—CW Laser Damage," *SPIE Proc.*, 6403: 640305, 2007.

CHAPTER 8
Zigzag Slab Lasers

Hagop Injeyan

Technical Fellow, Northrop Grumman Aerospace Systems, Redondo Beach, California

Gregory D. Goodno

Senior Scientist, Northrop Grumman Aerospace Systems, Redondo Beach, California

8.1 Introduction

The invention of zigzag slabs in the early 1970s by Bill Martin and Joe Chernock[1] launched a new paradigm in the development of solid-state lasers (SSLs). The idea of propagating laser beams in a direction that averages the temperature gradients in the gain medium has been the cornerstone of power scaling of SSLs, be it in the form of thin disks, zigzag slabs, or Brewster-plate amplifiers. Although zigzag slabs have been the most common architecture for SSL power scaling in the past 15 years, there has been significant evolution in the implementation of zigzag slabs by numerous groups. This chapter reviews the principles of zigzag slab propagation, its scaling laws, and various adaptations of this approach to optimize performance.

8.2 Zigzag Slab Principle and Advantages

8.2.1 Zigzag Geometry

The zigzag slab geometry is shown schematically in Fig. 8.1. Typically, a rectangular cross-section slab is cut to have angled input faces and polished sides. The slab is, in general, cooled from the polished faces. The laser beam is injected into the slab so that it will allow the beam to make multiple total internal reflections (TIRs) from the polished sides as it propagates down the slab. The main purpose of the zigzag

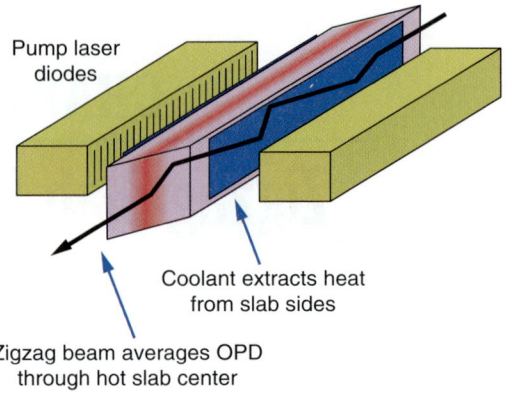

FIGURE 8.1 Schematic of a traditional side-pumped zigzag slab. OPD: optical path difference.

propagation is to average over the temperature gradients in the thin dimension of the slab.

Figure 8.2 shows two different injection schemes into a YAG (yttrium aluminum garnet) slab with refractive index $n = 1.82$. The first scheme uses a near-Brewster cut that favors linearly p-polarized light on the input face and that is often used in oscillators. At a slab cut of 30.9° (Brewster is 28.8°), the losses for p-polarized light are minimal, and the refraction angle is such that the beam reflects from the TIR surface parallel to the input face, optimally filling the slab. The second approach uses near-normal incidence and is polarization indifferent. The latter approach is best suited for two-pass amplifier designs, in which the first pass may be p-polarized and the second pass s-polarized. This approach creates small unextracted regions, called *dead zones*, that reduce extraction efficiency by a small amount (discussed later in this chapter) but that can also help provide areas for mounting and sealing the slab.

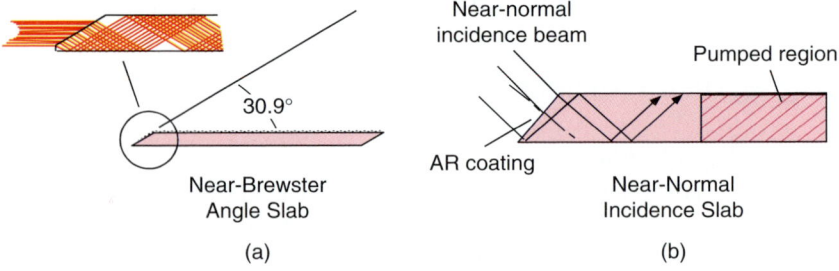

FIGURE 8.2 Propagation through a zigzag slab. (*a*) Near-Brewster cut for *p*-polarized light and (*b*) near-normal incidence for polarization-independent propagation. AR: antireflective

8.2.2 Scaling Laws

Under steady-state operating conditions, in which the gain medium is volumetrically pumped and simultaneously cooled from the surface, the temperature gradients in the gain medium are the ultimate limitation to power scaling. Figure 8.3 shows a simplified graphical representation of the cooling geometry for a slab and a cylindrical rod of thickness t and diameter d, respectively. The functional dependence of ΔT under uniform heat deposition for a slab is given by

$$\Delta T = Qt^2/8k \tag{8.1}$$

For a rod it is

$$\Delta T = Qd^2/16k \tag{8.2}$$

where Q is the volumetric heat density and k is the thermal conductivity. For propagation down the axis of a gain medium of length L, this center-to-edge temperature difference results in optical path difference (OPD) Δz across the aperture of the gain medium:

$$\Delta z = L\Delta T \frac{dn}{dT} \tag{8.3}$$

where dn/dT is the coefficient of index change with temperature.

To the first order, the parabolic wavefront curvature introduced by this OPD can be approximated as a thermally induced lens of focal length:

$$f = d^2/(8\Delta z) \tag{8.4}$$

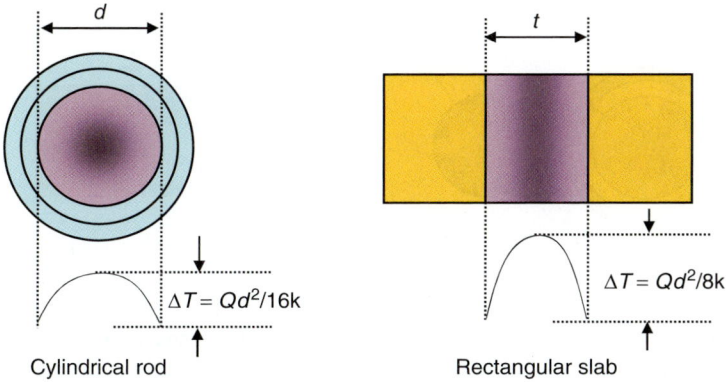

FIGURE 8.3 Temperature gradients in a uniformly heated cylindrical rod and slab.

Combining Eqs. (8.2), (8.3), and (8.4), we find that for a rod at constant heat loading, the focal length of the thermal lens is independent of the rod diameter:

$$f = 2k / \left(LQ \frac{dn}{dT} \right) \tag{8.5}$$

As a result, thermal lensing is the main limitation to power scaling in rod-based devices.

In slabs, however, zigzag propagation between the two cooled surfaces averages over the temperature gradients and results in virtually no thermal lensing to the first order. Thus, the main limitation in early side-pumped slabs was thermally induced stress, which can lead to slab fracture. Figure 8.4 shows a profile of the stress in rods and slabs and the functional dependence of the stress under uniform heat deposition. Note that the surfaces are under tensile stress (i.e., they are being stretched), which can lead to fracture.

The slab's fracture strength depends not only on the lasing material but also on the surface characteristics of the slab. Thus, a slab polished by one vendor may have higher fracture strength than another. This is not an unexpected result, because fracture begins from microcracks on the slab's surface. The number and depth of these microcracks depend on the quality and method for polishing the slab. A YAG slab with a high-quality optical polish will have a fracture limit on the order of 300 MPa; however, because of the uncertainty of the surface characteristic due to handling and mounting of the slab, a fracture safety margin of 3–4 is suggested in designing a high-power slab.

The ability of slabs to scale in power far beyond that achievable with rods is possible because, unlike a cylindrical rod, slab geometry

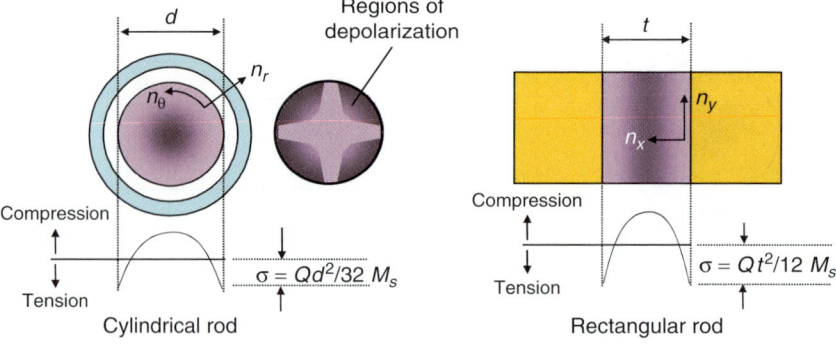

FIGURE 8.4 Stress distribution in a uniformly heated cylindrical rod and slab. $M_S = (1 - v)k/\alpha E$, where v is Poisson's ratio, α is the CTE, and E is the Young's modulus.

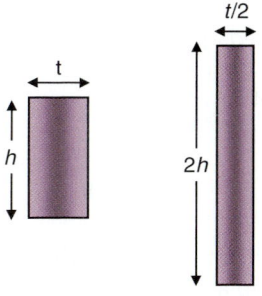

Parameter	Factor Change	
Thickness, t	1	1/2
Height, h	1	2
Heat density, Q	1	1
ΔT	1	1/4
Index change, Δn	1	1/4
Focal length	1	1
Stress, σ	1	1/4
Gain, g_0	1	1

FIGURE 8.5 The table highlights the advantages of aspect-ratio scaling of a slab.

offers two degrees of freedom when it comes to slab sizing. By making a slab taller and thinner, its center-to-edge temperature difference ΔT and the stresses can be reduced, enabling the slab to scale to higher powers. Figure 8.5 shows a comparison of two slabs with the same area and overall output power. The first one has a cross-sectional aspect ratio of 2:1, while the second one is a factor of 2 thinner and taller with an aspect ratio of 8:1. According to equations in Figs. 8.3 and 8.4, both ΔT and the stress decrease by a factor of 4, enabling a factor of 4 scaling in power.

Aspect-ratio scaling also has limitations. For traditional side-pumped slabs, as the slab becomes taller and thinner, pump absorption efficiency begins to suffer. In addition, losses due to diffraction within the slab become a factor that limits the length of the slab, and maintaining fabrication tolerances on TIR surface figures becomes increasingly more difficult. Finally, for crystalline slabs, there may be growth limits on the height of the slab. A brief discussion for each of these limitations follows.

Pump Absorption

Pump absorption efficiency for traditional side-pumped thin slabs is an issue primarily for Nd:YAG, because the Nd-doping concentrations are limited to ~1 percent. Higher concentrations of Nd in YAG result in low-optical-quality crystals and a rapid degradation in the fluorescence lifetime. For diode arrays centered on the 807-nm absorption band of Nd:YAG with a bandwidth of ~4 nm, the required slab thickness for greater than 80 percent absorption efficiency is ~6 mm. This efficiency follows Beer's law and degrades for thinner slabs. To overcome this problem, alternate edge-pumping and end-pumping techniques have been developed (discussed later in this section) that provide much longer absorption distances.

Diffraction

In most applications, the slab represents a distributed aperture as the propagating beam enters and exits. To minimize diffractive losses, the slab's Fresnel number must be on the order of 10 or larger. The Fresnel number is given by

$$N = a^2/\lambda L_{eff} \tag{8.6}$$

where a is the half-thickness of the slab and L_{eff} is the effective length of slab in zigzag propagation. Thus, for $\lambda = 1$ µm and a slab thickness of ~2 mm, the slab length is limited to ~10 cm. To overcome this limitation, slab architectures have been developed in which the beam propagates across, instead of within, a thin gain medium. A recently developed architecture that uses this approach is the Thinzag™ architecture (Chap.9).

Slab Fabrication

Because the beam typically makes many bounces from the TIR faces as it zigzags down the slab, the flatness (figure) and parallelism of these surfaces is critical for the laser's ultimate beam quality. Typical polishing specification for these surfaces is $\lambda/10$ in zigzag transmission. Holding this type of specification for large aspect ratio/thin slabs becomes very difficult. The polishing process stresses the slab, and when released from the polishing fixture, YAG slabs can change their shape in a phenomenon known as *springing*. A reasonable aspect ratio of slab height to thickness that maintains the slab shape is on the order of 20.

Slab Size

Before the development of ceramic laser host materials (c.f. Chap. 7), crystalline host material sizes were limited by the crystal's growth process. For Nd:YAG, the largest commercially available boules yield slabs that are ~3 cm tall. Ceramic Nd:YAG has increased this dimension by a factor of 5, and further increases will be possible in the near future. Another method of overcoming the size limitation of the crystalline host material is diffusion bonding, which was used on the Diode-Array Pumped Kilowatt Laser (DAPKL) laser in the mid-1990s and is described later in this chapter.

8.3 Traditional Side-Pumped Slabs

8.3.1 Architecture and Technical Issues

Figure 8.1 showed a schematic diagram of a traditional side-pumped slab. This type of slab is usually pumped by close-coupled diode arrays through a coolant that flows over the slab's TIR faces. The coolant is typically confined by a pair of windows that are sealed against the slab's TIR faces. This slab architecture has several design issues that have been addressed over the years using various techniques.

Non-Zigzag Axis Temperature Nonuniformity

The slab coolant is usually sealed using an O-ring or gasket that is positioned near the slab edges. This technique usually results in a cold region of unpumped material at the top and bottom edges of the slab, leading to OPD and wavefront distortion. To address this problem, scientists at Lawrence Livermore National Laboratory (LLNL) in the early 1980s introduced the concept of edge bars. Edge bars are typically metallic bars attached to the edges of the slab; depending on the need, these bars can cool the edges using coolant flow or heat the edges via embedded resistive-heating elements. This allows the user to control the slab's edge temperature and reduce or eliminate the OPD near the slab's edges (Fig. 8.6a).

Another design characteristic that can produce OPD in the non-zigzag (vertical) axis is the direction of coolant flow. Although it is tempting to design a slab in which the coolant flow is along the vertical dimension of the slab, this can lead to OPD, because the water at the slab's inlet edge will always be cooler than the exit edge. This can be mitigated by flowing the coolant in the slab's longitudinal (beam propagation) direction, as shown in Fig. 8.6b. Although the temperature increase in the coolant is typically higher with this geometry, the change in coolant temperature does not cause OPD, because there are no temperature gradients in the non-zigzag direction.

Optical Damage at Seal Contact Areas Near the Slab's Input and Exit Faces

The beam zigzagging down the slab can be apertured in the vertical direction to avoid the seals at the slab's top and bottom edges. However, if the slab has near-unity fill factor, the beam footprint may

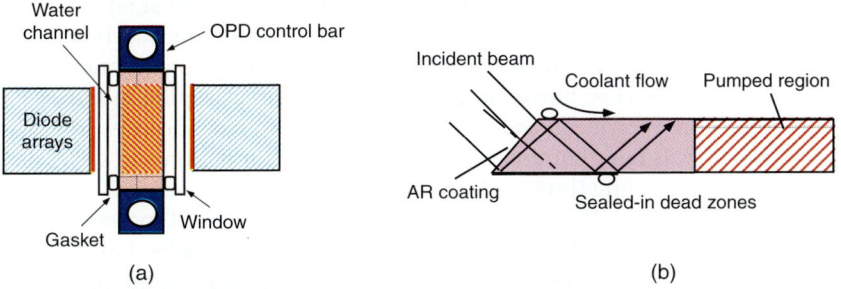

FIGURE 8.6 Schematic of cooling approach for side-pumped slabs. (a) End view showing edge bars and (b) top view showing positioning of seals and direction of coolant flow.

overlap the seals near the entrance and exit areas of the slab. This can easily lead to damage, because the evanescent waves penetrate into the medium outside the slab a distance comparable to the wavelength of light. Thus, if the O-ring material is even slightly absorbing, it can char and damage the slab. This problem can be mitigated by slab and beam injection geometry, which creates small dead zones along the TIR face where seals may be placed without risk of damage, as shown in Fig. 8.6b. The dead zones are areas along the TIR face where the beam does not touch the slab faces. This, in turn, leads to small, unextracted volumes in the slab. If we define the fill factor F as the ratio of the beam footprint on the TIR surface to the total zigzag footprint, the fraction of volume extracted η is given[2] as

$$\eta = F(2 - F) \tag{8.7}$$

This equation indicates that even if the dead zone is as large as 20 percent of the beam footprint, the reduction in extraction efficiency is proportional to the fraction of unextracted volume, which, in this example, is only 4 percent.

An alternative to creating dead zones is the use of an evanescent wave coating. Recent advances in coating technology have enabled the deposition of thick coatings (2 to 3 μm) of low-index material, such as fused silica or MgF_2. If the slab is designed such that the total internal reflection is at the slab-coating interface, this type of coating can isolate the beam from anything that may be on the outside of the TIR face.

Scaling Limitations

As mentioned earlier, traditional side-pumped slabs rely on the slab thickness for absorption of 808-nm diode light. For efficient absorption (> 70 percent) in Nd:YAG at a typical 1.1 percent doping level, the slab thickness must be ~4 mm or thicker. For a slab height of around 3 cm (i.e., the maximum available for monolithic crystalline slabs), the 4-mm thickness limits the slab to an extracted power of about 1 kW before the stress level reaches a significant fraction of the fracture stress. Thus, for further scaling using the side-pumped geometry, the user must either accept lower efficiency or devise a method for using the unabsorbed diode light. The latter can be achieved by stacking two or more thinner slabs side by side. This, however, creates nonuniform pumping in the thin dimension, which can cause the slab to bow. Recent advances in ceramic materials have eliminated the ~3-cm height limitation and can provide further scaling through taller slabs.

For Yb:YAG, the doping level can be higher—up to 100 percent (stoichiometric).[3] In principle, this would enable thinner slabs. However, recent work on thin-disk lasers and slabs has shown anomalous loss mechanisms for highly pumped slabs at doping levels beyond 7 to 8 percent.[4] This forces the user to either compromise efficiency or use multiple absorption passes with low-doped material as described in Chap. 10.

8.3.2 Performance

An example of the traditional side-pumped slab lasers was the Defense Advanced Research Projects Agency (DARPA)–sponsored DP25 precision laser machining (PLM) laser. The DP25 laser was able to scale the average power to greater than 5 kW with good beam quality of 2.4 times the diffraction limit (DL).[5] The laser used a power oscillator, power-amplifier approach with five identical gain modules (Fig. 8.7). Two of the gain modules were used inside an unstable resonator to produce approximately 2 kW of power, and three gain modules were used as single-pass amplifiers, each delivering power on the order of 1 kW.

Figure 8.8 shows one of the slab gain modules with 15 diode arrays pumping from each side of a 5 × 33 × 170 mm slab. Each diode

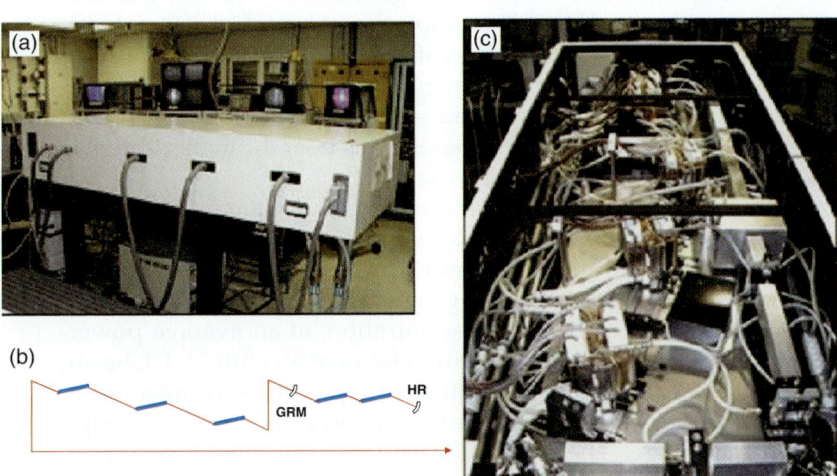

FIGURE 8.7 (a) DP25 laser in a 45 × 105 × 265 cm box, (b) schematic of the optical layout in the box, and (c) laser with the top removed. GRM: graded reflectivity mirror; HR: high reflector.

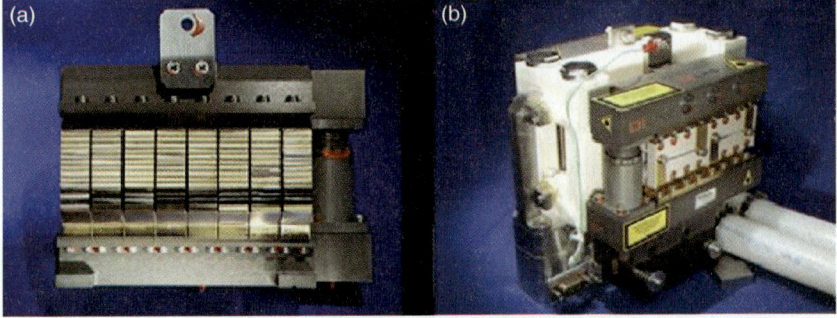

FIGURE 8.8 One of five DP25 gain modules. (a) View of diode arrays and (b) assembled.

Solid-State Lasers

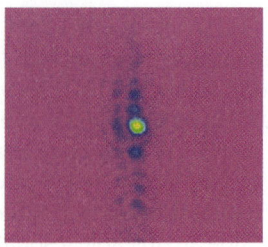

FIGURE 8.9
DP25 far-field intensity distribution.

array consists of 16 quasi–continuous wave (QCW) bars operating at 20 percent duty cycle and 50-W peak power per bar. Operating the device in a QCW mode increased the gain and enabled the use of a higher-magnification ($M = 1.5$) unstable resonator, which generated a robust mode despite several waves of slab OPD.

Figure 8.9 shows the far-field intensity distribution of the beam at a 5.4-kW power level. The measured beam quality based on the power in the central lobe was 2.4 × DL, making this laser the highest-brightness solid state laser for its time (year 2000).

Another interesting laser that used the traditional side-pumped approach was the DARPA Diode-Array Pumped Kilowatt Laser (DAPKL). A key feature of this laser was its simultaneous achievement of high pulse energy and high average power. The laser emitted 10 J per pulse with 7-ns pulse duration, at an average power level of 1 kW (100-Hz pulse repetition frequency) with 2 × DL beam quality.[6] This combination of high energy per pulse, coupled with high brightness, was, at the time (1997) and even currently, a significant challenge. The DAPKL laser used the master oscillator power amplifier (MOPA) approach and phase conjugation via stimulated Brillouin scattering (SBS) to provide good beam quality. Figure 8.10 shows a schematic layout of the laser that used three different sizes of amplifiers to achieve the required output. The largest amplifier aperture was sized based on optical damage considerations and had a cross-sectional area of 4 × 1.4 cm. Because Nd:YAG crystal growth does not support a monolithic slab with such an aperture, the slab was fabricated by diffusion bonding three smaller $1.5 \times 1.5 \times 18$ cm^3 slabs (Fig. 8.11). Although diffusion bonding of glasses was common at the time, diffusion bonding of YAG was very rare. It has since become an important tool for laser design and power scaling.[7]

The DAPKL program also advanced the state-of-the-art of SBS phase conjugation as an important tool for wavefront control in high-power pulsed solid-state lasers. Energy scaling of greater than 1.5 J with average powers greater than 150 W at the SBS cell was achieved in a simple focus geometry, using liquid Freon 113 as the SBS medium with good fidelity and without optical breakdown.

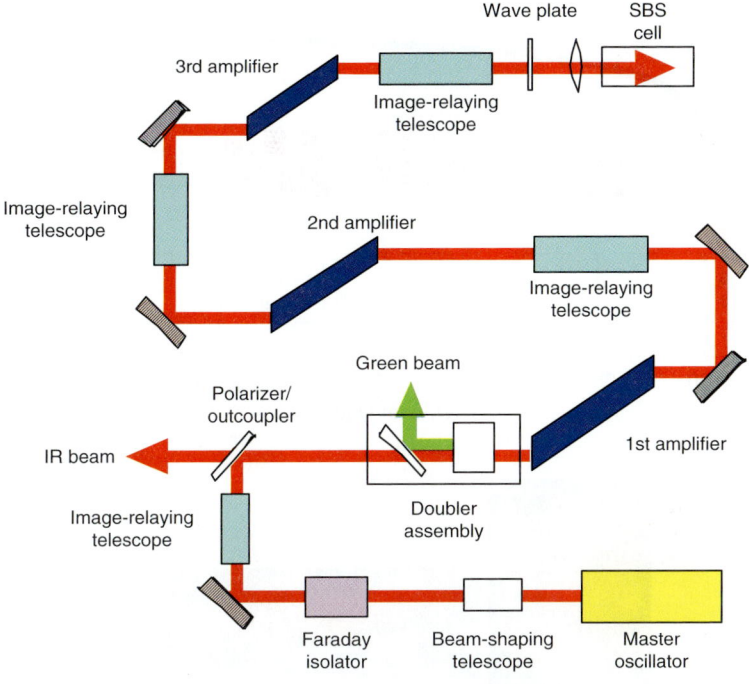

FIGURE 8.10 Schematic of Diode-Array Pumped Kilowatt Laser (DAPKL) optical layout.

FIGURE 8.11 DAPKL's largest amplifier composite slab. The 4.5 × 1.5 × 18 cm slab was fabricated by diffusion bonding three slabs.

Figure 8.12a shows the extracted power and the beam quality as a function of master oscillator power produced by DAPKL. This, too, was the brightest solid-state laser at the time. Figure 8.12b shows the near-field and far-field intensity distributions of the output beam. The near field shows the bond lines of the largest amplifier slab.

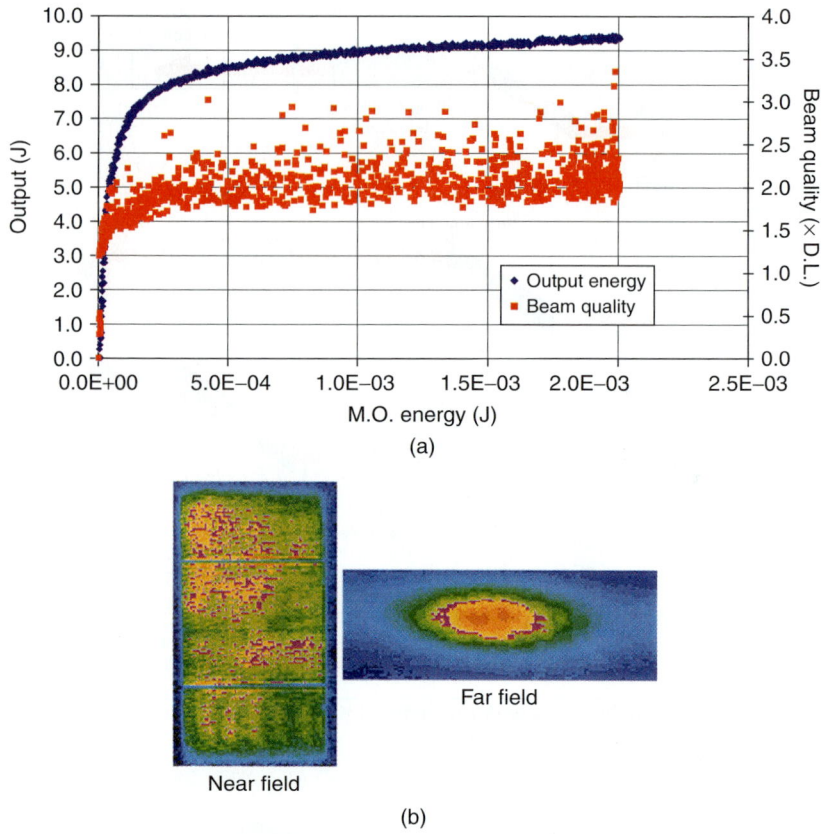

FIGURE 8.12 DAPKL performance. (*a*) Extracted power and beam quality and (*b*) near-field and far-field intensity distribution.

8.4 End-Pumped Slabs

8.4.1 Architecture and Technical Issues

End-pumped slab architectures decouple the slab absorption length from the traditional cooling geometry of the slabs, thus providing scalability that comes with using thinner slabs. In addition, end pumping offers higher pump intensities, which are important in quasi-three-level lasers, such as Yb:YAG. However, this advantage comes with the added complexity of coupling diode light through the end of the slabs. The conduction-cooled, end-pumped slab (CCEPS) is one example of such an architecture that has enabled power scaling of solid-state lasers beyond 15 kW from a single aperture with good beam quality.

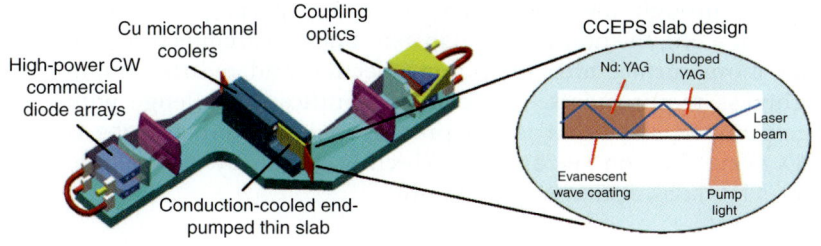

FIGURE 8.13 Conduction-cooled, end-pumped slab (CCEPS) laser concept.

Figure 8.13 shows the main features of a CCEPS gain module.[8] A thin slab is sandwiched between two microchannel coolers with a low thermal impedance interface for efficient conduction cooling. The ends of the slab are cut at 45° angles, and diode light is coupled into the slab through its side edge. The diode light makes a total internal reflection from the slab's input face and propagates axially down the slab. The diode light is coupled into the slab either by a set of lenses or by a lens duct. Unlike the side-pumped geometry, the diode arrays must have high brightness along the thin direction of the slab to allow efficient coupling into the slab aperture. This high brightness is achieved by using microlenses on the diode arrays to collimate the fast axis of the diode emitters. Because the slab ends protrude beyond the coolers, they have undoped, diffusion-bonded end caps that do not absorb light and therefore remain cool. Finally, a 2 to 3 µm SiO_2 evanescent wave coating on the slab's TIR faces ensures near lossless zigzag propagation of the high-power beam down the slab. The high-power beam is injected into the slab at angles that are 20° to 30° from the normal to the 45°-cut input face; this ensures that TIR occurs at the YAG-evanescent wave coating interface.

Although the CCEPS architecture overcomes some of the scaling limitations of the side-pumped approach, it does have issues common to all zigzag slabs. Foremost among these is the temperature uniformity in the non-zigzag direction. Temperature uniformity in the non-zigzag direction requires both uniform pumping and uniform cooling in this direction. For uniform pumping, a beam homogenizer, such as a lens duct, may be used; alternatively, a set of lenses can provide adequate uniformity by imaging the bars onto the slab aperture. For uniform cooling, the slab coolers should be designed to minimize internal temperature gradients, and the thermal interface between the slab and the cooler must be uniformly thin with low thermal impedance.

A similar architecture that shares with CCEPS the common feature of decoupling the slab thickness and pump absorption length is the edge-pumped slab.[9] In this architecture, pump light is coupled through the slab edges so that it propagates and is absorbed along the non-zigzag slab axis. Such an architecture provides more surface area for pump injection and reduces pump brightness requirements. However, a significant challenge with edge-pumped slabs that is not present in end-pumped slabs is OPD along the non-zigzag axis, which is driven by exponential (Beer's law) absorption of pump light.

8.4.2 Performance

The CCEPS architecture was first used to demonstrate a 250-W class Yb:YAG laser.[10] The laser used a $3 \times 2 \times 60$ mm^3 (height × thickness × length) slab, with the central 36-mm consisting of 1 percent Yb:YAG, and 12-mm-long diffusion-bonded undoped end sections. The slab was pumped from each end by a 15-bar, 700-W array of microlensed 940-nm diode bars, with an emitting area of 25×10 mm^2. A solid fused silica lens duct with 93 percent throughput concentrated the pump light to ~20 kW/cm^2 at the slab. Approximately 80 percent of the total pump light was absorbed in the slab. More than 415 W of multimode power was extracted, for an optical efficiency of 30 percent. Figure 8.14 shows the TEM$_{00}$ output and beam quality; 250 W was extracted with an average M^2 beam quality of 1.45.

Shortly thereafter, the CCEPS architecture was used with a Nd:YAG slab and demonstrated even higher optical efficiency.[11] A $5.6 \times 1.7 \times 67$ mm^3 composite slab with a central 49-mm section of 0.2 percent doped Nd:YAG was used to demonstrate 430 W of multimode output power with an optical efficiency of 34 percent,

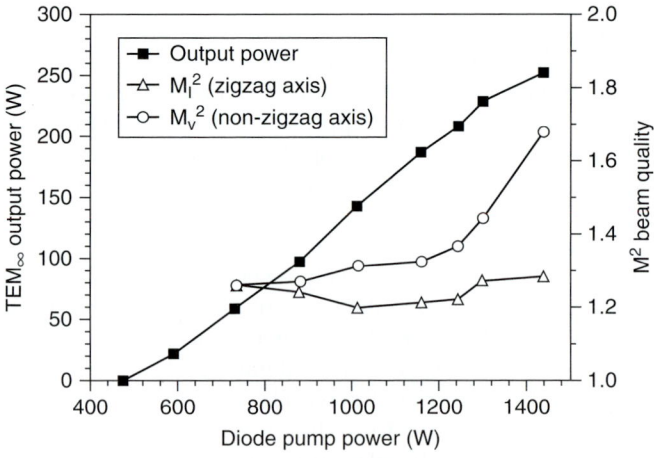

FIGURE 8.14 Performance of a Yb:YAG laser using the CCEPS concept.

Zigzag Slab Lasers

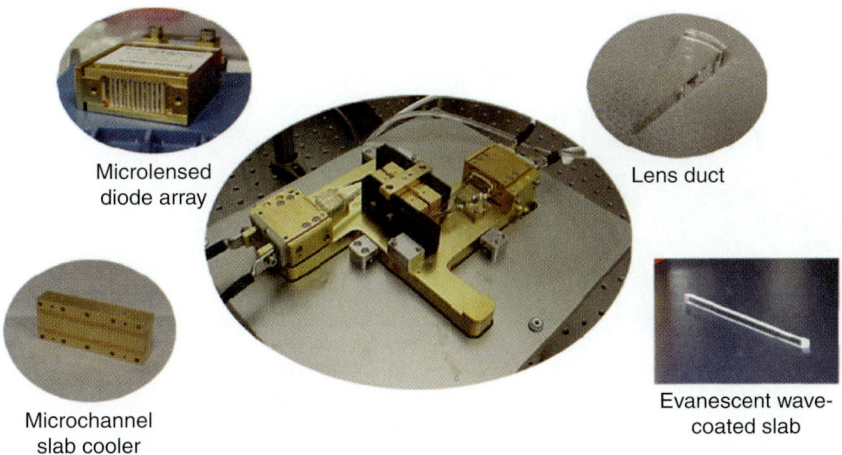

FIGURE 8.15 400-W Nd:YAG CCEPS gain module and key components.

and 380 W of linearly polarized output with an M^2 beam quality of 1.8, from a hybrid stable-unstable resonator (stable in the 1.7-mm dimension; unstable in the 5.6 mm dimension). Figure 8.15 shows the gain module with the key components, while Figure 8.16 shows the output power as a function of diode power, as well as the far-field intensity distribution of the output beam.

8.4.3 Power Scaling

The initial results were the foundation for scaling individual CCEPS-based gain modules to 4 times higher power than was achievable,

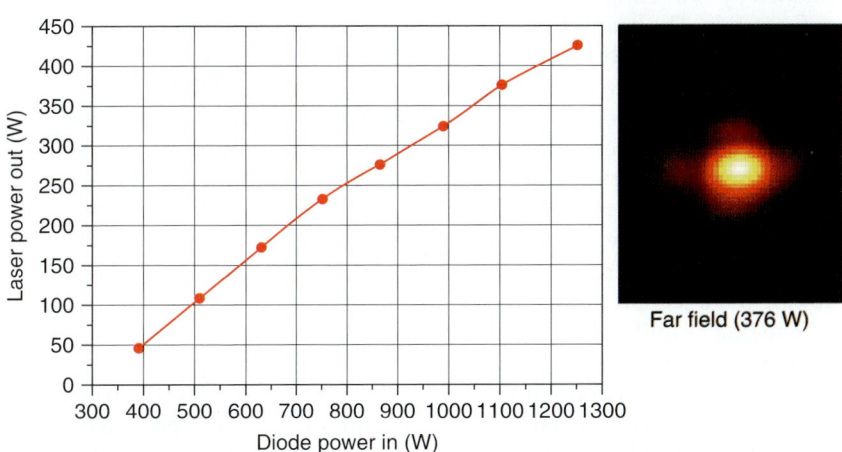

FIGURE 8.16 Nd:YAG CCEPS laser performance with a hybrid stable-unstable resonator.

Solid-State Lasers

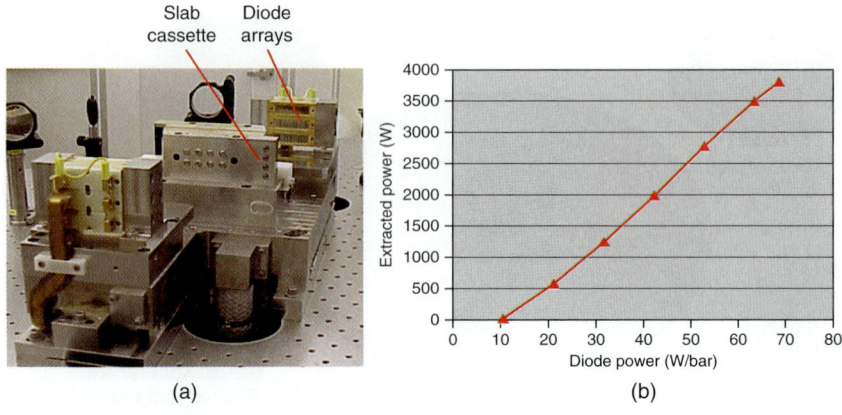

FIGURE 8.17 (a) Scaled CCEPS gain module, used for the Joint High-Power Solid-State Laser (JHPSSL) program and (b) representative extracted power in a multimode oscillator configuration.

using the side-pumped approach. Figure 8.17 shows the scaled gain module and representative power extracted from a multimode stable resonator. The gain module was scaled by using a taller slab and a stack of three diode arrays pumping from each end of the slab. Typical power levels of 3.5 to 4 kW were extracted from these gain modules with a stable multimode resonator configuration. The wavefront uniformity under full power loading and extraction was two to three waves.

The performance of these gain modules set the stage to demonstrate 100-kW CW power with good beam quality on the Joint High Power Solid-State Laser (JHPSSL) program.[12] High beam quality was obtained by extracting a chain of four CCEPS modules in a serial MOPA configuration. The beams from seven of these four-slab MOPAs (28 slabs total) were coherently combined for further parallel power scaling, using a technique developed for phase locking fiber amplifiers.[13]

The JHPSSL system architecture is shown in Fig. 8.18.[14] The output from a low-power, single-frequency master oscillator (MO) progresses through a network of 1-W Yb-doped fiber amplifiers (YDFAs) and splitters to form multiple low-power seed channels. One of these channels is frequency shifted by an acousto-optic modulator to serve as a heterodyne reference for coherent phasing. The other channels provide the injection inputs to each MOPA chain.

The first amplifiers in each chain consist of a mutistage YDFA that boosts the channel power to 200 W. Faraday isolators guard against feedback at the input and output of each YDFA stage. The output from the final YDFA is collimated into a beam and injected into the power amplifier stage, which consists of a series of four identical, 4-kW CCEPS modules (Fig. 8.19). The beam is image relayed from slab to slab to minimize geometric coupling losses. Double passing each slab

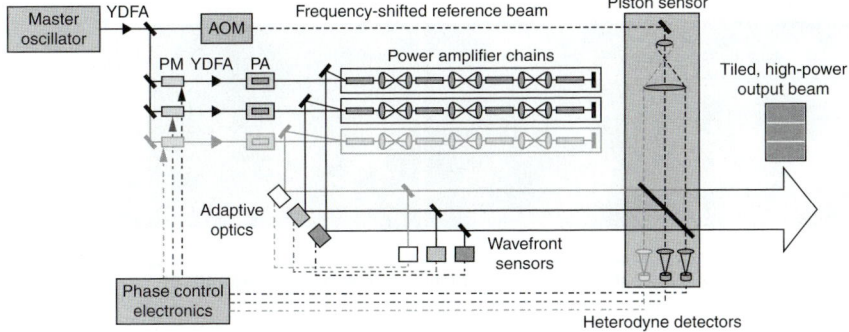

FIGURE 8.18 Schematic of the laser system. The grayed-out components indicate hardware duplication to scale past two chains. YDFA: Yb-doped fiber amplifier; PM: phase modulator; AOM: acousto-optic modulator; PA: preamplifier.

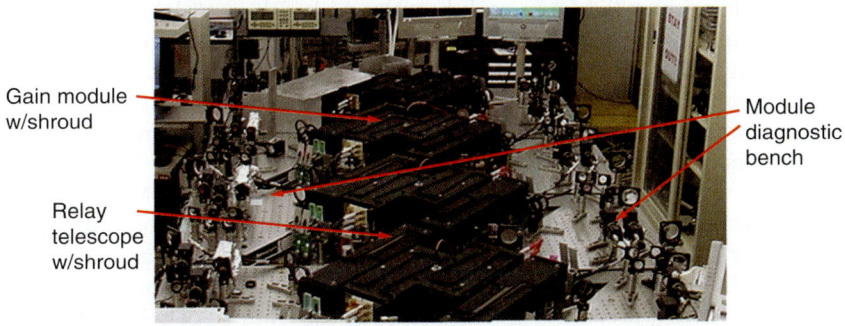

FIGURE 8.19 One of the JHPSSL MOPA chains.

via angular multiplexing enables good staturation and 30 percent optical extraction efficiency. Angular multiplexing of the slabs is made straightforward by choosing different integral numbers of zigzag reflections on each pass.[15] After all eight amplification passes, the beamlet powers are amplified to their final levels of 15 kW.

The slab amplifiers impose multiple waves of OPD on each beamlet, due to thermo-optic effects in the slabs that arise from spatial inhomogeneities in the heat deposition and removal and which are thus not removed by zigzagging. Figure 8.20 shows OPD imposed by a pass

FIGURE 8.20 Typical 4-kW slab gain module OPD meausured using a Mach-Zehnder inteferometer operating at 658 nm. The zigzag axis is vertical, and the non-zigzag axis is horizontal.

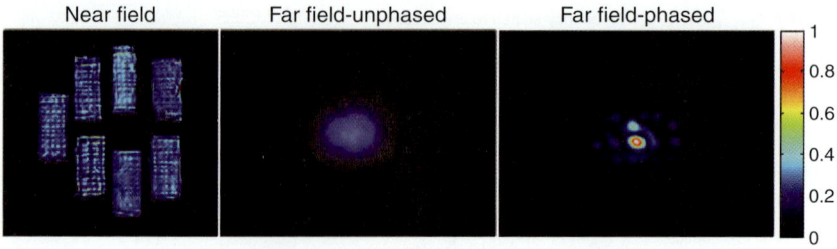

FIGURE 8.21 Near-field and far-field intensity profiles of a 100-kW slab laser system.

through one slab at full power, representing net thermal variations of ~4 percent across the slab aperture. This OPD is corrected using adaptive optics to generate good beam quality. The aberrated, high-power beamlets are expanded to fill the active area of continuous-facesheet deformable mirrors (DMs) in each beamlet path. Tilt is off-loaded to steering mirrors (SMs) to conserve DM stroke. High-reflectivity dielectric coatings on the DMs and SMs enable use of these elements in the 15-kW beamlet paths. A sample of each output beamlet is directed to a Shack–Hartmann wavefront sensor, which generates error signals to drive the active elements in a closed-loop configuration.

After wavefront correction, the beams from all seven MOPA chains are tiled together in a close-packed array configuration and coherently phased together to form a less than 3 times diffraction-limited, 100-kW composite output beam (Fig. 8.21). The far-field beam profiles displayed in Fig. 8.21 illustrate the features of coherent beam combination.[14] Disabling the phase controller results in only a linear increase of the far-field peak intensity with the number of beamlets $N = 7$. Enabling the phase controller would theoretically increase the far-field intensity by another factor of N. Because the beams exhibit some residual wavefront aberrations and jitter, the observed far-field brightness increases by a factor of ~4 times due to imperfect constructive interference among the beams. Nevertheless, this represents the brightest SSL ever demonstrated.

Finally, this laser architecture provides a vehicle for brightness scaling well beyond 100 kW. Because the phase of individual chains is controlled relative to a common reference, there are no cumulative errors as the number of chains is increased; in addition, brightness can, in principle, be scaled indefinitely in this architecture by adding more chains. The general topic of beam combining is explored in greater depth in Chap. 19.

References

1. Martin, W. S., and Chernoch, J. P., "Multiple Internal Reflection Face-Pumped Laser," U.S. Patent 3,633,126; 1972.

2. Eggleston, J. M., Frantz, L. M., and Injeyan, H., "Derivation of the Frantz-Nodvik Equation for Zig-zag Optical Path, Slab Geometry Laser Amplifiers," *IEEE J. Quantum Electron.*, 25: 1855, 1989.
3. Patel, F. D., Honea, E. C., Speth, J., Payne, S. A., Hutcheson, R., and Equall, R., "Laser Demonstration of $Yb_3Al_5O_{12}$ (YbAG) and Materials Properties of Highly Doped Yb:YAG," *IEEE J. Quantum Electron.*, 37: 135, 2001.
4. Larionov, M., Schumann, K., Speiser, J., Stolzenburg, C., and Giesen, A., "Nonlinear Decay of the Exited State in Yb:YAG," *Proc. Advanced Solid State Photonics Conf.* 18–20, 2005.
5. Machan, J. P., Long, W. H., Zamel, J., and Marabella, L., "5.4 kW Diode-Pumped, 2.4x Diffraction-Limited Nd:YAG Laser for Material Processing," *Proc. Advanced Solid State Laser Conf.*, 549, 2002.
6. St. Pierre, R., Mordaunt, D., Injeyan, H., Berg, J. G., Hilyard, R. C., Weber, M. E., Wickham, M. G., Harpole, G. M., and Senn, R., "Diode Array Pumped Kilowatt Laser," *IEEE J. Selected Top. Quantum Electron.*, 3: 53, 1997.
7. Meissner, H., "Composites Made from Single Crystal Substances," U. S. Patent 5,441,803; July 29, 1992.
8. Injeyan, H., and Hoefer, C. S., "End Pumped Zigzag Slab Laser Gain Medium," U.S. patent 6,094,297; July 5, 2000.
9. Rutherford, T. S., Tulloch, W. M., Gustafson, E. K., and Byer, R. L., "Edge-Pumped Quasi-Three-Level Slab Lasers: Design and Power Scaling," *IEEE J. Quantum Electron.*, 36: 205, 2000.
10. Goodno, G. D., Palese, S., Harkenrider, J., and Injeyan, H., "YbYAG Power Oscillator with High Brightness and Linear Polarization," *Opt. Lett.*, 26: 1672, 2001.
11. Palese, S., Harkenrider, J., Long, W., Chui, F., Hoffmaster, D., Burt, W., Injeyan, H., Conway, G., and Tapos, F., *Proc. Advanced Solid State Lasers Conf.*, 41–46, 2001.
12. McNaught, S. J., et al., "100-kW Coherently Combined Nd:YAG MOPA Laser Array," *Frontiers in Optics*, paper FThD2, 2009.
13. Anderegg, J., Brosnan, S., Weber, M., Komine, H., and Wickham, M., "8-W coherently phased 4-element fiber array," *Proc. SPIE*, 4974: 1, 2003.
14. Goodno, G. D., Komine, H., McNaught, S. J., Weiss, S. B., Redmond, S., Long, W., Simpson, R., Cheung, E. C., Howland, D., Epp, P., Weber, M., McClellan, M., Sollee, J., and Injeyan, H., "Coherent Combination of High-Power, Zigzag Slab Lasers," *Opt. Lett.*, 31: 1247, 2006.
15. Kane, T. J., Kozlovsky, W. J., and Byer, R. L., "62-dB-Gain Multiple-Pass Slab Geometry Nd:YAG Amplifier," *Opt. Lett.*, 11: 216, 1986.

CHAPTER 9

Nd:YAG Ceramic ThinZag® High-Power Laser Development

Daniel E. Klimek

Principal Research Scientist, Textron Defense Systems, Wilmington, Massachusetts

Alexander Mandl

Principal Research Scientist, Textron Defense Systems, Wilmington, Massachusetts

9.1 Introduction and ThinZag Concept Development

Over the past decade, solid-state lasers have demonstrated remarkable power in scaling. To a large extent, the emergence of solid-state lasers as competitive high-power devices is due to the availability of highly efficient (~60 percent), high-power (> 100 W), low-cost (< \$10/W mounted) laser diode bars.

As a laser gain material, Nd:YAG is by far the most commonly used in solid-state lasers, due to a combination of properties that uniquely favor high-power laser performance. The YAG host is a robust, fracture-resistant material with high thermal conductivity. Nd:YAG also has a narrow fluorescent line width, which results in high gain. There has also been a revolutionary development in laser gain material. Cubic structure materials like YAG can now be fabricated as ceramics with optical uniformity that is better than found in YAG crystals (for both dopant uniformity and variations in index of refraction), with scattering loss coefficients comparable to YAG crystals

Solid-State Lasers

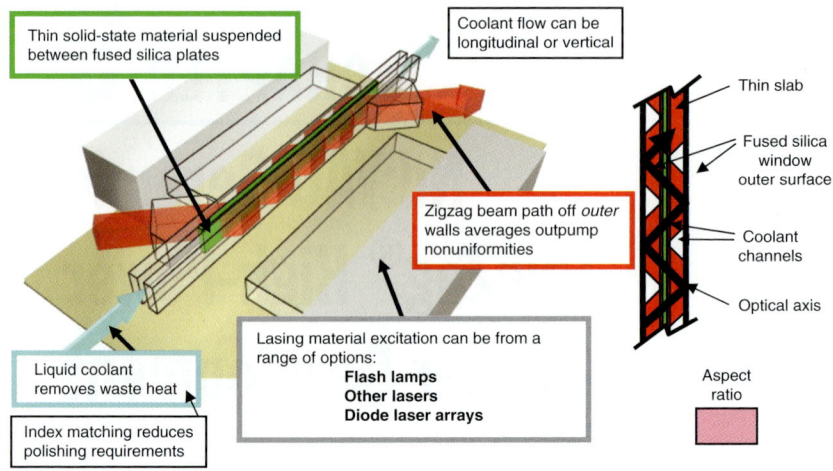

FIGURE 9.1 A schematic drawing of ThinZag configuration, including the key features of the laser and laser beam optical path within the cell.

($< 0.15\%/$cm). These materials can also be produced in sizes that YAG crystals cannot achieve (e.g., 400×400 mm^2 slabs).[1–3]

The unique properties of Nd:YAG ceramic combined with the ThinZag laser configuration, developed by scientists and engineers at Textron Defense Systems, have allowed scaling of these lasers to more than 16 kW average power from a single laser module. Higher power configurations involve a single-aperture power oscillator configuration consisting of a number of identical modules operating in series.

Figure 9.1 shows a schematic diagram of the ThinZag configuration. With this configuration, improved methods of thermal management for high-power diode-pumped, solid-state slab lasers have been demonstrated. This unique optical arrangement uses thin slabs of solid-state gain material immersed in a flowing cooling fluid and sandwiched between a pair of fused silica windows. The laser flux zigzags through the gain medium in a nontraditional manner—that is, it reflects off the outer surfaces of the fused silica windows rather than off the outer surfaces of the lasing material. The ThinZag configuration allows the use of thin slabs for good thermal control of the laser medium using a near-field beam that has a near-unity aspect ratio that is independent of the laser slab's thickness.

Many features of this design can be varied almost independently to allow optimization of key input parameters to improve performance. This design's orthogonal nature allows for independent variation of parameters such as slab thickness, diode pump intensity, diode pump distribution, thermal cooling rate, number of slabs, and so on.

In addition to the recent development of ceramic Nd:YAG-based devices, tests on a variety of laser gain media have been conducted over the years at Textron Defense Systems' laser laboratories using the ThinZag configuration. These tests have included flash lamp- and laser-pumped laser arrangements using liquid dye,[4,5] dye-impregnated plastics,[6] and Yb/Er:Glass, Nd:YLF, and Cr:LiSAF crystals.[7-9]

This section describes the progression of ThinZag laser designs from a 1-kW single-slab device (TZ-1) to a 5-kW two-slab device (TZ-2) to a larger-area two-slab nominal 15-kW device (TZ-3). The TZ-3 laser module is the basic building block for achieving higher-power (100-kW) output. Initial tests consist of coupling three TZ-3 modules as a single-aperture power oscillator. The Joint High Power Solid-State Laser (JHPSSL) 100-kW laser consists of six similar modules operating as a single-aperture power oscillator.

9.1.1 TZ-1 Module Development

The first diode-pumped Nd:YAG ThinZag laser (designated TZ-1) was a single-slab design with nominal output ~1 kW. ThinZag lasers at that time used short-pulse lasers or short-pulse flash lamp pumping (~1 µs) as an excitation source. The highest power achieved was about 80 W from a Cr:LiSAF laser, which operated at up to 10 Hz with output up to 8 J/pulse.[4,10,11] The thermal loads for the diode-pumped high-power devices are larger by more than 2 orders of magnitude and call for much greater attention to thermal control of the laser components.

The TZ-1 consisted of a single slab of Nd:YAG (either ceramic or crystal) that is pumped from both sides by high-power 808-nm continuous wave (CW) laser diode arrays. The TZ-1 laser achieved high-power output for extended runs, as shown in Fig. 9.2.

In comparing crystalline and ceramic Nd:YAG samples, it was found that the ceramic samples were generally optically superior to the crystalline samples. Nd:YAG ceramic also displayed better [Nd] uniformity compared to crystal. Typical measurements using a

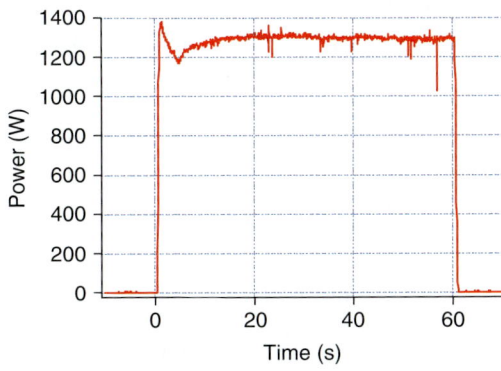

FIGURE 9.2 Demonstrated steady-state performance of TZ-1 laser using ceramic slab.

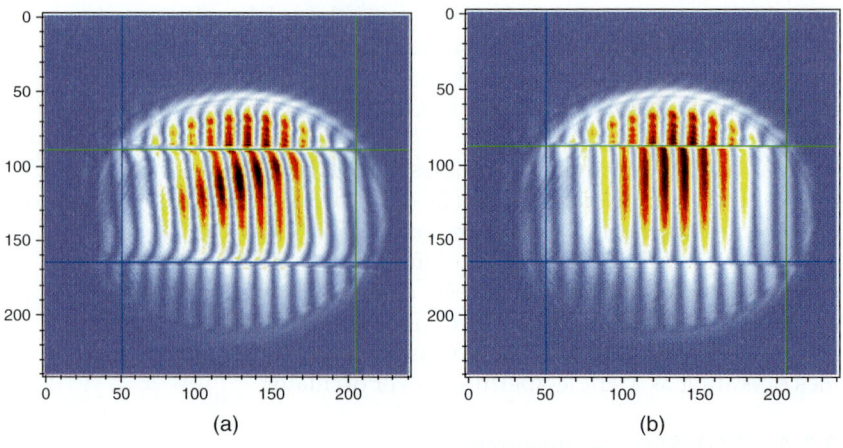

FIGURE 9.3 Double-pass interferometer measurements show that Nd:YAG ceramic slabs are intrinsically more uniform (b) than typical crystals (a).

two-pass interferometer comparing Nd:YAG crystal with Nd:YAG ceramic slabs are shown in Fig. 9.3. The ceramic slab's index of refraction uniformity was found to be generally superior. Although some crystalline slabs were close in quality to the ceramic slabs, the index variations from one crystal sample to another were significant and unpredictable.

Measurements of laser output using a stable optical cavity verified that ceramic and crystalline slabs of Nd:YAG produced the same output power, within experimental error (~1.2 kW), thus confirming that ceramic gain material was suitable for high-power laser operation. Typical output power measurements are shown in Fig. 9.4.

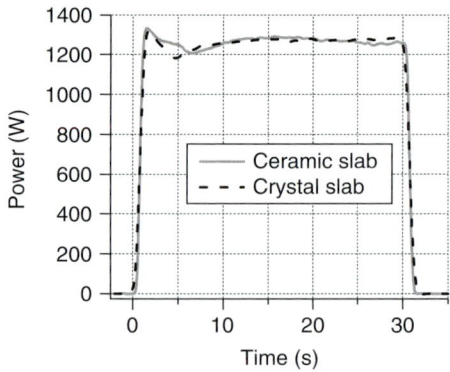

FIGURE 9.4 Comparison of ceramic and crystal slab laser performance.

9.1.2 TZ-2 Module Development

The next scaling path to higher average power was to incorporate a second gain slab into the basic ThinZag laser design. The volume was increased by increasing the slab length by a factor of 1.5, as well as by doubling the number of slabs from one slab to two, for a factor of 3 total increase in excited volume. The pump power was also increased by a factor of 1.5 for an overall deposited pump power increase of 4.5 ($1.5 \times 2 \times 1.5$), thereby projecting an increase in output power to over 5 kW (1.2 kW $\times$ 4.5 = 5.4 kW).

The two-slab ThinZag laser (designated TZ-2) was also pumped from both sides by Nuvonyx diode pump sources, each consisting of 80-W diode bars in series oriented with the fast axis in the horizontal plane. To ensure uniform pumping, the pump light was optically mixed in an optical "scrambler" before reaching the laser module. These scramblers were made of single blocks of fused silica, which acted as a light guide by confining the 808-nm pump radiation using total internal reflection (TIR). Measurements of the pump deposition profile were made using a single scrambler with one of the ThinZag windows. CCD images were taken of the pump light after it passed through the optical scrambler–window combination and impinged onto a Lambertian scattering surface. The measured profiles showed excellent deposition uniformity.

In the center of Fig. 9.5, which shows the TZ-2, the two-slab Thin-Zag laser head is positioned to show some of the gold-coated metal

FIGURE 9.5 TZ-2 laser, showing diode pump source and optical scramblers pumping device from two sides.

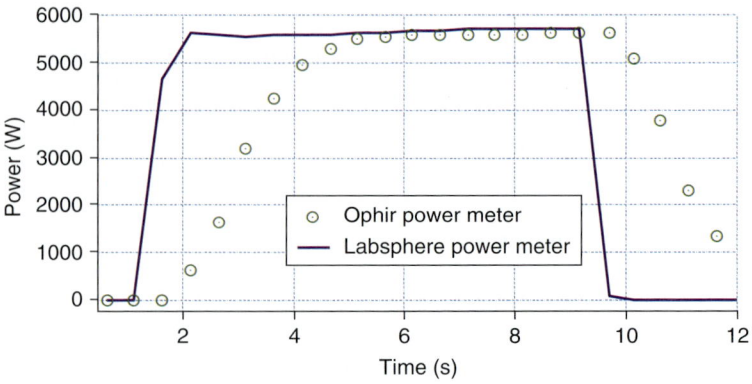

FIGURE 9.6 Measured laser output power from TZ-2 using a stable optical cavity. Laser output was simultaneously measured by an independently calibrated Ophir power meter (~3-s response time) and an independently calibrated Labsphere power meter (~1-s response time). The diode pump flux incident from each side of the laser slabs was 405 W/cm^2. The calculated optical efficiency (laser output/pump diode output) was 30 percent.

parts. Also shown are the laboratory optical components used for extracting laser light, for making diagnostic measurements, and for recording average power. A trace of the output power versus temporal profile greater than 5 kW is displayed in Fig. 9.6.

The TZ-2 laser was typically operated with a stable optical cavity, using a 4-m radius of curvature primary and a 70 percent reflective feedback flat-output coupler. Figure 9.6 displays two different measurements of laser output: an Ophir power meter, which has a response time of a few seconds, and a Labsphere integrating sphere power meter, which has a response time of about a second. Both instruments are independently calibrated by their manufacturers, and very good agreement was evident. The measured output was about 5.6 kW, which is in good agreement with scaling based on the TZ-1 measurements and the increase in system gain projected from our design changes. These data show an 8-s run with apparent steady-state output. The TZ-2 laser was operated using this stable optical cavity for various operating conditions and runtimes. A 30-s run is shown in Fig. 9.7. No real-time corrections were introduced to handle any thermally induced distortions, such as tilt and focus during this longer run, resulting in a gradual decrease in output with time.

For most applications, lasers must have good beam quality. To evaluate the potential of a ThinZag laser to produce a good-quality optical beam, the laser was placed in one arm of an interferometer, as shown in Fig. 9.8. These measurements were used to provide information on how one might modify the laser module to achieve improved performance. Throughout these tests, the distortions of the

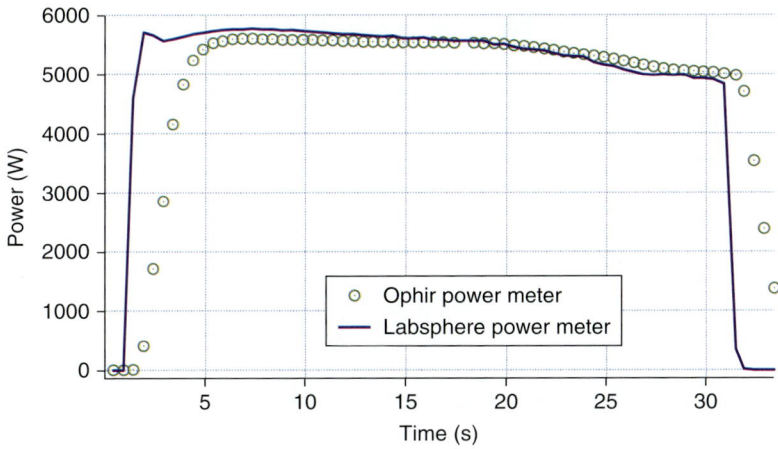

FIGURE 9.7 A TZ-2 laser output 30-s run using a stable optical cavity has no dynamic correction.

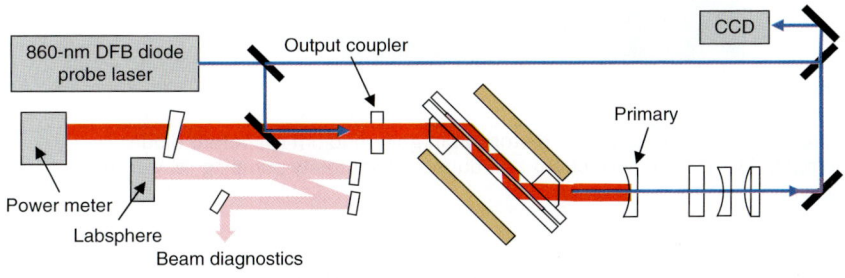

FIGURE 9.8 Schematic diagram of the setup used to measure interferometric distortions in the laser medium under full diode pump load during laser operation. CCD: charge-coupled device; DFB: distributed feedback.

laser medium were observed to be low order (e.g., focus, astigmatism, tilt, etc.) and slowly varying (time scales of seconds).

The probe laser can be used to measure the medium while the laser operates with a stable optical resonator or, by removing the output coupler, to measure the medium with only the diode pumping. The probe wavelength is in a spectral region in which Nd:YAG has very low absorption. Also shown in Fig. 9.8 are three lenses, which are placed in the optical path to remove low-order static phase errors that result from cell fabrication. These lenses consist of two orthogonal cylindrical lenses and one spherical lens.

Typical interferograms obtained with the TZ-2 laser with and without the resonator mirrors are shown in Fig. 9.9. The top interferogram (Fig. 9.9*a*) is before pumping starts and includes the static

Solid-State Lasers

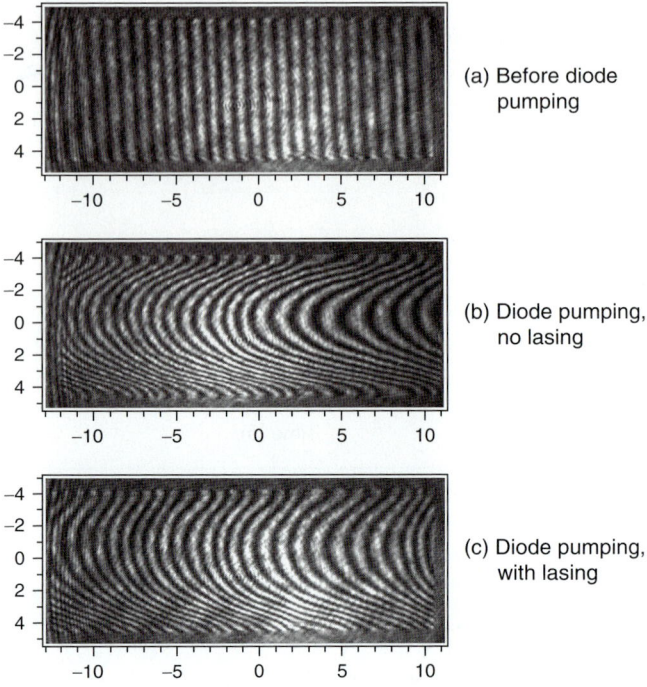

FIGURE 9.9 Measured phase distortion in gain medium (*a*) before diode pumping (with static correction) and during full intensity pumping (*c*) with and (*b*) without lasing.

correctors mentioned above. The data show that only simple distortions of cylinder exist in the medium before pumping and are easily corrected with planocylindrical optics. The next interferogram (Fig. 9.9*b*) shows that about eight waves of primarily vertical cylinder develop as a consequence of the diode pumping. The bottom interferogram (Fig. 9.9*c*), which was taken while the cavity was lasing, shows that the vertical distortion is reduced to about five waves. The distortions shown in Fig. 9.9 develop in about 1 s, with very little change thereafter, with the exception being a slowly growing tilt of about 0.1 waves per second.

Examination of Fig. 9.9*b* and 9.9*c* shows that there are distortions beyond pure cylinder and tilt. Figure 9.10 shows the residual phasefront distortion of the probe beam after both horizontal and vertical cylinders have been mathematically removed. As seen, the variation in the residual is only +1 to −1 waves, most of which is located near the top and bottom edges. As with the cylindrical distortion, there is very little change after about 1 s of operation.

An unstable optical cavity was then set up to perform beam quality measurements on the TZ-2 laser. In order to set up an unstable cavity

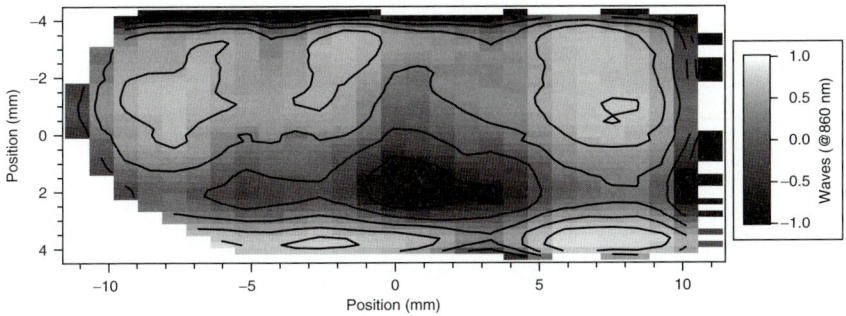

FIGURE 9.10 Residual phase-front distortion after the horizontal and vertical cylinders are removed.

on the TZ-2 module, the optical path was folded to double the gain length. Figure 9.11 is a schematic diagram of the folded TZ-2 cavity. The deformable mirror was added to the cavity to remove the residual distortions depicted in Fig. 9.10.

The TZ-2 laser, when operated with stable optics, has a near-field beam profile of roughly 1×2 cm. With a folded cavity, the beam profile in near field is approximately 1×1 cm. A graded reflectivity mirror (GRM) with a super-Gaussian square profile was designed and subsequently fabricated by INO (National Optics Institute, Quebec, Canada). Laser experiments were performed with this GRM as an output coupler. The measured laser output for this folded cavity was

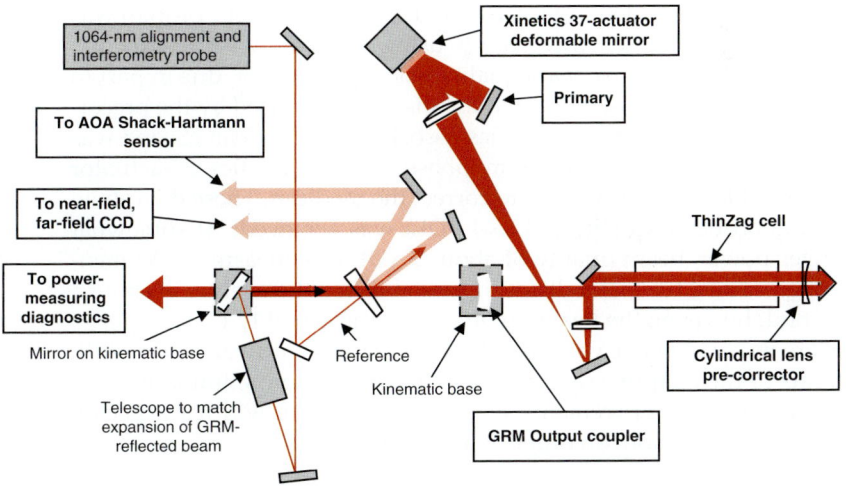

FIGURE 9.11 Folded cavity with increased gain length set up with unstable optics for beam quality measurements. AOA: adaptive optics associates; CCD: charge-coupled device; GRM: graded reflectivity mirror.

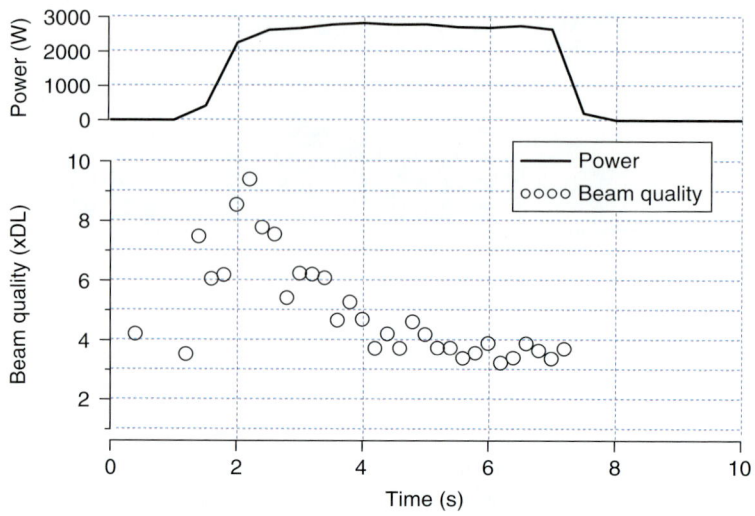

FIGURE 9.12 Measured beam quality (power in a 1 xDL bucket method) is plotted as a function of time for the TZ-2 cell operation at 3-kW output power.

reduced to about 3 kW, which is substantially below that measured using a stable optical cavity; this reduction was due mainly to the nonoptimum output coupling.

Beam quality measurements for the configuration of Fig. 9.11 are shown in Fig. 9.12, in which beam quality was measured using a CCD camera. The power into the central 1 times diffraction limited (xDL) spot (dimensions determined by the 1 × 1 cm near-field profile) was measured. The laser was set up using intracavity "precorrector" cylindrical lenses. The beam quality was initially poor, due in part to the medium itself and in part to the distortions caused by the precorrector lenses. An adaptive optics associates (AOA) WaveScope was used to measure the medium phase, and a Xinetics 37-actuator deformable mirror was used to correct the medium. These data show that good beam quality (~3 to 4 xDL) was achieved. At somewhat lower power, beam quality of about 2 xDL was measured. As mentioned earlier, the major distortions established themselves in about a second; however, the beam quality shown in Fig. 9.12 took longer to get to its steady-state value due to software and hardware bandwidth limits of the adaptive optics system used in these preliminary trials. With planned improvements, the AO system's bandwidth is expected to improve by about an order of magnitude.

9.1.3 TZ-3 Module Development

As described earlier, the TZ-3 and TZ-2 lasers have the same footprint and flow manifolds. The key difference between the two devices is

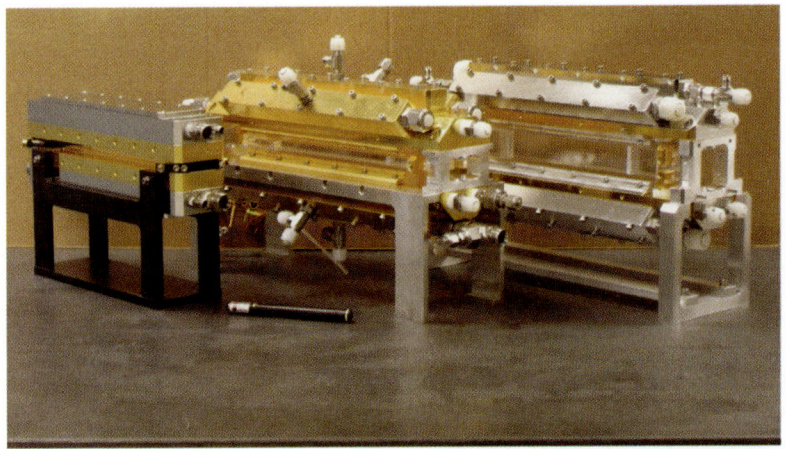

FIGURE 9.13   The three ThinZag modules: TZ-1, TZ-2, and TZ-3 (from left to right). Interestingly, there is not a significant size change in the modules as the power increases from about 1 kW laser output (TZ-1) to more than 15 kW laser output (TZ-3).

the height of the Nd:YAG slabs. The TZ-2 device uses 1-cm-high slabs, while the TZ-3 uses 3-cm-high slabs. Because the pump intensity in both devices is the same, the output from the TZ-3 module, compared with the TZ-2 module, is expected to be greater by a factor of 3. Since the TZ-2 module produced about 5.6 kW, the TZ-3 is expected to produce about 16.8 kW. Figure 9.13 shows the TZ-1, TZ-2, and TZ-3 lasers. Note the small change in the devices' overall dimensions, which produces more than an order of magnitude higher power when scaling from the TZ-1 to the TZ-3. Initial short-pulse measurements performed on the TZ-3 demonstrated outputs to 16.8 kW output using a stable cavity, as shown in Fig. 9.14.

The TZ-3 laser module operates, as did all the previous ThinZag laser modules, with laser medium distortions that are low order (mainly cylinder) and slowly varying (time scales of seconds). Modifications to the laser module continue to improve the device's thermal control, which in turn influences the medium quality when under full-power extraction.

9.1.4   Coupling Three TZ-3 Modules

Three TZ-3 modules were coupled in series as a single-aperture power oscillator. (Three is the minimum number of modules needed to operate with an unstable cavity for good beam quality.) The laser model calculations shown in Fig. 9.15 indicate that with three modules, optimum feedback for good extraction occurs at a little over 40 percent reflectivity. For graded reflectivity output couplers, the

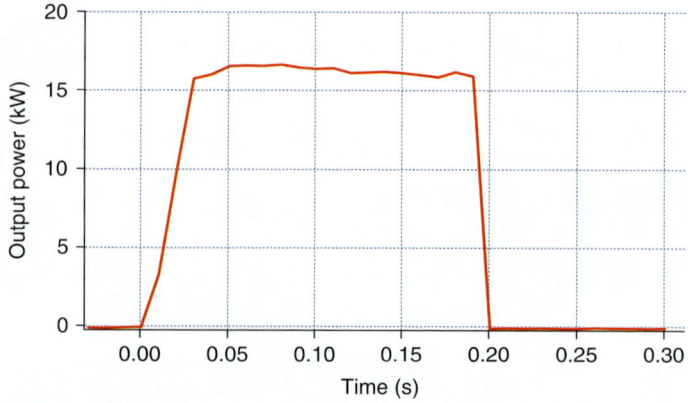

FIGURE 9.14 Short-pulse measurements were made using the TZ-3 module. Output of 16.8 kW was achieved in 200-ms pulses. The calculated optical efficiency (laser output/pump diode output) is 25 percent.

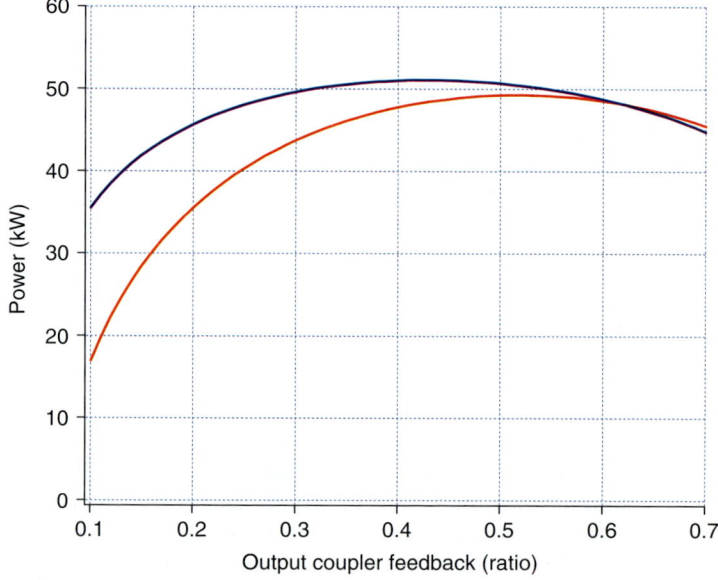

FIGURE 9.15 Model calculations of expected power extraction from three TZ-3 modules. The red curve predicts CW and the blue curve low duty cycle (LDC) performance.

highest feedback that can be achieved is about 40 percent; thus, operating with less than three modules would not allow good extraction. If the laser medium is not operating with good extraction, other loss mechanisms (i.e., amplified spontaneous emission [ASE] or parasitic losses) can further reduce the optical extraction.

The TZ-3 was designed to operate at 100 kW as six modules in a series power oscillator. Each TZ-3 module was designed to be low gain, so that ASE loss along the length of the slab would be minimal when operated at full power. To increase the gain of the modules for more efficient operation, an alternative mode of operation—namely, low duty cycle (LDC)—was tested. In LDC mode, the current is pulsed on for, say, 30 ms and off for 30 ms. The on times and off times are somewhat arbitrary, though the pulse time should be long compared with the kinetic lifetime of Nd:YAG (0.25 ms) and short compared with the slab's thermalization time (~1 s).

During the on time, the current is set much higher than the nominal 80-A full pump current used for CW operation; therefore, the instantaneous gain is much higher. In this case, the laser operates at the maximum allowed current of the Osram diodes. At the higher-diode pump currents, the laser operates at higher gain, with more efficient extraction and consequently lower thermal heating of the slab. The LDC current is chosen such that the laser output essentially doubles for the on time compared with the current for CW operation. For LDC mode, the average laser output is the same as for CW current; however, because of the more efficient extraction, the overall heating of the slabs is reduced, as are thermal distortions. Figure 9.15 shows the calculated improved extraction for lower-output coupling when operating in the LDC mode.

A series of measurements, shown in Fig. 9.16, was performed to test LDC mode operation. Measurements made with stable (S) and unstable (U) optical cavities gave essentially the same average output, as one would expect. The individual 30-ms pulses, measured using a fast-response Labsphere, were twice the average power, as predicted. Kinetic code calculations of the expected output also agreed. The stable optical cavity output power measured was 44 kW, compared with code calculations that predicted 42 kW average power

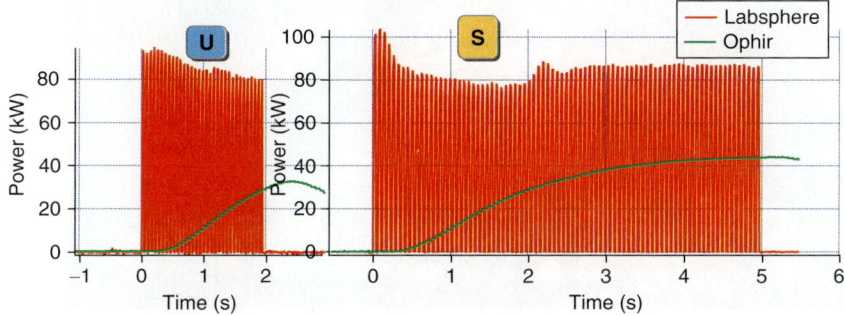

FIGURE 9.16 Three TZ-3 modules operating with unstable (U) and stable (S) optical cavities. Note that the instantaneous power output is essentially double the average power. Output from stable and unstable cavities is comparable.

for this case. Note that the instantaneous power of the pulses is close to 88 kW, close to calculation.

Several issues must be addressed when using LDC mode. The diode current's rise time must be fast as compared with the several-millisecond pulse on time. The Sorenson SFA-150 power supplies used have current rise times less than 0.25 ms, which is significantly less than the pulse on time. Diode wavelength measurements indicate that the steady-state diode temperature is achieved on the scale of the pulse rise time. A key issue that is yet to be addressed is optimized control of the deformable mirrors in pulsed mode for good beam quality. Controlling deformable mirrors is more straightforward in CW than in LDC mode. The downside is that thermal heating of the slabs is greater in CW mode, so the intrinsic medium distortion is greater.

Figure 9.17 shows the three modules set up as a single-aperture power oscillator. The design of the optical cavity has provisions for operation with a deformable mirror coupled to each cell to correct medium distortion. The distortions are sensed using a second color probe laser; a Shack-Hartmann detector measures phase and controls a deformable mirror to correct the phase of each module. Initial measurements were made without the internal deformable mirrors in the optical cavity. An external deformable mirror built by MZA Associates was used to correct beam quality.

A series of measurements was made at diode pump currents of 50, 60, and 70 A. The optical cavity used a GRM output coupler with a nominal magnification of 1.4. The GRM feedback was still

FIGURE 9.17 Three TZ-3 modules operating as a single-aperture power oscillator.

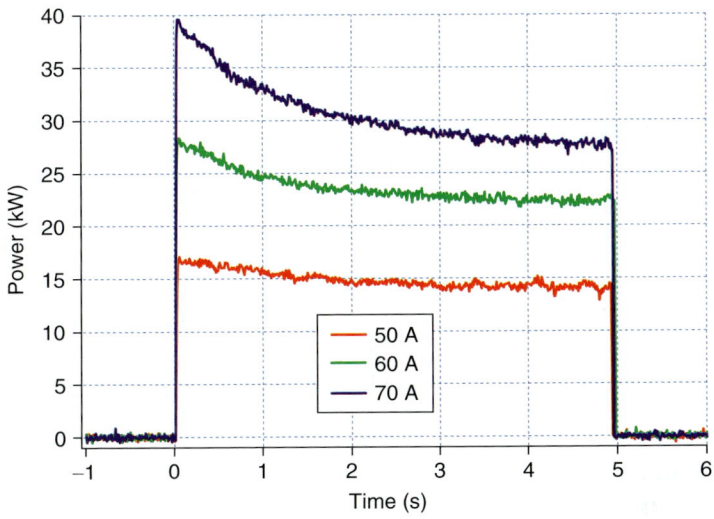

FIGURE 9.18 Characteristic output power from the three-module ThinZag configuration.

somewhat below the optimum feedback needed for the three-module configuration tested.

The laser was operated in 5-s bursts to ensure that steady-state operation was achieved by the end of the run. Characteristic output powers for 5-s runs are shown in Fig. 9.18. Laser output powers of between 15 and 30 kW were achieved with beam qualities of 2.4 xDL at the lower powers and 3.3 xDL at the highest powers. The diode pump source is capable of higher current operation, which should result in increased laser output power.

The next scaling of the device was to couple six TZ-3 modules into a single-aperture laser to produce 100 kW laser output. An engineer's drawing showing six TZ-3 modules on two coupled 5 × 10 ft optical benches is shown in Fig. 9.19. This device recently achieved

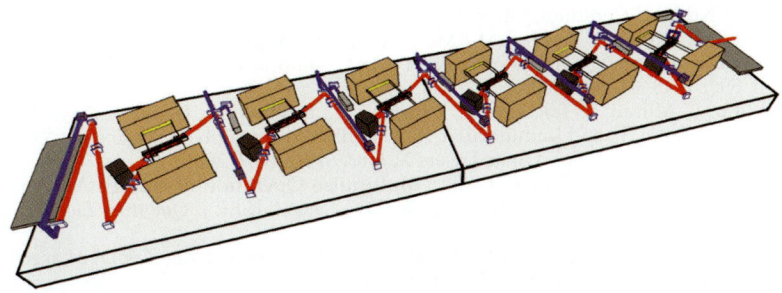

FIGURE 9.19 Layout design for six TZ-3 modules for 100-kW ThinZag laser on two coupled 5 × 10 ft optical benches.

average powers in excess of 100 kW in final testing on the JHPSSL program.

9.2 Summary

This chapter presented an overview of the approach, history, and current state of scaling Nd:YAG ceramic ThinZag lasers to significant power levels using a single-aperture power oscillator architecture. These lasers are compact and scalable to 100 kW and higher power levels. Recently average power levels in excess of 100-kW output were achieved in final government testing of the JHPSSL program. A critical issue, as with all very high power solid-state lasers, is achieving excellent beam quality at the highest powers.

Acknowledgments

This work was supported by the U.S. Army Space and Missile Defense Technical Center, SMDC-RDTC-TDD, in Huntsville, Alabama, under contract W9113M-05-C-0217, with funding from the Department of Defense (DOD) High Energy Laser Joint Technology Office, in Albuquerque, New Mexico, and from the office of the Secretary of the Army for Acquisition, Logistics, and Technology (ASA-ALT).

The authors wish to acknowledge the excellent technical assistance of R. Hayes for his creative design contributions to the ThinZag device. We also thank R. Budny and M. Trainor, for their invaluable operational support during the course of this research, and M. Foote, for his excellent insights into phase control of this device. We also acknowledge the support of W. Russell and S. Flintoff in the assembly of the ThinZag device. We also thank I. Sadovnik, J. Moran, and C. vonRosenberg for their support on thermal analysis of the laser module.

References

1. Lu, J., Prabhu, M., Song, J., Li, C., Xu, J., Ueda, K., Kaminskii, A. A., Yagi, H., and Yanagitani, T., "Optical Properties and Highly Efficient Laser Oscillation of Nd:YAG Ceramics," *Appl. Phys. B*, 71: 469–473, 2000.
2. Lu, J., Song, J., Prabhu, M., Xu, J., Ueda, K., Yagi, H., Yanagitani, T., and Kudryashov, A., "High Power Nd:$Y_3Al_5O_{12}$ Ceramic Laser," *Jpn. J. Appl. Phys.*, 39: L1048–L1050, 2000.
3. Lu, J., Murai, T., Takaichi, K., Umeatsu, T., Misawa, K., Ueda, K., Yagi, H., Yanagitani, T., and Kaminskii, A. A., "Highly Efficient Polycrystalline Nd:YAG Ceramic Laser," Solid State Lasers X, *Proc. SPIE*, 4267: 2001.
4. Mandl, A., and Klimek, D. E., "Multipulse Operation of a High Average Power, Good Beam Quality Zig-Zag Dye Laser," *IEEE J. Quantum Electron.*, 32: 378–382, 1996.
5. Mandl, A., and Klimek, D. E., "Single Mode Operation of a Zig-Zag Dye Laser," *IEEE J. Quantum Electron.*, 31: 916–922, 1995.
6. Mandl, A., Zavriyev, A., and Klimek, D. E., "Energy Scaling and Beam Quality Improvement of a Zig-Zag Solid-State Plastic Dye Laser," *IEEE J. Quantum Electron.*, 32: 1723–1726, 1996.

7. Mandl, A., Zavriyev, A., Klimek, D. E., and Ewing, J. J., "Cr:LiSAF Thin Slab Zigzag Laser," *IEEE J. Quantum Electron.*, 33: 1864–1868, 1997.
8. Mandl, A., Zavriyev, A., and Klimek, D. E., "Flashlamp Pumped Cr:LiSAF Thin-Slab Zig-Zag Laser," *IEEE J. Quantum Electron.*, 34: 1992–1995, 1998.
9. Klimek, D. E., and Mandl, A., "Power Scaling of Flashlamp-Pumped Cr:LiSAF Thin-Slab Zig-Zag Laser," *IEEE J. Quantum Electron.*, 38: 1607–1613, 2002.
10. Lu, J., Murai, T., Takaichi, K., Umeatsu, T., Misawa, K., Ueda, K., Yagi, H., and Yanagitani, T., "72 W Nd:YAG Ceramic Laser," *Appl. Phys. Lett.*, 78: 3586–3588, 2001.
11. Heller, A. "Transparent Ceramics Spark Laser Advances," Livermore National Research Laboratory Science and Technology Review, *S&TR,* Apr. 2006.

CHAPTER 10
Thin-Disc Lasers

Adolf Giesen

Head, Institute of Technical Physics, German Aerospace Center (DLR), Stuttgart, Germany

Jochen Speiser

Head, Solid State Lasers & Nonlinear Optics, Institute of Technical Physics, German Aerospace Center (DLR), Stuttgart, Germany

10.1   Introduction

Thin-disc lasers are diode-pumped, solid-state lasers scalable to very high output power and energy with very high wall-plug efficiency and good beam quality. These design properties have been demonstrated during the past decade, and therefore, many companies are offering thin-disc lasers for various applications ranging from lasers for eye surgery and other medical applications to metal cutting and welding applications, with output powers up to 16 kW. This chapter reviews the results for continuous wave (CW) operation and for pulsed operation with pulse durations ranging from 100 femtoseconds (fs) to several µs. In addition, the scaling laws are discussed, showing that the physical limits for CW and pulsed operation of thin-disc lasers are far beyond the state of the art today.

10.2   History

Since the late 1980s many groups have worked on diode-pumped solid-state lasers, replacing the lamps for pumping with laser diodes. One goal of this work was to increase the wall-plug efficiency of the laser systems; another was to improve the beam quality of such lasers. Nearly all groups worked on the classical rod or slab design. However, some groups studied the properties of other laser active materials that could not be pumped using ordinary lamps.[1,2]

In November 1991, during the Lasers and Electro-Optics Society (LEOS) conference, Adolf Giesen listened to a talk about diode pumped Yb:YAG lasers given by T. Y. Fan of the Massachusetts Institute of Technology (MIT). Fan explained the advantages of Yb:YAG for diode laser pumping in detail, but he also stated that it would be very difficult to build a high-power Yb:YAG laser using classical designs due to Yb:YAG's quasi-three-level nature. At that time, the laser output power was only a few watts.

After this talk, Giesen made some initial calculations, which showed that it would be possible to power scale Yb:YAG if the material were simply a very thin sheet of material cooled from one or two sides, so that the heat flux length to the cooling device were minimum. At the University of Stuttgart in Germany, Giesen convinced his colleagues of his idea. In January 1992, a small group (Uwe Brauch from the German Aerospace Center [DLR], Adolf Giesen, Klaus Wittig and Andreas Voss from the University of Stuttgart [IFSW]) started to develop the details of such a laser design, using a thin sheet of laser active material. The primary thin-disc laser design was developed at the end of March 1992, and in late spring 1993, the first demonstration was realized—first with 2 W output power and later with 4 W.[3, 4] Also in 1993, the group applied for the first patent for this design, which has since been successfully licensed to more than 20 companies. During the following years, Giesen's group demonstrated power scaling of thin-disc lasers, pulsed operation also with subpicosecond pulse duration, and the applicability of this design to many other laser active materials.

Fortunately, during the 1980s the German federal ministry of Research and Technology (BMFT) identified laser technology as a key technology for materials processing. Consequently, during these years, many projects were initiated and funded between research institutes and companies, which led to increased funds for thin-disc laser work. Later, companies like Trumpf Laser, Rofin-Sinar and Jenoptik started working on thin-disc lasers, generously supporting the institute's work. As a result, within just one decade, German companies became very strong in the field of laser technology for materials processing, eventually taking the leadership role in this industrial area.

10.3 Principles of Thin-Disc Lasers

The core concept of the principle behind thin-disc lasers is the use of a thin, disc-shaped active medium that is cooled through one of the flat faces of the disc; simultaneously, the cooled face is used as a folding or end mirror of the resonator. This face cooling minimizes the transversal temperature gradient, as well as the phase distortion transversal to the direction of the beam propagation, and it accounts

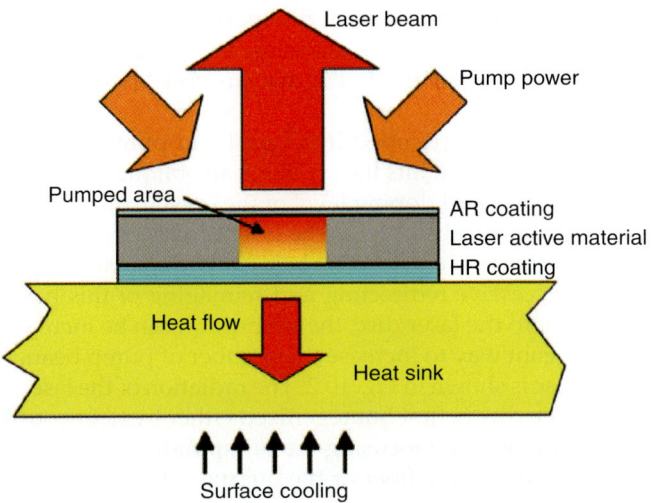

FIGURE 10.1 Thin-disc laser design. AR: antireflective; HR: highly reflective. (*Courtesy of Dausinger und Giesen GmbH*)

for one of the outstanding features of the thin-disc laser—that is, its excellent beam quality.

Figure 10.1 shows the principle of the thin-disc laser design.[3-8] The laser crystal has a diameter of several millimeters (depending on the output power or energy) and a thickness of 100 to 200 μm (depending on the laser active material, the doping concentration, and the pump design). The disc has a highly reflective (HR) coating on its back side for both the laser and the pump wavelengths and an antireflective (AR) coating on the front side for both wavelengths. This disc is mounted with its back on a water-cooled heat sink, using indium-tin or gold-tin solder, which allows for a very stiff fixation of the disc to the heat sink without disc deformation. To reduce the stress during and after the soldering process as much as possible, the heat sink is made from a heat-expansion-matched material (e.g., copper-tungsten metal matrix material [CuW]). The heat sink is water cooled by impingement cooling using several nozzles inside the heat sink.

As mentioned earlier, the temperature gradients inside the laser crystal are mainly coaxial to the disc axis and the laser beam axis due to this mounting and cooling technique. The temperature in the radial direction is nearly uniform within the disc's homogeneously pumped central area. Therefore, these temperature gradients only slightly influence the laser beam propagation through the disc. All the thermal lens effects and aspherical parts of the index of refraction profile are reduced by more than 1 order of magnitude as compared with rod laser systems. The stress-induced birefringence is even more reduced and can be neglected for real laser systems. In addition, due to the large surface-to-volume ratio, heat dissipation from the disc into the heat sink is very efficient, thus allowing for operation at extremely high volume power densities in the disc (up to 1 MW/cm^3 absorbed pump power density).

The crystal can be pumped in a quasi-end-pumped scheme. In this case, the pump beam hits the crystal at an oblique angle. Depending on the thickness and doping level of the crystal, only a fraction of the pump radiation is absorbed in the laser disc. Most of the incident pump power leaves the crystal after being reflected at the back side of the disc. By successive redirecting and reimaging of this part of the pump power onto the laser disc, the absorption can be increased.

A very elegant way to increase the number of pump beam passes through the disc is shown in Fig. 10.2. The radiation of the laser diodes for pumping the disc is first homogenized either by fiber coupling of the pump radiation or by focusing the pump radiation into a quartz-rod. The end of either the fiber or the quartz rod is the source of the pump radiation, which is imaged onto the disc using a collimation optic and the parabolic mirror. In this way a very homogeneous pump profile with the appropriate power density in the disc can be achieved, which is necessary for good beam quality. The unabsorbed part of the pump radiation is collimated again at the opposite side of the parabolic mirror. This beam is redirected via two mirrors to another part of the parabolic mirror, where the pump beam is focused again onto the disc, but this time from a different direction. This reimaging can be repeated until all the (virtual) positions of the parabolic mirror have been used. At the end, the pump beam is redirected back to the source, thereby doubling the number of pump beam passes through the disc. In this way, up to 32 passes of the pump radiation through the disc have been realized and more than 90 percent of the pump power will be absorbed in the disc.

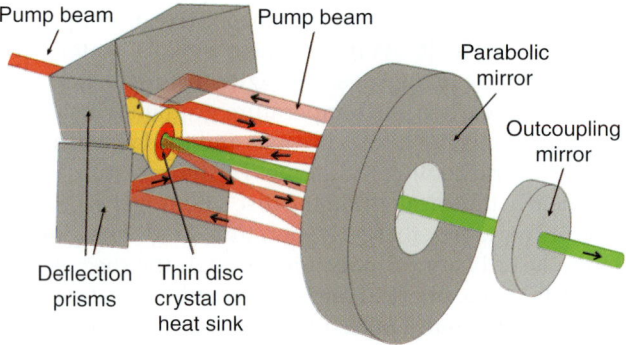

FIGURE 10.2 Pump design of the thin-disc laser with 24 pump beam passes. (*Courtesy of Institute of Laser Physics, University of Hamburg*)

Using multiple pump beam passes through the disc results in a thinner disc or a lower doping concentration, thus reducing such thermal effects like thermal lensing and stress in the disc. Another advantage of this system is that it increases the effective pump power density (nearly 4 times for 16 pump beam passes); thus, on the one hand, the demands on the pump diode's power density (beam quality) are reduced, while on the other hand, quasi-three-level laser materials (e.g., ytterbium-doped materials) can be used with this design.

Quasi-three-level materials offer the ability to build lasers with the highest efficiency. However, they are hard to operate, because the energy difference between the lower laser level and the ground level is small, leading to a significant thermal population of the lower laser level. Some amount of pump power density is necessary simply to reach transparency at the laser wavelength, making it necessary to pump the material with high pump power density in order to reach threshold without increasing the crystal's temperature too much. Using multiple pump beam passes through the crystal is thus the key to achieving low threshold and high efficiency, because it simultaneously helps reduce the thickness of the crystal and the doping concentration. This decoupling of laser and pump beam absorption is essential for operating quasi-three-level systems. The limit for the possible number of pump beam passes through the disc is given by the beam quality of the laser diodes, which determines the beam diameter on the parabolic mirror and, hence, the number of positions on the mirror that can be used. The better the beam quality of the pump laser diodes, the higher the number of pump beam passes and the higher the total efficiency of the thin-disc laser.

When operating the disc in this setup, it is easy to scale the output power simply by increasing the pump spot diameter, keeping the pump power density constant. In addition, there is no need to increase the brightness of the pump laser diodes.

10.4 Possible Laser Materials

Nearly all classical laser materials can be operated in the thin-disc design, especially if the absorption of the pump radiation is rather high and the lifetime of the excited state is not too short. The first material used with the thin-disc laser was Yb:YAG; with this material most of the high power or high energy results were reached. Yb^{3+} has two important benefits: a small quantum defect and no parasitic effects such as upconversion, cross relaxation, excited-state absorption and so on. Laser operation of Yb^{3+} with the thin-disc laser has been demonstrated in a large variety of host materials and also other active ions were successfully operated in the thin-disc laser setup. Table 10.1 gives an overview of successful combinations without intending to be exhaustive.

Host Material	
YAG	Yb^{3+}, $Nd^{(3+)}$ [9–11], $Tm^{(3+)}$ [12,13], $Ho^{(3+)}$ [14]
YVO_4	$Yb^{(3+)}$ [15–17], $Nd^{(3+)}$ [18–21]
Sc_2O_3	$Yb^{(3+)}$ [22]
Lu_2O_3	$Yb^{(3+)}$ [22,23]
$KY(WO_4)_2$	$Yb^{(3+)}$ [22]
$KGd(WO_4)_2$	$Yb^{(3+)}$ [22]
$NaGd(WO_4)_2$	$Yb^{(3+)}$ [15,17]
$LaSc_3(BO_3)_4$	$Yb^{(3+)}$ [24]
$Ca_4YO(BO_3)_3$	$Yb^{(3+)}$ [25]
$GdVO_4$	$Nd^{(3+)}$ [21]
ZnSe	$Cr^{(2+)}$ [26]

TABLE 10.1 Examples for Successful Combinations of Host and Active Ions in the Thin-Disc Laser Setup

With neodymium-doped materials, not only the four-level transitions could be used but also the quasi-three-level transitions, resulting in 5.8 W laser power at 914 nm with Nd:YVO[20] and 25 W laser power at 938 nm and 946 nm with Nd:YAG.[11]

10.5 Numerical Modeling and Scaling

10.5.1 Average Temperature

Because the disc is very thin and the pump spot is large, one can assume one-dimensional heat conduction. If we apply a pump power P_{pump} on a pump spot with radius r_p, absorption efficiency η_{abs}, and heat generation η_{heat} to a disc with thickness h, that is made of a material with thermal conductivity λ_{th}, we will get as heat load per area:

$$I_{heat} = \frac{P_{pump} \eta_{abs} \eta_{heat}}{\pi r_p^2} \tag{10.1}$$

This heat load will result in a parabolic temperature profile along the axis inside the disc of

$$T(z) = T_0 + I_{heat} R_{th,disk} \left(\frac{z}{h} - \frac{1}{2} \frac{z^2}{h^2} \right) \tag{10.2}$$

with $R_{th,disk} = h/\lambda_{th}$ being the heat resistance of the disc and T_0 being the temperature at the disc's cooled face. In particular, the maximum temperature will be

$$T_{max} = T_0 + \frac{1}{2} I_{heat} R_{th,disk} \qquad (10.3)$$

and the average temperature will be

$$T_{av} = T_0 + \frac{1}{3} I_{heat} R_{th,disk} \qquad (10.4)$$

For most thin-disc host materials, the thermal conductivity depends on the doping concentration and the material temperature. For YAG, a thermal conductivity of 6 W m^{-1} K^{-1} is a good approximation for low doping (~7 percent) and temperatures of ~100°C. For a disc of 180 µm thickness, this will result in a thermal resistance $R_{th,disk} = 30 \frac{\text{Kmm}^2}{\text{W}}$. Typically, the disc is not directly cooled. Instead, it is coated at the cooled face with an HR coating and this coating is mounted on a heat sink. The heat sink is then cooled with a cooling fluid of temperature T_{cool}. The thermal resistance of the HR coating is determined not only by the materials used, but also by the quality of the coating and the coating process used. From experimental results and numerical calculations, a thermal resistance $R_{th,HR} = 10 \frac{\text{Kmm}^2}{\text{W}}$ seems reasonable. The heat sink may consist of a large variety of materials, including a copper-tungsten (CuW) metal matrix material ($\lambda_{th} = 180$ W m^{-1} K^{-1}) or a chemical vapor deposition (CVD) diamond ($\lambda_{th} \approx 1000$ W m^{-1} K^{-1}), that have a typical thickness of 1 mm. The thermal resistance of the "mounting" itself can either be nearly neglected (e.g., for a soldering layer of 10 to 50 µm thickness, resulting in less than 1 Kmm²/W thermal resistance) or it can have a strong influence on the performance, as with glued discs, wherein the glue layer creates a thermal resistance of about 10 Kmm²/W due to its poor thermal conductivity. The heat transfer to the cooling fluid is also strongly design-dependent, with the best cooling reached via a highly turbulent flow of the cooling fluid. With water and a so-called impingement cooling, an effective thermal resistance of this transfer of 3 Kmm²/W was demonstrated. The resulting total effective thermal resistance with respect to the average temperature of the disc can therefore be expected to be about 30 to 35 Kmm²/W.

For high-purity Yb:YAG, the heat generation inside the disc is only due to the quantum defect, given by

$$\eta_{heat} = 1 - \frac{\lambda_p}{\lambda_l} \approx 8.7\% \qquad (10.5)$$

We can expect an average temperature in the disc of about 200°C if an absorbed pump power density of 60 W/mm² and a cooling fluid temperature of 15°C were used. We can also calculate an ultimate limit of the absorbed pump power density, because we must avoid boiling of the cooling fluid. With 300 W/mm² absorbed pump power density the resulting temperature at the back side of the heat sink would be 96°C.

From these calculations we can also derive that the maximum temperature difference ΔT in the disc will keep constant for a given material, as long as the ratio of absorbed pump power density and thickness of the disc is constant. Figure 10.3 illustrates this relation.

It is useful to introduce a thermal load parameter C which is the maximum allowed product of disc thickness and (absorbed) pump power density to keep the maximum temperature rise inside the disc below a given value of ΔT:

$$C = \frac{2\Delta T \eta_{heat}}{\lambda_{th}} \quad (10.6)$$

A similar parameter, the "thermal shock parameter", is often used in the context of slab lasers or active mirror lasers without additional supporting structures. It is motivated by the limitations given by the maximum thermally induced tensile stress.

10.5.2 Influence of Fluorescence

Up to now, only the heat generated inside the disc from the quantum defect was used for the temperature estimations. However, if we look

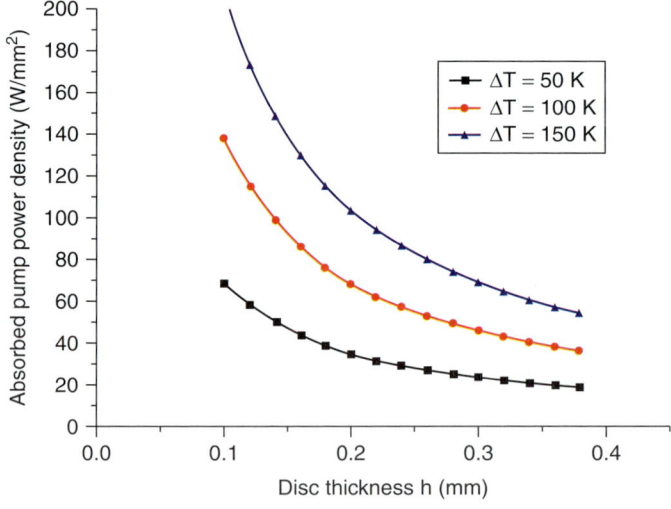

FIGURE 10.3 Absorbed pump power density to reach a temperature rise of 50 K, 100 K and 150 K as function of the thickness of the disc (assuming a heat generation η_{heat} = 8.7% and a thermal conductivity λ_{th} = W m⁻¹ K⁻¹).

at the system's "energy balance", roughly 9 percent of the pump power is transformed to heat, and, for the highly efficient thin-disc laser, about 60 percent is transformed to laser power. The remaining 31 percent is emitted as fluorescence radiation. We can expect that all fluorescence that is emitted at angles smaller than the critical angle will leave the disc through the AR-coated front face, either directly or after one reflection at the HR-coated face. For YAG the refractive index is 1.83, and the critical angle is therefore about 33°; therefore about 16 percent of the fluorescence will leave the disc through the AR face. If we sum these results, about 26 percent of the absorbed pump power will be transformed to fluorescence that is "captured" inside the disc.

Neglecting any further interactions of this fluorescence with the disc material, the HR coating design will determine whether this fluorescence is emitted or transformed to heat. A coating that is highly reflective at all angles and wavelengths will simply guide the fluorescence to the disc's lateral surface, where it will be reflected, scattered, "extracted", or perhaps transformed to heat. Neither back reflection nor back scattering is favorable due to the problems of amplified spontaneous emission (ASE) discussed later; in addition, extraction of several kilowatts of power at the lateral surface is technologically challenging. The contrary possibility is a coating which is highly transparent at all wavelengths and all angles larger than the critical angle, including a layer between the coating and the glue or solder (for mounting), which is highly absorbing. With this coating, nearly all "captured" fluorescence will be transformed to heat that must pass through the heat sink. Because the combination of heat sink, solder/glue and cooling has an effective thermal resistance of ~10 Kmm²/W, the 60 W/mm² absorbed pump power discussed above would create an additional temperature rise of 150°C.

A compromise between the reduction of fluorescence reaching the lateral surface and heat generation would be a partially transparent coating; such a coating design would also be closer to designs that are technically possible. For simplicity, assuming a transparency of 25 percent for all angles larger than the critical angle, the absorbed fluorescence would create an additional temperature rise of only 40°C. This additional heat generation would also reduce the limitation of absorbed pump power density to avoid boiling of the cooling fluid to about 175 W/mm².

The additional temperature rise due to fluorescence absorption would be much bigger if there were no lasing; in this case ~76 percent of the absorbed pump power would be transformed to "captured" fluorescence. With 25 percent transparency, the additional temperature rise would be ~110°C, and the "boiling limit" would be 95 W/mm².

Figure 10.4 shows the results for different values of the absorbed pump power. All these results are calculated without any heat spreading. In the disc, the heat spreading will have only a very small influence

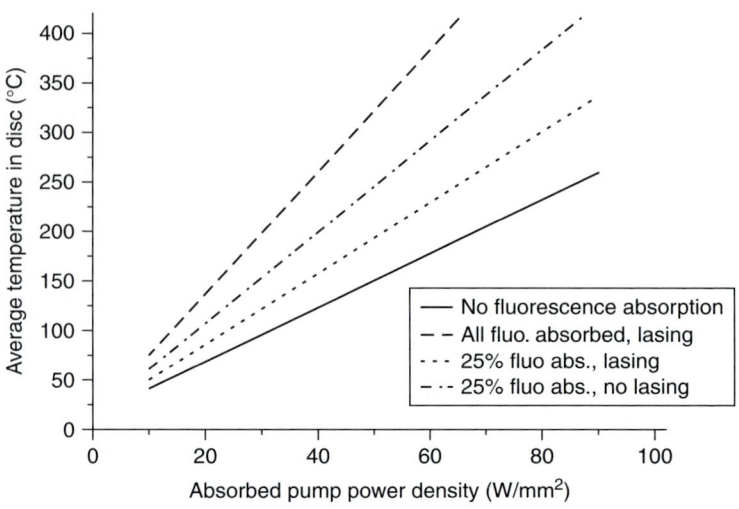

FIGURE 10.4 Average temperature in the thin-disc for idealized coating design as a function of the absorbed pump power density.

even for pump spots of a few mm in diameter; for the heat sink, the heat spreading is stronger and can especially strongly reduce the influence of the absorbed fluorescence in a real medium power (i.e., up to a few kilowatts) thin-disc laser.

10.5.3 Thermally Induced Stress

The temperature rise in the pumped region will lead to a thermal expansion of the thin-disc. Because the outer part of the disc will essentially be at the cooling temperature, this thermal expansion will lead to thermally induced stress within the disc. Most critical is the tensile stress with the highest tensile stress being generated at the boundary of the pumped region in azimuthal direction. In an idealized situation, the whole pump spot has the temperature T_{av}, the not-pumped part of the disc has temperature T_{cool}, the disc is not supported by any heat sink and there is no bending of the disc. In this case, we can use analytical results from elasticity theory: For the azimuthal stress $\sigma_{\phi,max}$ at the pump boundary spot we will get

$$\sigma_{\phi,max} = \frac{1}{2}\frac{\alpha_{th}E_{elast}}{1-\nu}(T_{av}-T_{cool})\left(1+\frac{r_p^2}{r_{disc}^2}\right) \qquad (10.7)$$

with disc radius r_{disc}, the thermal expansion (~7e-6 K^{-1}) α_{th}, Young's modulus (284 GPa) E_{elast} and Poisson's ratio (0.25) ν for YAG. The worst case is reached when the pump spot nearly fills the complete disc; thus we can use

$$\sigma_{\phi,\max} \leq \frac{\alpha_{th} E_{elast}}{1-\nu}(T_{av} - T_{cool}) \qquad (10.8)$$

The tensile strength of YAG is 130 MPa,[26] and the temperature difference between the disc and the cooling can be calculated with the effective thermal resistance derived above, but now neglecting the heat sink. With an effective thermal resistance of 23 Kmm²/W, the maximum heat density per area is 2.1 W/mm² (i.e., ~24 W/mm² absorbed pump power density if we only take into account the quantum defect as heat source).

The azimuthal stress inside the disc can be significantly reduced by mounting the disc on a heat sink with adequate stiffness. A detailed analysis of the stress inside the disc must also include the effects of bending; this will be done in the next section based on finite element analysis (see Sec. 10.5.4).

10.5.4 Deformation, Stress, and Thermal Lensing

A radially symmetric model of the disc mounted on the heat sink (or alternative supporting structures) was generated using the commercial finite element software COMSOL. Multiphysics and a uniform heat source distribution was applied inside the pump spot. Figure 10.5 shows the calculated temperatures for the situation discussed in the previous section though now for a large pump spot. The assumed 60-W/mm² absorbed pump power density and 7.5-mm pump spot radius are equivalent to ~10 kW absorbed pump power, which is sufficient for 6 kW laser power.

The main problem to be answered by finite element analysis (FEA) calculations is the amount of tensile stress inside the disc. This stress can be controlled even for high pump power densities by choosing an appropriate mounting design. Figure 10.6 shows that this stress is limited by the mounting on the CuW heat sink.

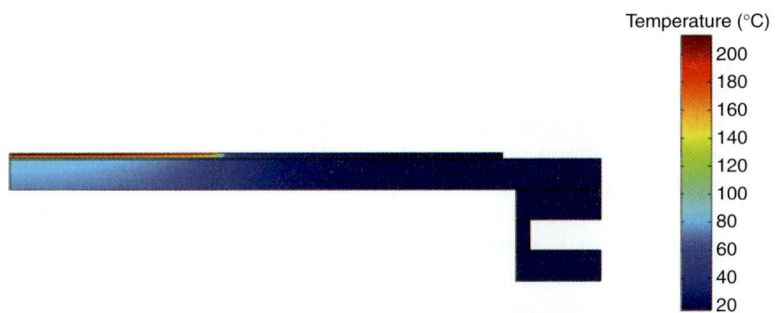

FIGURE 10.5 Calculated temperature for a Yb:YAG thin-disc; thickness 180 μm, mounted on a CuW heat sink with thickness 1 mm, pump spot radius 7.5 mm, and heat source per area 5.4 W/mm², equivalent to 60 W/mm² absorbed power density.

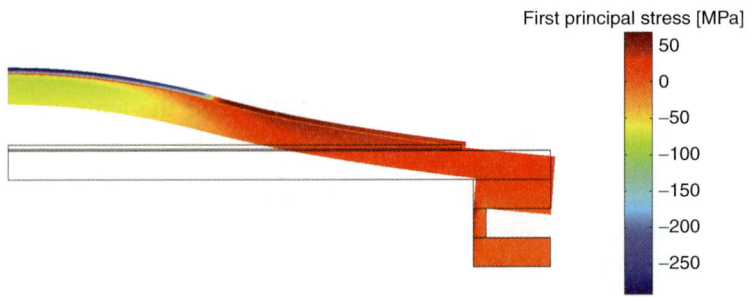

FIGURE 10.6 Calculated first principal stress and deformation (scaled by 100) for a Yb:YAG thin-disc of thickness 150 μm, mounted on a CuW heat sink with thickness of 1 mm, pump spot radius 7.5 mm, and heat source per area 5.4 W/mm², equivalent to 60 W/mm² absorbed power density.

From the results of the finite element software, the optical phase distortion (OPD) $\Phi(r)$ can be calculated. The OPD consists of two contributions: first, the change of the optical path length through the disc due to the thermal expansion and thermally induced change of the refractive index (what is typically considered as thermal lens for rod lasers) and second, the change of shape of the whole system. The change of shape is described by the displacement of the HR face of the disc. Both contributions are depicted in Fig. 10.7, including also the resulting OPD as sum of both.

The main part of the OPD is parabolic, caused by the bending due to the temperature gradients in the system. This parabolic part is equivalent to a curvature or a spherical contribution which can be expressed as a refractive power. The remaining aspherical part will cause diffraction losses. To determine the curvature or refractive power, the calculated OPD is separated into spherical and an aspherical part:

$$\Phi(r) = -2\pi r^2/(\lambda R_L) + \Delta\Phi(r) \tag{10.9}$$

The optimum value of the curvature R_L is determined by calculating the diffraction losses of the remaining $\Delta\Phi$ for different values of R_L. This calculation of diffraction loss is done by applying the phase distortion $\Delta\Phi$ to a plane wave (fundamental mode with a mode radius of typically 70 percent of the pump spot radius) and then determining which amount of the distorted mode is still fundamental. The data presented in Fig. 10.7 will result in a curvature of 2.98 m if a fundamental mode radius of 5.25 mm is assumed. Figure 10.8 shows the remaining phase distortion $\Delta\Phi$ of this analysis. Two non-parabolic contributions can be distinguished: a step-like structure (~500 nm) at the edge of the pump spot due to the temperature distribution in the disc and the non-parabolic part of the deformation due to the clamping.

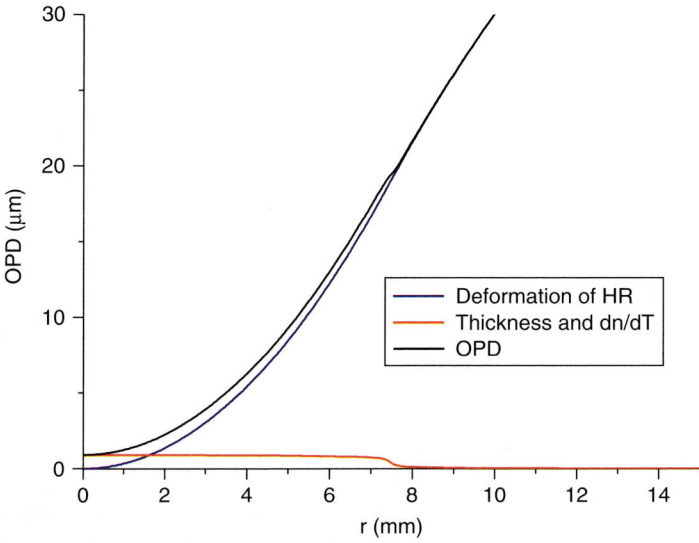

Figure 10.7 Contributions to the optical phase distortion OPD for a Yb:YAG thin-disc of thickness 150 μm, mounted on a CuW heat sink with thickness of 1 mm, pump spot radius 7.5 mm, and heat source per area 5.4 W/mm², equivalent to 60 W/mm² absorbed power density.

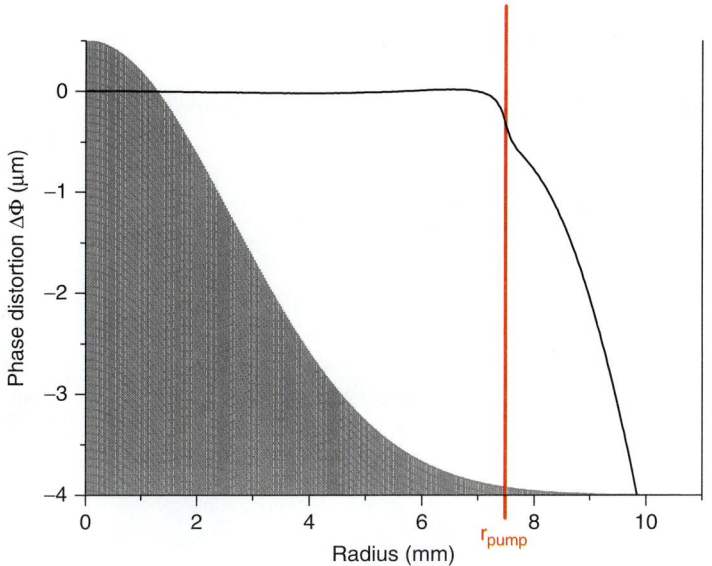

Figure 10.8 Aspherical part of the phase distortion of a Yb:YAG thin-disc of thickness 150 μm, mounted on a CuW heat sink with thickness of 1 mm, pump spot radius r_{pump} 7.5 mm, and heat source per area 5.4 W/mm².

10.5.5 Design Study for High-Power Thin-Disc Lasers

For the design of high power thin-disc lasers, two mounting designs seem to be promising. The "classical" design is to solder the disc onto a heat sink, the alternative is to use a transparent (e.g., undoped YAG) supporting structure on top of the disc and apply the cooling directly to the disc. Both concepts are sketched in Fig. 10.9. In this section, they will be compared based on their mechanical behavior and their thermal lens. In both cases, the Yb:YAG has a thickness of 180 µm and a diameter of 60 mm. It is either soldered on CuW (thickness 1.5 mm) or bonded to undoped YAG for direct cooling. The pump spot radius is 11 mm, and the pump power is varied between 6.4 kW and 25.6 kW. Based on the quasi-static model (c.f. Sec. 10.5.8, Fig. 10.12) this would be sufficient for 14 kW of laser power with one disc.

In Fig. 10.10, the results of FEA calculations are given. The mechanical behavior of both designs is quite different. Because the support from the heat sink is missing, the directly cooled design is less stiff and it shows tensile stress in radial direction. The classical design provides better compensation of the azimuthal stress. Nevertheless, the total temperature rise in the directly cooled design is smaller as there is no additional thermal resistance of the heat sink and the solder layer. Due to these lower temperatures, also the thermally induced stress inside the disc is below the critical value of 130 MPa.

The heat sink does not only provide stiffness to the system, it does also contribute to the deformation due to the temperature gradient.

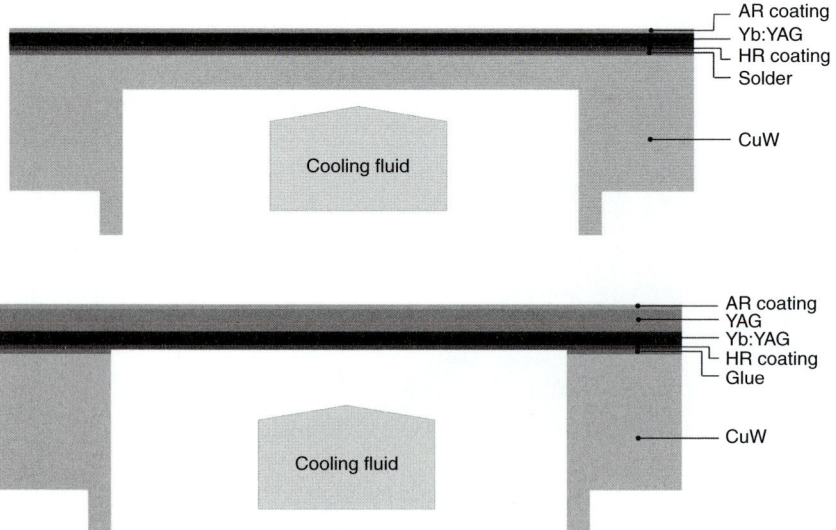

FIGURE 10.9 Different mounting designs, top: Yb:YAG soldered on CuW ("classical" thin-disc design); bottom: composite disc, directly cooled.

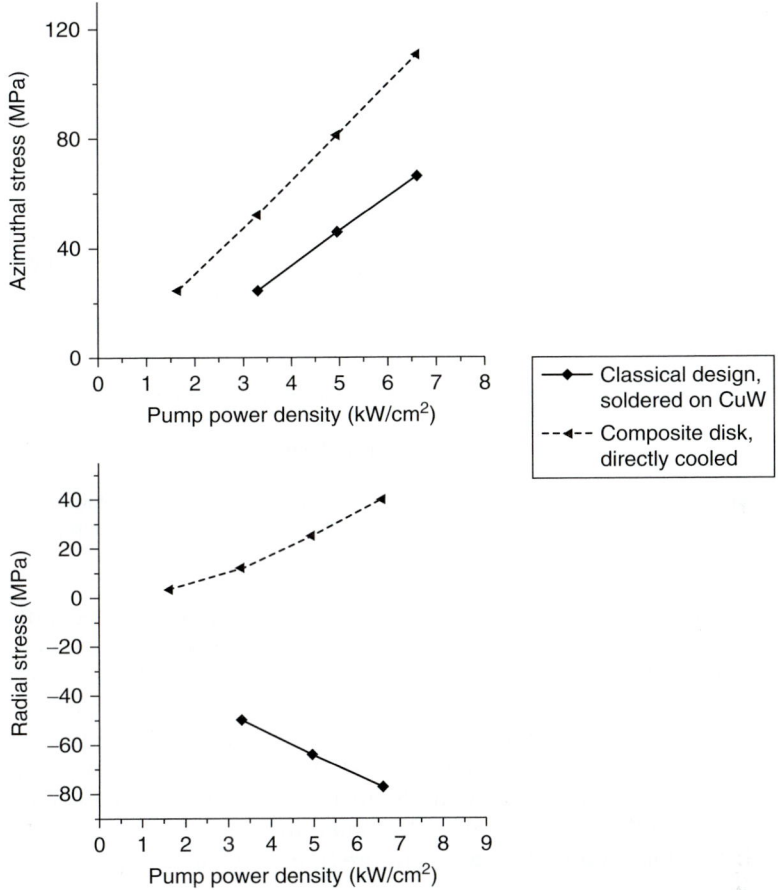

FIGURE 10.10 Comparison of the maximum radial and azimuthal stress inside the disc for two different mounting designs.

The transparent layer on top of the disc has nearly no axial temperature gradient and the directly cooled design shows a significantly smaller deformation. This results in a reduced refractive power of the thin-disc as can be seen in Fig. 10.11. But the remaining aspherical contributions of the OPD are stronger, as there is more hot material inside the resonator.

10.5.6 Numerical Modeling of Gain and Excitation

Yb:YAG shows a significant temperature-dependent reabsorption of the laser radiation. In the thin-disc design it is operated at a comparable high inversion level, resulting in a significant reduction of the pump absorption. Consequently, the coupling between the differential equations of pump absorption, laser amplification, inversion, and

Solid-State Lasers

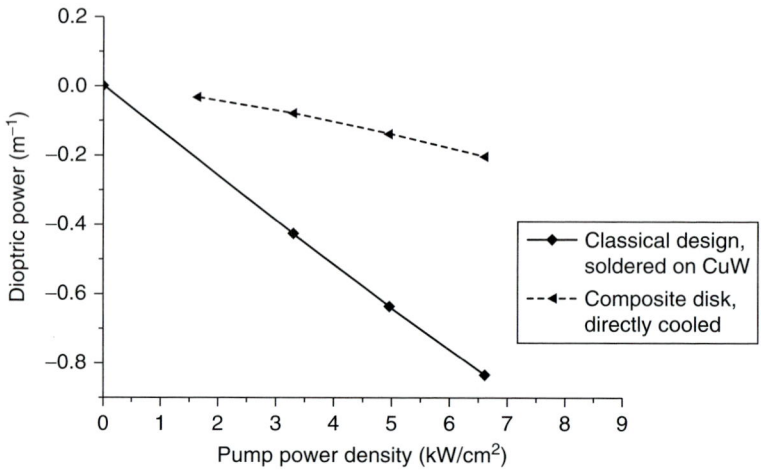

FIGURE 10.11 Calculated spherical part of the thermal lens of a soldered disc and a directly liquid cooled disc (pump diameter 22 mm, max. pump power 25 kW).

temperature cannot be neglected. A numerical model that accounts for this coupling was developed during the past decade.[27–31] A very important design feature of the thin-disc laser was developed based on this model: the immense benefit for the system performance from higher numbers of pump beam passes. As illustration, it was proven numerically and experimentally that doubling the number of pump beam passes from 8 to 16 (with optimized thickness and doping concentration in both cases) gives the same increase in efficiency like reducing the cooling fluid temperature from 15°C to −25°C.[29]

10.5.7 Equation of Motion

The fundamental equation of motion for the density of excited ions is

$$\dot{N}_2 = Q - \frac{N_2}{\tau} - \gamma_{laser} \Phi_r \quad (10.10)$$

with Q some kind of source (e.g., the absorbed pump photons per volume and time) and τ the fluorescence lifetime, γ_{laser} the gain per length at the laser wavelength (gain coefficient) and Φ_r the number of laser photons per area and time (photon flux density).

The typical laser material for the thin-disc setup is Yb:YAG, at room temperature a quasi-three-level laser material—that is, the lower laser level is thermally populated. For a given density of excited ions N_2, a density of laser ions N_0, and an emission cross section $\sigma_{em}(\lambda, T)$ at a wavelength λ, we get as gain coefficient:

$$\gamma_\lambda = \sigma_{em}(\lambda, T)(1 + f_{abs}(\lambda, T))N_2 - \sigma_{em}(\lambda, T)f_{abs}(\lambda, T)N_0 \quad (10.11)$$

with

$$f_{abs}(\lambda, T) = \frac{Z_2(T)}{Z_1(T)} \exp\left(\frac{2\pi\hbar c_{vac}}{\lambda k_B T}\right) \quad (10.12)$$

and Z_1, Z_2 the partition functions of the lower and upper laser level, respectively.

Similarly, one can calculate for a given absorption cross section $\sigma_{abs}(T)$ at pump wavelength λ_p the absorption coefficient

$$\alpha = \sigma_{abs}(T)N_0 - \sigma_{abs}(T)(1 + f_{em}(T))N_2 \quad (10.13)$$

with

$$f_{em}(T) = \frac{Z_1(T)}{Z_2(T)} \exp\left(-\frac{2\pi\hbar c_{vac}}{\lambda_p k_B T}\right) \quad (10.14)$$

With this absorption coefficient, a thickness of the disc h, and a pump power density E_p, we can calculate the number of absorbed pump photons per volume and time:

$$Q = \frac{E_p \lambda_p}{2\pi\hbar c_{vac}} \frac{[1 - \exp(-\alpha h M_p)]}{h} \quad (10.15)$$

if we use M_p pump beam passes through the disc.

We can also calculate the gain at the laser wavelength for one pass through the disc:

$$g = h[\sigma_{em,laser}(1 + f_{abs}(T))N_2 - \sigma_{em,laser} f_{abs}(T)N_0] \quad (10.16)$$

Because energy extraction is only possible if $g > 0$, we can similarly define the maximum extractable energy per area as:

$$H_{extractable} = \frac{2\pi\hbar c_{vac}}{\lambda_{laser}} h[(1 + f_{abs}(T))N_2 - f_{abs}(T)N_0] \quad (10.17)$$

These formulas ensure the correct handling of thermal population, bleaching, and saturation effects.

10.5.8 Coupled Quasi-Static Numerical Model

For the coupled model, the disc is discretized in finite elements in radial, azimuthal and axial direction. From the equation of motion (Eq. [10.10]), we can derive the formula for the density N_2 of the excited Yb^{3+} ions in the quasi-static limit in each element:

$$\frac{\lambda_p P_V}{2\pi\hbar c_{vac}} + \frac{\lambda_l M_r E_r}{2\pi\hbar c_{vac}} \sigma_{em,laser}[N_0 f_{abs} - N_2(1 + f_{abs})] - \frac{N_2}{\tau} - \Delta N_{ASE} = 0 \quad (10.18)$$

with P_V the absorbed pump power in the element and E_r the laser power density inside the thin-disc, N_0 the density of Yb^{3+} ions, $E_{r,r}$ the laser power density inside the thin-disc, $\sigma_{em,laser}$ the emission cross sections at the laser wavelength, τ the radiative lifetime, and N_{ASE} the difference between the number of emitted and the number of absorbed ASE-photons in the finite element. A Monte Carlo ray tracing method is used to calculate the absorbed pump power in each element, following each photon from the source through the complete system.

Calculation of the temperature distribution within the disc is based on the steady-state heat conduction equation with the Stokes Defect and the power transmitted through the HR coating as heat sources. This partial differential equation is solved by a finite volume method. Initial values for N_2 and E_r are derived analytically (with averaged crystal temperature and absorbed pump power density). In an iterative procedure, the laser power density and the excitation density are calculated.

To calculate ΔN_{ASE}, a Monte Carlo ray tracing method is used. A set of photons with a statistical distribution of wavelength, starting coordinates, and propagation vectors are traced through the crystal. Absorption and amplification are computed, as are reflection and transmission at the crystal boundaries.

Simulations [30-32] show that scaling of the output power of a single disc is only limited by ASE as the pump spot diameter becomes larger and larger. Fortunately, the gain of low-doped Yb:YAG is rather small, so ASE occurs only at very high pump power levels. With this numerical model, it was shown that an output power of more than 40 kW with one disc is possible.[31]

Figure 10.12 shows some scaling results to more than 10 kW output power that result from changing the pumped diameter, thus demonstrating the scalability by increasing the pumped area for high-power operation.

10.5.9 Influence of ASE

As we have seen, the temperature and the thermally induced stress are limitations which can be handled for the thin-disc design. The remaining possible limit is the amplified spontaneous emission (ASE). Increasing the output power of a thin-disc by increasing the size of the active region and keeping the thickness constant will lead to an increasing transversal gain. As consequence, this will lead to a reduction of the possible excitation in the disc, reducing signal gain and efficiency. To discuss this more in detail, we will look at the interaction of excitation, gain, pump absorption and ASE in more detail.

The quasi-static model is principally suitable to analyze the influence of ASE on the performance of a thin-disc laser. The calculation of ASE with a Monte Carlo ray tracing is very flexible and can also handle

Thin-Disc Lasers

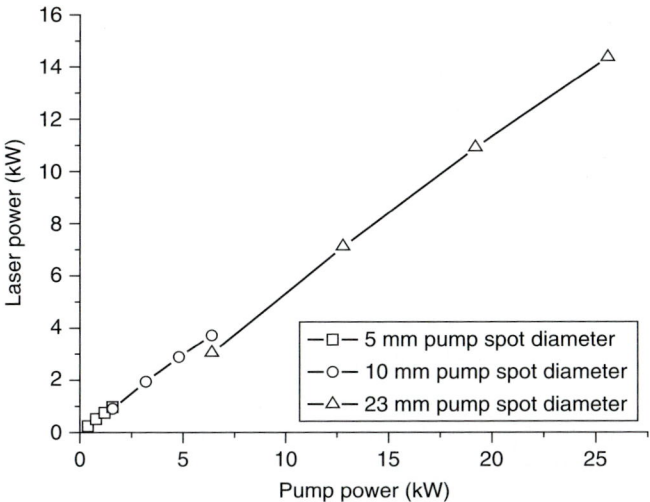

FIGURE 10.12 Calculated laser output power of a Yb:YAG thin-disc laser with doping concentration 9 percent and thickness 180 µm.

spatial variations of gain or temperature. Even spatial variations of the reflectivity of the faces of the thin-disc could be handled. But nevertheless, the iterative quasi-static approach represented by Eq. (10.18) limits the validity of the model to situations where the influence of the ASE is a "small" perturbation. The influence of the ASE can be approximated by $\Delta N_{ASE} \sim N_2^{2.33}$, therefore the assumption of a small perturbation is not suitable for situations with a high density of excited ions in large volume and high transversal gain—like thin-disc lasers for high energy pulse extraction. The convergence problems of the quasi-steady state iterative model are well known,[31] limiting the predictable output power in cw operation to roughly 50 kW and the predictable energy to 2.5 J. Replacing the quasi-static approach with a time-resolved model provides a solution.

10.5.10 Interaction of ASE and Excitation

The fundamental equation of motion (c.f. Eq. [10.10]) for the density of excited ions N_2 in a pumped active medium without resonator, including ASE but no additional effects such as upconversion, is

$$\dot{N}_2 = Q - \frac{N_2}{\tau} - \int \gamma_\lambda \Phi_{\lambda,\Omega} d\lambda d\Omega \qquad (10.19)$$

with Q some kind of source (e.g., the absorbed pump photons per volume and time) and τ the fluorescence lifetime, γ_λ the gain per length at the wavelength λ (gain coefficient) and $\Phi_{\lambda,\Omega}$ the number of (amplified) fluorescence photons per area and time (photon flux density) coming from the solid angle Ω.

First of all, it is necessary to calculate the photon flux density. With an excitation density $N_2(\vec{s})$ and a gain coefficient $\gamma_\lambda(\vec{s})$ the photon flux density arriving at the point $\vec{s}=0$, coming from a volume element dV at a distance $s=|\vec{s}|$ in the direction $\hat{s}=\vec{s}/s$, is

$$d\Phi_\lambda(\vec{s}) = \beta_\lambda \frac{N_2(\vec{s})}{\tau} \frac{1}{4\pi s^2} g_\lambda(\vec{s}) dV \qquad (10.20)$$

with the spectral distribution of the fluorescence β_λ, fulfilling $\int \beta_\lambda d\lambda = 1$ and with an amplification of the photon flux density of

$$g_\lambda(\vec{s}) = \exp\left[\int_0^s \gamma_\lambda(\tilde{s}\hat{s}) d\tilde{s}\right] \qquad (10.21)$$

The entire photon flux density at wavelength λ from direction $\hat{s}$ can be calculated as

$$d\Phi_\lambda(\hat{s}) = d\Omega \frac{\beta_\lambda}{\tau} \int_0^{s_{max}} N_2(\hat{s}s) g_\lambda(\hat{s}s) ds \qquad (10.22)$$

using $dV = s^2 d\Omega$.

The maximum integration distance s_{max} depends on the analyzed geometry. The thin-disc is a cylindrical volume of height (thickness) h and radius R, with the faces of the cylinder orientated horizontally (cf. Fig. 10.13). No reflection from the lateral surface is taken into account; with reflections from the lateral surface, no maximum integration distance could be defined. The reflectivity of the faces of the cylinder will be given by the functions $AR(\lambda, \vartheta)$ (antireflective)

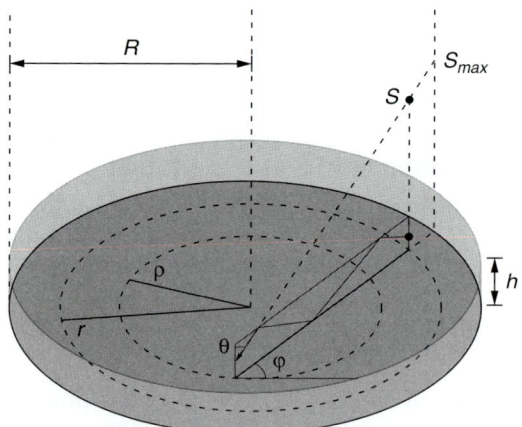

Figure 10.13 Geometry of the thin-disc with radius R and thickness h, illustrating the relations between the maximum integration distance S_{max} and the radial coordinates r and ρ.

and $HR(\lambda, \vartheta)$ (highly reflective). The AR face is on the top and the HR face is at the bottom.

The AR face is typically the interface between the crystalline laser medium and air (or vacuum); it is antireflective for normal incidence at the laser and pump wavelengths. Thus it is reasonable to assume that the AR face is nonreflective for angles smaller than the critical angle of total reflection $\vartheta_{tr} = \arcsin(1/n)$, with n the index of refraction of the active medium, and that it is ideally reflecting for larger angles. In addition, we can assume $HR(\lambda, \vartheta) \neq 0$ for all ϑ due to technical limitations of the coating design.

In spherical coordinates, Eq. (10.22) transforms to:

$$d\Phi_\lambda(\phi, \vartheta) = \sin\vartheta d\vartheta d\phi \frac{B_\lambda}{4\pi\tau} \int_0^{s_{max}} N_2(s, \phi, \vartheta) g_\lambda(s, \phi, \vartheta) ds \qquad (10.23)$$

Taking into account the multiple reflections at the faces of the cylinder, s_{max} can be expressed by:

$$s_{max} = \frac{\sqrt{R^2 - \rho\sin^2\phi} - \rho\cos\phi}{\sin\vartheta} \qquad (10.24)$$

if $AR(\lambda, \vartheta) HR(\lambda, \vartheta) \neq 0$.

To account for losses at the AR and HR faces, the gain coefficient can be modified to:

$$\tilde{\gamma}_{\lambda,\vartheta} = \gamma_\lambda + \frac{\ln(AR(\lambda, \vartheta) HR(\lambda, \vartheta))}{2h}\cos\vartheta \qquad (10.25)$$

if $AR(\lambda, \vartheta) HR(\lambda, \vartheta) \neq 0$.

10.5.11 Time Resolved Numerical Model

Based on these considerations, it is possible to develop a numerical model of the interaction of amplified spontaneous emission and excitation, a more detailed description can be found in literature[34]. For this, we discretize the problem in ρ, ϑ, ϕ, and λ. Neglecting the variation of excitation and gain in axial direction and assuming rotational symmetry, only the radial variation of $\tilde{N}_2(r)$ and $\tilde{\gamma}_\lambda(r)$ remains.

The temporal development of the excitation can easily be calculated by integrating the differential Eq. (10.19) with implicit methods. For each time step, the source Q and the photon flux densities $d\Phi_\lambda(\hat{s})$ are calculated based on the distribution of excitation from the previous time step. Because the typical time constant of the excitation is in the order of the spontaneous lifetime (several hundreds of microseconds), this is adequate for time steps of a few microseconds.

An instructive question is the extractable energy of a quasi-CW pumped thin-disc. The reduced duty cycle reduces the need of thickness optimization and the non-lasing condition facilitates the numerical handling of the differential equation.

The "model system" is a Yb:YAG disc with a thickness of 600 μm, Yb concentration 4.5 percent and a pump power of 16 kW (pump power density 5 to 6 kW/cm²), but with a duty cycle of only 10 percent (for the calculation of the average temperature in the active area). As a first result, Fig. 10.14 shows that in this case, the ASE will strongly reduce the achievable gain, and the gain will be saturated after less then 1 ms.

As the system is intended for energy extraction, it is also useful to look at the extractable energy. In Fig. 10.15, this is done for discs with different thicknesses. The product of doping concentration and thickness was kept constant to facilitate the comparability of the results. The influence of the thickness on the temperature is small due to the low duty cycle. Obviously, the classical strategy of making the disc very thin is no longer suitable at this energy level; the thickest disc reaches the highest gain and also the highest extractable energy.

Up to now, all calculations were done with a HR coating which is totally reflecting at all angles and wavelengths. Besides the technical difficulties to realize such a coating, it is also beneficial to use a coating with some loss for the ASE. Figure 10.16 shows results obtained with a so-called "ideal" coating with a reflectivity of only 75 percent for angles

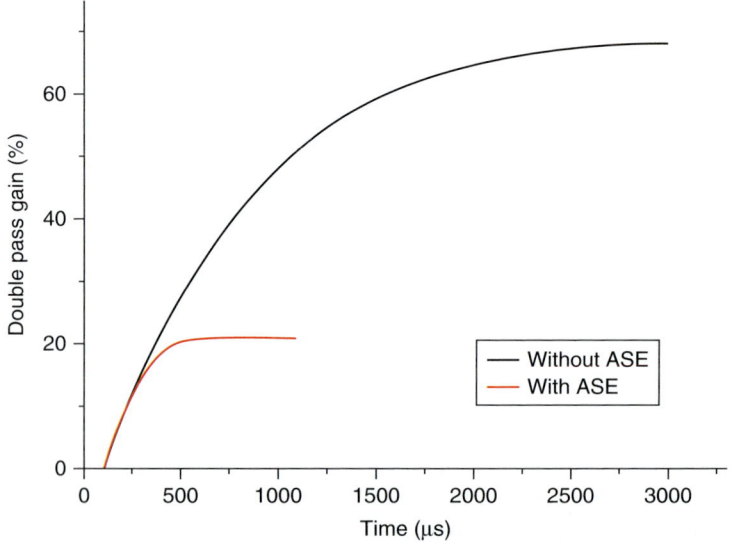

Figure 10.14 Temporal development of the gain in a thin-disc at quasi-cw pumping (10 percent duty cycle), with and without ASE. Doping concentration 4.5 percent, thickness 600 μm, pump power 16 kW, pump spot radius 9.8 mm.

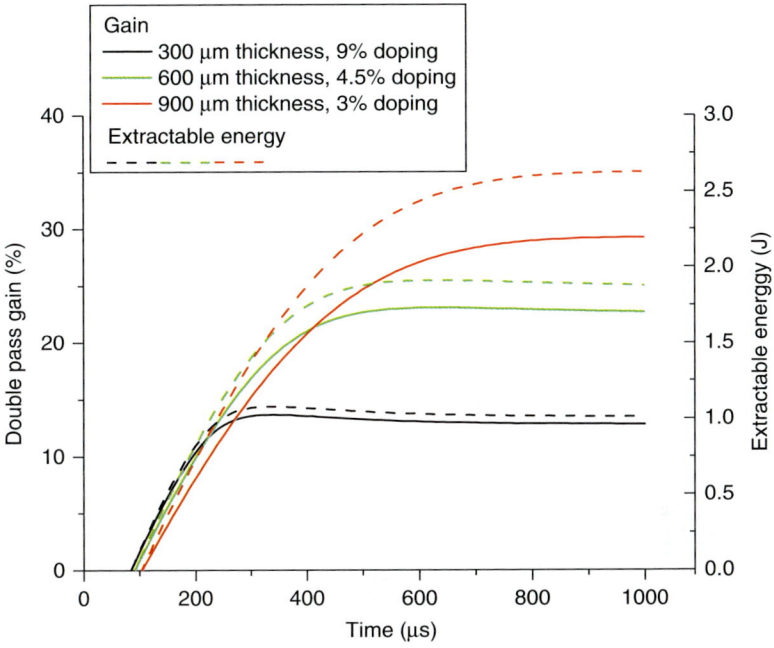

Figure 10.15 Temporal development of gain (solid lines) and extractable energy (dashed lines) with different thicknesses and doping concentrations. Applied pump power 16 kW, pump spot radius 9.2 mm.

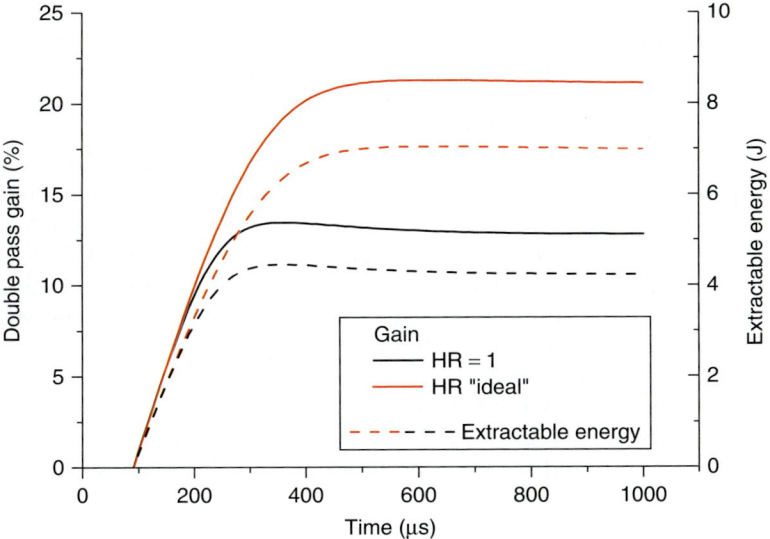

Figure 10.16 Temporal development of gain (solid lines) and extractable energy (dashed lines) with partial transmission of the ASE through the HR coating. Doping concentration 4.5 percent, thickness 600 μm, pump power 64 kW, pump spot radius 18.4 mm.

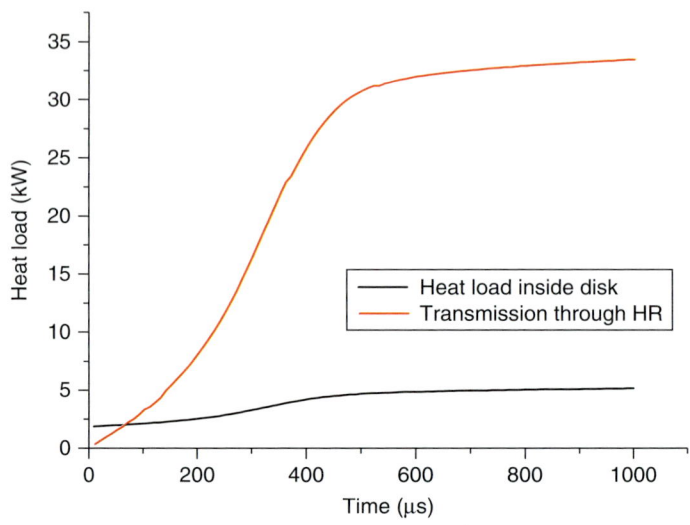

FIGURE 10.17 Heat load inside the disc and additional heat load at the HR coating due to partial ASE absorption and transmission. Doping concentration 4.5 percent, thickness 600 μm, pump power 16 kW, pump spot radius 9.8 mm.

larger than the critical angle (the technical realization would be also difficult, but it is possible to design coatings with modulated reflection spectra, reaching a similar average reflectivity). To enhance the visibility of the differences, for these calculations 64 kW peak pump power and a larger pump spot were used. The extractable energy is enhanced from 4 J to 7 J with this coating and the gain reaches a similar magnitude as in the low-power configuration in Fig. 10.15.

But there is also a drawback of this design, as the "loss" of 25 percent must either be absorbed in the coating or transmitted and then, depending of the mounting and cooling design, converted to heat somewhere in the contact layers. As shown in Fig. 10.17, this will produce an additional heat load of about 35 kW. In Sec. 10.5.2, we assumed that about 76 percent of the absorbed pump power will be captured inside the disc as fluorescence—and more than half of it will be eventually transformed into heat with this "ideal" coating. This is even more than what we have assumed in Sec. 10.5.2 for first calculations. In this case it can be beneficial to use a different mounting design for the disc, avoiding the heat sink and instead directly cooling the disc with water.

10.5.12 ASE-Limit

The method above can be used for numerical modeling of the influence of ASE on the performance of thin-disc lasers. To find a scaling limit without extensive numerical calculations, some simplifications are useful. We will assume for the further calculations that the density of excited ions N_2 and the temperature T are constant in the whole active region. Additionally, we will assume that all fluorescence is emitted at the laser wavelength. Therefore also γ_λ is constant and we can calculate

$$d\Phi(\phi,\vartheta) = \sin\vartheta d\vartheta d\phi \frac{N_2}{4\pi\tau} \frac{\exp(\gamma s_{max})-1}{\gamma} \qquad (10.26)$$

If we further assume that we are in the center of the active region, the dependency to ϕ will vanish and we get

$$\Phi = \frac{N_2}{2\tau\gamma} \int_0^\pi (\exp(\gamma s_{max})-1)\sin\vartheta d\vartheta \qquad (10.27)$$

The integration will not produce an analytic result, but it is useful to assume a constant s_{max} and we will get

$$\Phi = \frac{N_2}{\tau\gamma}(\exp(\gamma s_{max})-1) \qquad (10.28)$$

and

$$\begin{aligned}\dot{N}_2 &= Q - \frac{N_2}{\tau} - \gamma\frac{N_2}{\tau\gamma}(\exp(\gamma s_{max})-1) \\ &= Q - \frac{N_2}{\tau}\exp(\gamma s_{max})\end{aligned} \qquad (10.29)$$

This effect can be expressed as a reduced lifetime τ_{ASE}:

$$\tau_{ASE} = \tau\exp(-\gamma s_{max}) \qquad (10.30)$$

This approach was used to find a scaling limit for the maximum output power P_{max}, assuming as maximum integration distance the diameter of the active region, that is, $s_{max} = 2r_p$. The reduced lifetime in this case can also be written as[35]

$$\tau_{ASE} = \tau\exp\left(-\frac{2r_p}{h}g\right) \qquad (10.31)$$

with r_p the radius of the pump spot and g the single pass gain in the disc. Based on these assumptions, the maximum power is:

$$P_{max} = \frac{\lambda_p^2}{\lambda_l} \cdot \frac{27}{64\exp(2)} \cdot \frac{\sigma_{em}(\lambda_l,T)\cdot(1-f_{abs}(\lambda_l,T)f_{em}(\lambda_p,T))}{2\pi\hbar c} \cdot \frac{C^2}{\beta^3}$$

$$(10.32)$$

Besides the results on maximum power or efficiencies which are presented in the mentioned paper[35], there is especially one important feature in Eq. (10.32): the strong influence of the thermal load parameter C (cf. Sec. 10.5.1) and of the internal loss β inside the resonator. This relation also holds for efficiency calculations.

Based on considerations similar to the ideas presented in Sec. 10.5.10 and Fig. 10.13, a slightly different expression for the reduced lifetime was found:[36]

$$\tau_{ASE} \approx \frac{\tau}{\exp(2g) + 2g\,\mathrm{Ei}(2gr_p/h) + 2g\,\mathrm{Ei}(2g)} \qquad (10.33)$$

This can be approximated by

$$\tau_{ASE} \sim \tau \frac{r_p}{h} \exp\left(-\frac{2r_p}{h} g\right) \qquad (10.34)$$

With the lifetime reduction from Eq. (10.33), a similar dependence of maximum output power from the internal loss and the thermal load parameter as in Eq. (10.32) can be derived. The results are clearly beyond actually possible or planned thin-disc designs, but even more than 20 MW from one disc seem feasible if the internal loss β inside the resonator would be reduced to 0.25 percent—but requiring a pump spot diameter of ~5.5 m.

The achievable efficiencies are small (less than 10%), but following both papers,[35,36] higher efficiencies are possible with a slightly different optimization. With an internal loss of 0.25 percent, 1 MW laser power will be possible with nearly 50 percent optical-optical efficiency. Only a pump spot diameter of 20 cm will be required for this laser power.

10.6 Thin-Disc Laser in Continuous-Wave Operation

10.6.1 High Average Power

Very high laser output power can be achieved from one single disc by increasing the pump spot diameter while keeping the pump power density constant.[37,38] The highest output power reported for a single disc is 6.5 kW.[39]

Figure 10.18 shows one example for high output power with high efficiency from a single disc (Trumpf Laser). More than 5.3 kW of power has been achieved with a maximum optical efficiency of more than 65 percent. This high efficiency of the thin-disc laser results in a very high electrical efficiency for the total laser system—greater than 25 percent for industrial lasers with 8-kW output power and a beam propagation factor M^2 of less than 24.

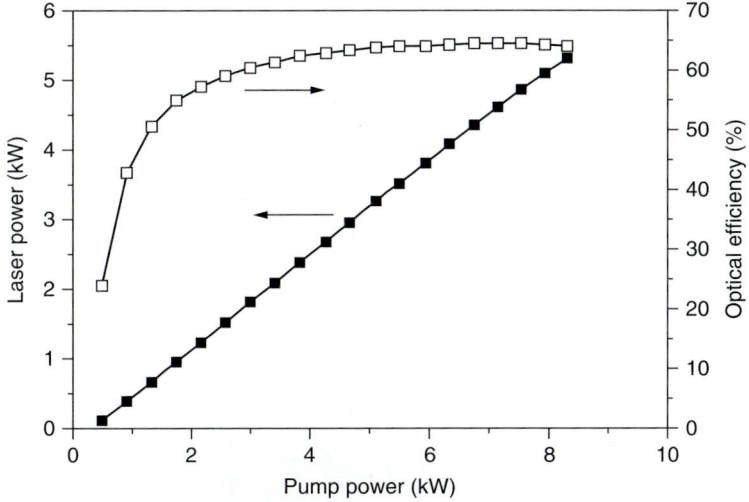

FIGURE 10.18 Output power and optical efficiency from a single disc. (*Courtesy of Trumpf Lase*)

An alternative way to scale the output power is to use several discs in one resonator. Figure 10.19 shows the design of a laboratory setup for high beam quality in which four discs are coupled together in one resonator. Figure 10.20 shows the output power and the optical efficiency of such a laser as function of the pump power. The high beam quality is made possible by the concept of neutral gain modules. For this concept, the discs are optically combined to modules which have a minimum effective optical length and refractive power.[40] Figure 10.21 shows a further example of power scaling by combination of several disks in one resonator, delivering more than 20 kW of output power, but with reduced beam quality.

10.6.2 Fundamental Mode, Single Frequency and Second Harmonic Generation (SHG)

High-power thin-disc lasers in the kilowatt-power range are typically operated with a beam propagation factor (beam quality) M^2 of about 20 (i.e., the laser beam's focusability is 20 times worse compared the theoretical limit $M^2 = 1$). This is sufficient for the typical demands of welding or cutting applications. Beyond this beam quality, the thin-disc laser design also offers the possibility to operate high-power lasers in the fundamental mode ($M^2 = 1$)[31,41–43] due to the disc's small thermal effects and small optical distortions.

Using a resonator design which has a stable fundamental mode diameter of 70 to 80 percent of the pump spot diameter, it is possible to achieve high laser output power with high optical efficiency. This

Solid-State Lasers

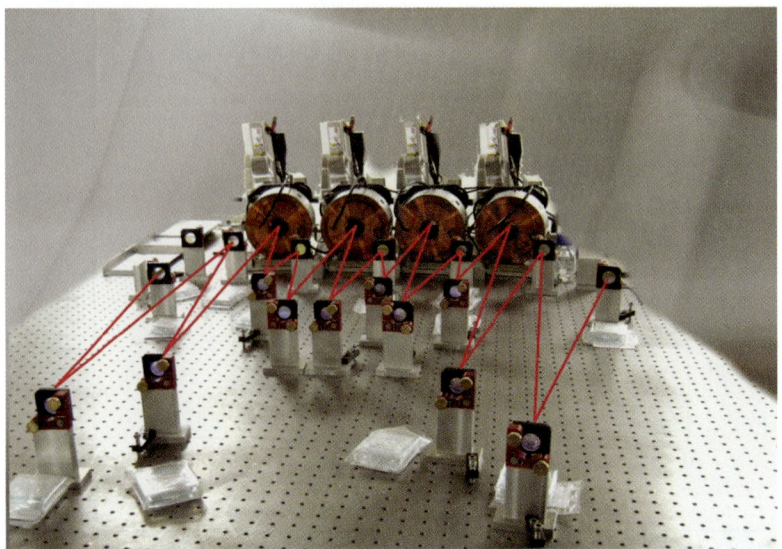

FIGURE 10.19 Artist's view of a setup for combining four discs in one resonator.

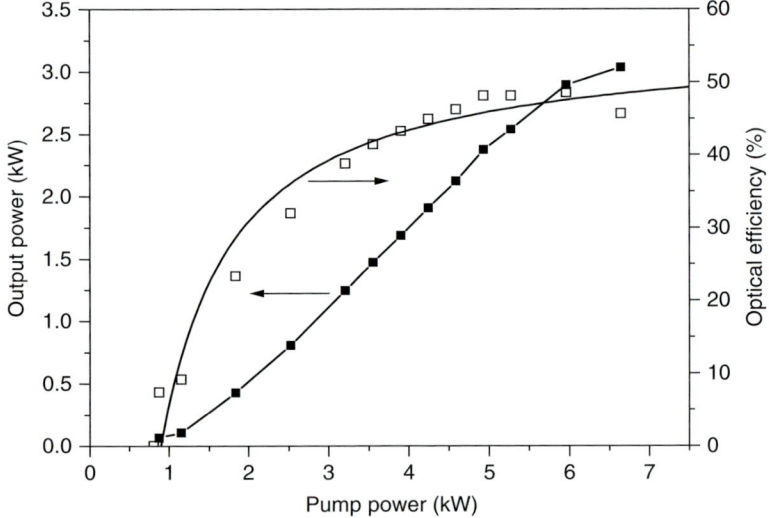

FIGURE 10.20 Output power and optical efficiency with 4 discs, beam quality $M^2 \approx 6$.

relationship between pump spot and fundamental mode is an optimization that concerns phase distortions and mode overlap. In Fig. 10.8, the intensity distribution of a fundamental mode with a mode diameter that is 70 percent of the pump spot diameter is sketched. Inside the pump spot, the remaining phase distortions are smaller than 400 nm and inside the mode diameter even less than

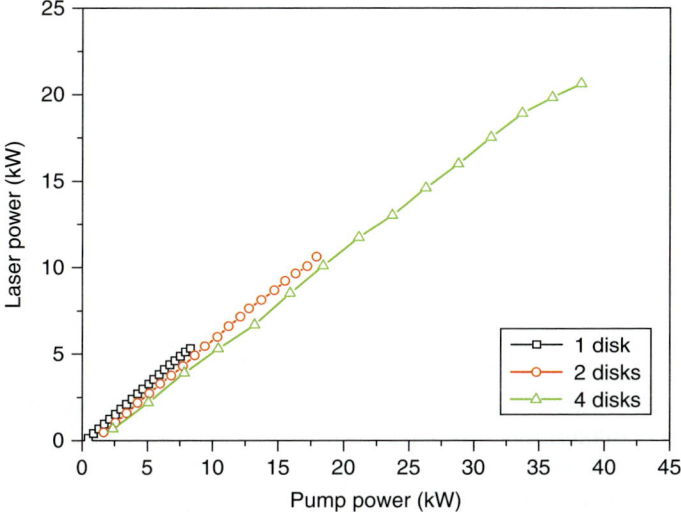

FIGURE 10.21 Output power with one, two, and four discs, beam quality $M^2 \approx 24$. (*Courtesy of Trumpf Lase*)

30 nm. Simultaneously, the losses due to absorption in the not-pumped region are negligible, and higher modes are effectively suppressed by the absorption in the not-pumped region ("gain aperture").

Figure 10.22 shows the result of a disc operated with more than 500 W laser power and an M^2 of better than 1.6. The optical efficiency of this laser was higher than 35 percent. Even higher

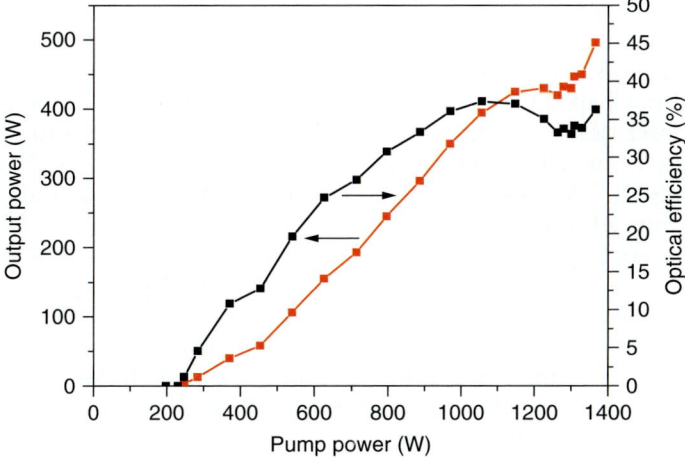

FIGURE 10.22 Output power and optical efficiency for operation close to fundamental mode operation, $M^2 \approx 1.6$.

laser power levels with nearly fundamental mode properties will be possible in future.

With the potential of very high output power levels for the fundamental mode, it is possible to operate thin-disc lasers also in single frequency operation.[43,44] To achieve this, it is necessary to use a birefringent filter and one or two uncoated etalons inside the fundamental mode resonator. With such resonators, up to 98-W single frequency power has been demonstrated.[43] In addition, the laser's wavelength can be tuned over a wide spectral range (1000–1060 nm for Yb:YAG) by tuning the birefringent filter.[5,43–46]

Another interesting feature is resonator's internal doubling of the laser frequency for covering the visible spectral range with high efficiency. This can be successfully demonstrated with different laser materials. With Yb:YAG the wavelength tunability between 500 and 530 nm, maximum power around 515 nm could be shown; 50-W green output power is commercially available. For Nd:YVO$_4$[18,19] more than 12 W can be demonstrated at 532-nm wavelength and more than 3 W can be demonstrated at 457 nm (doubling of the quasi-three-level transition at 914 nm). For Nd:YAG more than 1 W at 660 nm was achieved when doubling the 1320 nm transition.

10.7 Thin-Disc Laser in Pulsed Operation

In addition to the outstanding properties of the thin-disc laser's design for CW operation, it is also well suited for pulsed laser systems, especially if high average output power is demanded. Until recently, pulsed thin-disc laser systems had been developed and demonstrated for the nanosecond-, picosecond-, and femtosecond-pulse duration regime. All systems showed an excellent beam quality and a high efficiency.

In Ursula Keller's group at the ETH Zurich high average power fs-oscillators have been developed, which are described in more detail in Chap. 13.[47–51] It has been demonstrated that with the thin-disc laser design, high output powers are possible down to pulse durations of 220 fs, especially with the use of Yb:Lu$_2$O$_3$[52] and Yb:LuScO$_3$.[53] An exhaustive overview of possible laser materials and the achieved results with mode-locked thin-disc lasers can be found in a recent paper.[54] The mode locking of thin-disc lasers with semiconductor saturable absorbers (SESAM) is a very elegant approach to exploit the scaling behavior of both concepts, as the SESAM concept also can be scaled by increasing the active area. This advantage is already transformed to an industrial product, a mode-locked thin-disc laser with 800 fs pulse duration and 50 W output power available from *Time–Bandwidth products*.

In the following sections the results for q-switched lasers, cavity dumped lasers and for pulse laser amplifiers are discussed in more detail.

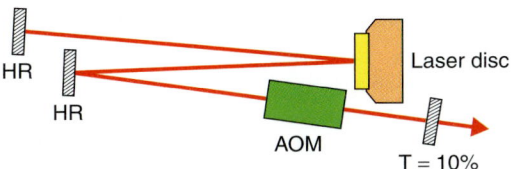

FIGURE 10.23 Resonator design of the q-switched laser. AOM: acousto-optic modulator. (*Courtesy of Dausinger und Giesen Gmb*)

10.7.1 Q-Switched Operation of the Thin-Disc Laser

An easy way to reach high pulse energies with the thin-disc concept is active q-switched operation.[55,56] Figure 10.23 shows a possible design for a q-switched thin-disc laser. The resonator is folded so that a short resonator length could be realized with a large mode area in the disc for fundamental mode operation. Q-switching is performed by a quartz acousto-optic modulator (AOM). Figure 10.24 shows the pulse energy as function of the repetition rate for the active laser material Yb:YAG. Stable operation can be achieved with repetition rates up to 13 kHz; for higher repetition rates bifurcations of the pulse energy may be observed. The maximum pulse energy is 18 mJ at 1-kHz repetition rate and the maximum average power is 64 W at 13 kHz, which corresponds to an optical efficiency of 34 percent. The beam propagation factor M^2 was better than 2 in all cases.[56]

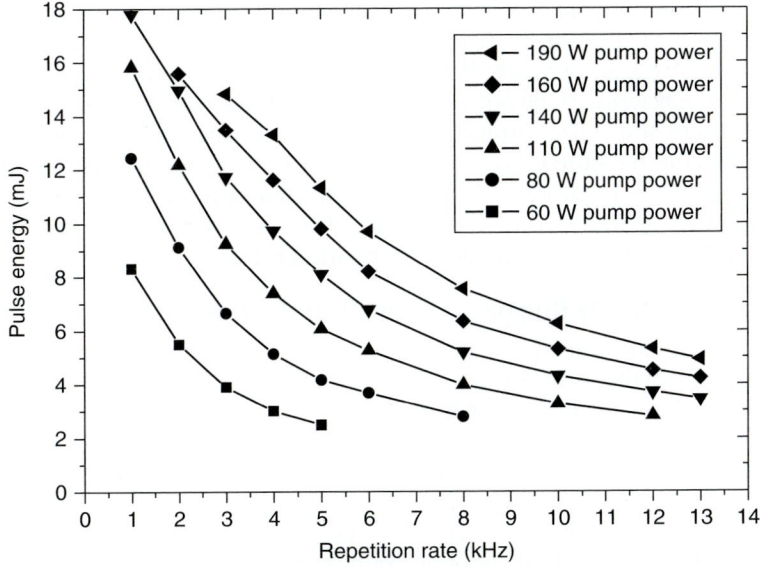

FIGURE 10.24 Pulse energy of the q-switched thin-disc laser as function of the repetition rate for different pump power levels.

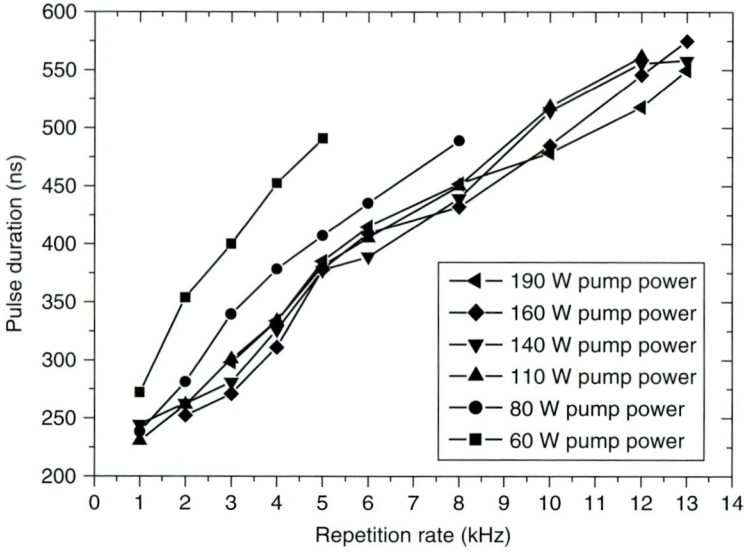

FIGURE 10.25 Pulse duration of the q-switched thin-disc laser as a function of the repetition rate for different pump power levels.

Figure 10.25 shows the pulse length of the pulses as a function of the pulse repetition rate for different pump power levels. At low repetition rates, the pulse duration is about 250 ns, whereas for higher repetition rates, the pulses become longer—up to 570 ns at a 13-kHz repetition rate. The reasons for these long pulses are the length of the resonator (840 mm for fundamental mode operation) and the relatively low gain per roundtrip of the disc—and, hence, the relatively high reflectance of the outcoupling mirror. These restrictions in repetition rate and pulse duration (limited to pulse durations longer than 200 ns for the setup used) could be overcome by using thin-disc amplifiers, which are described in Sec. 10.7.3. Alternatively, also the cavity dumped operation described in Sec. 10.7.2 is a very flexible scheme concerning pulse durations and repetition rates.

10.7.2 Cavity-Dumped Operation of the Thin-Disc Laser

Several possibilities exist for extracting the energy that is stored inside a cavity. In the setup shown in Fig. 10.26 either the thin film polarizer can be used as outcoupling mirror or the second harmonic generation (SHG) in the SHG crystal can be used to extract the energy from the cavity.[57,58]

Applying the full quarter-wave voltage to the Pockels cell, the outcoupling can be switched to 100 percent, creating pulses of some tens of nanoseconds. By applying only a small voltage, one can reach a kind of "cavity leaking" instead of cavity dumping with longer pulses. In this case, the pulse duration and pulse energy can be controlled very

Thin-Disc Lasers

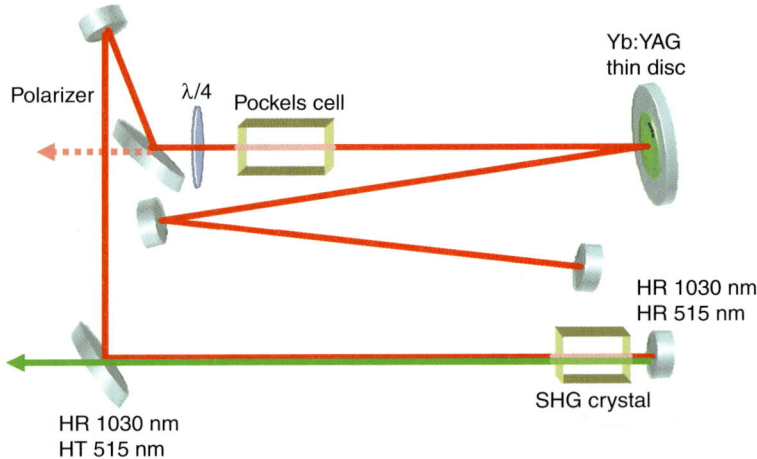

FIGURE 10.26 Concept of a pulsed thin-disc laser with SHG at 515 nm. SHG: second harmonic generation, HT: high transmission. (*Courtesy of Trump*)

precisely. Additionally, the intracavity power can be monitored with a photodiode behind an HR mirror to control the amplification time. This combination of outcoupling control and amplification control enables a versatile optimization of pulse duration, repetition rate and efficiency. The pulse duration can be varied between a little bit more than the cavity roundtrip time (~10 ns) and few µs. Additionally, it is possible to suppress instabilities at high repetition rates.

This system can be either optimized for SHG or for the fundamental wavelength. With an optimization for SHG, the maximum SHG output power at a repetition rate of 100 kHz was 700 W with a pulse duration of 300 ns (cf. Fig. 10.27). In this case, the duration of the SHG pulse is controlled by dumping the IR energy inside the cavity.

It is also possible to optimize such a cavity-dumped system for highest IR energy. With a similar concept, omitting the SHG crystal, 280 mJ at a repetition rate of 100 Hz and with a pulse duration of 25 ns have been demonstrated with $M^2 < 1.3$, using quasi-CW pumping.[58]

10.7.3 Amplification of Nanosecond, Picosecond, and Femtosecond Pulses

To produce shorter pulses with high pulse energy, a setup consisting of a seed oscillator followed by a regenerative amplifier is used.[58–61] The scheme of such a setup is shown in Fig. 10.28. The oscillator generates pulses with the desired properties (pulse length and wavelength), which are amplified to the desired energy in the thin-disc amplifier. The thin-disc amplifier in this scheme operates independently of the seed laser and is able to amplify any incoming pulse that

258 Solid-State Lasers

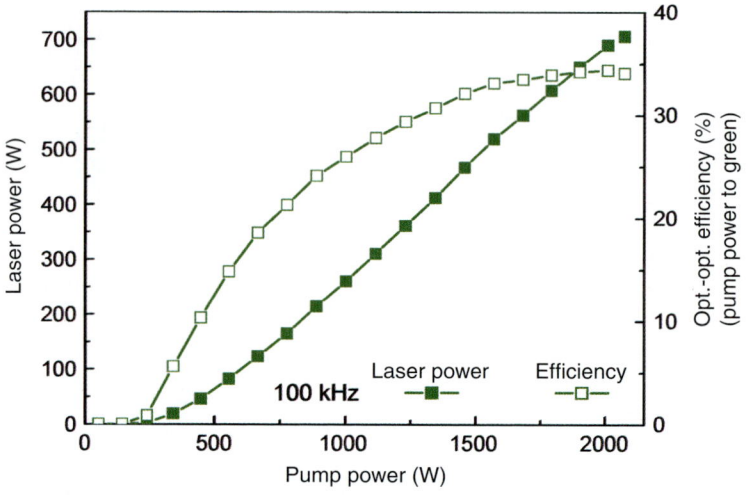

FIGURE 10.27 Output power of a cavity-dumped system with internal frequency conversion at 515 nm in multimode operation. (*Courtesy of Trumpf*)

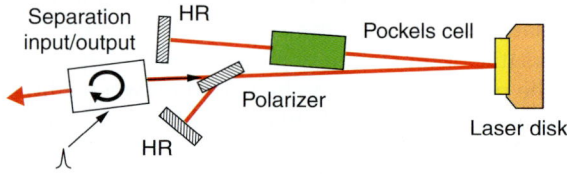

FIGURE 10.28 Schematic setup of an oscillator-amplifier system for pulse generation and amplification. (*Courtesy of Dausinger und Giesen Gmbh*)

has the right wavelength and a pulse duration shorter than the round-trip time of the amplifier resonator. To amplify the picosecond or femtosecond pulses, a seed oscillator with the appropriate pulse length (i.e., slightly shorter than the desired pulse duration after amplification) is used.

The key components of the regenerative amplifier are the disc as amplifying medium and the polarizer in combination with the Pockels cell as polarization switch for switching in and out the seed pulses and the amplified pulses, respectively. Additionally, for amplifying fs-pulses, Gires-Tournois interferometer (GTI) mirrors can be implemented to compensate for the positive group-velocity dispersion (GVD) of the Pockels cell and the optical elements during one round trip of the pulse inside the resonator.

Because of the small gain of the thin-disc in the amplifier (10–40 percent per roundtrip, depending on the operational conditions), the pulses need between 50 and 200 amplifying roundtrips in the amplifier to achieve the desired pulse energy. Therefore, it is very important to design the resonator in a way that the resonator's internal losses are as small as possible. Otherwise the efficiency is reduced dramatically.

Using a picosecond oscillator (pulse duration 1.8 ps) as a seed laser allows for the amplification of picosecond pulses up to nearly 5 mJ energy at 1 kHz repetition rate and up to 1 mJ at 20 kHz.[62–64] Due to gain narrowing in the amplifier, the pulse duration of the amplified pulses is extended to 4 ps. The beam is also for these pulses nearly diffraction limited. Using higher repetition rates, about 0.3 mJ at a repetition rate of 200 kHz were reached.[65] The average laser power is approximately independent of the repetition rate at such high repetition rates.

Also Yb:K(WO$_4$)$_2$ is used as the laser active medium in thin-disc pulse amplifiers. One advantage of this material is its much broader gain spectrum compared to Yb:YAG. Therefore even shorter pulses can be generated and amplified. In one experiment[66] GTI mirrors were used for keeping the pulses short during the amplification. As a seed laser, a Yb:Glass oscillator that delivered pulses with a pulse length of about 500 fs and an energy of about 1 nJ was used. These pulses were amplified to more than 100 µJ with a pulse duration of less than 900 fs.

It is also possible to omit the GTI mirrors. Using seed pulses with a duration of 270 fs and no compensation of the increasing pulse duration in the amplifier, but compressing the pulses after amplification with a grating compressor, a pulse length of 250 fs was demonstrated at an output energy of 116 µJ and at a repetition rate of 40 kHz, using Yb:KY(WO$_4$)$_2$.[67] It is remarkable that this result was achieved without chirped pulse amplification (CPA). Due to the large beam diameter inside the amplifier (2 mm in the Pockels cell) the pulse lengthening by non-linear effects could be limited to pulse duration values below 1 ps.

This result was further scaled with an Yb:KLu(WO$_4$)$_2$ thin-disc. Figure 10.29 shows the pulse energies for different pump powers at 50 kHz repetition rate. Up to 395 µJ could be reached with this setup. Figure 10.30 shows the autocorrelation traces for different pulse energies. With a pulse energy of 315 µJ, the pulse duration after amplification and compression is significantly shorter than the seed pulse duration. This is due to the spectral broadening of the pulse in the BBO crystal used for the Pockels cell.

10.7.4 High Pulse Energy Thin-Disc Lasers

It is also possible to generate higher pulse energies with thin-disc amplifiers. As already mentioned, 280 mJ at 100 Hz and 25 ns were reached with cavity dumping. With a regenerative amplifier, also 240 mJ at 100 Hz and 8 ns were realized.[58]

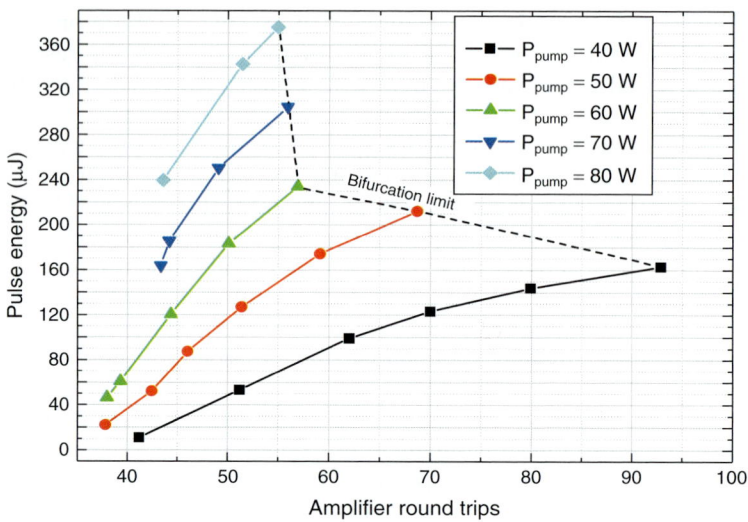

FIGURE 10.29 Pulse energy of the amplified pulses, 50 kHz repetition rate. (*Courtesy of Dausinger und Giesen Gmbh*)

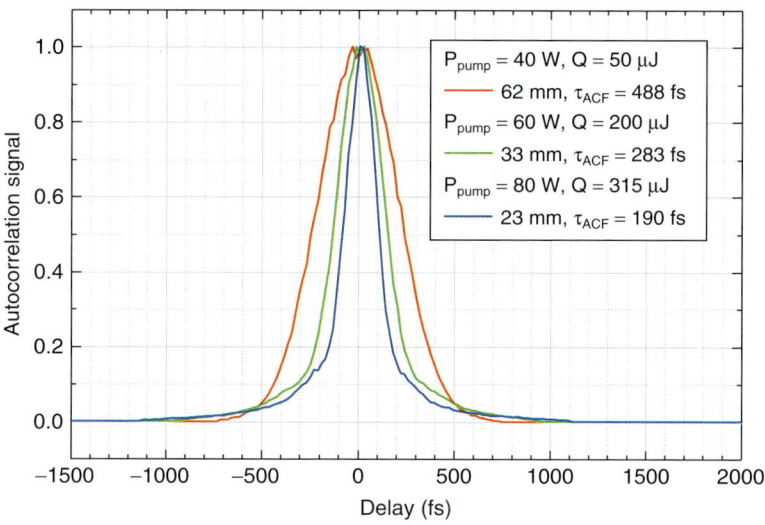

FIGURE 10.30 Autocorrelation trace of the amplified pulses at a repetition rate of 50 kHz with optimized adjustment of compressor gratings. (*Courtesy of Dausinger und Giesen Gmbh*)

Further scaling of the pulse energy is possible by combining regenerative amplifiers with geometrical multipass amplifiers. Such a multipass concept was developed to realize a system which is able to deliver pulses "on demand" with a very short reaction time. Starting with a specially designed q-switch thin-disc laser, it uses a multipass with 12 double passes through the disc, achieving an amplification of 4 to 6, depending on the starting pulse energy. The system delivers up to 48 mJ with a maximum delay of 400 ns after an electronic trigger signal.[68]

A similar geometrical multipass amplifier is used in a high energy CPA thin-disc system actually under development. The goal is to produce a pulse energy of about 1 J with a repetition rate of 100 Hz and a pulse duration of less than 5 ps. Actually, pulses stretched to 2 ns were amplified by a combination of a regenerative thin-disc amplifier and a multipass thin-disc amplifier to a pulse energy of 320 mJ; the compressibility to ~ 2 ps was demonstrated with pulse energies of ~180 mJ.[69]

From these results it can be seen that the thin-disc laser design is able to generate and amplify pulses up to high energies for all pulse durations between femtoseconds and nanoseconds.

10.8 Industrial Realizations

To date, 20 licenses of the thin-disc laser design have been sold to companies worldwide. Some of these companies already offer thin-disc lasers on the market.

Rofin-Sinar Laser GmbH offers materials processing lasers with output powers from 750 W to 4 kW; these lasers are coupled into an optical fiber with 200 µm core diameter (numerical aperture [N.A.] = 0.2). Trumpf Laser GmbH + Co KG offers thin-disc lasers from 250 W to 16 kW output power, which are also coupled into fibers with 200-µm core diameter. Because of the good beam quality of these lasers (comparable to that of CO_2 lasers in the same power range), these lasers are excellent for high power materials processing.

Jenoptik Laser, Optik, Systeme GmbH is selling high-power thin-disc lasers in the green, red, and blue spectral range (up to 8 W green, 0.8 W blue, and 2 W red). It is also selling q-switched thin-disc lasers with up to 40 W for marking and drilling.

These examples show that the thin-disc laser has successfully found its way from the laboratory to real applications in industry. More companies will enter the market within the next few years.

10.9 Summary

All the results in this chapter show the potential of the thin-disc laser design for high power lasers that are easy to build and operate. In particular, the simple scaling laws of the thin-disc laser design and the easy adaptation to many laser active materials will support the implementation of the thin-disc laser design to available industrial laser systems.

The thin-disc laser is an innovative laser concept that allows to build diode-pumped solid-state lasers with highest output powers, highest efficiency and excellent beam quality, simultaneously. Nearly all operational modes of solid-state lasers, including continuous wave, pulsed operation with pulse durations between femtoseconds and nanoseconds, as well as laser amplifiers can be built with this design with better properties as compared with other designs.

In future, new materials will be investigated with the goal of further increasing the power, energy, and beam quality. Laser output powers of more than 10 kW and energies of more than 1 J will be possible. New materials will open new markets for new wavelengths.

Acknowledgments

Our special thanks go to some 20 scientists, the technicians and numerous students who have worked on the thin-disc laser design since 1992 making possible the results summarized in this paper. Without their enormous commitment these results were not possible.

References

1. Lacovara, P., Choi, H. K., Wang, C. A., Aggarwal, R. L., and Fan, T. Y., "Room-temperature diode-pumped Yb:YAG laser," *Opt. Lett.*, 16: 1089–1091, 1991.
2. DeLoach, L. D., Payne, S. A., Krupke, W. F., Smith, L. K., Kway, W. L., Tassano, J. B, and Chai, B. H. T., "Laser and Spectroscopic Properties of Yb-Doped Apatite Crystals" *Advanced Solid State Lasers*, Washington, DC, OSA, 1993.
3. Giesen, A., Hügel, H., Voss, A., Wittig, K., Brauch, U., and Opower, H., "Scalable Concept for Diode-Pumped High-Power Solid-State Lasers," *Appl. Physics B*, 58: 365, 1994.
4. Giesen, A., Hügel, H., Voss, A., Wittig, K., Brauch, U., and Opower, H., "Diode-Pumped High-Power Solid-State Laser: Concept and First Results with Yb:YAG," *Advanced Solid State Lasers*, Washington, DC, OSA, 1994T.
5. Brauch, U., Giesen, A., Karszewski, M., Stewen, C., and Voss, A., "Multi Watt Diode-Pumped Yb:YAG Thin Disc Laser Continuously Tunable Between 1018 nm and 1053 nm," *Opt. Lett.*, 20 (7): 713, 1995.
6. Erhard, S., Giesen, A., Karszewski, M., Rupp, T., Stewen, C., Johannsen, I., and Contag, K., "Novel Pump Design of Yb:YAG Thin-Disc Laser for Operation at Room Temperature with Improved Efficiency," *OSA Trends in Optics and Photonics* (Advanced Solid-State Laser, vol. 34), eds. M. M. Fejer, H. Injeyan, and U. Keller, Washington, DC, OSA, 2000.
7. Erhard, S., Karszewski, M., Stewen, C., Contag, K., Voss, A., and Giesen, A., "Pumping Schemes for Multi-kW Thin Disc Lasers," *OSA Trends in Optics and Photonics* (Advanced Solid-State Laser, vol. 34), eds. M. M. Fejer, H. Injeyan, and U. Keller, Washington, DC, OSA, 2000.
8. Tünnermann, A., Zellmer, H., Schöne, W., Giesen, A., and Contag, K., "New Concepts for Diode-Pumped Solid-State Lasers," *High Power Diode Lasers: Fundamentals, Technology, Applications* (Topics in Applied Physics, vol. 78), ed. R. Diehl, Springer Verlag, Berlin, Heidelberg, 2000.
9. Giesen, A., Hollemann, G., Johannsen, I. "Diode-pumped Nd:YAG thin-disc laser," *Conference on Lasers and Electro-Optics* (CLEO '99), Washington, DC, OSA, 1999.
10. Johannsen, I., Erhard, S., Müller, D., Stewen, C., Giesen, A., and Contag, K., "Nd:YAG Thin Disc Laser," *OSA Trends in Optics and Photonics* (Advanced Solid-State Laser, vol. 34), eds. M. M. Fejer, H. Injeyan, and U. Keller, Washington, DC, OSA, 2000.

11. Gao, J., Speiser, J., and Giesen, A., "25-W Diode-Pumped Continuous-Wave Quasi-Three-Level Nd:YAG Thin Disc Laser," *Advanced Solid-State Photonics Technical Digest*, Washington, DC, OSA, 2005.
12. A. Diening, B.-M. Dicks, E. Heumann, G. Huber, A. Voß, M. Karszewski, A. Giesen, "High Power Tm:YAG Thin-Disc Laser," *OSA Technical Digest Series, Conference on Lasers and Electro Optics* (CLEO '98). Washington, DC, OSA, 1998.
13. N. Berner, A. Diening, E. Heumann, G. Huber, A. Voss, M. Karszewski, A. Giesen, "Tm:YAG: A Comparison between endpumped Laser-rods and the 'Thin- Disc'-Setup," *OSA Trends in Optics and Photonics* (Advanced Solid-State Laser, vol. 26), eds. M.M. Fejer, H. Injeyan, and U. Keller, Washington, DC, OSA, 1999.
14. Schellhorn, M., "Performance of a Ho:YAG Thin-Disc Laser Pumped by a Diode-Pumped 1.9 μm Thulium Laser," *Applied Physics B*, 85 (4): 549–552, 2006.
15. Kränkel, C., Peters, R., Petermann, K., and Huber, G., "High Power Operation of Yb:LuVO4 and Yb:YVO4 Crystals in the Thin Disc Laser Setup," *Advanced Solid-State Photonics 2007 Technical Digest*, Washington, DC, OSA, 2006.
16. Peters, R., Kränkel, C., Petermann, K., Huber, G., "Power Scaling Potential of Yb:NGW in Thin-Disc Laser Configuration," *Applied Physics B*, 91 (1): 25–28, 2008.
17. Peters, R., Kränkel, C., Petermann, K., and Huber, G., "Thin Disc Laser Operation of Yb^{3+} Doped $NaGd(WO_4)_2$," *Advanced Solid-State Photonics 2007 Technical Digest*, Washington, DC, OSA, 2006.
18. Koch, R., Hollemann, G., Clemens, R., Voelckel, H., Giesen, A., Voss, A., Karszewski, M., and Stewen, C., "Effective Near Diffraction Limited Diode Pumped Thin Disc $Nd:YVO_4$ Laser," *OSA Technical Digest Series, Conference on Lasers and Electro Optics* (CLEO '97), Washington, DC, OSA, 1997.
19. R. Koch, G. Hollemann, R. Clemens, H. Voelckel, A. Giesen, "Near Diffraction Limited Diode Pumped Thin Disc $Nd:YVO_4$ Laser," *LASER '97* (SPIE Proc. Vol. 3097), Bellingham, WA, SPIE, 1997.
20. Gao, J., Larionov, M., Speiser, J., Giesen, A., Douillet, A., Keupp, J., Rasel, E. M., and Ertmer, W., "$Nd:YVO_4$ Thin Disc Laser with 5.8 Watts Output Power at 914 nm," *OSA Trends in Optics and Photonics (TOPS), vol. 73*, Conference on Lasers and Electro-Optics (CLEO 2002) Technical Digest, Postconference Edition, Washington, DC, OSA, 2002.
21. Pavel, N., Kränkel, C., Peters, R., Petermann, K., Huber, G., "In-band pumping of Nd-vanadate thin-disc lasers," *Applied Physics B*, 91 (3): 415–419, 2008.
22. Larionov, M., Gao, J., Erhard, S., Giesen, A., Contag, K., Peters, V., Mix, E., Fornasiero, L., Petermann, K., Huber, G., Aus der Au, J., Spühler, G. J., Brunner, F., Pascotta, R., Keller, U., Lagatsky, A. A., Abdolvand, A., and Kuleshov, N. V., "Thin Disc Laser Operation and Spectroscopic Characterization of Yb-Doped Sesquioxides," *Advanced Solid-State Lasers* (Trends in Optics and Photonics, vol. 50), ed. C. Marshall, Washington, DC, OSA, 2001.
23. Peters, R., Kränkel, C., Petermann, K., Huber, G., "Broadly Tunable High-Power $Yb:Lu_2O_3$ Thin-Disc Laser with 80% Slope Efficiency," *OPTICS EXPRESS*, 15 (11): 7075–7082, 2007.
24. Kränkel, C., Johannsen, J., Peters, R., Petermann, K., Huber, G., "Continuous-wave High Power Laser Operation and Tunability of $Yb:LaSc_3(BO_3)_4$ in Thin-Disc Configuration," *Applied Physics B*, 87 (2): 217–220, 2007.
25. Kränkel, C., Peters, R., Petermann, K., Loiseau, P., Aka, G., Huber, G., "Efficient Continuous-wave Thin-Disc Laser Operation of $Yb:Ca4YO(BO_3)_3$ in E parallel to Z and E parallel to X Orientations with 26 W Output Power," *J. Opt. Soc. Am. B*, 26 (7): 1310–1314, 2009.
26. Schepler, K. L., Peterson, R. D., Berry, P. A., and McKay, J. B., "Thermal Effects in Cr^{2+}:ZnSe Thin-Disc lasers," *IEEE J. Sel. Top. Quant. Elec.*, 11 (3): 713–720, 2005.
27. Marion, J., "Strengthened Solid-State Laser Materials," *Appl. Phys. Lett.*, 47 (7): 694, 1985.

28. Contag, K., Brauch, U., Erhard, S., Giesen, A., Johannsen, I., Karszewski, M., Stewen, C., and Voss, A., "Simulations of the Lasing Properties of a Thin Disc Laser Combining High Output Powers with Good Beam Quality," *Modeling and Simulation of Higher-Power Laser Systems IV* (SPIE Proc., vol. 2989), eds. U. O. Farrukh and S. Basu, Bellingham, WA, SPIE, 1997.
29. Contag, K., Karszewski, M., Stewen, C., Giesen, A., and Hügel, H., "Theoretical Modeling and Experimental Investigations of the Diode-Pumped Thin-Disc Yb:YAG Laser," *Quant. Electron.*, 29 (8): 697, 1999.
30. Contag, K., Erhard, S., and Giesen, A., "Calculations of Optimum Design Parameters for Yb:YAG Thin Disc Lasers," *OSA Trends in Optics and Photonics* (Advanced Solid-State Laser, vol. 34), eds. M. M. Fejer, H. Injeyan, and U. Keller, Washington, DC, OSA, 2000.
31. Contag, K., *Modellierung und numerische Auslegung des Yb:YAG-Scheibenlasers*, Herbert Utz Verlag, München, 2002.
32. Speiser, J., and Giesen, A., "Numerical Modeling of High Power Continuous Wave Yb:YAG Thin Disc Lasers, Scaling to 14 kW," *Advanced Solid-State Photonics 2007 Technical Digest*, Washington, DC, OSA, 2006.
33. Barnes, N.P., and Walsh, B.M., "Amplified spontaneous emission - application to Nd:YAG lasers," *IEEE J. Quantum Electron.*, 35: 101–110, 1999.
34. Speiser, J., "Thin-Disc Laser—Energy Scaling," *LASER PHYSICS*, 19 (2), 2009.
35. Kouznetsov, D., Bisson, J. F., Dong, J., Ueda, K. I., "Surface loss limit of the power scaling of a thin-disc laser," *J. Opt. Soc. Am. B*, 23: 1074, 2006.
36. Speiser, J., "Scaling of Thin-Disc Lasers—Influence of Amplified Spontaneous Emission," *J. Opt. Soc. Am. B*, 26 (1): 26–35, 2009.
37. Contag, K., Brauch, U., Giesen, A., Johannsen, I., Karszewski, M., Schiegg, U., Stewen, C., and Voss, A., "Multi-Hundred Watt CW Diode Pumped Yb:YAG Thin-Disc Laser," *Solid State Lasers VI* (SPIE Proc., vol. 2986), ed. R. Scheps, Bellingham, WA, SPIE, 1997.
38. Stewen, C., Contag, K., Larionov, M., Giesen, A., and Hügel, H., "A 1-kW CW Thin-Disc Laser," *IEEE J. Sel. Top. Quant. Electron.*, 6 (4): S650, 2000.
39. Lobad, A., Newell, T., and Latham, W., "6.5 kW, Yb:YAG Ceramic Thin-Disc Laser," presented at the Solid State Lasers XIX: Technology and Devices, San Francisco, CA, USA, 24 January 2010.
40. Mende, J., Spindler, G., Speiser, J., and Giesen, A., "Concept of Neutral Gain Modules for Power Scaling of Thin-Disc Lasers" *Applied Physics B*, 97 (2): 307–315, 2009.
41. Karszewski, M., Brauch, U., Contag, K., Erhard, S., Giesen, A,. Johannsen, I., Stewen, C., and Voss, A., "100 W TEM_{00} Operation of Yb:YAG Thin-Disc Laser with High Efficiency," *OSA Trends in Optics and Photonics* (Advanced Solid-State Laser, vol. 19), eds. W. R. Bosenberg and M. M. Fejer, Washington, DC, OSA, 1998.
42. Karszewski, M., Erhard, S., Rupp, T., and Giesen, A., "Efficient High-Power TEM_{00} Mode Operation of Diode-Pumped Yb:YAG Thin Disc Lasers," *OSA Trends in Optics and Photonics* (Advanced Solid-State Laser, vol. 34), eds. M. M. Fejer, H. Injeyan, and U. Keller, Washington, DC, OSA, 2000.
43. Stolzenburg, C., Larionov, M., Giesen, A., and Butze, F., "Power Scalable Single-Frequency Thin Disc Oscillator," *Advanced Solid-State Photonics 2005 Technical Digest*, Washington, DC, OSA, 2005.
44. Giesen, A., Brauch, U., Karszewski, M., Stewen, C., and Voss, A., "High Power Near-Diffraction-Limited and Single Frequency Operation of Yb:YAG Thin-Disc Laser," *OSA Trends in Optics and Photonics* (Advanced Solid-State Laser, vol. 1), eds. S. A. Payne and C. R. Pollock, Washington, DC, OSA, 1996.
45. Giesen, A., Brauch, U., Johannsen, I., Karszewski, M., Schiegg, U., Stewen, C., and Voss, A., "Advanced Tunability and High-Power TEM_{00}-Operation of the Yb:YAG Thin-Disc Laser," *OSA Trends in Optics and Photonics* (Advanced Solid-State Laser, vol. 10), eds. C. R. Pollock and W. R. Bosenberg, Washington, DC, OSA, 1997.
46. Karszewski, M., Brauch, U., Contag, K., Giesen, A., Johannsen, I., Stewen, C., and Voss, A., "Multiwatt Diode-Pumped Yb:YAG Thin Disc Laser Tunable

Between 1016 nm and 1062 nm," *Proc. 2nd International Conference on Tunable Solid State Lasers, Wroclaw, Poland, 1996* (SPIE Proc., vol. 3176), eds. W. Strek, E. Tukowiak, and B. Nissen-Sobocinska, Bellingham, WA, SPIE, 1996.

47. Hönninger, C., Zhang, G., Keller, U., and Giesen, A., "Femtosecond Yb:YAG Laser Using Semiconductor Saturable Absorbers," *Opt. Lett.*, 20 (23): 2402, 1995.
48. Paschotta, R., Aus der Au, J., Spühler, G. J., Morier-Genoud, F., Hövel, R., Moser, M., Erhard, S., Karszewski, M., Giesen, A., and Keller, U., "Diode-Pumped Passively Mode-Locked Lasers with High Average Power," *Appl. Phys. B.*, 70: S25, 2000.
49. Spühler, G. J., Aus der Au, J., Paschotta, R., Keller, U., Moser, M., Erhard, S., Karszewski, M., and Giesen, A., "High-Power Femtosecond Yb:YAG Laser Based on a Power-Scalable Concept," *OSA Trends in Optics and Photonics* (Advanced Solid-State Laser, vol. 34), eds. M.M. Fejer, H. Injeyan, and U. Keller, Washington, DC, OSA, 2000.
50. Aus der Au, J., Spühler, G., J., Südmeyer, T., Paschotta, R., Hövel, R., Moser, M., Erhard, S., Karszewski, M., Giesen, A., and Keller, U., "16.2 W Average Power from a Diode-Pumped Femtosecond Yb:YAG Thin Disc Laser," *Opt. Lett.*, 25: 859, 2000.
51. Brunner, F., Südmeyer, T., Innhofer, E., Paschotta, R., Morier-Genoud, F., Keller, U., Gao, J., Contag, K., Giesen, A., Kisel, V. E., Shcherbitsky, V. G., and Kuleshov, N. G., "240-fs Pulses with 22-W Average Power from a Passively Mode-Locked Thin-Disc Yb:KY(WO$_4$)$_2$ Laser," *OSA Trends in Optics and Photonics (TOPS), vol. 73*, Conference on Lasers and Electro-Optics (CLEO 2002), Technical Digest, Postconference Edition, Washington, DC, OSA, 2002.
52. Baer, C. R. E., Kränkel, C., Saraceno, C. J., Heckl, O. H., Golling, M., Sudmeyer, T., Peters, R., Petermann, K., Huber, G., and Keller, U., "Femtosecond Yb:Lu$_2$O$_3$ Thin-Disc Laser with 63 W of Average Power," *OPTICS LETTERS*, 34 (18): 2823–2825, 2009.
53. Baer, C. R. E., Kränkel, C., Heckl, O. H., Golling, M., Sudmeyer, T., Peters, R., Petermann, K., Huber, G., and Keller, U., "227-fs Pulses from a Mode-locked Yb:LuScO$_3$ Thin-Disc Laser," *OPTICS EXPRESS*, 17 (13): 10725–10730, 2009.
54. Südmeyer, T., Kränkel, C., Baer, C. R. E., Heckl, O. H., Saraceno, C. J., Golling, M., Peters, R., Petermann, K., Huber, G., and Keller, U., "High-power ultrafast thin-disc laser oscillators and their potential for sub-100-femtosecond pulse generation," *Applied Physics B*, 97 (2): 281–295, 2009.
55. Johannsen, I., Erhard, S., Giesen, A., "Q-switched Yb:YAG thin-disc laser," *OSA Trends in Optics and Photonics (TOPS), vol. 50*, Advanced Solid-State Lasers, Washington, DC, OSA, 2001.
56. Butze, F., Larionov, M., Schuhmann, K., Stolzenburg, C., and Giesen, A., "Nanosecond Pulsed Thin Disc Yb:YAG Lasers," *Advanced Solid-State Photonics 2004*, Technical Digest, Washington, DC, OSA, 2004.
57. Stolzenburg, C., Giesen, A., Butze, F., Heist, P., and Hollemann, G., "Cavity Dumped Intracavity Frequency Doubled Yb:YAG Thin Disc Laser at 100 kHz Repetition Rate," *Advanced Solid-State Photonics 2007 Technical Digest*, Washington, DC, OSA, 2006.
58. Stolzenburg, C., Voss, A., Graf, T., Larionov, M., and Giesen, A., "Advanced Pulsed Thin-Disc Laser Sources," *Proc. SPIE 6871*, 2008.
59. Hönninger, C., Johannsen, I., Moser, M., Zhang, G., Giesen, A., and Keller, U., "Diode Pumped Thin Disc Yb:YAG Regenerative Amplifier," *Appl. Phys. B*, 65: 423, 1997.
60. Hönninger, C., Zhang, G., Moser, M., Keller, U., Johannsen, I., and Giesen, A., "Diode Pumped Thin Disc Yb:YAG Regenerative Amplifier" *OSA Trends in Optics and Photonics* (Advanced Solid-State Laser, vol. 19), eds. W. R. Bosenberg and M. M. Fejer, Washington, DC, OSA, 1998.
61. Hönninger, C., Paschotta, R., Graf, M., Morier-Genoud, F., Zhang, G., Moser, M., Biswal, S., Nees, J., Mourou, G. A., Johannsen, I., Giesen, A., Seeber, W., and Keller, U., "Ultrafast Ytterbium-Doped Bulk Lasers and Laser Amplifiers," *Appl. Phys. B*, 69 (1): 3, 1999.

62. Müller, D., Erhard, S., and Giesen, A., "High Power Thin Disc Yb:YAG Regenerative Amplifier," *OSA Trends in Optics and Photonics* (Advanced Solid-State Lasers, vol. 50), ed. C. Marshall, Washington, DC, OSA, 2001.
63. Müller, D., Erhard, S., and Giesen, A., "Nd:YVO$_4$ and Yb:YAG Thin Disc Regenerative Amplifier," *OSA Trends in Optics and Photonics (TOPS), vol. 56*, Conference on Lasers and Electro-Optics (CLEO 2001), Technical Digest, Postconference Edition, Washington, DC, OSA, 2001.
64. Müller, D., Giesen, A., and Hügel, H., "Picosecond Thin Disc Regenerative Amplifier," XIV International Symposium on Gas Flow, Chemical Lasers, and High-Power Lasers, *Proc. SPIE,* 5120: 281–286, 2003.
65. Stolzenburg, C., and Giesen, A., "Picosecond Regenerative Yb:YAG Thin Disc Amplifier at 200 kHz Repetition Rate and 62 W Output Power," *Advanced Solid-State Photonics 2007 Technical Digest*, Washington, DC, OSA, 2006.
66. Beyertt, A., Müller, D., Nickel, D., and Giesen, A., "CPA-Free Femtosecond Thin Disc Yb:KYW Regenerative Amplifier with High Repetition Rate," *Advanced Solid-State Photonics 2004 Technical Digest*, Washington, DC, OSA, 2004.
67. Larionov, M., Butze, F., Nickel, D., and Giesen, A., "Femtosecond Thin Disc Yb:KYW Regenerative Amplifier with Astigmatism Compensation," *Advanced Solid-State Photonics 2007 Technical Digest*, Washington, DC, OSA, 2006.
68. Antognini, A., Schuhmann, K., Amaro, FD., Biraben, F., Dax, A., Giesen, A., Graf, T., Hansch, T.W., et al. "Thin-Disc Yb:YAG Oscillator-Amplifier Laser, ASE, and Effective Yb:YAG Lifetime," *IEEE J. Quant. Electron.,* 45 (8): 983–995, 2009.
69. Tümmler, J., Jung, R., Stiel, H., Nickles, PV., and Sandner, W., "High-repetition-rate chirped-pulse-amplification thin-disc laser system with joule-level pulse energy," *OPTICS LETTERS,* 34 (9): 1378–1380, 2009.

CHAPTER 11
Heat-Capacity Lasers

Robert M. Yamamoto

Principal Investigator, Lawrence Livermore National Laboratory, Livermore, California

Mark D. Rotter

Member of the Technical Staff, Lawrence Livermore National Laboratory, Livermore, California

11.1 Introduction

Over the past decade, scientists and engineers have actively engaged in developing the key technologies required to realize the performance potential of the heat-capacity laser (HCL). Several scientific institutions around the world, most notably the Chinese Academy of Sciences in Beijing[1] and Shanghai[2] and the Lawrence Livermore National Laboratory (LLNL) in California,[3] have been developers of this type of solid-state laser architecture. The fundamental feature of the heat-capacity laser that makes it unique from other solid-state lasers is the distinct separation of the lasing action from the cooling required of the laser gain media. Heat is stored in the laser gain media during the lasing process and is then cooled off-line, away from the laser beam line. This allows aggressive cooling methods of the laser gain media to be realized, because the cooling does not interfere with, and is independent of, the lasing process.

11.2 System Architecture

An important attribute of the HCL is that it lends itself to an extremely simple design of the laser cavity, utilizing a single-aperture architecture comprised of large laser gain media (slabs) pumped by arrays of high-power diode bars. Figure 11.1 shows the latest configuration of the heat-capacity laser used at LLNL.[4]

268 Solid-State Lasers

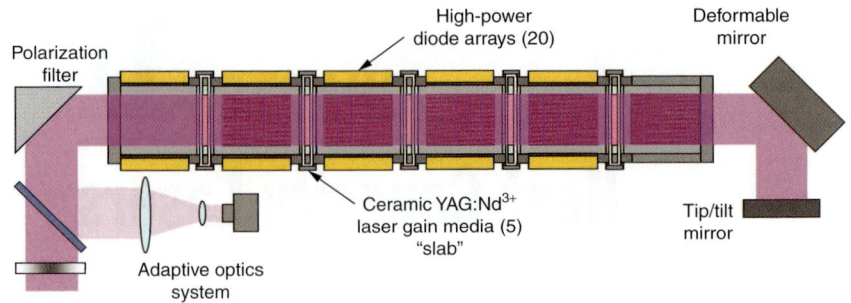

FIGURE 11.1 Heat-capacity laser architecture in use at Lawrence Livermore National Laboratory.

The HCL's basic building block is the laser gain module, which consists of a single slab pumped by four high-powered diode arrays, two on either side of the slab. Each diode array pumps the slab's adjacent face at a defined angle, providing uniform pump light intensity across the entire slab face. In this particular example, the laser gain media is a transparent ceramic Nd:YAG slab that is edge cladded with cobalt-doped gadolinium gallium garnet (GGG) to suppress amplified spontaneous emission. Figure 11.1 shows five gain modules, interlocked to form a compact cavity from which energy is extracted as a free-running resonator. An intracavity adaptive optic (AO) system, consisting of a wavefront sensor, a deformable mirror, a tip-tilt mirror, and a controller, maintains wavefront phase uniformity. The output laser beam wavelength for this HCL configuration is 1064 nanometers, and the diode light pump wavelength is 808 nm.

Two critical hardware components make up the HCL's gain module. The first is the high-powered diode arrays, which are used to pump the laser gain media. Each diode array comprises hundreds of relatively small, but very high-power, diode bars, which are carefully aligned and precision assembled to form a homogeneous diode array. Figure 11.2 shows a state-of-the-art high-powered diode array manufactured by Simmtec, Inc.[5] This diode array comprises 560 individual diode bars (seven rows of eight 10-bar tiles per row) and is capable of producing conservatively 84 kW of peak power. Electrical-to-optical conversion efficiency for a diode array is approximately 40 to 50 percent, giving rise to a significant cooling requirement to dissipate the waste heat generated (a cooling water flow rate of approximately 10 gallons per minute per diode array). In addition, the temperature of the cooling water must be maintained to within a few degrees to ensure that the wavelength of light being emitted by the diode bars is centered on the optimum absorption wavelength of the laser gain media. As a frame of reference, each high-powered diode array is about the size of a small loaf of bread, yet weighs about twice that of a standard bowling ball.

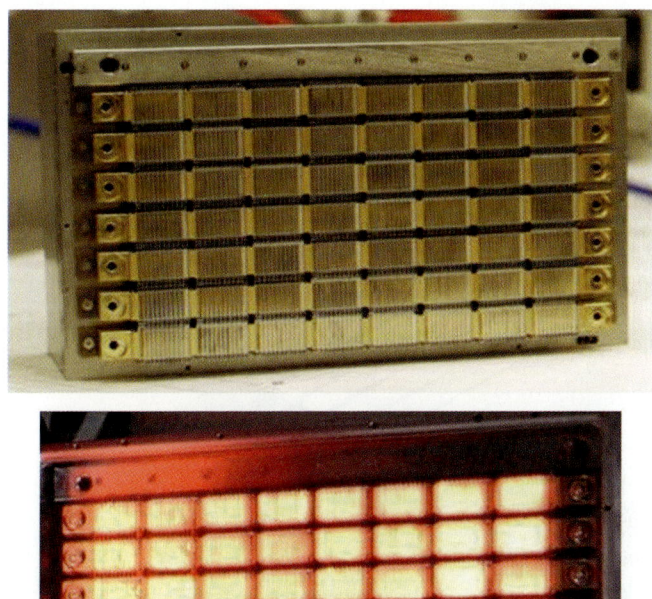

FIGURE 11.2 84 kW, 560-bar diode array manufactured by Simmtec.[5]

A detailed view of the structure of a 10-bar tile is shown in Fig. 11.3, which shows the relative relationship among all the major components of a diode package, as well as the sophistication of the design. Many processing steps, from etching the coolant passages in the glass block to accurately positioning and soldering the actual diode bars to the etched silicon submount, are required to make this sophisticated device.

Figure 11.4 shows a graph of output power versus current into the diode bars for quasi-continuous wave (QCW) operation. Approximately 2 kW of peak output power is achieved at a drive current of 140 amps, with an electrical-to-optical efficiency of 55 percent. These 10-bar tiles are "burned in" for a period of 12 continuous hours at 165 amps and a 20 percent duty cycle, which is well in excess of the nominal operating parameters and which ensures the desired performance under safe and robust operation.

The second critical hardware component of the HCL is the laser gain medium. The advent of large, transparent ceramics as laser gain media is a key technological advancement in development of the

Solid-State Lasers

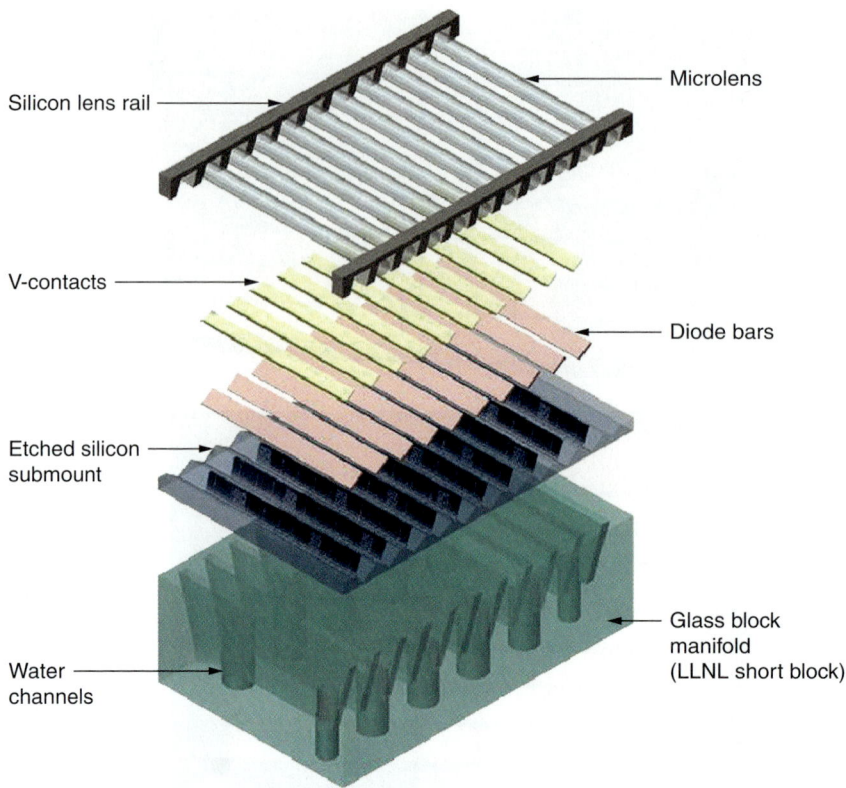

FIGURE 11.3 Ten-bar tile architecture as manufactured by Simmtec.[5]

heat-capacity laser, primarily because the HCL's power scales linearly with the size of the laser gain media. That is, the larger the slab's cross-sectional area, the more power is produced by the laser. The ability for transparent ceramics to be made in very large sizes—on the order of tens of centimeters—contributes significantly to the practicality of the HCL's architecture and its power scaling potential. (Characteristics and other advantages of transparent ceramics were presented in Chap. 7.)

In Fig. 11.5, the object on the left is a 10 × 10 × 2 cm transparent ceramic neodymium-doped yttrium aluminum garnet (Nd:YAG) laser gain medium (slab) designed for use in the HCL at LLNL. The object on the right is a transparent ceramic Nd:YAG slab, integrally framed (edge cladded) with transparent ceramic samarium-doped YAG (Sm:YAG), which is used to suppress amplified spontaneous emission (ASE). Both transparent ceramics were produced by Konoshima Chemical Company, Ltd.,[6] and Baikowski Japan Company, Ltd.[7]

Figure 11.6 shows an end view of the LLNL's heat-capacity laser, showing a 10 × 10 cm square transparent ceramic laser gain media being pumped by high-average-power diode arrays.

Heat-Capacity Lasers

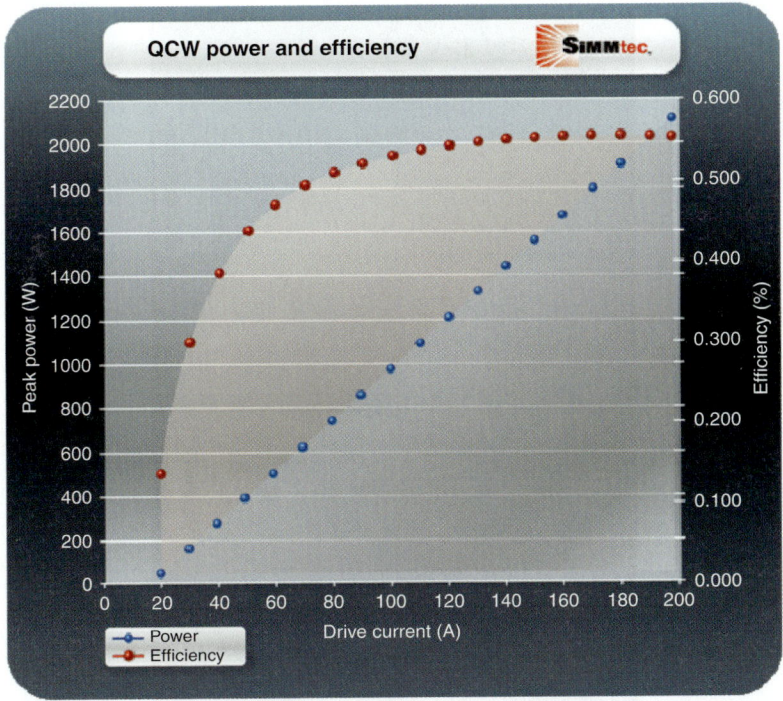

FIGURE 11.4 Ten-bar tile performance as manufactured by Simmtec.[5]

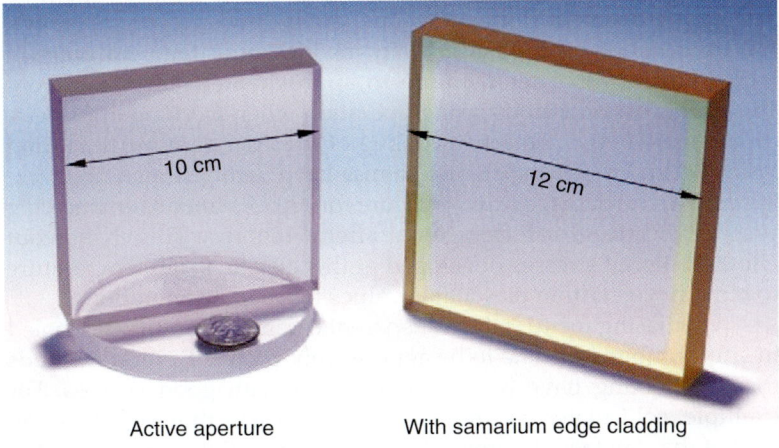

Active aperture   With samarium edge cladding

FIGURE 11.5 Transparent ceramic laser gain media manufactured by Konoshima Chemical Company, Ltd. and Baikowski Japan Company, Ltd.

FIGURE 11.6 Energized diode arrays "pumping" a transparent ceramic laser gain medium.

In a typical HCL operation with the architecture as described earlier, at 25-kW average laser output power, 200-Hz pulse repetition rate, and 10 percent duty cycle for the high-powered diode arrays, the transparent ceramic laser gain media's bulk temperature will increase from room temperature to approximately 130°C in about 10 seconds of continuous lasing operation when working in its heat-capacity mode (i.e., no active cooling of the laser gain media during lasing). Although transparent ceramics have demonstrated no cracking or physical damage at temperatures of 200°C, standard engineering practice dictates a slab change-out after a 10-s run. This change-out allows material stresses developed at this elevated slab temperature to remain well within reasonable values for reliable operation.

After lasing for 10 s, the transparent ceramic can either be cooled in situ or moved off-line to be aggressively cooled. Several methods for slab cooling have been demonstrated with good success. For example, cold plates are moved to within a few thousandths of an inch of the slab faces to provide a conduction path for the heat out of the slab; cold gas is flown across each slab face; or a fine mist that absorbs the heat is applied to the slab face surface. The requisite cooldown time to get the hot slabs from their elevated temperature down to room temperature can vary from tens of seconds to several minutes, depending on the type of cooldown system used. The actual application of the laser and how it will be required to operate (e.g., duty cycle, etc.) will dictate the most suitable type of cooling system. The HCL's flexibility with regard to how it is cooled is another key attribute for using this type of laser architecture, particularly for use in real-world applications.

11.3 Laser Performance Modeling

11.3.1 Pump Absorption, Gain, and Extraction

The geometry of the HCL as implemented in our model is shown in Fig. 11.7. The one-dimensional laser medium is composed of m slabs, each with a thickness of ℓ, between two mirrors of reflectivity R_1 and R_2. The cavity defined by the mirrors has length L_{cav}, and the total length of the active medium is $L_{slab} = m\ell$. Each slab surface is pumped at intensity $i_p(\lambda, t)$ W/cm²-nm. For simplicity, we'll assume the laser output is on a single line. Within the medium are circulating intensities $I_L^\pm$ W/cm², and the medium has bulk loss α cm^{-1}.

The absorption cross section of Nd:YAG in the vicinity of the 808-nm pump band is shown in Fig. 11.8, along with the time-integrated pump spectrum. As may be seen, the spectrum has numerous peaks and valleys. In addition, the pump laser diodes have a time-dependent center wavelength and spectral width. It is therefore important to use the more general expression for the pump rate into the upper laser level—that is,

$$R_p(z, t) \propto \int \lambda \sigma_a(\lambda) i_p(\lambda, t) [\exp(-N_0 \sigma_a(\lambda) z) \\ + \exp(-N_0 \sigma_a(\lambda)(\ell - z))] d\lambda \tag{11.1}$$

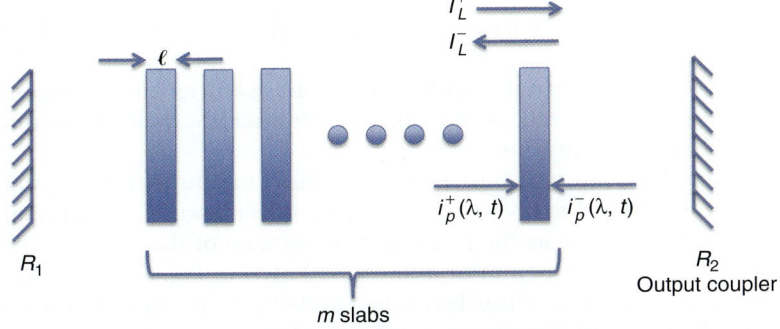

FIGURE 11.7 Geometry of the HCL used in the model.

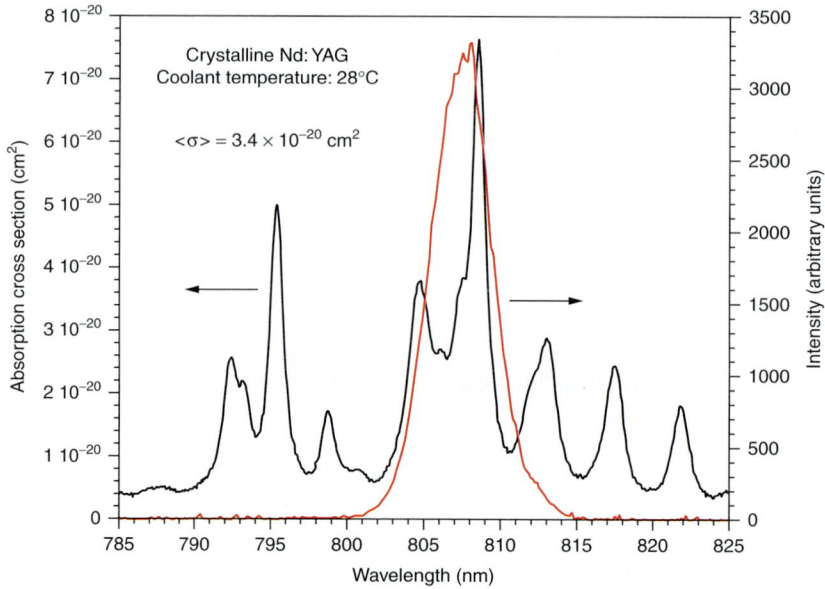

FIGURE 11.8 The 808-nm absorption band of Nd:YAG. Superimposed in red is the time-integrated laser diode pump spectrum.

where $\sigma_a(\lambda)$ is the absorption cross section, N_0 is the neodymium concentration in the ground state, and the integral is done over the pump band (typically 800–820 nm). Implicit in Eq. (11.1) is the assumption that both surfaces of the slab see the same pump intensity.

The instantaneous pump spectral intensity $i_p(\lambda, t)$ may be further written as

$$i_p(\lambda, t) = f_p(t)\exp[-4\ln 2(\lambda - \lambda_c(t))^2/\Delta\lambda^2(t)] \quad (11.2)$$

where $f_p(t)$ describes the overall pump profile; where

$$\lambda_c(t) = A\left[1 - e^{Bt}\mathrm{erfc}\left(\sqrt{Bt}\right)\right] \quad (11.3)$$

is the center wavelength, with A and B fit coefficients; and where $\Delta\lambda(t) = 2.7 + t/235$ nm is the spectral full-width, half-maximum (FWHM) with time in µs.

Because there will be considerable heat buildup in the lasing medium, we need to take into account the lower laser level's thermal population, as well as the thermal depopulation of the upper level. The energy level diagram of interest is shown in Fig. 11.9. Because we are dealing with a four-level laser system, the decay out of level 3 into level 2 (and likewise for level 1 into level 0) is extremely rapid on the time scales of interest; levels 3 and 2 (and levels 1 and 0) are

Heat-Capacity Lasers

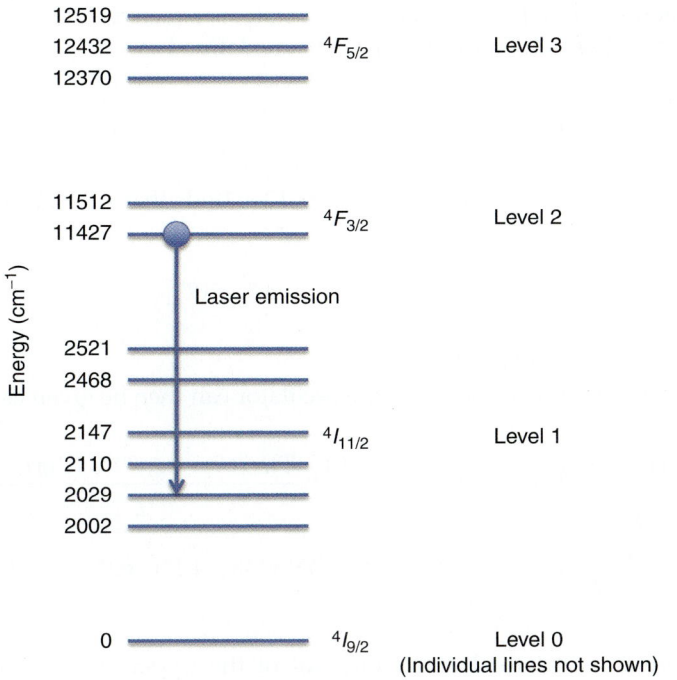

FIGURE 11.9 Energy-level diagram of Nd:YAG, showing the levels considered in the model. Pumping is into the $^4F_{5/2}$ band.

in thermodynamic equilibrium. Consequently, if ΔE_{10} and ΔE_{32} are the energy differences between levels 1, 0 and 3, 2 respectively, we may write the populations of level 1 at temperature T as

$$N_1(T) = N_0 e^{-\Delta E_{10}/kT} \tag{11.4}$$

where $\Delta E_{10} = 0.26$ eV, and k is Boltzmann's constant. Likewise the population of level 2 at temperature T is

$$N_2^0(T) = \frac{Z_{32}(T)}{Z_{32}(T_0)} N_2^0(T_0) \tag{11.5}$$

where T_0 is the reference temperature and where

$$Z_{32}(T) = \frac{Z_2(T)}{Z_2(T) + Z_3(T)} \tag{11.6}$$

is the fraction of atoms in level 2. The population $N_2^0(T)$ refers to the population of level 2 by the pump process at temperature T. It is distinguished from the purely thermal population $N_2(T)$, which, because of its position relative to the ground state, is effectively 0. The partition functions Z_2 and Z_3 in Eq. (11.6) are defined as

$$Z_i(T) = \sum_\alpha e^{-E_{i\alpha}/kT} \qquad (11.7)$$

where α labels the sublevels of level i. One finds that Z_{32} is, to a good approximation,

$$Z_{32}(T) = 1/[1 + 4\exp(-\Delta E_{32}/kT)] \qquad (11.8)$$

where $\Delta E_{32} = 0.13$ eV.

The equations describing the oscillator can then be given as

$$\frac{\partial N_2^0(T)}{\partial t} = \frac{N_0}{hc} R_p(z,t) - k_T N_2^0(T) - \frac{(\sigma_{21}(T)N_2^0(T) - \sigma_{12}(T)N_1(T))(I_L^+ + I_L^-)}{h\nu_L}$$

$$\pm \frac{\partial I_L^\pm}{\partial z} + \frac{n}{c}\frac{\partial I_L^\pm}{\partial t} = \{[\sigma_{21}(T)N_2^0(T) - \sigma_{12}(T)N_1(T)](I_L^\pm + I_n^\pm) - \alpha I_L^\pm\}(L_{slab}/L_{cav})$$

(11.9)

where k_T is the total decay rate out of the upper laser level (i.e., $k_T = k_F + k_{ASE}$, where k_F is the fluorescence decay rate and k_{ASE} is the decay rate due to ASE, see Sec. 11.3.2). The noise term that initiates the lasing process is given by $I_n^\pm$, and any distributed loss in the system is given by the parameter α. The above set of equations is closed by noting that $N_0 + N_2 = N$, where N is the total Nd concentration. The initial/boundary conditions are

$$N_2^0(t=0) = I_L^\pm(t=0) = 0$$

$$I_L^+(z=0,t) = R_1 I_L^-(z=0,t) \quad \text{and} \quad I_L^-(z=L_{cav},t) = R_2 I_L^+(z=L_{cav},t)$$

(11.10)

where R_1 is the high-reflector reflectivity and R_2 is the output-coupler reflectivity.

As an example of the results obtained with these calculations, we show in Fig. 11.10 the spatially averaged gain coefficient, the output intensity, and the output fluence as a function of time for a four-slab Nd:YAG system.

The presence of relaxation oscillations is readily apparent in Fig. 11.10, as is the clamping of the gain at threshold after a steady state has been reached. For this case, the output fluence is approximately 1 J/cm². For an active region 100 cm² in area, this represents an output energy/pulse of about 100 J, or an average power of 20 kW at a 200-Hz repetition rate.

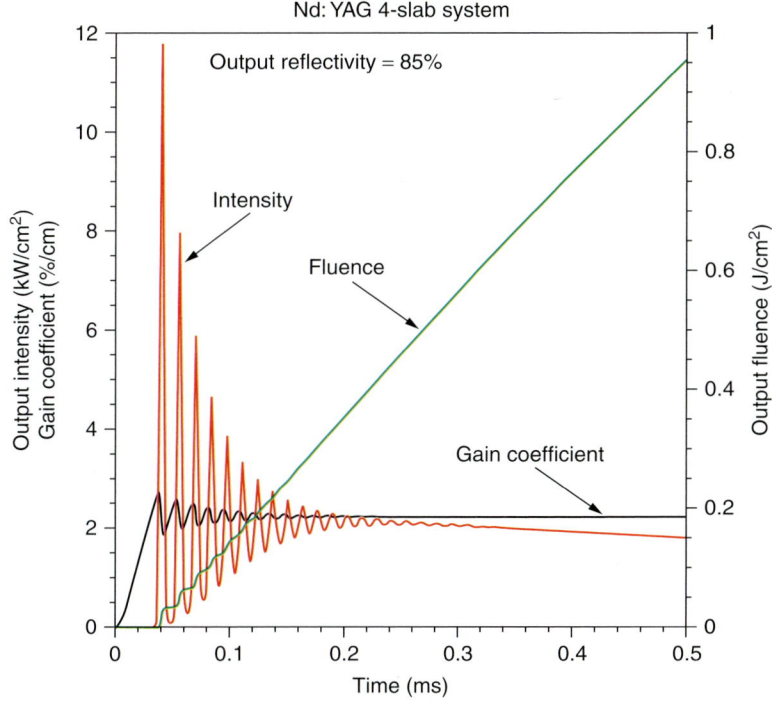

FIGURE 11.10 Calculated output variables for a four-slab Nd:YAG oscillator.

As an example of the type of parameter studies that may be done, in Fig. 11.11 we show the output power as a function of slab count and output-coupler reflectivity. The equivalent unstable resonator magnification ($M = 1/\sqrt{R_{oc}}$), as well as the measured output power for the $M = 1.5$ case, is shown. As expected, systems with higher slab count tend to optimize at higher values of magnification due to the increased amount of gain in the propagation direction.

Finally, we show in Fig. 11.12 the dependence of output power on the temperature of the slab. As mentioned earlier, the output power is reduced as the temperature is increased, due both to thermal population of the lower laser level and to thermal depopulation of the upper laser level. The calculation shown is for a seven-slab system, producing roughly 75-kW output power at the initial temperature of 300°K. For relatively limited temperature increases of 100°K, the output power at the end of the burst is about 80 percent of the initial power. We have found that a typical temperature rise/pulse is approximately 0.05°K. Thus, a 10-s burst at 200 Hz raises the temperature on the order of 100°K.

278 Solid-State Lasers

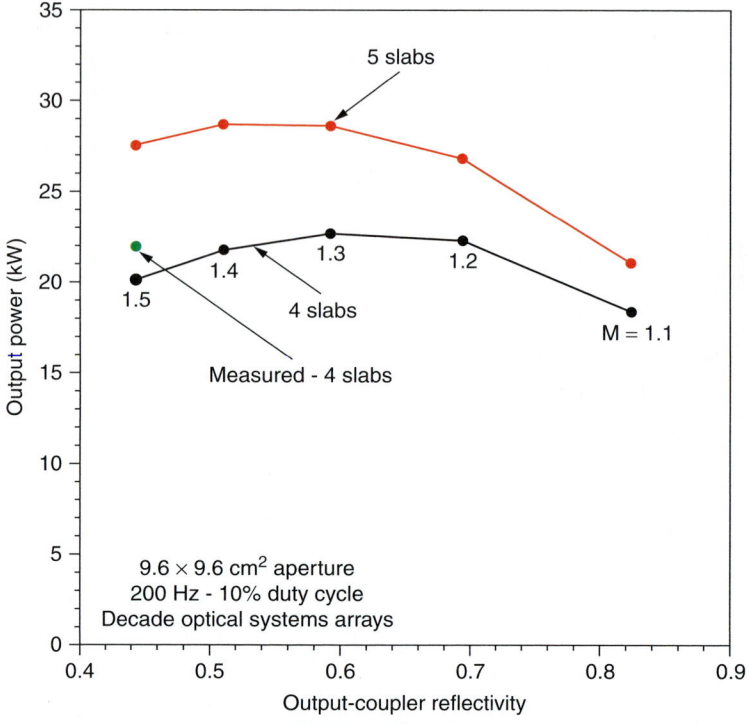

FIGURE 11.11 Output power as a function of output-coupler reflectivity and slab count for a Nd:YAG heat-capacity laser.

11.3.2 The Effects of Amplified Spontaneous Emission

In a solid-state laser medium, a large fraction of the spontaneous emission is trapped due to total internal reflection. To absorb this radiation, and thus prevent internal parasitics from forming, edge claddings are placed around the perimeter of the material, as shown in Fig. 11.13.

The effect of ASE on stored energy may be modeled through an artifice called the *ASE multiplier*, or M_{ASE}. If no ASE were present in the slab, the upper laser level would decay at the fluorescence decay rate $k_F = 1/t_F$, where t_F is the fluorescence lifetime. In the presence of ASE, the upper state will decay at rate $k_{ASE} = k_F(M_{ASE} - 1)$, where $M_{ase} \geq 1$. When $M_{ASE} = 1$, no ASE is present. The ASE multiplier may be parameterized by the gain-width product, or the product of the gain coefficient (in cm^{-1}) with the width of the clear aperture (in cm) of the slab.

We used a Monte Carlo three-dimensional ray tracing code[8] to calculate the ASE multiplier as a function of the gain-width product for a given slab geometry. The code launches rays at random positions

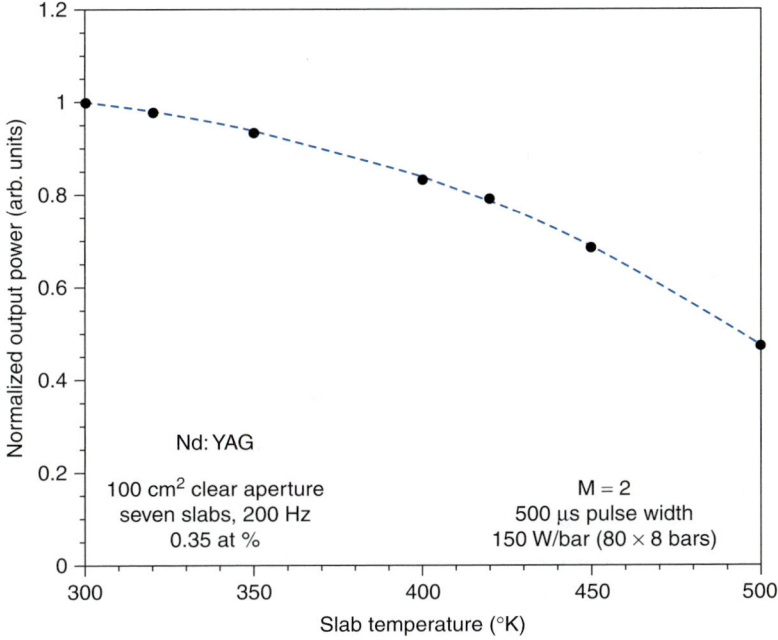

FIGURE 11.12 Output power as a function of slab temperature. Nominal output power at 300°K is 75 kW.

FIGURE 11.13 Ceramic Nd:YAG slab with cobalt-doped, epoxy-bonded edge cladding.

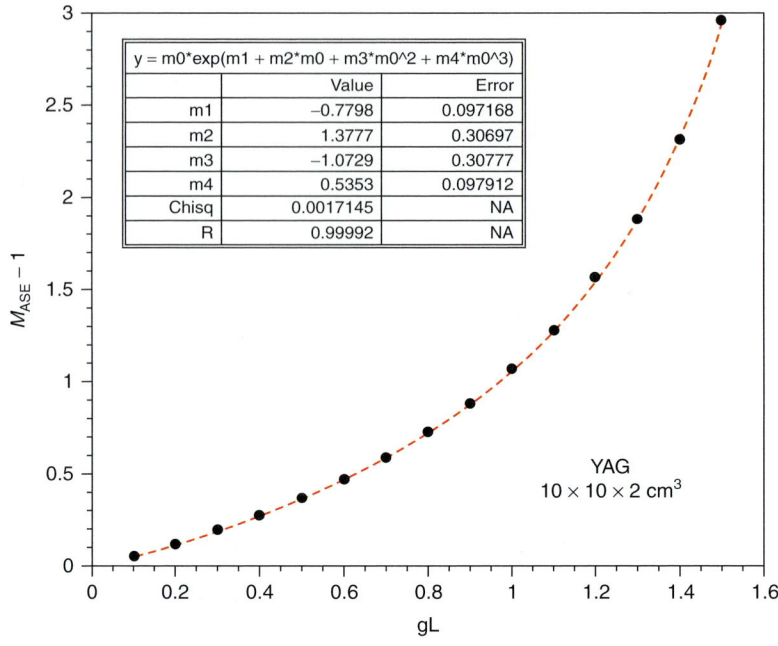

FIGURE 11.14 Variation of ASE multiplier with gain coefficient-width product and analytical fit.

and directions within the slab, keeping track of the gain (or loss) as the ray propagates through the slab. A parasitic condition is noted when $M_{ase} \to \infty$.[1]

As mentioned earlier, we can parameterize the ASE multiplier in terms of the variable $\beta = gL$, where L is the width of the pumped region. In particular,

$$M_{ASE} - 1 = \beta \exp(m_1 + m_2\beta + m_3\beta^2 + m_4\beta^3) \qquad (11.11)$$

where the m_i are curve-fit coefficients. An example of this type of calculation is shown in Fig. 11.14 for a YAG slab of dimensions $10 \times 10 \times 2$ cm³, assuming no index mismatch between the slab and the edge cladding.

In the presence of ASE, the rate equation for the gain coefficient g (or, equivalently, the stored energy density) may be written as

$$\frac{dg}{dt} = P(t) - g M_{ase}(g) k_F \qquad (11.12)$$

where $P(t)$ is the pump rate; it has been explicitly noted that the ASE multiplier is a function of the gain coefficient. From Eq. (11.12), we see that large values of the ASE multiplier lead to a rapid (in time) reduction of the gain coefficient and, consequently, of the stored energy density. Physically, the gain coefficient "clamps" at a given value; at this point, the pump energy goes into generating more ASE as opposed to increasing the gain coefficient.

Under ideal conditions, there would be no mismatch in refractive index between the slab and the edge cladding. These ideal conditions can be achieved by diffusion bonding the edge cladding to the slab or by co-sintering the edge cladding to the slab in the case of ceramic media. However, both of these approaches have proved to be time consuming and not very repeatable in terms of yield. Another approach is to use a bonding agent (such as epoxy) between the slab and the edge cladding. Unfortunately, most epoxies have a refractive index significantly lower than that of YAG. Thus, the Fresnel reflection of spontaneous emission at the slab/epoxy interface lowers the threshold for parasitic oscillations within the slab. If, however, the slab edge is roughened before bonding, the diffuse scattering that results raises the threshold and acts to inhibit the formation of parasitics.

The roughened surface in the ASE model can be treated as follows: The surface is characterized by $\zeta(x, y)$, which represents the difference in height (in the z-direction) at any point (x, y) from the mean z-value of the surface. We assume ζ is a normally distributed, stationary, random variable with 0 mean and variance σ^2. The random distribution is further described by correlation distance.

In the limit where one assumes that the surface is quite rough, so that $\sigma/\lambda \gg 1$, where λ is the wavelength of the light, the probability density for normally incident light to be scattered into angle θ may be written as[9]

$$p(\theta; \xi) = \frac{c\xi}{1+\cos\theta} \exp\left[-\frac{\xi^2 \sin^2\theta}{8(1+\cos\theta)^2}\right] \quad (11.13)$$

where

$$c = \left[\sqrt{2\pi} \operatorname{erf}\left(\frac{\xi}{2\sqrt{2}}\right)\right]^{-1}$$

and $\xi = T/\sigma$ characterizes the surface. A rough surface is given by $\xi \to 0$; a smooth surface, by $\xi \to \infty$. Each time a ray hits the slab edge, its new (reflected) direction is randomized according to the probability distribution shown in Eq. (11.13). If U represents a uniformly distributed random number on (0, 1), the scattering angle θ, as given by Eq. (11.13), may be generated from[10]

$$\theta(U; \xi) = 2\tan^{-1}\left\{\frac{2\sqrt{2}}{\xi} \operatorname{erf}^{-1}\left[U \operatorname{erf}\left(\frac{\xi}{2\sqrt{2}}\right)\right]\right\} \quad (11.14)$$

where erf^{-1} is the inverse error function.

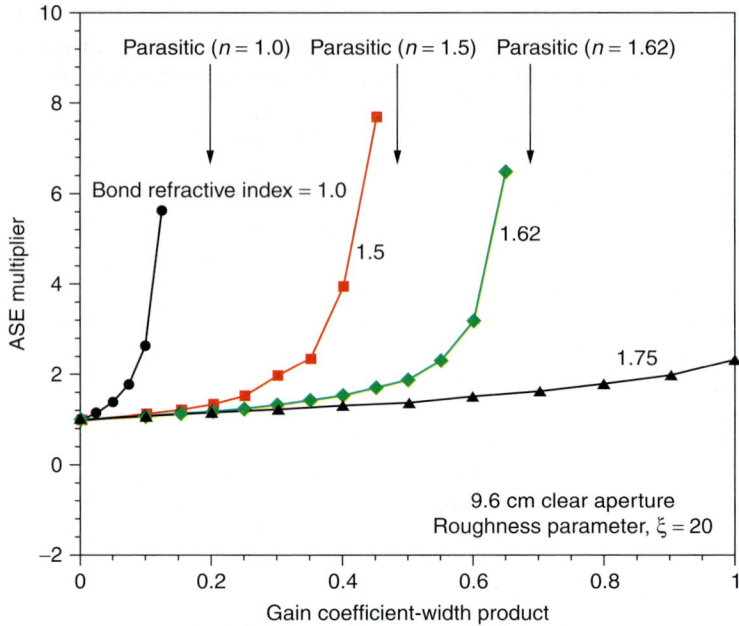

FIGURE 11.15 Effect of bond refractive index on ASE multiplier.

Figure 11.15 shows the effect of epoxy refractive index on the ASE multiplier as a function of gain coefficient-width product. As expected, the closer the refractive index of the epoxy approaches that of the YAG slab (refractive index ~1.82), the less of an effect the epoxy has on the multiplier. The arrows indicate at what value of gain-width product parasitics begin to occur. Figure 11.16 shows the measured and calculated gain coefficient of an epoxy-bonded edge cladding where the epoxy refractive index is 1.62. Eventually the slab develops parasitics, as noted by the clamping of the gain coefficient at 0.11 cm^{-1}; however, the operating point of the heat-capacity laser is well below this, as indicated by the dashed line.

11.3.3 Wavefront Distortion and Depolarization

Even though the HCL was designed to minimize thermal gradients, and hence thermally induced wavefront distortion, gradients still exist transverse to the propagation direction due to nonuniformities in the pump illumination. This section provides the approach for calculating these effects and shows how these effects limit the system's performance.

The modeled finite-element geometry is shown in Fig. 11.17. The central region in the figure represents the ceramic Nd:YAG slab, with the surrounding region denoting the Co:GGG edge cladding. Between the two materials is a 125-μm-thick epoxy bond. Due to the bond's

Heat-Capacity Lasers 283

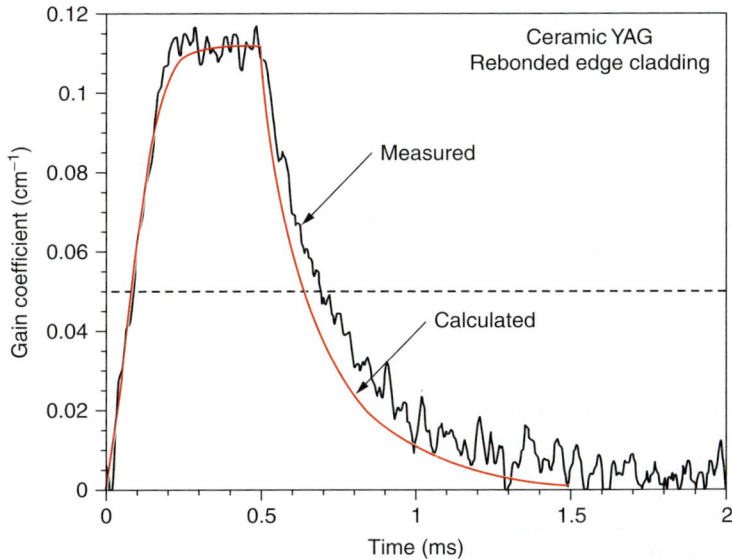

FIGURE 11.16 Measured and calculated gain coefficient for a ceramic YAG slab. Epoxy refractive index is 1.62. The dashed line indicates the operating point for a four-slab system with a magnification of 1.5 unstable resonator.

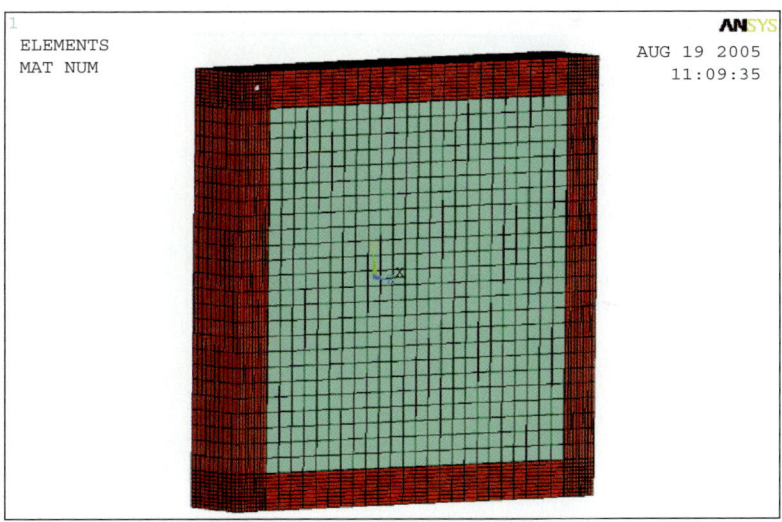

FIGURE 11.17 Geometry of modeled slab. The central region is ceramic Nd:YAG, while the surrounding region is the Co:GGG edge cladding. Between the two regions is a thin (125 μm) epoxy layer, not visible in this figure.

very thin dimension, this region is not visible in the figure. Also modeled is a small air gap between two adjacent pieces of edge cladding. Again, because of the thin dimension of the air gap, this region is not visible in the figure.

Temperature and Stress Calculations

The thermoelastic calculation begins with the specification of the thermal source function. To this end, for the YAG region, the actual measured laser diode array intensity profiles at the plane of the slab were used. For the edge cladding region, it was assumed that the unextracted energy was deposited uniformly around the active region's perimeter. The calculation was run for each slab in the laser individually, because the diode array profiles were different for each slab. For both the YAG and the GGG, the temperature dependence of the thermal constants (notably, thermal conductivity and specific heat) was taken into account. Time dependence was included for both the temperature and stress parts of the calculations.

The temperature distribution for slab 4 after 5 s is shown in Fig. 11.18. The scale on the bottom is in degrees Celsius, with an initial (uniform) temperature of 20°C. To a good approximation, the slab surface heats up at a rate of about 11 to 12°C/s. The diode light's nonuniformity is readily apparent in this figure.

A thermal camera enabled the temperature at the slab's surface to be measured and thus compared to the model predictions. Figure 11.19 shows the comparison at $t = 1$ s and $t = 5$ s. The YAG slab is located

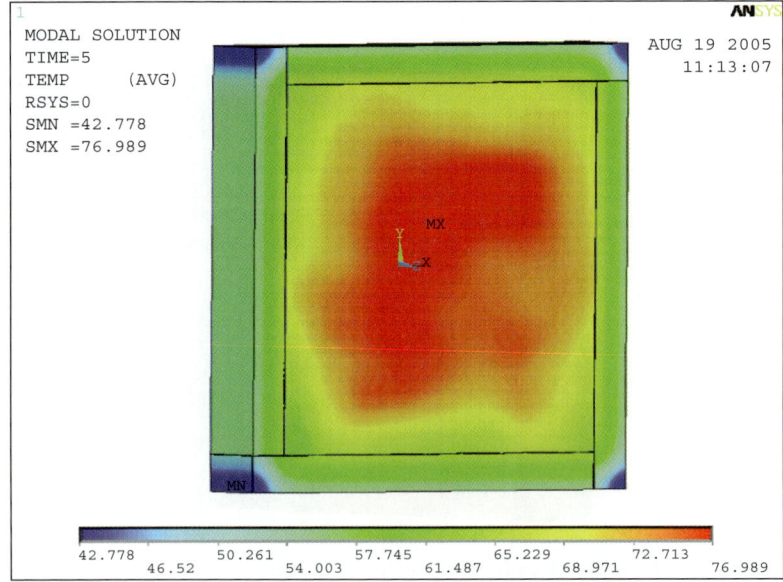

FIGURE 11.18 Temperature contours after 5 s. Scale is in degrees Celsius.

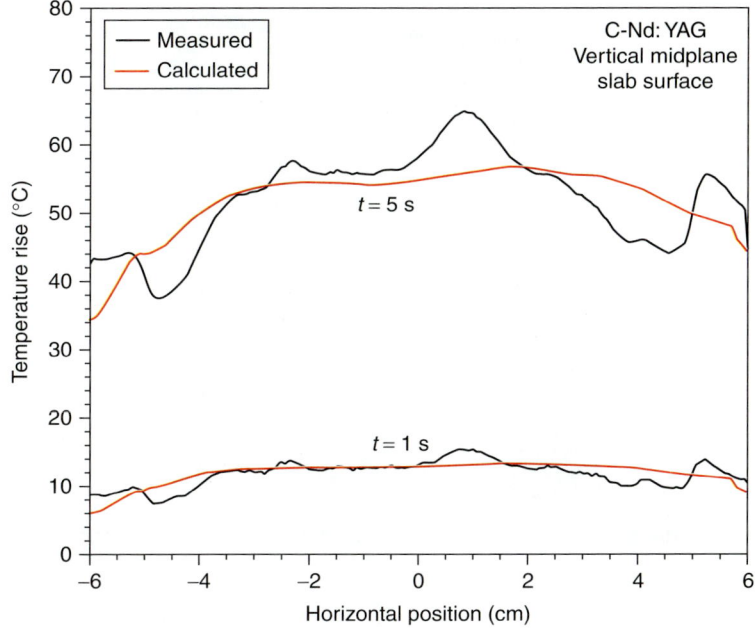

FIGURE 11.19 Measured and calculated temperature rises for slab 4.

between –5 and 5 cm on the graph. Although some of the fine structure is lacking, the model tracks the overall temperature rise rather well.

One of the main drivers in generating depolarization is the x-y shear stress, which is shown in Fig. 11.20 and corresponds to the temperature distribution in Fig. 11.18. As expected, the greatest shear stress occurs in the corners of the slab; thus, this is where one would expect to see the greatest amount of depolarization.

Wavefront Calculations

Given the temperature and stress distributions in the laser slabs, one can calculate the amount of wavefront distortion expected. In general, wavefront distortions come from three sources: (1) The temperature dependence of the refractive index, (2) mechanical deformation, and (3) stress-induced birefringence. Stress-induced birefringence also leads to depolarization of an initially linearly polarized beam. Figure 11.21 presents the total wavefront phase error for slab 1 at $t = 5$ s due to all effects, displacement, dn/dT, and stress. It should be noted that the vast majority of the wavefront is due to dn/dT and displacement effects; stress effects play a minor role insofar as they contribute to the amount of wavefront distortion. Units for the graph are waves at 1 μm. The peak-to-valley (P–V) wavefront distortion does not grow linearly during the 5 s; rather, the P–V value grows

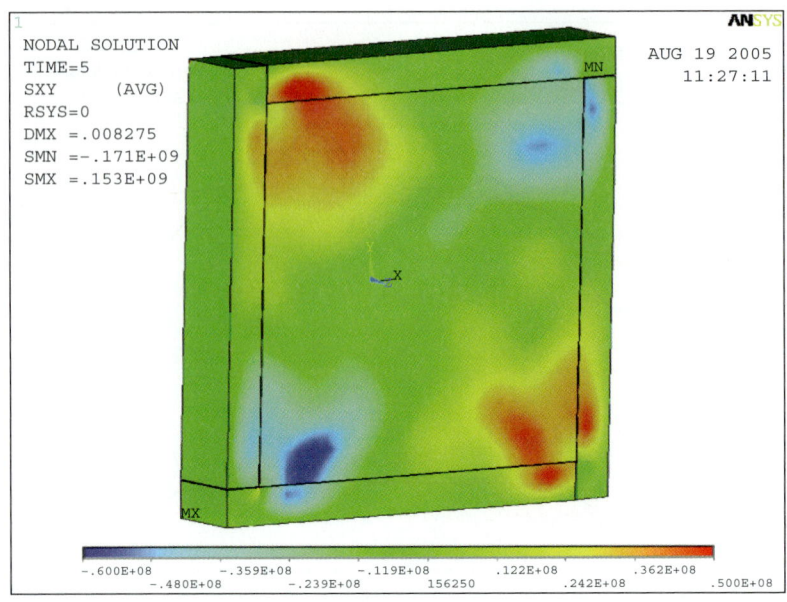

FIGURE 11.20 Contours of the x–y shear stress for the temperature distribution in Fig. 11.18. Scale units are dynes per centimeter squared; divide by 10^7 to get megapascals.

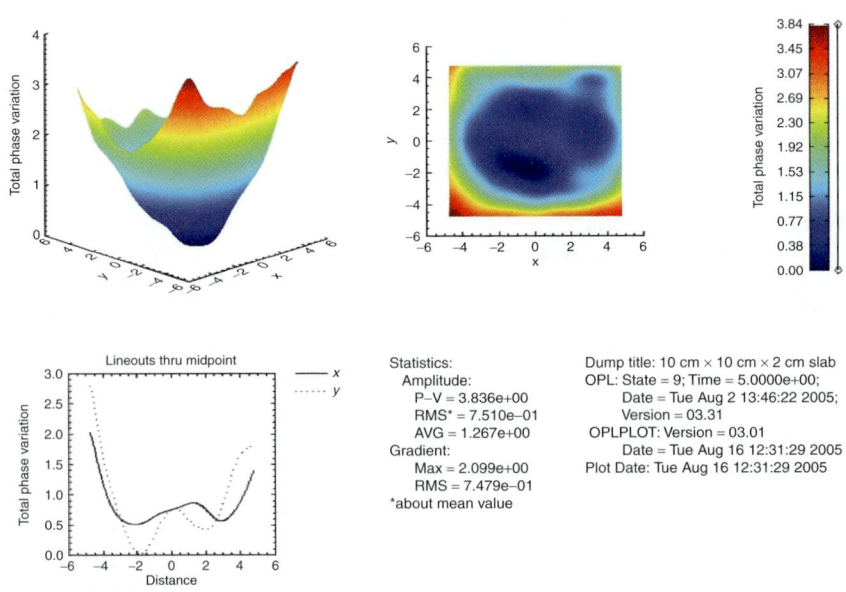

FIGURE 11.21 Total wavefront phase for slab 1 at $t = 5$ s.

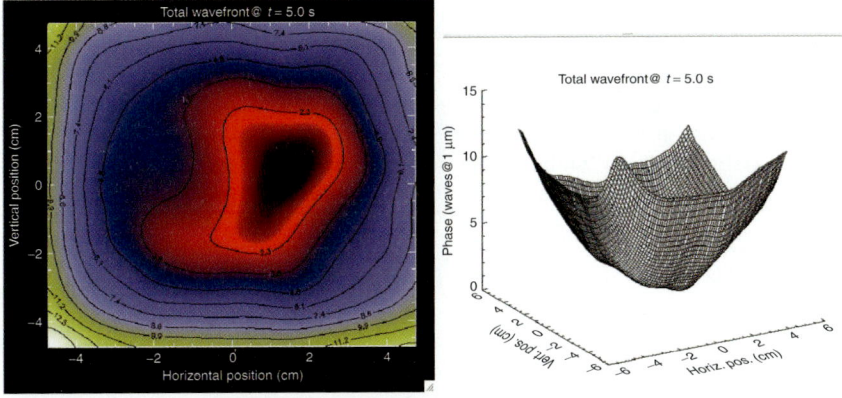

FIGURE 11.22 Total wavefront due to four slabs (single pass) at $t = 5$ s.

linearly during the first second, but then grows sublinearly—that is, the P–V value at 1 s is one wave, whereas at 5 s, it is 3.8 waves.

The total wavefront due to all four slabs is found to be a coherent addition of the individual slabs. Figure 11.22 shows the total wavefront for all four slabs at $t = 5$ s. Notice the substantial amount of curvature to the wavefront. By correcting the spherical wavefront error ($x^2 + y^2$) with another optic(s) (analogous to what is currently done for tip-tilt), the deformable mirror (DM) stroke would be saved to correct any higher-order wavefront error. Figure 11.23 shows the wavefront with the sphere removed, showing that the amount of P–V wavefront has been cut in haf.

Table 11.1 summarizes the total wavefront P–V values found by our calculations. The amount of phase aberration (and gradient) seen at the DM plane is twice the above values, due to double passing of

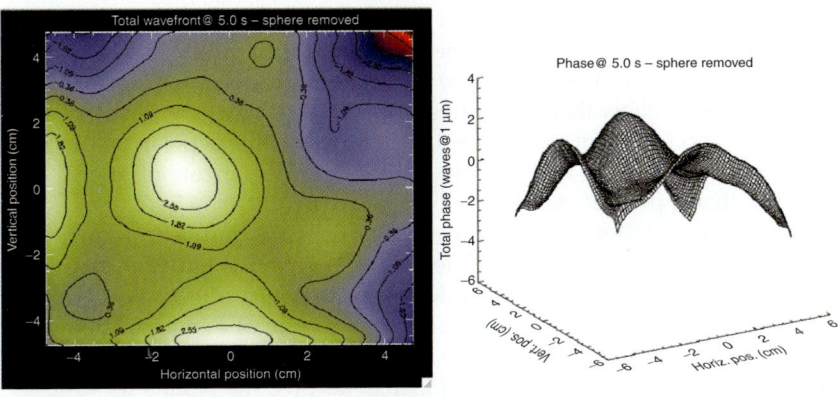

FIGURE 11.23 Total wavefront due to four slabs (single pass), with sphere removed, at $t = 5$ s.

Time (s)	Total phase P–V (waves @ 1 μm–four slabs single pass)	Total phase P–V, sphere removed (waves @ 1 μm–four slabs single pass)	Maximum phase gradient (waves/cm)	Maximum phase gradient – sphere removed (waves/cm)
0.25	1.0	0.5	0.5	0.3
0.50	2.0	1.0	1.0	0.6
1.00	3.7	1.8	1.8	1.1
5.00	14.0	8.0	5.3	4.6

TABLE 11.1 Calculated Total Wavefront Error and Gradients Due to Four Slabs.

the slabs. It should be noted that the magnitude of the above aberrations is well within the correction capability of the DM, which is up to 16 waves. However, what limit the runtime are the phase gradients. The DM will allow a maximum of ±2 μm of relative motion between actuators. Because there is approximately 1 cm between actuators, a gradient of 2 waves/cm will have reached this limit. From Table 11.1, we see that this occurs at approximately 1 s runtime without spherical error subtraction, or as much as 2 s with spherical error corrected by another optic.

Depolarization

For light that is linearly polarized along a given direction, the depolarization value gives the percentage of light that is rotated into the orthogonal polarization. For example, a value of 80 percent indicates that at a given point in the aperture, light that is linearly p-polarized emerges from the slab elliptically polarized, with 80 percent of the intensity s-polarized and 20 percent remaining p-polarized.

As mentioned earlier, the x-y shear component of the stress drives the depolarization. Consequently, the spatial distribution of the depolarization tends to follow that of the stress. Figure 11.24 shows the depolarization for slab 1 at $t = 5$ s. As expected, the majority of the depolarization occurs in the corners of the slab. The amount of depolarization ranges from less than 1 percent at $t = 0.25$ s to about 80 percent at $t = 5$ s.

The depolarization results for the individual slabs cannot be added in a simple way to obtain the total depolarization for the four-slab system. The reason is that because the depolarization *intensity* is given, all "phase" information is lost. To calculate the amount of depolarization for four slabs, the actual Jones matrices for a given slab must be used. These matrices may be multiplied together to give the results for an arbitrary number of slabs. The results of this calculation for four slabs at $t = 5$ s (single pass) are shown in Fig. 11.25. Peak depolarization values

Heat-Capacity Lasers

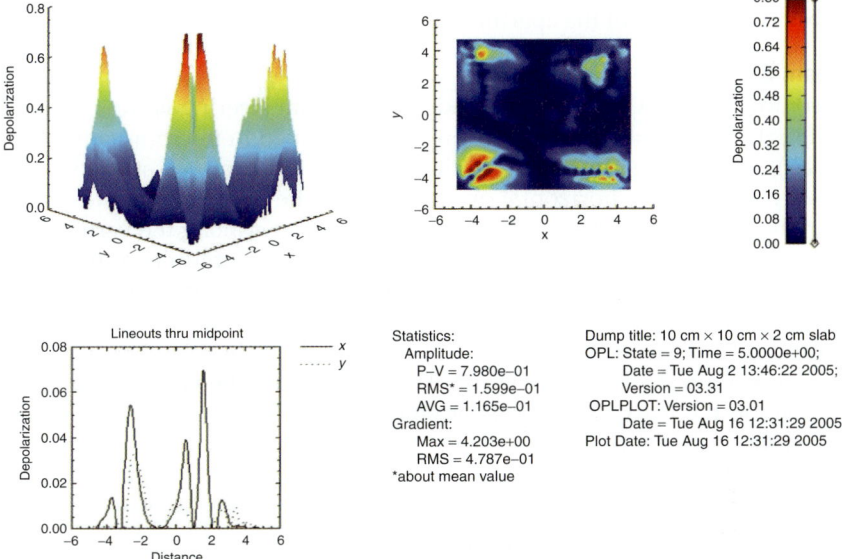

FIGURE 11.24 Slab 1 depolarization at $t = 5.0$ s.

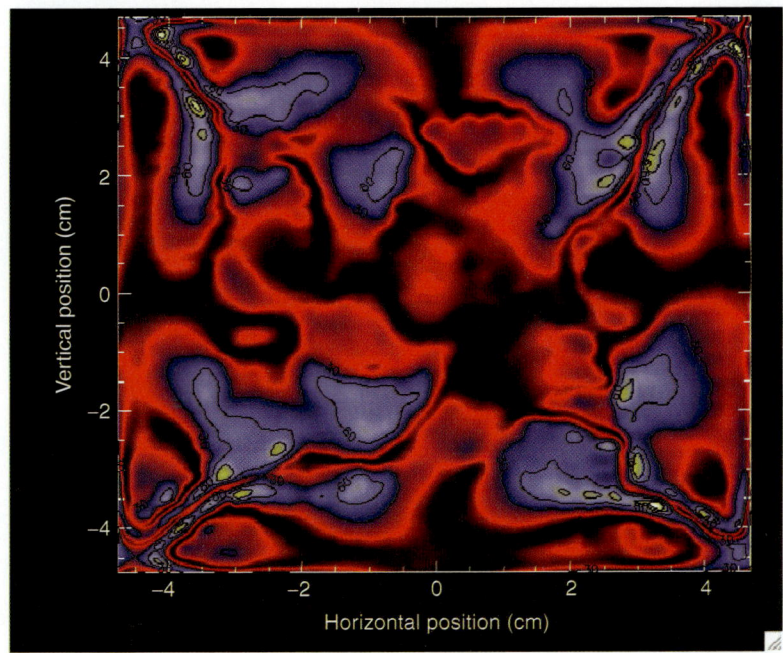

FIGURE 11.25 Four-slab (single-pass) depolarization at $t = 5$ s.

range from about 10 percent at $t = 0.25$ s to 100 percent at $t = 5$ s, with a substantial amount of the aperture depolarized.

Beam Steering

A contour plot of the horizontal and vertical beam steering is shown in Fig. 11.26 for $t = 5.0$ s. The steering angle is given in microradians (µrad), with a positive value indicating that the beam is steered toward the positive horizontal or vertical axis (the origin of the axes is in the center of the aperture). Cross-sectional views along the vertical midplane (for horizontal steering) and horizontal midplane (for vertical steering) are given in Fig. 11.27. After 1 s, the maximum steering angle is about 200 µrad (four slabs, single pass) for both horizontal and vertical steering. A double pass through the slabs would result in a maximum steering of 400 µrad. This value could then be used to determine the actual linear displacement of the beam on the DM, given the path length in the cavity.

11.4 Current State of the Art

11.4.1 Power Extraction

In January 2006, the heat-capacity laser at LLNL achieved 67 kW of average output laser power for short-fire durations consisting of 335 J/pulse at a 200-Hz pulse repetition rate,[4] setting a world record for pulsed, diode-pumped, solid-state lasers. The pulsed HCL had a 500-µs pulse width and used up to a 20 percent duty cycle from the high-powered diode arrays. This power level was accomplished by pumping five transparent ceramic YAG:Nd^{3+} slabs in series, each having an active lasing region of $10 \times 10 \times 2$ cm in thickness. Figure 11.28 shows an end-view and side-view photograph of this HCL system.

11.4.2 Wavefront Control

To control the amount of wavefront distortion in the HCL, a number of techniques were used. Figure 11.29 shows an optical layout schematic of the HCL. One of the turning mirrors—and the main method of controlling wavefront—is the intracavity deformable mirror (DM). Tip-tilt corrections are applied to the high reflector, and a quartz rotator midway through the optical chain acts as a birefringence compensator. Not shown in the schematic is the beam sampling plate (placed before the output coupler) and the Hartmann sensor which provides the measurement of the wavefront as well as the signals necessary to control the DM.

As mentioned earlier, the output beam quality depends very strongly on phase distortions in the resonator. Some of the sources of these distortions include (1) pump-induced thermal gradients in the gain medium, (2) heating of resonator optics by absorbing some of

Heat-Capacity Lasers 291

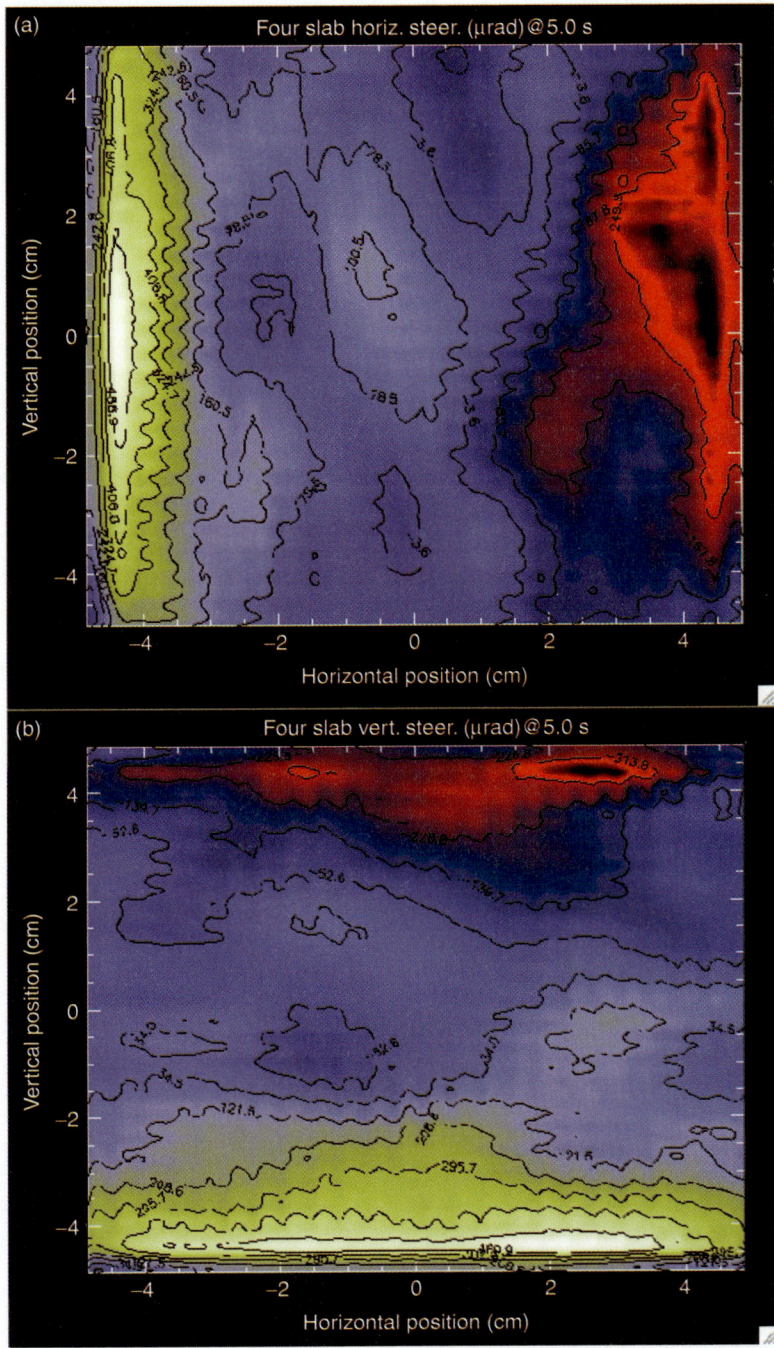

FIGURE 11.26 Four-slab (single-pass) steering (mrad) at $t = 5.0$ s. (*a*) Horizontal and (*b*) vertical.

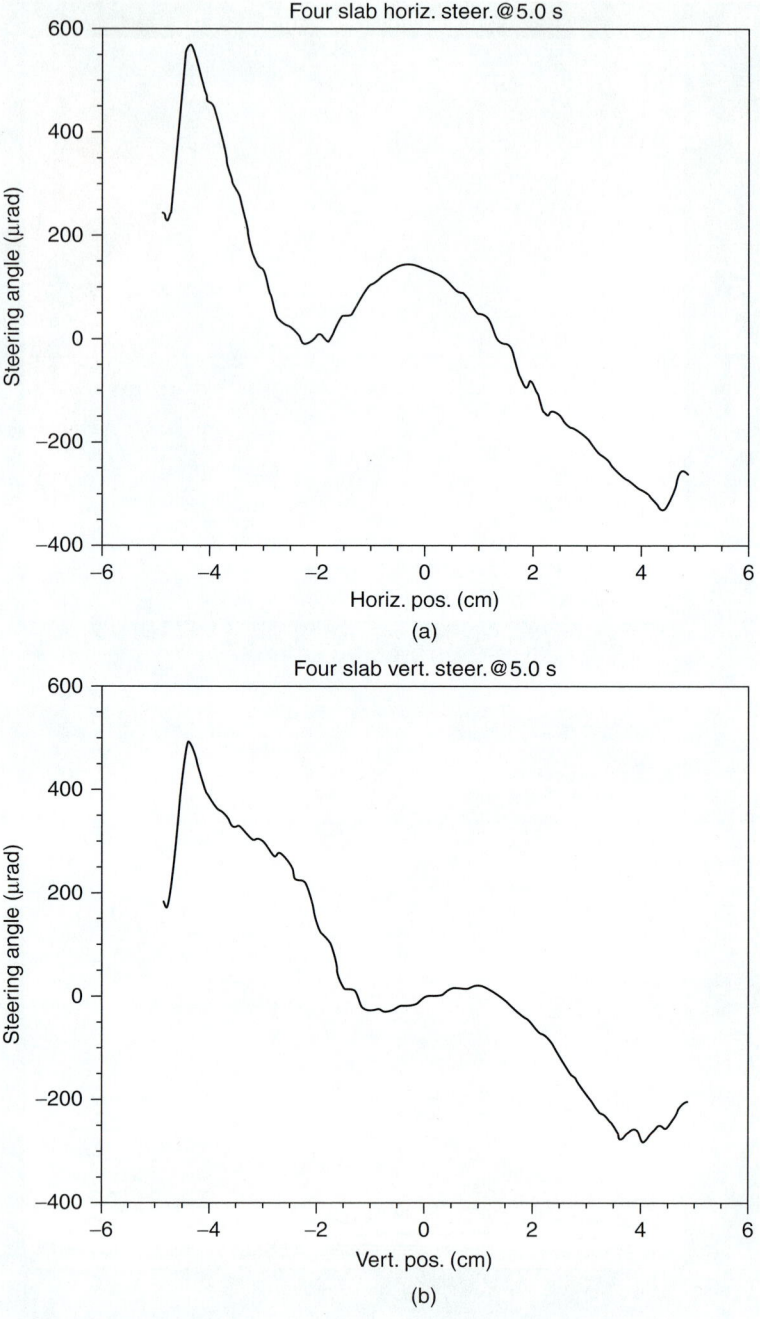

FIGURE 11.27 Four-slab (single-pass) steering (mrad) at $t = 5.0$ s. (*a*) Horizontal steering at vertical midplane and (*b*) vertical steering at horizontal midplane.

Heat-Capacity Lasers

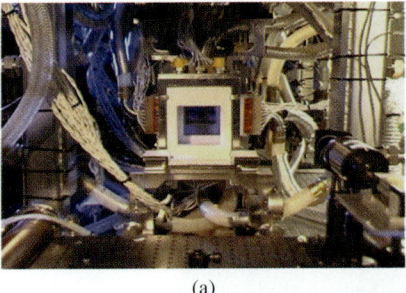
(a)

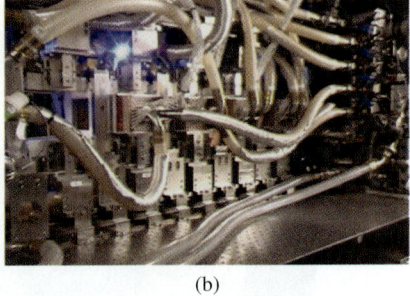
(b)

FIGURE 11.28 Current configuration of the heat-capacity laser at Lawrence Livermore National Laboratory. (a) End view and (b) side view.

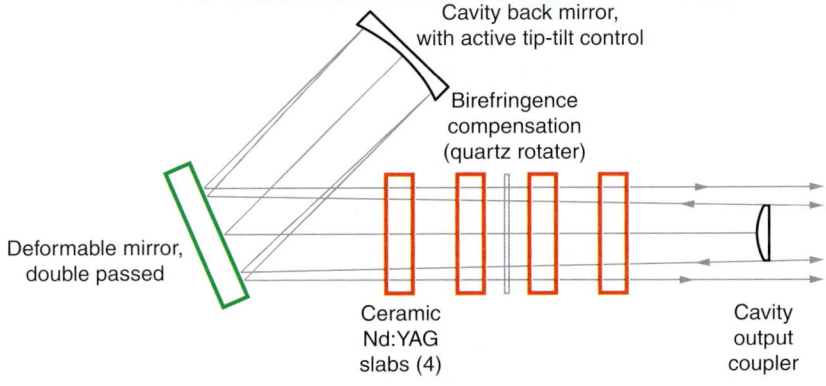

FIGURE 11.29 Optical layout schematic of the HCL. Not shown are the beam sampling optic (placed before the output coupler) and the Hartmann sensor.

the laser power, and (3) heating of the environment (i.e., the surrounding structures and, subsequently, the atmosphere). To avoid degradation in beam quality, the wavefront errors must be kept below ~35 nm RMS (~$\lambda/30$).

The DM is the primary method of aberration control in the HCL. Figure 11.30 shows the face of the DM. The optic in front of the DM not only protects the DM but also provides a channel for the air column, which is used to keep dust off the face. The DM, built by Xinetics, uses 206 discrete actuators on a pseudohex pattern with approximately 1 cm spacing. The total stroke limit is ±4 µm, with a maximum interactuator stroke limit of ±2 µm. The DM uses push-pull actuation and is amenable to zonal or modal AO correction schemes, while being susceptible to "print-through," which is the residual phase aberration after correction. Because the DM is used in a double-pass configuration, the total amount of correction possible is up to 16 waves at 1 µm (low spatial frequency).

FIGURE 11.30 The front face of the intracavity deformable mirror. The actuators can be seen through the front faceplate.

The main source of phase distortions is pump-induced thermal gradients in the gain medium (see Fig. 11.18 for the calculated temperature distribution in the slab). The source of these gradients is primarily nonuniform pump-light deposition on the face of the slab. These nonuniformities get directly imprinted on the wavefront.

Even though pump nonuniformities produce the greatest effect on wavefront, other effects, such as heating of optics or thermally induced air currents, also play a role. For example, the window on the face of the DM was initially BK7 glass in which there was unacceptable absorption of the laser light by the window, causing large amounts of distortion. These distortions were sufficient to be visible in the near-field intensity pattern. We also detected the presence of convection cells via the AO control loop. These cells resulted in large tilts that had to be applied to the rear mirror for mitigation. Operating the laser in a helium atmosphere would be one way to reduce the impact of these cells.

Because of pump nonuniformities and absorption by the DM window, the laser's initial runtime was limited to about 1 s before the level of aberrations was too large to be corrected by the DM. Note this runtime is in good agreement with the calculations presented earlier. By replacing the DM window and using a holographic diffuser to homogenize the pump arrays, the runtime was extended to 5 s at less than two times the diffraction limit, as shown in Fig. 11.31. The "Early fall 2005" graph represents the laser's initial condition. In late fall 2005, the BK7 window in front of the DM was replaced with a fused-silica version. In spring 2006, holographic diffusers were added to the pump arrays. The final result was a beam quality of no more than two times the diffraction limit at the end of the 5-s run.

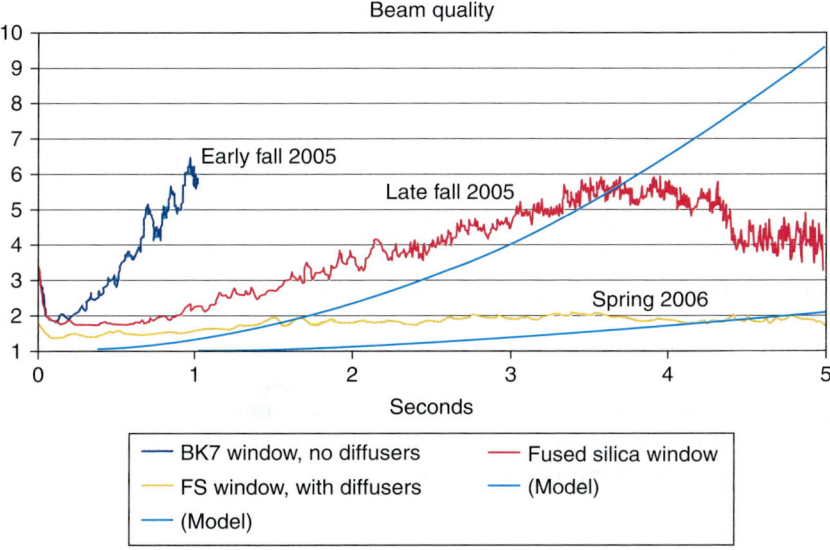

FIGURE 11.31 Improvements in beam quality (times the diffraction limit) as a result of changes made to the laser.

11.5 Scaling Approaches

The HCL's architecture described in this chapter has confirmed that significant amounts of laser output power (67 kW of average output power) can be produced in a very small volumetric footprint via an extremely simple, straightforward laser cavity design. The five laser gain module HCL shown in Fig. 11.28 could fit on a typical dining room table. The power of the HCL scales linearly in each of the following three independent methods:

- Adding more inline laser gain media (slabs)
- Increasing the cross-sectional area of the laser gain media (and a corresponding increase in diode pump light)
- Increasing the duty cycle of the high-powered diode arrays

The simplicity of increasing any or all of these three parameters makes for a very straightforward, practical approach to increasing the HCL's output power.

Looking ahead to the next level of power, a concept design for a megawatt-class HCL is as follows:

- 16 transparent ceramic Nd:YAG laser gain media arranged in series
- Each gain medium equipped with a $20 \times 20 \times 4$ cm–thick slab

296 Solid-State Lasers

- 64 high-powered diode arrays at 84 kW of average output power per array and a duty cycle of 20 percent

These parameters follow the aforementioned general power scalability formula for a heat-capacity laser: increasing the number of laser gain media, increasing the size of the laser gain media, and increasing the duty cycle of the high-powered diode arrays. Although many details of the entire laser system's architecture remain to be resolved, there is no fundamental reason that the HCL cannot attain megawatt-class output power using the same nominal architecture that is currently used today. In addition, the only technology that has not been physically demonstrated to date is the 20-cm transparent ceramic laser gain media. Thus, the "leap" to significantly higher laser output power levels is more of an evolutionary engineering process, rather than a wait for a significant technological breakthrough to occur.

11.6 Applications and Related Experimental Results

Because of its large output power capability, as well as its simple architecture resulting from the ease of operation and compact footprint, the heat-capacity laser is often used to conduct a variety of laser-material interaction experiments. Several investigations using the HCL at Lawrence Livermore National Laboratory are cited[11] to provide examples of the various capabilities of the heat-capacity laser.

11.6.1 Rapid Material Removal (Boring/Ablation)

Experiments have been conducted that showed the laser interaction on steel targets, initially in a static configuration. The collected data are often represented by the term Q* (Q star), or the amount of energy required to remove 1 g of material. In this particular experiment, a 25-kW beam produced by the heat-capacity laser, with a laser spot size of approximately 2.5 × 2.5 cm and a pulse frequency of 200 Hz, is impinged on a 1-in-thick block of carbon steel. The results of the laser-target interaction after 10 s of continuous laser operation are shown in Fig. 11.32. The initial hole through the steel block was generated after just 6 s of runtime.

A significant amount of material was removed during this laser-target interaction. This type of experimental data can be useful in determining machining rates for laser cutting tools, as well as in estimating burn-through times for targets of military interest.

11.6.2 Aerodynamic Imbalance Due to Airflow Interaction

The sequence shown in Fig. 11.33 shows an experimental simulation of a laser beam interacting with a thin aluminum structure in flight. The laser beam heats the material surface (13 × 13 cm spot

FIGURE 11.32   25 kW on a 1-in-thick carbon steel target for 10 s.

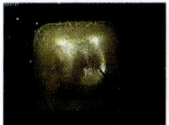 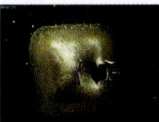 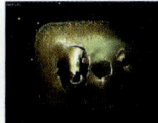 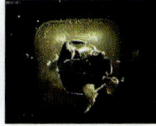 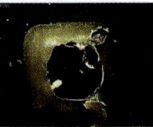

FIGURE 11.33   Laser interaction of a thin-walled aluminum sheet with airflow; 0.07 s total elapsed time.

size), softening it to the point at which initiation of a crack and the ultimate rupture of the material occur. A high rate of airflow is directed across the surface to simulate flying through the atmosphere. This experiment demonstrates that well before melting of the aluminum sheet, the material softens and bulges outward due to the low-pressure region formed by the flowing air. The hydrodynamic force generated by the stream of flowing air is sufficient to rip away the thin aluminum skin. This aerodynamic imbalance either destroys the structural integrity of the target or sends it off its desired flight path.

11.6.3   Laser Used for Humanitarian Mine Clearing

In 2004, the Lawrence Livermore National Laboratory received a Research and Development 100 award[12] for developing a heat-capacity laser for use in humanitarian mine clearing. Experiments showed that due to the laser system's pulsed format, the laser beam could bore through soil at a very fast rate, heat up a buried land mine in seconds, and raise the temperature of the high explosive within the land mine sufficiently for deflagration to occur.

The physics of the digging phenomenon is as follows: because of the HCL's pulse format, the peak power generated in each laser pulse is many times that of the laser's average power. This high peak power per pulse corresponds to a high peak temperature increase in the soil substrate. All soils contain some residual amount of moisture; the high peak power pulses of laser light generated by the HCL impinge on the soil in a very focused area and vaporize the moisture in the soil. This vaporization creates a microexplosion of the moisture on a pulse-by-pulse basis. This microexplosion generates the force required to displace the soil, allowing the laser to penetrate to the intended target. Each laser pulse vaporizes more moisture, thus creating more explosions, which allows the laser to penetrate deeper and deeper into the soil. Once the laser hits the outer casing of the mine, it rapidly begins to heat the material. Within a few seconds, the temperature of the high explosive within the mine is significantly raised (a few hundred degrees Celsius) to initiate deflagration.

Figure 11.34 provides a concept drawing of the HCL system used for humanitarian mine clearing. The system can be used for both buried and surface mines and can be operated at significant standoff distances to reduce the amount of human exposure within the blast-affected zone. In addition, the HCL's power can be easily modulated such that

FIGURE 11.34   Concept of the heat-capacity laser system used for humanitarian mine clearing.

Heat-Capacity Lasers

FIGURE 11.35 400-kW heat-capacity laser system on a mine-resistant, ambush-protected vehicle.

the highly explosive material in the mine does not explode and instead simply "sizzles" in place, thus further reducing the risk of human exposure to material fragmentation and the resulting shrapnel.

11.6.4 Self-Contained 400-kW Heat-Capacity Laser on a Military Vehicle

It is obvious that the HCL's many attributes will lend themselves to military applications. The heat-capacity laser, as described above, can be scaled to very high powers while still maintaining a very low weight and a compact footprint. In addition, due to its simple laser architecture, it is extremely compatible with military requirements for being robust, reliable, and easy to maintain. Figure 11.35 shows an artist's conception of a 400-kW heat-capacity laser system on a mine-resistant, ambush-protected (MRAP) vehicle. The system, as designed, is fully self-contained, including the laser, power management system, thermal management system, beam director, and computer control system. Initial targets could include rockets, artillery, and mortars (RAMs), as well as improvised explosive devices (IEDs).

11.7 Summary

The heat-capacity laser's simple architecture, including the separation of the lasing action from the cooling of the laser gain media, has demonstrated that it can be used for practical applications. Key components, such as the high-power diode arrays and the transparent ceramic laser gain media, are available from the industry, providing additional support for the maturity and practicality of the laser design. Experimental results using HCLs not only show conclusively its power scalability, but also demonstrate the many uses for the laser system. The near future will see the HCL transformed from a laboratory device to an established product geared for a variety of real-world applications, providing solutions to a variety of situations.

References

1. Yang, X., et al., "2277-W Continuous-Wave Diode-Pumped Heat Capacity Laser," *Chinese Optics Letters*, 5(4), April 2007.
2. Guo, M.-X., et al., "A Kilowatt Diode-Pumped Solid-State Heat-Capacity Double-Slab Laser," *Chinese Physics Letters*, 23(9), May 2006.
3. Yamamoto, R., et al., "Evolution of a Solid State Laser," SPIE Defense & Security Symposium, UCRL-ABS-229142, April 2007.
4. Yamamoto, R. M., et al., "The Use of Large Transparent Ceramics in a High Powered, Diode Pumped Solid State Laser," Advanced Solid State Photonics Conference, UCRL-CONF-235413, January 2008.
5. Simmtec, Allison Park, Pennsylvania: http://www.simm-tec.com.
6. Konoshima Chemical Company, Takuma-cho, Mitoyo-gun, Kagawa, Japan: http://www.konoshima.co.jp.
7. Baikowski Japan Company, Ltd., Chiba-ken, Japan: http://www.baikowski.com.
8. Jancaitis, K. S., *Laser Program Annual Report*, UCRL 50021–87 (p. 5-3), Livermore, CA: Lawrence Livermore National Lab, 1987.
9. Beckmann, P., and Spizzichino, A., *The Scattering of Electromagnetic Waves from Rough Surfaces*, New York: Pergamon Press, 1963.
10. Devroye, L., *Non-Uniform Random Variate Generation* (Chap. 2), New York: Springer-Verlag, 1986.
11. Yamamoto, R., et al., "Laser-Material Interaction Studies Utilizing the Solid-State Heat Capacity Laser," 20th Annual Solid State and Diode Laser Technology Review, UCRL-CONF-230816, June 2007.
12. "Laser Burrows into the Earth to Destroy Land Mines," *Science & Technology*, October 2004 (https://www.llnl.gov/str/October04/Rotter.html).

CHAPTER 12
Ultrafast Solid-State Lasers

Sterling Backus

Vice President, Research and Development, Kapteyn-Murnane Laboratories, Inc., Boulder, Colorado

12.1 Introduction

Over the past 15 years, ultrafast laser technology and its applications have progressed by leaps and bounds, ever since the widespread introduction of solid-state ultrafast laser materials in the early 1990s.[1] In 1990, the state-of-the-art femtosecond (fs) laser used dye laser media and could generate output powers in the tens of milliwatt (mW) range, with pulse durations of ~100 fs. The successful application of titanium-doped sapphire (Ti:sapphire) to ultrafast lasers immediately resulted in an order of magnitude increase in average power (to ~1 W), as well as in the ability to easily and reliably generate pulses of less than 10 fs.[2] This technological advance has since led to a tremendous broadening of the field of ultrafast science, and more applications could be successfully implemented with the new generation of lasers. For example, the use of ultrafast lasers for machining and materials ablation began in the mid-1980s, with the realization that the high-intensity laser–matter interaction is fundamentally different on femtosecond (compared with picosecond or nanosecond) timescales, allowing for a much more precise and well-controlled ablation.[3] Peak powers into the petawatt (PW) regime have been realized, owing to ultrafast pulses.[4–6] This high-peak-power capability has also defined many other applications throughout physics, chemistry, and biology. However, the "real world" applications of femtosecond lasers only became practical with the development of high-power solid-state (predominantly Ti:sapphire) lasers. Femtosecond lasers are now used in a few industrial and medical settings, such as the precision machining of explosives without detonation[7] and the cutting of the corneal flap for refractive corrective surgery, just to name a few.[8]

Ti:sapphire, however, has its limitations. For example, until recently it was not directly diode pumpable. Although it can be pumped with 4XXnm diode lasers in an oscillator,[9] output power is limited because these pumps are low power, and the nonlinear absorption effects are quite severe. New directly diode-pumped materials have become more widespread. Ytterbium-, chromium-, and erbium-doped materials can have broad emission bands and low quantum defect (reduced thermal problems) and are directly pumped by high-power laser diodes. The first years of the 21st century have seen continued rapid progress in the development of higher average-power ultrafast lasers, with the introduction of widespread thermoelectric and cryogenic cooling technology for ultrafast laser amplifiers to mitigate large thermal effects. These effects plague laser systems across the board and are not unique to femtosecond lasers; however, they can have a dramatic effect on the generation of short pulses.

This chapter describes ultrafast sources, amplification methods, thermal mitigation, and ways to measure the fastest events ever recorded in human history.

12.2 Ultrafast Laser Sources and Oscillators

Modern ultrafast sources are predominantly solid state and passively mode locked. Two specific types of mode locking are used today—Kerr lens mode locking and mode locking from saturable absorbers, specifically semiconductor saturable absorber mirrors (SESAMs).[10]

12.2.1 Kerr Effect

In 1990, the modern solid-state ultrafast laser was developed by Wilson Sibbett at the University of St. Andrews.[11] This laser used a new, passive mode-locking mechanism and a third-order effect, known as the Kerr effect, that was given by a change in the index of refraction in Ti:sapphire:

$$n(\omega) = n(\omega_0) + \frac{3\chi^{(3)}}{8n(\omega_0)} I(\omega); \quad n_2(\omega_0) = \frac{3\chi^{(3)}}{8n(\omega_0)} \quad (12.1)$$

where $n(\omega)$ is the index of refraction, $\chi^{(3)}$ is the third-order susceptibility tensor component magnitude, and $I(\omega)$ is the intensity of light. The nonlinear index $n_2(\omega_0)$ gives rise to a lensing effect at the very peak of the intensity profile, with a value of ~2×10^{-16} cm^2/W for Ti:sapphire. If an optical cavity is designed with the lens shown in Eq. (12.2) in mind, passive mode locking can occur.

$$f_{\text{Kerr}}^{-1} = \frac{4n_2(\omega_0)L_m}{\pi w^4} P \qquad (12.2)$$

Here L_m is the material length, w is the beam radius, and P is the beam power. From this expression for a typical Ti:sapphire oscillator, we get a focal length of ~1 m.

12.2.2 Ultrafast Oscillators

A typical ultrafast oscillator has some distinguishing characteristics. First, it needs a pump source, whether diodes or another laser. Second, it needs some form of dispersion compensation—either prisms, chirped mirrors, or both, depending on the desired result.[12] Finally, some sort of starting mechanism, such as a shock (prism jog is typical), to induce an intensity modulation to start the Kerr effect, or a SESAM, which induces lower loss for a given intensity. Figure 12.1 shows a standard Ti:sapphire laser. Note that if the cavity is set just right, self-mode locking can occur.

Many other femtosecond lasers have since been developed and are widely used today. Table 12.1 gives a sampling of available femtosecond laser sources. These sources can cover a wide range of pulse durations, from less than 10 fs to 1 ps. An advantage of some of these ultrafast laser sources is their ability to be directly laser diode pumped, which can reduce cost and complexity. Ti:sapphire, which has the potential for the shortest pulses, still must be pumped by complex intracavity-doubled Nd:YVO (neodymium-doped yttrium orthovanadate) lasers. Although new laser diodes in the 4XXnm regime, and potentially in the 5XXnm regime, may help this problem, this technology has a long way to go to reach usable powers of around 1 to 5 W at 532 nm.[13] New optically pumped semiconductor

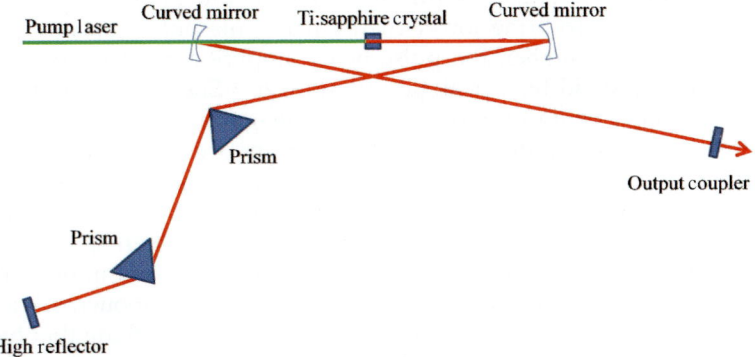

FIGURE 12.1 Diagram of a standard Ti:sapphire oscillator with prisms used for phase compensation.

Material	Center Wavelength	Pulse Duration	Pump Laser	Typical Average Power
Ti:sapphire	800 nm	< 10 fs	Nd:YVO, 532 nm	100–2000 mW
Yb:KGW/KYW	1050 nm	< 200 fs	Diodes, 980 nm	1–3 W
Yb:YAG	1030 nm	< 200 fs	Diodes, 940 nm	1–10 W
Cr:LiSAF	840 nm	< 50 fs	Diodes, 670 nm	100 mW
Cr:Forsterite	1235 nm	< 100 fs	Nd:YAG, 1064 nm	100 mW
Cr:ZnSe	2500 nm	< 100 fs	Tm:Fiber, 1900 nm	50–100 mW
Er:Fiber	1550 nm	< 50 fs	Diodes, 940 nm	50 mW
Yb:Fiber	1030 nm	< 200 fs	Diodes, 980 nm	100–1000 mW

TABLE 12.1 Sample of Femtosecond Sources (List is Not Meant to be Comprehensive.)

lasers (OPSLs) have been introduced as a new source for pumping Ti:sapphire.[14] In addition, frequency-doubled fiber lasers are an attractive low-cost alternative to Nd:YVO systems.[15]

12.3 Ultrafast Amplification Techniques

Ultrafast laser systems suffer from complexity due to their high peak power nature. To bring lower-energy nanojoule pulses up to millijoule pulses or higher, the pulse being amplified must increase in duration to avoid high peak powers (Power = Energy/Duration) in the amplifier chain to avoid causing damage. In 1985, the idea of chirped pulse amplification (CPA) was introduced as a method for bringing low-energy, ultrafast pulses to energies of less than 1 J.[16] The broad-bandwidth nature of ultrafast pulses can also be challenging. Because bandwidths can be rather large (oscillators can span more than an octave), managing all the different frequencies can be difficult. Care must be taken when choosing ultrafast components, such as waveplates, polarizers, Brewster windows, and anything that has a frequency-dependent result. In particular, strongly dispersive elements, such as gratings and prisms, have a propensity to introduce aberrations by coupling the spatial and spectral content of the beams.

12.3.1 Chirped Pulse Amplification

CPA starts by "stretching" the low-energy pulse from an oscillator by passing it through a 1:1 imaging system that contains a frequency-separating element, such as a grating or prism. This imaging system is then moved out of the imaging plane, leading to a different path length for each frequency in the ultrafast pulse. This technique effectively "chirps" the pulse and can add 1×10^5 in stretch, taking a 10-fs pulse to 100 to 1000 ps. After this stretching, amplification can be safely done to greater than 10^9, or from 1 nJ to 1 J (Fig. 12.2).

After amplification, recompression is done by a compressor, which is typically a grating pair. The grating pair undoes the stretch originally put on the pulses by the stretcher. In theory, the stretch put on by a stretcher is given by[17,18]

$$\varphi_s(\omega) = -\frac{8\omega L}{c}\left[1 - \left(\frac{2\pi c}{\omega d} - \sin\gamma\right)^2\right]^{1/2} \quad (12.3)$$

where $\varphi_s(\omega)$ is the phase delay between the frequencies of light in the pulse denoted by ω, L is the length that the stretcher is detuned from the focal plane, d is the grating groove spacing, and γ is the grating's incident angle. Conversely, the grating pair compressor is related simply by a change in sign and a factor of 2; for the stretcher in Fig. 12.3, L is defined as deviation from the focal plane, whereas in the compressor (Fig. 12.4), it is defined as the distance between the gratings:

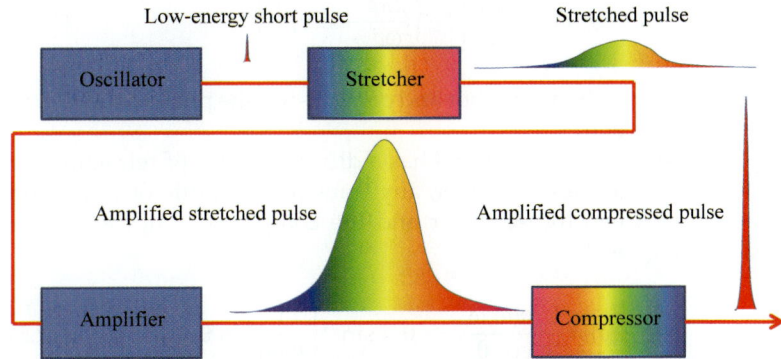

FIGURE 12.2 Diagram of chirped pulse amplification used to avoid damage in ultrafast laser amplifier systems.

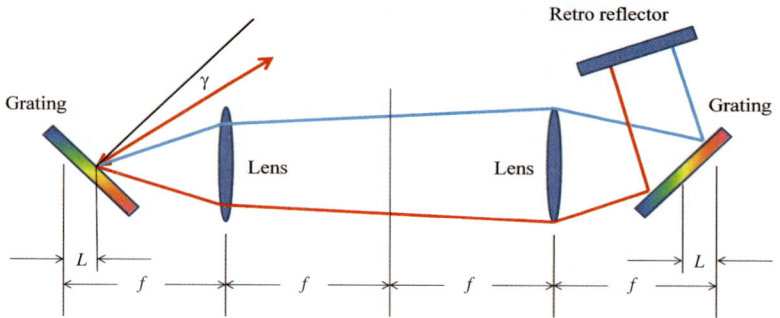

FIGURE 12.3 Diagram of 1:1 telescope-style pulse stretcher. Shown is a double-pass system (4f), with lenses for clarity. Mirrors may be used to avoid chromatic aberrations in the stretcher.

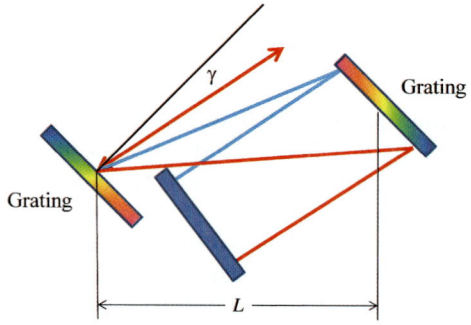

FIGURE 12.4 Diagram of Treacy-style grating compressor.[19]

$$\varphi_c(\omega) = \frac{4\omega L}{c}\left[1 - \left(\frac{2\pi c}{\omega d} - \sin\gamma\right)^2\right]^{1/2} \quad (12.4)$$

With such a matched stretcher and compressor, the phase delay for all frequencies is zero.

Because refractive material has a different index of refraction for different frequencies, however, any transmissive optic or laser gain material between the stretcher and the compressor imposes phase distortion, given by

$$\varphi_m(\omega) = \frac{L_m n(\omega)\omega}{c\,\cos\theta} \qquad \theta = \sin^{-1}\left[\frac{1}{n(\omega)}\sin\theta_i\right] \quad (12.5)$$

where $\varphi_m(\omega)$ is the material phase delay, L_m is the material length, $n(\omega)$ is the index of refraction, θ is the refracted angle inside the material, and θ_i is incident angle.

The beam layout of the stretcher can be understood as follows: The beam from the oscillator is directed onto the first diffraction grating and is then imaged via the focusing optics onto a second grating. The beam is then retroreflected back through the grating pair, which returns all the frequencies to a single spatial mode. It is critical that the focusing optics are separated by 2f. If they are not, the output beam will not return to the same spatial mode, leading to a condition known as *spatial chirp*, which is a very undesirable frequency sweep across the laser beam. When the gratings are placed at points other than the object and image planes, the path lengths for the lower and higher frequencies are different, which causes a temporally chirped pulse to emerge from the stretcher. The degree of chirp depends on L [see Eq. (12.3)], or the distance from the gratings to the image and object planes. In the stretcher, the gratings are placed inside the image and object planes, resulting in a shorter path length for the redder wavelengths, and thus a positive chirp. Typically, a stretcher is aligned to stretch a 15- to 20-fs pulse to a 150- to 400-ps pulse.

Dispersion is necessarily introduced whenever the beam passes through any material; this is due to variations in refractive index over the beam's wavelength range [Eq. (12.5)]. Dispersion further positively chirps the pulse. However, higher-order terms of this dispersion are difficult to counteract when recompressing the pulse. Therefore, curved mirrors, rather than lenses, are typically used as the focusing optics in the stretcher design. After amplification, the pulses will accumulate a certain amount of phase distortion, which is defined as high-order phase terms that cannot be compensated for by the stretcher and compressor. However, a slight mismatch in the incident angle and L of the compressor can compensate terms up to the third order in the Taylor expansion of the total phase of the system (stretcher, amplifier material, and compressor), given by

$$\varphi_{sys}(\omega) = \varphi(\omega_0) + \varphi'(\omega_0)(\omega - \omega_0) + \frac{1}{2!}\varphi''(\omega_0)(\omega - \omega_0)^2 \qquad (12.6)$$
$$+ \frac{1}{3!}\varphi'''(\omega_0)(\omega - \omega_0)^3 + \cdots$$

where $\varphi_{sys}(\omega)$ is the total system phase delay and ω_0 is the pulse center frequency. The first two terms are constants related to the absolute time delay of the pulses, and the φ'' and φ''' terms are the group-velocity dispersion (GVD) and third-order dispersion (TOD), respectively. These terms, and their effect on the resulting output pulses, will be discussed in Sec. 12.5.

12.3.2 Aberrations

Misalignment of stretcher and compressor optics can have deleterious effects on ultrafast pulses. The main effects to watch for are spherical aberration, chromatic aberration (when using lenses), thermal distortion, and spatial chirp. One way to address spherical aberrations is to use a ray-tracing software package when designing a stretcher. Chromatic aberrations can either be eliminated by removing any lenses in the system or be greatly reduced by using F-numbers (Focal length/Beam diameter) greater than ~20 for 40 nm of bandwidth. Spatial chirp can be greatly reduced by making sure that (1) in stretchers, the spread-out spectrum does not receive any tilt, and (2) in compressors, the gratings, as well as their lines, are parallel face to face. (For thermal distortions, see Sec. 12.4.) More on these and other aberrations can be found in Muller et al.[20]

12.3.3 Amplifier Schemes

The main goal of amplification is to bring low-energy pulses in the nanojoule regime to high-energy pulses in the millijoule to joule regime for high-intensity experiments. At these levels, with say 20-fs pulses, intensities greater than 1×10^{19} W/cm² can be obtained, which is extremely useful in high-field physics and materials processing. To efficiently extract the stored energy from the amplifier, one must reach the material's saturation fluence. For a four-level laser, this is given by

$$F_{sat} = \frac{h\omega}{2\pi\sigma(\omega)} \qquad (12.7)$$

where h is Plank's constant, and $\sigma(\omega)$ is the stimulated emission cross section as a function of frequency. In the case of Ti:sapphire, the saturation fluence is ~1 J/cm², and working at $2F_{sat}$ will typically give the best extraction efficiency. However, one must be careful, because Yb:KGW (ytterbium-doped potassium gadolinium tungstate) has a saturation fluence of ~10 J/cm², and $2F_{sat}$ will exceed the material's damage threshold, making energy extraction very difficult, though not impossible.

Two types of amplifier schemes are used in amplifying ultrafast pulses (at least where a storage medium is concerned): regenerative amplification and multipass amplification. This section illustrates the advantages and disadvantages of both schemes. Regardless of which scheme is used, the effect of B integral, gain narrowing, and frequency pulling prevent ultrafast amplifiers from producing pulses as short as those that come from the oscillator. Gain narrowing is a result of a finite gain bandwidth in the amplifying medium:

$$n(t, \omega) = n_i(0, \omega)e^{\sigma(\omega)\Delta N} \qquad (12.8)$$

where $n(t, \omega)$ is the total amplification factor, and ΔN is the excited state population.[17] Because the small signal gain is exponential, the frequencies at the edges of the gain bandwidth will see less gain than will the center frequency. This effectively narrows the amplified spectrum, which, in turn, increases the compressed pulse duration. Other factors, such as finite bandwidth mirror sets and other optical elements, can also reduce the overall bandwidth.

In high-intensity lasers, a nonlinear process that arises from the amplified beam's gaussian intensity distribution leads to a lensing effect known as B integral. This effect is a nonlinear phase shift across the beam profile:

$$\varphi(t) = \frac{\omega_0}{c} n_2 \int I(t, l) \, dl \qquad (12.9)$$

where n_2 is the nonlinear index for a given material, and $I(t, l)$ is the beam intensity. B is the peak value of Eq. (12.9); in practice, B should be kept to a minimum in the amplifier. Large amounts of B (much greater than 1 rad) can lead to self-focusing and damage in the amplifier or to filamentation outside the amplifier after compression.

Frequency pulling happens when the amplifier reaches saturation. Because the red frequencies lead the blue in a positively chirped pulse, it sees higher gain in saturation. This causes the peak of the spectrum to red shift, which can be undesirable.

12.3.4 Regenerative Amplification

A regenerative amplifier (also known as a *regen*) is basically a stable optical cavity with either an acousto-optic modulator (AOM) or an electro-optic modulator (EOM) that switches pulses for a number of gain passes and then extracts the amplified pulse out of the cavity. Figure 12.5 shows a typical regen amplifier.[21]

Two major advantages of the regen amplifier are its simplicity and the fact that it is an optical cavity, which gives out superior beam

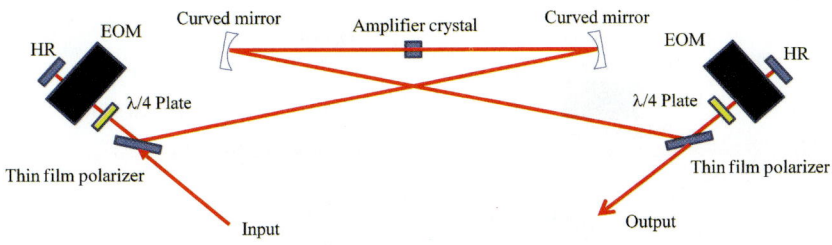

FIGURE 12.5 Regenerative amplifier diagram. The electro-optic modulator can be replaced with an acousto-optic modulator. (The pump laser input is not shown.) HR: high reflector; EOM: electro-optic modulator.

quality. This can also be independent of the beam quality going into the amplifier. Regen operation is quite simple: The stretched pulses are injected through a thin film polarizer (TFP), where the EOM traps the pulse in the cavity. The pulses amplify; when they reach their peak, the other EOM switches the pulses out through a TFP. Typically in a millijoule regen in Ti:sapphire, it takes about 20 to 40 passes to amplify. Alternatively, a regen can be run with only one EOM and one TFP, which then requires a Faraday isolator to prevent the output from destroying back-stream optics.

Due to the large number of passes in the amplifier system and the amount of refractive material, this scheme suffers from large phase distortion. Therefore, it is difficult to recompress the pulses to less than 50-fs durations without adverse effects. In addition, as the gain changes (i.e., the pump laser power), the number of passes changes; therefore, the compressor must adjust to compensate both angle and separation. Although large bandwidths and short pulses have been obtained by regen amplifiers, these pulses can "breathe," due to small environmental changes because the pulse spectrum is highly confined by the phase distortion.[22,23]

12.3.5 Multipass Amplification

Another way to amplify is to pass the beams through the gain medium and have each pass spatially separated (Fig. 12.6).[24] The major advantage of this scheme is that it moves the EOM outside the amplifier, thus dramatically reducing the overall refractive material. This type of amplifier is also run at single-pass gains of ~10, rather than at the regen's ~2, which means there are fewer actual amplifier passes overall. Due to the lack of high phase distortion, pulses can be compressed to shorter durations with the multipass amplifier using standard techniques. Pulses as short as 15 fs at 1 mJ have been realized in a multipass amplifier.[25] Another advantage comes in the form of mitigating gain narrowing. Applying a filter (i.e., transmissive optic) in the first five or so passes to suppress the peak of the gain

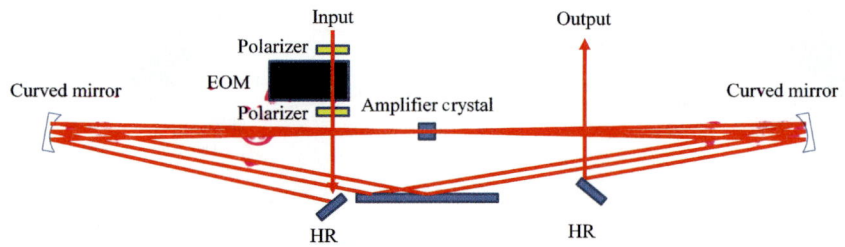

FIGURE 12.6 Multipass amplifier diagram. The EOM has been moved outside the amplifier, which greatly reduces the refractive material in the chain.

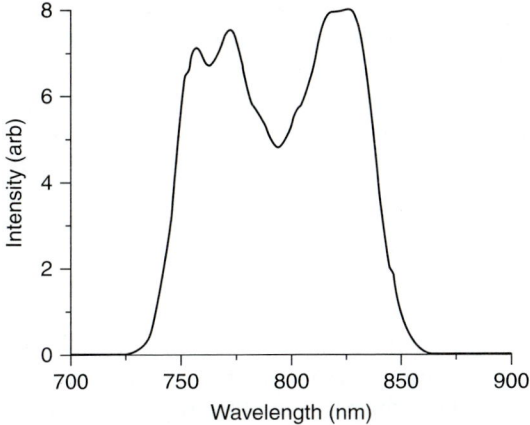

FIGURE 12.7 Spectrum from a multipass amplifier producing 16.8-fs pulses at 2 mJ and 10 kHz.

curve gives a flatter gain curve, resulting in a spectrum greater than 90 nm (Fig. 12.7).

Although this technique can also apply to regenerative amplifiers, the gain-flattening device must be in all passes. Although 90-nm spectra have been obtained, the wings of the spectrum are very sharp, giving rise to significant prepulse structure on the pulse output. One drawback of the multipass amplifier is that the beam quality can suffer if the amplifier is overdriven or if the energy extraction is too high.

12.3.6 Downchirped Pulse Amplification

As seen in Sec. 12.3.1, the CPA technique is a well-established method for generating high peak-power pulses with 10-fs to 10-ps duration. However, the CPA scheme has significant limitations, primarily associated with the construction and alignment of the pulse compressor. If even slightly misaligned, pulse compressors tend to exhibit "spatial chirp," or a physical separation of the colors of a pulse (see Sec. 12.3.3). Furthermore, pulse compressors tend to exhibit high loss (> 30%), as is discussed below. Thus, an alternative technique is needed for compression of chirped optical pulses emerging from an ultrafast laser amplifier or other optical device.

Past CPA implementations have used a configuration in which the pulse being amplified has a positive chirp (see Sec. 12.3.1). The pulse stretcher is configured such that the redder components of the pulse emerge from it earlier than the bluer components. After amplification, the compressor then undoes this by providing a "negative" dispersion—that is, in the compressor, the redder components have a longer optical path length than the bluer components. In a properly designed system, the entire optical system's net dispersion, including the stretcher, the amplifier components, and the compressor, is designed to be as nearly zero as possible. Typically, the pulse compressor consists of a pair of diffraction gratings (Fig. 12.4) or an equivalent configuration. In some past work, prisms, or a combination of prisms and chirped mirrors, have been used for the compression process.[26] The use of prisms, rather than gratings, has also been employed to avoid some of these limitations. However, prisms do not avoid spatial dispersion effects; furthermore, prisms typically need to use specially designed mirrors to compensate for residual higher-order dispersion.[27]

With downchirped pulse amplification (DPA), the pulse is stretched using negative dispersion. The pulse injected into the amplifier is thus negatively chirped—in other words, the blue colors come first in the pulse, and the red colors come later. This pulse stretching can be accomplished using a grating or prism pair, which is the same type of negative-dispersion element that is normally employed for recompressing the pulse. Other possible optical elements that might be included are specially designed mirrors, which compensate for dispersion or which correct for high-order dispersion errors introduced by other optical elements, or pulse shapers, which use adaptive-optics devices to adjust pulse dispersion in either a predetermined or a programmable manner. The use of grisms (grating-prism combinations) has also been successfully made (as is discussed later in this section).

Compression of the optical pulses after amplification is accomplished using positive dispersion. Perhaps the most advantageous way of doing this is by using material dispersion, or propagating the pulse through a block of glass or other transparent material. Other devices, such as the positive-dispersion grating arrangement used for pulse stretchers in CPA systems, could also be used. However, the use of a simple, transparent optical element has a number of significant advantages over past pulse compressor designs. First, a transparent material can be virtually lossless, thus avoiding the 30 to 50 percent loss in average power typical of a grating pulse compressor. Furthermore, it also helps avoid thermal distortion effects. Second, a simple block of glass is alignment-insensitive, making alignment of the pulse compressor, as well as accurate dispersion compensation, much simpler to obtain.

Unlike conventional CPA, the fully compressed femtosecond-duration pulse will emerge from a material, such as a block of glass or similar, that compresses the pulse. Thus, the possibility exists for nonlinear distortion of the pulse due to Kerr self-phase modulation, or the B integral [Eq. (12.9)].[17] However, this problem is not fundamental and unavoidable. After amplification, the pulse beam will typically be expanded to a larger physical cross section. By expanding the beam, the peak power inside the compressor can be kept low enough to avoid nonlinear distortions. The necessity for expanding the beam is not a major disadvantage over conventional CPA, because

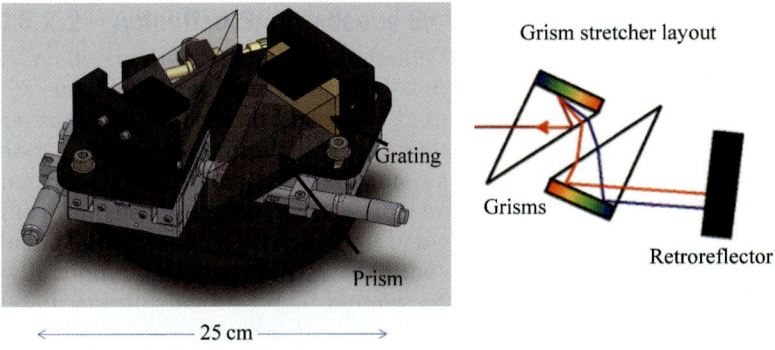

FIGURE 12.8 Grism stretcher for downchirped pulse amplification (DPA) ultrafast laser system. The stretcher is capable of chirping a pulse from 15 fs to greater than 40 ps.

the beam from a CPA laser must also typically be expanded to avoid damage to the gratings.

One advantage of DPA is the ability to use grisms as the stretcher for the amplifier system. Grisms, a combination of prisms and gratings, have a very high dispersion. (Figure 12.8 shows a diagram of a commercially available grism stretcher.) The GVD:TOD ratio of a grism pair is also an exact match for most bulk materials, which means that phase distortion in the system can be corrected for up to TOD.[28] For a full description of grism pairs, see Durfee, Squier, and Kane.[29]

Although DPA is attractive because it is simple and highly efficient, its main drawback is the large amounts of material in the system, which can lead to a substantial B integral. However, this technique has been very useful for ultrafast systems in the hundreds of microjoules to 1 millijoule range of energies at very high repetition frequencies.[30,31] A DPA system's compressor usually consists of a pair of mirrors and a block of glass, preferably one of the short flint glasses available from Schott. Figure 12.9 shows the layout of such a compressor,

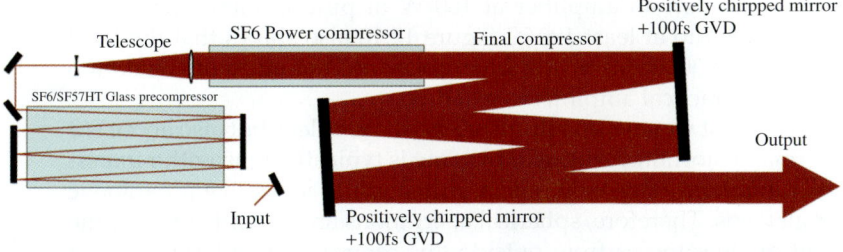

FIGURE 12.9 Glass compressor for DPA ultrafast laser system. Compressors can handle up to ~1 mJ, while keeping the B integral to ~1. GVD: group-velocity dispersion.

which has a precompressor, and a final compressor, in which the beam is expanded and the final compression step is with chirped mirrors.

12.4 Thermal Mitigation

Whether in a regenerative or a multipass amplifier, the first stage is by far the most sensitive to the deleterious effects of thermal lensing, thermal astigmatism, and spherical aberrations. This is because of the small mode size and the large number of passes through the gain material. Although it is possible to stabilize a first amplification stage under ~20-W pump powers with conventional water or thermoelectric cooling near room temperature, the system is then restricted to operate only at a single power level (i.e., a single energy and repetition rate), which makes it very inflexible in operation. Cryogenic cooling can extend this operation range to high average powers and high energies, minimizing aberrations. In the case of near-room temperature cooling, higher-order aberrations remain, drastically limiting beam quality. Although spatial filtering can restore beam quality, it is at the cost of laser efficiency and, therefore, of maximum operating power. The thermal lens is given by Koechner[32] as

$$f_{therm}^{-1} = \frac{dn}{dT}\frac{1}{2\kappa A}P \qquad (12.10)$$

where f_{therm} is the dioptric power, dn/dT is the refractive index change with temperature, κ is the thermal conductivity at a given temperature, A is the area in which the power is deposited, and P is the total power deposited. If we plug in some numbers for Ti:sapphire, we can see a factor of 250 reduction in the thermal lens power and, thus, a reduction in distortions in the pumped crystal as the temperature is reduced from 300 K to 77 K due to the drop in dn/dT and the increase in κ (Fig. 12.10).

If we look at the focal length of the thermal lens as a function of pump power, we can see that it would be difficult to make multiple passes through an amplifier at 100 W of pump, unless the crystal were cooled to at least 100 K. Figure 12.10 also shows that the focal length from 300 to 233 K only changes from 1 to 3 cm, which is far too short for practical amplifiers.

We must worry not only about the thermal lens but also about the thermal distortions. Because pumping is typically done with a gaussian mode, only the central part of that mode looks like a parabolic singlet lens. Therefore, spherical aberrations are present any time the seed mode samples from outside this central pumped region. As a rule, keeping the pump intensity below 7 kW/cm^2 has been somewhat successful with ultrafast lasers in the range of 300 to 233 K. In this case, spherical aberrations can be considered as a loss mechanism

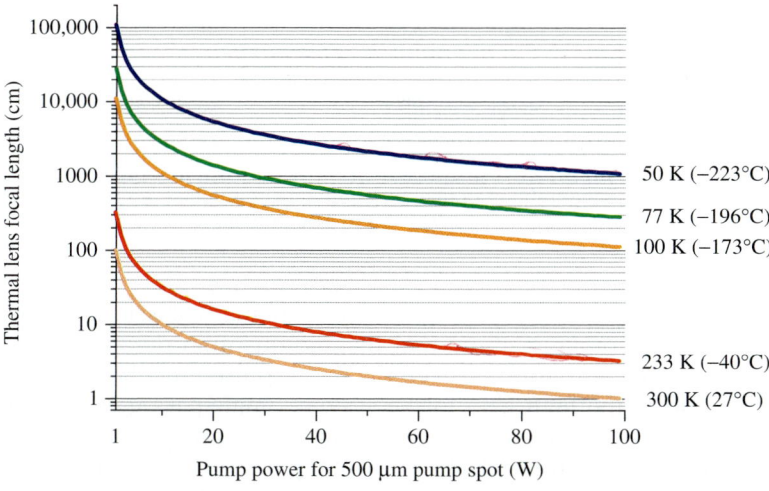

FIGURE 12.10 Thermal lens focal length as a function of deposited power in a Ti:sapphire rod from 300 to 50 K.

for a regenerative amplifier, because they act as a strong spatial filter. However, more loss leads to higher overall gain to reach the desired output and, therefore, to more phase distortion and gain narrowing. For high-power applications with greater than 20 W of pump, cryo-cooling is preferable.

12.4.1 Optical Parametric Chirped Pulse Amplification

Classically near- and midinfrared ultrafast pulses have been generated using Ti:sapphire-amplified laser systems in conjunction with an optical parametric amplifier (OPA). This system can generate very short (< 50 fs) pulses in the OPA idler around 2 µm. However, in this scheme, the Ti:sapphire laser (though a rugged technology) can have a large footprint and require laboratory-like conditions. These systems also tend to be quite expensive (~$300,000). In addition, in terms of reliability, Ti:sapphire systems require bulky, frequency-doubled Nd laser systems. Although fiber-based green pump lasers are now available and have been used to pump high-power Ti:sapphire oscillators, they are a very new technology with energy scalability issues.

Optical parametric chirped pulse amplification (OPCPA) provides an alternative to laser amplification.[33] It uses the nonlinear process of parametric generation (Fig. 12.11), which splits a pump photon into two parts: the signal (the high-energy photon) and the idler (the low-energy photon). It also has the advantage of being able to use standard stretching and compression techniques, such as CPA and DPA.

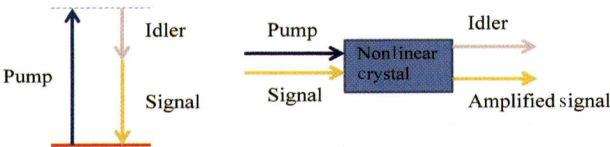

FIGURE 12.11 Schematic of optical parametric chirped pulse amplification (OPCPA). The pump laser is usually a 10 to 100-ps source at ~1 μm.

This process may seem simple and free of thermal issues, because there is no storage medium and thus no quantum defect. For this process to be efficient, however, the pump pulse must be square in time due to the high gain in the system. Single-pass gains can be greater than 1000; therefore, if we want to amplify a chirped pulse, the gain (which is now related to the pump pulse shape) must be flat; otherwise, gain narrowing can be quite severe. In addition, if the pump pulse is gaussian, we can only amplify in the narrow central region of the gaussian intensity profile, which leaves the temporal wings of the pump laser unconverted, reducing the efficiency (see Fig. 12.12).

A spatially flattop or super-gaussian mode profile is also desired to avoid massive mode reshaping of the amplified beam. An OPCPA system's gain bandwidth can be very large, in some cases supporting less than 10-fs pulses. This gain bandwidth is a direct result of phase matching in the crystal used. In the case of OPCPA, cryocooling is not necessary; however, single-mode, high-beam-quality picosecond pump lasers must be used. The major advantage of this technology is the wavelength tunability for the entire system. The same architecture may be used for many different wavelengths, from the ultraviolet into the midinfrared. Figure 12.13 shows a scaled version of a recently demonstrated 3.0-μm OPCPA system.[34] This system uses a fiber oscillator (Er:Fiber), which is split and

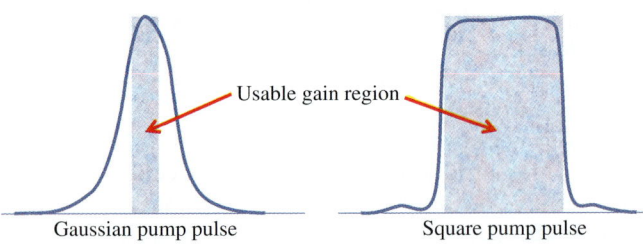

FIGURE 12.12 Pump pulse temporal profile for efficient OPCPA. The super-gaussian, or "square," pulse leaves less energy behind, greatly improving efficiency.

Ultrafast Solid-State Lasers

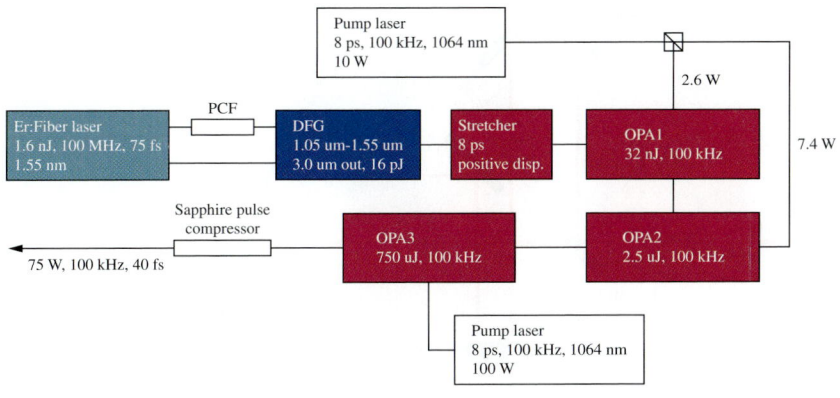

FIGURE 12.13 Scaled OPCPA laser system.

amplified in two channels. One channel is spectrally broadened to create wavelengths at 1.05 μm. The 1.55 and 1.05 μm beams are then converted using difference frequency generation (DFG) to 3.0 μm. The beam is then stretched to match the pump laser pulse width, is amplified in three OPA stages, and is finally compressed in a sapphire block. The pumps are Nd:YAG mode-locked lasers in either a master oscillator power amplifier (MOPA) or a regenerative amplifier configuration. This OPCPA scheme has also led to very high peak powers of greater than 1 PW.[35] Gaul, Ditmire, et al. constructed a tabletop petawatt laser system that is capable of 1.1 PW in 167 fs and 186 J of energy.

12.5 Pulse Measurement

Measuring femtosecond pulses can be tricky, because electronic methods can only measure ~10-ps pulses. Therefore, optical techniques must be used to determine the pulse duration of a femtosecond pulse. One advantage of using optical techniques is that short pulses can more easily drive nonlinear processes, which are intensity dependent. The first method used was the process of autocorrelation,[36] which is essentially a Mach-Zehnder interferometer in which a nonlinear crystal (usually KDP [potassium dihydrogen phosphate] or beta barium borate [BBO] for Ti:sapphire wavelengths) is placed at the focus of the output.

The delay line is oscillated, and the detector's output can be read on an oscilloscope (Fig. 12.14). Although the measured autocorrelation width gives approximately the actual pulse width multiplied by the autocorrelation factor (1.55 for $sech^2$ and 1.41 for gaussian spectra), it does not give the shape or the phase of the pulse. A new

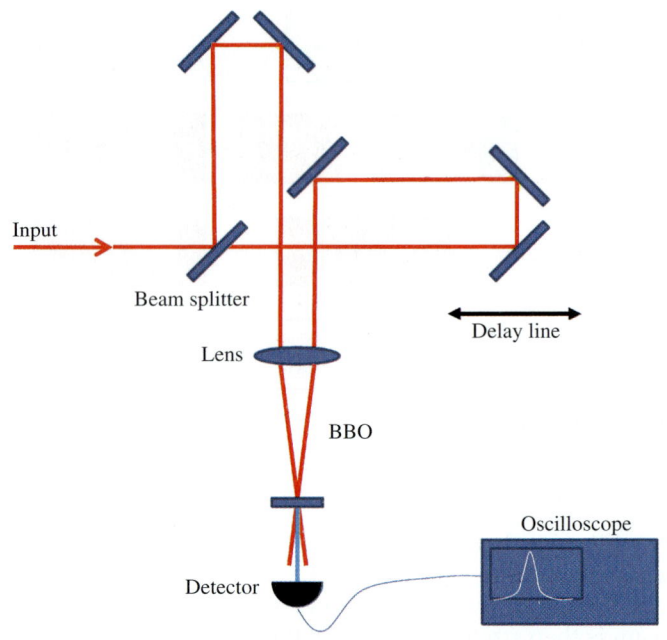

FIGURE 12.14 Autocorrelation setup.

technique, pioneered by Dan Kane and Rick Trebino, allows both the shape and phase of the pulse to be retrieved; in addition, this method will indicate whether the measurement is being done properly.[37] In this new method, which is called frequency resolved optical gating (FROG), the autocorrelator setup basically stays the same, while the detector is replaced with a spectrometer, so that a spectrum can be taken at each time delay. From this, a two-dimensional spectrogram is measured; this spectrogram carries all the amplitude and phase information. A simple algorithm is applied to the data to retrieve the pulse shape, temporal phase, spectrum, and spectral phase.

Other methods have since been developed that make the measurement faster and easier. On the FROG side, the GRENOUILLE (Swamp Optics, Inc.) is a real-time device that displays the spectrogram and the retrieved information.[38] Another device, called a Scan FROG, uses the standard Mach-Zehnder interferometer in addition to a voice coil to perform the time delay very quickly and an algorithm that can update at ~2 Hz. Another widely used method is the Spectral Phase Interferometry for Direct Electric-field Reconstruction (SPIDER).[39] Many other techniques may be found for a wide variety of wavelengths, from the extreme ultraviolet to the midinfrared.

12.6 Applications

12.6.1 Filaments

When focused into air, terawatt-level (10^{12} W) femtosecond laser pulses can, under the right circumstances, generate a tightly focused filament that can propagate, without diffraction, over extended distances.[40] These self-trapped filaments are formed by the balance of high-intensity, self-focusing of light with the generation ionization of the air, which defocuses the light. The result is a filament that keeps light focused at high intensity ($> 10^{15}$ W/cm^2) over extended (> 100 m) propagation lengths.

Because this light, when incident on a solid target, is intense enough to cause ablation, filamentation has attracted recent attention for military applications. Although a single filament is not sufficient to directly cause disabling damage to an enemy missile or aircraft, a large number of co-propagating filaments could cause significant damage in a way that is exceedingly difficult to protect against, because no material can sustain greater than 10^{15} W/cm^2 without damage. This can prepare the target's surface for efficient absorption of high-energy, longer-duration pulses that might otherwise simply be reflected without harm. The disruption of optical and imaging sensors is another obvious potential application. Furthermore, the target composition could be determined by a "remote" version of laser-induced breakdown spectroscopy (LIBS). Finally, emission of an electromagnetic pulse from the laser-matter interaction may also provide opportunities for disruption of sensors and electronic systems.

Another use of these high-intensity filaments is as a backlight source for measuring atmospheric composition. This possibility has been demonstrated, in dramatic fashion, in Europe, where a terawatt (TW) laser system built into a cargo container, called the "teramobile" (www.teramobile.org), has been used for a variety of atmospheric studies.[41] These studies were made possible by the findings that a white-light filament can be generated in the upper atmosphere at altitudes up to 20 km and that the white light generated preferentially scatters in the backward direction.[42] These characteristics essentially provide a multispectral "lightbulb" source that can be placed anywhere within the range of the laser, giving simultaneous spectral and light detection and ranging (lidar) information, while also making it possible to measure atmospheric absorption and identify pollutants and contaminants, such as atmospheric aerosols.

12.6.2 Precision Machining with Minimum Collateral Damage

In recent years, micromachining with femtosecond lasers has received considerable attention from researchers because the dynamics of

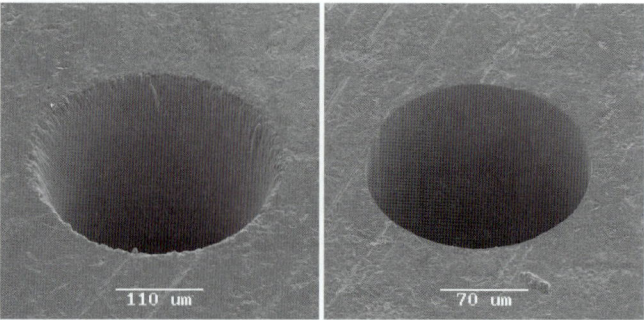

FIGURE 12.15   Entrance (left) and exit (right) micrographs of hole drilled in 1-mm mild steel stock, using a cryogenically cooled, high-average-power Ti:sapphire amplifier system.[44] Using this laser resulted in greater than 10X reduction in drill time compared with previous efforts—1.5 seconds was required to drill this hole.

material removal can be substantially different from that of longer pulses, going from melt expulsion for microsecond and nanosecond pulses to vaporization or sublimation for femtosecond pulses. These dynamical differences produce concrete differences in the results obtained during laser machining of surfaces (Fig. 12.15). Machining with femtosecond laser pulses generally reduces the amount of debris and surface contamination compared with that produced by longer pulses, a feature that is partially responsible for making femtosecond lasers the preferred tools for repairing photolithographic masks. Femtosecond lasers also have advantages for micromachining inside transparent materials.[43]

When a femtosecond laser pulse is focused inside the bulk of a transparent material, the intensity in the focal volume can become high enough to cause absorption through nonlinear processes, leading to optical breakdown in the material. Because the absorption is strongly nonlinear, this breakdown is localized to only the regions of highest irradiance in the focal volume, without affecting the surface. The energy deposited in the bulk material then produces permanent structural changes in the sample, which can be used to micromachine three-dimensional structures inside the bulk of the material. Moreover, the threshold nature of a femtosecond pulse interacting with a material allows ultrashort pulse machining with feature sizes below the diffraction limit. Although micromachining in glasses and crystals has many uses, another means of producing microscopic structures is to use light-induced polymerization, in which light initiates a polymerization reaction to produce a solid polymeric object. Other researchers have already used single-photon polymerization to fabricate microrotors only 5 µm in diameter and to produce light-powered micromachinery. Femtosecond lasers are also used for microfabrication by light-induced polymerization, because two-photon absorption may be used to initiate the polymerization reaction. As in their use for micromachining, femtosecond laser pulses allow significant two-photon absorption in localized volumes and can produce small, high-resolution spatial features.

12.6.3 Laser-Based Photon and Particle Sources

Recent experiments have demonstrated that an intense femtosecond laser pulse focused into a gas-puff target can drive a strong plasma wake field, which can accelerate electrons to tens of mega-electronvolt energy in a propagation distance of just a few millimeters. Recent experiments have shown that under the right conditions, the emitted electron bunch can be *monochromatic*, with an energy bandwidth of a few percent at electron energies as high as 80 MeV.[45] In this recent work, ultrafast laser pulses of peak power ~1 to 10 TW are needed to effect relativistic self-focusing of the pulse at intensities greater than 10^{18} W/cm^2, which is required to generate the wake field. Further work in ultrafast laser development, as well as continued progress in optimizing parameters for plasma generation, will allow this type of electron accelerator to work reliably at higher 100 to 1000 Hz repetition rates. This would make laser-plasma-based electron sources practical for such applications as radioisotope production and, eventually, for use as much-brighter photoinjector electron sources for compact free-electron lasers.

An important set of new, high-resolution radiological applications may soon be possible using these intense lasers. It is well known that such intense laser pulses can produce copious amounts of radiation of various sorts. These radiation sources range from photons with energy of 10 eV to greater than 1 MeV to a host of particles, including neutrons, protons, and electrons. Moreover, the fluxes of these radiation sources can be quite substantial, even though their source sizes are small (microns to tens of microns). Therefore, these new photon and particle radiation sources would have unique applications in high-resolution probing of materials—for example, to detect buried voids and cracks in aircraft surfaces or to understand fuel flow and combustion. Thus, this field has the potential for these sources of extreme photons and particles to be combined with novel diagnostic techniques to realize enabling technologies that will have a major impact on diverse areas of the defense, manufacturing, environmental and medical industries.

12.6.4 High Harmonic Generation

One of the major thrusts in modern science and technology has been to understand and make use of electromagnetic (EM) radiation. Understanding the interaction of EM radiation with matter led to the development of quantum theory and, subsequently, of solid-state physics, electronics, and lasers. However, one region of the EM spectrum has been relatively underutilized to date: the extreme ultraviolet (EUV) and soft x-ray range, corresponding to wavelengths 10 to 100 times shorter than visible light and with photon energies in the range of tens to hundreds of electron volts. EUV light is both useful and difficult to exploit for the same reason—that is, it is ionizing radiation that interacts strongly with matter. This strong interaction makes EUV light difficult to generate and severely restricts the types of optics that can be used. However, the development of EUV optical technologies has strong motivation. With EUV light, it is possible to make microscopes that can resolve smaller features than is possible using visible light. In addition, with EUV lithography, it is possible to write smaller patterns. Furthermore, these wavelengths are well matched to the primary atomic resonances of most elements, making possible many element- and chemical-specific spectroscopies and spectromicroscopies.

The compelling scientific applications of EUV light have led to the development of several dozen large-scale synchrotron radiation sources, with more than 10,000 users worldwide. However, synchrotron light sources have major disadvantages, especially when uses for EUV light move from the research lab into manufacturing or analytical applications. The most obvious disadvantage is the large size and cost of these sources. Experiments must be constructed at the facility itself, and any samples must be brought to the facility. Furthermore, a number of emerging applications of EUV and soft x rays, such as soft x-ray holography, require *coherent* light. This need has prompted the development of large-scale "fourth generation" free-electron lasers. However, these sources are even larger and often more costly than synchrotron light sources. The need for *small-scale* coherent light sources has motivated research in both x-ray lasers and upconversion of coherent light from a laser to very short wavelengths. During the past decade, both types of light sources have been successfully used for a variety of application experiments, such as nanoscale imaging and studies of molecular dynamics.

In particular, the process of high-order harmonic generation (HHG) has proven to be a very useful coherent tabletop x-ray laser source that can be used for a variety of applications in basic and applied science (Fig. 12.16).[46,47] In HHG, a very intense femtosecond laser focused into an atomic gas is upconverted into the EUV or soft x-ray regions of the spectrum. The HHG process results from a complex laser-atom interaction, in which the light from an incident intense laser pulse first pulls an electron from an atom through a process of field ionization and then drives this electron back into its parent ion. The resulting recollision process coherently emits a short-wavelength photon whose energy is given by

$$E_{max} = I_p + 3.2 U_p \qquad (12.11)$$

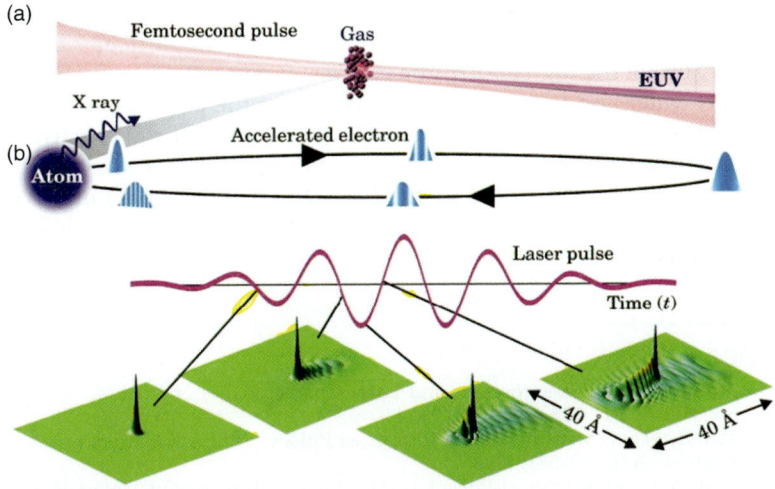

FIGURE 12.16 In the process of high-harmonic generation, coherent x-ray beams are generated through a coherent electron ionization and recollision process. (*a*) The classical picture of strong-field ionization. (*b*) A representation of the quantum equivalent.

where I_p is the ionization potential of the atom, and $U_p \propto I\lambda^2$ is the ponderomotive potential or energy gained in the driving field. The dynamics of the recollision process occur on attosecond timescales; an understanding of this process has led to the birth of the field of attosecond science.[48–53] The HHG radiation is actually emitted as a sequence of attosecond bursts; under the correct conditions, single isolated attosecond pulses can result.[54] The extremely short duration of the EUV and soft x-ray light emitted by HHG makes it possible to observe extremely fast processes in atomic, molecular, and solid-state systems.

The HHG process is powered by high-power ultrashort pulse lasers. Although the required intensities of up to 10^{15} W/cm² are comparable to those used in laser fusion, the pulse energy required to obtain this intensity is modest because femtosecond duration pulses are used. The high-power laser used to drive the HHG process can easily fit into a fraction of a standard optical table, essentially providing a robust and practical way of implementing a tabletop EUV or soft x-ray laser. Much of the recent rapid progress in the use of HHG has been due to the development of a new generation of tabletop-scale, solid-state, ultrashort-pulse lasers capable of generating femtosecond pulses with very high peak and average power. In the longer term, the further development of HHG-based light sources at shorter wavelengths in the "water window" region of the soft x-ray spectrum (corresponding to photon energies of greater than 300 eV) will allow for ultrahigh-resolution biological and materials imaging.[55,56] The water window region of the soft x-ray spectrum is of particular interest because of the high absorption contrast between (relatively transparent) water and (opaque) carbon, as well as the presence of many absorption edges throughout the soft-x-ray region of the spectrum. Several soft x-ray microscopes have been implemented at synchrotron facilities worldwide and have provided new capabilities for biology—for example, in three-dimensional tomographic imaging of single cells.

References

1. Moulton, P. F., "Spectroscopic and Laser Characteristics of Ti:Al$_2$O$_3$," *J. Opt. Soc. Am. B*, 3(1): 125–133, 1986.
2. Taft, G., et al., "Measurement of 10-fs Laser Pulses." *IEEE J. Select Topics Quant. Electron.*, 2(3): 575–585, 1996.
3. Tien, A. C., et al., "Short-Pulse Laser Damage in Transparent Materials as a Function of Pulse Duration," *Phys. Rev. Lett.*, 82(19): 3883–3886, 1999.
4. Perry, M. D., and Mourou, G., "Terawatt to Petawatt Subpicosecond Lasers," *Science*, 264: 917–923, 1994.
5. Perry, M. D., et al., "Petawatt Laser Pulses," *Opt. Lett.*, 24(3): 160–162, 1999.
6. Pennington, D. M., et al., "Petawatt Laser System and Experiments," *IEEE J. Select Topics Quant. Electron.*, 6(4): 676–688, 1994.
7. Roeske, F., et al., "Cutting and Machining Energetic Materials with a Femtosecond Laser," *Propellants Explosives Pyrotechnics*, 28(2): 53–57, 2003.
8. Juhasz, T., et al., "Corneal Refractive Surgery with Femtosecond Lasers," *IEEE J. Select Topics Quant. Electron.*, 5(4): 902–910, 1999.
9. Roth, P. W., et al., "Directly Diode-Laser-Pumped Ti:sapphire Laser," *Opt. Lett.*, 34(21): 3334–3336, 2009.
10. Keller, U., et al., "Semiconductor Saturable Absorber Mirrors (SESAMS) for Femtosecond to Nanosecond Pulse Generation in Solid-State Lasers," *IEEE J. Select Topics Quant. Electron.*, 2: 435–453, 1996.
11. Spence, D. E., et al., "Regeneratively Initiated Self-Mode-Locked Ti:sapphire Laser," *Opt. Lett.*, 16(22): 1762–1764, 1991.
12. Asaki, M. T., et al., "Generation of 11-fs Pulses from a Modelocked Ti:sapphire Laser," *Opt. Lett.*, 18: 977, 1993.
13. Sharma, T. K., and Towe, E., "Application-Oriented Nitride Substrates: The Key to Long-Wavelength Nitride Lasers Beyond 500 nm," *J. App. Phys.*, 107(2): 2010.
14. Hunziker, L. E., Ihli, C., and Steingrube, D. S., "Miniaturization and Power Scaling of Fundamental Mode Optically Pumped Semiconductor Lasers," *IEEE J. Select Topics Quant. Electron.*, 13(3): 610–618, 2007.
15. IPG Photonics. http://www.ipgphotonics.com/Collateral/Documents/EnglishUS/Green_CW_Fiber_Laser_IPG%20web.pdf.
16. Strickland, D., and Mourou, G., "Compression of Amplified Chirped Optical Pulses," *Opt. Comm.*, 56(3): 219–221.
17. Backus, S., et al., "High Power Ultrafast Lasers," *Review of Scientific Instruments*, 69 (3): 1207–1223, 1998.
18. Martinez, O. E., "Design of High-Power Ultrashort Pulse Amplifiers by Expansion and Recompression," *IEEE J. Quant. Electron.*, QE-23(8): 1385–1387, 1987.
19. Treacy, E. B., "Optical Pulse Compression with Diffraction Gratings," *IEEE J. Quant. Electron.*, QE-5(9): 454–458, 1969.
20. Muller, D., et al., "Cryogenic Cooling Multiplies Output of Ti:sapphire Output," *Laser Focus World*, 41(10): 65–68, 2005.

21. Pessot, M., et al., "Chirped-Pulse Amplification of 100-fs Pulses," *Opt. Lett.*, 14(15): 797–799, 1989.
22. Huang, C.-P., et al. "Amplification of 26 fs, 2 TW Pulses in Ti:sapphire," *Generation, Amplification and Measurement of Ultrashort Laser Pulses II*. San Jose, CA: SPIE, 1995.
23. Yamakawa, K., et al., "Generation of 16 fs, 10 TW Pulses at a 10 Hz Repetition Rate with Efficient Ti:sapphire Amplifiers," *Opt. Lett.*, 23(7): 525–527, 1998.
24. Backus, S., et al., "Ti:Sapphire Amplifier Producing Millijoule-Level, 21 fs Pulses at 1 kHz," *Opt. Lett.*, 20(19): 2000, 1995.
25. Zeek, E., et al., "Adaptive Pulse Compression for Transform-Limited 15-fs High-Energy Pulse Generation," *Opt. Lett.*, 25(8): 587–589, 2000.
26. Spielmann, C., et al., "Compact, High-Throughput Expansion–Compression Scheme for Chirped Pulse Amplification in the 10 Fs Range," *Opt. Comm.*, 120(5–6): 321–324, 1995.
27. Lenzner, M., et al., "Sub-20 fs, Kilohertz-Repetition-Rate Ti:sapphire Amplifier," *Opt. Lett.*, 20(12): 1397, 1995.
28. Kane, S., and Squier, J., "Grism-Pair Stretcher-Compressor System for Simultaneous Second- and Third-Order Dispersion Compensation in Chirped Pulse Amplification," *J. Opt. Soc. Am. B*, 14(3): 661–665, 1997.
29. Durfee, C. G., Squier, J. A., and Kane, S., "A Modular Approach to the Analytic Calculation of Spectral Phase for Grisms and Other Refractive/Diffractive Structures," *Opt. Express*, 16(22): 18004–18016, 2008.
30. Backus, S., "100 kHz Ultrafast Laser System for OPA/NOPA Frequency Conversion," *ASSP 2008 Proceedings*. Japan: 2008.
31. Gaudiosi, D., et al., "Multi-Kilohertz Repetition Rate Ti:sapphire Amplifier Based on Down-Chirped Pulse Amplification," *Opt. Express*, 14(20): 9277–9283, 2006.
32. Koechner, W., *Solid-State Laser Engineering*, Heidelberg, Germany: Springer-Verlag, 1996.
33. Matousek, P., Rus, B., and Ross, I. N., "Design of a Multi-Petawatt Optical Parametric Chirped Pulse Amplifier for the Iodine Laser ASTERIX IV," *IEEE J. Quant. Electron.*, 36(2): 158–163, 2000.
34. Chalus, O., et al., "Mid-IR Short-Pulse OPCPA with Microjoule Energy at 100 kHz," *Opt. Express*, 17(5): 3587–3594, 2009.
35. Gaul, E. W., et al., "Demonstration of a 1.1 Petawatt Laser Based on a Hybrid Optical Parametric Chirped Pulse Amplification/Mixed Nd:glass Amplifier," *Appl. Opt.*, 49(9): 1676–1681, 2010.
36. Braun, A., et al., "Characterization of Short-Pulse Oscillators by Means of a High-Dynamic-Range Autocorrelation Measurement," *Opt. Lett.*, 20(18): 1889–1891, 1995.
37. Trebino, R., et al., "Measuring Ultrashort Laser Pulses in the Time-Frequency Domain Using Frequency-Resolved Optical Gating," *Rev. Sci. Instrum.*, 68(9): 3277–3295, 1997.
38. O'Shea, P., Kimmel, M., and Trebino, R., "Increased Phase-Matching Bandwidth in Simple Ultrashort-Laser-Pulse Measurements," *J. Opt. B: Quant. Semiclassical Opt.*, 4(1): 44–48, 2002.
39. Iaconis, C., and Walmsley, I. A., "Spectral Phase Interferometry for Direct Electric-Field Reconstruction of Ultrashort Optical Pulses," *Opt. Lett.*, 23(10): 792–794, 1998.
40. Kasparian, J., Sauerbrey, R., and Chin, S. L., "The Critical Laser Intensity of Self-Guided Light Filaments in Air," *Appl. Phys. B: Lasers Opt.*, 71(6): 877–879, 2000.
41. Kasparian, J., et al., "White-Light Filaments for Atmospheric Analysis," *Science*, 301(5629): 61–64, 2003.
42. Mejean, G., et al., "Remote Detection and Identification of Biological Aerosols Using a Femtosecond Terawatt Lidar System," *Appl. Phys. B: Lasers Opt.*, 78(5): 535–537, 2004.
43. Tien, A., et al., "Short Pulse Laser Damage in Transparent Materials as a Function of Laser Pulse Duration," *Phys. Rev. Lett.*, 82: 3883–3886, 1999.

44. Backus, S., et al., "High-Efficiency, Single-Stage 7-kHz High-Average-Power Ultrafast Laser System," *Opt. Lett.*, 26(7): 465–467, 2001.
45. Glinec, Y., et al., "High-Resolution Gamma-Ray Radiography Produced by a Laser-Plasma Driven Electron Source," *Phys. Rev. Lett.*, 94(2): 025003, 2005.
46. Ferray, M., et al., "Multiple-Harmonic Conversion of 1064 nm Radiation in Rare Gasses," *J. Phys. B*, 21: L31, 1987.
47. Zhou, J., et al., "Enhanced High Harmonic Generation Using 25 Femtosecond Laser Pulses," *Phys. Rev. Lett.*, 76(5): 752–755, 1996.
48. Bartels, R., et al., "Shaped-Pulse Optimization of Coherent Emission of High-Harmonic Soft X-Rays," *Nature*, 406(6792): 164–166, 2000.
49. Miaja-Avila, L., et al., "Ultrafast Studies of Electronic Processes at Surfaces Using the Laser-Assisted Photoelectric Effect with Long-Wavelength Dressing Light," *Phys. Rev. A*, 79(3): 4, 2009.
50. Zhang, X., et al., "Quasi Phase Matching of High Harmonic Generation in Waveguides Using Counter-Propagating Beams," *Ultrafast Phenomena XV*, Asilomar, CA: Springer-Verlag, 2006.
51. Drescher, M., et al., "X-Ray Pulses Approaching the Attosecond Frontier," *Science*, 291(5510): 1923–1927, 2001.
52. Dombi, P., Krausz, F., and Farkas, G., "Ultrafast Dynamics and Carrier-Envelope Phase Sensitivity of Multiphoton Photoemission from Metal Surfaces," *J. Mod. Opt.*, 53(1–2): 163–172, 2006.
53. Baltuska, A., et al., "Attosecond Control of Electronic Processes by Intense Light Fields," *Nature*, 421(6923): 611–615, 2003.
54. Krausz, F., and Ivanov, M., "Attosecond Physics," *Rev. Mod. Phys.*, 81(1): 163–234, 2009.
55. Popmintchev, T., et al., "Phase Matching of High Harmonic Generation in the Soft and Hard X-Ray Regions of the Spectrum," *Proceedings of the National Academy of Sciences of the United States of America*, 106(26): 10516–10521, 2009.
56. Popmintchev, T., et al., "Extended Phase Matching of High Harmonics Driven by Mid-Infrared Light," *Opt. Lett.*, 33(18): 2128–2130, 2008.

CHAPTER 13
Ultrafast Lasers in Thin-Disk Geometry

Christian Kränkel

Institute of Quantum Electronics, Physics Department, Swiss Federal Institute of Technology (ETH Zurich), Switzerland

Deran J. H. C. Maas

Institute of Quantum Electronics, Physics Department, Swiss Federal Institute of Technology (ETH Zurich), Switzerland

Thomas Südmeyer

Institute of Quantum Electronics, Physics Department, Swiss Federal Institute of Technology (ETH Zurich), Switzerland

Ursula Keller

Institute of Quantum Electronics, Physics Department, Swiss Federal Institute of Technology (ETH Zurich), Switzerland

13.1 Introduction

The tremendous progress in the research and development of femtosecond and picosecond lasers, typically referred to as *ultrafast lasers*, has enabled many breakthroughs in science and technology. Ultrafast lasers were a crucial contributor to two recent Nobel prizes: one in femtochemistry by A. Zewail in 1999, and the other in frequency metrology by J. L. Hall and T. W. Hänsch in 2005. Femtosecond lasers have also enabled many other new technologies in areas as diverse as biology, medicine, and material science. A highly attractive commercial application is precision materials processing. The short duration of a few picosecond or even femtosecond pulses can ablate material before its temperature increases from the absorbed energy. This non-thermal "cold" ablation enables precise materials processing with negligible secondary damage effects from heating and melting.[1-3] To date, however, ultrafast laser technology has not found widespread use in industry. The main challenges have been low average power, high costs, and limited reliability of typical femtosecond laser systems. The pico- to femtosecond materials processing application is a representative example for many other industrial applications, for which, in principle, excellent improvements and even new opportunities have been demonstrated in research laboratories.

We believe that novel ultrafast lasers in the thin-disk geometry based on either diode-pumped ytterbium (Yb)-doped solid-state lasers or semiconductor lasers can offer a solution for many applications. Stable ultrafast pulses are obtained with semiconductor saturable absorber mirrors (SESAMs).[4,5] Such SESAM mode-locked thin-disk lasers offer reduced complexity and cost, with improved reliability and average power.[6,7]

SESAM mode-locked ultrafast laser oscillators in the thin-disk geometry are very promising. The gain material's geometry is an important factor for a laser's efficient thermal management. For average power scaling, the gain medium must be efficiently cooled, which is achieved through a large surface-to-volume ratio. Possible options are fiber, slab, and thin-disk geometries. In thin-disk geometry, the active medium has the shape of a thin-disk with an aperture much larger than its thickness. Applying this concept to diode-pumped solid-state lasers led to the development of the thin-disk laser (TDL),[8] which initially used the crystalline material $Yb:Y_3Al_5O_{12}$ (Yb:YAG) as the active medium. Today, multikilowatt continuous-wave (CW) Yb:YAG TDLs have successfully been established in the automotive industry and have demonstrated excellent reliability, high efficiency, and good beam quality.[9] In addition, CW semiconductor TDLs can generate greater than 20 W of output power in fundamental transverse mode,[10] which is significantly higher than any other semiconductor laser. Such lasers were initially referred to as vertical external cavity surface-emitting lasers (VECSELs)[11] or optically pumped semiconductor lasers (OPSLs); because of their similarity to solid state thin-disk lasers, however, they are more recently also referred to as semiconductor disk lasers (SDLs).

Because both VECSELs and TDLs use the same thin-disk geometry of the gain material, they share many common features. Both lasers produce state-of-the-art performance and are ideally suited for ultrafast passive mode locking with a SESAM.[4,5] Even though both rely on SESAM mode locking, it is important to realize that their basic mode-locking mechanisms are significantly different. In addition, their ideal operation parameters, with 10 to 100 µJ pulse energies at megahertz repetition rates for TDLs and pico- to nanojoule pulse energies at gigahertz repetition rates for VECSELs, are very different, even though they share the high-average-power scaling benefits of their respective operation regimes. Ultrafast TDLs can generate 141 W of average power in femtosecond pulses,[12,13] which is higher than any other mode-locked laser oscillator. They also generate the highest pulse energies, with up to 25 µJ at a pulse repetition rate of 2.93 MHz, which is sufficient for high-speed micromachining applications.[14] Ultrafast VECSELs access a different operation regime than TDLs, generating pulse energies in the pico- to nanojoule regime at gigahertz pulse repetition rates but with relatively high average power in the 100 mW to multiwatt regime, which is the highest in comparison to any other gigahertz laser oscillator. Ultrafast VECSELs have a number of compelling advantages, including compactness and their ability to operate in wavelength regions that are not easily accessible with established ion-doped, solid-state laser materials. Furthermore, it is possible to combine gain and saturable absorber in one semiconductor structure, enabling mode locking in a simple, straight cavity. These devices are referred to as mode-locked integrated external-cavity surface emitting lasers (MIXSELs).[15] Their good mode-locking performance, in combination with the potential for cost-efficient mass production, makes MIXSELs a promising alternative for many applications that currently rely on more bulky and expensive laser systems.

This chapter describes the differences between and common features of passively mode-locked high-power laser oscillators in the thin-disk geometry using either diode-pumped solid-state lasers or optically pumped semiconductor lasers. The chapter starts with a brief introduction of the pump concepts of solid-state TDLs and VECSELs, including a discussion of their thermal management. We then explain why the fundamental laser material parameters lead to different pulse formation mechanisms and to different operation regimes, though with the same power scaling benefits. The chapter closes with a brief summary and an outlook toward further improvement of the performance of passively mode-locked solid-state TDLs and VECSELs.

13.2 Pump Geometry

In the thin-disk geometry (c.f., Chap. 10), the disk-shaped active medium has a highly reflective (HR) coating on the back and an antireflection (AR) coating on the front for both the pump and laser wavelength. In the simplest case, the resonator can be formed by the disk, which then acts as an end mirror, and only one additional output coupler (Fig. 13.1a), which is why it is also known as the active mirror concept.[16] Especially for diode-pumped solid-state TDLs, the pump absorption length is significantly larger than the disk thickness. Therefore, the pump light is launched onto the disk under a

330   Solid-State Lasers

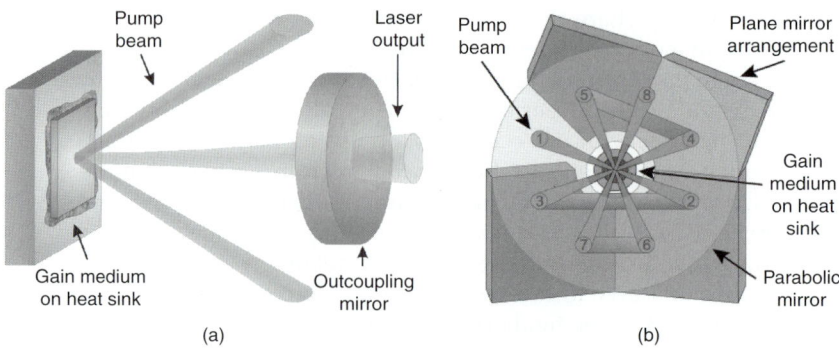

FIGURE 13.1 (*a*) Simplest resonator scheme with two passes of the pump beam, which is typically applied for barrier-pumped VECSELs. (*b*) Schematic of a more sophisticated solid-state TDL pump module for 16 passes through the disk. The numbers correspond to the number of passes through the gain medium. After 8 passes, the pump light is back reflected at the rooftop formed between two plane mirrors.

certain angle of incidence (Fig. 13.1*a*), which supports a stable multi-path pump concept with a high pump absorption (Fig. 13.1*b*). The HR coating on the backside of the thin disk reflects the nonabsorbed pump light after every path through the active medium.

The gain in a standard VECSEL is based on several quantum wells embedded between nonactive barrier layers. The barrier-pumped VECSELs are pumped with a higher pump energy than the barrier material band gap, thus providing efficient pump absorption within a single or double pass through the gain region. This is in contrast to in-well pumped VECSELs (see Sec. 13.3) and to Yb^{3+}-doped solid-state TDLs, for which additional passes of the pump light through the gain medium are necessary. As an example, four passes through the crystal can easily be achieved by a simple back reflection of the nonabsorbed pump light along the initial path. In typical commercial TDLs, the pump beam passes up to 32 times through the crystal, using a more sophisticated arrangement of one parabolic mirror and four plane mirrors. Here, the nonabsorbed pump light is reflected back onto its initial path, which doubles the number of passes through the thin disk.

An example for such a pumping scheme with 16 pump passes through the active medium is shown in Fig. 13.1*b*. This multipass concept of pump light through the gain medium allows for an excellent absorption of more than 99 percent of the incident pump. Moreover, it reduces the demands on the beam quality and brightness of the pump diodes and leads to a lower laser threshold in Yb^{3+}-doped three-level laser systems. Scaling of output power in all kinds of disk lasers is realized with an increasing pump and laser mode area on the disk, while keeping the maximum intensities and the deposited heat per volume constant.

13.3 Thermal Management in Thin-Disk Geometry

In Yb^{3+}-based solid-state lasers, as well as in semiconductor lasers, the performance is sensitive to an increase in the temperature of the gain material. Yb^{3+}-doped lasers exhibit a quasi-three-level laser scheme,[17] with a thermal population of the lower laser level according to the Boltzmann distribution. The lower laser level's population increases with rising temperature, which lowers the achievable gain for a given pump intensity. A comparable behavior can be found in semiconductor lasers, where the carrier distribution in the valence and conduction band is described by the Fermi-Dirac distribution. In this case, a rising temperature leads to a broader energy distribution of the carriers and, consequently, a lower maximum occupation number, which also affects the gain.[18] In both cases, an elevated gain temperature requires a higher density of excited states to achieve the same gain as is reached in a "cold" laser and leads to a nonlinear increase of processes that are detrimental for the laser performance. These processes mainly result from different types of interactions between exited states. In solid-state lasers, this effect is known as "quenching," and the dominating processes are migration to impurities[19] and upconversion (which is not present in Yb^{3+}-doped lasers due to the lack of suitable higher energy levels). The corresponding processes in semiconductor lasers are Auger recombination[20] and thermally excited escape of the carriers over the confining potentials into the barrier regions. It is also important to note that the semiconductor band gap decreases with rising temperature, leading to a typical red shift of the central emission wavelength of ~0.3 nm/K.

Furthermore, in both material classes, the index of refraction n and the length l exhibit a dependency on temperature T. Whereas the dn/dT causes the formation of a thermal lens with rising temperature, the dl/dT causes stress in the laser material and can induce depolarization. Both effects have a detrimental influence on beam quality, which deteriorates for strong temperature gradients in different directions.

The strong thermal sensitivity of laser performance and beam quality in Yb^{3+}-based solid-state lasers and semiconductor lasers thus require efficient heat removal for power scaling. The thin-disk geometry is ideally suited for this task. The disks are usually mounted onto an actively cooled heat sink with their backside HR coated. The thin disk supports efficient heat removal due to the large ratio of cooled surface to pumped volume. Solving the corresponding heat equations shows that more than 90 percent of the heat is extracted via the back face of the cylinder-shaped pumped region for beam radii that are about six times larger than the disk thickness.[21,22] Therefore, even if it is not possible to totally avoid a temperature gradient in the disk, the remaining gradient is mainly one dimensional and perpendicular to the faces of the disk. This maintains good beam quality, because the resulting thermal lens is isotropic and can be compensated by a standard resonator design.

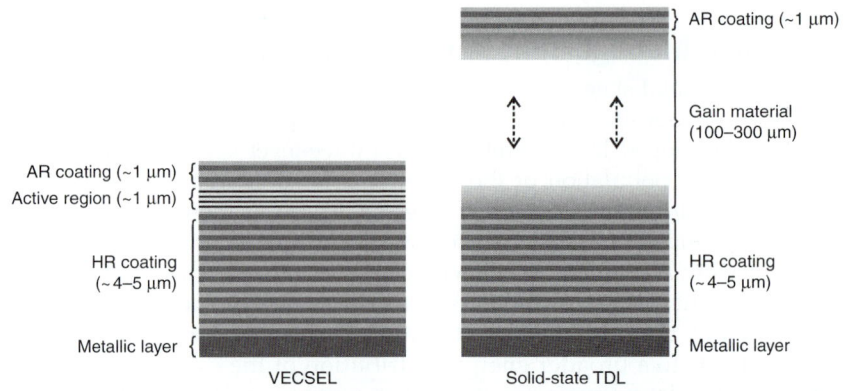

FIGURE 13.2 Comparison of the composition of a VECSEL and a TDL disk. Although the coatings are of comparable thickness, the active region in the VECSEL is roughly two orders of magnitude thinner than the TDL crystal.

Despite the similarities outlined above, there are basic differences in terms of thermal management between solid-state TDLs and VECSELs. The most obvious difference is the thickness of the active region (see Fig. 13.2), which in both cases is sandwiched between a roughly 1-μm thick AR coating and a 4- to 5-μm thick HR distributed Bragg reflector (DBR) coating. Although typical active regions of VECSELs exhibit a thickness of around 1 μm, the significantly lower absorption efficiency of Yb^{3+}-doped materials requires a thickness on the order of 100 μm to achieve a good absorption efficiency, even for the multipass pump concept described previously. Furthermore, the thermal conductivity for semiconductors is much higher than for suitable crystalline insulator host materials of Yb^{3+} ions (e.g., YAG). Consequently, the normalized thermal resistance of a semiconductor disk is much lower than that of a YAG disk (see Table 13.1), which allows for significantly higher pump power densities of more than 30 kW/cm² in VECSELs, even in single-mode operation.[10] In contrast, the pump intensity in a solid-state TDL is typically below 15 kW/cm² and is even lower for fundamental mode operation[23] (see Table 13.1). However, the typical pump beam diameters in VECSELs are smaller, leading to a higher absolute thermal resistance—that is, to a larger temperature increase for the same heating power. The high thermal conductivity of the semiconductors requires a heat sink material with an even higher thermal conductivity. According to the scaling law for lasers in the thin-disk geometry, the output power increases linearly with the pump and laser mode area if the pump density is kept constant and the heat flow is dominated by a one-dimensional propagation into the heat sink.

As an example, we consider a 5-μm thick AlGaAs VECSEL structure directly mounted on a copper heat sink. Numerical calculations

	VECSEL* $In_{0.13}Ga_{0.87}As/GaAs$	Solid-State TDL 10% Yb:YAG
Output Parameters		
Average output power	2.1 W	80 W
Repetition rate	4 GHz	57 MHz
Pulse duration	4.7 ps	703 fs
Pulse energy	0.53 pJ	1.4 µJ
Pump power	18.9 W	360 W
Opt.-to-opt. efficiency	11%	22%
Setup Parameters		
Total disk thickness	~6 µm	~105 nm
Active region thickness	~1 µm	100 µm
Pump spot diameter	350 µm	2.8 mm
Pump density	19.6 kW/cm^2	5.8 kW/cm^2
Outcoupling rate	2.5%	8.5%
Pump wavelength	808 nm	941 nm
Laser wavelength	957 nm	1030 nm
Quantum defect	15.6%	8.6%
Material Parameters		
Upper state lifetime	~1 ns	~1 ms
Absorption coefficient	10^4 cm^{-1}	10 cm^{-1}
Absolute thermal resistance*	8 K/W	3.3 K/W
Normalized thermal resistance**	0.77 Kmm^2/W	20.6 Kmm^2/W
Gain cross sections	~10^{-16} cm^2	~10^{-21} cm^2

*The data for the VECSEL were taken from Refs. 21, 38–41; those for the solid-state TDL are from Refs. 42–44.

**The normalized thermal resistance is independent of the pumped area, while the absolute thermal resistance value is obtained by dividing the normalized thermal resistance by the pumped area, which is much smaller in the case of a typical VECSEL.

TABLE 13.1 Comparison of Output, Setup, and Material Parameters of the VECSEL and of the Yb:YAG-Based TDL with the Currently Highest-Average Output Power in Mode-Locked Operation

reveal that the main thermal impedance is determined by the heat sink for pump spots larger than ~450 μm in diameter and pump power densities around 10 kW/cm².[7,21] Although this does not make it impossible to further scale the pumped area, and thus the output power, the performance will be affected by the temperature increase, and the loss in efficiency will ultimately cancel the benefits of a larger size. However, further scaling of the pump spot size is feasible using heat sink materials with a better thermal conductivity, such as diamond.

As an example, we discuss the thermal management of the VECSEL currently generating the highest CW power in the fundamental transverse mode[10] (Fig. 13.3). The structure's GaAs wafer has been removed, and mounted on a diamond heat sink. In Fig. 13.3a, the output power is shown as a function of the pump power. Already at a pump power of 30 W, an output power of 12.6 W is generated at 42 percent total optical-to-optical efficiency. The maximum output power of 20 W is achieved for 50 W of pump radiation at 40 percent efficiency. At 30 W pump power, the incident pump power density is 16.6 kW/cm²; at 50 W, it is 27.6 kW/cm². Figure 13.3b shows the calculated temperature difference between the maximum temperature in the gain region and that of the heat sink using a standard finite-element simulation. The temperature increase as a function of the pump mode radius is given for the two discussed pump intensities (27.6 kW/cm²: solid gray line; 16.6 kW/cm²: dashed gray line). The vertical line indicates the 240-μm pump radius used in the experiment. At the highest pump intensity, we obtain a temperature increase of 40 K for the 240-μm pump radius. A comparison with an unprocessed gain structure on a 600-μm thick GaAs wafer shows the importance of thermal management for power scaling (black curve).

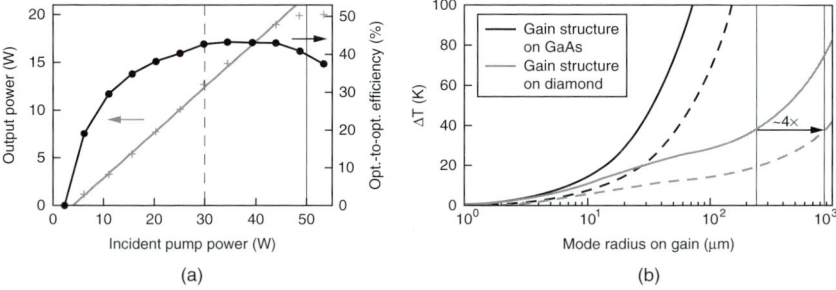

FIGURE 13.3 (a) Output power of the currently highest fundamental mode continuous-wave VECSEL versus the incident pump power. The mode radius is 240 μm, and the gain structure is mounted on a diamond head spreader. (b) Finite-element simulation of the heating of the gain structure versus the mode radius on the gain at a fixed pump and heating intensity. The dashed lines correspond to an incident pump power density of 16.6 kW/cm², or 30 W of incident pump power in (a); the solid lines correspond to 27.6 kW/cm², or 50 W of incident pump power.[10,24]

A 40-K temperature increase is already reached at a pump radius below 30 µm at the higher pump intensity (or 60 µm at the lower pump intensity). On the other hand, the diamond-mounted structure is suitable for further power increase by enlarging the pump diameter: Operating the laser at the 40 percent lower pump intensity of 16.6 kW/cm² should allow for an increase of the pump spot radius by roughly a factor of four, while maintaining the same 40-K temperature increase. Considering this 16-fold increase of the pump area and the slightly higher efficiency at the lower pump intensity, it should be possible to increase the output power by nearly an order of magnitude to well above 100 W.

It is currently not clear which effects will finally limit the power scaling in VECSELs. Additional challenges will arise at very large pump radii, such as inversion depletion due to amplified spontaneous emission (ASE) inside the disk, which can strongly affect the laser's efficiency.[25]

In solid-state TDL materials, the pump and laser mode diameters can be scaled to several millimeters and even more than 1 cm, due to the lower amount of generated heat per volume thanks to the lower quantum defect of Yb^{3+} lasers and the lower pump power density. Furthermore, the ratio of the total thermal impedance of the disk and the heat sink, which is usually made from copper or copper tungsten, is larger. Therefore, the heat will not accumulate in the heat sink. An approach for overcoming the thermal limitations is to reduce the quantity of generated heat. The main contribution results from the quantum defect, which is the energetic difference between the pump and laser photons. If the quantum defect is reduced, higher pump powers can be applied. For Yb^{3+}-based solid-state TDLs, the quantum defect is already very low. Yb:YAG is typically pumped at 941 nm, and the laser wavelength is 1030 nm, resulting in a quantum defect of less than 9 percent. However, rapid progress has been made in recent years in developing new laser materials that are pumped directly into the zero-phonon line of the Yb^{3+} ion.[26–34] Pump wavelengths around 975 nm reduce the quantum defect and thus the total generated heat by nearly a factor of two.

In VECSELs, the quantum defect and the thermal load can be reduced via in-well pumping. In this case, the pump wavelength is chosen such that the incoming photons are only absorbed in the quantum wells.[35] The interaction of the pump light with the quantum wells takes place in a small region a few nanometers in length, which is much shorter than in barrier pumping, where the typical interaction length is ~1 µm. Therefore, the fraction of absorbed pump light in a single pass is significantly lower. The absorption efficiency can thus be increased with the established multipass pump scheme used for solid-state TDLs (see Fig. 13.1b). Another approach for improving the absorption efficiency is based on resonant VECSEL structures. Typically, the pump radiation's internal angle of incidence is chosen in such a way that the antinodes of the pump and laser light are brought into alignment, which makes the structure resonant for both the pump and the laser wavelength.[36] Initial experiments indicate that in-well pumping bears the potential for further scaling of the output powers of VECSELs.[36,37]

13.4 SESAM Mode Locking

A SESAM acts as an intracavity loss modulator with an intensity-dependent reflectivity. Its macroscopic nonlinear optical properties are mainly determined by modulation depth ΔR, or the difference in reflectivity between a fully saturated and an unsaturated SESAM, as well as by the saturation fluence F_{sat}, which is the pulse fluence needed to reduce the losses by $1/e$ of the initial value (neglecting the nonsaturable losses R_{ns}). An example of a nonlinear reflectivity measurement of a SESAM is shown in Fig. 13.4a. As mentioned earlier, the power scaling principle of disk lasers relies on increasing the mode area on the active region. Analogous arguments apply for the SESAM, such that a fixed set of parameters can be used in different average power regimes. Hence, among the various techniques that can force a laser into mode-locked operation,[45–47] passive mode locking with a semiconductor saturable absorber mirror (SESAM)[4,5] is ideally suited for ultrafast disk lasers.

13.4.1 Pulse Formation Mechanisms

Another crucial parameter describing the dynamics of a SESAM is the recovery time $\tau_{1/e}$ (see Fig. 13.4b), which is defined as the exponential time constant of the return to the unsaturated reflectivity after

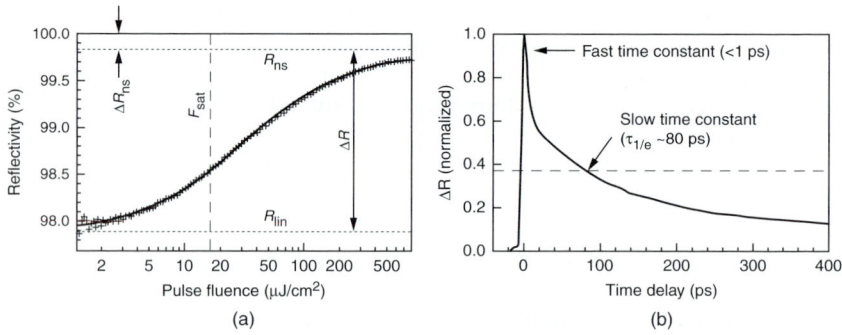

FIGURE 13.4 (a) Example of a measurement of the nonlinear reflectivity of a SESAM (crosses) as a function of the incident pulse fluence. The theoretical fit (solid) results in F_{sat} = 16.6 µJ/cm², ΔR = 1.95%, and ΔR_{ns} = 0.16%. (b) Example of the temporal response of a SESAM. The measurement was performed with 2.7-ps pulses, which were too long to resolve the fast recombination time constant (the SESAM is described in Ref. 42).

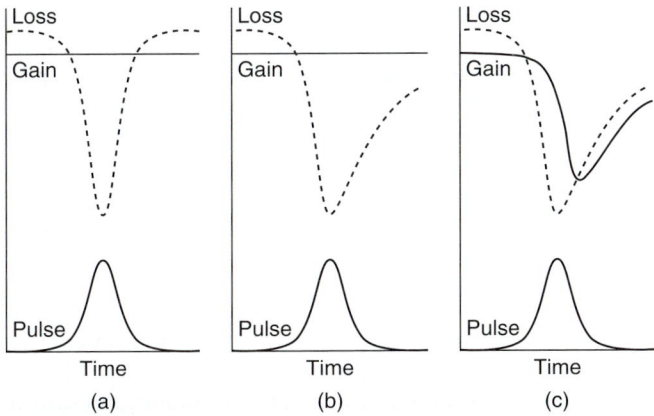

FIGURE 13.5 Mechanisms of pulse shaping and stabilization using saturable absorbers. (a) Fast saturable absorber: Net gain window is opened and closed by the saturable absorber. (b) Slow saturable absorber and no dynamic gain saturation: Long net gain window is opened by the saturable absorber, and amplified background noise behind the pulse is swallowed with a timing delay of the pulse caused by an attenuation of the leading edge by the absorber in each roundtrip. (c) Slow saturable absorber with dynamic gain saturation: Net gain window is opened by the saturable absorber and closed by gain saturation.

a pulse. Depending on the recovery time of the SESAM and the saturation dynamics of the gain, we can distinguish among three distinct mechanisms of stable pulse formation.

The first mechanism, shown in Fig. 13.5a, relies on a fast, saturable absorber with a recovery time shorter than the pulse duration. Standard SESAMs, which are based on absorption bleaching in the valence band to conduction band transition using quantum wells or bulk materials, usually do not have such a dominant short recovery time in the femtosecond-pulse-width regime. Thus, this mechanism is particularly important for Kerr lens mode locking,[48–50] a technique that has not yet been applied to disk lasers.

In the case of a slow saturable absorber, the recovery time is longer than the pulse duration, leading to two further mechanisms of pulse formation. For gain materials with small gain cross sections, we do not observe any significant change of the saturation of the gain for each pulse—the so-called dynamic gain saturation. In this case, the gain remains constant and is only saturated with the laser's CW power (Fig. 13.5b). This case is of particular importance for rare earth–doped solid-state gain materials, because the 4f transitions, which are relevant for most laser processes in this material group, are only weakly allowed. They exhibit long lifetimes in the order of hundreds of microseconds to several milliseconds. According to basic laser equations,[51] they consequently have low-gain cross sections in the order of 10^{-18} to 10^{-21} cm². Semiconductor lasers, on the other hand, rely on allowed transitions with upper-state lifetimes of a few nanoseconds or less and thus have cross sections higher by several orders of magnitude (see Table 13.1). Semiconductor lasers exhibit significantly lower saturation energy E_{sat}, according to

$$E_{sat} = A \cdot F_{sat} = A \cdot \frac{h\nu}{\sigma_{em} + \sigma_{abs}} \tag{13.1}$$

with mode area A and saturation fluence F_{sat}, as well as photon energy $h\nu$ and absorption and emission cross sections σ_{abs} and σ_{em}, respectively, at the signal wavelength. This lower saturation energy gives rise to dynamic gain saturation during propagation of the pulse through the active region (Fig. 13.5c). This dynamic gain saturation is then partially or fully recovered between two consecutive pulses.

As shown in Fig. 13.5b, the SESAM's slow recovery time, in combination with a constant gain saturation, leads to a long net gain window after the pulse. Intuitively, one would expect background noise behind the pulse to be amplified as well. However, because the pulse's leading edge is absorbed each time the pulse hits the absorber, the center of the pulse is shifted backward each roundtrip, and amplified noise is swallowed by the pulse after some roundtrips. Numerical simulations reveal that the pulse duration can be even more than one order of magnitude shorter than the SESAM recovery time.[52] To achieve shorter pulse durations in the femtosecond regime, further pulse stabilization mechanisms are necessary. This can be achieved by introducing a well-balanced amount of self-phase modulation (SPM) and group delay dispersion (GDD) into the cavity. The resulting pulses can be considered as solitons; therefore, the typical mode-locking process in solid-state TDLs is known as *soliton mode locking*.[53–55]

In most VECSELs, the following dynamic gain saturation process, depicted in Fig. 13.5c, takes place: A net gain window is opened by saturating the SESAM. This temporal window is closed by the dynamic saturation of the gain induced by the pulse in the active region. This mechanism is typical for lasers with larger gain cross sections, such as semiconductor and dye lasers.[56] To enable stable mode locking with an open net gain window, it is obvious that the SESAM's saturation energy must be smaller than that of the gain, so that the absorber saturates first. In addition, a shorter net gain window can be obtained when the absorber recovers faster than the gain.

13.4.2 Different Operation Regimes

Continuous wave (CW) TDLs deliver several kilowatts of output power[91] and several 100 W in diffraction-limited beam quality.[57] Such performance enables mode-locked operation with average output powers and pulse energy levels that cannot be obtained directly from any other mode-locked laser oscillator technology. Because the average power is the product of the pulse energy and the repetition rate, decreasing the repetition rate at constant pulse duration leads to higher pulse energy and peak power. Ultrafast TDLs can generate greater than 10-μJ pulses directly out of the oscillator, which makes them highly attractive for applications in material structuring and high-field science. On the other hand, semiconductor lasers are ideally suited for cost-efficient mass production due to their epitaxial growth on wafers. Therefore, they are attractive for such applications as telecommunication or optical clocking and interconnects, in which a high repetition rate is more important than high pulse energy. Furthermore, mode-locked VECSELs cover a wide range of the infrared spectral range—between 0.95 μm and 2.01 μm, depending on the semiconductor gain material.[7] This range is in strong contrast to that of mode-locked solid-state TDLs, which up to now were only demonstrated with Yb^{3+}-doped gain materials at a laser wavelength around (1035 ± 10) nm. This section highlights the fundamental reasons for the different operation regimes of VECSELs and TDLs, which are clearly distinguishable in Fig. 13.6.

Average Power

The upper limit for the average power in mode-locked operation is the maximum fundamental mode output power in CW operation. The presence of higher-order modes tends to introduce destabilizing

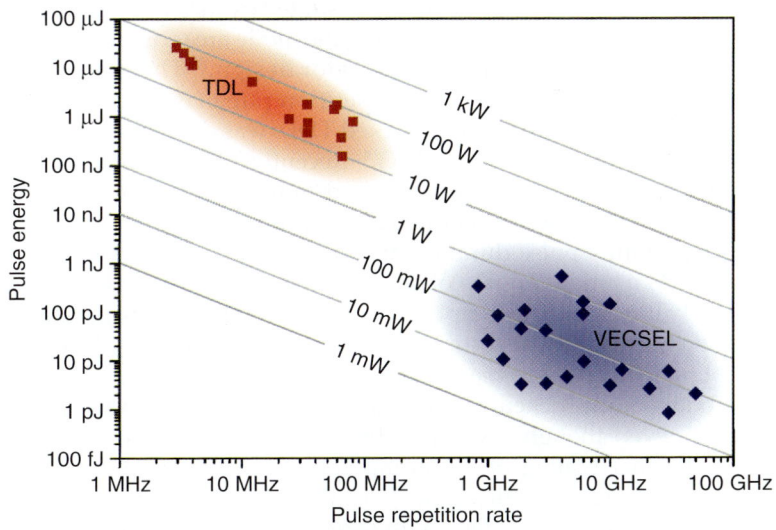

FIGURE 13.6 Graphical representation of the different output parameter regimes for SESAM mode-locked VECSELs and TDLs. The diagonal lines correspond to constant average output power. The data for the TDL results were taken from Refs. 12, 14, 33, 43, 58–67; the VECSEL results were published in Refs. 21, 38, 68–83.

effects and prevents the laser from stable mode locking. As already mentioned in Sec. 13.3, good thermal control and one-dimensional heat removal in the disk geometry allow for very high output power in TEM_{00} operation. However, it has also been shown that in VECSELs, the general scaling law for TDLs is no longer valid when the pump spot reaches a size at which the major part of the thermal impedance is caused by the heat sink and no longer by the semiconductor device itself.[21] At a certain point, the increased temperature will affect the threshold intensity as well as the slope efficiency; thus, the total loss in efficiency will dominate the benefit of a larger gain mode size. Furthermore, as mentioned in Sec. 13.3, the onset of ASE can also be a challenge for further scaling the mode areas and thus the output power.[25] So far, the highest TEM_{00} output power that has been achieved with a VECSEL is 20.2 W.[10] In contrast, the maximum fundamental mode output power that has been obtained with solid-state TDLs is in the order of several 100 W—for example, 360 W were demonstrated by Killi et al.[57] Even in this power regime, ASE was not a limiting factor.[84]

In addition, a much higher average power was obtained in mode-locked operation with solid-state TDLs. The first SESAM mode-locked solid-state TDL, which was demonstrated in 2000, already delivered an average output power of 16.2 W in 730-fs pulses.[64] This laser used Yb:YAG as the gain material, which was also the first gain material for solid-state TDLs. The pump spot had a diameter of 1.2 mm. Power scaling to 60 W was later demonstrated by increasing the pump spot diameter to 2.8 mm and adapting the mode areas on the disk and the SESAM by an appropriate resonator design.[85] The pulse duration of this laser was 810 fs at a repetition rate of 34 MHz. Further power scaling of SESAM mode-locked TDLs was demonstrated up to 80 W average power using Yb:YAG.[43] Recently a 100-W breakthrough result was achieved with the new and more efficient gain material $Yb:Lu_2O_3$.[12,13] This laser generated a maximum average output power of 141 W with a pulse duration of 738 fs and a high efficiency of 37 percent. At a repetition rate of 58 MHz, the pulse energy is 2.1 µJ. At a slightly lower average output power of 103 W, the laser operated at an even higher optical-to-optical efficiency of 42 percent,[12,13] nearly doubling the typical efficiency of Yb:YAG.

Today, mode-locked output powers in the kilowatt range seem feasible, even though several additional issues must first be overcome. Further scaling of the pump spot diameter to the necessary centimeter range is challenging, because it requires a uniform thermal lensing effect in the disk. A uniform thermal lens can be compensated for by a standard resonator design without the need for complicated adaptive optics. However, this compensation becomes more difficult at larger pump spot diameters, because stability range narrows with an increasing pump spot diameter.[86]

Although today's ultrafast solid-state TDLs typically operate in the average output power regime of several tens of watts, the output power of mode-locked VECSELs is often in the order of 100 mW. The first passively mode-locked VECSEL was also realized in 2000 and had an average output power of 21.6 mW in 22-ps pulses at a repetition rate of 4 GHz.[76] Since then, the average power has been scaled to 2.1 W with a pulse duration of 4.7 ps and a similar repetition rate.[38] Further scaling above the 10-W regime seems feasible, as nearly an order of magnitude higher output power in fundamental mode operation has recently been demonstrated.[10]

Table 13.1 shows a comparison of the output data and other parameters of the VECSEL and the Yb:YAG-based TDL with the highest average output powers in mode-locked operation.

Repetition Rate

Section 13.4.1 explained that VECSELs and solid-state TDLs rely on different pulse formation mechanisms: Whereas mode-locked VECSELs rely on dynamic gain saturation, the gain in TDLs is saturated to a constant level according to the laser's average power. In addition, TDLs, with their low-gain cross sections and long upper-state lifetimes, carry the risk of unwanted Q-switching instabilities. The reason for these instabilities is that pulses with higher energies within typical noise fluctuations can saturate the SESAM more strongly and therefore reduce their losses. The gain then must respond fast enough with additional saturation; otherwise, there will be positive feedback, and the noise pulse will be able to increase its energy further. This would destabilize the damping of relaxation oscillations and could lead to stable Q-switched mode locking (QML) or only Q-switching instabilities. In a stable QML regime of operation, the laser output consists of mode-locked pulses underneath a Q-switched envelope. (Typically, the mode-locked pulse repetition rate is in the order of 100 MHz, determined by the laser cavity length, whereas Q-switching modulations usually have frequencies in the kilohertz region, similar to the typical relaxation oscillations of TDLs.) One way to protect a laser from QML is to decrease the repetition rate, thus increasing the pulse energy in such a way that the gain saturation sets in faster. Hönninger et al.[87] developed a stability criterion for the minimum required pulse energy E_p in comparison to the saturation energies of the gain material $E_{sat,gain}$ and the saturable absorber $E_{sat,abs}$:

$$E_p^2 > E_{sat,gain} E_{sat,abs} \Delta R \qquad (13.2)$$

This equation shows that low saturation energies and a low modulation depth ΔR are also beneficial to prevent the laser from Q-switching instabilities. For the femtosecond laser, the QML threshold is typically five times lower, because the soliton effect, together with the spectral filtering due to the gain bandwidth, stabilizes additional pulse generation against Q-switching instabilities.[87] The basic idea for the additional stabilization in the femtosecond region is as follows: If the energy of a pulse increases by relaxation oscillations, the spectrum of the pulse is broadened by SPM. A broader spectrum, however, will experience a smaller average gain due to the laser material's finite bandwidth. This effect has a much smaller influence on picosecond lasers, because SPM is much weaker. In addition, inverse saturable absorption of the absorber further reduces the QML threshold.[88,89] Again, the basic idea is simple: An inverse saturable absorption causes a rollover in the nonlinear reflectivity, which increases the losses for pulses with higher energy, thus damping relaxation oscillations. Therefore, SESAMs are ideal for mode-locking diode-pumped solid-state lasers, because semiconductor saturable absorbers inherently have a large absorption cross section (a low absorber saturation energy), and their nonlinear reflectivity dynamics can be designed over a wide parameter range. In contrast, the low-gain saturation energy of VECSELs makes them immune to such Q-switching instabilities, moreover their short upper-state lifetime (typically in the nanosecond regime) tends to restrict the pulse repetition rate to the gigahertz range for stable CW mode locking. High-power TDLs typically operate in the 1- to 100-MHz regime, because at much higher pulse repetition rates, QML instabilities become more severe. This is, however, not a problem for most applications, because the larger pulse energy at lower pulse repetition rates is advantageous for many applications, such as precision micromachining.

Stable mode-locked operation with dynamic gain saturation requires the absorber to saturate at lower intracavity energy than the gain in order to obtain stable pulse formation. For mode-locked VECSELs, the active region typically consists of the same material as the absorber. Therefore, a large ratio between the mode areas on the gain and on the absorber material is necessary to achieve saturation of the absorber before the gain saturates. In this case (see Fig. 13.7a), geometrical issues for the laser cavity give an upper limit for the repetition rate of a fundamentally mode-locked VECSEL. One way to overcome these issues is mode locking with similar mode areas on the gain and the absorber (see Fig. 13.7b), which is referred to as 1:1 mode locking. In this case, SESAMs with low saturation fluence are required, which can be achieved using quantum dot (QD) SESAMs instead of conventional quantum well (QW) SESAMs. In QW-SESAMs, the product of the saturation fluence F_{sat} and the modulation depth ΔR is proportional to the energy needed to completely saturate the absorber. This means that these two parameters cannot be adapted independently. However, for 1:1 mode locking with ultrahigh repetition rates and very low pulse energies, a low F_{sat} and ΔR are required, according to the QML criterion. This problem could be overcome by

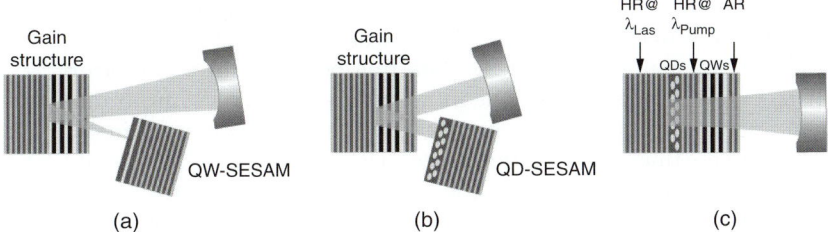

FIGURE 13.7 Integration scheme, progressing from conventional VECSEL SESAM mode locking with (*a*) large mode area ratios, thus limited to large cavities; (*b*) identical mode areas on gain structure and SESAM, making high repetition rate and integration possible; and (*c*) absorber-gain integration in a single device. The MIXSEL contains two high reflectors (HRs), a quantum dot (QD) saturable absorber, a quantum well (QW) gain, and an antireflection (AR) coating. The intermediate HR prevents the pump light from bleaching the saturable absorber.

the use of QD-SESAMs, which have a lower density of states and thus a lower total saturation energy. QD-SESAMs enabled the first demonstration of 1:1 mode locking in VECSELs, with 25-GHz repetition rate[81] and further scaling to 50 GHz.[83] Recently, even the integration of the saturable absorber into the gain structure was realized. This novel type of ultrafast semiconductor laser—the mode-locked integrated external-cavity surface emitting laser (MIXSEL)[15] (see Fig. 13.7*c*)—is a technology that opens the way to cost-efficient mass production for widespread applications. Furthermore, the concept is easily scalable to even higher repetition rates. More details on this topic can be found in Refs. 90 and 91.

The QML instabilities limit the scalability of TDLs to gigahertz repetition rates. However, this topic has not yet been fully explored, and the highest reported repetition rate of 81 MHz[67] is presumably far below the highest achievable repetition rate with this technology. As an example, an Er,Yb:glass laser with a similarly low-gain cross section already allows for a repetition rate of up to 100 GHz at 35-mW average output power.[92] However, the practical use of gigahertz TDLs at tens of watts of average output power is limited at this time.

On the other hand, it is often desirable to operate at low repetition rates to increase the available pulse energy. Scaling mode-locked VECSELs to lower repetition rates is limited by the onset of harmonic mode locking. The typical carrier lifetime in a QW-VECSEL is in the order of nanoseconds. If the cavity roundtrip time becomes longer, two subsequent pulses with lower energy will have a gain advantage over a single pulse with higher energy and will therefore be favored.[93] The threshold for harmonic mode locking strongly depends on the laser's operation parameters. However, the lowest repetition rates that have been demonstrated up to now are around 1 GHz, with an average output power of 275 mW.[68]

Solid-state lasers can be operated at repetition rates that are several orders of magnitude lower. The lower limit of the pulse repetition rate is usually set by the self-starting behavior. In the buildup phase of CW mode locking, the intracavity pulse evolves from random fluctuations of the CW operation, with much lower peak intensities than the final pulse. As discussed in Refs. 94 and 95, the intensity of these random fluctuations in the case of too-long resonators might not be sufficient to ensure self-starting mode locking. So far, stable mode locking has not yet been reported for repetition rates below 1 MHz; the lowest repetition rate from a mode-locked bulk solid-state laser is 1.2 MHz,[96] which corresponds to a cavity length of 121 m. A sophisticated cavity design and several additional mirror bounces were used to keep a reasonable footprint and a sufficient overall mechanical stability. So far, the lowest repetition rate of ultrafast TDLs has been 4 MHz[60] when using the passive multipass cell concept[97] or 2.93 MHz when using an active multipass cell with multiple passes through the gain disk.[14] In both cases, the operation was self starting, the average output power was in the tens of watts range, and the pulse energy exceeded 10 µJ.

Pulse Energy

The average output power of a mode-locked laser is the product of the pulse energy and the repetition rate. Therefore, the typical higher-average powers and lower repetition rates in solid-state TDLs result in pulse energies about five orders of magnitude higher than in the case of VECSELs (see Fig. 13.6).

Pulse-energy scaling of TDLs into the 100-µJ regime requires further considerations, such as the nonlinearity of the atmosphere in the resonator.[61] In a typical mode-locked TDL, the nonlinearity in the cavity is controlled by moving a few-millimeter-thick fused silica plate that is inserted at Brewster's angle along the axis of the diverging beam near the output coupler. The amount of SPM scales inversely proportionally to the cross section of the laser beam in the plate. The first intuition one might have is that the nonlinearity of air (3×10^{-19} cm^2/W[98]) is negligible compared with the nonlinearity of fused silica (2.46×10^{-16} cm^2/W[99]), because air's nonlinear coefficient is roughly three orders of magnitude lower than silica's. However, the cavity length of a TDL with high pulse energies can be as large as several tens of meters; therefore, the total SPM introduced by the air can easily dominate over the SPM introduced by the thin fused silica plate.

Another challenging point for generating pulse energies exceeding 10 µJ in the standard TDL configuration is the large amount of dispersion needed to balance the total SPM introduced by the air atmosphere in these very long cavities in order to obtain stable soliton mode locking. Here, the dispersion needed becomes too large to be balanced by a reasonable amount of bounces on dispersive mirrors.

Possibilities for overcoming this problem and for achieving pulse energies in the 10 to 100 µJ regime are to reduce the nonlinearity in the resonator by operating it in vacuum or in helium or to reduce the intracavity pulse energy by increasing the laser cavity's output coupler transmission. Under helium atmosphere, pulse energies of up to 11 µJ were obtained,[60] with an intracavity pulse energy exceeding 100 µJ. The second approach requires an increased gain per cavity roundtrip, which can be achieved in a cavity setup with multiple passes through the gain medium. With this concept, pulse energies of 26 µJ and an output coupler of 78 percent, and thus an intracavity pulse energy of only 34 µJ, were demonstrated in air atmosphere.[14]

The available pulse energy for VECSELs is limited by the carrier lifetime, which hinders operation at megahertz repetition rates for CW pumping. As discussed in the previous section, the maximum cavity length is limited by the onset of multiple pulsing instabilities or harmonic mode locking.[93] The highest pulse energies reported for mode-locked VECSELs are only in the order of several 100 pJ,[38,68] which is orders of magnitude lower than in solid-state TDLs.

Pulse Duration

For both ultrafast TDLs and semiconductor disk lasers, the full potential to achieve extremely short pulse durations has not yet been exploited. Currently, ultrafast TDLs are restricted to pulse durations of more than 220 fs. To date, all ultrafast solid-state TDLs have been based on Yb^{3+}-doped gain materials. In the Yb^{3+} ion, the so-called lanthanide contraction leads to a lowered distance of the $5s$ and $5d$ shells from the atom core. Therefore, the $4f$ shell, in which the optical transitions take place, is less shielded from the surrounding crystal field than is the case in other rare earth ions. This situation leads to a stronger coupling to the host's phonons and thus to broad absorption and emission spectra, which have been shown to support the generation of pulse durations less than 60 fs in longitudinally pumped low-power bulk lasers (e.g., in Yb:glass,[100] $Yb:LuVO_4$,[101] $Yb:CaGdAlO_4$,[102] $Yb:LaSc_3(BO_3)_4$,[103] or $Yb:NaY(WO_4)_2$[104]). Such short pulses require a gain bandwidth Δf_g of about 20 nm in the spectral range of ~1 µm. However, the most common gain material for the solid-state TDL is Yb:YAG, which was chosen for its beneficial CW properties, even though its gain bandwidth is narrower than for many other Yb^{3+}-doped gain materials. Consequently, the shortest pulse durations obtained with Yb:YAG TDLs were around 700 fs,[42,64] whereas $Yb:Lu_2O_3$ TDLs enabled the generation of 535-fs and 329-fs pulses at 63 W and 40 W of average output power, respectively.[67] Investigation of new gain materials with even broader emission bandwidths has enabled the generation of pulses as short as 240 fs at 22 W of average output power with Yb:KYW ($Yb:KY(WO_4)_2$)[62] and 227 fs with an average output power of 7.2 W with $Yb:LuScO_3$.[66] The differences in the gain bandwidths of some of these materials can be seen in Fig. 13.8a. Other Yb-doped materials may have the potential to push the high-power TDLs into the sub-100-femtosecond regime.[105] On the other hand, longer pulse durations can easily be achieved by inserting a spectral filter into the thin-disk laser cavity, thus limiting the available gain bandwidth.[8]

The pulse durations in high-power TDLs are significantly longer than the pulse durations achievable by low-power SESAM mode-locked lasers, which use a bulk crystal as gain material. For example, pulses as short as 340 fs were obtained with Yb:YAG in such a setup,[106] while a Yb:LuScO$_3$ delivered 111-fs pulses.[107] This difference occurs because the pulse duration is not only determined by the gain bandwidth but also depends on other parameters. Detailed investigations[52,53] on stable soliton mode locking with a SESAM revealed that according to

$$\tau_p \approx 0.2 \left(\frac{1}{\Delta f_g}\right)^{3/4} \left(\frac{\tau_a}{\Delta R}\right)^{1/4} \frac{g^{3/8}}{\Phi_0^{1/8}} \tag{13.3}$$

the pulse duration τ_p is also strongly influenced by the gain saturation g, even more so than by the SESAM parameters recovery time τ_a and modulation depth ΔR or the soliton phase shift Φ_0. High-power solid-state TDLs usually use a larger output coupler transmission than do low-power mode-locked lasers, because high intracavity pulse energies lead to an unwanted SPM contribution of the ambient atmosphere (compare Sec. 13.4.2) or even to damage of the optical components. Thus, these lasers are operated at a significantly higher saturated gain. Moreover, the short length of the active medium requires a comparably high inversion level, which often narrows the gain bandwidth that can be used for generating the pulses. Shorter pulse durations may be achieved with the concept of the active multipass cell, with multiple passes through the gain material during one resonator roundtrip[14] (see Sec. 13.4.2). With a lower output coupler, one would obtain low saturated gain and inversion, which may enable the generation of shorter pulses in the future.

Typical VECSELs exhibit a broad gain spectrum that is comparable to that of broadband Yb^{3+}-doped materials (see Fig. 13.8a). Furthermore, the emission bandwidth can be easily engineered by appropriate design of the gain structure. The overall gain spectrum depends on the intrinsic emission properties of the QW layers, as well as on the wavelength-dependent field strength at the position of the QW layers. The latter can be influenced by the design of a resonant or antiresonant structure for the standing wave pattern inside the gain medium, which is referred to as *field enhancement*. Typically, several QW layers are employed, and the overall bandwidth can even be larger than the intrinsic bandwidth of one QW layer (the VECSEL

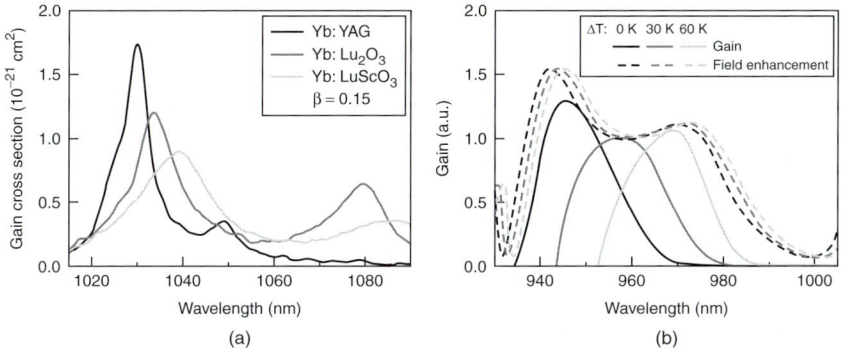

FIGURE 13.8 (a) Gain spectra of Yb:YAG, Yb:Lu$_2$O$_3$, and Yb:LuScO$_3$ for an inversion level β of 0.15. (b) Field enhancement and resulting gain spectra for a typical VECSEL structure at different temperatures.[91] ΔT represents the temperature difference to the designated operation temperature.

used for Fig. 13.8 uses seven QW layers in successive maxima of the standing wave pattern[91]). However, the generation of transform-limited pulses in the femtosecond regime, which exploit a significant fraction of the bandwidth, is challenging. Most SESAM mode-locked VECSELs operate at few-picosecond pulse durations, with an optical bandwidth below 1 nm in the slow saturable absorber regime.

The first subpicosecond pulses from a VECSEL[70] were obtained with a special SESAM, utilizing the AC Stark effect.[108] In this device, the strong electromagnetic field during the pulse leads to a blue shift of the absorption. Hence, for wavelengths longer than the peak absorption wavelength, the absorption decreases. Because no carriers are involved in this process, the recovery time is comparable to the pulse duration and is much faster than in conventional SESAMs. To further decrease the SESAM's recovery time, the single QW was placed near the surface to enable fast recombination for carriers by tunneling into surface states. By applying such an AC Stark SESAM, pulses as short as 477 fs with 100-mW average output power at a repetition rate of 1.21 GHz were realized as early as 2002.[70] Six years later, an improved AC Stark SESAM and a carefully tailored gain spectrum of the VECSEL resulted in 260-fs pulses with 25 mW of average output power.[69]

Even shorter pulses of only 190 fs could be obtained by optimizing the spectral position of the SESAM's absorption maximum in relation to the VECSEL's gain maximum by varying the temperature of both devices. However, in this sensitive operation regime, bandwidth-limited pulses could be observed only in a temperature range of about 10°C, which limited the applicable pump power. Therefore, the average output power did not exceed 5 mW for the shortest observed pulses.[109] A recent breakthrough was the demonstration of a mode-locked VECSEL with only 60-fs pulses; however, the output power of 35 mW and the quality of the pulse train are not sufficient for many applications, in particular because multiple pulses were circulating in the cavity.[110] An important aspect for improving the performance will be the optimization of GDD in the cavity;[109] this idea is supported by recent experiments confirming the quasi-soliton-theory,[111] which predicts the shortest pulse durations for slightly positive cavity GDD.[112] In any case, a careful control of the cavity GDD is regarded as being crucial for achieving femtosecond high-average-power operation of mode-locked VECSELs.

13.5 Conclusion and Outlook

Ultrafast SESAM mode-locked thin-disk lasers based on either Yb-doped solid-state gain materials or semiconductors offer a robust and power-scalable solution to the challenges of ultrashort pulse generation at high power levels. The key for this performance is efficient heat removal, which minimizes thermal lensing and aberrations, thus enabling high power levels in a fundamental transverse mode. The SESAM is an ideal device for mode locking at high power levels due to its large design flexibility. The concept of the SESAM mode-locked thin-disk laser has the essential advantage of power scalability: The output power can be scaled up by increasing pump power and mode areas on both gain medium and SESAM. For high-power ion-doped solid-state as well as for semiconductor lasers, this technology has enabled new power records.

Femtosecond ion-doped solid-state TDLs achieved pulse energies beyond the 10-mJ level at megahertz repetition rates directly from the oscillator. The average power level was increased to the 100-W level, which is particularly attractive for materials processing applications at high throughput. The first thin disk gain material was Yb:YAG, which until recently delivered the highest average output powers and pulse energies. However, the impressive advances in the research and development of new Yb-doped hosts and the availability of suitable pump diodes operating in the 980-nm spectral region have both led to new power records by applying Yb-doped sesquioxides. In particular, $Yb:Lu_2O_3$ is a promising material, achieving a mode-locked average output power of 141 W in 738-fs pulses.[12,13] Further scaling toward several hundred watts of average power and pulse energies of more than 50 µJ appear to be within reach. A critical challenge will be the demonstration of similar power levels and pulse energies from systems operating at pulse durations below 100 fs, which will require gain materials with larger-emission bandwidth than the dominant Yb:YAG gain material. Such systems will be useful for numerous industrial and scientific applications—for example, in the area of high-field science and high harmonic generation.[6]

Ultrafast semiconductor disk lasers operate at multiwatt power levels, which is higher than any other ultrafast semiconductor laser technology. Even average output power levels exceeding 10 W can be expected in the near future. Recently a MIXSEL already achieved an average output power of 6.4 W.[113] In comparison to the mode-locked solid-state TDLs, these semiconductor disk lasers access substantially higher repetition rates in the gigahertz regime. An important future research task is the demonstration of femtosecond-pulse durations at high power levels. Although sub-100-fs pulses have already been demonstrated, it will be challenging to achieve such performance at the watt level, which is a requirement for many applications. Ultrafast VECSELs have a large potential for the realization of robust, cost-efficient, ultracompact sources. The simple, straight MIXSEL cavity geometry should allow a further increase in repetition rates to the 10 to 100-GHz regime. Ultrafast VECSELs and MIXSELs appear well suited for replacing more complex solid-state lasers for many applications in areas as diverse as telecommunications, optical clocking, frequency metrology, and microscopy.

References

1. Liu, X., Du, D., and Mourou, G., "Laser Ablation and Micromachining with Ultrashort Laser Pulses," *IEEE J. Quantum Electron.*, 33: 1706–1716, 1997.
2. Nolte, S., Momma, C., Jacobs, H., Tünnermann, A., Chichkov, B. N., Wellegehausen, B., and Welling, H., "Ablation of Metals by Ultrashort Laser Pulses," *J. Opt. Soc. Am. B*, 14: 2716–2722, 1997.
3. von der Linde, D., Sokolowski-Tinten, K., and Bialkowski, J., "Laser-Solid Interactions in the Femtosecond Time Regime," *Appl. Surf. Sci.*, 109/110: 1–10, 1997.
4. Keller, U., Miller, D. A. B., Boyd, G. D., Chiu, T. H., Ferguson, J. F., and Asom, M. T., "Solid-State Low-Loss Intracavity Saturable Absorber for Nd:YLF Lasers: An Antiresonant Semiconductor Fabry-Perot Saturable Absorber," *Opt. Lett.*, 17: 505–507, 1992.
5. Keller, U., Weingarten, K. J., Kärtner, F. X., Kopf, D., Braun, B., Jung, I. D., Fluck, R., et al., "Semiconductor Saturable Absorber Mirrors (Sesams) for Femtosecond to Nanosecond Pulse Generation in Solid-State Lasers," *IEEE J. Sel. Top. Quantum Electron.*, 2: 435–453, 1996.
6. Südmeyer, T., Marchese, S.V., Hashimoto, S., Baer, C. R. E., Gingras, G., Witzel, B., and Keller, U., "Femtosecond Laser Oscillators for High-Field Science," *Nature Photonics*, 2: 599–604, 2008.
7. Keller, U., and Tropper, A. C., "Passively Modelocked Surface-Emitting Semiconductor Lasers," *Phys. Rep.*, 429: 67–120, 2006.
8. Giesen, A., Hügel, H., Voss, A., Wittig, K., Brauch, U., and Opower, H., "Scalable Concept for Diode-Pumped High-Power Solid-State Lasers," *Appl. Phys. B*, 58: 365–372, 1994.
9. Giesen, A., and Speiser, J., "Fifteen Years of Work on Thin-Disk Lasers: Results and Scaling Laws," *IEEE J. Sel. Top. Quantum Electron.*, 13: 598–609, 2007.
10. Rudin, B., Rutz, A., Hoffmann, M., Maas, D. J. H. C., Bellancourt, A.-R., Gini, E., Südmeyer, T., and Keller, U., "Highly Efficient Optically Pumped Vertical Emitting Semiconductor Laser with More Than 20-W Average Output Power in a Fundamental Transverse Mode," *Opt. Lett.*, 33: 2719–2721, 2008.
11. Kuznetsov, M., Hakimi, F., Sprague, R., and Mooradian, A., "High-Power (>0.5-W CW) Diode-Pumped Vertical-External-Cavity Surface-Emitting Semiconductor Lasers with Circular TEM00 Beams," *IEEE Photon. Technol. Lett.*, 9: 1063–1065, 1997.

12. Baer, C. R. E., Kränkel, C., Saraceno, C. J., Heckl, O. H., Golling, M., Südmeyer, T., Keller, U., et al., "Efficient Mode-Locked Yb:Lu$_2$O$_3$ Thin Disk Laser with an Average Power of 103 W" (talk AMD2), *Advanced Solid State Photonics (ASSP)*, San Diego, CA: 2010.
13. Baer, C. R. E., Kränkel, C, Saraceno, C. J. ; Heckl, O. H., Golling, M., Peters, R., Petermann, et al., "Femtosecond Thin-disk Laser with 141 W of Average Power," *Opt. Lett.*, 35, 2719–2721, 2010.
14. Neuhaus, J., Bauer, D., Zhang, J., Killi, A., Kleinbauer, J., Kumkar, M., Weiler, S., et al., "Subpicosecond Thin-Disk Laser Oscillator with Pulse Energies of up to 25.9 Microjoules by Use of an Active Multipass Geometry," *Opt. Express*, 16: 20530–20539, 2008.
15. Maas, D. J. H. C., Bellancourt, A.-R., Rudin, B., Golling, M., Unold, H. J., Südmeyer, T., and Keller, U., "Vertical Integration of Ultrafast Semiconductor Lasers," *Appl. Phys. B*, 88: 493–497, 2007.
16. Abate, J. A., Lund, L., Brown, D., Jacobs, S., Refermat, S., Kelly, J., Gavin, M., et al., "Active Mirror: A Large-Aperture Medium-Repetition Rate Nd:Glass Amplifier," *Appl. Opt.*, 20: 351–361, 1981.
17. Fan, T. Y., and Byer, R. L., "Modelling and CW Operation of a Quasi-Three-Level 946 nm Nd:YAG Laser," *IEEE J. Quantum Elect.*, 23: 605–612, 1987.
18. Chow, W. W., and Koch, S. W., *Semiconductor: Laser Fundamentals. Physics of the Gain Materials*. Berlin, Germany: Springer, 1999.
19. Forster, T., "Zwischenmolekulare Energiewanderung und Fluoreszenz," *Ann. Phys.* Berlin, 437: 55–75, 1948.
20. Beattie, A. R., and Landsberg, P. T., "Auger Effect in Semiconductors," *Proceedings of the Royal Society of London Series A: Mathematical and Physical Sciences*, 249: 16–29, 1959.
21. Häring, R., Paschotta, R., Aschwanden, A., Gini, E., Morier-Genoud, F., and Keller, U., "High-power passively mode-locked semiconductor lasers," *IEEE J. Quantum Electron.*, 38: 1268–1275, 2002.
22. Larionov, M., "Kontaktierung und Charakterisierung von Kristallen für Scheibenlaser," *Institut für Strahlwerkzeuge*, Stuttgart, Germany: Universität Stuttgart, 2009.
23. Killi, A., Zawischa, I., Sutter, D., Kleinbauer, J., Schad, S., Neuhaus, J., and Schmitz, C., "Current Status and Development Trends of Disk Laser Technology," *Conference on Solid State Lasers XVII*, eds. W. A. Clarkson, N. H. Hodgson, and R. K. Shori, San Jose, CA: SPIE-Int. Soc. Optical Engineering, L8710–L8710, 2008.
24. Rudin, B., Rutz, A., Maas, D. J. H. C., Bellancourt, A. R., Gini, E., Südmeyer, T., and Keller, U., "Efficient High-Power VECSEL Generates 20 W Continuouswave Radiation in a Fundamental Transverse Mode" (paper ME2), *Advanced Solid-State Photonics (ASSP)*, Denver, USA C 2009.
25. Bedford, R. G., Kolesik, M., Chilla, J. L. A., Reed, M. K., Nelson, T. R., and Moloney, J. V., "Power-Limiting Mechanisms in VECSELs," *Conference on Enabling Photonics Technologies for Defense, Security, and Aerospace Applications*, ed. A. R. Pirich, Orlando, FL: SPIE-Int. Soc. Optical Engineering, 199–208, 2005.
26. Brunner, F., Südmeyer, T., Innerhofer, E., Paschotta, R., Morier-Genoud, F., Keller, U., Gao, J., et al., "240-fs Pulses with 22-W Average Power from a Passively Mode-Locked Thin-Disk Yb:KY(WO$_4$)$_2$ Laser" (talk CME3), *Conference on Laser and Electro-Optics CLEO 2002*, Long Beach, CA, 2002.
27. Kränkel, C., Johannsen, J., Peters, R., Petermann, K., and Huber, G., "Continuous-Wave High Power Laser Operation and Tunability of Yb:LaSc$_3$(BO$_3$)$_4$ in Thin Disk Configuration," *Appl. Phys. B*, 87: 217–220, 2007.
28. Kränkel, C., Peters, R., Petermann, K., Loiseau, P., Aka, G., and Huber, G., "Efficient Continuous-Wave Thin Disk Laser Operation of Yb:Ca$_4$YO(BO$_3$)$_3$ in EIIZ and EIIX Orientations with 26 W Output Power," *J. Opt Soc. Am. B*, 26: 1310–1314, 2009.
29. Kränkel, C., Peters, R., Petermann, K., and Huber, G., "High Power Operation of Yb:LuVO$_4$ and Yb:YVO$_4$ Crystals in the Thin-Disk Laser Setup" (paper MA 3), *Advanced Solid-State Photonics (ASSP)*, Vancouver, Canada, 2007.

30. Peters, R., Kränkel, C., Petermann, K., and Huber, G., "Power Scaling Potential of Yb:NGW in Thin Disk Laser Configuration," *Appl. Phys. B,* 91: 25–28, 2008.
31. Peters, R., Kränkel, C., Petermann, K., and Huber, G., "Broadly Tunable High-Power Yb:Lu$_2$O$_3$ Thin Disk Laser with 80% Slope Efficiency," *Opt. Express,* 15: 7075–7082, 2007.
32. Peters, R., Kränkel, C., Petermann, K., and Huber, G., "High Power Laser Operation of Sesquioxides Yb:Lu$_2$O$_3$ and Yb:Sc$_2$O$_3$" (paper CTuKK4), *Conference on Lasers and Electro-Optics,* San Jose, CA, 2008.
33. Palmer, G., Schultze, M., Siegel, M., Emons, M., Bünting, U., and Morgner, U., "Passively Mode-Locked Yb:KLu(WO$_4$)$_2$ Thin-Disk Oscillator Operated in the Positive and Negative Dispersion Regime," *Opt. Lett.,* 33: 1608–1610, 2008.
34. Giesen, A., Speiser, J., Peters, R., Krankel, C., and Petermann, K., "Thin-Disk Lasers Come of Age," *Photonics Spectra,* 41: 52, 2007.
35. Schmid, M., Benchabane, S., Torabi-Goudarzi, F., Abram, R., Ferguson, A. I., and Riis, E., "Optical In-Well Pumping of a Vertical-External-Cavity Surface-Emitting Laser," *Appl. Phys. Lett.,* 84: 4860–4862, 2004.
36. Beyertt, S. S., Zorn, M., Kubler, T., Wenzel, H., Weyers, M., Giesen, A., Trankle, G., and Brauch, U., "Optical In-Well Pumping of a Semiconductor Disk Laser with High Optical Efficiency," *IEEE J. Quantum Electron.,* 41: 1439–1449, 2005.
37. Beyertt, S.-S., Brauch, U., Demaria, F., Dhidah, N., Giesen, A., Kübler, T., Lorch, S., et al., "Efficient Gallium–Arsenide Disk Laser," *IEEE J. Quantum Electron.,* 43: 869–875, 2007.
38. Aschwanden, A., Lorenser, D., Unold, H. J., Paschotta, R., Gini, E., and Keller, U., "2.1-W Picosecond Passively Mode-Locked External-Cavity Semiconductor Laser," *Opt. Lett.,* 30: 272–274, 2005.
39. Kneubühl, F. K., and Sigrist, M. W., *Laser,* Stuttgart, Germany: B. G. Teubner, 1991.
40. Corzine, S. W., Yan, R. H., and Coldren, L. A., "Theoretical Gain in Strained InGaAs/AlGaAs Quantum Wells Including Valence-Band Mixing Effects," *Appl. Phys. Lett.,* 57: 2835–2837, 1990.
41. Casey, H. C., Sell, D. D., and Wecht, K. W., "Concentration Depedence of the Absorption Coefficient for n- and p-Type GaAs Between 1.3 and 1.6 eV," *J. Appl. Phys.,* 46: 250–257, 1974.
42. Marchese, S., *Towards High Field Physics with High Power Thin Disk Laser Oscillators,* Dissertation at ETH Zurich, Nr. 17583, Hartung-Gorre Verlag, Konstanz, 2008.
43. Brunner, F., Innerhofer, E., Marchese, S. V., Südmeyer, T., Paschotta, R., Usami, T., Ito, H., et al., "Powerful Red-Green-Blue Laser Source Pumped with a Mode-Locked Thin Disk Laser," *Opt. Lett.,* 29: 1921–1923, 2004.
44. Contag, J., "*Modellierung und numerische Auslegung des Yb:YAG Scheibenlasers,*" *Institut für Strahlwerkzeuge,* Stuttgart, Germany: Universität Stuttgart, 2002.
45. Keller, U., "Ultrafast Solid-State Lasers," *Landolt-Börnstein. Laser Physics and Applications. Subvolume B: Laser Systems. Part I.,* eds. G. Herziger, H. Weber, and R. Proprawe, Heidelberg, Germany: Springer Verlag, 33–167, 2007.
46. Keller, U., "Ultrafast Solid-State Lasers," *Progress in Optics,* 46: 1–115, 2004.
47. Keller, U., "Recent Developments in Compact Ultrafast Lasers," *Nature,* 424: 831–838, 2003.
48. Spence, D. E., Kean, P. N., and Sibbett, W., "60-fsec Pulse Generation from a Self-Mode-Locked Ti:sapphire Laser," *Opt. Lett.,* 16: 42–44, 1991.
49. Keller, U., 'tHooft, G. W., Knox, W. H., and Cunningham, J. E., "Femtosecond Pulses from a Continuously Self-Starting Passively Mode-Locked Ti:Sapphire Laser," *Opt. Lett.,* 16: 1022–1024, 1991.
50. Brabec, T., Spielmann, C., Curley, P. F., and Krausz, F., "Kerr Lens Mode Locking," *Opt. Lett.,* 17: 1292–1294, 1992.
51. McCumber, D. W., "Einstein Relations Connecting Broadband Emission and Absorption Spectra," *Phys. Rev.,* 136: 954–957, 1964.
52. Paschotta, R., and Keller, U., "Passive Mode Locking with Slow Saturable Absorbers," *Appl. Phys. B,* 73: 653–662, 2001.

53. Kärtner, F. X., and Keller, U., "Stabilization of Soliton-Like Pulses with a Slow Saturable Absorber," *Opt. Lett.*, 20: 16–18, 1995.
54. Kärtner, F. X., Jung, I. D., and Keller, U., "Soliton Mode-Locking with Saturable Absorbers," *IEEE J. Sel. Top. Quant.*, 2: 540–556, 1996.
55. Jung, I. D., Kärtner, F. X., Brovelli, L. R., Kamp, M., and Keller, U., "Experimental Verification of Soliton Modelocking Using Only a Slow Saturable Absorber," *Opt. Lett.*, 20: 1892–1894, 1995.
56. New, G. H. C., "Pulse Evolution in Mode-Locked Quasi-Continuous Lasers," *IEEE J. Quantum Electron.*, 10: 115–124, 1974.
57. Killi, A., Zawischa, I., Sutter, D., Kleinbauer, J., Schad, S., Neuhaus, J., and Schmitz, C., "Current Status and Development Trends of Disk Laser Technology" (art. no. 68710L), *Conference on Solid State Lasers XVII*, eds. W. A. Clarkson, N. H. Hodgson, and R. K. Shori, San Jose, CA: SPIE-Int. Soc. Optical Engineering, L8710–L8710, 2008.
58. Neuhaus, J., Bauer, D., Kleinbauer, J., Killi, A., Weiler, S., Sutter, D. H., and Dekorsy, T., "Pulse Energies Exceeding 20 µJ Directly from a Femtosecond Yb:YAG Oscillator," *Proceedings of the Ultrafast Phenomena XVI*, eds. P. B. Corkum, S. D. Silvestri, K. A. Nelson, E. Riedle, and R. W. Schoenlein, Heidelberg, Germany: Springer, 2008.
59. Neuhaus, J., Kleinbauer, J., Killi, A., Weiler, S., Sutter, D., and Dekorsy, T., "Passively Mode-Locked Yb:YAG Thin-Disk Laser with Pulse Energies Exceeding 13 µJ by Use of an Active Multipass Geometry," *Opt. Lett.*, 33: 726–728, 2008.
60. Marchese, S. V., Baer, C. R. E., Engqvist, A. G., Hashimoto, S., Maas, D. J. H. C., Golling, M., Südmeyer, T., and Keller, U., "Femtosecond Thin Disk Laser Oscillator with Pulse Energy Beyond the 10-Microjoule Level," *Opt. Express*, 16: 6397–6407, 2008.
61. Marchese, S. V., Südmeyer, T., Golling, M., Grange, R., and Keller, U., "Pulse Energy Scaling to 5 µJ from a Femtosecond Thin Disk Laser," *Opt. Lett.*, 31: 2728–2730, 2006.
62. Brunner, F., Südmeyer, T., Innerhofer, E., Paschotta, R., Morier-Genoud, F., Gao, J., Contag, K., et al., "240-fs Pulses with 22-W Average Power from a Mode-Locked Thin-Disk Yb:KY(WO$_4$)$_2$ Laser," *Opt. Lett.*, 27: 1162–1164, 2002.
63. Innerhofer, E., Südmeyer, T., Brunner, F., Häring, R., Aschwanden, A., Paschotta, R., Keller, U., et al., "60 W Average Power in 810-fs Pulses from a Thin-Disk Yb:YAG Laser," *Opt. Lett.*, 28: 367–369, 2003.
64. Aus der Au, J., Spühler, G. J., Südmeyer, T., Paschotta, R., Hövel, R., Moser, M., Erhard, S., et al., "16.2 W Average Power from a Diode-Pumped Femtosecond Yb:YAG Thin Disk Laser," *Opt. Lett.*, 25: 859–861, 2000.
65. Marchese, S. V., Baer, C. R. E., Peters, R., Kränkel, C., Engqvist, A. G., Golling, M., Maas, D. J. H. C., et al., "Efficient Femtosecond High Power Yb:Lu$_2$O$_3$ Thin Disk Laser," *Opt. Express*, 15: 16966–16971, 2007.
66. Baer, C. R. E., Kränkel, C., Heckl, O. H., Golling, M., Südmeyer, T., Peters, R., Petermann, K., et al., "227-fs Pulses from a Mode-Locked Yb:LuScO$_3$ Thin Disk Laser," *Opt. Express*, 17: 10725–10730, 2009.
67. Baer, C. R. E., Kränkel, C., Saraceno, C. J., Heckl, O. H., Golling, M., Südmeyer, T., Peters, R., et al., "Femtosecond Yb:Lu$_2$O$_3$ Thin Disk Laser with 63 W of Average Power," *Opt. Lett.*, 34: 2823–2825, 2009.
68. Rautiainen, J., Korpijärvi, V.-M., Puustinen, J., Guina, M., and Okhotnikov, O. G., "Passively Mode-Locked GaInNAs Disk Laser Operating at 1220 nm," *Opt. Express*, 16: 2008.
69. Wilcox, K. G., Mihoubi, Z., Daniell, G. J., Elsmere, S., Quarterman, A., Farrer, I., Ritchie, D. A., and Tropper, A., "Ultrafast Optical Stark Mode-Locked Semiconductor Laser," *Opt. Lett.*, 33: 2797–2799, 2008.
70. Garnache, A., Hoogland, S., Tropper, A. C., Sagnes, I., Saint-Girons, G., and Roberts, J. S., "Sub-500-fs Soliton Pulse in a Passively Mode-Locked Broadband Surface-Emitting Laser with 100-mW average power," *Appl. Phys. Lett.*, 80: 3892–3894, 2002.
71. Hoogland, S., Paldus, B., Garnache, A., Weingarten, K. J., Grange, R., Haiml, M., Paschotta, R., et al., "Picosecond Pulse Generation with a 1.5 µM Passively

Modelocked Surface Emitting Semiconductor Laser," *Electron. Lett.*, 39: 846, 2003.
72. Casel, O., Woll, D., Tremont, M. A., Fuchs, H., Wallenstein, R., Gerster, E., Unger, P., et al., "Blue 489-nm Picosecond Pulses Generated by Intracavity Frequency Doubling in a Passively Mode-Locked Optically Pumped Semiconductor Disk Laser," *Appl. Phys. B*, 81: 443–446, 2005.
73. Häring, R., Paschotta, R., Gini, E., Morier-Genoud, F., Melchior, H., Martin, D., and Keller, U., "Picosecond Surface-Emitting Semiconductor Laser with >200 mW Average Power," *Electron. Lett.*, 37: 766–767, 2001.
74. Lindberg, H., Sadeghi, M., Westlund, M., Wang, S., Larsson, A., Strassner, M., and Marcinkevicius, S., "Mode Locking a 1550 nm Semiconductor Disk Laser by Using a GaInNAs Saturable Absorber, *Opt. Lett.*, 30: 2793–2795, 2005.
75. Klopp, P., Saas, F., Zorn, M., Weyers, M., and Griebner, U., "290-fs Pulses from a Semiconductor Disk Laser," *Opt. Express*, 16: 5770–5775, 2008.
76. Hoogland, S., Dhanjal, S., Tropper, A. C., Roberts, S. J., Häring, R., Paschotta, R., and Keller, U., "Passively Mode-Locked Diode-Pumped Surface-Emitting Semiconductor Laser," *IEEE Photonics Tech. Lett.*, 12: 1135–1138, 2000.
77. Rutz, A., Liverini, V., Maas, D. J. H. C., Rudin, B., Bellancourt, A.-R., Schön, S., and Keller, U., "Passively Modelocked GaInNAs VECSEL at Centre Wavelength Around 1.3," *Electron. Lett.*, 42: 926, 2006.
78. Aschwanden, A., Lorenser, D., Unold, H. J., Paschotta, R., Gini, E., and Keller, U., "10-GHz Passively Mode-Locked Surface-Emitting Semiconductor Laser with 1.4 W Output Power," *Conference on Lasers and Electro-Optics/International Quantum Electronics Conference (CLEO/IQEC)*, San Francisco, CA: Optical Society of America, 2004.
79. Hoogland, S., Garnache, A., Sagnes, I., Roberts, J. S., and Tropper, A. C., "10-GHz Train of Sub-500-fs Optical Soliton-Like Pulses From a Surface-Emitting Semiconductor Laser," *IEEE Photonics Tech. Lett.*, 17: 267–269, 2005.
80. Härkönen, A., Rautiainen, J., Orsila, L., Guina, M., Rößner, K., Hümmer, M., Lehnhardt, T., et al., "2-μm Mode-Locked Semiconductor Disk Laser Synchronously Pumped Using an Amplified Diode Laser," *IEEE Photonics Technol. Lett.*, 20: 1332–1334, 2008.
81. Lorenser, D., Unold, H. J., Maas, D. J. H. C., Aschwanden, A., Grange, R., Paschotta, R., Ebling, E., et al., "Towards Wafer-Scale Integration of High Repetition Rate Passively Mode-Locked Surface-Emitting Semiconductor Lasers," *Appl. Phys. B*, 79: 927–932, 2004.
82. Rudin, B., Maas, D. J. H. C., Lorenser, D., Bellancourt, A.-R., Unold, H. J., and Keller, U., "High-Performance Mode-Locking with Up to 50 GHz Repetition Rate from Integrable VECSELs," *Conference on Lasers and Electro-Optics (CLEO)*, Long Beach, CA, 2006.
83. Lorenser, D., Maas, D. J. H. C., Unold, H. J., Bellancourt, A.-R., Rudin, B., Gini, E., Ebling, D., and Keller, U., "50-GHz Passively Mode-Locked Surface-Emitting Semiconductor Laser with 100 MW Average Output Power," *IEEE J. Quantum Electron.*, 42: 838–847, 2006.
84. Speiser, J., "Scaling of Thin-Disk Lasers-Influence of Amplified Spontaneous Emission," *J. Optic. Soc. Am. B-Opt. Phys.*, 26: 26–35, 2009.
85. Innerhofer, E., Südmeyer, T., Brunner, F., Häring, R., Aschwanden, A., Paschotta, R., Keller, U., et al., "60 W Average Power in Picosecond Pulses from a Passively Mode-Locked Yb:YAG Thin-Disk Laser" (talk CTuD4), *Conference on Laser and Electro-Optics CLEO 2002*, Long Beach, CA, 2002.
86. Magni, V., "Multielement Stable Resonators Containing a Variable Lens," *J. Opt. Soc. Am. A*, 4: 1962–1969, 1987.
87. Hönninger, C., Paschotta, R., Morier-Genoud, F., Moser, M., and Keller, U., "Q-Switching Stability Limits of Continuous-Wave Passive Mode Locking," *J. Opt. Soc. Am. B*, 16: 46–56, 1999.
88. Grange, R., Haiml, M., Paschotta, R., Spuhler, G. J., Krainer, L., Golling, M., Ostinelli, O., and Keller, U., "New Regime of Inverse Saturable Absorption for Self-Stabilizing Passively Mode-Locked Lasers," *Appl. Phys. B*, 80: 151–158, 2005.

89. Schibli, T. R., Thoen, E. R., Kärtner, F. X., and Ippen, E. P., "Suppression of Q-Switched Mode Locking and Break-Up into Multiple Pulses by Inverse Saturable Absorption," *Appl. Phys. B*, 70: S41–S49, 2000.
90. Südmeyer, T., Maas, D. J. H. C., and Keller, U., "Mode-Locked Semiconductor Disk Lasers," *Semiconductor Disk Lasers:Physics and Technology*, ed. O. Okhotnikov, Wiley-VCH Verlag KGaA, 2010.
91. Maas, D., *MIXSELs: A New Class of Ultrafast Semiconductor Lasers*. Dissertation at ETH Zurich, Nr. 18121, Hartung-Gorre Verlag, Konstanz, 2009.
92. Oehler, A. E. H., Südmeyer, T., Weingarten, K. J., and Keller, U., "100 GHz Passively Mode-Locked Er:Yb:glass Laser at 1.5 µm with 1.6-ps Pulses," *Opt. Express*, 16: 21930–21935, 2008.
93. Saarinen, E. J., Harkonen, A., Herda, R., Suomalainen, S., Orsila, L., Hakulinen, T., Guina, M., and Okhotnikov, O. G., "Harmonically Mode-Locked VECSELs for Multi-GHz Pulse Train Generation," *Opt. Express*, 15: 955–964, 2007.
94. Ippen, E. P., Liu, L. Y., and Haus, H. A., "Self-Starting Condition for Additive-Pulse Modelocked Lasers," *Opt. Lett.*, 15: 183–185, 1990.
95. Haus, H. A., and Ippen, E. P., "Self-Starting of Passively Mode-Locked Lasers," *Opt. Lett.*, 16: 1331–1333, 1991.
96. Papadopoulos, D. N., Forget, S., Delaigue, M., Druon, F., Balembois, F., and Georges, P., "Passively Mode-Locked Diode-Pumped Nd:YVO$_4$ Oscillator Operating at an Ultralow Repetition Rate," *Opt. Lett.* 28: 1838–1840, 2003.
97. Herriott, D., Kogelnik, H., and Kompfner, R., "Off-Axis Paths in Spherical Mirror Interferometers," *Appl. Opt.*, 3: 523–526, 1964.
98. Nibbering, E. T. J., Grillon, G., Franco, M. A., Prade, B. S., and Mysyrowicz, A., "Determination of the Inertial Contribution to the Nonlinear Refractive Index of Air, N_2, and O_2 by Use of Unfocused High-Intensity Femtosecond Laser Pulses," *J. Opt. Soc. Am. B*, 14: 650–660, 1997.
99. Adair, R., Chase, L. L., and Payne, S. A., "Nonlinear Refractive Index of Optical Crystals," *Phys. Rev. B*, 39: 3337–3350, 1989.
100. Hönninger, C., Morier-Genoud, F., Moser, M., Keller, U., Brovelli, L. R., and Harder, C., "Efficient and Tunable Diode-Pumped Femtosecond Yb:glass Lasers," *Opt. Lett.*, 23: 126–128, 1998.
101. Rivier, S., Mateos, X., Liu, J., Petrov, V., Griebner, U., Zorn, M., Weyers, M., Zhang, H., et al., "Passively Mode-Locked Yb:LuVO$_4$ Oscillator," *Opt. Express*, 14: 11668–11671, 2006.
102. Zaouter, Y., Didierjean, J., Balembois, F., Lucas Leclin, G., Druon, F., Georges, P., Petit, J., et al., "47-fs Diode-Pumped Yb^{3+}:CaGdAlO$_4$ Laser," *Opt. Lett.*, 31: 119–121, 2006.
103. Rivier, S., Schmidt, A., Kränkel, C., Peters, R., Petermann, K., Huber, G., Zorn, M., et al., "Ultrashort Pulse Yb:LaSc$_3$(BO$_3$)$_4$ Mode-Locked Oscillator," *Opt. Express*, 15: 15539–15544, 2007.
104. García-Cortés, A., Cano-Torres, J. M., Serrano, M. D., Cascales, C., Zaldo, C., Rivier, S., Mateos, X., et al., "Spectroscopy and Lasing of Yb-doped NaY(WO$_4$)$_2$: Tunable and Femtosecond Mode-Locked Laser Operation," *IEEE J. Quantum Elect.*, 43: 758–764, 2007.
105. Südmeyer, T., Kränkel, C., Baer, C. R. E., Heckl, O. H., Saraceno, C. J., Golling, M., Peters, R., et al., "High-Power Ultrafast Thin Disk Laser Oscillators and Their Potential for Sub-100-Femtosecond Pulse Generation," *Appl. Phys. B*, 97: 281–295, 2009.
106. Hönninger, C., Paschotta, R., Graf, M., Morier-Genoud, F., Zhang, G., Moser, M., Biswal, S., et al., "Ultrafast Ytterbium-Doped Bulk Lasers and Laser Amplifiers," *Appl. Phys. B*, 69: 3–17, 1999.
107. Schmidt, A., Mateos, X., Petrov, V., Griebner, U., Peters, R., Petermann, K., Huber, G., et al., "Passively Mode-Locked Yb:LuScO$_3$ Oscillator" (paper MB12), *Advanced Solid-State Photonics (ASSP)*, Denver, CO: 2009.
108. Mysyrowicz, A., Hulin, D., Antonetti, A., and Migus, A., "Dressed Excitons in a Multiple-Quantum-Well Structure: Evidence for an Optical Stark-Effect with Femtosecond Response-Time," *Phys. Rev. Lett.*, 56: 2748–2751, 1986.

109. Klopp, P., Griebner, U., Zorn, M., Klehr, A., Liero, A., Weyers, M., and Erbert, G., "Mode-Locked InGaAs-AlGaAs Disk Laser Generating Sub-200-fs Pulses, Pulse Picking and Amplification by a Tapered Diode Amplifier," *Opt. Express*, 17: 10820–10834, 2009.
110. Quarterman, A. H., Wilcox, K. G., Apostolopoulos, V., Mihoubi, Z., Elsmere, S. P., Farrer, I., Ritchie, D. A., and Tropper, A., "A Passively Mode-Locked External-Cavity Semiconductor Laser Emitting 60-fs Pulses," *Nat. Photonics*, 3: 729–731, 2009.
111. Paschotta, R., Häring, R., Keller, U., Garnache, A., Hoogland, S., and Tropper, A. C., "Soliton-Like Pulse-Shaping Mechanism in Passively Mode-Locked Surface-Emitting Semiconductor Lasers," *Appl. Phys. B*, 75: 445–451, 2002.
112. Hoffmann, M., Sieber, O. D., Maas, D. J. H. C., Wittwer, V. J., Golling, M., Sudmeyer, T., and Keller, U., "Experimental Verification of Soliton-like Pulse-shaping Mechanisms in Passively Mode-locked VECSELs," *Opt. Express*, 18, 10143–10153, 2010.
113. Wittwer, V. J., Rudin, B. Maas, D. J. H. C., Hoffmann, M., Sieber, O., Barbarin, Y., Golling, et al.,"An Integrated Passively Modelocked External-Cavity Semiconductor Laser with 6.4 W Average Power" (talk ThD1), *4th EPS-QEOD Europhoton Conference*, Hamburg, Germany, 2010.

CHAPTER 14
The National Ignition Facility Laser
High-Pulse Energy Fusion Laser

Richard A. Sacks

Senior Scientist and Technical Lead, ICF and HED Science Program (NIF), Lawrence Livermore National Laboratory, Livermore, California

Christopher A. Haynam

Associate Program Leader, ICF and HED Science Program (NIF), Lawrence Livermore National Laboratory, Livermore, California

14.1 Introduction

The 192-beam National Ignition Facility (NIF) laser is the world's largest, most complex optical system. To meet its goal of achieving energy gain (ignition) in a deuterium-tritium (DT) nuclear fusion target, laser design criteria include the ability to generate pulses of up to 1.8 megajoules (MJ) total energy, with peak power as high as 500 terawatts (TW) and temporal pulse shapes spanning 2 orders of magnitude at the third harmonic (351 nm or 3ω) of the laser wavelength. The focal spot fluence distribution of these pulses is carefully controlled through a combination of special optics in the 1ω (1053-nm) portion of the laser (continuous phase plates), smoothing by spectral dispersion (SSD), and overlapping of multiple beams with orthogonal polarization (polarization smoothing). The NIF laser has been successfully tested and verified to meet its laser performance design criteria, as well as the temporal pulse shaping, focal spot conditioning, and peak power requirements for two candidate indirect-drive ignition designs.

We have structured this chapter as follows: Section 14.2 summarizes the development history of high-energy solid-state lasers that are intended to probe thermonuclear fusion physics. Section 14.3 provides a brief overview of the NIF facility and laser design. This is followed in Secs. 14.4 to 14.6 with a detailed description of each major laser subsystem, along with performance validation experiments carried out in 2006. (These sections are largely excerpts from a review article we wrote.[1]) These experiments demonstrated that the NIF laser would meet both its original design specifications, as laid out in 1994, and the ignition campaign requirements that evolved as progress in target physics modeling, fabrication, and understanding was made. The results presented in Secs. 14.4 to 14.6 cover the predicted and measured performance of the laser obtained during the final stages of the activation or commissioning of the first of NIF's 24 bundles. The performance envelope of the laser's 1ω portion was explored by a series of shots at progressively higher 1ω energies. In Sec. 14.4, we compare model predictions of each of the eight beamlines with measured energies, report the shot-to-shot energy reproducibility, and show the 1ω power and energy operating envelopes for NIF. Section 14.5 details how a shaped pulse is created, diagnosed, and amplified as it traverses the NIF laser and how its frequency is converted to the third harmonic by nonlinear crystals in the final optics assembly (FOA). It also describes a series of laser shots that validated NIF's capability of meeting its energy, power, and temporal contrast design goals. These performance qualification (PQ) shots were taken with an entire bundle operating at 1ω. Section 14.6 describes the addition of focal spot beam conditioning to the laser. It also details the generation of two shaped pulses that had all three beam-conditioning methods applied and that simultaneously generated the single-beam 3ω powers and energies planned for the first ignition campaigns on NIF. We conclude in Sec. 14.7 with a description of the present state as the completed machine carries out initial plasma physics and target compression experiments preparatory to a fusion ignition campaign in late 2010 to mid-2012.

14.2 Historical Background

The laser era was born on May 16, 1960, when Theodore Maiman of Hughes Research Laboratory first exposed a 1-cm ruby crystal, polished on two parallel faces, to a high-power pulsed flash lamp and observed a marked narrowing of the emission spectrum.[2] Within days of the publication of this event, Stirling Colgate, Ray Kidder, and John Nuckolls of the Livermore branch of the Lawrence Radiation Laboratory in Livermore (now the Lawrence Livermore National Laboratory [LLNL]) separately proposed investigating whether devices based on this phenomenon could be used to drive thermonuclear fusion in a controlled laboratory environment.[3] In 1962, a small laser-fusion project, under the leadership of Kidder, was established in the Livermore lab's physics department to explore this possibility.

Over the following decade, this group produced a number of advances, including the development of the Lasnex laser-fusion simulation code[4] and the seminal first open-literature publication of the physics behind inertial fusion.[5] In this publication, the authors estimated that thermonuclear burn in a compressed hot spot might be observed with laser irradiation of about 10 kilojoules (kJ), while significant fuel burnup and high gain would require ~1 MJ in a 10-ns temporally shaped pulse.

In 1973, the first Livermore inertial confinement fusion (ICF) laser—the single-arm Cyclops laser—was commissioned. Cyclops generated several hundred joules in a few hundred-picosecond pulse and was used for laser research and development (R&D), especially for developing techniques for controlling optical self-focusing. Cyclops pioneered the use of specially engineered low-nonlinear-index glass, of Brewster-angle amplifier slabs, and of spatial filtering. The first experiments to generate x rays by irradiating the interior of a hohlraum were carried out on Cyclops by Lindl, Manes, and Brooks in 1976.[6]

The two-beam Janus laser, built in 1974, was a 40-J, 100-ps laser that used many of Cyclops's component designs. This laser, which was used for target irradiation experiments, was the first Livermore facility to demonstrate target compression and the production of thermonuclear neutrons. In 1976, the Argus laser built on both of these successes to push the performance envelope. Argus's two beams had 20-cm output apertures and a series of five groups of amplifiers and spatial filters. Because spatial filtering was built into the design, the telescopes were longer, thus improving the beam smoothing achieved. Argus could deliver as much as ~2 kJ in a 1-ns pulse into a 100-μm spot, generating as many as 10^9 neutrons per shot on direct-drive exploding-pusher targets. It also pioneered the use of nonlinear crystals to convert light to the second or third harmonic, and significant improvement in coupling the light into the target was noted.

The next step along the path to ICF was taken in 1977, when the 20-beam Shiva laser was commissioned. Compared with previous ICF lasers, Shiva was a giant—about 100 m × 50 m. Shiva was able to deliver as much as 20 TW in short (100-ps) pulses and up to 10 kJ at nanosecond pulse lengths, approximately fivefold increases in both energy and power over Argus. It is arguable that Shiva's greatest success was its failure to accomplish all that had been hoped for. Experiments with Shiva were able to achieve capsule compressions of about 100 times, which is in the right ballpark for ignition targets; however, both hohlraum temperature and capsule compression fell below expectations. These effects were traced to laser-plasma instabilities ($2\omega_{pe}$ and forward stimulated Raman scattering), which coupled laser energy into high-energy electrons. These instabilities both decreased x-ray generation and preheated the ablator and the fuel.[7] It had been previously demonstrated that shorter-wavelength lasers would couple more effectively to targets.[8] The Shiva results, along with the improved simulation and analysis that accompanied them,[8] firmly established that achieving DT ignition requires both more energy and shorter wavelength. For neodymium-doped glass lasers, this means that harmonic conversion is essential.

By June 1979, when the Shiva compression experiments were completed, design work for Shiva's successor was already well advanced. Nova was envisioned as a 20-beam, 200-kJ, 100 ps to 10 ns infrared (IR) laser that would achieve the long-sought goal of fusion burn at laboratory scale. The Shiva results showed that coupling at 1-μm wavelength could not be coaxed to be good enough to efficiently drive the capsule. In addition, reported results from Campbell et al., École Polytechnique, the University of Rochester, and KMS Fusion, Inc., all showed that conversion to 351-nm wavelength could be carried out with efficiency in excess of 50 percent.[9] Based on this information, a review chaired by John Foster[10] confirmed that Nova should not be expected to reach ignition and recommended that it be reconfigured as a 10-beam, 100-kJ IR device with frequency conversion to the third harmonic.

Even before Nova construction began, it broke new ground in a number of ways. In 1976, Bliss et al.[11] reported measurements of the rate of nonlinear growth of beam-intensity fluctuations (filamentation) versus spatial frequency. In the same year, Trenholme and Goodwin[12] developed and made available computational tools that quantitatively explained these measurements, demonstrated the efficacy of spatial filtering at controlling filamentation, and enabled examination of alternative Nova architectures to assess their relative filamentation risk. Also in 1976, Hunt et al.[13] invented the use of relay imaging to allow high spatial fill factor. Both of these techniques were built into the Nova design. The Nova laser was the first whose design was guided by numerical modeling and optimization[14] and the first whose construction was preceded by the building of a prototype (the two-beam Novette, commissioned in 1983). When Nova was commissioned in 1985, it could deliver as much as 100 kJ of IR light or 40 to 50 kJ at 351 nm, with flexible pulse-shaping capability ranging from ~100-ps impulses to ~10-ns multistep ramps for target implosions.

For more than a decade, Nova was the premier fusion laser in the world. Among the accomplishments achieved on Nova were:

- First quantitative measurements of beam-breakup threshold (due to small-angle forward-rotational Raman scattering) in long air-path beam propagation[15]

- Discovery of, theoretical description of, and development of countermeasures for optical damage due to transverse Brillouin scattering in large-aperture fused-silica optics[16]
- Neutron yield greater than 10^{13} in 1986[17]
- First petawatt laser (1.3 PW in 500-fs pulse) in 1996[18]
- First x-ray laser (213 Å) in 1985[19]
- Compression of an ICF target by a factor of 35 (linear dimension), which was close to that needed for gain, in 1987[20]

A number of lasers at other facilities throughout the world have also made important contributions to ICF research and have helped lay the foundation for the NIF design and specifications. Omega, at the University of Rochester's Laboratory for Laser Energetics, was commissioned in 1995. This 60-beam Nd:glass laser is capable of delivering as much as 30 kJ at up to 60 TW at 351 nm. Omega was the world's highest-energy laser from 1999, when Nova was decommissioned, until 2005, when the first eight beams of NIF demonstrated output approximately double Omega's. Omega's record of 10^{14} neutrons in a single shot, set in 1999,[21] still stands, as does its record for compressed fuel density (100 g/cm^3), set in 2008.[22] Since 1999, much of the experimental science preparatory to the ignition campaign has been developed and prototyped on Omega. In 2008, the Omega EP (extended performance) laser system became operational. This addition to Omega consists of four beamlines that are near-clones of NIF beams, a new target chamber, and a vacuum pulse-compression chamber for achieving petawatt pulses at ~1 ps. In addition to basic science experiments, the goal of this combined facility will be to carry out fully integrated cryogenic fast-ignition experiments.[23]

Gekko XII, a 12-beam, Nd:glass laser at Osaka University's Institute of Laser Engineering, was completed in 1983. It delivered as much as 10 kJ in 1 to 2-ns pulses and was initially used to study direct-drive implosion symmetry and exploding-pusher target yields. In 1996–1997, Gekko was upgraded by the addition of a 400-J, ~100-fs short-pulse beam and has been used for experiments in fast-ignition physics. In 2002, it demonstrated a factor of 10^3 yield increase by using its original 12 beams to achieve capsule compression and then heating with its petawatt beam. Gekko is currently being upgraded with the addition of a 10-kJ, 10-ps beamline.

Phebus, a two-beam laser capable of 20-kJ IR, 5-kJ ultraviolet (UV) in 1-ns pulses, is part of the Laboratoire pour l'Utilisation des Lasers Intenses (LULI) at the École Polytechnique (Palaiseau, France). It has been a European center for research into high-energy density physics and has been responsible for important advances in plasma diagnostics and understanding of the initiation and growth of optical damage.

The Ligne d'Integration Laser (LIL) was completed in 2002 at Cesta, France, by the French nuclear science directorate Commissariat à l'Énergie Atomique (CEA). This four-beam prototype of the 240-beam Laser Mégajoule (LMJ) was designed, like NIF, to be capable of demonstrating inertial fusion ignition and gain. There has been very active collaboration between NIF and LMJ designers, and we have learned much from each other. A notable example is the development at CEA of programmable spatial light modulators for detailed tailoring of the light that is launched into the large amplifiers.[24] As of this writing, NIF is in the process of installing these devices in the front end of all beamlines.

NIF is the culmination of the experience gained at all of these facilities over the years. Its fundamental requirements in terms of energy, power, pulse-shaping finesse, far-field spot-size control, power balance, and shot-to-shot reproducibility, among other criteria, were laid out in 1994.[25] Its goal is to achieve fusion breakeven, with more thermonuclear energy released than is delivered in the form of laser energy, based on the understanding that has been developed by the plasma-physics and target-coupling data gathered in previous facilities.

Ceremonial groundbreaking occurred in May 1997, and by June 1999, the stadium-sized facility was sufficiently completed that the 10-m, 264,000-lb target chamber could be installed in the target bay. The conventional facility was completed in September 2001, and in December 2002, its first four beams were fired at 43-kJ IR in a 5-ns pulse. On May 30, 2003, NIF set the first of its world records by firing a single-beam 10.4-kJ, 3.5-ns UV pulse into its precision diagnostics system, thus meeting its primary criteria for beam energy, uniformity, and pulse-shaping capability. The first plasma-coupling experiments were carried out in August 2003. On February 24 and 25, 2009, a special subcommittee of the National Ignition Campaign (NIC) Review Committee met with NIF scientists for "a formal performance review of the status of the Laser System of NIF and the readiness of the Laser System for its role in the National Ignition Campaign."[26] Their conclusion was "that each and every one of the laser performance Completion Criteria, established under the NIF Project Completion Criteria, has been met or exceeded."[26]

14.3 NIF Facility and Laser Overview

For more than a decade, up to 1000 engineers, scientists, technicians, and skilled laborers, as well as more than 2300 vendors, have worked on NIF. NIF's 192-laser beamlines are housed in a building with a volume of about 350,000 m^3. Figure 14.1 shows an aerial photograph of the NIF site, doctored to remove the roofs and to show the internal structure. In the upper left is the optical assembly building, where final assembly,

The National Ignition Facility Laser

FIGURE 14.1 The National Ignition Facility (NIF) is approximately 150 m × 90 m and seven stories tall. The roof has been "removed" from this aerial photo to show an engineering rendering of the laser. The two laser bays are shown on the upper left. The switchyard (in red) is shown on the lower right, as is the spherical target chamber (in silver) into which the 192 beamlines converge.

cleaning, and preparation for clean transport of all large optics are carried out. Near the bottom right, one can see the spiderweb of beam tubes that separate the close-packed beams and direct them to the beam ports distributed around the 10-m-diameter target chamber. In between are the two large laser bays, each holding 12 bundles of eight 40 cm × 40 cm beamlines. Altogether, the building is approximately 150 m × 90 m by seven stories high, about the size of a large sports arena but filled with high-precision optical components.

Each of the 192-laser beamlines is composed of 36 to 38 large-scale optics, depending on beamline configuration (see Fig. 14.2), plus hundreds of smaller optics, yielding a total area of ~3600 m² for all of NIF optics. The total near-field area of all 192 laser beams is about 22 m². For indirect-drive fusion studies, all 192 beams are focused into a cylindrical hohlraum through two circular entrance holes that are each about 2.5 mm in diameter (see Fig. 14.3). The conditions created in the hohlraum or in other targets will provide the necessary environment to explore a wide range of high-energy-density physics experiments, including laboratory-scale thermonuclear ignition and burn.

We summarize the laser design very briefly in this section. See Refs. 27 to 34 and the further references in those papers for detailed

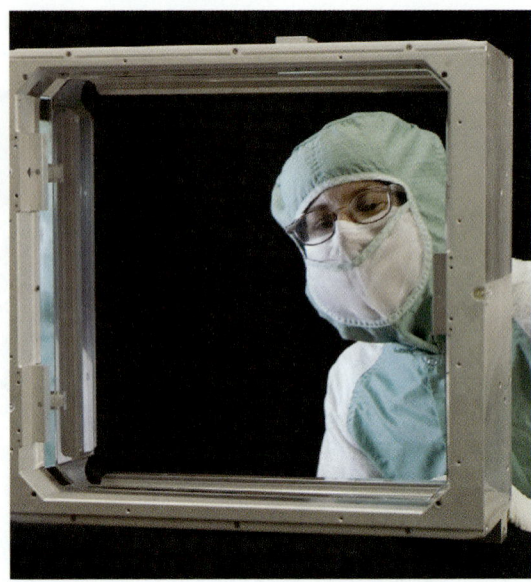

FIGURE 14.2 NIF's large optics each have an area of approximately 40 cm × 40 cm. The optic shown here—a 7.7-m focal length wedged lens used to focus one beam onto the target—was used during the 2006 campaigns discussed in Secs. 14.4 to 14.6. This photograph was taken after the lens was exposed to 11 shots with 8 to 9.4 kJ of 351-nm light (equivalent to 1.6 to 1.8 MJ of 351-nm light for the full 192-beam NIF).

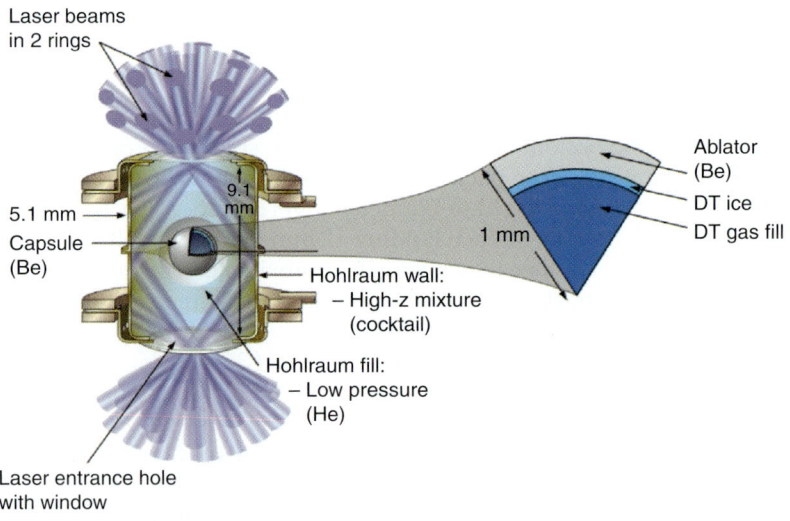

FIGURE 14.3 All of the 192 NIF laser beams are schematically shown focused into a single cylindrical hohlraum. Each cone comprises four individual beams. The hohlraum is approximately 10 mm × 5 mm in diameter. The laser entrance hole is about 2.5 mm in diameter. Each laser beam will be pointed to a precise location on the hohlraum wall and will generate x rays that will then drive the implosion of the central 1-mm radius spherical fusion capsule. The ensuing nuclear reaction is expected to release about 10 MJ of energy.

The National Ignition Facility Laser

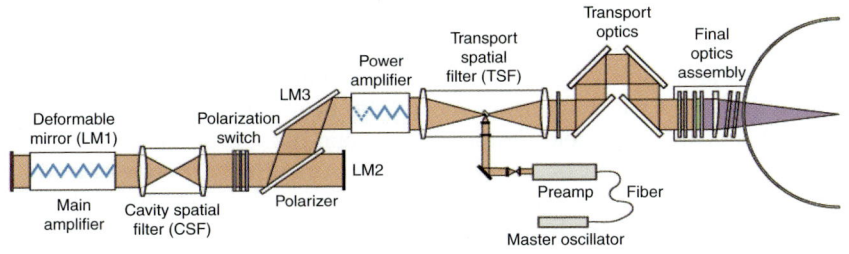

FIGURE 14.4 Schematic layout of one of NIF's 192 beamlines. The laser's path through the optics is discussed in the text.

information. The performance of Beamlet, a physics prototype for the NIF laser, was described several years ago.[35] References 36 to 39 discuss the laser energies and pulse shapes required for various ignition targets.

The NIF laser pulse starts in a continuous wave (CW) Yb:fiber master oscillator. From there, it passes through an array of fiber optical components to provide temporal amplitude and bandwidth control and is split to drive 48 preamplifier modules located under the main laser's transport spatial filter (see Fig. 14.4). This injection laser system (ILS) will be discussed in more detail in Sec. 14.5.2.

Immediately following the ILS, about 1 percent of the laser energy is diverted to a diagnostic suite known as the input sensor package (ISP). Here, the total energy, temporal shape, and near-field spatial shape from each preamplifier module (PAM) is measured.[30] The ILS can fire roughly one shot every 20 minutes. ISP measurements are important both for validating and normalizing numerical models of the laser performance and for ensuring that the ILS is properly configured prior to a main laser shot.

Pulses from the ILS are split four ways, supplying each of four main beamlines with energy that is adjustable from millijoules to more than a joule. Figure 14.4 shows a schematic of a single beamline of the main laser system. The pulse from the ILS is injected near the focal plane of the transport spatial filter (TSF). It expands to the full beam size of 37.2 cm × 37.2 cm (at the level of 0.1 percent of the peak fluence) and is collimated by the spatial filter lens. It then passes through the power amplifier (PA), reflects from a mirror and polarizer, and enters the cavity spatial filter (CSF). It traverses the main amplifier (MA), reflects off a deformable mirror that is used to correct wavefront distortions, and then goes through the MA and CSF again. By the time it makes this second pass through the CSF, a plasma-electrode Pockels cell (PEPC) switch has been fired to rotate the beam polarization by 90 degrees, allowing it to pass through the polarizer and be reflected back for another double pass through the CSF and MA. When the beam returns to the PEPC, the cell has switched off, so the beam now reflects from the polarizer and passes a second time through the PA and TSF. After the TSF, a beam splitter reflects a small sample of the output pulse back to the central TSF area, where it is collimated and directed to an output sensor package (OSP) located under the TSF. OSP diagnostics record the beam energy, temporal pulse shape, and near-field profiles.[30] The main pulse proceeds to the switchyard, where four or five transport mirrors direct it to one of a number of final optics assemblies (FOAs) symmetrically located about and mounted on the target chamber. Each FOA contains a 1ω vacuum window, focal spot beam-conditioning optics, two frequency-conversion crystals to reach 351-nm wavelength, a focusing lens, a main debris shield that also serves as a beam diagnostic pickoff to measure energy and power, and a 3-mm-thick disposable debris shield. The debris shields protect the upstream optics from target debris.

For the experiments reported in Secs. 14.4 to 14.6, the beam was not transported to the target chamber. Instead, an array of either seven or eight calorimeters was inserted at the TSF output to both measure and absorb the 1ω laser energy. When the eighth calorimeter was absent, it was replaced by a pickoff that routed that beam to an extensive suite of diagnostics for 1ω, 2ω, and 3ω light called the precision diagnostic system (PDS). The PDS instruments can diagnose one beamline in great detail, whereas the OSP diagnostics can acquire 1ω data on all 192 beams during a shot. In PDS, the laser was frequency converted to the second or third harmonic using typical NIF final optics, and detailed studies of the 1ω beam entering the FOA, as well as the 1ω, 2ω, and 3ω beams exiting it were performed.[1]

The MA contains eleven Nd:glass laser slabs. The PA is configured to have as many as seven slabs, though it typically contains only five. Some NIF shots have had one, three, or seven slabs in the PA to explore the full range of operating conditions. As an indication of scale, the CSF is 22 m long, the TSF is 60 m, the path length from the TSF output to the target chamber is 60 to 75 m, and the target chamber is 5 m in radius.

14.4 1ω Bundle Performance and $1\omega/3\omega$ NIF Operating Envelopes

Each of NIF's eight-beam laser bundles undergoes 1ω operational and performance qualification (OQ and PQ) before being used in any experiments. The OQ-PQ consists of firing 8 to 10 shots, using all 8 beams, into a bank of full-aperture calorimeters. These calorimeters measure absolute beam energy and calibrate a system of diodes in the OSP that serve as energy diagnostics during routine operations. Beam energies at 1ω for these shots range from 1 to 19 kJ, and pulse shapes are either flat in time (FIT) at the output or shaped to match user specifications. In addition to verifying the bundle performance, these shots are used to calibrate and validate the laser performance operations model (LPOM) description of these beamlines. LPOM is then used to predict laser performance and to set up the ILS for all NIF shots.

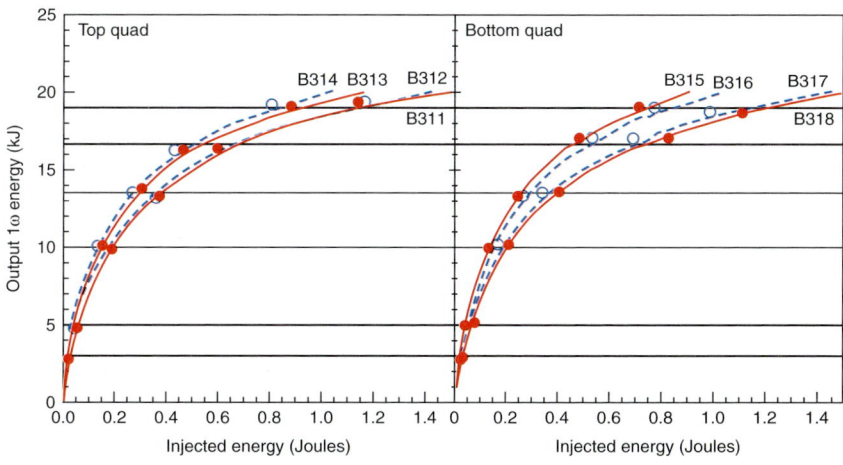

FIGURE 14.5 Comparison of modeled (dashed and solid lines) and measured (open and solid points) energies for eight shots on NIF's first operational bundle (Bundle 31: beamlines 311 through 318). The output energy is measured by the full-aperture calorimeters.

14.4.1 Energetics and the Laser Performance Operations Model Calibration Results

Figure 14.5 shows the comparison between modeled performance using LPOM and energy measurements for eight shots on the first bundle of the NIF laser. In this figure, output 1ω energy refers to the energy measured at the output of the main laser with the full-aperture calorimeters. The OSP was calibrated to these calorimeters. The injected energy is inferred from the ISP measurements, the known four-way ILS beam split ratios, and the known transmission from the ISP to the injection at the TSF. LPOM's predictions differ from the measurements by no more than 1.2 percent, demonstrating that LPOM can be used to set the desired energy from each beamline accurately over an extended range of operations.

The laser 1ω output energy is required to be reproducible to within 2 percent root mean square (rms) from shot to shot for proper ignition target performance. To test this performance criterion, the 19.2-kJ shot in Fig. 14.5 was repeated three times. After the first shot, no adjustments were made to either the injected pulse shape or the injected energy. As Table 14.1 shows, agreement with the target energy, the rms spread in total energy among the four shots, and the standard deviation of the eight beamline energies in each shot were all better than 1 percent. The estimated error in the 19.2-kJ energy measurement is 1.4 percent, or 0.27 kJ. This error estimate is a root sum of squares (rss) of the observed random component (1.3 percent) and the known systematic uncertainty (0.42 percent) of the calibration standard from the National Institute of Standards and Technology (NIST) that was used in calibrating NIF's calorimeters.

Solid-State Lasers

Shot Sequence Number	Desired 1ω Energy (kJ)	Average Measured Beamline 1ω Energy (kJ)	Total Bundle 1ω Energy (kJ)	Deviation of Average from Desired (%)	St Dev of Beamline Energies from Mean (%)
1	19.02	19.20	153.6	+0.98	0.84
2	"	19.15	153.2	+0.68	0.94
3	"	19.11	152.9	+0.50	0.67
4	"	19.10	152.8	+0.43	0.89

TABLE 14.1 Analysis of the 1ω Beam Energetics of Four Identical 19-kJ Shots

14.4.2 Power versus Energy Operating Envelopes for NIF

A *system shot* is defined as any event in which the flash lamps are fired in a bundle with all of the bundle's main laser (1ω) optics installed. From commissioning the first four NIF beamlines in April 2001 to the time of the initial tests in 2006, NIF had fired more than 600 system shots. Figure 14.6 shows a summary of the 1ω shots fired, together with the NIF standard 1ω operating envelope, as set in LPOM. This envelope does not represent the absolute limits of

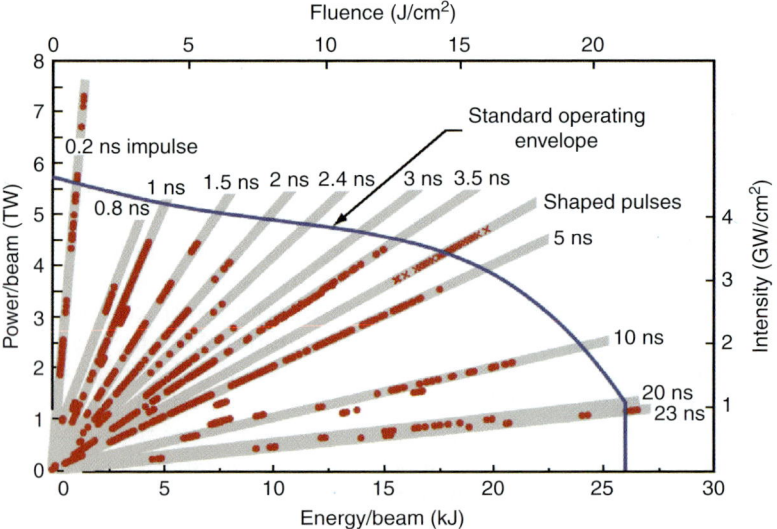

FIGURE 14.6 Plot of 1ω peak power per beam versus 1ω energy per beam for initial NIF shots. The thin solid line is the laser performance operations model (LPOM) "equipment protection" operating limit.

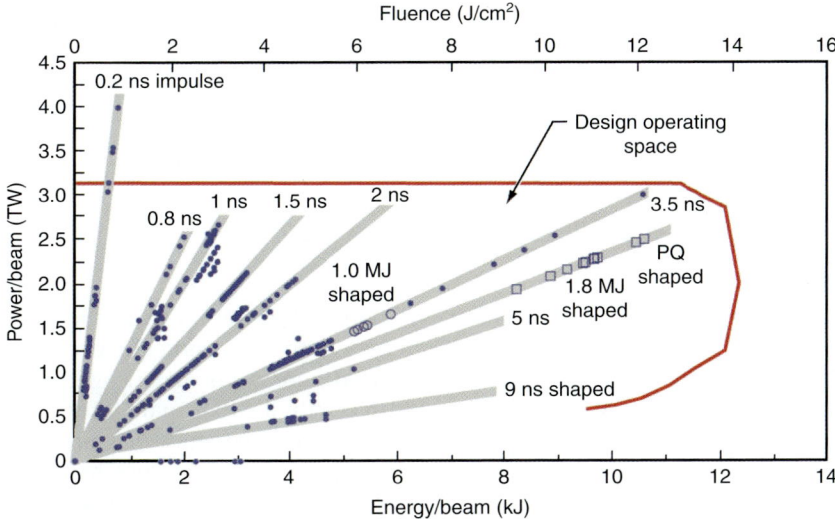

FIGURE 14.7 Plot of 3ω beam power versus 3ω beam energy for initial NIF shots.

operation, as one can see from the several shots that lie above the limit; rather, it is a guide for routine operations. In general, the limit for high-power operation is set by the growth of small-scale intensity irregularities due to the nonlinear index in glass. For high-energy operation, the limit is determined by the injected energy available from the ILS.

Figure 14.7 similarly summarizes all 3ω shots from 2001 to 2006. This 3ω performance space includes shaped pulses that meet or exceed the energy and power levels required for the current ignition target design. The NIF design operating range predicted[33] in 1994 is also plotted on this figure. These initial 3ω shots, combined with the validation of LPOM projections over the range of shots shown, indicate that we can achieve the design power versus energy range described in 1994.

High-power operation of previous LLNL Nd:glass laser systems was limited by small-scale beam breakup,[35,40] which, in turn, was driven by the nonlinear index of the transmissive optics in the beam path. Small-scale contaminants or optics imperfections lead to beam intensity modulations. At high intensity, these modulations are amplified and focused by the nonlinear index effect. An early sign of the development of this instability is growth in the beam contrast, which is defined as the standard deviation of the fluence divided by its mean value. Contrast is measured by taking a sample of the near-field beam, projecting it on a camera, and calculating the fluence variation as recorded in the $m \times n$ camera image.

Solid-State Lasers

$$\text{Fluence beam contrast} \equiv \sqrt{\frac{1}{nm}\sum_{i=1}^{m}\sum_{j=1}^{n}\left(\frac{F(x_i, y_j) - \overline{F}}{\overline{F}}\right)^2} \quad (14.1)$$

$F(x_i, y_j)$ = Pixelated fluence from "near-field" camera image

$\overline{F}$ = Average fluence of image

Figure 14.8 demonstrates that NIF contrast at the input to the frequency converter consistently decreases with increasing fluence and energy per beam. The contrast reported here is calculated over the central 27×27 cm² of the laser, measured with the PDS main laser output camera. The decrease seen is a simple consequence of gain saturation: high-fluence regions in the beam experience less net gain than do low-fluence regions, tending to decrease any intensity modulation. The data in Fig. 14.8 span the NIF design's operating space, indicating that the careful attention we have paid to optical quality throughout the beamlines[28] has successfully controlled beam breakup.

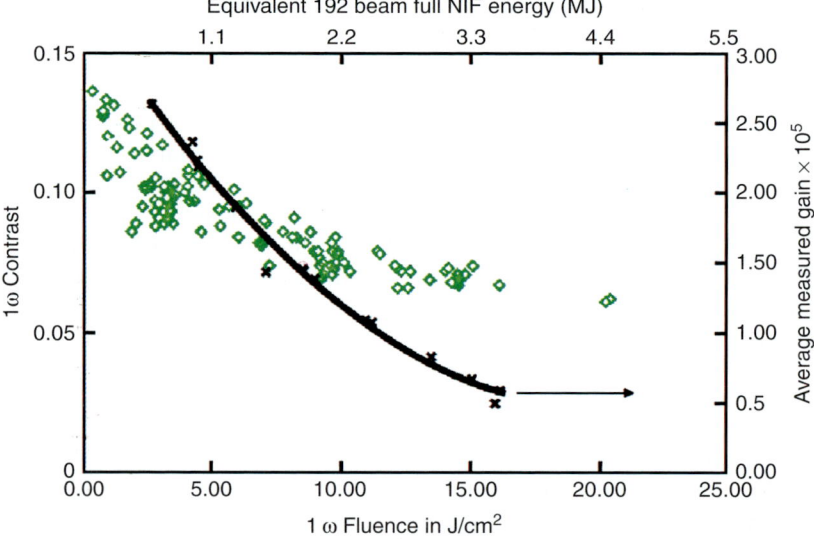

FIGURE 14.8 1ω near-field fluence beam contrast (small diamonds) versus 1ω fluence, measured at the converter input in the precision diagnostic system (PDS). These points represent shots covering the 1ω operating range and pulse lengths, as shown in Fig. 14.6. The solid line represents the measured amplifier gain versus fluence, showing that the contrast drops as the gain saturates.

14.5 Performance Qualification Shots for Ignition Target Pulse Shapes

In March 2006, we fired two 1ω PQ shots separated by three hours and eighteen minutes, an interval that is significantly shorter than the NIF design requirement of less than or equal to 8 hours between system shots. Shot intervals of less than 4 hours have been repeated on a regular basis during the commissioning of the first 40 NIF beamlines, with no discernable degradation in either beam wavefront or near-field modulation.

These PQ shots were taken to validate NIF's capability to meet its energy, power, and temporal contrast design goals. One beam from each shot was routed to the PDS. The other seven beams were measured in the 1ω calorimeters. We will follow the performance of the laser, as measured by the diagnostics, through the four sections of the laser, starting with the 1ω sections (master oscillator room, preamplifier module, and main laser) and finishing with the 3ω diagnostics following the FOA. A detailed discussion of the PDS diagnostics, main laser diagnostics, and calorimeters can be found in the appendices to Haynam et al.[1]

14.5.1 Master Oscillator and Pulse Shaping System

The master oscillator and pulse shaping system, referred to by the acronym MOR (master oscillator room) (Fig. 14.9) creates the temporal pulse shape specified by LPOM. The MOR temporal pulse shape compensates for gain saturation in the rest of the 1ω laser and for the power dependence of the frequency converter efficiency, so that the desired 3ω pulse shape is achieved.

The pulse begins in a CW Yb:fiber master oscillator tuned to 1.053-μm wavelength. The CW signal from the oscillator's output is chopped by an acousto-optic modulator to a pulse width of 100 ns at a pulse repetition rate of 960 Hz. The light is phase modulated at a frequency of 3 GHz to a total bandwidth of 30 GHz in order to suppress stimulated Brillouin scattering (SBS) in the main laser optics.[41] A high-reliability fail-safe system is in place to guarantee that the pulse cannot proceed beyond the MOR unless adequate modulation has been applied to ensure that SBS will be suppressed.[34] A separate modulator operating at 17 GHz can apply more than 150 GHz of additional bandwidth at 1ω (450 GHz at 3ω) for beam smoothing by spectral dispersion (SSD), as will be discussed in Sec. 14.6.2. The pulse then transits a cascade of fiber splitters and Yb:fiber amplifiers, culminating in 48 fiber outputs, each of about 1-nJ energy. Each output goes into an amplitude modulator chassis (AMC) that sets the pulse shape for injection into a preamplifier module (PAM).

In order to account for varying gain/loss characteristics among the beamlines, and to afford operational flexibility to fire a variety of

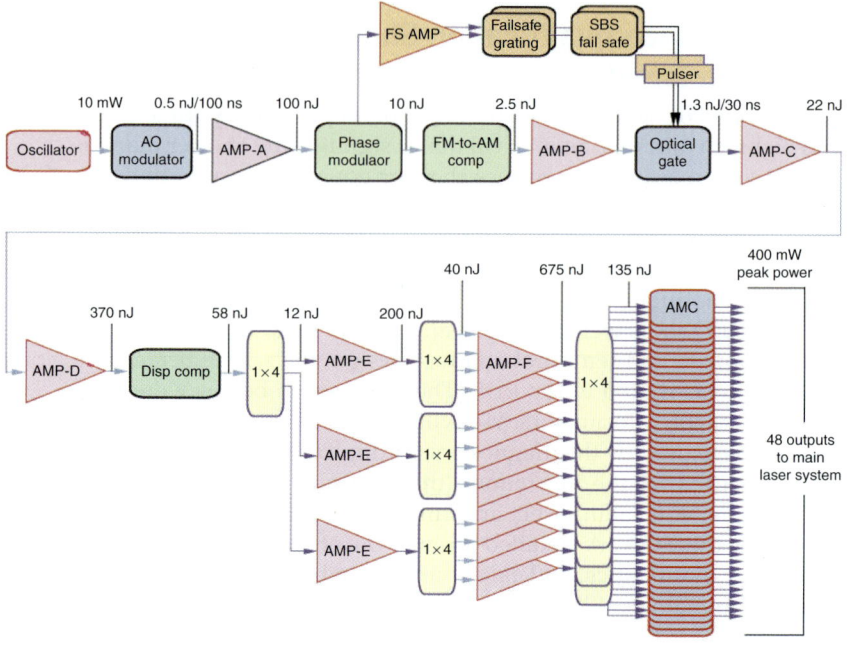

FIGURE 14.9 Schematic of the master oscillator (MOR) and NIF pulse-shaping system, including power and energy levels at each stage. Fiber amplifiers (triangles) are used to compensate for optical losses as the initial continuous wave (CW) beam is chopped by the acousto-optic modulator, frequency broadened to 30-GHz bandwidth by the phase modulator, precompensated by the frequency modulation to amplitude modulation compensator (to minimize amplitude modulation of the high-power beam), corrected for group velocity dispersion in the dispersion compensator, then split, and finally temporally shaped in the amplitude modulator chassis (AMC). The components shown produce the shaped pulse for all of NIF's 48 preamplifier modules (PAMs).

pulse shapes on a single shot, NIF has 48 AMCs, each of which independently provides the pulse to drive the corresponding PAM and its associated four main beamlines. A digital oscilloscope immediately following each AMC records its pulse shape. The AMC controller averages a few hundred individual pulses, calculates the deviation of that average from the requested pulse shapes, and then uses a negative feedback loop to minimize this deviation. Figure 14.10 compares the requested and measured pulse shapes for the two PQ shots.

14.5.2 Preamplifier Module Description and Performance

Each of the 48 pulses from the MOR enters the main laser bay on an optical fiber and is injected into a PAM, where it is amplified first by a regenerative amplifier system and then by a four-pass rod amplifier (shown schematically in Fig. 14.11). The pulse makes approximately 30 round-trips in the regenerative amplifier, experiencing a gain that raises

The National Ignition Facility Laser 373

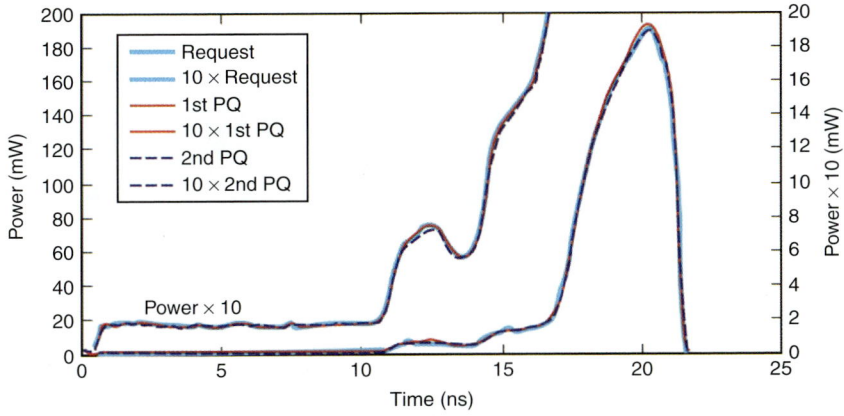

FIGURE 14.10 The temporal pulse shape at the output of the MOR for the two performance qualification (PQ) shots, designated as 1st PQ (N060329-002-999) and 2nd PQ (N060329-003-999). The pulse shape was measured with a 1-GHz transient digitizer.

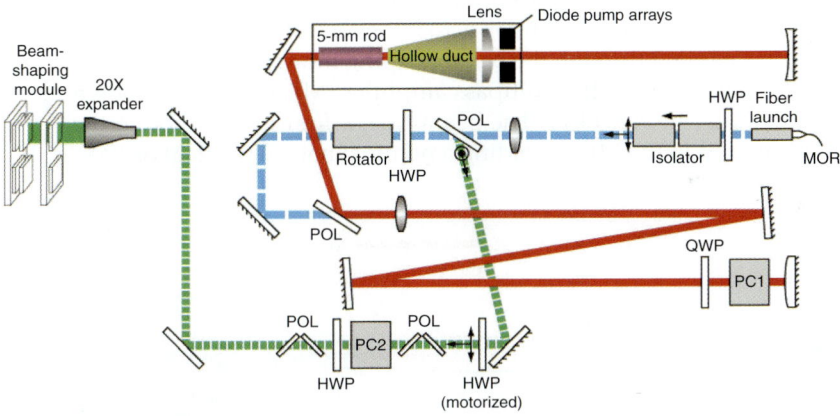

FIGURE 14.11 Schematic of an injection laser system's (ILS's) regenerative amplifier. Light enters the amplifier from the MOR fiber launch at the right of the figure (dashed blue line). It is collimated, passed through an optical isolator, and injected through a polarizer (POL) into the main regenerative amplifier cavity (solid red line). After the beam passes through the Pockels cell (PC1) once, the PC is switched on, trapping the pulse in the cavity for approximately 30 round-trips. During each round-trip, the pulse passes twice through a diode-pumped rod amplifier. Before the final pass, the PC is switched off, and the light exits through a second polarizer (dashed green line). A motorized half-wave plate (HWP), in combination with a set of polarizers, controls the energy transmitted to the next stage of amplification. A second Pockels cell (PC2) can be used to clip off a trailing portion of the pulse that is meant to saturate the regenerative amplifier for energy stability but is not required in the rest of the laser. A 20X beam expander, in combination with a beam-shaping module, sculpts the beam to the desired spatial shape (solid green line).

Solid-State Lasers

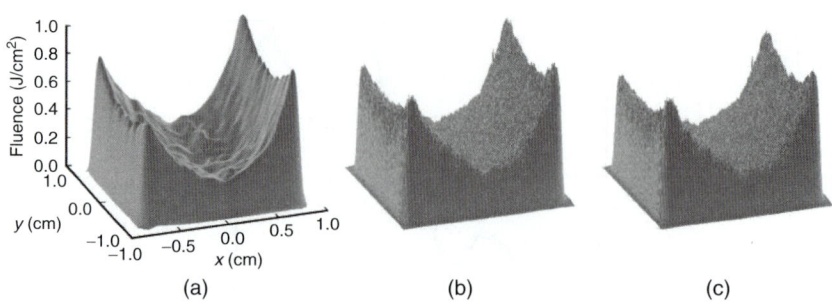

FIGURE 14.12 Predicted (*a*) and measured, near-field profiles for the first (*b*) and second (*c*) PQ shots.

its energy from ~1 nJ to approximately 20 mJ, as appropriate for each PAM. After being switched out of the regenerative amplifier, the pulse traverses a spatial shaping module that transforms the gaussian spatial shape to a profile that is designed to compensate for the gain's spatial nonuniformity throughout the rest of the laser. Figure 14.12 compares the predicted spatial profile measured at the ISP location with measurements from the two PQ shots. The ability to accurately shape the spatial profile allows NIF to produce beams at the output of the system that have a flat irradiance distribution across the central part of the beam.

After passing through the beam-shaping module in the PAM, the pulse is injected into the multipass amplifier (MPA), which is shown schematically in Fig. 14.13. The beam makes four passes though the amplifier rod in the MPA, yielding a nominal net energy gain of 1000.

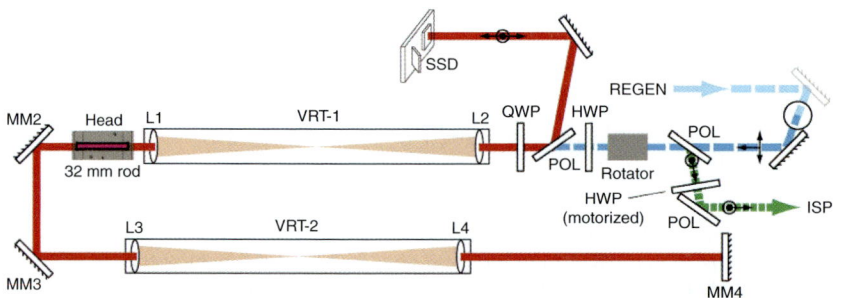

FIGURE 14.13 Schematic of the MPA system. Light enters from the regenerative amplifier (REGEN) at the right of the figure (dashed blue line) and transmits through the polarizer (POL). The polarization is rotated by a series of half-wave plates (HWPs) and quarter-wave plates (QWPs), so that the pulse passes four times through the 32-mm flash lamp-pumped rod amplifier (solid line) before exiting through the polarizer. Each pass is optically relayed using a set of two vacuum relay telescopes (VRTs), which are evacuated to prevent air breakdown at the telescope's central focus. As the pulse exits the cavity (dashed green line), it passes through a combination of a motorized half-wave plate and a polarization-sensitive mirror to allow control of the energy transmitted to the preamplifier beam transport system (PABTS) and the main laser.

The National Ignition Facility Laser

	MPA Input Energy (mJ)	MPA Output Energy (J)
Requested	1.41	1.11
First PQ shot	1.40	1.09
First PQ deviation	−0.7%	−2%
Second PQ shot	1.40	1.02
Second PQ deviation	−0.7%	−8%

TABLE 14.2 Requested and Measured Energies at the Input and Output of the MPA

The ILS's overall energy gain is of the order 10^9. LPOM uses both off-line and online data analysis to maintain ILS models that have the accuracy needed to predict the energetics of this high-gain system. Table 14.2 shows a comparison of the modeled and measured energies at the input and output of the MPA for both PQ shots.

The input energy to the MPA is monitored at 1 Hz. It is maintained at the requested value by a closed-loop control, using attenuation provided by the combined action of an adjustable half-wave plate and a polarizer used in transmission. The closed-loop control mechanism produces energies within ±2 percent of the request. The LPOM's MPA model is accurate to within ±5 percent for injected energies ranging from 0.5 to 10 mJ. Figure 14.14 shows a comparison of the predicted and measured ISP power sensor traces for the two PQ shots.

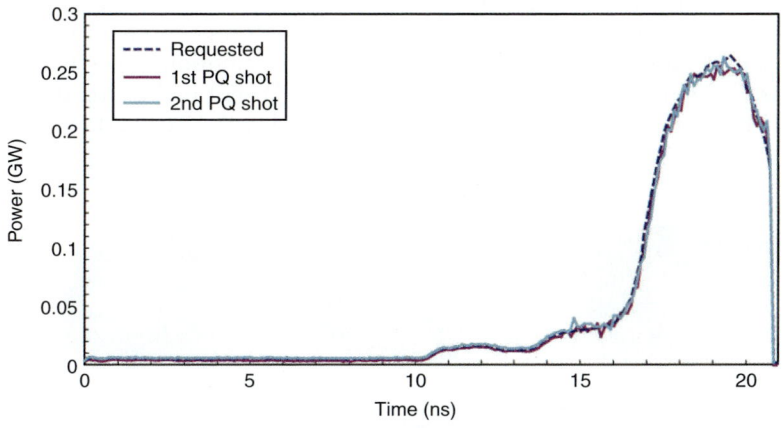

FIGURE 14.14 Requested and measured temporal profiles at the output of the preamplifier module (prior to injection into the main laser), as measured by the ISP for the two PQ shots.

14.5.3 Main Laser 1ω Performance

After the ISP, the pulse is injected into the main laser, the part of the laser system that contains the full-aperture (40-cm) components. The near-field and far-field spatial and temporal profiles at the 1ω output of the main laser are modeled using the NIF virtual beamline (VBL) propagation code, which has been incorporated into LPOM. LPOM contains detailed information regarding sources of wavefront distortion. All large optics undergo full-aperture, high-resolution interferometer measurements during their manufacture. This interferometry data is used directly in the LPOM description for each optic at the position in the chain where the optic is located. The distortion that is induced as the laser slabs are deformed by nonuniform flash lamp heating has been both calculated and measured; calculated aberrations are used in LPOM. Calculated estimates for distortions due to mounting stresses and a contribution for air turbulence in the amplifier cavities are also included. Finally, a model of the 39-actuator, full-aperture deformable mirror, using measured influence functions for each actuator, is also used to represent the correction done online in the Hartmann sensor/deformable mirror loop.

High-spatial-frequency wavefront errors generate corresponding high-spatial-frequency intensity variations in the measured beam profile. Lower-spatial-frequency wavefront errors (less than about 0.1/mm) affect spot size but not near-field intensity, because laser propagation distances are insufficient for them to diffract into intensity variations. The lower-spatial-frequency variations in the near-field measurements are caused primarily by the input spatial shape, the gain spatial profiles, and aberrations in the laser's front end.

Figure 14.15 compares the measured and modeled near fields at the 1ω PDS near-field camera position for both PQ shots. These shots had a 1.8-MJ ignition-target pulse shape (discussed in Sec. 14.6.4) and 1ω energy of ~18 kJ per beam. Figure 14.16 shows an overlap of the measured and modeled fluence probability distributions over the central 27 cm × 27 cm of the beam. The first PQ shot had a slightly higher energy than the second (18.0 kJ compared with 17.6 kJ), due to

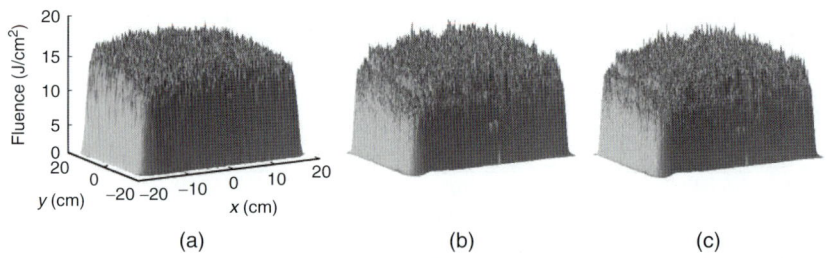

FIGURE 14.15 Comparison of modeled (a) and measured near-field 1ω fluence distributions at PDS for the first (b) and second (c) PQ shots, respectively.

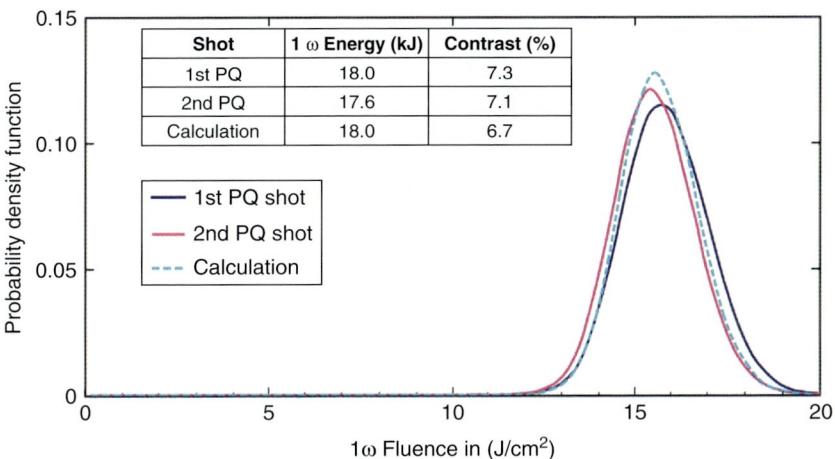

FIGURE 14.16 Comparison of modeled and measured fluence probability distributions at the PDS 1ω diagnostic over the central 27 cm × 27 cm of the beam for the two PQ shots. The small shifts in mean 1ω fluence are due to differing total energies in the two PQ shots. The calculation is reported at the mean fluence of the two PQ shots over the central 27 cm × 27 cm of the beam. Measured contrast is nearly identical for both shots, is in reasonable agreement with prediction and is well under our design goal of 10 percent.

an adjustment to the injected energy. Agreement between the measured and modeled contrast is sufficient for LPOM to specify laser energetics and pulse shapes, protecting against equipment damage caused by off-normal laser operation. The less than 0.5 percent absolute discrepancy in contrast may arise from such sources as small inaccuracies in the modeled gain spatial shape (overall flatness of the beam), approximations made in the statistical modeling of front-end optic aberrations, or the calculational estimate made of the contrast added by the diagnostic optics. The measured values of 1ω contrast are well below the NIF design goal of less than or equal to 10 percent.

Figure 14.17 displays plots of the enclosed fraction of the focal spot energy as a function of radius, starting at the centroid of the spot. Two measurements are shown for each PQ shot. The first measurement was taken directly from the PDS 1ω far-field camera. The second used the measured wavefront from the 1ω radial shear interferometer and fluence from the near-field camera. From these two inputs, the beam field was numerically reconstructed, and a far field was predicted. Both the LPOM and radial shear predictions are at paraxial focus (simple Fourier transform of the field) and are in good agreement. Both, however, predict somewhat smaller focal spots than the direct measurements. The most likely explanation is that our diagnostic imaged a location that was slightly displaced from best focus (1 to 2 mm out of 7700 mm). Figure 14.18 shows the spatial fluence

378 Solid-State Lasers

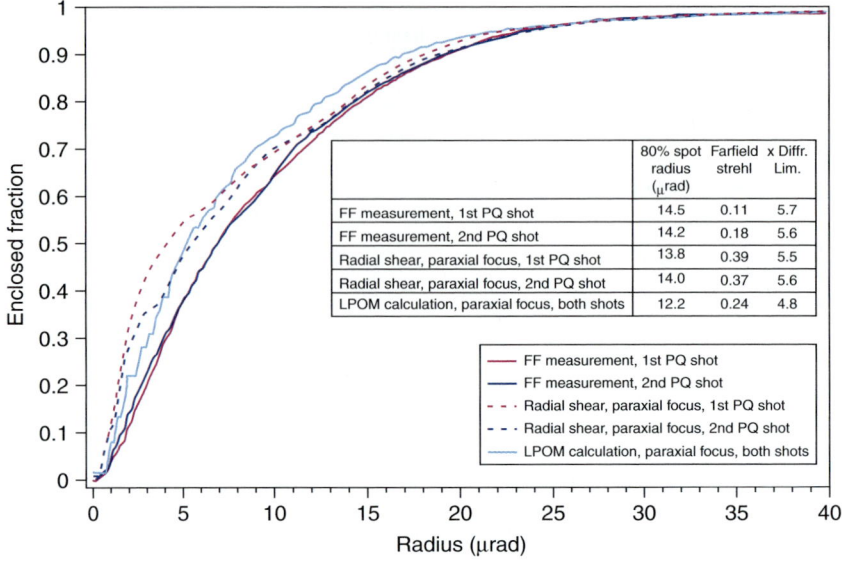

FIGURE 14.17 Enclosed 1ω focal spot energy fractions for the two PQ shots. Direct far-field measurements as well as predictions based on reconstruction of the field using the radial shear and near-field diagnostics are shown. The calculated far field applies to both shots.

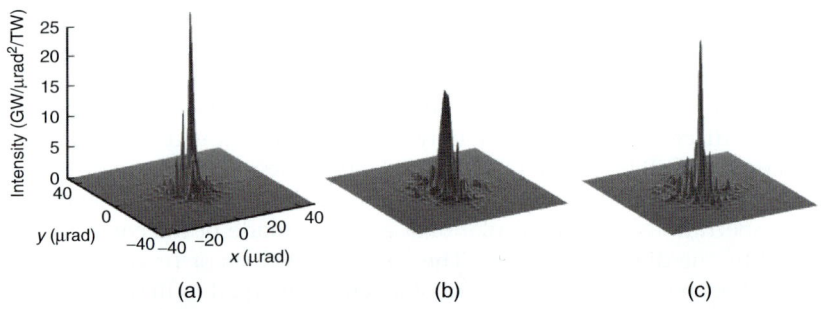

FIGURE 14.18 Calculated (*a*) and directly measured 1ω focal spots for the first (*b*) and second (*c*) PQ shots. All plots have a common set of axes, which is shown on the left. The change in peak fluence between the first and second shots is attributed to turbulence in the beam path.

profile of the calculated and measured focal spots. The shot-to-shot variability is minor, as demonstrated by the small change in the 80 percent spot radius (see Fig. 14.17).

14.5.4 Frequency Conversion Performance

The target must be irradiated with 351-nm light. NIF converts the main laser output pulse to the third harmonic using a pair of potassium

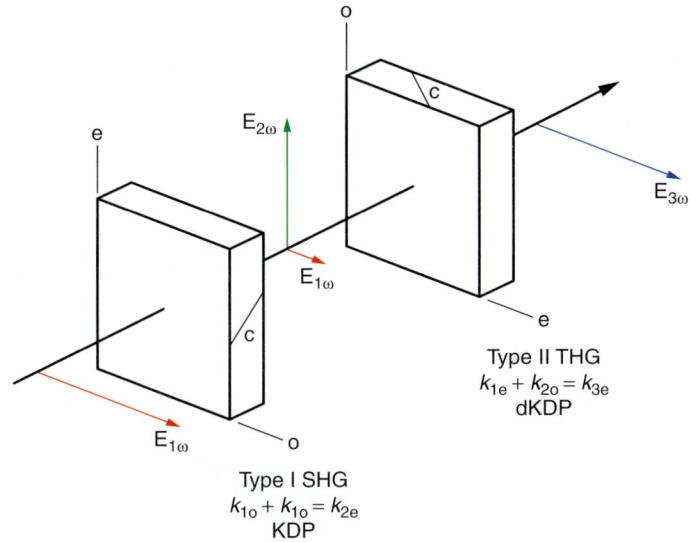

FIGURE 14.19 Illustration of a Type I–Type II converter scheme. The NIF doubler (second harmonic generator [SHG]) thicknesses range from 11 to 14 mm, and the tripler (third harmonic generator [THG]) thicknesses range from 9 to 10 mm. The measurements described here were primarily performed with a 14-mm SHG and a 10-mm THG.

dihydrogen phosphate (KDP) frequency conversion crystals[42,43], as illustrated in Fig. 14.19. The first crystal, or doubler, converts approximately two-thirds of the incident laser energy to the second harmonic via Type I phase-matched degenerate sum-frequency mixing: $1\omega(o) + 1\omega(o) \rightarrow 2\omega(e)$. The copropagating second harmonic and residual fundamental beams are then passed through a deuterated KDP (dKDP) tripler, where the third harmonic beam is created by Type II phase-matched sum-frequency mixing: $2\omega(o) + 1\omega(e) \rightarrow 3\omega(e)$. We set the critical 2:1 mix ratio of 2ω to 1ω energy needed for efficient mixing in the tripler by angularly biasing the Type I doubler a few hundred microradians from exact phase matching. The optimum bias angle depends both on crystal thickness and drive irradiance. The sensitivity of conversion efficiency to this optimum bias angle is shown in Fig. 14.20.

Figure 14.21 shows measured 3ω energy produced as a function of 1ω energy into the converter for flat-in-time (FIT) pulses. The figure compares two different converter configurations, one with crystal thicknesses $L_1/L_2 = 11$ mm/9 mm, and a second with $L_1/L_2 = 14$ mm/10 mm. The data for the 11/9 configuration was obtained from shots with a 3.5-ns pulse length, with the doubler operating at a bias of 220 ± 5 μrad, and with the tripler tuned for phase matching to

Solid-State Lasers

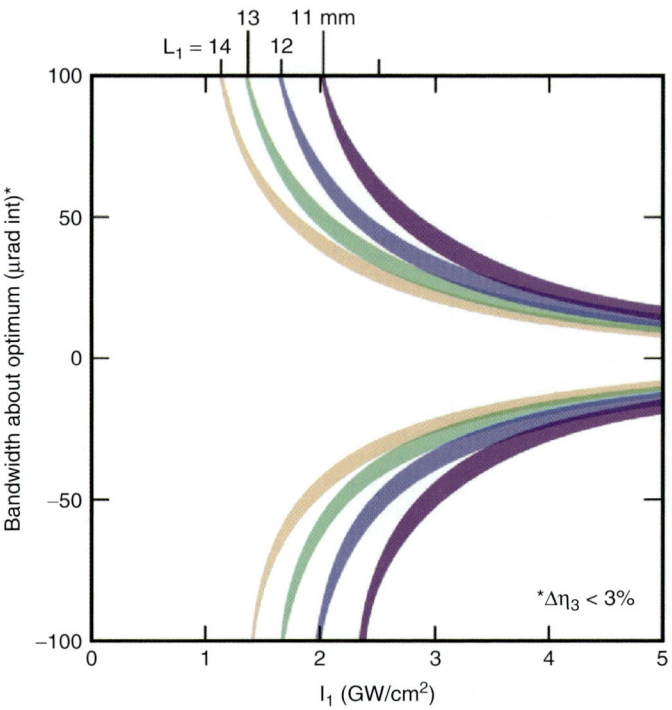

FIGURE 14.20 Angular bandwidth of the Type I–Type II 3ω conversion scheme versus drive irradiance for different choices of crystal thickness. The curves depict the angle away from exact phase matching at which conversion efficiency is decreased by 3 percent, with the bands at each SHG thickness (L_1) spanning the THG thickness range of 9 to 10 mm.

within ±15 μrad (all angle values are internal to the crystal). At the highest input energy tested (12.9 kJ), this configuration produced an output 3ω energy of 10.6 kJ—that is, the energy conversion efficiency across the converter was greater than 80 percent. The data for the 14/10 configuration was obtained from shots with a 5-ns pulse length and with the doubler at a bias of 195 μrad. As Fig. 14.21 indicates, the thicker crystals have better low-irradiance performance than the 11/9 configuration, because a similar conversion efficiency is achieved at approximately two-thirds (3.5 ns/5 ns) the drive irradiance. The increased efficiency at low drive is an advantage for converting high-contrast ignition pulses, provided the reduced angular bandwidth of the thicker crystals is manageable (see Fig. 14.20). Results on NIF demonstrate that the crystal alignment system is precise enough to allow accurate alignment of the thicker crystals. All 3ω performance data discussed in the remainder of this chapter were obtained using the 14/10 configuration.

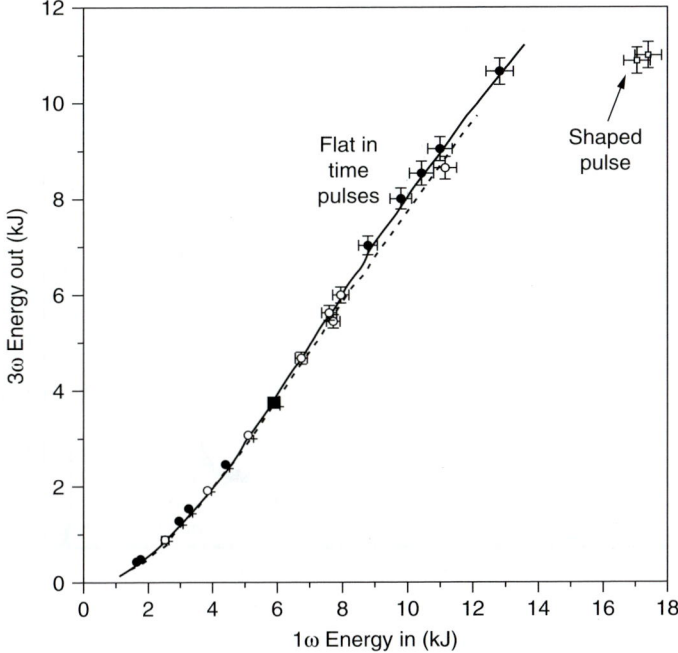

FIGURE 14.21 Measured 3ω energy out of the converter versus measured 1ω energy into the converter for three illustrative cases: an 11/9 converter with 3.5-ns flat-in-time (FIT) pulses (filled circles); a 14/10 converter with 5.0-ns FIT pulses (open circles); and a 14/10 converter with a 1.8 MJ/500 TW (FNE) shaped pulse (open squares). The model (solid line for FIT 11/9; dashed for FIT 14/10) is described in the text.

The measured third harmonic performance of the laser under PQ conditions is summarized in Figs. 14.22 to 14.24. Figure 14.22 plots the harmonic energies and pulse shapes for a 17.1-kJ input pulse with a peak power of 3.65 TW and a temporal contrast of 17:1 that was frequency converted to 10.9 kJ of 3ω with a peak power of 2.90 TW and a temporal contrast of 150:1 at the output of the converter. The measurements are in good agreement with simulations employing a three-dimensional (x, y, z) time-slice model. The model uses the paraxial formulation of the coupled wave equations and accounts for diffraction, phase matching, Poynting vector walk-off, linear absorption, nonlinear refractive index, cross-phase modulation, and two-photon absorption at the third harmonic.[43] It incorporates representative measured crystal data for surface aberrations[44] and spatial birefringence variations,[45] as well as measured data for the spatial profile of the electric-field amplitude, phase, temporal shape of the input pulse (see previous section), and Fresnel losses (Table 14.3). The first two rows of Table 14.3 give the Fresnel losses in the converter components

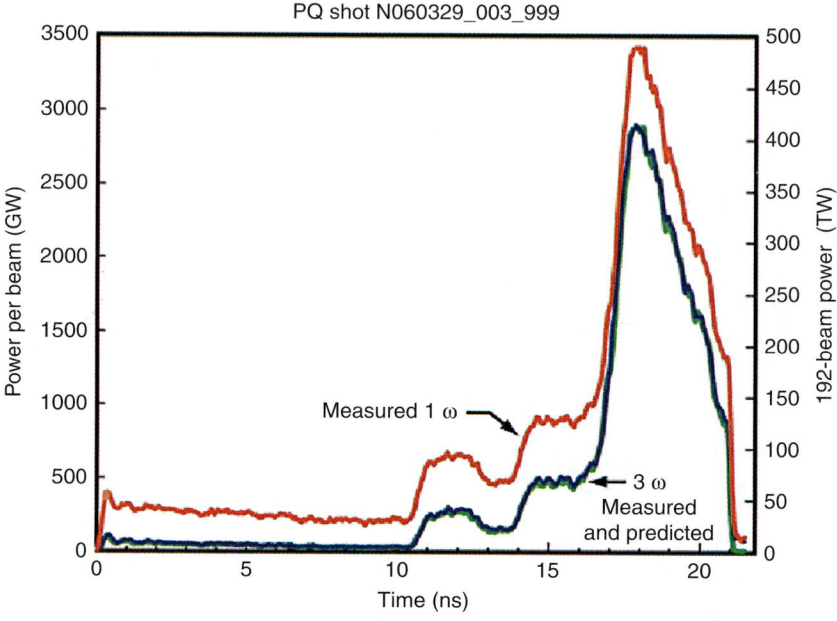

FIGURE 14.22 Comparison of measured and predicted pulse shape for the 3ω PQ pulse, along with the input 1ω pulse shape.

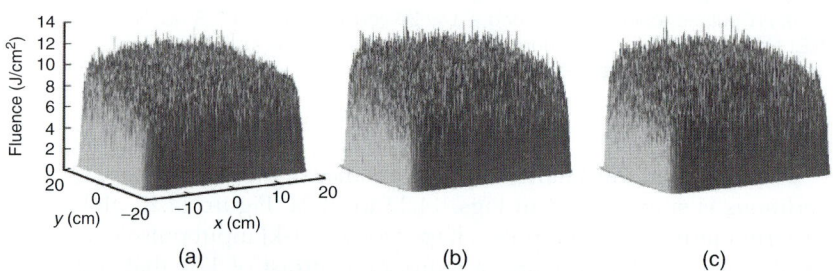

FIGURE 14.23 Comparison of modeled (a) and measured near-field 3ω fluence distributions at PDS for the first (b) and second (c) PQ shots, respectively.

(second and third harmonic generators). The final row summarizes the remaining transport losses to the target. The majority of this remaining loss occurs at a grating that is etched onto a silica debris shield in order to direct a portion of the light to the drive diagnostic. The simulations were performed on a 512 × 512 transverse spatial grid with 1-mm resolution, using the split-operator method[46] and fast Fourier transforms for the field propagation, with 15 z steps per crystal. The temporal pulse shape was modeled with discrete time slices (typically 50). The effect of temporal bandwidth on the input pulse

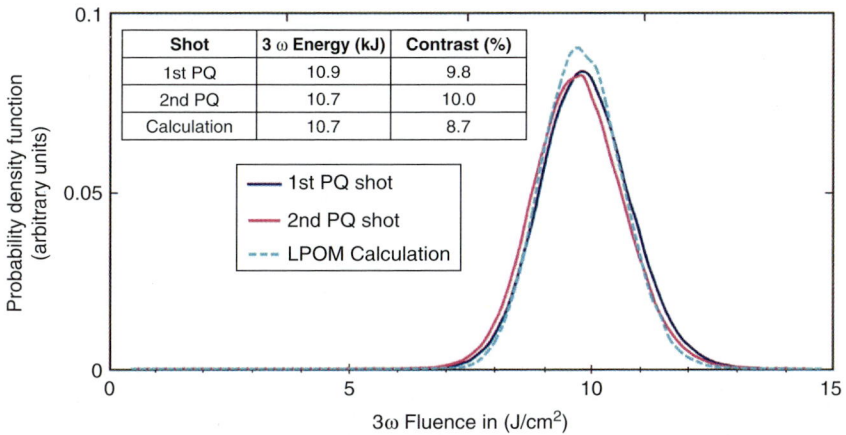

FIGURE 14.24 Comparison of modeled and measured fluence probability distributions at 3ω PDS over the central 27 cm × 27 cm of the beam. The calculation is reported at the mean fluence over this aperture for the two shots. Measured contrast is nearly identical for both shots—~1 percent higher than the model and well under our design goal of 15 percent.

	Transmission		
Component	1ω (1.053 μm)	2ω (0.532 μm)	3ω (0.351 μm)
SHG	0.9900	0.9925	NA
THG	0.9607	0.9766	0.9975
To TCC	0.8995	0.9184	0.9545

TABLE 14.3 Transmission of Final Optics Assembly (FOA) Optics as a Function of Wavelength

was modeled as an effective tripler detuning of 1.9 μrad/GHz. The model for Fig. 14.22 uses as field inputs the PDS 1ω measured near-field fluence, radial shear wavefront, and temporal pulse shape. It confirms the ability of the frequency conversion model to match measurements.

As discussed in Sec. 14.3.3, LPOM uses the frequency conversion model to predict both energetics and near- and far-field profiles at 3ω. Figure 14.23 compares the near-field prediction at the output of the final focusing lens with measurements for the two PQ shots. Figure 14.24 similarly compares the near-field fluence probability distributions over the center 27 cm × 27 cm of the beam. LPOM predicts a beam contrast of 8.7 percent, which is slightly lower than the measured values of ~10 percent. As in the 1ω section, the calculation includes an estimate of the contrast added by the 3ω diagnostic optics. The measured value is substantially below the 15 percent contrast design goal.

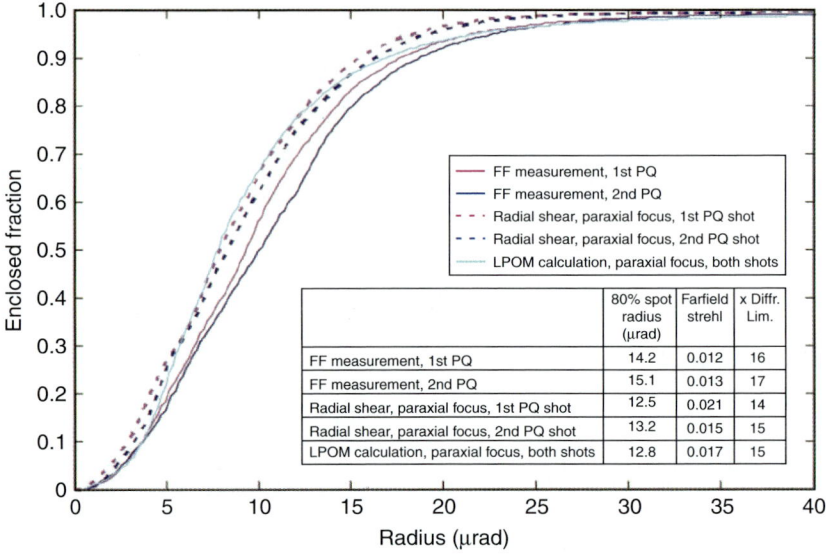

FIGURE 14.25 Enclosed 3ω focal spot energy fractions for measurement and calculation of the two PQ shots. Both direct far-field measurements and predictions based on reconstruction of the field using the radial shear and near-field diagnostics are shown.

Figure 14.25 shows the enclosed energy fraction of measured and modeled 3ω focal spots as a function of radius, starting at the centroid of the spot. As with the 1ω spots in Fig. 14.17, two measurements are indicated for each PQ shot: one directly from the PDS far-field camera, the other a reconstruction from the measured 1ω near-field fluence and wavefront. The LPOM model agrees reasonably well with the reconstructed-field prediction. Both yield ~10 percent smaller spots than the far-field camera measurement. Figure 14.26 shows the LPOM-modeled far field and the directly measured far fields, demonstrating the good qualitative agreement and shot-to-shot repeatability.

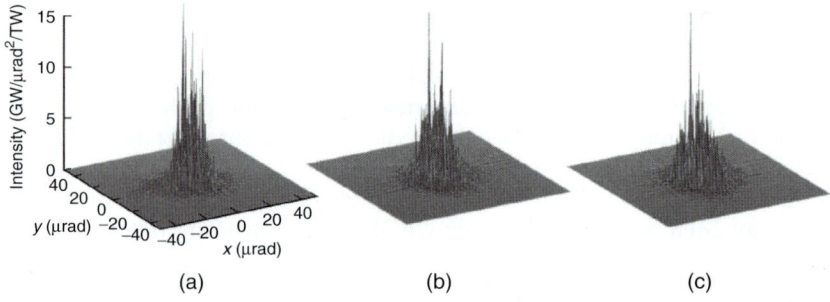

FIGURE 14.26 LPOM-calculated (*a*) and directly measured 3ω focal spots for the first (*b*) and second (*c*) PQ shots. All plots have common axes, shown on the left.

14.6 Focal Spot Beam Conditioning and Precision Pulse Shaping for Ignition Experiments

The PQ demonstrations discussed to this point were shot without focal spot beam conditioning in order to study the fine-scale characteristics of the NIF focal spots. NIF ignition targets, however, require spatial and temporal beam conditioning, both to tailor the irradiance profile in the focal plane and to reduce hot spots that might seed laser-plasma instabilities.[36,37] Spatial beam conditioning is provided by phase plates designed to produce elliptical speckle patterns with about 1- to 1.3-mm average diameter and ellipticity that varies from beam to beam, depending on the angle of incidence at the target. The laser speckle contrast is then reduced, both instantaneously and in a time-averaged sense, by the application of polarization smoothing (PS)[47] and smoothing by spectral dispersion in one dimension (1D SSD).[48,49] Polarization smoothing is limited to a maximum reduction in contrast of $1/\sqrt{2}$.[50] SSD, as implemented on NIF, achieves an additional ~5 times reduction in speckle contrast on a time scale of a few tens of picoseconds.

The tests described in this section also used precisely shaped pulses with high temporal contrast (~150:1), single-beam 3ω peak powers in the range of 1.9 to 2.6 TW, and energies of 5.2 to 9.4 kJ (370 to 500 TW; 1 to 1.8 MJ full NIF equivalents). The two ignition pulse shapes used in these experiments are shown in Fig. 14.27. The 1-MJ shape is the Rev. 1 baseline[39] for the first ignition campaign on NIF. The 1.8-MJ shape is a slightly updated version of the reference ignition pulse shape that we assumed for the NIF laser design. For further discussion of pulse-shaping requirements for the ignition point design, see Refs. 36 to 39.

Table 14.4 summarizes results for two shots: a 1-MJ pulse with a 0.50 mm × 0.95 mm (diameter) elliptical focal spot and 270 GHz of

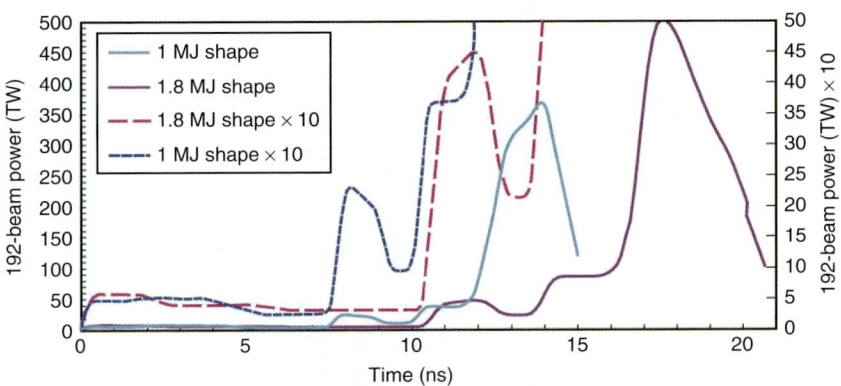

FIGURE 14.27 The two shaped pulses used in these experiments, scaled to their 192-beam equivalents. The temporal contrasts are 158:1 and 176:1 for the 1- and 1.8-MJ pulses, respectively.

Campaign Description			Beam Energy and Power			Beam Smoothing		
Campaign	Pulse shape	Pulse length (ns)	3ω energy per beam (J)	3ω energy full NIF (MJ)	Peak power (TW/beam)	CPP (mm) [FWHM]	Polarization Rotation	SSD (GHz 3ω)
1.0 MJ ignition Design	Ignition	15.4	5208	1.00	1.85	.95 × .5	Yes	270
Demonstrated: 1.0 MJ	Ignition	15.4	5316	1.02	1.9	.95 × .5	Yes	270
1.8 MJ ignition Design	Ignition	20.4	9375	1.80	2.6	1.3 × 1.16	Yes	90
Demonstrated: 1.8 MJ	Ignition	20.4	9438	1.81	2.6	1.3 × 1.16	Yes	120

TABLE 14.4 Three Methods of Beam Conditioning Demonstrated Simultaneously on Two Candidate Ignition Temporal Pulse Shapes

SSD, and a 1.8-MJ pulse with a 1.2 mm × 1.3 mm (diameter) focal spot and 90 GHz of SSD. (In this section, all SSD bandwidths are specified at 3ω, unless otherwise indicated. To good accuracy, the frequency converter triples the imposed bandwidth, along with the fundamental laser frequency.) These fully integrated tests include all three of NIF's beam-conditioning techniques simultaneously: phase plates, SSD, and PS. Table 14.4 shows that the energies, peak powers, focal spot sizes obtained, and SSD bandwidths for the two candidate ignition temporal pulse shapes agree with expectations and meet or exceed the campaign goals. Polarization smoothing will be accomplished on NIF by rotating the polarization in two of the four apertures in each final optics assembly by 90 degrees, and by then overlapping all four beams at the target. Consistent with this strategy, the tests described here were conducted with a prototype dKDP 1ω half-wave plate and a rotated set of frequency conversion crystals installed in the PDS final optics (Fig. 14.28). The average polarization

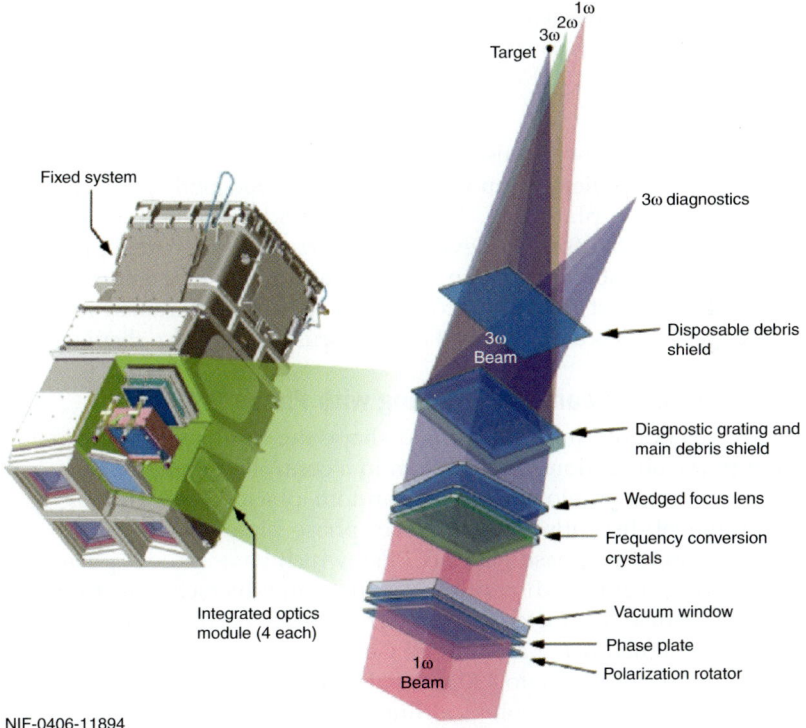

NIF-0406-11894

FIGURE 14.28   (a) The schematic layout of the final optics assembly on NIF: This mechanical system mounts to the NIF target chamber and contains the final set of optics for four NIF beamlines. (b) The suite of optics for one of these beamlines: the same mechanical, optical, and beam control components that are used in the FOA at the target chamber are reproduced for a single beamline in the PDS.

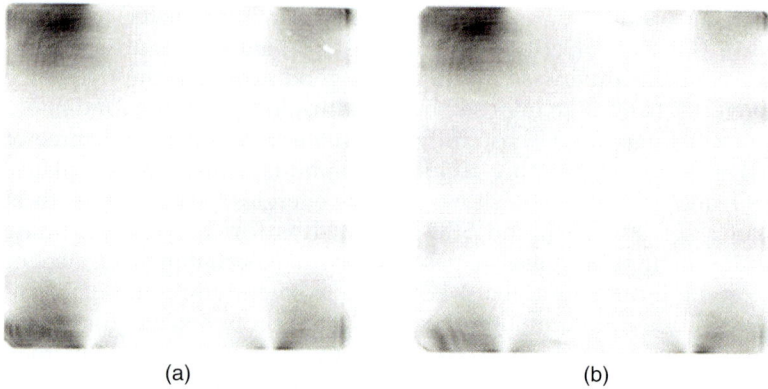

FIGURE 14.29   Measured depolarization on the NIF beam without (a) and with (b) the polarization rotator crystal. The linear grayscale varies from 0 percent (white) to 2 percent (black) depolarization. The spatial extent of the image is 38 cm on each side. The small variations of beam polarization are due to the stress-induced birefringence in the vacuum-loaded spatial filter lenses. The average depolarization is 0.11 percent for each case, which results in a frequency conversion loss that is both small when compared with the 1ω and 3ω FOA transmission losses shown previously in Table 14.3 and negligible in an absolute sense.

impurity of a low-power pulsed 1ω beam (generated by leaving the rod and slab amplifiers unpumped) was measured both with and without the wave plate (Fig. 14.29) and found to be better than 0.11 percent in each case. This level of depolarization has a negligible impact on frequency conversion. Phase-plate divergence and SSD bandwidth do affect frequency conversion and must be taken into account. These effects are addressed in the discussion on pulse shaping.

14.6.1 Spatial Beam Conditioning with Phase Plates

Phase plates (kinoforms) enlarge and shape the focal spot by introducing phase aberrations on the beam in a controlled manner. Early implementations employed binary random phase plates (RPPs)[51] and multilevel discontinuous kinoform phase plates (KPPs).[52] NIF employs continuous phase plates (CPPs), which have smooth phase profiles with no abrupt discontinuities that can adversely affect the beam's near-field characteristics.[53,54] The phase profiles for these plates are designed using a modified Gerchberg-Saxton algorithm,[53] and they are imprinted onto 430 mm × 430 mm × 10 mm fused silica plates using a magnetorheological finishing (MRF) process.[55] These CPPs are achromatic, affording flexibility in their placement relative to the frequency conversion crystals. For the tests described here, the plates were sol-gel antireflection coated for 1ω operation (less than 0.2 percent Fresnel loss per surface) and installed in the PDS final optics, as shown in Fig. 14.28.

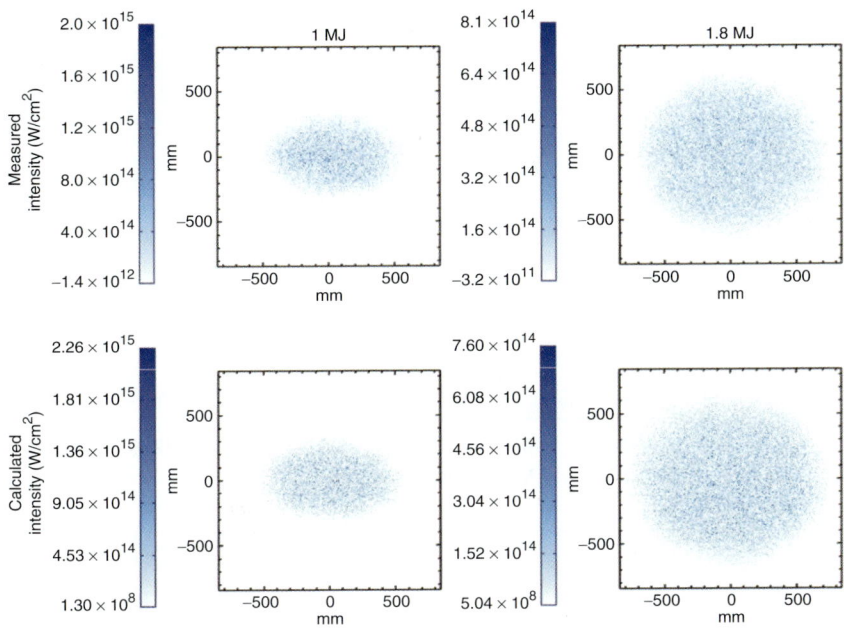

FIGURE 14.30 Comparison of measured (top) and calculated (bottom) NIF focal spots with no applied SSD. The images on the left are from a 1-MJ shot with a 0.50 mm × 0.95 mm full-width, half-maximum (FWHM) spot size CPP. Images on the right are from a 1.8-MJ shot with a 1.16 mm × 1.3 mm FWHM spot size CPP. The measured data are from the shots described in Table 14.4. Measured (time-integrated) images and calculated (time-dependent) images are both normalized to an input power of 1 TW. See text for discussion.

Figure 14.30 compares measured and modeled focal spots obtained with the appropriate CPPs for the 1.0- and 1.8-MJ pulses with no SSD present. We also compare (Fig. 14.31) the encircled energy for these spots and the fractional power above intensity (FOPAI), defined as

$$\mathrm{FOPAI}(I_0) = \frac{\int_{\mathrm{beam\,area\,where\,}I(x,y)<I_0} I(x,y)\,dx\,dy}{\int_{\mathrm{total\,beam\,area}} I(x,y)\,dx\,dy} \qquad (14.2)$$

The model starts with the measured 1ω near-field fluence, temporal shape, and phase profiles (from the PDS radial shear interferometer). It then adds the measured CPP phase to construct the complex 1ω electric field. It then calculates the frequency conversion of this beam and the propagation of the resulting 3ω beam through the final optics and to focus. The modeled FOPAI is evaluated at the time of peak power. We derive the measured FOPAI by assuming that the intensity

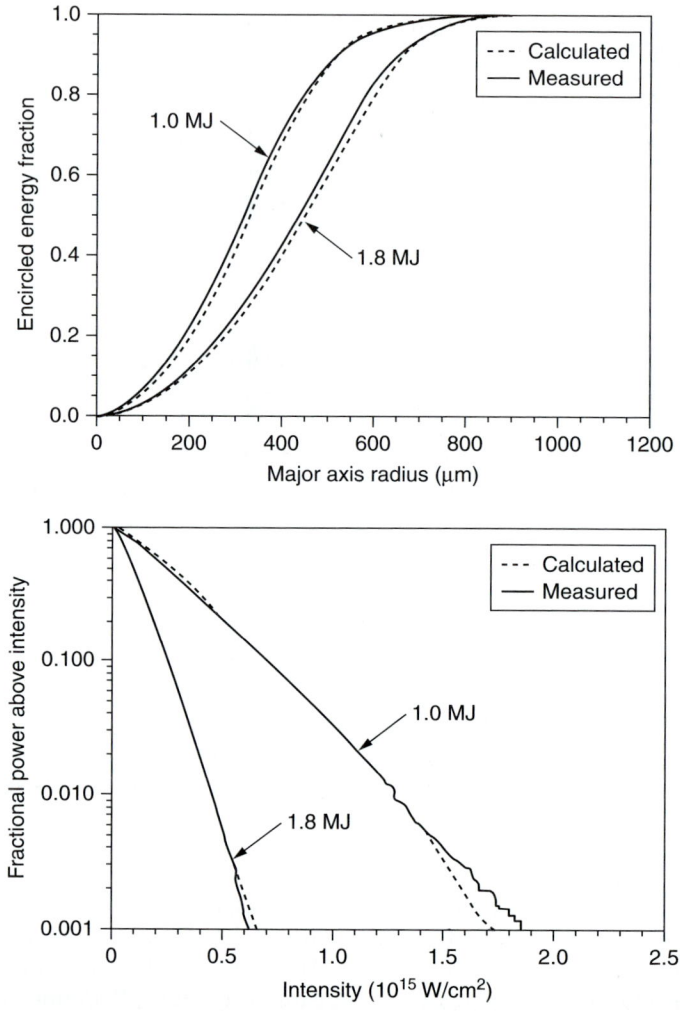

FIGURE 14.31 Comparison of the encircled energy fraction and the fractional power above intensity (FOPAI) for the 1-MJ and the 1.8-MJ shots described in Fig. 14.30. The encircled energy was calculated in elliptical coordinates, with eccentricity of 0.55 for the 1-MJ case and 0.88 for the 1.8-MJ case. The total power is normalized to 1 TW for each case.

is separable in time and space: $I(x, y, t) = F(x, y)*P(t)/E$, where F is the measured near-field fluence, P the time-dependent whole-beam power, and E the total energy. Equation (14.2) is, again, evaluated at the time of peak power. Our models indicate that for these pulses, intensity-dependent effects, such as nonlinear refractive index and frequency conversion, do not cause significant changes in the focal spot characteristics as a function of time, thus justifying the separability assumption. Figures 14.30 and 14.31 show that the modeled and measured focal spots are in good agreement. The encircled energies and the FOPAI also agree well for both the 1-MJ and the 1.8-MJ CPP spots.

The smallest speckle size in the patterns seen in Fig. 14.30 is the diffraction limit of the final focusing lens: $2\lambda f/D = 15.4$ µm. Although the contrast of an ideal speckle pattern is unity, the measured focal spots show contrast of 0.79 ± 0.02. We account for this lower value by noting the presence of the SBS-suppression modulation (3-GHz modulation frequency, 30-GHz full-width, half-maximum [FWHM] bandwidth at 1ω, 90 GHz at 3ω) and the chromatic dispersion in the wedged final focusing lens. The lateral displacement in the focal plane due to the lens chromatic dispersion is about 0.045 µm/GHz at the third harmonic. When averaged over the pulse length, the shifted speckle patterns add incoherently and reduce the contrast, in a process analogous to SSD. This effect predicts a decrease in contrast to 0.84, which is in reasonably good agreement with the measurement.

14.6.2 Temporal Beam Conditioning with One-Dimensional SSD

SSD consists of phase modulating the laser pulse and angularly dispersing its spectral content sufficiently to displace individual FM side bands in the focal plane by at least half the speckle size, a condition generally referred to as *critical dispersion*.[48,49] On NIF, the SSD modulator runs at 17 GHz (v_{mod}), and the 3ω lateral spectral displacement at the target is 0.58 µm/GHz, which is comfortably beyond the critical dispersion value of 0.45 µm/GHz. This dispersion is provided by a Littrow grating in the PAM, which is oriented so that the dispersion direction is aligned along the short axis of the elliptical focal spot. SSD bandwidths of up to ~150 GHz (1ω) can be produced by adjusting the modulation index (δ) of the modulator ($\Delta v_{1\omega} = 2\delta v_{mod}$). The maximum 1ω bandwidth in the tests reported here was measured to be 95 ± 5 GHz.

Figure 14.32 compares the measured and calculated focal spots with both CPP and SSD for the 1-MJ PQ shot and for one of the 1.8-MJ ignition pulses. The time-averaged SSD focal spot was calculated by performing a spectrally weighted incoherent sum of spatially translated non-SSD focal spots (Fig. 14.30), using the measured 3ω spectrum (Fig. 14.33). This spectrum includes both the 3-GHz-modulated SBS-suppression bandwidth and the 17-GHz-modulated SSD bandwidth. The calculations include both the lens chromatic dispersion (horizontal in the figure) and the grating SSD dispersion (vertical). The observed reduction in speckle contrast from 0.79 to 0.19 for the 1-MJ focal spot is equivalent to an incoherent average of ~28 speckles. The speckle contrast for the 1.8-MJ spot decreased from 0.79 to 0.24, which compares well with the calculated decrease to 0.26 and

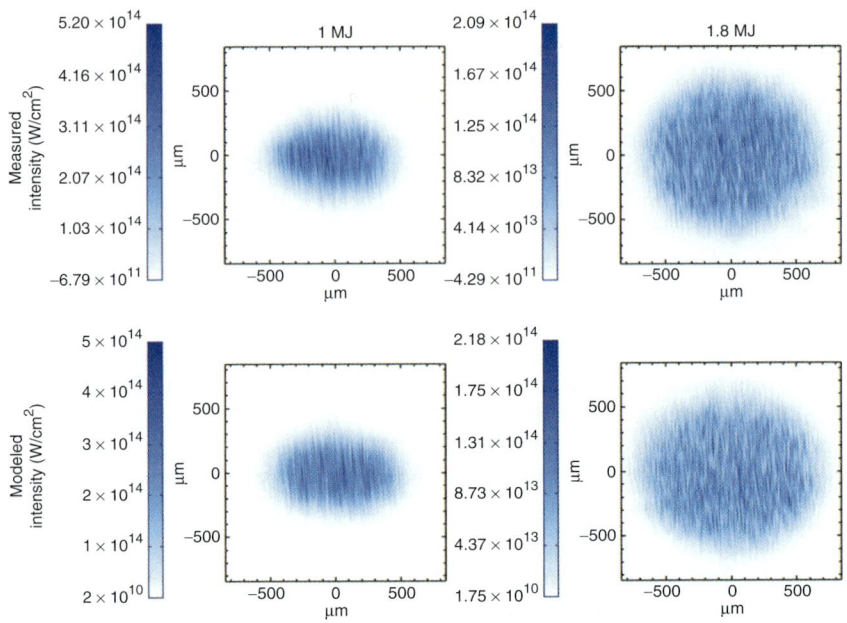

Figure 14.32 Comparison of measured (top) and calculated (bottom) focal spots with both CPP and SSD. The images on the left are from a 1-MJ shot with a 0.50 mm × 0.95 mm FWHM spot size CPP. The images on the right are from a 1.8-MJ shot with a 1.16 mm × 1.3 mm FWHM spot size CPP. The measured data are from the shots described in Table 14.4. The 3ω spectra used to generate the predictions are shown in Fig. 14.33. The intensity scales are normalized for 1-TW total power.

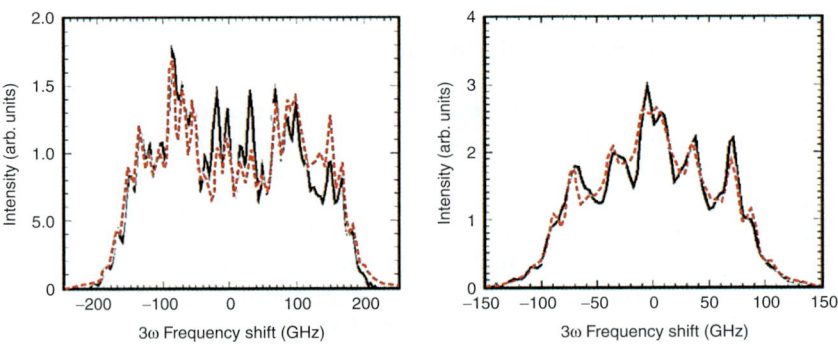

Figure 14.33 Measured (solid black lines) and fitted (dashed red lines) spectra for the 1-MJ (left) and 1.8-MJ shots described in Table 14.4. The fit assumes a sum of 3-GHz and 17-GHz FM components and yields 3ω SBS bandwidths of 90 GHz for both cases, and 3ω SSD bandwidths of 270 GHz and 120 GHz for the 1-MJ and 1.8-MJ shots, respectively.

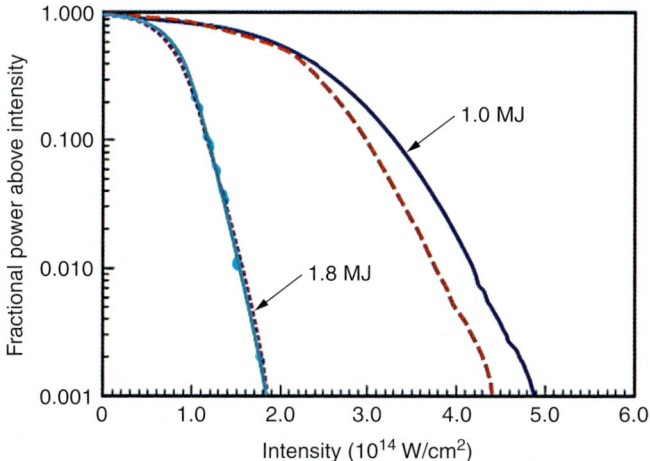

FIGURE 14.34 FOPAI comparisons for CPP-generated, SSD-smoothed focal spots. All curves are normalized to 1-TW total power. Solid lines are measurements; dashed lines are model. The 1-MJ curves are on the right; 1.8-MJ curves are on the left.

corresponds to averaging of ~16 speckles. The speckle averaging effect can be seen by comparing the images of Fig. 14.30 (no SSD) with those of Fig. 14.32 (with SSD). FOPAI plots for both the 1-MJ and the 1.8-MJ focal spots (Fig. 14.34) demonstrate both the reduction of the intensity of the hot spots by SSD and the agreement between prediction and measurement of that decrease.

14.6.3 Frequency Conversion of Spatially and Temporally Conditioned Pulses

Precision pulse shaping requires an accurate laser energetics model that, among other things, correctly accounts for conversion efficiency losses associated with beam conditioning. Table 14.5 summarizes the results of a series of high-power shots conducted with 1-ns FIT pulses for the purposes of validating our converter model. Data were obtained at a drive irradiance of ~3 GW/cm² for three different 1ω CPP configurations (none, 0.50 μm × 0.95 μm, and 1.16 μm × 1.3 μm), each with varying amounts of SSD bandwidth. The two CPP spot sizes correspond to the point designs for the 1-MJ and 1.8-MJ ignition experiments, respectively. The relevant energies are shown at the input to the second harmonic generator (1ω at SHG input) and the output of the third harmonic generator (3ω at THG output). For each shot, conversion efficiency was calculated using the model described in Sec. 14.5, including measured phase profiles for the phase plates. Bandwidth was simulated as

Shot	CPP	SSD (GHz)	SHG input (kJ)	THG output (kJ)		
				Measured	Model	Delta (%)
N060214-002	–	0	3.759	2.920	2.9072	–0.44
N060216-003	–	65.8	3.644	2.668	2.7122	1.66
N060216-002	–	95.2	3.721	2.477	2.4640	–0.52
N060224-001	1MJ	0	3.672	2.925	2.9325	0.26
N060224-002	1MJ	96.7	3.668	2.462	2.4551	–0.28
N060313-001	1.8MJ	0	3.553	2.656	2.7120	2.11
N060314-002	1.8MJ	37.2	3.757	2.667	2.7336	2.50
N060314-001	1.8MJ	94.8	3.766	2.367	2.3871	0.85

TABLE 14.5 Comparison Between Modeled and Measured Frequency Converter Performance

an effective tripler detuning of 1.9 μrad/GHz, assuming quadrature addition of the 3- and 17-GHz spectra. Off-line time-dependent plane-wave calculations have validated that this treatment is accurate over a wide range of input power. In all cases, the model is within 2.5 percent of measurement.

14.6.4 Temporal Pulse Shaping

The ignition campaign plan calls for a high-contrast, frequency-tripled temporal pulse shape, with all beam-conditioning techniques in place, to be specified and controlled to an rms deviation over the 48 NIF quads of less than or equal to 15 percent in the foot of the pulse and to less than or equal to 3 percent at the peak of the pulse. A precision pulse-shaping sequence was performed to test how well the current NIF hardware can generate the requested pulses and to develop a strategy for routinely matching them with high accuracy. Figure 14.27 shows the requested pulse shapes for the current 1-MJ and the 1.8-MJ baseline target drives.

The LPOM code is the first and primary tool used to determine the required setup pulse shape at the MOR. It uses its calibrated model of the state of all individual components, along with a solver capability built into its propagation/extraction code (VBL) to perform a first-principles numerical solution. As a side benefit of this solution, LPOM predicts the expected pulse shape that will be measured at the ISP and the OSP. For FIT pulse shapes, we have found this solution to be very accurate. For precise control of high amplitude-contrast pulses, we have developed an iterative operational

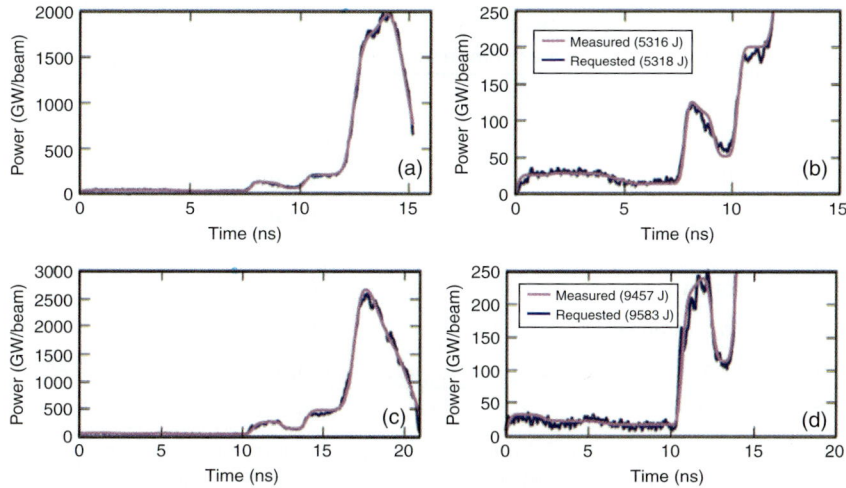

FIGURE 14.35 Comparison of measurement to request for 1-MJ (*a* and *b*) and 1.8-MJ (*c* and *d*) pulses, showing the peak (*a* and *c*) and foot (*b* and *d*) for both.

procedure to refine LPOM's results and to adjust for minor discrepancies between the model and measurements.

As Fig. 14.35 demonstrates, this iterative procedure led to an accurate match to the requested pulse shapes at both 1 MJ and 1.8 MJ. Once minor corrections to the drive prescription have been derived, the results can be incorporated into the general LPOM description. LPOM can then be relied on to make the desired pulse modifications required for optimizing the drive to ignition capsules.

14.7   2010 NIF Status and Experiments

Sections 14.4 through 14.6 describe detailed measurements made on eight arms of the NIF laser in 2005–2006, including single-beam measurements in the PDS. These measurements were the first end-to-end verification, using typical production hardware, that the NIF design would be able to meet the functional requirements that had been laid out in 1994. At the time of those measurements, 40 beamlines had been operated in the infrared at greater than or equal to 19 kJ per beamline. In the succeeding three and a half years, the remainder of the laser construction and commissioning has been completed, and NIF has demonstrated that it could fire at 3ω energies in excess of 1 MJ, with all required temporal pulse shaping, pointing, beam synchronization, and focal spot conditioning (beam smoothing) in place. As of mid-June 2010, 1795 full-system shots had been fired, for a cumulative 102 MJ of 1ω energy and 42 MJ at 3ω. More than 90 shots included targets, yielding important laser-plasma coupling and

Solid-State Lasers

	96 Beam Performance	Single Bundle Performance
Peak energy	500 kJ	75 kJ
Peak power	200 TW	21 TW
Wavelength	.35 μm	.35 μm
Positioning accuracy	100 μm rms at target plane	100 μm
Pulse duration	20 ns	20 ns
Pulse dynamic range	>25:1	50:1
Pulse spot size	600 μm	600 μm
Prepulse power	$< 10^8$ W/cm^2	$< 4 \times 10^6$ W/cm^2
Cycle time	8 hours max between full system shots	8 hours max between full system shots

TABLE 14.6 NIF Project Completion Criteria (PCC). A Single Eight-Beam Bundle Needed to Demonstrate the Ability to Operate at Full NIF Equivalent (FNE) of 1.8 MJ/500 TW, Meeting All Requirements Set Out in the 1994 Description of NIF Primary Criteria;[25] Half of NIF (96 beams) Was Required to Demonstrate Operation at 1 MJ/400 TW FNE

capsule implosion data. The program is on track to begin inertial fusion ignition experiments by the end of 2010.

On January 14, 2009, the last of the shots necessary to satisfy the project completion criteria was fired. Although this did not mark the end of the commissioning of laser hardware, it did mark the transition to a functioning scientific facility capable of carrying out important target experiments while commissioning activities continued. As can be seen in Table 14.6, "project completion" formally required that eight beamlines be able to operate at the full per-beam energy, power, positioning accuracy, pulse-shape contrast, spot size, and shot rate specified in 1994. For this interim goal, it was also required that 96 beams could be fired simultaneously at the 1 MJ/400 TW full-NIF-equivalent level with the same precision. On February 24–25, 2010, NIF data and progress were reviewed in detail by a special subcommittee of the National Ignition Campaign (NIC) Review Committee, which affirmed that "all laser performance Completion Criteria have been met or exceeded."[26] Tables 14.7 and 14.8 support this conclusion.

To date, target physics experiments have centered on hohlraum energetics, on the closely related topics of laser-plasma interactions and backscatter physics, and on capsule implosion symmetry. Data from these campaigns can be found in Glenzer et al.[56] As one example

3ω Laser Parameter	Single Bundle PCC (in context)	Achieved Values	Shot Number
Pulse energy	≥75 kJ	78 kJ for 8 beams FNE 1.87 MJ	N081010-002-999
Peak power	≥21 TW	21.3 TW 511 TW FNE	N081011-001-999
Wavelength	0.35 μm	0.35 μm	Entire campaign
Positioning accuracy	<100 μm RMS	70 μm RMS 61 μm RMS	N081205-001-999 N090114-002-999
Pulse duration	Up to 20 ns	Up to 21 ns	N081102-002-999
Pulse dynamic range	≥50:1	>90:1	N081011-001-999
Pulse spot size	≤600 μm	500 TW in 330 μm 530 TW in 600 μm	N060329-002-999 (B318 on PDS)
Prepulse power	$<4 \times 10^6$ W/cm^2	$<10^6$ ω/cm^2, Measured at PAM or 1ω main laser output, analysis to TCC	Many
Cycle time	8 hours max between full system shots	4:15 hours between shots (Two Rev 3.1 Be shots)	N081013-003-999 N081014-001-999

TABLE 14.7 Comparison of Single-Bundle Performance with PCC Single-Bundle Requirements. The Shot Numbers Encode the Shot Date (N081102 = November 11, 2008), Sequence Number (002 Means the Second Shot Fired That Day), And a Code (999) That Signifies a Full-System Shot.

of these results, Fig. 14.36 is an x-ray framing camera image of an imploded 1-mm-radius gas-filled capsule. Linear compression of about 20 times is seen, with spherical contour maintained to within ~15 percent. Prior to the onset of the implosion campaign, theory and modeling had predicted that stimulated scattering in the laser entrance hole would act to transfer energy between the inner-cone (low angle) and outer-cone (high angle) beamlines and that this effect would be sensitive to the wavelength separation between the two sets.[57] Experiments have confirmed this effect and have demonstrated that the imploded contour can be tuned predictably from oblate to round to prolate by modifying the wavelength separation.[58]

On December 4 and 5 of 2009, NIF fired 192-beam shots of 1.04 and 1.2 MJ into gas-filled hohlraums with gas-filled capsules. These shots were part of a series that established that laser light can be efficiently coupled into the hohlraum (90 percent or better coupling into

3ω Laser Parameter	96 Beam PCC	Achieved Values	Shot Number
Pulse energy	≥500 kJ	542kJ for 96 beams 564kJ for 96 beams	N081219-002-999 N081221-003-999
Peak power	≥200 TW	215 TW Peak 205 TW Average over 2 ns	N081222-001-999 N081221-003-999
Wavelength	0.35 µm	0.35 µm	Entire campaign
Positioning accuracy	<100 µm RMS	64 µm RMS	N090114-002-999
Pulse duration	Up to 20 ns	Up to 20.9 ns	N081222-002-999
Pulse dynamic range	≥25:1	>90:1	N081221-003-999
Pulse spot size (without smoothing)	≤600 µm	500 TW in 330 µm 530 TW in 600 µm	N060329-002-999 (B318 on PDS)
Prepulse power	$<10^8$ W/cm^2	$<10^8$ W/cm^2, Measure at PAM or 1ω main laser output, analysis to TCC	Many
Cycle time	8 hours max between full system shots	5:10, 6:02, and 6:12 hours between shots during campaign	Shortest: N081218-003-999 N081219-001-999

TABLE 14.8 Comparison of 96-Beam Performance with PCC Requirements

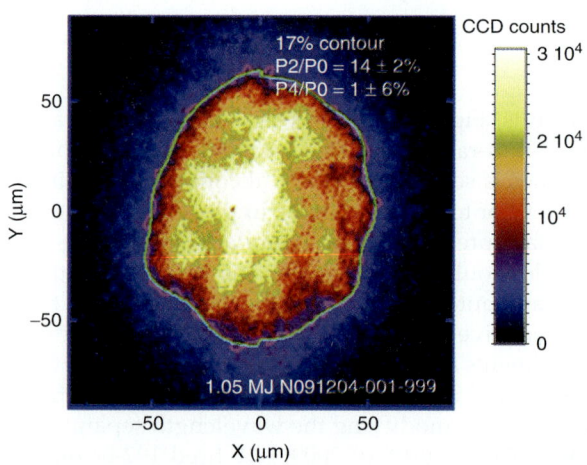

FIGURE 14.36 X-ray framing camera image of an imploded 2-mm-diameter gas-filled capsule. NIF scientists have demonstrated the ability to achieve high capsule compression while controlling the imploded-configuration symmetry.

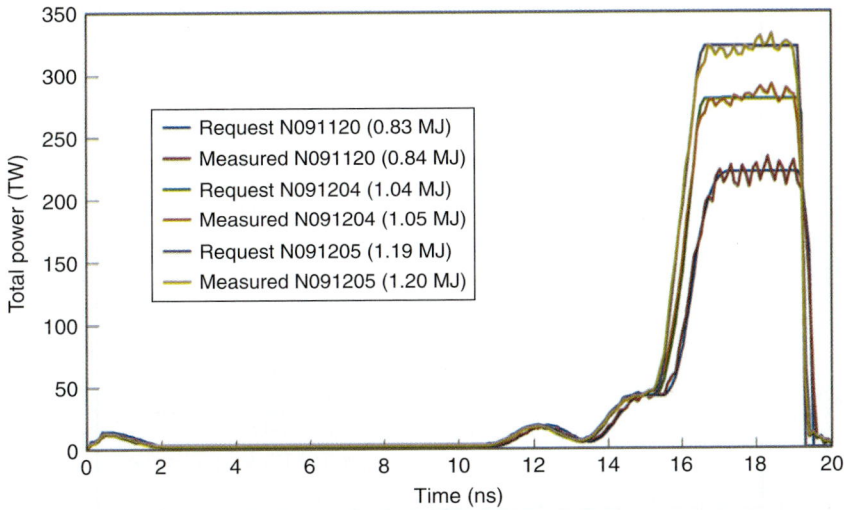

FIGURE 14.37 Requested (smooth lines) and measured (fluctuating lines) temporal pulse shapes for three shots fired at different energies on November 20, December 4, and December 5, 2009. The small periodic variation in the measured power is a result of the 3-GHz phase modulation applied in the front end to suppress transverse Brillouin scattering in the large fused silica optics; it has no effect on target performance. Other than this, achieved power history shows excellent agreement with the request.

soft x rays) and hence can generate sufficiently high hohlraum temperature to meet the NIF ignition design requirements.[56] Figure 14.37 compares the pulse shapes from those two shots, along with that of a previous shot fired on November 20, with the requests. In all three cases, the measured temporal pulse agrees with the request to within a few percent.

A perennial difficulty that has plagued previous ICF lasers is blooming of the far-field spot size as the power is raised. This blooming results from self-focusing of localized high-intensity regions in the beam near field; at the high power at which these lasers operate, the refractive index of the transmissive optics is modified by the light. Self-focusing, in turn, introduces high angular content to the beam, which is reflected in a growing focal spot size. The NIF design was developed with strict consideration of this effect and with strict limits both on sources of the fluctuations that seed the effect and on the B integral that measures its growth rate.[59] The success of this effort is demonstrated in Fig. 14.38, which plots single-beam far-field spot radius (90 percent enclosed energy), as measured on PDS, against the single-beam peak power (displayed as equivalent full-NIF power). The small increase between 0.5 TW per beam and 2.8 TW per beam is close to the measurement accuracy.

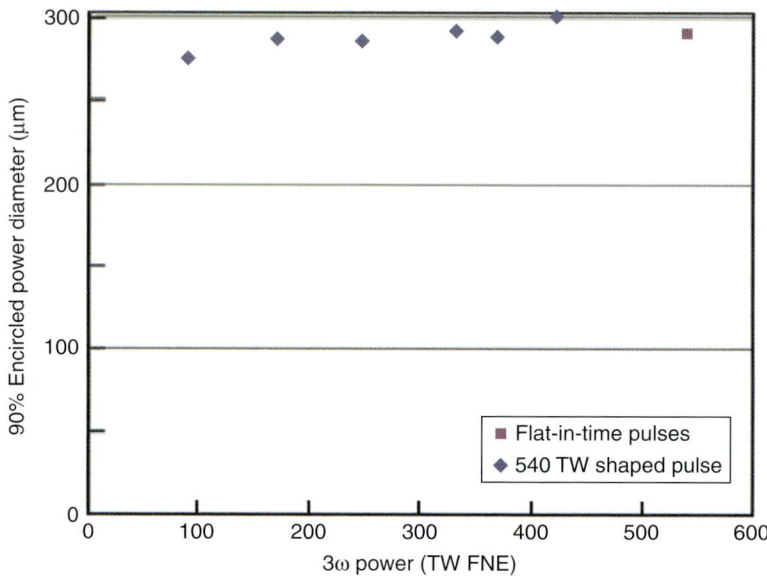

FIGURE 14.38 Unconditioned (no beam smoothing applied) far-field spot size as a function of peak beam power, measured on PDS. No significant spot size increase is observed for beam powers up to about 10 percent above the maximum design power. The spot size is comfortably less than the design requirement of 80 percent of the pulse energy within a 600-μm diameter.

A viable inertial fusion campaign has a number of stringent requirements, laid out both in the 1994 primary criteria and in the succeeding capsule, hohlraum, target positioning, and laser-performance specification roll-ups that have come to be known as Rev. 1 through Rev. 5 point designs. Beyond global numbers like energy, peak power, and temporal-pulse contrast, these point designs specify details such as time-dependent power balance, pointing accuracy, beam synchronization, shot-to-shot pulse reproducibility, and finesse in making small adjustments to the pulse shape to accommodate the results of pulse-tuning experiments. Figures 14.39 to 14.43 address the current state of NIF's demonstrated ability to achieve these requirements.

Power balance and pulse-shape reproducibility are illustrated in Figs. 14.39 and 14.40. Figure 14.39 is single-beam PDS data. A sequence of 16 shots was fired, holding constant the nanojoule pulse shape requested from the MOR. The desired 1.3 MJ Rev. 2 3ω pulse shape is shown, along with the time-dependent power balance specification and the time-dependent rms variation actually achieved among the 16 shots. Treating the variation among multiple shots of a single beamline as a surrogate for the variation among multiple beamlines

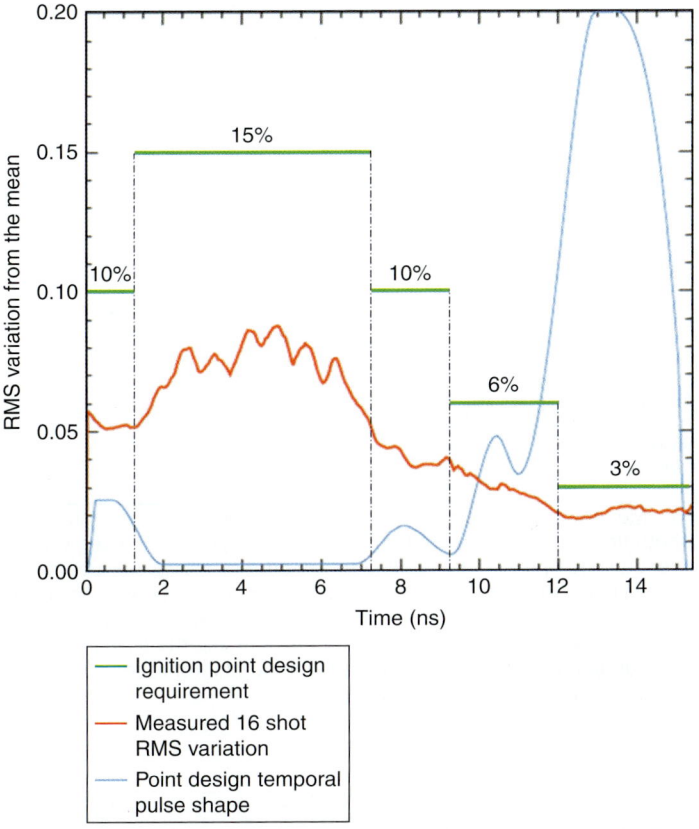

FIGURE 14.39 The time-dependent root mean square variation among a set of 16 nominally identical shots (red), compared with the Rev. 2 power-balance requirement (green); the temporal pulse shape (blue) is for reference.

on a single shot, this test predicted that NIF would meet its power-balance specifications with a comfortable margin. Further, assuming that shot-to-shot power variations are correlated within a quad, but uncorrelated between quads, we would expect the maximum shot-to-shot variation for the full NIF to be $\sim 9\%/\sqrt{48} = 1.3\%$, which is also significantly less than the 3 percent Rev. 2 requirement.[60] Two direct measurements of the achievable full-system power balance are represented in Fig. 14.40. Data from the 96-beam, 500-kJ project-completion shot on December 21, 2008 (Fig. 14.40a), verify that the quad-to-quad rms power variation is well below the Rev. 3.1 requirement throughout the pulse, including the 3 percent requirement at the peak of the pulse (compare with 8 percent in the 1994 Functional Requirements and Primary Criteria[25]). The 1-MJ, 192-beam shot fired on March 9, 2009 (Fig. 14.40b), had a known error in the request sent to the MOR

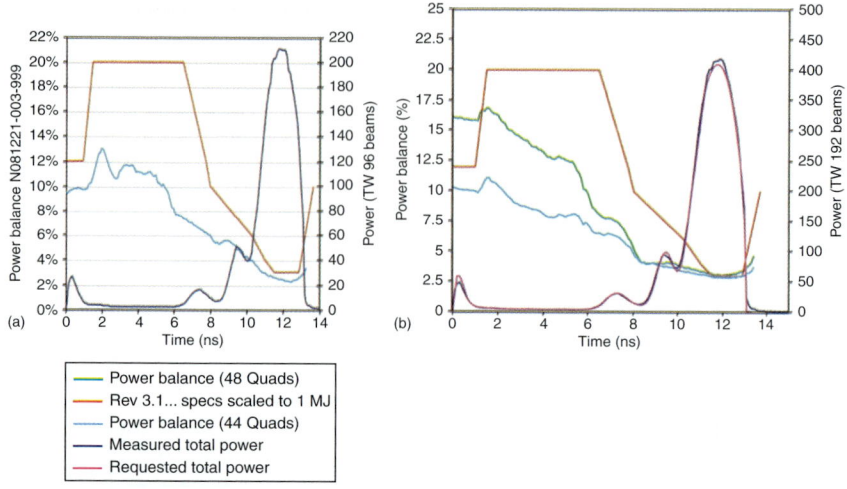

FIGURE 14.40 Measured rms power balance on (*a*) 500-kJ, 96-beam and (*b*) 1-MJ, 192-beam shots fired in January and March of 2009, respectively, along with the Rev. 3.1 power-balance specification scaled to the correct energy per beamline and the requested 3ω pulse shapes (for reference). The 96-beam test was below the specification throughout the pulse. On the 192-beam test, an error was made in communicating the front-end request on four quads, causing the power balance to be out of spec during the first 2 ns. Excluding those quads from the analysis, the rms of the remaining 44 quads met the specification.

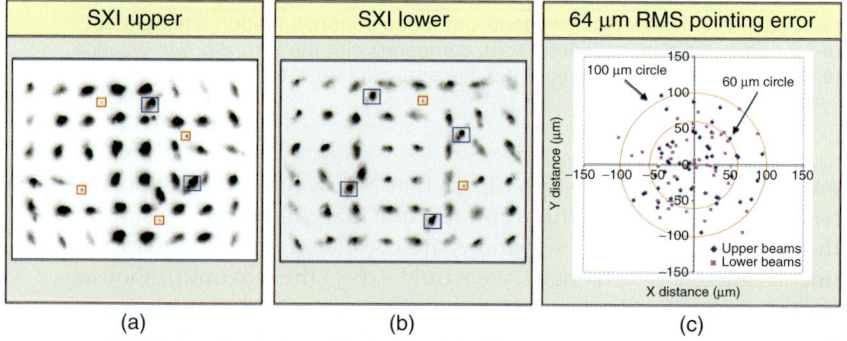

FIGURE 14.41 96-beam pointing experiment (shot N090114-002-999): Beams are intentionally aimed at a rectangular array of locations, 800 μm between focal spots, and their actual locations are measured with two static x-ray imagers (SXI-upper and SXI-lower). Six additional beams supply fiducials to position the two images with respect to each other. The difference between the measured spot centroids and their corresponding aim points (*c*) has an rms value of 64 μm, and the worst single-beam deviation is 120 μm. Both values are smaller than either the 1994 NIF primary criteria or the current ignition specification.

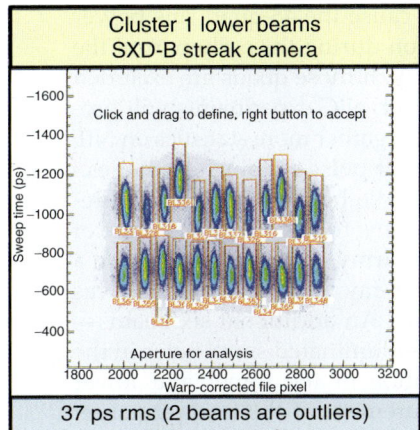 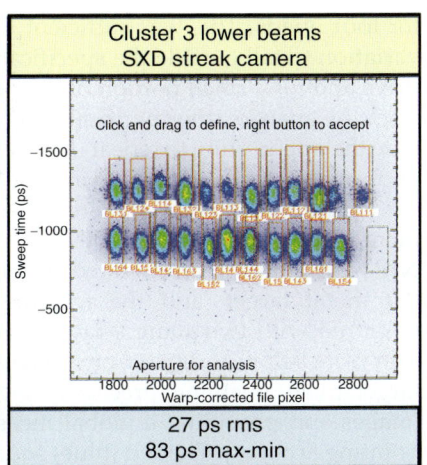

FIGURE 14.42 Streak camera images of 47-beam, 88-ps gaussian pulses. Target points are separated by 700 μm, and some quads are delayed in the MOR by 333 ps to get more beams per record. Using this method, all of NIF was synchronized to 64 ps rms. Developments in 2010 have enabled us to decrease this to less than 30 ps rms.

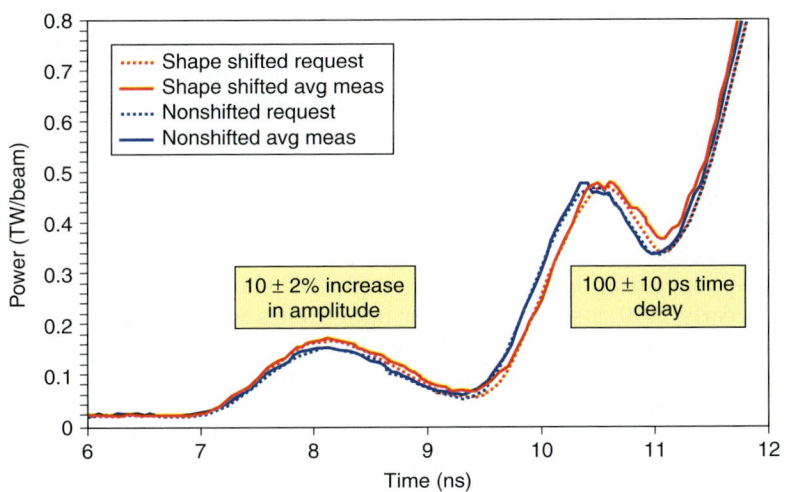

FIGURE 14.43 PDS data from July 2007 illustrating NIF's ability to reliably generate subtle modifications to pulse shape in order to accommodate target requirements. Each "bump" in the pulse shape generates a weak shock in the capsule ablator. To achieve compression to high density, it is necessary to time the propagation of these shocks through the shell precisely—a process that can be accomplished by adjusting either the shock strengths (laser power) or the shock-initiation times (pulse timing). These data quantify NIF's ability to control both.

for four quads. This was sufficient to cause the 48-quad rms power variation to fall outside the specification during the early part of the drive and for a small portion of the peak. If those quads are excluded from the analysis, the specification is met. NIC experiments will provide the opportunity to repeat these tests, gather more statistics on NIF performance, and continue to improve our pulse-shaping techniques.

Figure 14.41 displays the results of a 96-beam pointing measurement made on January 14, 2009.[61] 48 beams from each hemisphere of NIF were aimed at an 8 × 6 rectangular array of target locations on a flat metal target, and the resulting x-ray emission pattern was observed with two static x-ray imagers. An additional six beams—four from below and two from above—illuminated small holes in the target plate, enabling precise collocation of the upper and lower images and providing a global measurement of the beam-to-target pointing accuracy. The red (blue) squares in each image surround the holes illuminated by beams hitting the holes from the far (near) side of the target. The x-ray emission centroids differed from their intended locations on the target by an rms error of 64 ± 4 μm, compared with the current Rev. 5 point design specification of less than or equal to 80 μm. The worst beam missed its mark by ~120 μm, twice as good as the Rev. 5 requirement of less than or equal to 250 μm.

NIF's ability to achieve target-chamber-center (TCC) pulse synchronization is illustrated in Fig. 14.42. Short (88-ps) impulses are fired at a flat target positioned at TCC, and the x-ray emission is measured with one of two streaked x-ray detectors (SXDs). The target can be oriented either normal to the SXD field of view, thus maximizing the number of beams that can be viewed simultaneously, or at an angle, allowing cross timing between upper and lower hemispheres. Spatially separating the beam target points and temporally staggering the impulse times in the MOR allowed as many as 96 beams to be observed on the two SXDs on a single shot. Using this method, the 192 NIF beams were timed to within 64 ps rms, which is worse than the 30-ps specification but sufficient for the 2009 target shots.

In early 2010, a set of four fiber optic cables was installed on a diagnostic manipulator that could be placed at the center of the NIF target chamber. Low-energy laser pulses (from the NIF regenerative amplifier) can be directed onto these fibers, allowing the relative timing of each beamline on NIF to be measured, four beamlines at a time, at a high repetition rate. Using this new capability has allowed us to adjust the timing of NIF to less than 30 ps rms, which meets both the NIF Functional Requirements and Primary Criteria and the NIC ignition requirements.

NIF's pulse-shaping capability is exquisite. A combination of model-based shot setup, an advanced arbitrary wave form generator pulse-forming network, and careful attention to stabilized front-end operation have given the laser the ability to generate pulses varying from 88 ps quasi-gaussians to 20$^+$-ns shaped pulses. The current target modeling indicates that ignition will require pulses with 3ω contrast (maximum:minimum power) slightly in excess of 200:1.[62] The first attempt at generating such a pulse with the full NIF facility, fired on August 21, 2010, actually delivered 1.04 MJ at a contrast of 187 and matched the requested peak power to within less than 5 percent.

Because only NIF target experiments will be able to determine the precise pulse shape needed for low-adiabat, high-convergence capsule implosion at the scale of interest, design requirements include the ability to make subtle pulse-shape adjustments. Figure 14.43 shows PDS measurements from 2007 that illustrate this flexibility. Two pulse-shape requests are shown by the dotted lines—a baseline, or "unshifted," pulse in red and a "shifted" pulse generated by increasing the power at 8 ns by 10 percent and delaying the point at 10 ns by 100 ps. In both cases, these were full 1.3 MJ, 385 TW FNE (7 kJ, 2 TW per beamline) pulse shapes; the scale has been expanded to emphasize the low-power portion, because that is where the changes were made. The two solid lines are averages of multiple single-beam measurements. The laser was first adjusted to give a good fit to the unshifted shape; then 16 shots were taken without changing the front-end setup. The result is the blue solid curve. The front-end setup was then changed, based solely on numerical prediction, and 12 shots with this modified setup were taken, producing the solid red curve. NIF's success in accurately achieving the requested pulse-shape modification is apparent.

Conclusion

The NIF laser is by far the largest and most complex optical system that has ever been built. It has about 3 times as many beams and 60 times the energy or power of its nearest competitor. It can do pointing, timing, pulse shaping, and power balancing with unprecedented precision. While work continues to improve the laser performance, important plasma physics experiments are progressing, and early results compare well with expectations. Backscatter has been measured at less than 10 percent.[56] Full-ignition-scale (1 cm length × 5.4 mm diameter) hohlraums have been heated to 285 eV with megajoule drive energies, in agreement with radiation-hydrodynamics simulation. High-convergence implosion of ignition-scale capsules has been achieved, and the ability to tune the symmetry has been demonstrated. As of the date of this writing (July, 2010) we are on track to begin the first-ever inertial fusion ignition campaign in late 2010. The first cryogenic layered tritium, hydrogen, deuterium shot was fired in October 2010, and the shock-timing campaign began in November 2010.

NIF has also begun to fulfill its other two missions: ensuring the safety and reliability of the U.S. nuclear deterrent and serving as an international user facility for basic research in high-energy-density physics. As of this writing, 18 shots have been devoted to the development of experimental platforms and initial data acquisition in the areas of radiation transport, equation of state, and x-ray sources for stockpile-stewardship applications. The first set of experiments by an external academic team—a collaboration with the University of Michigan's Center for Radiative Shock Hydrodynamics to study hydrodynamic instabilities influenced by strong radiative shocks that are relevant to core-collapsing supernovae[62]—has already begun. In addition, a recent invitation for proposals for shot time drew 44 letters of intent and 40 full proposals from 20 institutions in 6 countries.

Acknowledgments

We would like to express our gratitude to J. Atherton, J. Auerbach, E. Bliss, M. Bowers, S. Burkhart, F. Chambers, P. Coyle, J. Di-Nicola, P. Di-Nicola, S. Dixit, G. Erbert, G. Guruangan, G. Heestand, M. Henesian, M. Hermann, R. House, M. Jackson, K. Jancaitis, D. Kalantar, R. Kirkwood, K. LaFortune, O. Landon, D. Larson, B. MacGowan, K. Manes, C. Marshall, J. Menapace, E. Moses, J. Murray, M. Nostrand, C. Orth, T. Parham, H. Park, R. Patterson, B. Raymond, T. Salmon, M. Schneider, M. Shaw, M. Spaeth, S. Sutton, P. Wegner, C. Widmayer, W. Williams, R. White, P. Whitman, S. Yang, and B. Van Wonterghem for their many important contributions to this work.

References

1. Haynam, C. A. Wegner, P. J., Auerbach, J. M., Bowers, M. W., Dixit, S. N., Erbert, G. V., Heestand, G. M., et al., "National Ignition Facility Laser Performance Status," *Appl. Opt.*, 46: 3276–3303, 2007.
2. Maiman, T. H., "Stimulated Optical Radiation in Ruby," *Nature*, 187: 493–494, 1960.
3. Nuckolls, J. H., *Laser Interactions and Related Plasma Phenomena*, vol. 20, eds. G. H. Miley and H. Hora, New York: Plenum, pp. 23–24, 1992.
4. Zimmerman, G. B., and Kruer, W. L., "Numerical Simulation of Laser-Initiated Fusion," *Commun. Plasma Phys. Control. Fusion*, 11: 51–61, 1975.
5. Nuckolls, J. H., Wood, L., Thiessen, A., and Zimmerman, G., "Laser Compression of Matter to Super-High Densities: Thermonuclear (CTR) Applications," *Nature*, 239: 139–142, 1972.
6. Lindl, J. D., Manes, K. R., and Brooks, K., "'Forerunner,' First Observation of Laser-Induced Thermonuclear Fusion and Radiation Implosion," UCRL-52202, Livermore, CA: Lawrence Livermore National Laboratory, pp. 88–98, 1976 (unpublished).
7. Phillion, D. W., and Banner, D. L., "Stimulated Raman Scattering in Large Plasmas," UCRL-84854, CONF-801119-7 (1980); presented at American Physical Society, Division of Plasma Physics, San Diego, CA, pp. 10–14, 1980.
8. Holzrichter, J., "Lasers and Inertial Fusion Experiments at Livermore," *Inertial Confinement Fusion: A Historical Approach by Its Pioneers*, eds., Guillermo Velarde and Natividad Carpintero Santamaria, London: Foxwell and Davies, p. 79, 2007.

9. Campbell, E. M., Turner, R. E., Griffith, L. V., Kornblum, H., McCauley, E. W., Mead, W. C., Lasinski, B. F., Phillion, D.W., and Pruett, B.L. et al., "Argus Scaling Experiments," *Laser Program Annual Report,* UCRL-50055-81/82, p. 4, 1982.
10. Foster Committee, "Final Report of the Ad Hoc Experts Group on Fusion," U.S. Department of Energy Report, Washington, DC: DOE, October 17, 1979.
11. Bliss, E. S., Hunt, J. T., Renard, P. A., Sommergren, G. E., and Weaver, H. J., "Effects of Nonlinear Propagation on Laser Focusing Properties," *IEEE J. Quant. Electron.,* QW-12: 402, 1976.
12. Trenholme, J. B., and Goodwin, E. J., "Fast Lumped-Element Computer Analysis of Laser Systems" and "Bespalov-Talanov Ripple Growth Calculations in Laser Systems," Laser Program Annual Report, UCRL-50021-76, pp. 2-333–2-244, 1976.
13. Hunt, J. T., Glaze, J. A., Simmons, N. W., and Renard, P. A., "Supression of Self-Focusing Through Low-Pass Spatial Filtering and Relay Imaging," *Appl. Opt.,* 17: 2053, 1976.
14. Holzrichter, J., "Lasers and Inertial Fusion Experiments at Livermore," *Inertial Confinement Fusion: A Historical Approach by Its Pioneers,* eds., Guillermo Velarde and Natividad Carpintero Santamaria, London: Foxwell and Davies, p. 81, 2007.
15. Henesian, M., Swift, C., and Murray, J. R., "Stimulated Rotational Raman Scattering in Long Air Paths," *Opt. Lett.,* 10: 565, 1985.
16. Murray, J. R., Smith, J. R., Ehrlich, R. B., Kyrazis, D. T., Thompson, C. E., Weiland, T. L., and Wilcox, R. B., "Experimental Observation and Suppression of Transverse Stimulated Brillouin Scattering in Large Optical Components," *J. Optic. Soc. Am. B,* 6: 2402–2411, 1989.
17. Lane, S. M., "High Yield Direct-Drive Implosions," *1986 Laser Program Annual Report,* UCRL-50021-86, Livermore, CA: Lawrence Livermore National Laboratory, 3-2–3-6, 1986.
18. Pennington, D. M., Perry, M. D., Stuart, B. C., Britten, J. A., Brown, C. G., Herman, S., Miller, J. L., et al., "The Petawatt Laser System," *ICF Quarterly,* UCRL-LR-105821-97-4, Livermore, CA: Lawrence Livermore National Laboratory, 4: 7, 1998.
19. Matthews, D. L., Hagelstein, P. L., Rosen, M. D., Eckart, M. J., Ceglio, N. M., Hazi, A. U., Medecki, H., et al., " Demonstration of a Soft X-Ray Amplifier," *Phys. Rev. Lett.,* 54: 110, 1985.
20. Storm, E., "Approach to High Compression in Inertial Fusion," *J. Fusion Energy,* 7(2): 131, 1988.
21. Soures, J. M., McCrory, R. L., Verdon, C. P., Babushkin, A., Bahr, R. E., Boehly, T. R., Boni, R., et al., "Direct-Drive Laser-Fusion Experiments with the OMEGA, 60-Beam, >40 kJ, Ultraviolet Laser System," *Phys. Plasmas,* 3: 2108, 1996.
22. Sangster, T. C., Goncharov, V. N., Radha, P. B., Smalyuk, V. A., Betti, R., Craxton, R. S., Delettrez, J. A., et al., "High-Areal-Density Fuel Assembly in Direct-Drive Cryogenic Implosions," *Phys. Rev. Lett.,* 100: 185006, 2008.
23. Tabak, M., Hammer, J., Glinski, M. E., Kruer, W. L., Wilks, S. C., Woodworth, J., Campbell, E. M., et al., "Ignition and High Gain with Ultrapowerful Lasers," *Phys. Plasmas,* 1: 1626–1634, 1994.
24. Di-Nicola, J. M., Fleurot, N., Lonjaret, T., Julien, X., Bordenave, E., Le Garrec, B., Mangeant, M., et al., "The LIL Facility Quadruplet Commissioning," *J. Phys. IV France,* 133: 595–600, 2006.
25. "National Ignition Facility Functional Requirements and Primary Criteria," revision 1.3, Report NIF-LLNL-93-058, Lawrence Livermore National Laboratory, 1994.
26. Dunne, M., Byer, R., Edwards, C., Grunder, H., Le Garrec, B., Kelley, J., and Wittenbury, C., letter to Scott L. Samuelson, Director National Ignition Facility Project Division, National Nuclear Security Administration, February 25, 2009.
27. Miller, G. H., Moses, E. I., and Wuest, C. R., "The National Ignition Facility," *Opt. Eng.,* 43: 2841–2853, 2004.

28. Spaeth, M. L., Manes, K. R., Widmayer, C. C., Williams, W. H., Whitman, P. K., Henesian, M. A., Stowers, I. F., and Honig, J., "National Ignition Facility Wavefront Requirements and Optical Architecture," *Opt. Eng.*, 43: 2954–2965, 2004.
29. Bonanno, R. E., "Assembling and Installing Line-Replaceable Units for the National Ignition Facility," *Opt. Eng.*, 43: 2866–2872, 2004.
30. Zacharias, R. A., Beer, N. R., Bliss, E. S., Burkhart, S. C., Cohen, S. J., Sutton, S. B., Van Atta, R. L., et al., "Alignment and Wavefront Control Systems of the National Ignition Facility," *Opt. Eng.*, 43: 2873–2884, 2004.
31. Shaw, M., Williams, W., House, R., and Haynam, C., "Laser Performance Operations Model," *Opt. Eng.*, 43: 2884–2895, 2004.
32. Moses, E. I., and Wuest, C. R., "The National Ignition Facility: Laser Performance and First Experiments," *Fusion Sci. Tech.*, 47(3): 314–322, 2005.
33. Hunt, J. T., Manes, K. R., Murray, J. R., Renard, P. A., Sawicki, R., Trenholme, J. B., and Williams, W., "Laser Design Basis for the National Ignition Facility," *Fusion Tech.*, 26: 767–771, 1994.
34. Wisoff, P. J., Bowers, M. W., Erbert, G. V., Browning, D. F., and Jedlovec, D. R., "NIF Injection Laser System," *Proc. SPIE*, 5341: 146–155, 2004.
35. Van Wonterghem, B. M., Murray, J. R., Campbell, J. H., Speck, D. R., Barker, C. E., Smith, I. C., Browning, D. F., and Behrendt, W. C., "Performance of a Prototype for a Large-Aperture Multipass Nd:glass Laser for Inertial Confinement Fusion," *Appl. Opt.*, 36:4932–4953, 1997.
36. Lindl, J. D., *Inertial Confinement Fusion*, New York: Springer, 1998, "Development of the Indirect-Drive Approach to Inertial Confinement Fusion and the Target Physics Basis for Ignition and Gain," *Phys. Plasmas*, 2: 3933–4024, 1995.
37. Lindl, J. D., Amendt, P., Berger, R. L., Glendenning, S. G., Glenzer, S. H., Haan, S. W., Kaufmann, R. L., et al., "The Physics Basis for Ignition Using Indirect-Drive Targets on the National Ignition Facility," *Phys. Plasmas*, 11: 339–491, 2004.
38. Hinkel, D. E., Haan, S. W., Langdon, A. B., Dittrich, T. R., Still, C. H., and Marinak, M. M., "National Ignition Facility Targets Driven at High Radiation Temperature: Ignition, Hydrodynamic Stability, and Laser-Plasma Interactions," *Phys. Plasmas*, 11: 1128–1144, 2004.
39. Haan, S. W., Herrmann, M. C., Amendt, P. A., Callahan, D. A., Dittrich, T. R., Edwards, M. J., Jones, O. S., et al., "Update on Specifications for NIF Ignition Targets, and Their Roll Up into an Error Budget," *Fusion Sci. Tech.*, 49: 553–557, 2006.
40. Manes, K. R., and Simmons, W. W., "Statistical Optics Applied to High-Power Glass Lasers," *J. Opt. Soc. Am. A*, 2: 528–538, 1984.
41. Murray, J. R., Smith, J. R., Ehrlich, R. B., Kyrazis, D. T., Thompson, C. W., and Wilcox, R. B., "Observation and Suppression of Transverse Stimulated Brillouin Scattering in Large Optics," *J. Opt. Soc. Am. B*, 6: 2402–2411, 1989.
42. Craxton, R., "High-Efficiency Tripling Schemes for High-Power Nd:glass Lasers," *IEEE J. Quantum Electron.*, QE-17: 1771–1782, 1989.
43. Eimerl, D., Auerbach, J. M., and Milonni, P. W., "Paraxial Wave Theory of Second and Third Harmonic Generation in Uniaxial Crystals," *J. Mod. Opt.*, 42(5): 1037–1067, 1995.
44. Williams, W. H., Auerbach, J. M., Henesian, M. A., Jancaitis, K. S., Manes, K. R., Mehta, N. C., Orth, C. D., et al., "Optical Propagation Modeling for the National Ignition Facility," *Proc. SPIE*, 5341: 277-78, 2004.
45. Auerbach, J. M., Wegner, P. J., Couture, S. A., Eimerl, D., Hibbard, R. L., Milam, D., Norton, M. A., et al., "Modeling of Frequency Doubling and Tripling with Measured Crystal Spatial Refractive-Index Nonuniformities," *Appl. Opt.*, 40(9): March 2001.
46. Hardin, R. H., and Tappert, F. D., "Application of the Split-Step Fourier Method to the Numerical Solution of Nonlinear and Variable Coefficient Wave Equations," *SIAM Rev.*, 15: 423, 1973; Cooley, P. M., and Tukey, J. W., "An Algorithm for the Machine Computation of Complex Fourier Series," *Mathematics of Computation*, 19: 297, 1965.
47. Munro, D. H., Dixit, S. N., Langdon, A. B., and Murray, J. R., "Polarization Smoothing in a Convergent Beam," *Appl. Opt.*, 43: 6639–6647, 2004.

48. Skupsky, S., Short, R. W., Kessler, T., Craxton, R. S., Letzring, S., and Soures, J. M., "Improved Laser Beam Uniformity Using the Angular Dispersion of Frequency-Modulated Light," *J. Appl. Phys.*, 66: 3456–3462, 1989.
49. Rothenberg, J. E., "Comparison of Beam-Smoothing Methods for Direct-Drive Inertial Confinement Fusion," *J. Opt. Soc. Am. B*, 14: 1664–1671, 1997.
50. Goodman, J., Chapter 2, in *Laser Speckle and Related Phenomena*, ed., J. C. Dainty, New York: Springer-Verlag, 1984.
51. Powell, H. T., Dixit, S. N., and Henesian, M. A., "Beam Smoothing Capability on the Nova Laser," Lawrence Livermore National Laboratory (LLNL) Report, UCRL-LR-105821-91-1: 28–38, 1990.
52. Dixit, S. N., Thomas, I. M., Woods, B. W., Morgan, A. J., Henesian, M. A., Wegner, P. J., and Powell, H. T., "Random Phase Plates for Beam Smoothing on the Nova Laser," *Appl. Opt.*, 32: 2543–2554, 1993.
53. Dixit, S. N., Thomas, I. M., Rushford, M. R., Merrill, R., Perry, M. D., Powell, H. T., and Nugent, K. A., "Kinoform Phase Plates for Tailoring Focal Plane Intensity Profiles," LLNL Report, UCRL-LR-105821-94-4: 152–159, 1994.
54. Dixit, S. N., Feit, M. D., Perry, M. D., and Powell, H. T., "Designing Fully Continuous Phase Plates for Tailoring Focal Plane Irradiance Profiles," *Opt. Lett.*, 21: 1715–1717, 1996.
55. Menapace, J. A., Dixit, S. N., Génin, F. Y., and Brocious, W. F., "Magnetorheological Finishing for Imprinting Continuous Phase Plate Structure onto Optical Surfaces," *Proc. SPIE*, 5273: 220–230, 2003.
56. Glenzer, S. H., MacGowan, B. J., Meezan, N. B., Adams, P., Alfonso, J., Alger, E., Alherz, Z., et al., "Demonstration of Ignition Radiation Temperatures in Indirect-Drive Inertial Confinement Fusion Hohlraums," *Phys. Rev. Lett.*: in press.
57. Lindl, J. D., Amendt, P., Berger, R. L., Glendinning, S. G., Glenzer, S. H., Haan, S. W., Kaufman, R. L., et al., "The Physics Basis for Ignition Using Indirect-Drive Targets on the National Ignition Facility," *Phys. Plasmas*, 11: 339, 2004.
58. Glenzer, S. H., MacGowan, B. J., Michel, P., Meezan, N. B., Suter, L. J., Dixit, S. N., Kline, J. L., et al., "Symmetric Inertial Confinement Fusion Implosions at Ultra-High Laser Energies," *Science*, 327: 1228, 2010.
59. Siegman, A. E., *Lasers*, Mill Valley, CA: University Science Books, 1986, 385–386.
60. Haynam, C. A., Sacks, R. A., Wegner, P. J., Bowers, M. W., Dixit, S. N., Erbert, G. V., Heestand, G. M., et al., "The National Ignition Facility 2007 Laser Performance Status," *J. Phys.: Conf. Ser.*, 112: 032004, 2008. Schneider, M., "Beam Pointing Results at the National Ignition Facility," Poster Session 1.10.044, Inertial Fusion Science and Applications Conference, San Francisco, CA, Sept. 6–11, 2009.
61. Haan, S. W., personal communication, August 12, 2010.
62. Kuranz, C. C., Park, H.-S., Remington, B. A., Drake, R. P., Miles, A. R., Robey, H. F., Kilkenny, J. D., et al., "Astrophysically Relevant Radiation Hydrodynamics Experiment at the National Ignition Facility," Proceedings of the 8th International Conference of High Energy Density Laboratory Astrophysics, March 15–18, 2010, Pasadena, CA, 2010.

PART 4

Fiber Lasers

CHAPTER 15
Introduction to Optical Fiber Lasers

CHAPTER 16
Pulsed Fiber Lasers

CHAPTER 17
High-Power Ultrafast Fiber Laser Systems

CHAPTER 18
High-Power Fiber Lasers for Industry and Defense

CHAPTER 15
Introduction to Optical Fiber Lasers

Liang Dong

Department of Electrical and Computer Engineering and Center for Optical Materials Science and Engineering Technologies (COMSET), Clemson University, Clemson, South Carolina

Martin E. Fermann

IMRA America, Ann Arbor, Michigan

15.1 Background

15.1.1 History

The benefit of using a rare-earth-doped single-mode optical fiber in a laser cavity to provide a robust single spatial mode at the laser output was realized as early as 1961 by E. Snitzer, who was then at America Optical Company.[1] High gain in neodymium-doped multimode silica optical fiber lasers pumped by flash lamps was demonstrated a few years later by Snitzer and C. Koester.[2] A diode-pumped neodymium multimode optical fiber laser was then demonstrated in the early 1970s by J. Stone and C. A. Burrus at Bell Laboratories.[3] Work on single-mode optical fiber lasers started in the mid-1980s, after the development of rare-earth-doping methods that used modern optical fiber fabrication processes based on vapor-phase deposition.[4–7] Soon thereafter, interest in fiber lasers and amplifiers took off, thanks to a convergence of two key technologies in the late 1980s. The first technology was the availability of erbium-doped single-mode fibers, which provided a unique gain medium for the very important telecommunication window at 1.55 µm; erbium-doped fiber amplifiers

(EDFAs) were the dominant fiber technology research topic in the late 1980s.[8,9] The second technology was the availability of pump diodes that could be used in EDFAs to provide robust, compact devices for use in demanding telecommunication applications, such as submarine systems.[10,11] The ability of optical fiber amplifiers to replace electronic repeaters and consequently to eliminate the need for optical-to-electronic and electronic-to-optical conversion were easy to see. This was especially true when the wavelength-transparent nature of optical fiber amplifiers was considered; wavelength-division-multiplexing transmission systems were seen as the path to meet the increasing bandwidth demands for the new digital data traffic beyond the traditional analogue traffic. Both single-mode erbium-doped fibers and pump diodes are derived from technologies that have been well known and field tested in the telecommunication industry. The maturity of the key technologies, in combination with an increasing demand in data traffic, was key to the rapid development and subsequent deployment of EDFAs in the early 1990s.

The development of EDFAs, which was spurred by important telecommunication market demands and backed by the immense resources of the telecommunication industry, quickly led to the wide availability of knowledge, components, technologies, and equipment relevant to the development of optical fiber lasers. This, in turn, led to an extensive amount of research being conducted on rare-earth-doped fiber lasers covering continuous wave (CW) lasers, Q-switched lasers, mode-locked lasers, upconversion lasers, and single-frequency lasers in the late 1980s and early 1990s. For the majority of the 1990s, the highest-power single-mode optical fiber lasers were usually pumped by gas lasers or solid-state lasers, which made them impractical for commercial applications. The average powers from most fiber lasers pumped by single-mode pump diodes were too low for serious industry applications.

Another important development in the late 1980s led to the eventual development of single-mode optical fiber lasers with average powers well above the subwatt levels available from single-mode pump diodes. Cladding pumping was initially studied to enable the use of much higher powers available from low-brightness multimode single-emitter diodes,[12,13] which were being developed for pumping solid-state lasers, leading to more efficient and reliable high-power solid-state lasers. A double-clad fiber used in cladding pumping has a small rare-earth-doped single-mode core embedded in a much larger multimode pump guide. This configuration effectively behaves like a brightness converter, which allows conversion of highly multimode low-brightness pump light to a single-mode high-brightness laser beam guided in the rare-earth-doped single-mode core. Because of the limited brightness and low packing density of multiemitter diodes or diode arrays, some form of beam-shaping optics is often used to further maximize the pump power that can be coupled into a double-clad fiber. The development of optical fiber lasers accelerated in the second half of the 1990s, leading to the first commercialization effort by JDS Uniphase Corporation (JDSU) to market watt-level fiber lasers for printing applications.

The burst of the telecom bubble around 2001 was a blessing for the development of high-power optical fiber lasers. Much of the research and development demands in telecommunications suddenly evaporated. The investments and people who had gathered to meet those needs were looking for new business directions. The research and development of multimode pump diodes led to significantly more powerful and reliable multimode pump diodes at much lower cost. Military-funded programs, in a drive for directed energy weapon systems and countermeasures, also played a significant role in pushing for higher powers from fiber lasers and the development of related technologies. It is worth noting that the U.S. directed energy weapons program started by working on solid-state lasers in the early 1960s. It went through gas lasers in the 1960s, chemical lasers in the 1970s, x-ray lasers in the 1980s, and eventually back to solid-state lasers in recent years.[14]

In the late 1990s, another technology development further advanced peak-power scaling of fiber lasers. The small core of a single-mode optical fiber leads to high optical intensities and, consequently, low nonlinear thresholds. The solution to this issue comes in the form of single-mode operation in multimode fibers with much larger cores, significantly improving nonlinear thresholds.[15,16] With the convergence of all these factors, a much accelerated rate of development took place, culminating in the demonstration of single-mode diffraction-limited multikilowatt-level fiber laser systems in recent years.

15.1.2 Advantages of Fiber Lasers

An optical fiber can be designed to support only the lowest-order mode—that is, fundamental HE_{11} mode or LP_{01} mode, which eliminates the need for spatial mode control in an optical fiber laser consisting of a rare-earth-doped single-mode core with two reflectors placed at each end. Another significant benefit originates from better heat dissipation, coming from the use of a long fiber with a large heat-dissipating surface positioned just a few tens to a few hundreds of micrometers (μm) away from the active region. In a rod laser, heat is generated within a small volume in the rod center, with a limited heat-dissipating surface some distance away. This limits power scaling due to rod fracture by the large temperature gradient from a high heat load. Disc lasers have been developed to combat this issue by introducing a limited heat-dissipating surface on one side of a thin disc, very close to the active region. The power scalability of disc lasers is still an area of intense research. An additional benefit of fiber lasers, especially from the perspective of practical high-power laser systems, is high efficiency, mostly coming from better confinement and efficient absorption of the pump once coupled into a double-clad optical fiber. A glass host also provides broader absorption and emission spectra for rare earth ions due to strong inhomogeneous broadening, leading to less constraint on pump wavelength stability, a wider range of lasing wavelengths, and a wide gain bandwidth, all of which are critical factors for ultrashort pulse lasers.

15.2 Rare-Earth-Doped Optical Fibers

15.2.1 Basics of Optical Fibers

Figure 15.1 shows a typical optical fiber. At the center of the fiber is a core of diameter 2ρ and refractive index n_{co}, surrounded by a cladding layer with refractive index n_{cl}. Both the core and the cladding are made mostly of silica glass. Germanium doping in the core is typically used to raise the core's refractive index. Phosphorous and aluminum doping can also be used to raise the refractive index. Fluorine and occasionally boron doping are used to lower the refractive index of silica glass. The core glass and at least the inner part of the cladding glass are typically made of very-high-purity glass through vapor-phase deposition processes, which minimize impurities, especially transition metal ions, to achieve very low transmission loss. The intrinsic loss of between 1 and 2 μm, a wavelength range relevant to most optical fiber lasers, is at most a few decibels per kilometer (dB/km). This loss is negligible in a fiber laser of a few meters long. A layer of acrylic coating with a higher refractive index is typically used both to protect the glass surface and to strip away any unwanted light propagating in the cladding glass.

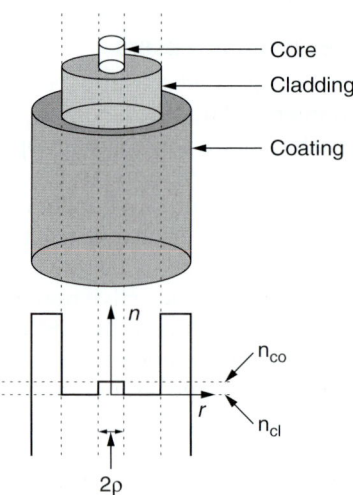

FIGURE 15.1 An optical fiber and its transverse refractive index distribution.

The relative refractive index difference of a step-index optical fiber is defined as

$$\Delta = \frac{n_{co} - n_{cl}}{n_{co}} \tag{15.1}$$

where Δ is typically below 2 percent. In the weakly guiding regime, reflected and transmitted fields around core and cladding interfaces are well approximated without considering field orientations. Linearly polarized modes can then be used to describe the modes in the optical fiber. By ignoring the vector nature of the fields, the weakly guiding approximation greatly simplifies the theoretical analysis of an optical fiber. The numerical aperture (NA) of an optical fiber is defined as

$$NA = \sqrt{n_{co}^2 - n_{cl}^2} \tag{15.2}$$

A very important parameter in an optical fiber is normalized frequency, which is defined as

$$V = \frac{2\pi \rho NA}{\lambda} \tag{15.3}$$

where λ is the vacuum wavelength. The guided modes of a waveguide can be obtained from the Helmholtz eigenvalue equation, which is derived from Maxwell's equations and which ensures all relevant field continuities at all boundaries. A guided mode can be seen as a robust fundamental spatial distribution that propagates at the propagation constant β and maintains a constant wavefront. It can be expressed as

$$E(r, \theta, z) = E_0(r, \theta) e^{-i\beta z} \tag{15.4}$$

where β is propagation constant, E_0 is transverse mode distribution, and z is the propagation distance. Due to the Helmholtz equation's unique scaling characteristics, a waveguide's mode properties are entirely determined once the normalized frequency V is known. It is worth noting that proportionally scaling both ρ and λ leads to modes with the same relative field distribution and propagation constant. The effective mode index n_{eff} can be obtained from the relation

$$\beta = \frac{2\pi n_{eff}}{\lambda} \tag{15.5}$$

In the weakly guiding regime, waveguide modes can be represented as LP_{lm}, where LP stands for linearly polarized and l and m are

Fiber Lasers

azimuthal and radial mode numbers, respectively. The effective mode index falls into the following range:

$$n_{co} > n_{eff} > n_{cl} \qquad (15.6)$$

Modes with $n_{eff} < n_{cl}$ have oscillating fields in the cladding and are no longer guided. They are sometimes referred to as *radiation modes* or, if discrete, *leaky modes*. A useful interpretation for leaky modes is that the total internal reflection condition is no longer met, resulting in power loss at each reflection at the core-cladding interface. The normalized propagation constant b, defined in Eq. (15.7), is a useful parameter for measuring how strongly a mode is guided. For guided modes, b falls between 0 and 1. A mode is no longer guided when $b \leq 0$.

$$b = \frac{n_{eff}^2 - n_{cl}^2}{n_{co}^2 - n_{cl}^2} \qquad (15.7)$$

The effective mode index of a single-mode optical fiber and an optical fiber that supports few modes is illustrated in Fig. 15.2a and 15.2b, respectively. The fundamental mode LP_{01} always has the highest effective mode index, followed by the second order mode LP_{11}. It is also worth noting that the effective mode index difference between modes can serve as a rough measure of how easily two modes can be phase matched for intermodal coupling, though a more rigorous analysis of intermodal coupling will also have to involve the spatial overlap integral between the modes and the external perturbation. Modes cannot couple to each other in the absence of external perturbations due to their orthogonal nature. As the number of guided modes increases in a waveguide, the effective mode index difference gets smaller, which means it is generally easier for intermodal coupling to occur.

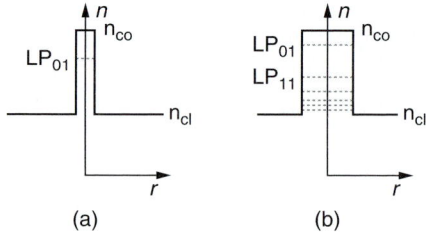

FIGURE 15.2 Effective mode index in (*a*) a single-mode fiber and (*b*) a fiber that supports few modes.

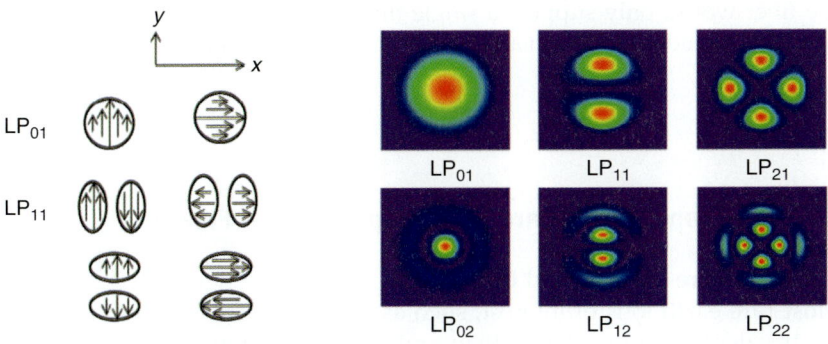

FIGURE 15.3 The first few modes of an optical fiber. The arrows represent the direction of the electric field.

The first two modes of an optical fiber (LP_{01} and LP_{11}) are illustrated in Fig. 15.3. The fundamental mode has two degenerate modes, and the second mode has four nearly degenerate modes in an optical fiber with perfect circular symmetry. This degeneracy of the two orthogonally polarized modes can lead to easy coupling between them, inducing depolarization after a short propagation length assuming a linearly polarized input. The degeneracy can be broken by introducing a noncircular symmetry, which is what happens in a polarization maintaining (PM) fiber, leading to much improved polarization preservation. The normalized propagation constant b of all LP modes in a step-index fiber are plotted in Fig. 15.4 for $V < 10$. It can be seen that

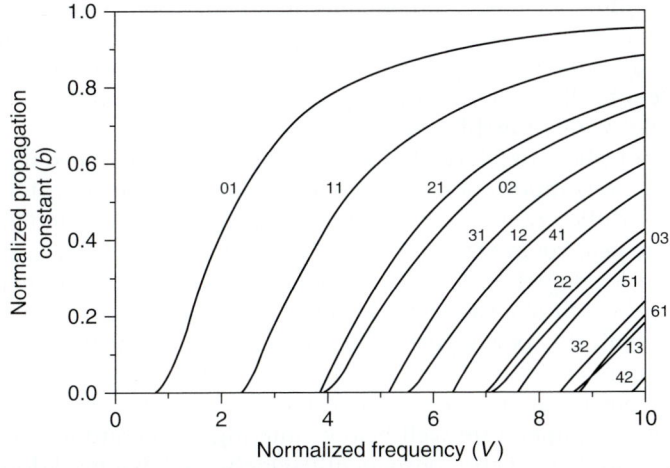

FIGURE 15.4 Normalized propagation constant for all LP modes with normalized frequency $V < 10$ for a step-index fiber.

the fiber would only support a single mode for $V < 2.405$. The total number of guided modes in a fiber can be estimated from

$$N = \frac{V^2}{2} \tag{15.8}$$

15.2.2 Properties of Rare-Earth-Doped Optical Fibers

Basics of Rare-Earth-Doped Glass

Most rare earth ions of interest, such as Yb^{3+}, Er^{3+}, Tm^{3+}, and Nd^{3+}, are trivalent and much larger than typical glass formers. This is particularly true in silica glass, which comprises the tetrahedral structures of SiO_2. Incorporation of rare earth ions requires a disruption of the regular glass network, which ultimately limits the level of possible rare earth doping before the onset of clustering due to phase separation. A high level of Al^{3+} ions can be added to a silica glass host to form a homogeneously modified glass network, which allows a much improved incorporation of rare earth ions. Incorporation of P^{5+} ions in silica glass can have a similar effect. In fact, phosphate glass, a glass made mostly of P_2O_5, is known to allow very high levels of rare earth doping to few tens of weight percent (wt%) levels before significant phase separation. In any case, much lower rare-earth-doping levels and gain per unit length in a silica-based glass host are expected in comparison with that in a crystal host. In an EDFA, typical erbium-doping levels are in the few tens of mole parts per million (mol ppm) to a few hundreds of mol ppm. Absorption and emission spectra of rare earth ions in a glass host are usually much broader in comparison with those in crystal hosts due to significant inhomogeneous broadening in an amorphous host. This is beneficial, as it relaxes the required wavelength control for diode-pumped lasers and allows for broadband amplification, such as in wavelength-division-multiplexing systems, ultrashort pulse generation, and widely tunable lasers.

Vapor-phase delivery of rare earth compounds has been used in early demonstrations of rare-earth-doped single-mode optical fibers. The key difficulty is a lack of compounds with high enough vapor pressure near room temperature. Elaborate heated delivery systems must be developed to incorporate both rare earth and aluminum ions. The key benefit of a vapor-phase delivery system is its compatibility with chemical vapor deposition (CVD) processes, which are widely used for optical fiber fabrication. The solution doping method, based on aqueous impregnation of soot formed by a CVD process at a low deposition temperature well below sintering temperature, is much easier to implement and consequently widely used. For most dopants of interest, soluble compounds are easily available for common laboratory solvents. The drawbacks are that a preform usually has to be

Introduction to Optical Fiber Lasers

taken off a CVD system for solution doping, and an additional drying process is required afterward.

Spectral Properties of Rare-Earth-Doped Fibers

Infrared transitions from trivalent rare earth ions originate from f-shell electrons, which are relatively shielded and consequently do not change significantly from one host to another. Upper-state lifetimes, precise transition spectra, and nonradiative processes relying on energy coupling to phonons of the surrounding lattice can, however, be modified by the host glass. Lower-level transitions for Yb^{3+}, Er^{3+}, and Tm^{3+} ions are illustrated in Fig. 15.5. Erbium-doped optical fibers are the most studied due to their ability to amplify around the telecommunication window of 1.55 µm. Initially, erbium-doped optical fibers were pumped in the visible by gas and dye lasers due to a lack of suitable pumps at longer wavelength. The first diode-pumped EDFAs were pumped around 1480 nm using diodes developed for Raman amplifiers for telecommunications.[10,11] EDFAs spurred the development of single-mode diode lasers around 980 nm, which allows more efficient inversion and lower amplifier noise. Correspondingly, high-power optical fiber lasers using these pump wavelengths were also developed. Even though Nd^{3+}-doped optical fiber lasers were the first to be studied due to the four-level nature of Nd^{3+} transitions, Yb^{3+} ions have recently become the dopant of choice for high-power fiber lasers. The first reason for this is the Yb^{3+} ion's simple system with low quantum defect. Low quantum defect equals less heat generation, which is a huge plus for high-power lasers. The simple system of ytterbium is also beneficial, because there is no need

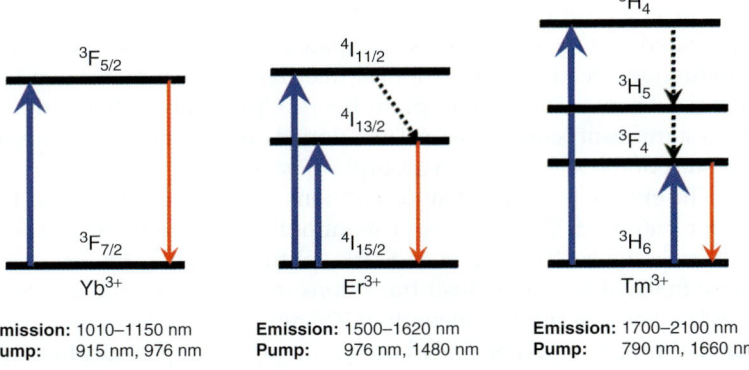

Figure 15.5 Low-level transitions of Yb^{3+}, Er^{3+}, and Tm^{3+} in silica glass host. Nonradiative processes are indicated by dotted lines. Pump transitions are indicated by upward-pointing arrows, and emission is indicated by downward-pointing arrows.

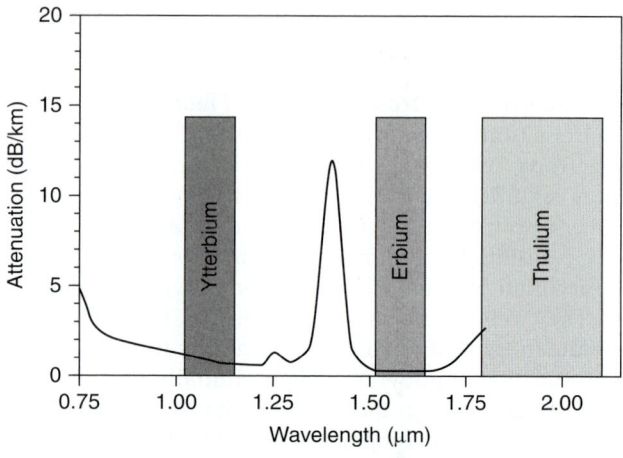

FIGURE 15.6 Emission wavelengths of optical fiber lasers plotted with a typical attenuation spectrum of a silica optical fiber.

to worry too much about excited state absorption and cooperative upconversion, both of which are channels for power loss. (Note that there is still some level of upconversion in Yb^{3+} system showing up as blue fluorescence.) The upconversion process can also lead to photodarkening, which compromises the long-term reliability of an optical fiber laser. Tm^{3+}-doped optical fiber lasers have recently attracted much interest for applications around 2 µm. The most preferred wavelength of pumping is 790 nm, which allows an efficient one-for-two process and access to higher-power pump diodes. The one-for-two process allows one ion excited to the $^3H_4$ level to generate two ions excited to the $^3F_4$ level.

The emission wavelengths of Yb^{3+}, Er^{3+}, and Tm^{3+} fiber lasers are illustrated in Fig. 15.6 and are compared with a typical silica fiber attenuation spectrum. Most optical fiber lasers use a meter to a few meters of rare-earth-doped optical fibers. Optical fiber attenuation is not a significant issue, even for Tm^{3+}-doped optical fiber lasers around the start of the silica phonon absorption band.

The energy level diagram of Yb^{3+} ions is shown in Fig. 15.7a. The upper manifold $^3F_{5/2}$ and lower manifold $^3F_{7/2}$ have three and four sublevels, respectively, due to Stark splitting. Unlike for Er^{3+} ions, these manifolds give distinct transitions in absorption and emission spectra. For a pump wavelength of 976 nm, $a \rightarrow e$ is used, especially for cladding pumped optical fibers due to high pump absorption. For pump wavelengths around 915 nm, $a \rightarrow f$ is sometimes used to provide more tolerance in pump wavelength control; this also allows higher gain per unit length due to the much higher inversion possible when pumped around 915 nm. Two transitions—one at 1025 nm, $e \rightarrow b$,

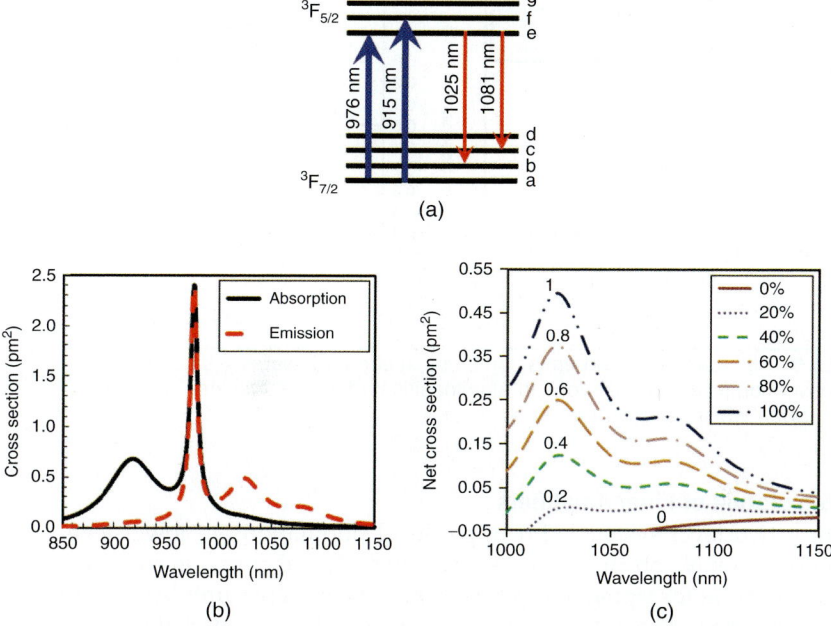

FIGURE 15.7 (a) Energy levels of Yb^{3+} ions. (b) Absorption and emission cross section of Yb^{3+} ions. (c) Net cross section for various inversion levels in aluminosilicate fibers.

and one at 1080 nm, $e \to c$—dominate the emission spectra. The absorption and emission cross section of Yb^{3+} in an aluminosilicate host are given in Fig. 15.7b. Net cross sections at various inversion levels are given in Fig. 15.7c. For comparison, similar curves for Er^{3+} and Tm^{3+} system are given in Figs. 15.8 and 15.9, respectively.

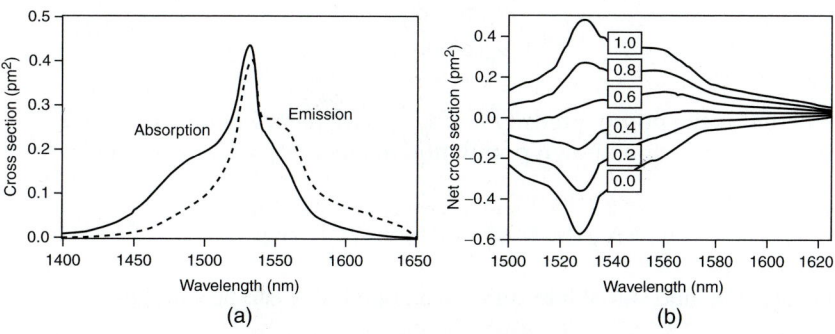

FIGURE 15.8 (a) Absorption and emission cross sections. (b) Net cross sections at various inversion levels of Er^{3+} ions in aluminosilicate fibers.

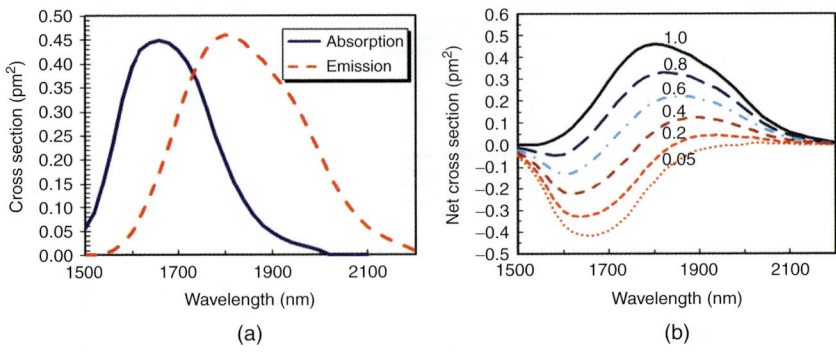

FIGURE 15.9 (a) Absorption and emission cross section. (b) Net cross sections at various inversion levels of Tm^{3+} ions in aluminosilicate fibers. (*Courtesy of Peter Moulton, Brian Walsh, and Nufern*)

Simulation of Fiber Amplifiers

In a two-level system with upper-level population N_2, lower-level population N_1, absorption cross section $\sigma_a^{(s,p)}$, and emission cross section $\sigma_e^{(s,p)}$ (s for signal and p for pump), the gain per unit length can be obtained from rate equations. Ignoring absorption at the signal wavelength,

$$g = \frac{g_0}{1+\dfrac{I_s}{I_{sat}}} \qquad (15.9)$$

where g_0 is small signal gain and I_s and I_{sat} are signal intensity and saturation intensity, respectively. This relation is plotted in Fig. 15.10, which shows a significant reduction of gain at higher signal intensities. For negligible amplified spontaneous emission (ASE), I_{sat} is given as

$$I_{sat} = \frac{\nu_s(\sigma_a^p + \sigma_e^p)}{\nu_p(\sigma_a^s + \sigma_e^s)} I_p \qquad (15.10)$$

where I_p is the pump intensity and $\nu_{(s,p)}$ is frequency (s for signal and p for pump). For a known population inversion $\eta = N_2/N$, with $N = N_1 + N_2$,

$$g_0 = N_2\sigma_e^s - N_1\sigma_a^s = N[\eta(\sigma_e^s + \sigma_a^s) - \sigma_a^s] \qquad (15.11)$$

Local pump and signal intensities in a doped fiber can be simulated as

$$\frac{dI_p(z)}{dz} = (N_2\sigma_e^p - N_1\sigma_a^p)I_p(z) \qquad (15.12a)$$

Introduction to Optical Fiber Lasers

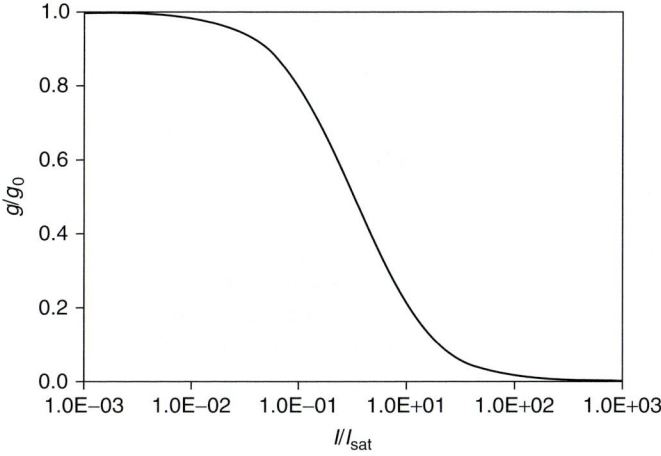

FIGURE 15.10 Normalized gain versus normalized signal intensity.

$$\frac{dI_s(z)}{dz} = (N_2 \sigma_e^s - N_1 \sigma_a^s) I_s(z) \qquad (15.12b)$$

The evolution of signal power, pump power, and population inversion is illustrated for a ytterbium fiber amplifier for copumping and counterpumping configurations in Fig. 15.11. For the copumping case, high inversion is achieved at the fiber's front end, which leads to a more rapid signal growth early on in the fiber. A higher inversion can be maintained at the output end of the fiber in a counterpumping configuration, leading to higher output powers. The counterpumping configuration also suffers less from nonlinear effects due to a reduction of the effective nonlinear length or effective amplifier length, as given later on by Eq. (15.17).

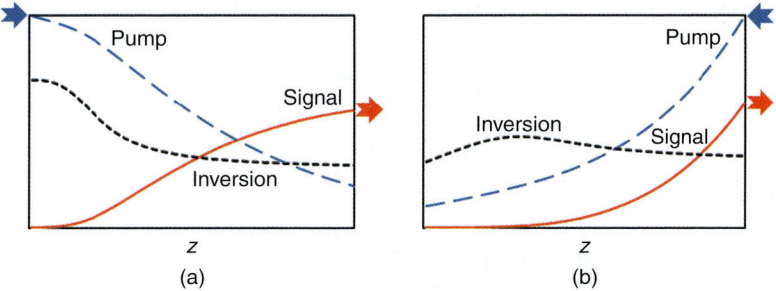

FIGURE 15.11 Signal power, pump power, and inversion in an ytterbium optical fiber amplifier for (*a*) copumping and (*b*) counterpumping configurations.

15.2.3 Power Scaling of Fiber Lasers

Nonlinear Limits

Stimulated Brillouin Scattering In stimulated Brillouin scattering (SBS), a pump photon is annihilated to produce a Stokes photon and an acoustic phonon through the electrostriction process.[17] This process is further stimulated by the presence of the generated Stokes photons and acoustic phonons in the fiber. In an optical fiber, momentum conservation requires that the scattered Stokes photons can only propagate in the opposite direction to the pump photons. Energy conservation and momentum conservation requires

$$v_a = v_p - v_s, \quad k_a = k_p - k_s \tag{15.13}$$

where v_p, v_s, and v_a are frequencies for pump, Stokes, and acoustic waves, respectively, and k_p, k_s, and k_a are wave vectors for pump, Stokes, and acoustic waves, respectively. The acoustic frequency v_a is around 11 GHz in silica fibers at 1.55 μm, and the acoustic velocity in silica is $v_a = 5.944$ km/s. The phonon lifetime is less than 10 ns, giving a phonon travel distance of less than 60 μm. The SBS threshold is a strong function of the spectral bandwidth of optical pulses:

$$P_{cr} = 21 \frac{A_{eff}}{g_B L_{eff}} \frac{\Delta v_s}{\Delta v_a} \tag{15.14}$$

where g_B is peak Brillouin gain, A_{eff} is the effective mode area, L_{eff} is the effective nonlinear length, Δv_a is the acoustic spectral bandwidth, and Δv_s is the signal spectral bandwidth. The peak Brillouin gain is typically $3-5 \times 10^{-11}$ m/W in a silica fiber and is almost independent of wavelength. The acoustic bandwidth Δv_a is typically around 10–1000 MHz. The effective mode area A_{eff} for a mode with a spatial electric field distribution of $E(x, y)$ is defined as

$$A_{eff} = \frac{\left(\iint_{-\infty}^{\infty} |E(x, y)|^2 \, dx \, dy \right)^2}{\iint_{-\infty}^{\infty} |E(x, y)|^4 \, dx \, dy} \tag{15.15}$$

The effective length for an amplifier with a maximum output power P_0 and power distribution $P(z) = P_0 f(z)$ is

$$L_{eff} = \int_0^L f(z) \, dz \tag{15.16}$$

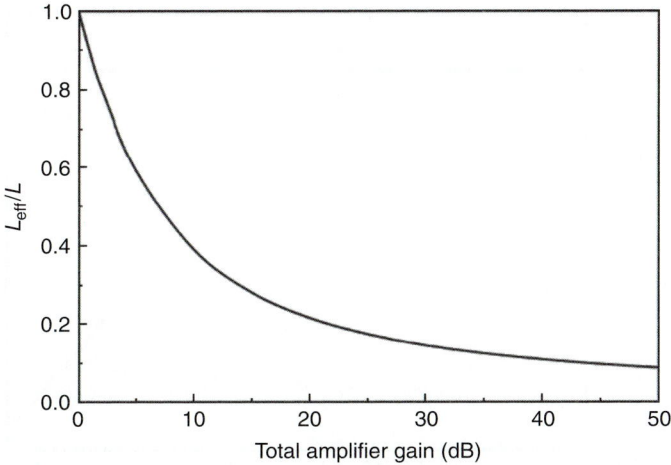

FIGURE 15.12 Effective amplifier length for a counterpumped fiber amplifier.

In a counterpumped fiber amplifier with length L where signal growth is more or less exponential as $f(z) = e^{gz}$, the effective nonlinear length can be estimated as

$$\frac{L_{\text{eff}}}{L} = \frac{1-e^{-gL}}{g} \approx \frac{4.343}{G(\text{dB})} \quad (15.17)$$

which is plotted in Fig. 15.12. $G(\text{dB})$ is the total gain expressed in decibels. The approximation applies when $G > 10$ dB. Because the total gain can be as high as 40–50 dB, the effective nonlinear length can be a small fraction of the total amplifier length.

Using the peak Brillouin gain $g_B = 5 \times 10^{-11}$ m/W and an acoustic spectral bandwidth $\Delta v_a = 100$ MHz for transform-limited gaussian pulses with $\Delta v_s \Delta \tau = 0.44$, where Δv_s and $\Delta \tau$ are the full-width, half-maximum (FWHM) pulse spectral and temporal widths, respectively, we obtain $P_{cr}{}^*L_{\text{eff}}/A_{\text{eff}} = 1.85 \times 10^{-3}/\Delta\tau$ MW/m, where pulse width is expressed in seconds. $P_{cr}{}^*L_{\text{eff}}/A_{\text{eff}}$ as a function of pulse duration is plotted for SBS in Fig. 15.13. For example, for $L_{\text{eff}} = 1$ m and $A_{\text{eff}} = 1000$ μm², $P_{cr} = 1.85 \times 10^{-6}/\Delta\tau$ W/m; the critical power is inversely proportional to the pulse width. For optical fibers with inhomogeneous transverse doping profile, which produces varying acoustic properties across the core, the overall Brillouin gain seen by the guided mode requires spectral and spatial integration over the core. This has been shown to lead to SBS suppression compared with a homogeneously doped fiber core, due to a reduction of overall peak Brillouin gain as a result of the spectrally varying Brillouin gain peak across the fiber core.[18,19]

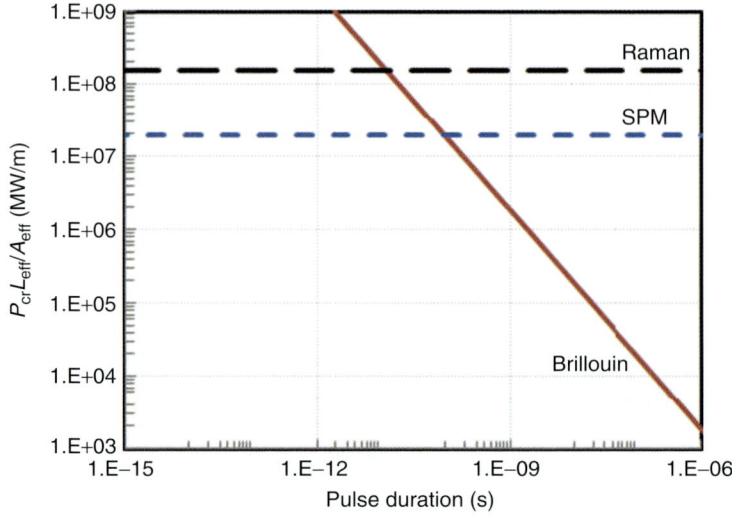

FIGURE 15.13 Nonlinear thresholds for stimulated Brillouin scattering (SBS), stimulated Raman scattering (SRS), and self-phase modulation (SPM).

Stimulated Raman Scattering In stimulated Raman scattering (SRS), a pump photon is annihilated, and a Stokes photon and an optical phonon are generated. This effect was discovered in 1928, well before the advent of lasers, by using focused sunlight and an optical grating to measure air-scattered light that is in the forward direction. Energy is conserved as in SBS; however, momentum is not conserved. Due to the amorphous nature of silica glass, a broad Stokes band of more than 40 THz and peaked at ~13 THz from the pump is typically seen in optical fibers. The peak Raman gain is around 1×10^{-13} m/W. Unlike SBS, the peak Raman gain g_R scales inversely with pump wavelength λ_p.

The SRS threshold is determined by

$$P_{cr} = 16 \frac{A_{eff}}{g_R L_{eff}} \tag{15.18}$$

which is plotted in Fig. 15.13 for $g_R = 1 \times 10^{-13}$ m/W at a wavelength of 1 μm, where $P_{cr} * L_{eff}/A_{eff} = 1.6 \times 10^8$ MW/m. For example, for $L_{eff} = 1$ m and $A_{eff} = 1000$ μm, we obtain $P_{cr} = 160$ kW; the critical power is independent of pulse width.

Self-Phase Modulation Self-phase modulation (SPM) arises from the nonlinear phase dependence on local intensity during pulse propagation. The time-dependent phase change leads to the generation of

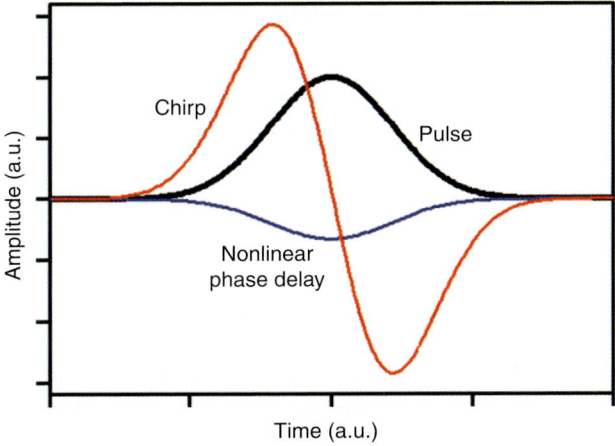

FIGURE 15.14   SPM and related chirp generated by an optical pulse.

new spectral components at the leading and trailing edge of an optical pulse. Continuous generation of red-shifted components happens at the leading edge of a pulse and blue-shifted components at the trailing edge of a pulse, resulting in a positive chirp (see Fig. 15.14 and Agrawal[17] for more theoretical details). This chirp is linear near the center of a gaussian pulse. The nonlinear phase ϕ_{NL} inside a fiber induced by an optical signal of power P_0 can be expressed as

$$\phi_{NL} = \gamma P_0 L_{eff} \quad (15.19)$$

The nonlinear coefficient γ is defined as

$$\gamma = \frac{2\pi n_2}{\lambda A_{eff}} \quad (15.20)$$

where λ is the vacuum wavelength and n_2 is a Kerr nonlinear coefficient that is typically around 2.6×10^{-20} m^2/W in silica fibers. The nonlinear phase induced by SPM is sometimes referred to as the B integral, which is particularly important in chirped pulse amplification (CPA) systems where an optical pulse is first stretched in the temporal domain to reduce its peak power,[20] amplified, and then compressed back to its original temporal shape. As a rule of thumb, to prevent SPM-induced pulse breakup in a CPA scheme, the B integral needs to be less than π, unless special measures are implemented to eliminate any small ripples in the temporal and spectral pulse profiles.

The nonlinear length L_{NL} is defined as the effective fiber length for which $\phi_{NL} = 1$:

$$L_{NL} = \frac{1}{\gamma P_0} \quad (15.21)$$

The nonlinear coefficient γ in silica fibers typically varies from 1×10^{-5} to 0.1 (mW)$^{-1}$. For a gaussian pulse, the maximum induced chirp is given in Agrawal[17] as

$$\delta\omega_{max} = 0.52 \Delta v_s \phi_{NL} \quad (15.22)$$

We obtain $P_{cr} * L_{eff} / A_{eff} = 1.92 \times 10^7$ MW/m for SPM. For example, for $L_{eff} = 1$ m and $A_{eff} = 1000$ μm, we have $P_{cr} = 19.2$ kW for SPM; the critical power is independent of pulse width for SPM.

Nonlinear Self-Focusing Although SPM can be viewed as the work of the Kerr nonlinearity in the temporal domain, nonlinear self-focusing arises from the Kerr nonlinearity operating in the spatial domain. For the fundamental mode of an optical fiber, the refractive index distribution of the waveguide confines the mode. In a bulk medium, there is no waveguide, and a propagating mode naturally diffracts. Self-focusing by Kerr nonlinearity reduces diffraction and eventually leads to a balance between the effects of nonlinear self-focusing and diffraction. Self-focusing collapses the beam and can produce optical damage when the effect of nonlinear self-focusing overcomes diffraction. A larger beam size equally reduces the nonlinearity coefficient and diffraction, leading to a critical power P_{cr}, not intensity, which determines the onset of nonlinear self-focusing.

In a waveguide, the effect of diffraction is already balanced by waveguiding. The presence of the Kerr nonlinearity below the critical power P_{cr} leads to a nonlinear guided stationary mode with a smaller mode size than the waveguide's original fundamental mode. Self-focusing takes place above P_{cr}[21]

$$P_{cr} = \frac{1.86}{k_0 n_0 n_2} \quad (15.23)$$

where $k_0 = 2\pi/\lambda$ is the vacuum wave number and n_0 is the fundamental mode index. The critical power for nonlinear self-focusing is estimated to be around 4–6 MW in optical fibers. The normalized spot sizes of nonlinear-guided stationary modes versus normalized power are plotted in Fig. 15.15a for various V values.[22] The evolution of the normalized spot size along an optical fiber at various normalized powers is illustrated in Fig. 15.15b.[22]

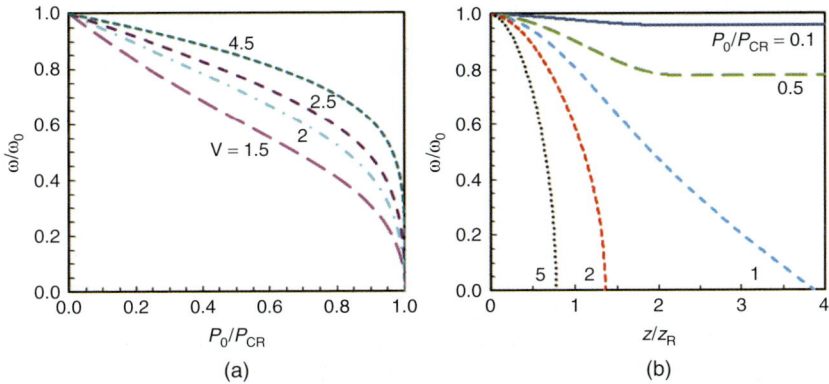

FIGURE 15.15 Effect of self-focusing in optical fibers—normalized mode spot size versus (a) normalized power at various V values and (b) z/z_R at various normalized powers, where z_R is Rayleigh range $z_R = \pi\omega_0^2/\lambda$ and ω_0 is $1/e^2$ beam radius.

Thermal Effects

Assuming only convection heat removal from the fiber surface[23] for a bare fiber with core radius ρ, outer radius R, homogenous thermal conductivity K_c, surface heat transfer coefficient h, heat density Q_0 generated only in the core, and ambient temperature T_s, the steady-state temperature distribution is given by

$$T(r) = T_s + \frac{Q_0\rho^2}{4K_c}\left[1 + 2\ln\left(\frac{R}{\rho}\right) + \frac{2K_c}{Rh} - \left(\frac{r}{\rho}\right)^2\right] \quad 0 \leq r \leq \rho$$

$$T(r) = T_s + \frac{Q_0\rho^2}{2K_c}\left[\frac{K_c}{Rh} - \ln\left(\frac{r}{R}\right)\right] \quad \rho \leq r \leq R \quad (15.24)$$

The temperature rise in the core is given by

$$\Delta T = T(0) - T_s = \frac{Q_0\rho^2}{4K_c}\left[1 + 2\ln\left(\frac{D}{\rho}\right) + \frac{2K_c}{Rh}\right] \quad (15.25)$$

For silica fibers, K_c is 1.38 W/(m·K), and the heat transfer coefficient h is both a function of fiber radius R and temperature difference between fiber surface and air $\Delta T = T(R) - T_s$. The heat transfer coefficient h increases with the temperature difference between fiber surface and air ΔT and can be estimated as $h = 0.2 K_a/R$, where $K_a = 2.54 \times 10^{-2}$ W/m²/KW/(m·K), at very small temperature difference ΔT.

In silica fibers, fast heat conduction at the timescale of less than 1 ms quickly leads to an almost uniform steady-state temperature distribution across the fiber; the fiber temperature rise ΔT is dominated by the heat transfer term [third term in the bracket in Eq. (15.25)]:

$$\Delta T \approx \frac{Q_0 \rho^2}{2Rh} \quad (15.26)$$

Equation (15.26) is plotted in Fig. 15.16 for various heat loads deposited in the core (W_h in units of watts per meter) for a fiber with diameter $2R = 250$ μm. Assuming 20 percent of the pump is turned into heat and a maximum fiber temperature of 200°C, this would limit the maximum convection heat removal to less than 70 W/m. Forced air or water cooling must be used for higher output powers. When the temperature rise is near 0, ΔT can be approximated as $\Delta T = 31.3\, W_h$, independent of fiber diameter.

Optical Damage

The energy threshold E_{th} for optical damage is given by

$$E_{th}(J) = 480\sqrt{\Delta \tau (ns)}\, A_{eff}(cm^2) \quad (15.27)$$

For a perfectly polished surface, the surface damage threshold is close to that of a bulk medium. In reality, however, the surface damage

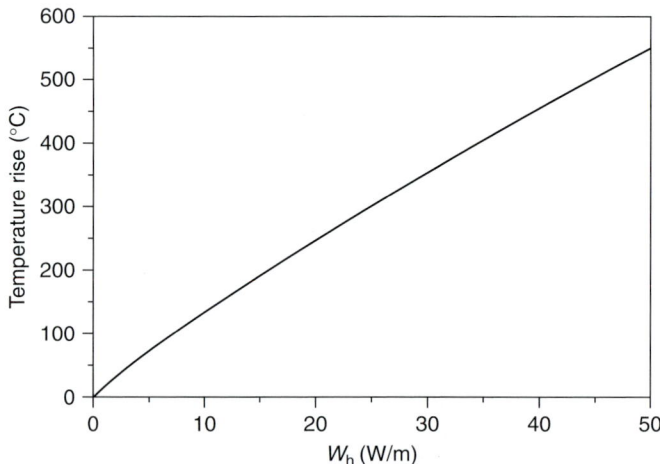

Figure 15.16 Temperature rise in a silica fiber with diameter $2R = 250$ μm ($h = 60.64 + 15.5\Delta T^{0.25}$) versus heat load deposited in the core of the fiber, assuming only convective heat removal from the fiber surface.

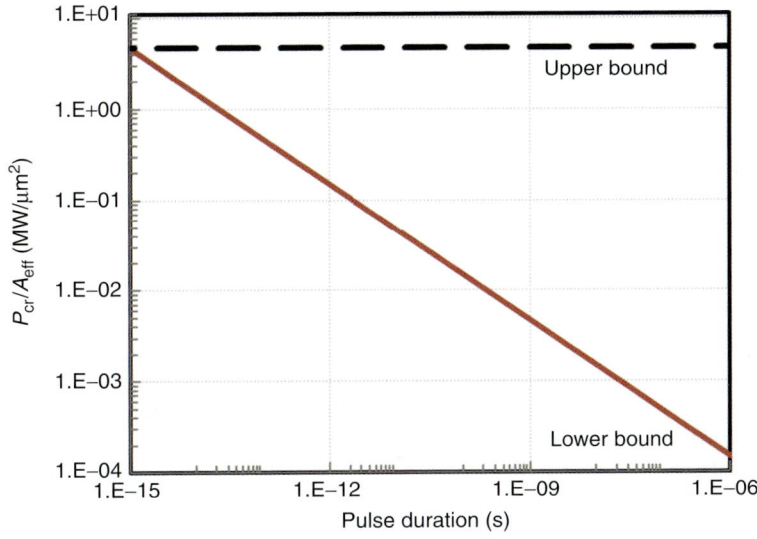

FIGURE 15.17 Damage threshold of silica optical fibers. Lower bound is from Eq. (15.27); upper bound is from Smith, Do, and Söderlund.[24]

threshold is much lower than that of the bulk. Recently, new observations indicated that the optical damage threshold is independent of pulse width for pulse durations above 50 ps and is close to 480 GW/cm², or 4.8 kW/µm²,[24] limited by electron avalanche. For shorter pulses (less than 50 ps), the avalanche process evolves more slowly than the pulse envelope, leading to a slightly higher threshold. Both limits are plotted in Fig. 15.17.

Energy Extraction

For amplifiers operating at repetition rates much greater than the inverse of upper-state lifetime—that is, for pulse repetition rates greater than 10 kHz—and for output pulse fluence much less than F_{sat}, where F_{sat} is the saturation fluence given by

$$F_{sat} = \frac{h\nu}{\sigma_e^s + \sigma_a^s} \tag{15.28}$$

we can generally ignore intrapulse saturation characteristics, and amplifier performance can be modeled using average values of signal and pump power. High repetition rates are typically used in micromachining applications. For counterpumped amplifiers, energy extraction efficiencies approaching 100 percent can be obtained even for high-gain amplifiers, which is one of the main advantages of fiber, compared with solid-state, amplifier technology. The main reason for the higher extraction efficiency in fiber amplifiers is the much larger available small signal gain enabled by the waveguide structure, which can exceed 50 dB. In fact, in high-power amplifiers, the small signal gain is limited by the onset of parasitic amplifier oscillations due to Rayleigh scattering, backscattering from optical fiber end faces, as well as external optical components. Therefore, it is often advisable to reduce the amplifier gain to values of 20 to 40 dB by providing a sufficiently strong seed signal. A ytterbium optical fiber amplifier operating at 100 W output power typically requires an injection power of 0.1 to 1.0 W for reasonably safe operation.

For operation at repetition rates less than 10 kHz or at a pulse fluence close to the saturation fluence, the time-dependent amplifier saturation characteristics must be considered. In a timeframe moving with the pulse, the amplified pulse and the evolution of the amplifier inversion during pulse propagation as a function of z can be obtained using

$$\frac{\partial I(z,t)}{\partial t} = \sigma_e^s \Delta N_{eff}(z,t) I_s(z,t) \tag{15.29a}$$

$$\frac{\partial \Delta N_{eff}(z,t)}{\partial t} = -\frac{1}{h\nu} \sigma_e^s \Delta N_{eff}(z,t) I_s(z,t) \tag{15.29b}$$

where $\Delta N_{eff} = N_2 - (\sigma_e^s/\sigma_a^s) N_1$ is the effective population inversion at position z as a function of time. Here we have neglected both ASE and the influence of signal-dependent changes in pump absorption. For operation at repetition rates less than 10 kHz, it is also beneficial to use pulsed pumping in order to minimize the buildup of ASE between pulses. Equations (15.29a) and (15.29b) can be readily extended to include pulsed pumping.[25] Analytic solutions to these equations for arbitrary pulse shapes can also be readily found.[26]

The total energy extracted from the amplifier is given by

$$E_{extr} = F_{sat} A_{eff} \ln(G_0/G_f) \tag{15.30}$$

where $G_{0,f}$ are the amplifier gain values before and after the pulse propagates through the amplifier and A_{eff} is the mode area. E_{extr} is maximized when the amplifier is totally saturated, such that $G_f = 1$.

An amplifier needs to be seeded at the saturation fluence to achieve significant energy extraction from an inverted system. $F_{sat} = 100$ J/cm², or 1 μJ/μm², for Yb³⁺-doped optical fibers at the gain peak of 1030 nm, where $\sigma_e^s = 0.2$ pm². Because peak pulse powers are limited by SRS, SPM, damage threshold, and nonlinear self-focusing, the minimum pulse duration to reach saturation fluence while staying below various nonlinear thresholds can be determined for known saturation fluence and effective mode area A_{eff}. Because the power

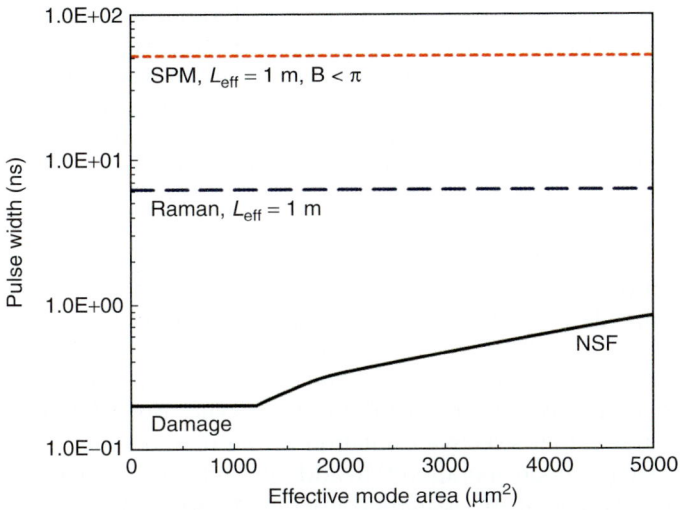

FIGURE 15.18 Lower bounds of pulse duration for meeting F_{sat} and various limits in Yb^{3+}-doped fibers. NSF: nonlinear self-focusing.

limits for SRS, SPM, and damage threshold all have the same dependence on A_{eff} as the saturation pulse energy does, A_{eff} cancels out in the calculation for minimum pulse duration. The minimum allowable pulse duration for reaching F_{sat} thus becomes independent of A_{eff} for SRS, SPS, and damage threshold (see Fig. 15.18). The optical damage limit, which is less than 480 GW/cm^2,[24] requires a pulse duration $\Delta\tau > 40$ ps; the Raman scattering limit, assuming $L_{eff} = 1$ m, requires $\Delta\tau > 6$ ns; and SPM, assuming $L_{eff} = 1$ m and $B < \pi$, requires $\Delta\tau > 50$ ns. The only exception is the nonlinear self-focusing (NSF) limit, which is independent of A_{eff}, and the minimum allowable pulse duration is consequently proportional to A_{eff}.

15.2.4 Fibers for High-Power Fiber Lasers

Conventional Large Mode Area Fibers

A conventional step-index fiber operates in the single-mode regime when normalized frequency $V < 2.405$. Single-mode fibers with core diameters in the 2 to 8 μm range are regularly made for telecommunication applications. For high-power fiber lasers, the large effective mode area of a large fiber core is desirable for reducing nonlinear and optical damage related effects. Single-mode operation is also required to maintain good mode quality. The easiest way to extend the core diameter, while remaining in the single-mode regime, is to minimize a fiber's NA. The consequences of this approach are weak guidance and high bend loss. As will become clear later, this is a general consequence of all large-core designs for fundamental mode operation. A small NA requires a small refractive index contrast between core and cladding. The minimum NA is also limited by how reliably the refractive index can be controlled in the CVD processes. An NA of 0.06 is considered to be the minimum NA possible with standard fiber fabrication processes and reasonably good controls.

Single-mode operation can be achieved in optical fibers that support a few modes, typically for $V < 12$. The fundamental mode can be excited by underfilling the optical fiber's acceptance cone, which is determined by the fiber NA. In an optical fiber with a few modes, the fundamental node's acceptance cone is much smaller than the maximum fiber acceptance cone (see Fig. 15.19a). Even though a few higher-order modes can propagate free of loss in such optical fibers, it has been shown that selective excitation of the fundamental mode can allow single-mode operation.[15] In addition, small coils can be used to introduce higher losses to higher-order modes to suppress their propagation.[16] Scaling up the core diameter with this approach is mainly limited by the increasing number of guided modes, which makes it increasingly difficult to achieve fundamental mode launching and to introduce differential modal losses by coiling. Commercial fibers with 50-μm core diameters are available with conventional designs for ytterbium-doped fiber amplifiers (see Fig. 15.19b), with an effective mode area of ~1000 μm². With core-pumped designs, mode areas up to ~1800 μm² are possible for erbium-doped fiber amplifiers.[27] Experimental results with 80-μm core diameters have been demonstrated for ytterbium-doped fiber amplifiers in cladding-pumped configurations.[28] However, most of the fiber was coiled to a very small diameter with a much smaller effective mode area than that in a straight fiber (more discussion on mode area compression in bent fibers later). Nevertheless, a straight section at the output end can be used for its large effective mode area, assuming all higher-order

FIGURE 15.19 (*a*) Excitation of a fundamental mode in a fiber that supports few modes. (*b*) Cross section of a commercial double-clad polarization-maintaining (PM) optical fiber of conventional design made by Nufern.

modes can be stripped off by the small coil diameters without significant intermodal coupling.

Many lasers use linearly polarized beams. In an optical fiber with perfect circular symmetry, the state of polarization can vary with propagation due to the birefringence caused by random bends and stress and phase-matched coupling between polarization modes. As a result, uncertainties in the orientation of polarization, as well as in the state of polarization, can arise. These uncertainties can be mitigated by PM fiber designs. Polarization-maintaining large-core optical fibers are fabricated by placing a pair of stress-inducing elements—typically, boron-doped silica, which has a large thermal expansion coefficient mismatch compared with that of silica—on either side of the core to create a polarization-dependent refractive index (see Fig. 15.19b). Normally, the two polarization modes of a circular fiber are degenerate; they become non-degenerate with effective mode indices n_x and n_y, respectively, in a PM fiber, providing birefringence $\Delta n = n_x - n_y$. The beat length $L_B = \lambda/(n_x - n_y)$ is defined as the period of polarization mode beating. A cross section of a commercial double-clad PM fiber is shown in Fig. 15.19b. The octagonal pump guide is designed to minimize skew rays, which do not pass the doped core. The fibers are usually coated with a low-index polymer to provide a pump NA as high as ~0.45.

Photonic Crystal Fibers

Photonic crystal fibers (PCFs) were first demonstrated in the late 1990s.[29,30] Its potential for achieving single-mode operation with large cores was realized very early on.[31] A PCF is drawn from a hexagonal stack of capillaries, with typically one to seven capillaries replaced by rods in the center. Pressurization of the airholes is typically used to keep the holes from collapsing from surface tension during drawing. The center rods form the core. The composite cladding material of glass and air makes it easy to achieve a very low refractive index contrast between the core and the composite cladding, consequently providing much better control at achieving fibers with low NAs.[32] Weak guidance, however, makes PCFs very bend sensitive, especially at large core diameters. In practice, they need to be kept straight beyond core diameters of 40 μm. Fibers with these large core diameters are made into rods with outer diameters anywhere from 1 to 2 mm and are referred to as *fiber rods*.[33,34] Fiber rods with core diameters as large as 100 μm and an effective mode area of ~4500 μm² have been demonstrated.[35] The ytterbium-doped core glass for large core fibers requires good uniformity and exceptional average refractive index control. For PCFs, this control is achieved by stacking together ytterbium-doped glass, which has a high refractive index, and fluorine-doped glass, which has a lower refractive index, in the correct area ratio to achieve a high level of average refractive index control. The process can be repeated many times in theory to achieve much improved refractive index uniformity.

It was shown early on that the composite cladding's effective refractive index can be represented by the effective mode index of the fundamental mode of an infinite medium, which consists of the cladding's basic unit cell extending to infinity in all directions.[30] This mode is referred to as *fundamental space-filling mode* with effective index n_{FSM}, which can be obtained through a mode solver. It took a few years until it was determined that the core radius is better approximated by $\Lambda/3^{1/2}$,[36] where Λ is the center-to-center spacing of the holes and d is usually used to represent the hole diameter. The unique scaling properties of the Helmholtz eigenvalue equation dictate that the modes of a waveguide are not fundamentally changed in terms of modal properties when all length-related properties (x, y, and λ) are scaled by the same amount. This property is useful because it means that such waveguides can be analyzed efficiently using normalized parameters. It is common to plot the solutions in normalized length scales of d/Λ and λ/Λ. Single-mode and multimode regimes of PCFs with one missing hole are plotted in Fig. 15.20 for two designs—one with air in the holes[37,38] and one with a glass with a refractive index $n_F = 1.2 \times 10^{-3}$ lower than that of the background glass, which has $n_B = 1.444$.[39] An effective V value can be obtained for a PCF using

$$V_{eff} = \frac{2\pi}{\lambda} \frac{\Lambda}{\sqrt{3}} \sqrt{n_{co}^2 - n_{SFM}^2} \qquad (15.31)$$

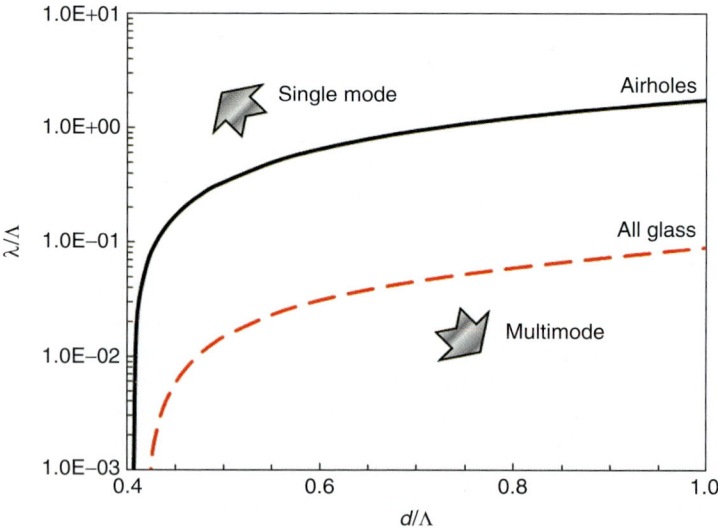

FIGURE 15.20 Single-mode and multimode regimes of photonic crystal fibers (PCFs) with airholes and all glass designs, with $n_B = 1.444$ and $n_F = 1.4428$.

Introduction to Optical Fiber Lasers

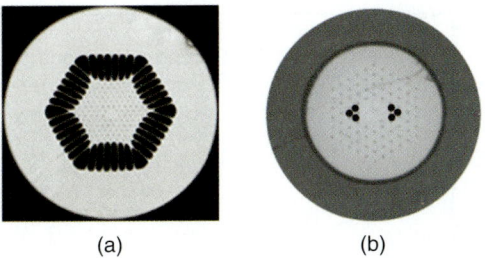

FIGURE 15.21 (*a*) Double-clad PCF with air-clad pump guide and (*b*) related PM design.

The pump cladding in PCFs is typically made with a layer of airholes with very thin glass webs (see Fig. 15.21). This structure has been demonstrated to be capable of providing an effective pump NA of ~1. In practice, the glass webs must be made thick enough to allow cleaving of the end face. This typically limits pump NAs to ~0.6, which is larger than what is possible with low-refractive index polymers. As a result, lower-brightness pump diodes can be used, as will be discussed later. The use of airhole pump guides also enables designs with much larger fiber diameters, while maintaining a relatively smaller pump guide. Large fiber diameters help keep the PCFs straight, thereby minimizing effects of macrobending and microbending. Microbending can lead to mode coupling in optical fibers that support more than one mode. Removal of the polymers from the pump path also improves the fiber's long-term reliability, especially for high-average-power lasers. Examples of commercial double-clad PCFs are shown in Fig. 15.21.

Leakage Channel Fibers

Leakage channel fibers (LCFs) are designed with built-in higher loss for higher-order modes to suppress their propagation in order to extend single-mode operation beyond conventional designs and PCFs.[40] The fully enclosed core and cladding boundary of a conventional optical fiber ensures total internal reflection everywhere once a mode is guided; consequently, in theory, all guided modes can propagate without loss. A series of channels are introduced in the immediate cladding material in an LCF to make the waveguide leaky in order to engineer the required differential modal loss. LCFs can be designed to be made entirely of glass, which significantly improves their ease of fabrication and use.[41] Fundamental and second-order mode losses are plotted in Fig. 15.22*a* for an LCF with 50-μm core diameter. Higher-order mode losses are much higher and are not shown here. A very high differential mode loss can be achieved. The design space for various core diameters is illustrated in Fig. 15.22*b* for fundamental

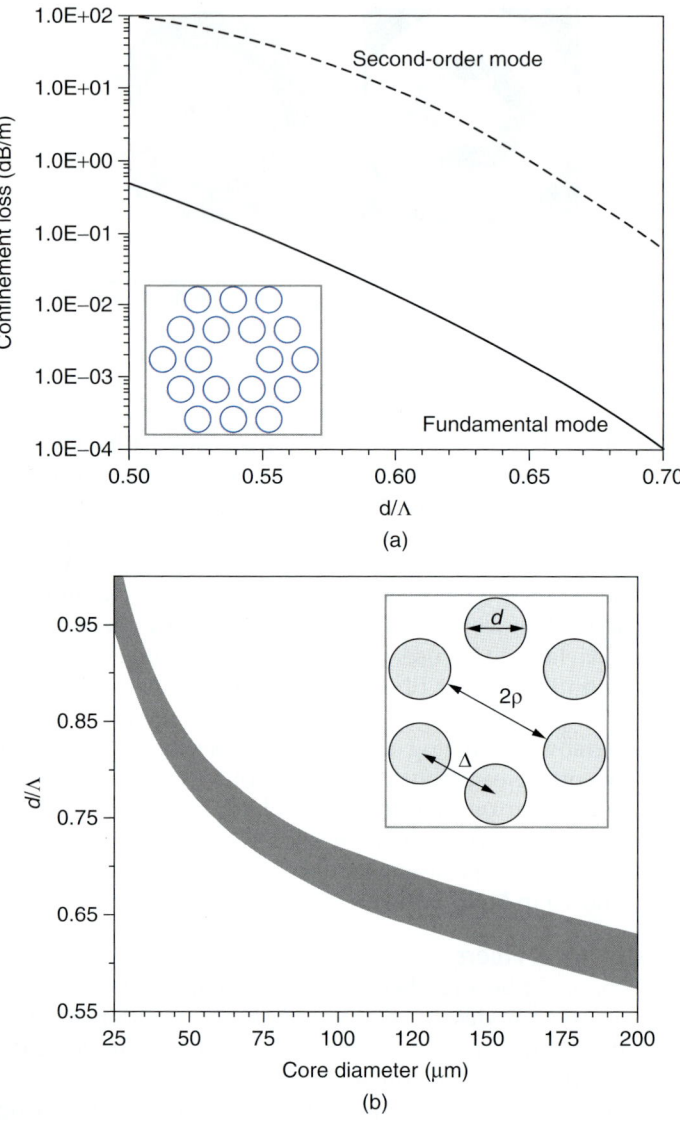

FIGURE 15.22 (*a*) Confinement losses for the fundamental and second-order mode in leakage channel fibers (LCFs) with a 50-μm core diameter. (*b*) Design space of LCFs constrained by a fundamental mode loss of less than 0.1 dB/m and second-order mode loss larger than 1 dB/m. (Designs are illustrated in the insets. $n_B = 1.444$, $n_F = 1.4428$, and $\lambda = 1$ μm.)

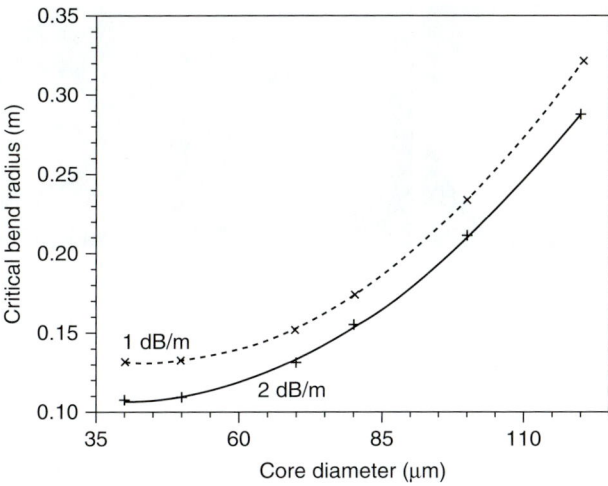

FIGURE 15.23 Critical bend radius for 1 dB/m and 2 dB/m of fundamental mode loss for LCFs with $n_B = 1.444$, $n_F = 1.4428$, $d/\Lambda = 0.9$, and $\lambda = 1$ μm. The simulation was performed by finite element method (FEM).

mode losses less than 0.1 dB/m and second-order mode losses larger than 1 dB/m, thus showing that core diameters as large as 200 μm are potentially possible. LCFs with core diameters up to 170 μm have been demonstrated.[42] For the simulations in Fig. 15.22, the background glass is silica with refractive index $n_B = 1.444$, and the holes are filled with a glass with refractive index $n_F = 1.4428$ and with wavelength of operation at 1 μm.

LCFs can be bent even at very large core diameters. The critical bend radii are shown in Fig. 15.23 at various core diameters for LCFs with $n_B = 1.444$, $n_F = 1.4428$, $d/\Lambda = 0.9$, and $\lambda = 1$ μm. Various types of LCFs have been fabricated, some of which are shown in Fig. 15.24.[43]

Other Advanced Large-Core Fiber Designs

One approach to increasing core area relies on the addition of one or more additional cores, which are designed to be resonant with higher-order modes of the primary core. The design concept is based on the use of mode coupling to channel higher-order mode power away from the primary core. This concept is equivalent to additional higher-order mode losses if the coupled power is dissipated in the second core before it gets coupled back to the primary core, because all coupling processes are bidirectional. Two implementations of this scheme have been proposed. In one case, the fiber is uniform along its length.[44] In a second case, an appropriately designed circular core is placed next to the primary core at the center, and the fiber is spun during the

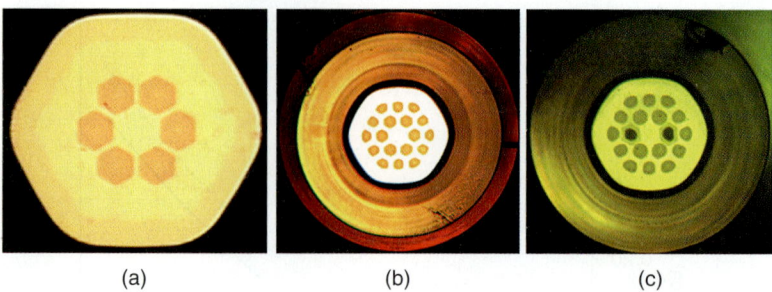

FIGURE 15.24 Cross section of double-clad LCFs with (a) low-index polymer pump guide, (b) fluorine-doped silica pump guide, and (c) fluorine-doped silica pump guide and a PM core. All cores are ytterbium doped.

drawing process. As a result, the second core follows a helical pattern around the primary core, which is sometimes referred to as *chirally coupled core fibers*.[45,46] One major issue with such designs is that all mode coupling requires modal spatial overlap as well as phase matching. As the primary core gets larger and becomes more multimode, all modes are increasingly confined to the primary core, which makes it difficult for coupling to take place over practical length scales due to diminished spatial modal overlap. A second issue is that the mode density in the modal index space gets increasingly larger as the core gets more multimode (see Fig. 15.2), which makes it hard to phase match only the unwanted modes without affecting the nearby desired mode. Optical fibers with overall performance that is significantly better than conventional fibers have not yet been demonstrated with these approaches.

A second approach is based on the propagation of a higher-order mode in a multimode fiber. It has been shown that a higher-order mode can be excited using a mode converter based on a long-period grating that is fabricated in a fiber designed to have a central single-mode core surrounded by a much larger multimode core. The long-period grating couples the fundamental mode input guided in the small single-mode core to a co-propagating higher-order mode guided mostly in the larger multimode core. It has been shown that higher-order modes can propagate with low sensitivity to external perturbations. Effective mode areas as high as 3200 to 2100 µm² have been demonstrated with LP_{04} to LP_{07} modes, respectively.[47] It was found that higher-order modes propagate more robustly than lower ones in the same fiber. (Note that a higher-order mode also has a smaller effective mode area than lower-order modes in the same fiber.) A second mode converter is required at the output end to convert the higher-order mode back to the fundamental mode. Recently, active fibers have been demonstrated with this approach.[48] ASE limits these fibers to a low gain of just more than

10 dB, though higher gain values can be obtained by propagating the pump in the same higher-order mode as the signal. The effective mode area of higher-order mode fibers has shown to be of smaller compression in bent fibers.

A third approach to increasing core area is an antiguiding optical fiber, which uses an active core doped with rare earth ions that have lower index than that of the cladding.[49] In this case, there are only a few very leaky modes guided in the core, with higher-order modes suffering higher losses. The concept is not fundamentally different from many other approaches addressed earlier, except that the fundamental mode loss can be very high in an antiguiding fiber and becomes only tolerable for core diameters of a few hundred micrometers (loss scales with the inverse of the cube of the core diameter). Lasers with fundamental mode output have been demonstrated using sufficient gain to overcome the excessive loss of the fundamental mode, while keeping the higher-order modes below threshold.[50] It is worth noting that these fibers are not likely to guide light through the real part of the refractive index being modified in the presence of gain—that is, the imaginary part of the refractive index—through the Kramers-Kronig relation. In practice, the increase of refractive index (due to gain) in rare-earth-doped glass is typically well below what is sufficient to modify waveguide behavior. Gain in these fibers is likely mostly used to compensate waveguide losses for the lower-order modes.

At large core diameters, the increasing size of the fundamental mode makes it behave more and more like a free space beam. One way of looking at how a guided mode navigates a bend is by imagining it continuously transforming itself into the local bent waveguide mode adiabatically while propagating around the curve. As the mode gets larger, it takes longer distances for the mode to make this transition adiabatically; consequently, larger coil diameters must be used to minimize these transition losses (see Fig. 15.23). The bent fundamental modes also have a much reduced effective mode area. Effective mode areas of bent fibers were first simulated by Fini et al. and are plotted in Fig. 15.25, which shows the increasing effective mode area compression at large core diameters.[51,52,53] This compression is a result of an equivalent linear refractive index gradient introduced by bending in an optical fiber. A summary of progress of large core fiber development in recent years is illustrated in Fig. 15.26.

Pumping of Optical Fiber Lasers and Beam Quality

Similar to high-power solid-state lasers, high-power optical fiber lasers constructed with large core fibers can also have limited beam quality and suffer from depolarization. Beam-quality limitations may arise due to the presence of higher-order modes in the fiber output, thermal stresses from fiber holders, or thermal lensing and thermal birefringence in external components, such as collimation lenses, polarizers, diffraction gratings, prisms, and isolators. At power levels

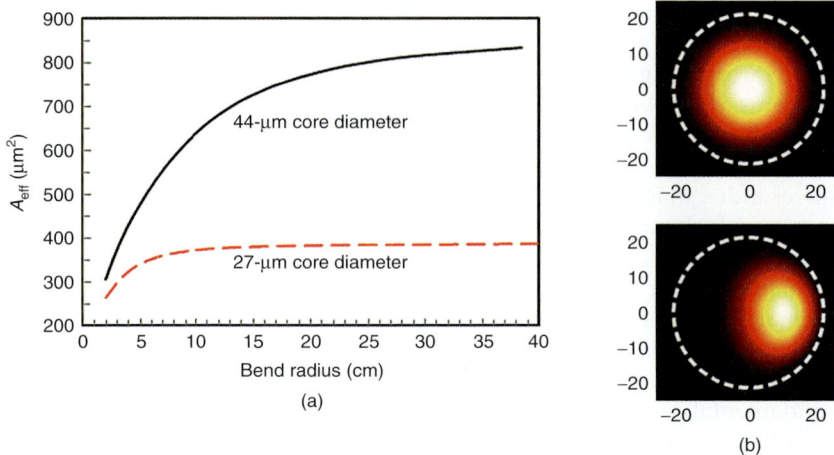

FIGURE 15.25 (a) Effective mode area compression in bent fibers. (b) Top: Mode in straight fiber with $A_{eff} = 862$ μm²; Bottom: Mode in bent fiber with a bend radius of 10 cm and $A_{eff} = 627$ μm² (core diameter is 44 μm). (*Courtesy of John Fini*)

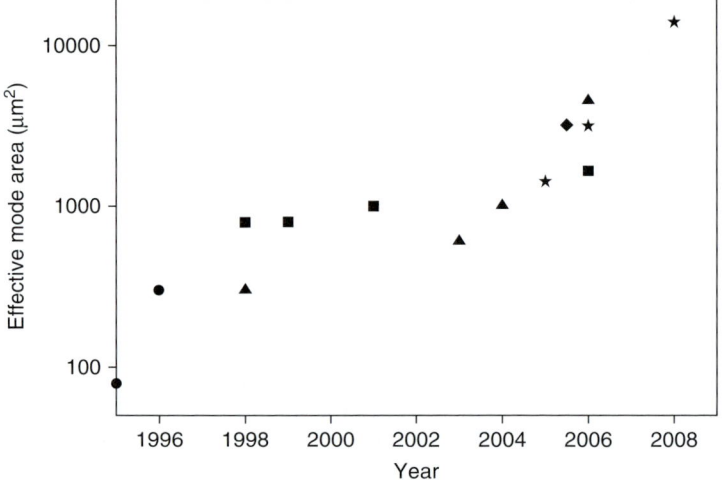

FIGURE 15.26 Development of fiber mode areas of a variety of diffraction-limited single-core optical fibers as a function of time. Circles: single-mode step-index fibers; squares: multimode step-index fibers; triangles: photonic crystal fibers and rods; diamonds: higher-order mode fiber; stars: leakage channel fibers.

greater than 50 W, it is typically advisable to use silica optics to minimize thermal beam distortions. In addition, water-cooled fiber holders and isolators may also be required at high powers.

Beam quality is usually measured by the M^2 parameter, which characterizes how much a beam's diffraction angle is increased relative to that of a gaussian beam. The diffraction half-angle θ_d can be written as

$$\theta_d = M^2 \left(\frac{\lambda}{\pi \omega} \right) \quad (15.32)$$

where ω is the $1/e^2$ intensity radius of a gaussian beam. Some manufacturers also use the beam parameter product (BPP), which is defined as

$$\text{BPP}(\text{mm} \times \text{mrad}) = \omega \times \theta_d \quad (15.33)$$

For a diffraction-limited beam, $M^2 = 1$ and BPP = 0.33 mm × mrad at 1 µm. Sometimes BPP is also defined as BPP = $2\omega \times 2\theta_d$, corresponding to a diffraction-limited BPP of 1.3 mm × mrad at 1 µm.

High-power fiber lasers are generally pumped with high-brightness fiber-coupled diodes, where brightness is defined as

$$B = \frac{4P}{(\pi d NA)^2} \quad (15.34)$$

where P is total available pump power, NA is the fiber NA, and d is beam diameter. Another parameter used is the etendue of a pump source, defined as $E = B/P$. Current state-of-the-art fiber-coupled broad-area diodes have a brightness in the range of 1.8–6.0 MW/cm², where a brightness of 1.5 MW/cm² corresponds to a pump power of 10 W in a 100-µm-diameter fiber with NA = 0.15. Efforts are ongoing to increase the brightness to 10 MW/cm², but such high values require simultaneous coupling of up to 10 broad-area diodes into a single 100-µm-diameter fiber, which can be accomplished with sophisticated beam-shaping techniques. Even higher brightness values can be obtained with beam shaping of diode arrays, where currently up to 400 W can be obtained from a 200-µm-diameter fiber with NA = 0.22, corresponding to a brightness of 8.4 MW/cm². Efforts are also ongoing to reach 1 kW from a 200-µm-diameter fiber. Even higher power levels can be obtained by implementing wavelength multiplexing. With present state-of-the-art diode lasers, several kilowatts of power can be coupled into a 600-µm-diameter gain fiber. Apart from enabling higher-power fiber laser operation, research in high-brightness diode lasers is motivated by machining applications that can use the direct output of semiconductor pump lasers without any intermediate fiber laser stage.

15.3 Optical Fiber Lasers

15.3.1 Continuous Wave Fiber Lasers

Background
Many configurations of fiber lasers have been studied since the mid-1980s. Most are core-pumped lasers, with little possibility of scaling up to higher power levels. Many of the early higher-power fiber

lasers were pumped by more powerful gas and solid-state lasers, with up to a few tens of watts of pump power, which is much more powerful than the subwatt level available from a single-mode pump diode. These optical fiber lasers are, however, impractical for commercial use. Cladding pumping by multimode pump diodes, first demonstrated by Po et al.[13] in 1988, was a major milestone that eventually led to high-power single-mode lasers using the much higher pump powers available from multimode pump diodes. We have focused our discussions on high-power optical fiber lasers based on cladding pumping configurations.

Typical cavity designs used commercially for higher-power applications are based on Fabry-Pérot cavity designs with fiber Bragg gratings (FBG) as reflectors. Photosensitivity in germanium-doped silica fibers was first discovered by Hill et al.[54] in 1978. It was the work by Meltz et al.[55] in 1989 that led to intensive development of FBG. The significance of the work by Meltz et al. is the first demonstration of a Bragg grating in an optical fiber core for wavelengths besides the wavelength of the writing laser by two interfering ultraviolet (UV) beams (see Fig. 15.27). The pitch of an FBG can be varied by adjusting the angle between the interfering beams. Even though the initial work was motivated by fiber sensors for temperature and strain measurements, FBGs received immense immediate interest mainly due to their application as filters in wavelength-division-multiplexing (WDM) systems for telecommunications. The application in optical fiber lasers as reflectors was also quite straightforward. Chirped FBGs are also used for intracavity dispersion management in mode-locked lasers, as will be discussed later. FBGs were, in fact, a critical ingredient in enabling the development and eventual commercial success of fiber lasers (see Ref. 56 for a review).

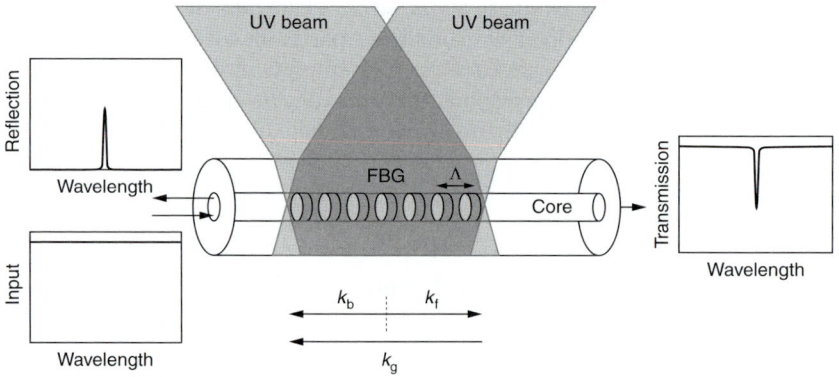

FIGURE 15.27 A fiber Bragg gratings (FBG) writing process.

Ultraviolet photosensitivity in the germanium-doped core of an optical fiber originates from germanium oxygen-deficient centers with well-defined absorption of around 240 nm, accessible by KrF excimer lasers, frequency-doubled argon ion lasers, or frequency-quadrupled Nd:YAG lasers. In addition to the depletion of the germanium oxygen-deficient centers upon UV exposure, there is also evidence of stress relaxation and sometimes compaction. All these effects are believed to contribute to the observed refractive index change. It was eventually found that hydrogen loading in a high-pressure cell near room temperature prior to UV exposure leads to a significant improvement of photosensitivity.[57] This discovery has enabled FBG writing in some germanium-free glasses. Large amounts of hydroxyls are found in the glass after UV exposure, and the significant composition modification is believed, in this case, to play a role in the refractive index change. FBGs are found to be highly stable at room temperature, though they can be erased at elevated temperatures.[58] Most FBGs will be erased at temperatures above 400°C, and FBGs with improved temperature stability can be made using special writing processes. One unique aspect that makes FBGs possible is that silica glass in the cladding is not photosensitive, which allows the UV beam to reach the core glass without being absorbed.

In an FBG, due to momentum conservation, we have

$$k_g = k_f - k_b \tag{15.35}$$

This leads to a relation between peak wavelength λ_0 and grating pitch Λ.

$$\lambda_0 = 2n_0 \Lambda \tag{15.36}$$

where n_0 is effective mode index. For a uniform grating, reflectivity can be analytically determined:

$$R = \frac{\kappa^2 \sinh^2(\kappa L)}{\delta^2 \sinh^2(\kappa L) + (\kappa^2 - \delta^2)\cosh^2(\kappa L)} \tag{15.37}$$

where L is grating length. The coupling coefficient is defined as

$$\kappa = \frac{\pi \Delta n_g}{\lambda_0} \eta_g \tag{15.38}$$

where η_g is the spatial modal overlap with the grating. It is also the proportion of modal power in the core if the FBG is written uniformly over the core; $2\Delta n_g$ is the peak-to-peak refractive index modulation.

Fiber Lasers

The detuning factor δ is given by $\delta = 2\pi n_0(1/\lambda - 1/\lambda_0)$. The peak reflectivity is given by

$$R = \tanh^2(\kappa L) \tag{15.39}$$

The FWHM bandwidth of a FBG is given by

$$\Delta\lambda = \lambda_0 \sqrt{\left(\frac{\Delta n_g}{2n_0}\right)^2 + \left(\frac{1}{N}\right)^2} \tag{15.40}$$

For weak gratings, FBG bandwidth is inversely proportional to the number of grating periods, $N = L/\Lambda$. For strong FBGs, it becomes proportional to the relative index modulation and is independent of FBG length. Reflections from weak FBGs of 10 percent peak reflection at various lengths are plotted in Fig. 15.28a, while those from FBGs of 10 mm length at various coupling coefficients κ are plotted in Fig. 15.28b. Frequency detuning from the peak frequency is used as the horizontal axis.

In a fiber laser with an output mirror of transmission T_0 and total round cavity loss α_t, the laser output can be written as[59]

$$P_{out} = \frac{T_o}{\alpha_t} \frac{f_s}{f_p} (P_{abs} - P_{th}) \tag{15.41}$$

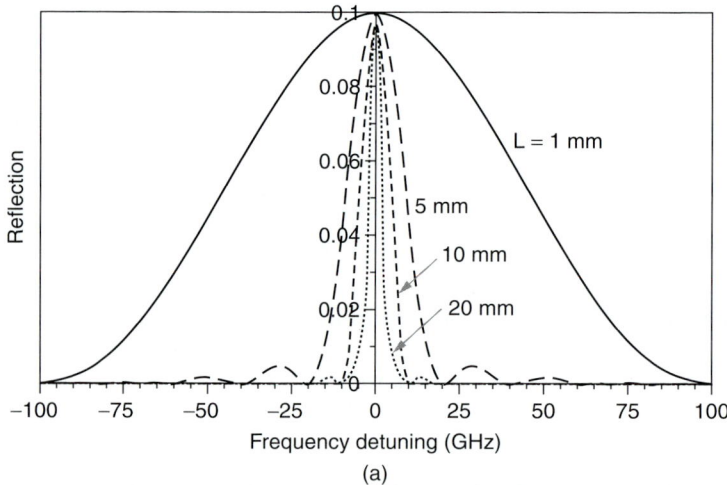

(a)

Figure 15.28 FBG reflection (a) at various lengths and 10% reflection and (b) at various κ and $L = 10$ mm.

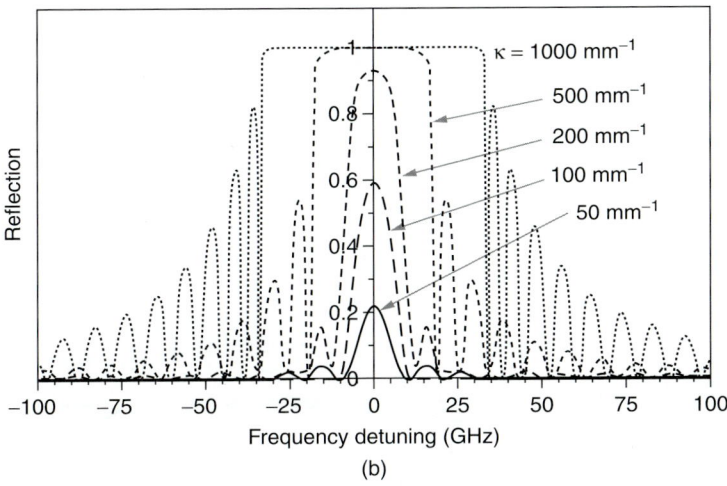

FIGURE 15.28 (*Continued*)

where f_s and f_p are signal and pump frequencies, respectively, and where P_{abs} and P_{th} are absorbed pump power and pump threshold, respectively. The slope efficiency η_s

$$\eta_s = \frac{T_0}{\alpha_t}\frac{f_s}{f_p} \qquad (15.42)$$

increases with output mirror transmission T_0. In cases of typical optical fiber lasers with FBGs as reflectors, in which excessive intracavity loss is low—that is, $\alpha_t \approx T_0$—the slope efficiency becomes $\eta_s = f_s/f_p$, which is the quantum defect, and it is independent of output mirror transmission. In Yb^{3+}-doped optical fiber lasers, a slope efficiency of ~80 percent is regularly achieved, making them extremely attractive for power scaling to high average powers.

One drawback of cladding pumping configurations is the much reduced pump overlap with the rare-earth-doped core. Consequently, a much longer fiber must be used for efficient pump absorption as compared with that for a direct core-pumped configuration. Most nonlinear effects scale with fiber length, which imposes lower nonlinear thresholds. This is especially detrimental for power scaling lasers with high peak powers limited by SPM or SRS and lasers with narrow spectral width limited by SBS. Because ytterbium can be doped in relatively higher concentrations in silica glass compared with other rare earth ions, it is a favorite candidate for cladding pumped optical fiber lasers.

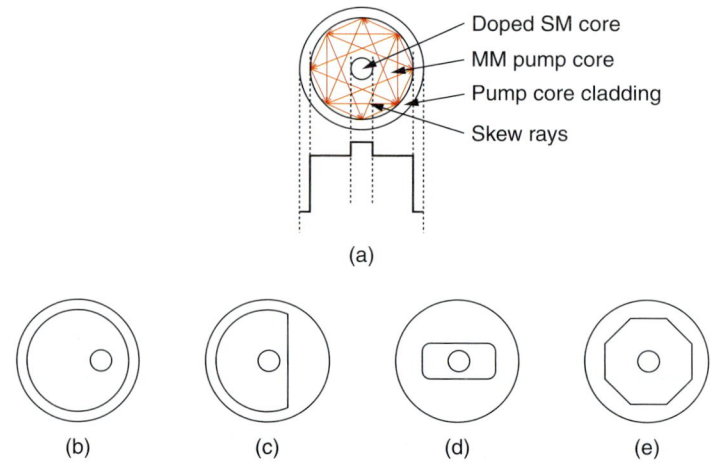

FIGURE 15.29 A double-clad fiber with (a) a circular pump guide, (b) an offset core, (c) a D-shaped pump guide, (d) a square pump guide and (e) an octagon pump guide. SM: single mode; MM: multimode.

Earlier double-clad fibers for cladding pumping used circular pump guides. The existence of skew rays in circular multimode pump guides—that is, rays propagating in helical patterns around the doped core—can substantially affect efficient pump absorption. Coiling, especially coiling with varying curvatures along the fiber, was found to be very useful for improving pump absorption. Eventually, it was found that pump absorption could be substantially improved by designs of offset signal core and, even more efficiently, by noncircular pump guides. Some examples of these are illustrated in Fig. 15.29. It was later found that the presence of low-index inclusions in the pump guides, such as stress elements, could also improve pump absorption.

Pump Coupling Schemes

The easiest way to inject multimode pump light into the pump core of a double-clad fiber is through the fiber end of the end-pumping scheme (see Fig. 15.30a). A dichroic mirror can be used between the two lenses to separate pump from signal light. Although an end-pumping scheme is easy to implement, the maximum pump injection points are limited to two. The thermal load from uncoupled pump light must be removed from the fiber ends, which can be a significant issue in very high power lasers. An alternative path is through the side-pumping scheme, illustrated in Fig. 15.30b,[60] in which pump light is reflected off a coated V groove machined into the side of a double-clad fiber. This scheme can be implemented multiple times along a fiber and, in theory, allows multiple injection points. Because the V groove also disrupts the propagating pump light, a laser must be carefully designed for multiple pump injections. No high-power

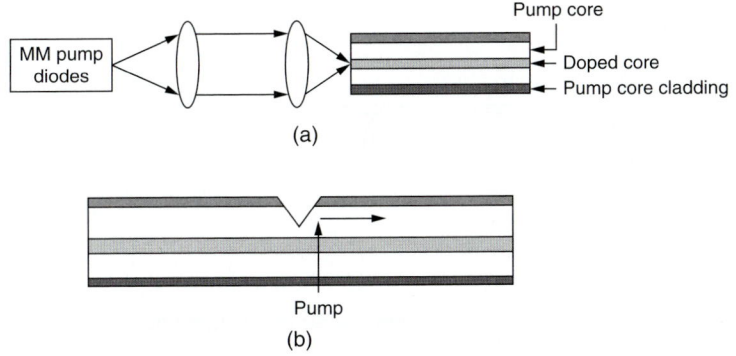

FIGURE 15.30 (a) End-pump scheme and (b) side-pumping scheme.

laser has yet been demonstrated with this scheme. Most commercial high-power lasers use fiber-based combiners, which enable multiple pump injections (see Fig. 15.31).

Power Scaling of Continuous Wave Fiber Lasers

A typical design of a CW fiber laser is shown in Fig. 15.32. The high reflector (HR) and output coupler (OC) are implemented with FBGs. For most high-power CW fiber lasers, optical nonlinear effects are not

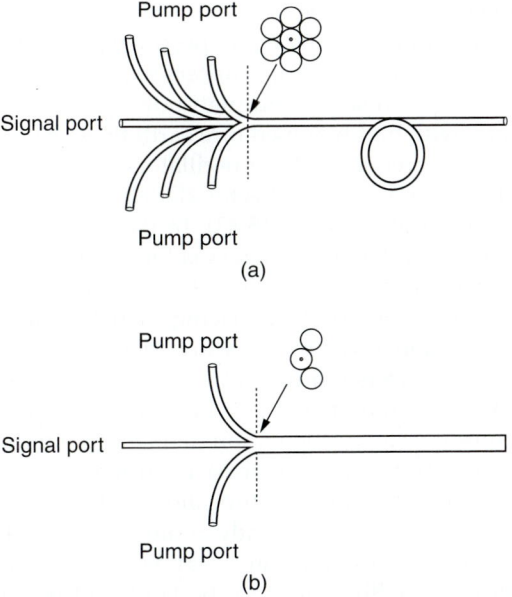

FIGURE 15.31 (a) Pump combiner with six pump ports and a signal port, and (b) GT-wave made by Southampton Photonics with two pump ports and a signal port.

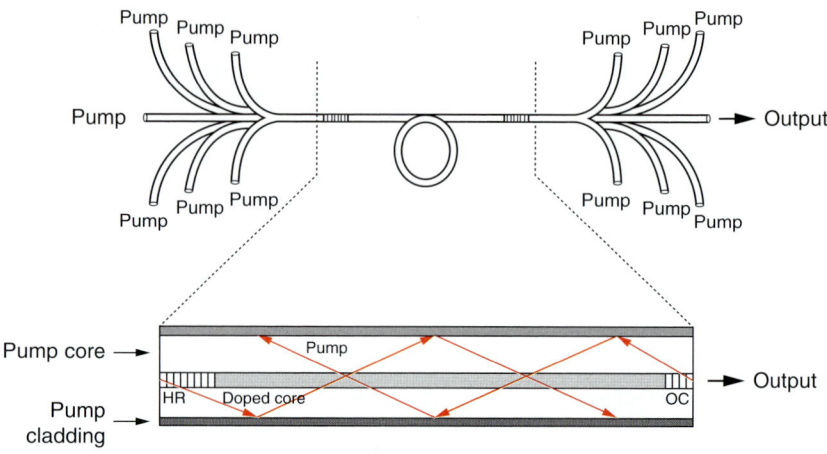

FIGURE 15.32 Schematic illustration of a high-power continuous wave fiber laser.

the major limit. Instead, pump coupling and heat management are the main design constraints. Bidirectional pumping can also be used to provide maximum pump powers. A number of CW fiber lasers with kilowatt power levels have been demonstrated in recent years with both end-pumping schemes and pump combiners.[61,62]

Single-Frequency Fiber Lasers
Single-frequency lasers offering laser spectral line widths of few tens of kilohertz are desirable in applications requiring long coherence lengths, such as optical sonar and high-power coherent combining. FBGs combined with highly doped rare earth fibers provide a perfect platform for single-frequency laser oscillators. Two designs are usually used—distributed Bragg reflector (DBR) and distributed feedback (DFB) configurations (Fig. 15.33). In the DBR configuration, a cavity length of just a few centimeters is formed between two FBGs with spectral line widths of a fraction of a nanometer. The short cavity length increases the spectral spacing of the cavity modes and allows single-frequency operation when the narrow spectral line width of the FBGs selects a single cavity mode to go beyond threshold. In a DFB configuration, a longer FBG with a built-in $\pi/2$ phase jump constitutes the cavity. Operation of a DFB is not different from a DBR design, except that precise phase control over the long FBG, in combination with the $\pi/2$ phase jump, allows much narrower overall spectral line width and, consequently, more robust single-frequency operation. The FGB phase jump in a DFB laser can either be implemented by moving the FBG's line position or by introducing an additional average index change after the FBG is written. In the latter case, the appropriate average index change can be achieved by exposure to

Introduction to Optical Fiber Lasers

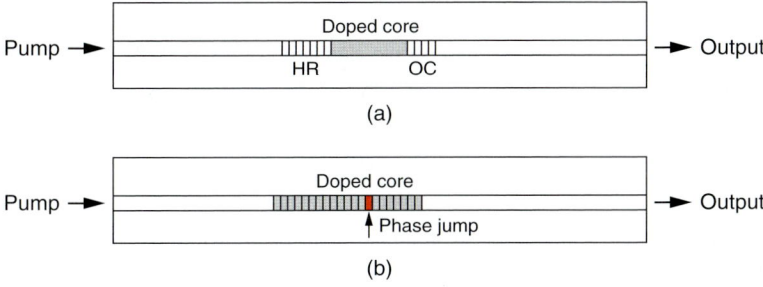

FIGURE 15.33 (a) Distributed Bragg reflector (DBR) design and (b) distributed feedback (DFB) design.

a UV beam, while monitoring the FBG's transmission. Single-polarization operation can also be achieved by introducing birefringence in the phase jump by having the polarization of the UV writing beam perpendicular to the fiber axis. DBR fiber lasers can be easily implemented by writing FBGs in photosensitive fibers and then splicing them to a section of highly rare-earth-doped fiber. A DFB fiber laser, however, requires the writing of FBGs directly in a highly rare-earth-doped fiber, which requires photosensitivity as well as high doping levels, which are conflicting requirements. This conflict can however be resolved by engineering photosensitive claddings with a highly rare-earth-doped core.[63]

A single-frequency fiber laser is usually pumped by a single-mode pump diode to provide an output power in the range of a few tens to a few hundred milliwatts.[64] The laser can be amplified in a double-clad fiber to a few hundred watts. Currently, SBS is the key limiting factor in amplification of a single-frequency seed to high-output powers. The reason for this is the low SBS threshold with narrow spectral line width operation. The use of large effective mode areas increases the SBS threshold. Recently, an amplified single-frequency output power as high as 500 W has been demonstrated.[65] It was also found that core temperature variation along the amplifier arising from pump decay raises the SBS threshold due to the temperature dependence of the SBS process, which leads to an effective SBS spectral broadening. Very recently, a 1.7-KW single-frequency fiber was demonstrated;[18] furthermore, it was found that ~10 dB SBS suppression can be obtained through SBS spectral broadening by dopant variations across the core.[18,19]

15.3.2 Q-Switched Lasers

In a pulsed rare-earth-doped optical fiber, the extractable energy is ultimately limited by ASE, which leads to a reduction of population inversion due to interpulse ASE (see Sec. 15.2.3). The extractable energy scales linearly with the effective mode area, as do the nonlinear

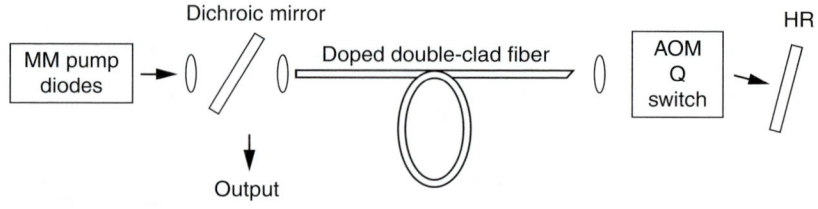

FIGURE 15.34 A typical scheme for a Q-switched fiber laser. AOM: acousto-optic modulator.

thresholds. Large core optical fibers also help minimize optical damage at the end faces. Although high interpulse gain is clearly desirable for high pulse energy, it is limited by potential interpulse lasing from spurious reflections, such as from Rayleigh scattering.[66,67] In general, the maximum extractable energy is limited to ten times saturation energy $F_{sat}A$, which is $A*10$ µJ/µm² for ytterbium-doped fibers. The pulse energy does not critically depend on fiber length. Longer pulses are typically generated in longer fibers in Q-switched lasers, while pulse breakup can happen if the pulse duration becomes much longer than the round-trip cavity time. Typically, the pulse buildup time is much shorter than the decay time in a high-gain system, leading to a steeper leading edge and a slower trailing edge. A typical Q-switched fiber laser is illustrated in Fig. 15.34.

15.3.3 Mode-Locked Fiber Lasers

The large bandwidth of optical fiber amplifiers, as well as their ease of construction, has created a considerable interest in the generation of ultrashort pulses, which remains an ongoing research area. Ytterbium, erbium and thulium optical fiber lasers support bandwidths of the order of 50, 45, and 200 nm, respectively, allowing for the generation of pulses of the order of 30 fs. Even shorter pulses can be obtained by pulse-compression techniques. The ultimate limit for the generation of the highest peak power inside an optical fiber is governed by self-focusing (as explained in Sec. 15.2.3) and is in the range of 4–6 MW.

To use optical fibers in ultrashort pulse generation, many different techniques have been tested over the years. An early review can be found in Fermann.[68] Here we limit our discussion to short-pulse generation via mode locking, which optimally utilizes the intrinsic cavity mode structure in an optical fiber oscillator. It is customary to distinguish active and passive mode-locking techniques.

Active Mode Locking

A fiber oscillator supports a range of cavity modes with frequencies f_k given by

$$f_k = k\frac{c}{n_p L} \tag{15.43}$$

where k is an integer, c is the velocity of light, n_p is the phase index at frequency f_k, and L is the cavity length. To a first order approximation, the cavity mode spacing δf can then be written as

$$\delta f = \frac{c}{n_g L} \tag{15.44}$$

Note that due to dispersion, the cavity mode spacing is slightly nonuniform in any cavity that contains an actual physical gain medium. Therefore the cavity mode spacing is governed by the group index n_g rather than by the phase index n_p.

In active mode locking, a modulator operating at the cavity round-trip time is introduced into the cavity. A subsection of the cavity modes then becomes phase locked through the generation of side bands from the modulator. Moreover, the applied modulation pulls the cavity modes into a precisely uniform frequency grid, leading to the generation of stable pulses at the repetition rate given by Eq. 15.44.

The cavity setup of an active mode-locked neodymium fiber laser generating 2.4-ps pulses at a repetition rate of 90 MHz is shown in Fig. 15.35. In this case, a grating pair is further introduced to provide negative cavity dispersion, and an amplitude modulator is implemented.

Following the analysis by A. E. Siegman,[26] active mode locking produces gaussian-shaped pulses of the form $A(t) = A_0 \exp[-(t/\tau)^2]$, in which the pulse width is given by

$$\Delta \tau = 0.315 \left(\frac{g}{\delta_a} \right)^{1/4} \left(\frac{1}{f_m \Delta f_a} \right)^{1/2} \tag{15.45}$$

where Δf_a is the bandwidth of the gain medium, f_m is the optical modulation frequency, g is the saturated amplitude gain in the gain medium, and $\delta_a \approx 1$ is the modulation depth of the modulator. The

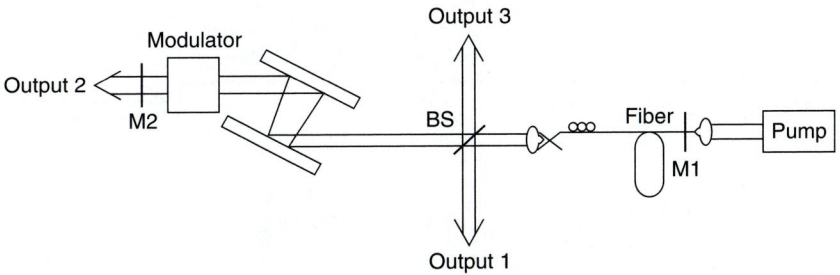

FIGURE 15.35 Early setup of an actively mode-locked neodymium fiber laser. BS: beam splitter.

generated pulses are bandwidth limited and have a time-bandwidth product $\Delta\nu\Delta\tau = 0.44$. When introducing a phase modulator, the modulation depth must be replaced with the exerted peak phase retardation in Eq. 15.45. The pulse width is then $\sqrt{2}$ longer than given by Eq. 15.45, and the pulses have a time-bandwidth product of $\Delta\nu\Delta\tau = 0.63$.

Active mode locking is mainly used in the generation of high-repetition-rate pulses. When using erbium fiber gain media, all-fiber systems can be constructed that operate at repetition rates of 10 GHz and higher by implementing modulation at a high harmonic of the cavity round-trip time. Typically, the minimum generated pulse widths are of the order of 1 ps.

Passive Mode Locking

The shortest possible pulses from an optical fiber oscillator can generally be obtained via passive mode locking, in which pulse widths with a bandwidth comparable to the gain bandwidth can be readily generated. The shortest pulses from a passively mode-locked oscillator to date are 28 fs.[69] Passive mode locking can produce such short pulses because the effective phase (or amplitude) modulation happens on a time scale comparable to the pulse width. As the pulses get shorter while oscillating in the cavity, the modulation strength increases, producing, in turn, shorter and shorter pulses. Eventually, this pulse width shortening is balanced by pulse width broadening from cavity dispersion and the limited bandwidth of the gain medium.

To induce passive mode locking, an appropriate passive amplitude modulator needs to be inserted into the cavity. Most convenient are semiconductor saturable absorbers, which exhibit power-dependent saturation characteristics. Alternatively, nonlinear interference between the two polarization modes of an optical fiber can be used to induce a power-dependent polarization state and passive amplitude modulation. A setup of a typical passively mode-locked laser used in industrial ultrafast fiber laser systems is shown in Fig. 15.36.

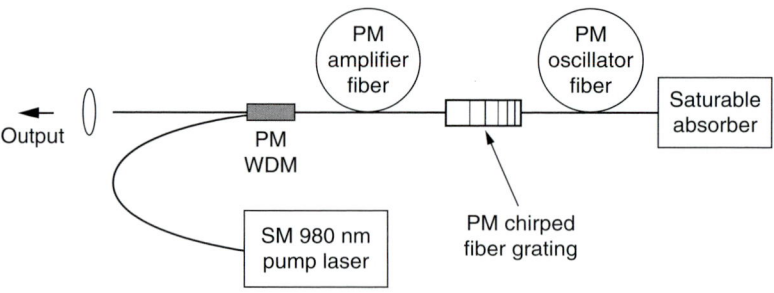

FIGURE 15.36 Setup of an industrial passively mode-locked ytterbium fiber laser. WDM: wavelength division multiplexing.

In Fig. 15.36, a semiconductor saturable absorber is used as a passive amplitude modulator. The use of PM fiber throughout the cavity allows for environmentally stable single-polarization operation. A chirped fiber grating is used for dispersion compensation, which allows for a very compact system construction. With the attached fiber amplifier, pulse energies up to 10 nJ can be generated, and the pulse widths can be as short as 100 fs. As with actively mode-locked lasers, the system from Fig. 15.36 can also produce gaussian-shaped pulses, in which, assuming optimum dispersion compensation, the obtainable pulse widths and pulse energies are given by

$$\Delta t_g = 0.66\sqrt{|D_2|} \tag{15.46}$$

$$E_g = 4.47\sqrt{|D_2|/\gamma L} \tag{15.47}$$

Here $|D_2|$ is the total amount of positive dispersion inside the cavity due mainly to the intracavity PM oscillator fiber. γ is the nonlinear fiber coefficient defined in Eq. (15.20).

Frequency Combs

Mode locking produces a pulse train with an optical frequency spectrum consisting of individual frequency modes f_k on a highly precise uniform frequency grid. This grid can be written as

$$f_k = k\delta f + f_{ceo} \tag{15.48}$$

where δf is the repetition rate of the laser given by Eq. (15.44), and the group index is evaluated at the carrier frequency (or central frequency) of the optical pulses. f_{ceo}, which is the carrier envelope offset frequency, remains uncontrolled; therefore, the absolute location of the frequency modes (or combs) in the optical frequency spectrum is undetermined and fluctuates slowly in the oscillator. As a result, the optical pulse phase at the output of the mode-locked laser also fluctuates from pulse to pulse. The value of f_{ceo} can be detected via a beat signal generated with a nonlinear "$f - 2f$" interferometer, which requires a coherent octave-spanning supercontinuum. This supercontinuum can be generated when coupling the oscillator output into a highly nonlinear fiber. The f_{ceo} beat signal is typically sensitive to the laser pump power; therefore, a control of the pump power in conjunction with a phase-locked loop can be used to phase lock f_{ceo} to an external radio frequency (RF) reference, thus stabilizing the phase of the output pulses.

Ultrafast Pulse Amplification

Most applications of mode-locked lasers require further amplification in external fiber amplifiers. These amplification schemes are subject to the fiber limitations discussed in Sec. 15.2.3. Efficient amplification of femtosecond pulses up to millijoule energy levels is enabled by the simultaneous implementation of the CPA technique and large core fibers, as discussed in Sec. 15.2.4. In CPA, the pulses are stretched temporally prior to amplification. After pulse amplification, the pulses are recompressed close to the bandwidth limit in a bulk grating pair. Because pulse stretching and compression ratios on the order of 1000–10000 can be implemented, the peak power inside the actual fiber amplifier can be greatly reduced. Because of the various nonlinear limitations of optical fibers, CPA schemes operate with typical pulse peak powers of the order of 0.1–1.0 MW inside the fiber. A more detailed discussion of ultrafast fiber amplification schemes is provided in Chap. 17.

References

1. Snitzer, E., "Proposed Fiber Cavities for Optical Masers," *J. Appl. Phys.*, 23: 36–39, 1961.
2. Koester, C. J., and Snitzer, E., "Amplification in a Fiber Laser," *Appl. Optics*, 3: 1182, 1964.
3. Stone, J., and Burrus, C. A., "Neodymium-Doped Fiber Lasers: Room Temperature CW Operation with an Injection Laser Pump," *Appl. Optics*, 13: 1256–1258, 1974.
4. Hegarty, J., Broer, M. M., Golding, B., Simpson, J. R., and MacChesney, J. B., "Photon Echoes Below 1 K in a Nd^{3+}-Doped Glass Fiber," *Phys. Rev. Lett.*, 51: 2033–2035, 1983.
5. Poole, S. B., Payne, D. N., and Fermann, M. E., "Fabrication of Low Loss Optical Fibers Containing Rare Earth Ions," *Electron. Lett.*, 21: 737–738, 1985.
6. Poole, S. B., Payne, D. N., Mears, R. J., Fermann, M. E., and Laming, R. I., "Fabrication and Characterization of Low Loss Optical Fibers Containing Rare Earth," *J. Lightwave Tech.*, LT-3: 870–876, 1986.
7. Mears, R. J., Reekie, L., Poole, S. B., and Payne, D. N., "Neodymium-Doped Silica Single-Mode Fiber Lasers," *Electron. Lett.*, 21: 737–738, 1985.
8. Mears, R. J., Reekie, L., Jauncie, I. M., and Payne, D. N., "Low Noise Erbium-Doped Amplifier Operating at 1.54 μm," *Electron. Lett.*, 23: 1026–1028, 1987.
9. Desurvire, E., Simpson, J. R., and Becker, P. C., "High Gain Erbium-Doped Traveling-Wave Fiber Amplifier," *Optics Lett.*, 12: 888–890, 1987.
10. Snitzer, E., Po, H., Hakimi, F., Tuminelli, R., and MaCollum, B. C., "Erbium Fiber Laser Amplifier at 1.55 μm with Pump at 1.49 μm and Yb Sensitized Er Oscillator," *OSA Tech. Digest*, 1: 218–221, 1988.
11. Nakazawa, M., Kimura, Y., and Suzuki, K., "Efficient Er^{3+}-doped Optical Fiber Amplifier Pumped by a 1.48 μm InGasP Laser Diode," *Appl. Phys. Lett.*, 54: 295, 1989.
12. Snitzer, E., Po, H., Hakimi, F., Tumminelli, R., and McCollum, B. C., "Double-Clad, Offset Core Nd Fiber Laser" (postdeadline paper PD5), Proceedings of Conference on Optical Fiber Sensors, 1988.
13. Po, H., Snitzer, E., Tumminelli, R., Zenteno, L., Hakimi, F., Cho, N. M., and Haw, T., "Double-Clad High Brightness Nd Fiber Laser Pumped by GaA/As Phased Array" (postdeadline paper PD7), Proceedings of Optical Fiber Communication 1989.
14. Hecht, J., "Half a Century of Laser Weapons," *Opt. Photon. News*, 20: 16–21, 2009.
15. Fermann, M. E., "Single-Mode Excitation of Multimode Fibers with Ultra-Short Pulses," *Opt. Lett.*, 23: 52–54, 1998.
16. Koplow, J. P., Kliner, D. A. V., and Goldberg, L., "Singe-Mode Operation of a Coiled Multimode Fiber Amplifier," *Opt. Lett.*, 25: 442–444, 2000.

17. Agrawal, G. P., *Nonlinear Fiber Optics,* Academic Press, 2007.
18. Nillson, J., "SBS Suppression at the Kilowatt Level" (paper 7195-50), *SPIE Photon. West,* San Jose, 2009.
19. Ward, B. G., and Spring, J. B., "Brillouin Gain in Optical Fibers with Inhomogeneous Acoustic Velocity" (paper 7195-54), *SPIE Photon. West,* San Jose, 2009.
20. Strickland, D., and Mourou, G., "Compression of Amplified Chirped Optical Pulses," *Opt. Commun.,* 56: 219–221, 1985.
21. Fibich, G., and Gaeta, A. L., "Critical Power for Self-Focusing in Bulk Media and in Hollow Waveguides," *Opt. Lett.,* 25: 335–337, 2000.
22. Dong, L., "Approximate Treatment of Nonlinear Waveguide Equation in the Regime of Nonlinear Self-Focus," *J. Lightwave Tech.,* 26: 3476–3485, 2008.
23. Davis, M. K., Digonnet, M. J. F., and Pantell, R. H., "Thermal Effects in Doped Fibers," *IEEE J. Lightwave Tech.,* 16: 1013–1023, 1998.
24. Smith, A. V., Do, B. T., and Söderlund, M. J., "Optical Damage Limits to Pulse Energy from Fibers," *IEEE J. Sel. Top. Quantum Electron.,* 15: 153–158, 2009.
25. Digonnet, M. J., *Rare-Earth-Doped Fiber Lasers and Amplifiers,* CRC Press, 2001.
26. Siegman, A. E., *Lasers,* University Science Books, 1986.
27. Jasapara, J. C., Andrejco, M. J., DeSantolo, A., Yablon, A. D., Várallyay, Z., Nicholson, J. W., Fini, J. M., et al., "Diffraction-Limited Fundamental Mode of Core-Pumped Very-Large-Mode-Area Er Fiber Amplifier," *IEEE Sel. Top. Quantum Electron.,* 15: 3–11, 2009.
28. Galvanauskas, A., Cheng, M. Y., Hou, K. C., and Liao, K. H., "High Peak Power Pulse Amplification in Large-Core Yb-Doped Fiber Amplifiers," *IEEE J. Sel. Top. Quantum Electron.,* 13: 559–566, 2007.
29. Knight, J. C., Birks, T. A., Russell, P. S.-J., and Atkin, D. M., "All-Silica Single-Mode Optical Fiber with Photonic Crystal Cladding," *Opt. Lett.,* 21: 1547–1549, 1996.
30. Birks, T. A., Knight, J. C., and Russell, P. S.-J., "Endless Single-Mode Photonic Crystal Fiber," *Opt. Lett.,* 22: 961–963, 1997.
31. Knight, J. C., Birks, T. A., Cregan, R. F., Russell, P. S.-J., and de Sandro, J. P., "Large Mode Area Photonic Crystal Fiber," *Electron. Lett.,* 34: 1347–1348, 1998.
32. Limpert, J., Liem, A., Reich, M., Schreiber, T., Nolte, S., Zellmer, H., Tünnermann, A., et al., "Low-Nonlinearity Single-Transverse-Mode Ytterbium-Doped Photonic Crystal Fiber Amplifier," *Opt. Express,* 12: 1313–1319, 2004.
33. J. Limpert, N. Deguil-Robin, I. Manek-Hönninger, F. Salin, F. Röser, A. Liem, T. Schreiber, et al., "High-Power Rod-Type Photonic Crystal Fiber Laser," *Opt. Express,* 13: 1055–1058, 2005.
34. Limpert, J., Schmidt, O., Rothhardt, J., Röser, F., Schreiber, T., and Tünnermann, A., "Extended Single-Mode Photonic Crystal Fiber," *Opt. Express,* 14: 2715–2719, 2006.
35. Brooks, C. D., and Di Teodoro, F., "Multi-Megawatt Peak Power, Single-Transverse-Mode Operation of a 100 μm Core Diameter, Yb-Doped Rod-Like Photonic Crystal Fiber Amplifier," *Appl. Phys. Lett.,* 89: 111119–111121, 2006.
36. Saitoh, K., and Koshiba, M., "Empirical Relations for Simple Design of Photonic Crystal Fibers," *Opt. Express,* 13: 267–274, 2004.
37. Kuhlmey, B. T., McPhedran, R. C., and de Sterke, C. M., "Modal Cutoff in Micro-Structured Optical Fibers," *Opt. Lett.,* 27: 1684–1686, 2002.
38. Kuhlmey, B. T., McPhedran, R. C., de Sterke, C. M., Robinson, P. A., Renversez, G., and Maystre, D., "Modal Cutoff in Micro-Structured Optical Fibers," *Opt. Lett.,* 27: 1684–1686, 2002.
39. Dong, L., McKay, H. A., and Fu, L., "All-Glass Endless Single-Mode Photonic Crystal Fibers," *Opt. Lett.,* 33: 2440–2442, 2008.
40. Dong, L., Peng, X., and Li, J., "Leakage Channel Optical Fibers with Large Effective Area," *J. Opt. Soc. Am. B,* 24:1689–1697, 2007.
41. Dong, L., Wu, T. W., McKay, H. A., Fu, L., Li, J., and Winful, H. G., "All-Glass Large-Core Leakage Channel Fibers," *IEEE J. Sel. Top. Quantum Electron.,* 15: 47–53, 2009.

42. Dong, L., Li, J., McKay, H. A., Marcinkevicius, A., Thomas, B. T., Moore, M., Fu, L. B., and Fermann, M. E., "Robust and Practical Optical Fibers for Single Mode Operation with Core Diameters up to 170 μm" (postdeadline paper CPDB6), Conference on Lasers and Electro Optics, San Jose, May 2008.
43. Fu, L., McKay, H. A., Suzuki, S., Ohta, M., and Dong, L., "All-Glass PM Leakage Channel Fibers with up to 80μm Core Diameters for High Gain and High Peak Power Fiber Amplifiers" (postdeadline paper MF3), Advanced Solid State Photonics, Denver, February 2009.
44. Fini, J. M., "Design of Solid and Microstructure Fibers for Suppression of Higher-Order Modes," *Opt. Express*, 13: 3477–3490, 2005.
45. Liu, C. H., Chang, G., Litchinitser, N., Galvanauskas, A., Guertin, D., Jabobson, N., and Tankala, K., "Effectively Single-Mode Chirally-Coupled Core Fiber" (paper ME2), *Advanced Solid-State Photonics*, OSA Technical Digest Series (CD), Optical Society of America, 2007.
46. Swan, M. C., Liu, C. H., Guertin, D., Jacobsen, N., Tankala, K., and Galvanauskas, A., "33 μm Core Effectively Single-Mode Chirally-Coupled-Core Fiber Laser at 1064-nm" (paper OWU2), Optical Fiber Communication/National Fiber Optic Engineers Conference, 2008.
47. Ramachandran, S., Nicholson, J. W., Ghalmi, S., Yan, M. F., Wisk, P., Monberg, E., and Dimarcello, F. V., "Light Propagation with Ultra Large Modal Areas in Optical Fibers," *Opt. Lett.*, 31: 1797–1799, 2006.
48. Ramachandran, S., "Spatially Structured Light in Optical Fibers, Applications to High Power Lasers" (paper MD1), Advanced Solid State Photonics, Denver, February 2009.
49. Siegman, A. E., "Gain-Guided, Index-Antiguided Fiber Lasers," *J. Opt. Soc. Am. B*, 24: 1677–1682, 2007.
50. Sims, R., Sudesh, V., McComb, T., Chen, Y., Bass, M., Richardson, M., James, A. G., et al., "Diode-Pumped Very Large Core, Gain Guided, Index Anti-Guided Single Mode Fiber Laser" (paper WB3), Advanced Solid State Photonics, Denver, February 2009.
51. Fini, J. M., "Bend-Resistant Design of Conventional and Microstructure Fibers with Very Large Mode Area," *Opt. Express*, 14: 69–81, 2006.
52. Fini, J. M., "Intuitive Modeling of Bend Distortion in Large-Mode-Area Fibers," *Opt. Lett.*, 32: 1632–1634, 2007.
53. Nicholson, J. W., Fini, J. M., Yablon, A. D., Westbrook, P. S., Feder, K., and Headley, C., "Demonstration of Bend-Induced Nonlinearities in Large-Mode-Area Fibers," *Opt. Lett.*, 32: 2562–2564, 2007.
54. Hill, K. O., Fujii, Y., Johnson, D. C., and Kawasaki, B. S., "Photosensitivity in Optical Fiber Waveguides: Application to Reflection Filter Fabrication," *Appl. Phys. Lett.*, 32: 647–649, 1978.
55. Meltz, G., Morey, W. W., and Glenn, W. H., "Formation of Bragg Gratings in Optical Fibers by a Transverse Holographic Method," *Opt. Lett.*, 14: 823–825, 1989.
56. Hill, K. O., and Meltz, G., "Fiber Bragg Grating Technology Fundamentals and Overview," *J. Lightwave Tech.*, 15: 1263–1276, 1997.
57. Lemaire, P. J., Atkins, R. M., Mizrahi, V., and Reed, W. A., "High Pressure H_2 Loading as a Technique for Achieving Ultrahigh UV Photosensitivity and Thermal Sensitivity in GeO_2 Doped Optical Fibers," *Electron. Lett.*, 29: 1191–1193, 1993.
58. Erdogan, T., Mizrahi, V., Lemaire, P. J., and Monroe, D., "Decay of Ultraviolet-Induced Fiber Bragg Gratings," *J. Appl. Phys.*, 76: 73–80, 1994.
59. Digonnet, M. J. F., *Rare-Earth-Doped Fiber Lasers and Amplifiers*, CRC Press, New York, 2001.
60. Goldberg, L., Cole, B., and Snitzer, E., "V-Groove Side-Pumped 1.5. m Fiber Amplifier," *Electron. Lett.*, 33: 2127–2129, 1997.
61. Jeong, Y., Sahu, J., Payne, D., and Nilsson, J., "Ytterbium-Doped Large-Core Fiber Laser with 1.36 kW Continuous-Wave Output Power," *Opt. Express*, 12: 6088–6092, 2004.

62. Gapontsev, V., Gapontsev, D., Platonov, N., Shkurikhin, D., Fomin, V., Mashkin, A., Abramov, M., and Ferin, S., "2 kW CW Ytterbium Fiber Laser with Record Diffraction-Limited Brightness" (paper CJ1-1-THU), Conference on Lasers and Electro-Optics Europe, Munich, Germany, 2005.
63. Dong, L., Loh, W. H., Caplen, J. E., Minelly, J. D., Hsu, K., and Reekie, L., "Efficient Single Frequency Fiber Lasers Using Novel Photosensitive Er/Yb Optical Fibers," *Opt. Lett.,* 22: 694–696, 2007.
64. Loh, W. H., Samson, B. N., Dong, L., Cowle, G. J., and Hsu, K., "High Performance Single Frequency Fiber Grating-Based Erbium: Ytterbium-Codoped Fiber Lasers," *IEEE J. Lightwave Tech.,* 16: 114–118, 1998.
65. Jeong, Y., Nilsson, J., Sahu, J. K., Payne, D. N., Horley, R., Hickey, L. M. B., and Turner, P. W., "Power Scaling of Single-Frequency Ytterbium-Doped Fiber Master Oscillator Power Amplifier Sources Up to 500W," *IEEE J. Sel. Topics Quantum Electron.,* 13: 546–551, 2007.
66. Offerhaus, H. L., Broderick, N. G., Richardson, D. J., Sammut, R., Caplen, J., and Dong, L., "High-Energy Single-Transverse-Mode Q-Switched Fiber Laser Based on a Multimode Large-Mode-Area Erbium-Doped Fiber," *Opt. Lett.,* 23: 1683–1685, 1998.
67. Renaud, C. C., Offerhaus, H. L., Alvarez-Charvez, J. A., Nilsson, J., Clarkson, W. A., Turner, P. W., Richardson, D. J., and Grudinin, A. B., "Characteristics of Q-Switched Cladding-Pumped Ytterbium-Doped Fiber Lasers with Different High-Energy Fiber Designs," *IEEE J. Quantum Electron.,* 37: 199–206, 2001.
68. Fermann, M. E., **"Ultrashort Pulse Sources Based on Single-Mode Rare-Earth-Doped Fibers,"** *J. Appl. Phys. B,* B58: 197, 1994.
69. Zhou, X., Yoshitomi, D., Kobayashi, Y., and Torizuka, K., "Generation of 28-fs Pulses from a Mode-Locked Ytterbium Fiber Oscillator," *Opt. Express,* 16: 7055, 2008.

CHAPTER 16
Pulsed Fiber Lasers

Fabio Di Teodoro

*Senior Scientist, Northrop Grumman Aerospace Systems,
Redondo Beach, California*

16.1 Introduction

The combination of nanosecond pulse durations and multikilohertz pulse repetition frequency (PRF) represents an operation regime of great interest for a variety of remote sensing applications, including ranging, altimetry, and terrain mapping. Thanks to subgigahertz transform-limited spectral bandwidths, such pulsed output is also applicable to close-range or standoff chemical detection and probing via resonant excitation. Finally, it is a key enabler for many materials processing tasks, primarily in the area of laser marking, engraving, and precision drilling. In many cases of practical interest, the generation of this pulse format must be accompanied by excellent spatial and spectral brightness and must afford packaging characteristics compatible with deployment in harsh, nonlaboratory environments.

Until recently, nanosecond-pulse bulk (non-waveguided) diode-pumped solid-state (DPSS) lasers have represented the only viable solution for the above applications. However, as the PRF, and hence the average power, scales up for a given pulse energy, DPSS lasers encounter well-known problems in maintaining high brightness due to the increasing magnitude of thermo-optically induced aberrations. These are conversely very manageable in fibers thanks to thermal load distribution and single-mode waveguiding. The higher optical efficiency of fiber laser sources, which stems from the larger overlap integral between optical pump beam and active medium, is also a critical discriminator for integration in platforms where electrical power may come at a premium, such as in space-borne and airborne payloads or industrial machines. Finally, fiber-based optical sources bear the promise of ultimate compactness and ruggedness, mainly due to the minimization of free-space alignment-sensitive optical paths and the potential for transition from the traditional optical bench concept toward distributed architectures of fluid, application-tailored form factor.

However, traditional pulsed fiber laser (PFL) systems have been confined to relatively low values of pulse energy and peak power, which only suit optical telecommunications and other niche applications. The primary obstacle to overcoming this limitation has been the onset of unwanted nonlinear optical effects, which stem from high in-fiber irradiance × length products and which result in spectral broadening. This issue directly impairs the PFL's ability to serve applications that require high spectral brightness, such as laser frequency tuning to material resonances, return signal discrimination against a broadband background, or optical efficiency maximization in nonlinear wavelength conversion. Indirectly, spectral broadening precludes ultimate power scaling through beam combining schemes that preserve spatial brightness—namely, coherent phasing and wavelength multiplexing. Other important challenges to overcome in pulse power scaling in the kilohertz PRF regime are energy storage limitations due to amplified spontaneous emission (ASE) and optical damage.

This chapter details the nature and extent of such challenges, with an emphasis on the specific cases of nanosecond pulse widths and high in-fiber pulse energy and peak power generation. It then reviews key enabling solutions that have permitted, in recent years, substantial power scaling for fiber-based lasers and amplifiers operating in this regime. This review focuses entirely on ~1-µm-wavelength Yb-doped fiber technologies, which dominate the high-power fiber laser scene in both the continuous wave and pulsed regimes. An outlook toward near-future developments concludes the analysis.

16.2 Challenges to Pulse Power Scaling

This section provides an overview of the issues that must be faced in order to boost the performance of PFLs and pulsed fiber amplifiers and to meet the requirements of demanding nontelecommunications applications. Three key problem areas are identified and discussed: nonlinear optical effects, amplified spontaneous emission, and optical damage.

16.2.1 Nonlinear Optical Effects

Nonlinear effects (NLEs) represent the principal limitation for in-fiber pulse peak power. This situation is a direct consequence of the tight optical confinement in the small guiding core of typical long fibers. The NLE's "strength," S_{NLE}, can be expressed as

$$S_{\text{NLE}} = \int_0^L \frac{P_{\text{peak}}(z)}{A} dz \qquad (16.1)$$

where L is the fiber length, $P_{\text{peak}}(z)$ is the pulse peak power at position z along the fiber, and A is the in-core guided-mode field area.

Fiber-supported NLEs are quite diverse and include forms of energy exchange between photons and lattice vibrations of the host material, as well as parametric processes, which are manifestations of the irradiance-dependent change in the material refractive index (optical Kerr effect). In this section, NLEs of special interest for near-nanosecond-pulse fiber sources will be reviewed, and their impact on laser and amplifier performance discussed.

Stimulated Brillouin Scattering

Stimulated Brillouin scattering (SBS) consists of the inelastic energy exchange between incident photons and acoustic phonons in the glass lattice, which leads to the generation of a phase-conjugated, red-shifted (Stokes), typically backward-propagating optical beam. Power transfer from the main beam to the Stokes beam is a major disruption in the operation of high-power fiber sources, because it reduces the output optical efficiency and yields a source of power-dependent optical feedback that can induce potentially destructive parasitic pulsing behaviors. In addition, SBS exhibits very high gain (more than 2 orders of magnitude larger than stimulated Raman scattering, addressed in the next subsection), which results in very-low-onset threshold, especially in the case of powerful and spectrally narrow input beams. Because of such low threshold and detrimental effects, SBS remains a subject of intense study, primarily in the context of high-power continuous wave (CW) fiber lasers, which must exhibit a high spectral quality in order to enable open-ended power scaling via spectral or coherent beam combining. SBS is also a primary limiting factor for multinanosecond PFLs and pulsed fiber amplifiers tasked with producing time-bandwidth products close to the Fourier transform limit—for example, for applications that involve high-resolution spectroscopic probing and discrimination of molecular species.[1,2]

The importance of SBS, however, is drastically reduced as the pulse duration is decreased to values less than 2–3 ns, a pulsed regime that is of primary interest for peak-power maximization. This is due to two concomitant causes. First, the SBS gain is substantially reduced when the input beam's spectral width exceeds the inverse acoustic phonon lifetime in fused silica, which is usually around 50 MHz. For comparison, the Fourier-transform-limited bandwidth of a 3-ns full-width, half-maximum (FWHM) Gaussian-shaped pulse is ~150 MHz. Second, the length of a sub-3-ns pulse in fiber is less than 1 m, which means that the spatial overlap (and interaction length) between such short input pulses and their corresponding counterpropagating, SBS-generated Stokes pulses is typically only a small fraction of the overall fiber length.

This analysis holds generally valid for the frequent, simple cases in which SBS is not "seeded," meaning that only the main beam is present in the fiber and that the Stokes components build from diminutively small parametric photon noise or from ASE occurring at the Stokes-shifted wavelength. However, due to the relatively small SBS frequency shift (e.g., ~15 GHz in Yb-doped fibers operated at around 1060 nm), it is possible that the input beam being amplified in (or delivered through) the fiber medium may fortuitously feature spectral components of appreciable power at the SBS Stokes frequency. This may occur, for example, if such a beam is generated by a short-cavity laser (e.g., certain types of Q-switched or semiconductor lasers) with longitudinal mode spacing sufficiently close to the SBS shift. In this case, SBS can set in at lower power and shorter pulses. Another case of interest is cumulative buildup of SBS in high-repetition-rate coherent pulse trains, in which the pulse period is significantly shorter than the acoustic phonon lifetime (~10 ns in fused silica). In such a scenario, significant SBS can occur even for short pulses, as successive pulses can constructively interact with the same Stokes wave.

Finally, due to its counterpropagating nature, SBS is strongly influenced by optical feedback, which may lead to the onset of chaotic pulsing behavior superimposed on the normal (CW or pulsed) operation. This parasitic effect is usually referred to as SBS-induced modulation instability[3] and is especially detrimental in linear-cavity (standing-wave) lasers.

Stimulated Raman Scattering
Stimulated Raman scattering (SRS) is another important form of inelastic nonlinear process in fiber, stemming from the interaction between a laser beam and laser-induced vibrations associated with intramolecular bonds. This interaction results in a variation of the electric dipole moment (optical phonons). Such vibrations are more energetic compared with those involved in SBS; in fused silica, they result in photon-to-phonon transferred energy of ~55 meV, corresponding to a Stokes red shift in excess of 13 THz. The ensuing large wavelength difference (e.g., greater than 60 nm at ~1 µm) is unacceptable for any applications that are required to maintain some degree of spectral control.

Because the peak Raman gain (~10^{-13} m/W in fused silica fibers) is approximately 2 orders of magnitude lower compared with SBS, SRS occurs at much higher power. Unlike SBS, however, the SRS Stokes beam can copropagate with the main beam, in which case significant SRS suppression can occur only when the Stokes pulse (which exhibits lower group velocity) accumulates a delay comparable to the pulse duration over a sufficiently short portion of the fiber. Such a walk-off effect becomes important in fused-silica fibers only for pulse widths less than 10 ps and is nearly negligible for nanosecond pulses. Moreover, the SRS gain bandwidth is much larger (~40 THz) than SBS; therefore, the effect strength remains virtually unchanged, even for very broad pulse spectra.

In light of the above characteristics, SRS is a dominant NLE for the near-nanosecond regime, and its mitigation appears possible only through a direct reduction of the S_{NLE} value or by means of specialty fiber designs, in which high propagation loss is selectively introduced for the SRS-generated field. Recently proposed SRS-suppressing solutions in the latter category include dual-hole,[4] W-shaped-core,[5] and photonic stop-band fibers.[6]

Nonlinear Phase Modulation

A direct manifestation of the optical Kerr effect is an intrapulse phase shift ϕ_{NL}, which, at each point z along the fiber and for spectrally narrow pulses of interest here,[7] is given by

$$\phi_{NL}(z,\tau) = \frac{2\pi n_2}{\lambda} \int_0^z \frac{P(z',\tau)}{A} dz' \qquad (16.2)$$

where λ is the pulse carrier wavelength, n_2 is the irradiance-dependent refractive index coefficient, P is the pulse instantaneous power, and τ denotes time in the reference frame that moves along the fiber at the pulse group velocity. Incidentally, in the case of negligible changes in the pulse temporal shape during in-fiber propagation, the time dependence of P can be factored out of the integral in Eq. (16.2), which, for $z = L$ (whole fiber length), becomes equal to S_{NLE} (Eq. (16.1). As a result of the phase shift ϕ_{NL}, which is referred to as self-phase modulation (SPM), the carrier frequency undergoes a corresponding intrapulse shift (a chirp) $\Delta \nu$, given by

$$\Delta \nu = -\frac{1}{2\pi} \frac{\partial \phi_{NL}}{\partial \tau} = -\frac{n_2}{\lambda} \int_0^z \frac{1}{A} \frac{\partial P}{\partial \tau} dz' \qquad (16.3)$$

For typical laser pulses, the larger the magnitude of such a frequency shift at the end of the fiber ($z = L$), the greater the pulse spectral broadening—hence, the departure from high spectral brightness conditions of interest for many applications. The nonlinear phase shift also represents a limiting factor for the generation of ultrashort pulses (see Chap. 17) and for the phase locking of multiple fiber-based amplifiers and lasers in coherently beam-combined schemes.

Based on Eq. (16.3), the SPM-induced spectral broadening depends on the pulse shape and is more pronounced in pulses of steep leading or trailing edges. This aspect is detrimental in pulse amplifiers that operate in the saturated regime (i.e., pulse energy ~ saturation energy; see Sec. 16.2.2), in which case the population inversion can be significantly depleted by the pulse's leading edge. As a result, the pulse becomes distorted upon amplification and "steepens," thus incurring greater SPM for a given peak power.

Another case that can further exacerbate nonlinear phase modulation effects is that of optical pulses exhibiting multiple frequencies in their spectrum (e.g., corresponding to several longitudinal modes of a laser cavity). In such cases, the nonlinear phase shifts of the distinct spectral components become coupled, which is referred to as *cross-phase modulation* (XPM) and which may lead to overall greater spectral broadening.[3]

As is the general case with NLEs, nonlinear phase-shift detriments are generally mitigated by reducing S_{NLE} through the use of large-core and short fibers. The choice of a seeding source that presents single-frequency spectral purity (for XPM avoidance) and a gently sloped temporal pulse profile is also instrumental for retaining high spectral brightness through amplification.

Further SPM and XPM mitigation can be obtained by proper temporal and spectral preconditioning of the input pulses, which may be possible, for example, when the seeding source is a gain-switched semiconductor laser or an externally modulated CW oscillator. For example, the pulse steepening effect described earlier can be countered by shaping the input pulse profile as a positive ramp, which results in a symmetric near-Gaussian-like profile upon amplification.[7,8]

Moreover, Eq. (16.3) shows that the SPM-induced $\Delta \nu$ is opposite in sign to the slope of the pulse and therefore always negative (positive) in the pulse leading (trailing) edge, which is referred to as *positive chirp*. Therefore, deliberately imparting a negative chirp on the input pulse (e.g., by means of an electro-optical phase modulator) can, in principle, compensate for such an effect and result in good containment of SPM-induced spectral broadening upon propagation or amplification through the fiber.[9]

Four-Wave Mixing

In fibers, the four-wave mixing (FWM) process consists of an energy conversion incurred by photon pairs in the main, high-irradiance beam (usually called the *FWM pump beam*) upon scattering off the host material (in this case, fused silica). The efficiency of this scattering event depends on the third-order term of the medium dielectric susceptibility—hence, ultimately on n_2.

Because the process does not involve material resonances, a necessary condition for significant FWM gain is that the total photon energy be preserved, which means that the initial FWM spectral signature is the appearance and ensuing power buildup of two side bands of equal frequency spacing with respect to the pump beam. Because of the induced power transfer from the main (pump) beam to such satellite spectral sidebands (often many nanometers away), FWM represents a major limitation when it occurs in PFLs and pulsed fiber amplifiers meant to retain high spectral brightness. The effectiveness of this power transfer can be further enhanced by SRS, when FWM side bands occur where significant Raman gain is available, as well as by nonlinear phase modulation. In fact, such nonlinearities can cascade in a runaway fashion if the peak in-core irradiance is increased indefinitely, which results in the cumulative effect of supercontinuum generation.

Another necessary condition for efficient FWM is that the total photon momentum be preserved, which is known as the *phase matching condition* and is quantified as

$$\Delta k = 2\phi_0 - \phi_1 + \phi_2 = 0 \tag{16.4}$$

where

$$\phi_i = \frac{2\pi n_{eff}(\lambda_i)}{\lambda_i} + \phi_{NL(i)} \quad i = 0, 1, 2 \tag{16.5}$$

Here the indices 0, 1, and 2 denote the pump and two scattered beams, respectively, whereas λ_i, $n_{eff}(\lambda_i)$, and $\phi_{NL(i)}$ are the wavelength, corresponding in-fiber effective refractive index, and nonlinear phase shift (defined by Eq. (16.2)), respectively, for each beam. Note that at FWM onset, the scattered beam power is low, and therefore only $\phi_{NL(0)}$ is appreciable. In phase-mismatched conditions ($\Delta \kappa \neq 0$), the scattered beam power simply oscillates along the fiber, without buildup, with a spatial period (called the *FWM coherence length*) of $\sim 2\pi/|\Delta \kappa|$.

Because the material dispersion parameter $D = d/d\lambda[n(\lambda)/\lambda]$ in typical silica-based fibers is positive for $\lambda < 1.3$ μm (zero-dispersion wavelength), FWM is generally phase mismatched; hence inefficient in this wavelength region (called *normal dispersion*) unless the pump and scattered beams propagate in different guided modes (only possible in multimode fibers) or, in some cases, exhibit different polarizations. Conversely, FWM can be phase matched at low pump powers even in single-mode fibers near the zero-dispersion wavelength (where $D = 0$), as well as in the anomalous dispersion region ($\lambda > 1.3$ μm) where D is negative; however, phase matching can be attained (at least over a fraction of the fiber length) due to the always positive contribution of the nonlinear phase shift ϕ_{NL}, provided that the pump is powerful enough.

Note that in fiber amplifiers, the FWM pump beam may experience significant, near-exponential gain and ensuing power growth over short distances, comparable to the FWM coherence length defined earlier. This occurrence breaks down the oscillatory nature of the power exchange between the pump and scattered beams and may lead to substantial FWM buildup even in non-phase-matched conditions in the normal dispersion wavelength region (~1 μm).[10,11]

Similar to the behavior of SBS and nonlinear phase modulation discussed earlier, the impact of FWM also becomes enhanced when a fraction of the main beam power is distributed over secondary, closely spaced spectral components, such as multiple longitudinal modes. In such cases, FWM side bands can become effectively "seeded" at these satellite frequencies (rather than building up from diminutive parametric noise) and, thus, set in at much lower peak powers, especially if the initial spectral features are separated by few tens of gigahertz or less.

This scenario, known in optical telecommunication applications as *quasi-phase-matched cross talk*, can lead to substantial spectral broadening, as shown in Fig. 16.1, which pertains to the Yb-doped fiber amplification of ~1-ns pulses emitted by a single-frequency Nd:YAG having side-longitudinal mode suppression of less than 10 dB. This important effect must be considered in the PFL design optimization, particularly in terms of the choice of seeding sources for amplifiers, as discussed in Sec. 16.3.

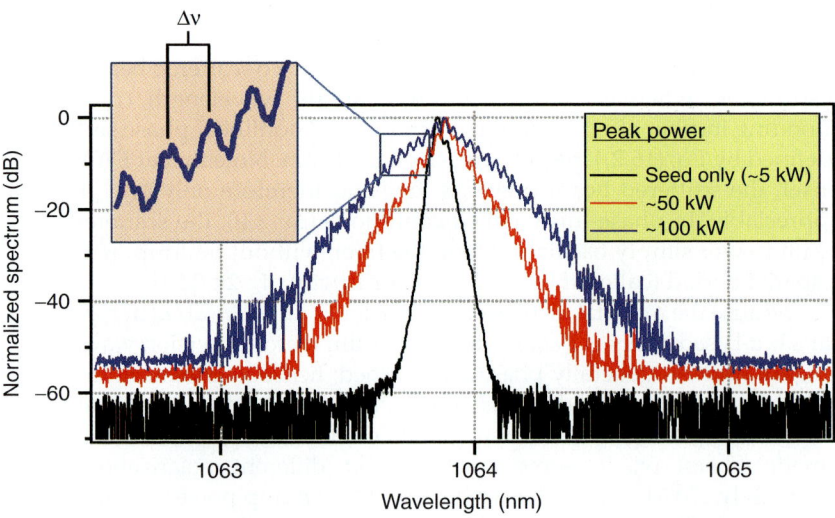

FIGURE 16.1 Peak-normalized logarithmic scale spectrum of the fiber-amplified output of an actively Q-switched Nd:YAG seed laser (pulse duration ~1 ns, pulse repetition frequency [PRF] = 10 kHz, output peak power ~5–10 kW). Black trace: Seed throughput only (unpumped fiber); Red trace: Amplified pulse peak power ~50 kW (~10 dB optical gain); Blue trace: Peak power ~100 kW. The gain fiber was a ~3-m long, 25-μm core, Yb-doped fiber. The frequency spacing of the cross-talk comb (Δν) is approximately equal to the longitudinal mode spacing in the seed laser.

16.2.2 Amplified Spontaneous Emission

A fraction of the spontaneous emission (isotropically distributed over the entire solid angle) from excited-state rare earth ions in the fiber is captured in the core and becomes bidirectionally guided through it and amplified via stimulated emission when in the presence of existing population inversion. In the interval between pulses, when only the pump beam is present in the fiber and the population inversion is being built up (i.e., when energy is being stored in the core), such ASE will continue to grow in power to the extent that, beyond a certain point, any further absorbed amount of pump power is simply rendered into ASE and does not lead to any additional increase in the excited-state population. As a general rule, this begins to happen when the stimulated and spontaneous emission rates become comparable. By limiting the population inversion, the ASE limits the achievable small-signal gain G and ultimately the energy extractable by the incoming pulse, which is given by

$$E_{ext} = GE_{sat} \qquad (16.6)$$

Here E_{sat} is the fiber saturation energy, which is, in turn, given by

$$E_{sat} = \frac{\varepsilon}{\sigma_a + \sigma_e} \frac{A}{\Gamma} \qquad (16.7)$$

where $\sigma_{a(e)}$ and ε are the absorption (or emission) cross section of the core dopant and the photon energy, respectively, which are both calculated at the pulse wavelength; A is the mode field area; and Γ is the transverse overlap integral between mode field and doping distribution.

Especially unfavorable are low PRF regimes, in which a longer time is available for ASE buildup, resulting in substantial degradation of the pulse energy versus pump power conversion efficiency. For example, in typical Yb-doped large mode area (LMA) fibers (e.g., ~25 µm/0.06 [core diameter/numerical aperture (NA)]) that are CW pumped at 975-nm wavelength, PRF values less than 10 kHz represent challenging pulse formats from the standpoint of efficiency.[12]

In addition to its energy-clamping consequences, the portion of ASE that copropagates with the signal also degrades the pulse contrast (i.e., the pulse-to-CW background power ratio), whereas backward-propagating ASE may be harmful for components located in the amplifier front end (such as in the master oscillator). All such detrimental effects are greatly exacerbated if back reflections are present, which results in ASE multipassing of the fiber gain medium. Ultimately, parasitic lasing at the ASE gain peak will occur. The threshold pump power for such an effect can be quite low if ASE and sources of unwanted optical feedback are left unmanaged. However, parasitic lasing eventually sets in within any fiber amplifier of high enough gain, because even with the implementation of optical isolators and successful suppression of all external feedback, in-fiber Rayleigh scattering still provides an intrinsic source of distributed feedback, especially in long fibers. In some cases, parasitic lasing may result in forward- or backward-propagating pulses that are energetic enough to cause damage.

16.2.3 Optical Damage

Pulse power scaling within fibers is ultimately limited by optical damage, which is primarily ascribed to dielectric breakdown triggered by photoionization. According to recent measurements in Yb-doped fibers,[13] failure occurs at constant irradiance I_f in excess of ~450 GW/cm² for pulses longer than ~50 ps. This value is consistent with irradiance values (~400 GW/cm²) inferred from some recent megawatt peak power demonstrations. Correspondingly, the maximum damage-free peak power scales as $I_f \times A$, with A being the mode field area, until a critical value P_c is attained. Beyond P_c, the in-fiber guided beam is expected to undergo self-focusing (SF), which is the spatial manifestation of the irradiance-proportional increase in the medium refractive index. Ultimately, SF results in spatial beam collapse and ensuing runaway irradiance divergence, which leads to damage. Recent analyses[14,15] suggest that P_c for the fundamental mode (LP_{01}) of silica fibers is independent of mode field area and can be approximated with that of a Gaussian beam in bulk silica. This is given by

$$P_c \quad \frac{\lambda^2}{2\pi n_0 n_2} \tag{16.8}$$

Here λ is the wavelength, n_0 is the refractive index, and n_2 is the second-order nonlinear refractive index coefficient, which is ~2.6 × 10⁻²⁰ m²/W for pulses greater than 1 ns (and as low as 2.2 × 10⁻²⁰ m²/W in shorter pulses, for which electrostrictive contributions to n_2 vanish). At λ = ~1.06 μm, which is a common wavelength for Yb-doped fibers, P_c for fused silica is in the ~5 MW range, which is actually higher than for many optical materials. For example, solid-state laser crystals of widespread use for high-power pulsed lasers, such as Nd:YAG and Ti:sapphire, exhibit greater values for n_2—namely, ~3 × 10⁻¹⁹ and ~3 × 10⁻²⁰ m²/W, respectively.[16] This corresponds to P_c = ~330 kW (Nd:YAG at 1.06 μm) and ~2 MW (Ti:sapphire at 0.8 μm). However, bulk lasers and amplifiers based on these and other crystals can be designed for damage-free operation in the SF regime, which is usually accomplished by ensuring that the crystal's physical length does not exceed the SF collapse length L_c, which for a Gaussian beam is given by[17]

$$L_e \quad \frac{A}{\lambda \sqrt{\frac{P}{P_c} - 1}} \tag{16.9}$$

Here A is beam area and P is peak power. Even in the most extreme single-mode Yb-doped fibers (e.g., $A = \sim5000\ \mu m^2$), corresponding to the LP_{01} mode in a ~100-μm core fiber, and for P exceeding P_c by a mere 10 percent, L_c is already less than 15 mm—that is, it is much shorter than any fiber length of practical usability. In bulk crystals, however, A can be much larger, and significant energy can be extracted even in a few centimeter lengths.

So far, the high-peak-power onset of SF in rare-earth-doped, as well as transport, fibers has proved a rather elusive phenomenon to conclusively observe. There still exists experimental evidence with which the theory has not been completely reconciled. One of them pertains to the SF behavior of high-order fiber modes, which are calculated to exhibit intrinsic P_c values that are several times higher than LP_{01}, but which are then predicted to undergo spatial instabilities during propagation and to incur collapse at about the same power level of 4 to 5 MW.[15] Further analysis is warranted, as such a description does not seem consistent with the reported delivery of greater than 20 MW peak power pulses with a highly multimode passive fiber.[18] Moreover, alternative analyses have recently been presented according to which the in-fiber P_c for LP_{01} does depend on mode field area, due to the onset of spatial beam filamentation, which is caused by spatial hole burning in fiber amplifiers.[19]

Fiber facets are normally subject to damage at much lower irradiance values than the fiber bulk. However, this difference is largely extrinsic and ascribed to the fiber facet's much lower optical quality, which is due to imperfections left behind by the cleaving and polishing processes. This issue can be circumvented by the use of beam-expanding end caps (Fig. 16.2), which are segments of nonguiding silica fused to the output fiber facet. The fiber-guided beam can then freely diffract within the end cap to attain much lower irradiance

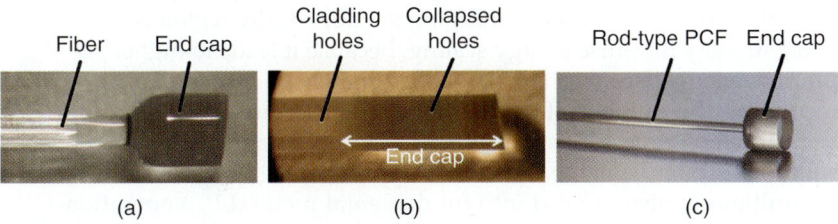

FIGURE 16.2 Side-view photographs of beam-expanding end caps. (*a*) End cap obtained by fusion splicing to the main fiber a piece of large, coreless fused silica fiber that is successively angle-polished. (*b*) Photonic crystal fiber (PCF), in which an end cap has been obtained by thermally collapsing the axial hollow channels (thus suppressing wave guidance) over a short (~2 mm) portion of the fiber end. (*c*) Rod-type PCF (greater than 1.5 mm outer diameter) fused to a large (~8 mm diameter), antireflection-coated end cap (bulk-fused silica) by means of a high-temperature torch.

values when reaching the vulnerable air-end cap interface. No disturbance to the beam quality is incurred, as long as the freely expanding beam does not overfill the refractive-index-homogeneous portion of the end cap's cross-sectional area. Fiber end capping is now ubiquitous and represents a key enabler for both pulsed and CW high-power lasers and amplifiers.

16.3 Fiber Laser Trades for High-Pulse-Power Operation

This section reviews design solutions that must be considered in the development of high pulse energy and peak power fiber-based laser sources. Such solutions are conceived to maximize performance while addressing and overcoming the challenges described in Sec. 16.2. Many of the traditional criteria for well-architected pulse fiber amplifiers originated in the early application of those amplifiers to optical telecommunications; thus, they emphasize gain maximization, noise management, low dispersion, and operation at wavelengths of ultra-low loss for fused silica (such as ~1.5 μm). Within this framework, fibers represented a very different medium as compared with bulk DPSS lasers. In the cases of interest in this chapter, however, such differences become less pronounced; therefore, a primary goal is to correctly negotiate the trades that enable power scaling without giving up key advantages, including efficiency and beam quality, of fibers over other solid-state sources. The following discussion spans three main areas: fiber designs, ASE management, and source architecture.

16.3.1 Type of Fiber

The selection criteria for the best fiber candidates to enable high peak power operation are, in principle, quite simple: Using a large core (and correspondingly a large mode field area A) and relatively short fiber is the most straightforward approach to minimizing the NLE strength given by Eq. (16.1). Enlarging the doped core region is also a valid strategy for pulse energy scaling, because it leads to higher saturation energy [see Eq. (16.7)]. Unfortunately, it also amounts to increasing the number of guided modes and, therefore, degrading beam quality, unless the core NA is reduced or additional special solutions are implemented. A classic approach involves LMA fibers, or multimode fibers forced into fundamental mode (LP_{01}) operation by matching the launched optical field to the fiber's LP_{01} mode and using selective bend loss to suppress higher-order modes. Unfortunately, the scalability is practically limited to mode field areas (MFAs) of ~700 μm², due to vanishing intermodal bend-loss discrimination. Moreover, tight bending distorts the guided mode field, thus greatly reducing its effective area,[20] which ultimately offsets the benefit of a large core. Finally, even in low-M^2 fibers, the spatial interference of residual higher-order modes causes beam-centroid wander and far-field beam-pointing instability, which especially hampers their use in remote sensors and materials processing.[21]

Another approach is to resort to waveguides designed to be effectively single-mode despite very large cores. An important example is photonic crystal fibers (PCFs). In these fibers, an air-clad multimode waveguide (designed for pump guiding) embeds a hexagonal array of axial holes in which a number of central holes are replaced with a solid hexagonal array of rare-earth-doped elements to form the core. The refractive indices of the pump waveguide and core are controlled by the axial-hole diameter-to-spacing ratio and the amount of index-lowering fluorine doping, respectively. They are also fine-tuned to provide a core-to-pump waveguide index contrast that supports single transverse mode operation. This microstructure is realized through a stack-and-draw fabrication process that offers significantly higher design precision compared with the chemical deposition and solution doping that are normally used for standard fibers. In particular, very-low-NA cores can be reproducibly obtained. This enables single-mode core size scaling (up to 100 μm diameter, to date), even though bend loss ultimately becomes significant, and thus the PCF must be enclosed in a rigidly straight overcladding that forms the main feature of the so-called rod-type PCF. Performance examples of this are reviewed in Sec. 16.4.

Similar to standard polarization-maintaining (PM) fibers, birefringence along a preferential PCF core axis can be elasto-optically imparted by the frozen-in stress resulting from incorporation of two sets of borosilicate axial rods thermally mismatched to the PCF silica-based cladding. Unlike in standard PM fibers, the stress rods contribute to light confinement and are located in close proximity of the fiber core, which affords high birefringence encompassing the large core area without the need for excessive (and hence structurally unsound) boron doping. Because of the exceedingly low core-to-inner cladding index contrast, the induced birefringence can be sufficient for light polarized along the PCF fast axis to experience an effective index lower than the inner cladding and, as a consequence, leak out of the core, which is known as *polarizing* behavior.

Alternative fiber concepts, such as leakage channel[22] and chirally coupled core fibers[23] (which are discussed in Chap. 15) have been proposed to foster MFA scaling with good beam quality. In the limit of very large cores, however, such diverse single-mode solutions seem inevitably to converge toward waveguides that must be kept nearly straight to escape bend loss or bend-induced mode field compression. Another adopted approach relies on fibers that selectively operate in a single, cylindrically symmetric large-area high-order mode (HOM), such as LP_{04} or LP_{07}. This mode is selected at the fiber input by means of a judiciously designed long-period grating.[24] The rationale is that HOMs are subject to lower bend-induced area shrinkage[24] and intermodal mixing,[25] which should enable tight bending with concomitant preservation of modal purity and low NLE. However, the HOM beam quality (BQ) factor is modest (e.g., M^2 for LP_{04} is greater than 6.7, according to Ref. 26), which, in high-power applications, requires a lossy bulk mode adaptor to recover good output BQ. In addition, the HOM peak irradiance is locally much higher than LP_{01} (e.g., a 2100-μm^2-area LP_{07} has the same peak irradiance as a 316-μm^2-area LP_{01}), which poses a bulk-damage hazard at high peak power. Finally, energy extraction efficiency and multimode behavior issues may arise due to spatial hole burning.

As for core enlargement, the need for minimization of the fiber length toward NLE containment poses important fiber design and laser architecture challenges. In some cases, shortening the fiber leads to optical efficiency degradation caused by incomplete pump absorption. This problem can be avoided in two ways: One is to increase the rare-earth-dopant density. However, this technique is hardly an open-ended solution due to the onset of doping-ion clustering at high enough doping concentrations, resulting in excited-state quenching and ensuing reduction of the excited-state lifetime. The second approach is to increase the pump and doped-core overlap, which can be done by design via the geometric increase of the fiber core:cladding area ratio in double-clad fibers. This approach may result in reduced brightness acceptance in the fiber pump cladding and has only recently become a truly viable option due to the ever-increasing brightness of pump diode lasers, which currently exceeds 10 MW/(cm^{2}sr) within commercially available, fiber-delivered, angle- or polarization-multiplexed single-stripe diodes.

Even assuming that effective pump absorption has been obtained from maximization of such core:cladding area ratio, any net reduction of the doped area volume (i.e., reduction in the total number of rare earth ions) ultimately results in a lower achievable small-signal gain, simply due to the reduced energy storage capacity. Accordingly, the extractable energy (proportional to the small-signal gain) will be degraded, which means that when used as an amplifier, such fiber must be seeded at high power for acceptably efficient operation.[27]

Yet another issue with short-fiber amplifiers is thermal management. Fibers are often referred to as a solid-state gain medium of extremely favorable thermal properties, as strikingly proved by the commercial availability of multikilowatt average power, single transverse mode, near-diffraction-limited lasers. However, this advantageous feature is inherently related to the possibility of distributing thermal loads over long stretches of fiber, resulting in negligible thermo-optical aberrations. Conversely, when the fiber must be shortened to avoid NLE, the thermal load increases at a given pump and output power. In such a situation, a very low core NA fiber may experience thermally induced refractive index changes that are of comparable magnitude with respect to the "cold" fiber core–cladding index step.

The net effect of such occurrence is that the in-core confinement of high-order modes may increase, resulting in BQ degradation.[28]

16.3.2 Amplified Spontaneous Emission Management

The possibility of suppressing or managing ASE by means of special characteristics engineered in gain fibers has attracted some interest over the years, though research efforts in this area have not yet yielded a practically viable and effective recipe. A relatively simple design solution is to lower the core NA, which (in the case of total internal reflection wave guidance) quadratically reduces the fraction of spontaneous emission captured in the fiber core.

Another interesting recent development has been the introduction of microstructured gain fibers, which can incorporate photonic stop bands for light that propagates in specific wavelength intervals, thus acting as a distributed spectral filter. This concept has been demonstrated by Goto et al.[29] in the suppression of ~1030-nm ASE within a Yb-doped solid-core photonic band-gap fiber.

A more common approach to managing ASE for high-pulse-energy generation entails proper fiber laser and amplifier architecture design. As shown in other laser sources, a widely applicable ASE mitigation strategy is gain staging. This solution is naturally supported by master oscillator power amplifier (MOPA) architectures and consists of distributing the desired optical gain over several pieces of doped fiber separated by band-pass spectral filters centered at the signal wavelength. The filters allow for cumulative amplification of the signal pulses across the entire amplifier chain, while rejecting, after each stage, all ASE generated outside their pass band. This concept, which requires the signal spectral bandwidth to be much narrower than the ASE spectrum, is frequently realized in gain media embedded within amorphous hosts, such as glass fibers, where energy transitions incur severe Stark broadening and often exhibit line widths of tens of nanometers.

Of course, spectral filtering is not effective for ASE that lies within the filter pass band (in-band ASE), which is best managed by time-domain techniques. A common solution is to insert an electro-optic or acousto-optic amplitude modulator at some point along the amplifier chain. The modulator is then used as a gating optical switch that provides high transmission in a short-time window centered at each pulse and low transmission at any other time. If the modulator exhibits sufficiently high on/off extinction (e.g., ~20 dB or higher) and the time window is a sufficiently small fraction of the pulse repetition period, then substantial suppression of in-band ASE is possible. A drawback to this approach is that it increases the system complexity and parts count. However, this drawback is somewhat mitigated by the fact that in a well-engineered MOPA source, most in-band ASE is generated in the high-gain, low-power first stages of the amplifier chain. At low-pulse power levels, monolithic fiber-pigtailed modulators are commercially available, which can be conveniently fusion spliced in the chain and which avoid alignment-sensitive components that might hinder ruggedness.

Finally, in cases of low-PRF operation, pulsed pumping can be effectively implemented. This solution, often adopted for bulk DPSS lasers, is especially viable due to the frequent use of single-emitter diodes (which can be easily modulated at high rates) for fiber pumping. This technique entails the judicious interruption of optical pumping between pulses to prevent ASE buildup. However, maximum improvement compared with CW pumping is anticipated for pulse periods longer than the rare earth inversion (i.e., excited-state) lifetime and for sufficiently long fibers, for which the ASE generated during one pump pulse ends up propagating through a portion of fiber gain medium that is no longer inverted and that corresponds to a significant fraction of the overall fiber length. For silica-based Yb- and Er-doped fibers, for example, this regime approximately corresponds to pulse repetition rates less than 1 kHz and less than 100 Hz, respectively.

16.3.3 MOPA versus Power Oscillators

In MOPA sources, a low-power laser acts as the seeder for a single- or multistage amplifier. As such, MOPAs enable function separation and independent optimization of the spectral and temporal aspects of the pulse formation and the generation of high power. In many cases documented in the literature, pulsed fiber-based MOPAs have featured a bulk solid-state laser as the master oscillator. A very common choice, borrowed from optical telecommunications and highly compatible with end-to-end all-fusion-spliced-fiber MOPA designs, is to use a fiber-coupled semiconductor laser. Pulsed operation from such lasers is obtained through gain switching with a pulse-driving current or by means of an external electronically controlled modulator. These techniques support nearly one-to-one mapping of arbitrary electrical-to-optical pulseforms and, therefore, can provide the highest degree of active control on the pulse format, including continuous and independent adjustability of pulse duration (from a few picoseconds to CW), shape and repetition rate, as well as minimal-jitter synchronization to a trigger signal. Moreover, high spectral purity with support for single-frequency and near-Fourier-transform-limited operation can be obtained by means of specially designed external-cavity or distributed-feedback architectures. A disadvantage of semiconductor lasers is that their limited energy storage results in nanojoule-level pulse energies for few-nanosecond pulses, which burdens the following fiber amplifiers with the task of supplying a very large gain. This problem has reportedly been circumvented by replacing the semiconductor lasers with microchip lasers. Such bulk

DPSS sources exhibit an extremely compact form factor that is very amenable to integration in a fiber-based system, as well as a short cavity that naturally yields pulses of ~1 ns or shorter at multi-kilohertz repetition rates, with pulse energies ~10 µJ, while providing support for single longitudinal mode operation. However, being passively Q switched, microchip lasers do not exhibit a dial-in electronically controllable repetition rate, and they incur substantial temporal jitter, unless some form of injection seeding is implemented.

In addition to pulse control, a fundamental advantage offered by MOPA architectures is the possibility of staging the gain, which is a key ASE mitigation strategy (see earlier discussion). This advantage also permits such architectures to optimally deploy different gain fibers for different segments of the amplifier, which can be leveraged for simultaneous maximization of efficiency and minimization of nonlinearities. However, gain-staged MOPAs require a plurality of components, including interstage isolators and spectral filters or active time gates, which may amount to high complexity and cost.

To provide a simpler alternative, considerable research and development has been devoted to the power scaling of actively Q-switched fiber lasers, which can, in principle, replace a MOPA with a mere power oscillator of significantly lower parts count. In these sources, a linear or ring fiber laser cavity incorporates an intracavity electro-optic (such as a Pockels cell) or acousto-optic switch, according to a design reminiscent of bulk DPSS lasers. For millijoule-pulse-energy operation, however, the intracavity irradiance usually exceeds the capabilities of traditional fiber-coupled telecommunications-type components; therefore, the laser cavity must include bulk free-space modulators and coupling optics.

An important limitation of Q-switched fiber lasers is that the fiber, and hence the laser cavity, length usually falls multi-meter range. Because the pulse duration in Q-switched lasers is proportional to the laser cavity length (for given cavity losses and gain medium inversion above transparency[30]), pulses in Q-switched fiber lasers are usually tens of nanoseconds long, which hampers applications seeking high peak powers. This problem has recently been addressed by the implementation of rod-type PCF as the gain medium, which provides ample energy storage in a significantly shorter length (see Sec. 16.4 for related results). Another issue with actively Q-switched lasers is that even though the pulse repetition rate is electronically controlled by the driving signal operating the intracavity switch, pulse duration (and corresponding peak power) and repetition rate are not independently settable. The pulse-PRF coupling stems from the fact that for a given cavity length, the pulse duration increases exponentially as the built-up population inversion decreases to values close to the gain medium inversion at transparency.[30] In other words, for a given amount of available pump power, the pulse duration will increase at higher PRF.

16.4 High Pulse Energy and Peak Power Fiber Amplifiers: Results

This section reviews milestones recently obtained in the pursuit of in-fiber pulse power scaling for the important regime of nanosecond-pulse durations and multikilohertz PRFs. Special emphasis is placed on work that aimed at, and clearly achieved, concurrent optimization of multiple performance parameters, including high-peak and average-power, beam quality, spectral brightness, polarization, pulse contrast, and pulse format control.

16.4.1 Single-Stage Fiber Amplifiers

Early demonstrations of high spatial and spectral brightness pulse fiber amplifiers trace back to the introduction of bend-loss mode filtering and input mode matching (discussed in Sec. 16.3), which enabled the essentially single transverse mode operation of lower-nonlinearity, rare-earth-doped LMA fibers. One of the first examples of LMA fiber usage for high-BQ pulse amplification is provided in Ref. 31, which refers to a fiber amplifier consisting of ~7-m-long, 25 µm/0.1 core diameter/NA, Yb-doped fiber coiled at a 1.67 cm diameter and seeded by a passively Q-switched microchip laser (780 ps/ 8.5 kHz pulse duration/PRF) and backward pumped (for maximum reduction of NLE interaction length) at 975 nm wavelength. As shown in Fig. 16.3, this amplifier produced a maximum pulse energy of ~255 µJ and corresponding peak and average power in excess of 300 kW and 2.2 W, respectively, within a diffraction-limited output beam ($M^2 < 1.1$). The pulse spectrum exhibited significant nonlinearities and primary SRS and FWM, but unappreciable SPM broadening. At the time, this performance represented the highest peak power obtained from a diffraction-limited BQ fiber amplifier and triggered significant interest in the development of LMA gain fibers for high pulse power applications.

This result was later surpassed by the introduction of larger-core, single transverse mode Yb-doped PCF. Figure 16.4 documents the performance of an air-clad, 40-µm-core (greater than three times larger MFA compared with Ref. 31), 2.5-m-long Yb-doped PCF amplifier, also seeded by a passively Q-switched Nd:YAG microchip laser (13.4 kHz PRF, subnanosecond-pulse duration, 1064-nm wavelength) and backward pumped at 975 nm.[32] A peak power in excess of 1.1 MW (pulse energy/average power = ~0.54 mJ/~7.2 W) was obtained (corresponding to optical gain greater than 20 dB), within a diffraction-limited beam of less than 10 GHz effective spectral bandwidth.

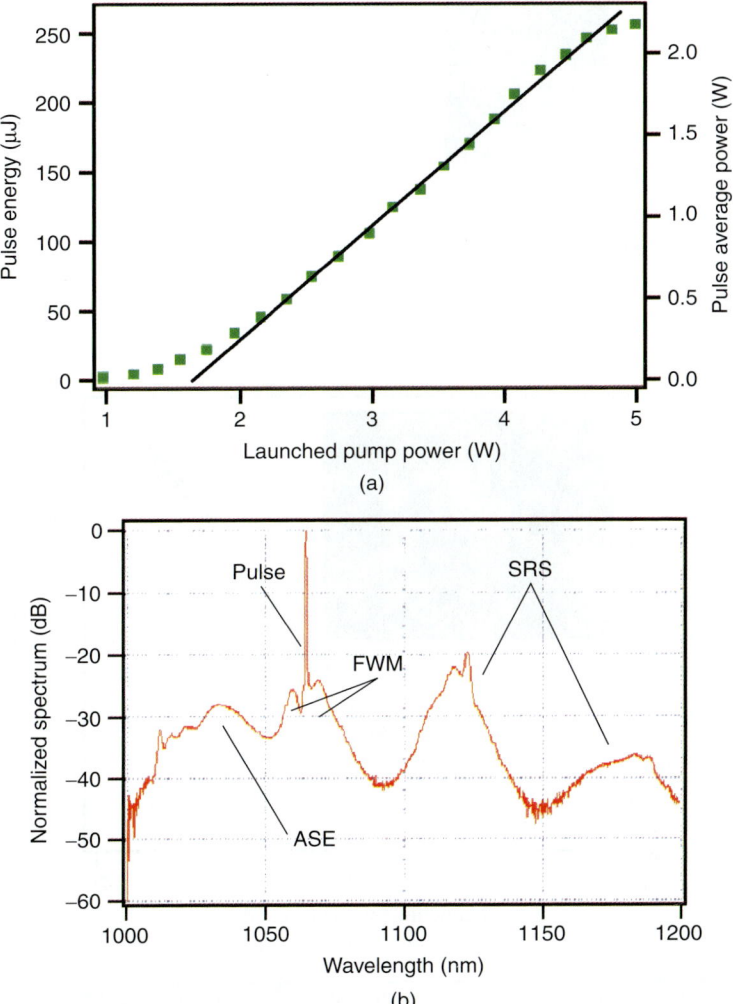

FIGURE 16.3 (*a*) Pulse energy and corresponding average power vs. pump power for a master-oscillator/power-amplifier (MOPA) architecture featuring 8.5 kHz pulse repetition frequency and 780 ps pulse duration microchip seed laser and 25μm-core bend-loss mode-filtered Yb-doped fiber amplifier. (*b*) Broadband spectrum of MOPA output recorded at maximum pulse energy (~255μJ). FWM: four-wave mixing.

As inferred from the broad-scale spectrum (Fig. 16.4*b*), further power scaling in such an amplifier was precluded by the significant buildup of ASE in the 1030–1040 nm region.

Similar results were obtained at about the same time by other groups experimenting with the same optical architecture, though terminated by smaller core LMA fibers. In such cases, greater nonlinearity was observed, in addition to similar levels of ASE.[33]

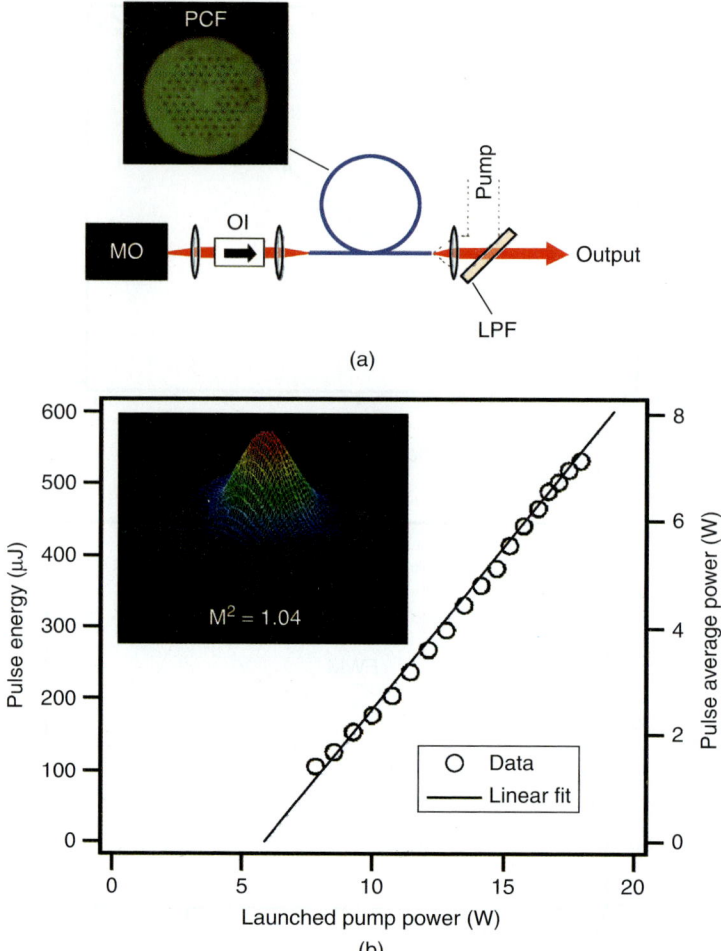

FIGURE 16.4 (a) Schematic architecture of MOPA source consisting of microchip laser seeder (pulse repetition frequency = 13.4 kHz, sub-nanosecond pulse duration), followed by Yb-doped 40 μm-core photonic crystal fiber (PCF) amplifier. The single-mode PCF was 2.5-m long and could be coiled at a ~35–40-cm diameter, without inducing appreciable loss. MO: master oscillator; OI: optical isolator; LPF: low pass filter. (b) Pulse energy and corresponding average power vs. pump power for this MOPA (Inset: Near-field intensity profile of MOPA output at maximum pulse energy). (c) Broadband logarithmic spectrum of MOPA output recorded at maximum pulse energy. ASE: amplified spontaneous emission; SRS: stimulated Raman scattering. (Inset: Corresponding pulse temporal profile.)

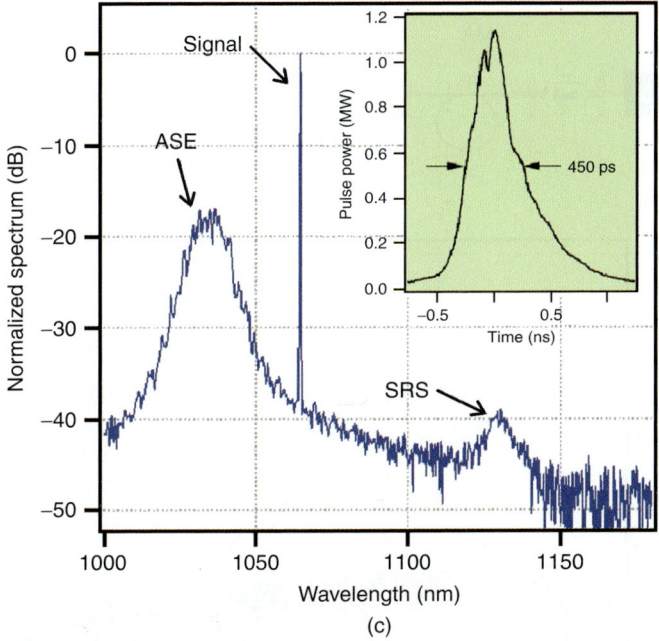

FIGURE 16.4 (*Continued*)

16.4.2 Gain-Staged MOPAs

The ASE issues that limited the performance of the fiber amplifiers described earlier have since been circumvented in a straightforward manner by resorting to gain-staged chains (see discussion in Sec. 16.3). Figure 16.5 documents, for example, the performance of a MOPA similar to that in Fig. 16.4, but now including a 30-μm-core Yb-doped LMA fiber preamplifier (followed by an ASE-rejecting spectral filter and optical isolator) between the same microchip laser seeder and the PCF power amplifier.[34] The preamplifier is used to boost the seed signal pulse energy to ~50 μJ, which allows the PCF final amplifier to achieve pulse energies even larger than those in Fig. 16.4, while supplying optical gain approximately 10 times lower. This solution allowed the PCF to be shortened (1.5 m vs. 2.5 m) and resulted in lower nonlinearity, as well as improved ASE containment and pulse-CW background contrast.

The same gain distribution strategy was adopted by Torruellas et al.[35] to achieve megawatt peak power levels by means of a final amplifier featuring a Yb-doped, flattened-mode LMA fiber.[36]

As evidenced by the observation of significant SRS, FWM, and phase modulation nonlinearities, peak power scaling beyond the ~1-MW level must require a substantial increase in MFA compared

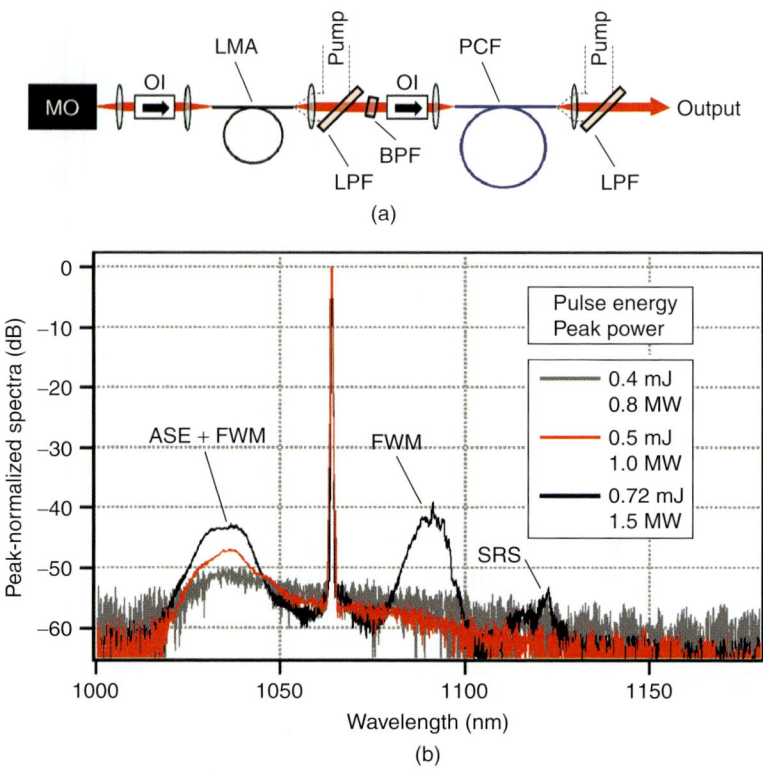

FIGURE 16.5 (a) Schematic view of 2-stage MOPA architecture, terminated by an Yb-doped 40-μm-core PCF amplifier (see text for details). (b) Broadband logarithmic spectrum of MOPA output at different pulse energies and corresponding peak powers (pulse repetition frequency = 13.4 kHz). LMA: large mode area; BPF: band pass filter.

with the fibers used in the work described earlier.[31–35] This development has been pursued either by pushing the technology of high-BQ, effectively single transverse mode fibers or by resorting to multimode fibers. Because of the inherent fabrication limits for standard fibers (see the discussion in Sec. 16.3), the single-mode route has been pursued primarily with the aid of microstructured, stack-and-draw fibers, such as PCF. A notable development in this area has been the introduction of very large-core, rod-type PCF, in which the issue of increased micro- and macrobending sensitivity induced by the core-cladding refractive index step reduction (which is necessary for single-mode operation in large cores) is circumvented by enclosing the fiber in a greater than 1-mm-diameter, rigidly straight glass overcladding. By practicing this concept, peak powers in excess of 2 MW were first reported in 2006.[37] As shown in Fig. 16.6, the result was obtained

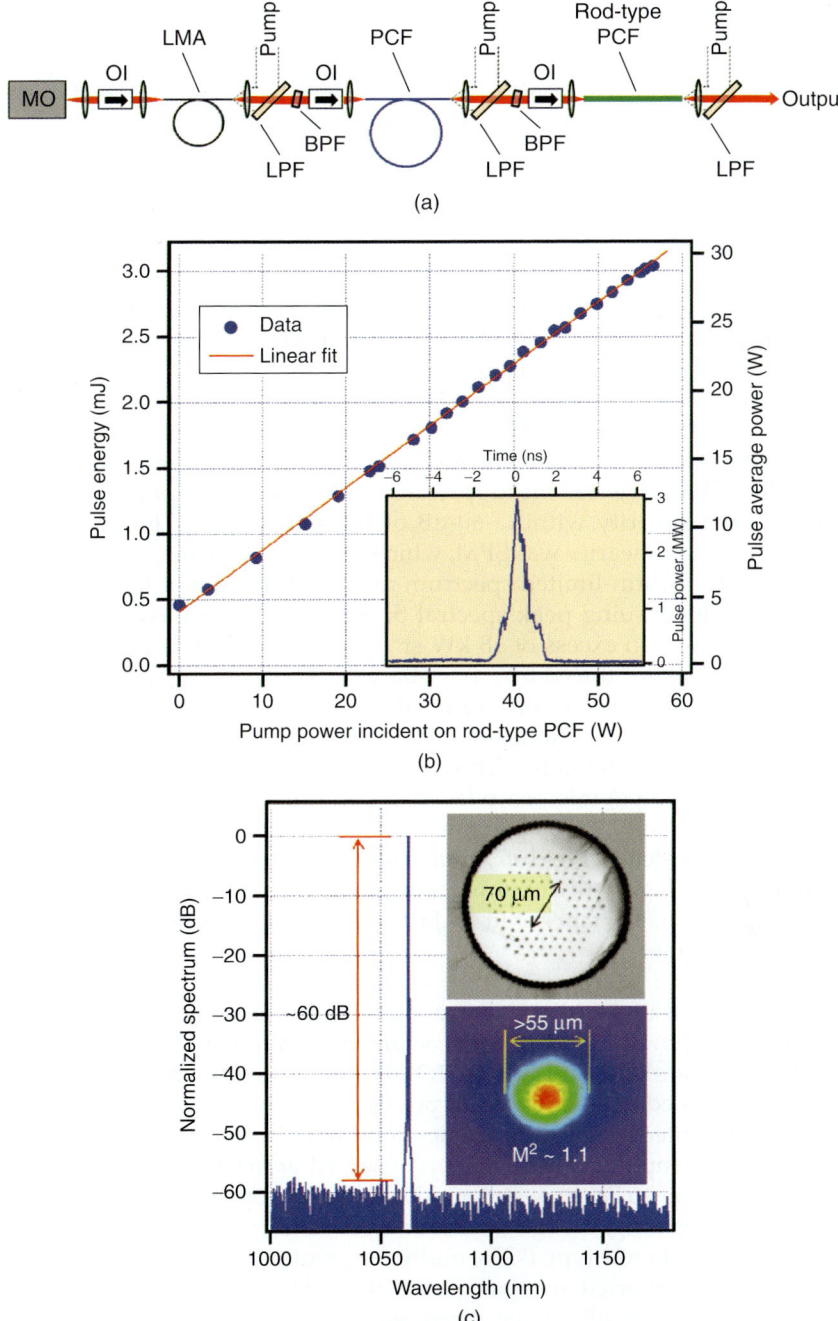

FIGURE 16.6 (*a*) Scheme of 3-stage MOPA architecture featuring Yb-doped rod-type PCF (70 μm-core diameter) as the final amplifier. (*b*) Pulse energy and corresponding average power vs. pump power from this MOPA (Inset: Pulse temporal profile recorded at maximum energy ~3 mJ). (*c*) Broadband logarithmic spectrum of MOPA output at maximum pulse energy; Insets: (top) Cross-sectional photograph of 70 μm-core diameter Yb-doped rod-type PCF used as the final amplifier and (bottom) near-field image of output beam as recorded at maximum pulse energy.

in a three-stage MOPA driven by a single-frequency 1062-nm, Nd-doped Lanthanum Scandium Borate (Nd:LSB) microchip laser (~1 ns pulse width, 9.6 kHz PRF), featuring interstage optical isolation and ASE rejection via band-pass filtering, and including a preamplifier (25/250 μm core/cladding solid-silica LMA Yb-doped fiber), booster amplifier (40/170 μm Yb-doped PCF), and power amplifier (70/200 μm Yb-doped rod-type PCF) stage, all backward pumped at 975 nm. Due to its very low-core NA design and the rod-type PCF, it emitted a near-diffraction-limited, fundamental-mode output of $M^2 = $ ~1.1, despite the very large core diameter (70 μm).

The ensuing MFA was in excess of 2000 μm^2. As shown in Fig. 16.6b, the MOPA generated ~1 ns pulses of pulse peak power P_{peak} in excess of 3 MW and corresponding energy/average power of 3 mJ/30 W, with excellent suppression of ASE and no appreciable off-band nonlinearity within ~60-dB of the signal peak. The only appreciable nonlinearity was SPM, which broadened the microchip-laser near-transform-limited spectrum to a FWHM bandwidth $\Delta\nu = $ ~20 GHz. The ensuing peak spectral brightness (defined as $P_{peak}/[\lambda^2(M^2)^2\Delta\nu]$) was in excess of 18 kW sr^{-1}Hz^{-1} cm^{-2}, which is reportedly the highest produced by a fiber-based optical source.[37] A similarly high-peak-power (~3 MW) result was reported a few years later by Schmidt et al.[38] in a two-stage MOPA featuring a slightly larger-core (80 μm diameter) Yb-doped rod-type PCF seeded by a shorter-pulse (85 ps) microchip laser and emitting an output beam of good BQ ($M^2 = $ ~1.4). An extreme realization of the rod-type PCF concept, illustrated in Fig. 16.7, entails a MOPA system nearly identical to that in Fig. 16.6, but now featuring a 100-μm core diameter (the largest core PCF fabricated to date), Yb-doped rod-type PCF as the final amplifier.

From such a source, ~10 kHz, subnanosecond pulses of peak power in excess of 4.5 MW (corresponding to pulse energy ~4.3 mJ and average power ~41 W) were obtained within a high-spectral brightness, high-BQ beam ($M^2 = $ ~1.3).[39] This result amounts to the highest combined pulse energy and peak power generated by a near-diffraction-limited, effectively single-mode fiber-based laser source to date. The temporal pulse breakup observed at maximum power (see inset in Fig. 16.7b) may be an indication of an operation regime close to self-focusing, though further analysis is warranted.

In addition to rod-type PCFs, multimegawatt peak power results have also been reported in large-core multimode fibers. In most cases, however, these results have been accompanied by somewhat degraded beam qualities. For example, Cheng et al.[40] reported peak power in excess of 2.4 MW (in ~4-ns pulses of ~9.6-mJ pulse energy), as well as pulse energy of ~82 mJ (the highest extracted from a rare-earth-doped fiber) within a beam of $M^2 = $ ~6.5 from a coiled, Yb-doped, solid-silica, 200-μm core fiber, which was used as the final stage in a

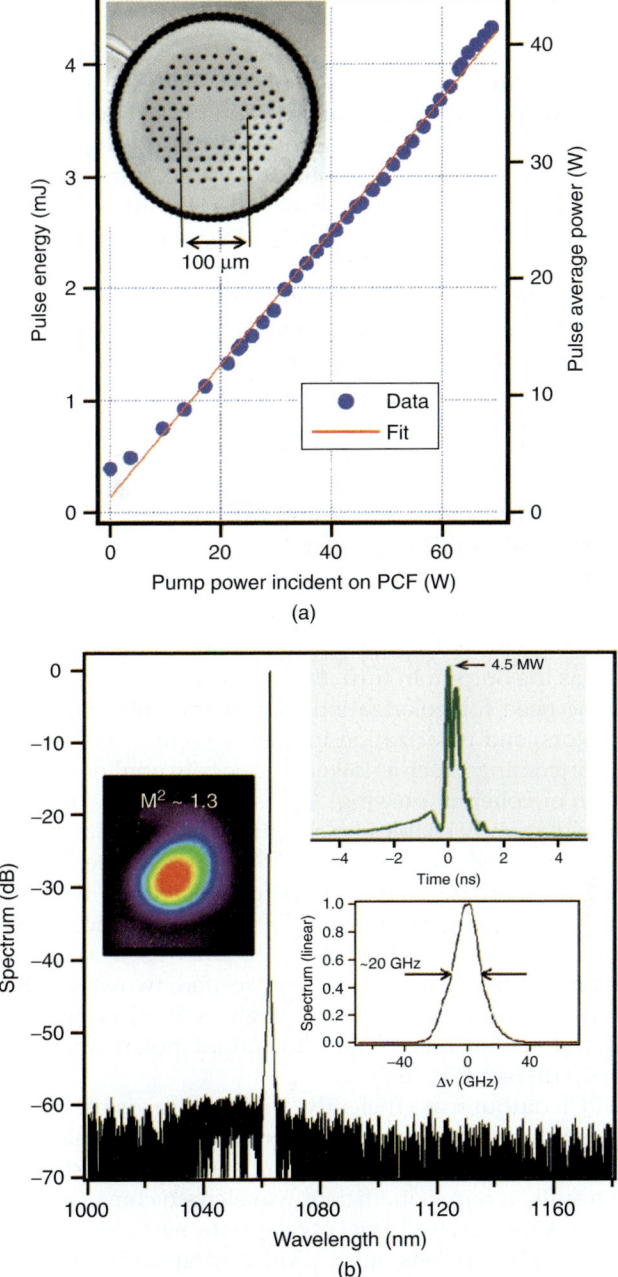

FIGURE 16.7 (a) Pulse energy and corresponding average power vs. pump power, generated by MOPA similar to that in Fig. 16.6, but featuring a Yb-doped 100 μm-core rod-type PCF as the final amplifier (Inset: Cross-sectional photograph of this PCF). (b) Broadband logarithmic spectrum of PCF output recorded at maximum pulse energy (~4.3 mJ). Insets: Near-field image of output beam, pulse temporal profile, and high-resolution linear spectrum of the PCF output, all recorded at maximum pulse energy.

pulse-pumped MOPA seeded by a sub-100-Hz PRF, relatively broadband (~1-nm spectral width) pulsed diode laser. Even higher peak power (~5.8 MW, corresponding to 6 mJ pulse energy and ~60 W average power) has been obtained from a 140-μm-core, Yb-doped multimode fiber amplifier seeded by ~10-kHz-PRF ~1-ns pulses and emitting an output beam of M^2 = ~10.[41] In one case,[42] a highly multimode fiber amplifier (80 μm/0.06 core diameter/NA) has been reported to generate peak powers in excess of 5 MW, while emitting a beam of M^2 = ~1.3 obtained via bend-loss mode filtering. However, the bend-induced mode discrimination in such an extreme case is so marginal that stably reproducing a similar result without exquisitely controlled laboratory and experimental conditions appears to be a formidable challenge.

16.4.3 Polarization-Maintaining MOPAs and Wavelength Conversion

The results described in the previous discussions are all based on non-PM fibers, which leads to an environmentally unstable polarization state in the pulsed signal traveling through the amplifier(s) and delivered as the output. In turn, this introduces design complication, such as the need for polarization controllers, polarization-independent isolators, and polarization-insensitive optics, as well as performance shortcomings, such as lower efficiency in nonlinear wavelength conversion or coherent phasing. As discussed in Sec. 16.3.2, the PCF technology naturally enables polarization control in large-core fibers and has represented an effective platform for pulse peak power scaling coupled to a robustly linearly polarized output. For example, a polarizing 40-μm-core Yb-doped PCF, designed in accordance with criteria detailed by Schreiber et al.,[43,44] was used as the terminal amplifier in an all-PM, microchip laser seeded, two-stage MOPA that generated 1062-nm peak power as high as 800 kW, with excellent spectral and spatial quality and an output polarization extinction ratio of ~20 dB (see Fig. 16.8).

The PCF output was efficiently converted to visible and ultraviolet (UV) by means of nonlinear external crystals, achieving record peak power for a fiber-based source at such wavelengths.[45,46] Other notable results in terms of efficient wavelength conversion to visible and UV and midinfrared (via optical parametric generation) with similar pulse format (1-ns pulse width and up to 50-kHz PRF) have also been reported for a single-stage (microchip laser seeded) MOPA featuring a ~30-μm-core PM, Yb-doped LMA fiber, although with somewhat lower peak power.[47] Other wavelengths of scientific or industrial interest have also been reached by leveraging the flexibility of pulsed-fiber MOPA architectures. For example, a PM Yb-doped LMA fiber similar to that in Ref. 47 was used as the terminal amplifier for a 0.7-ns pulse/100-kHz-PRF, Yb-fiber-based multistage MOPA

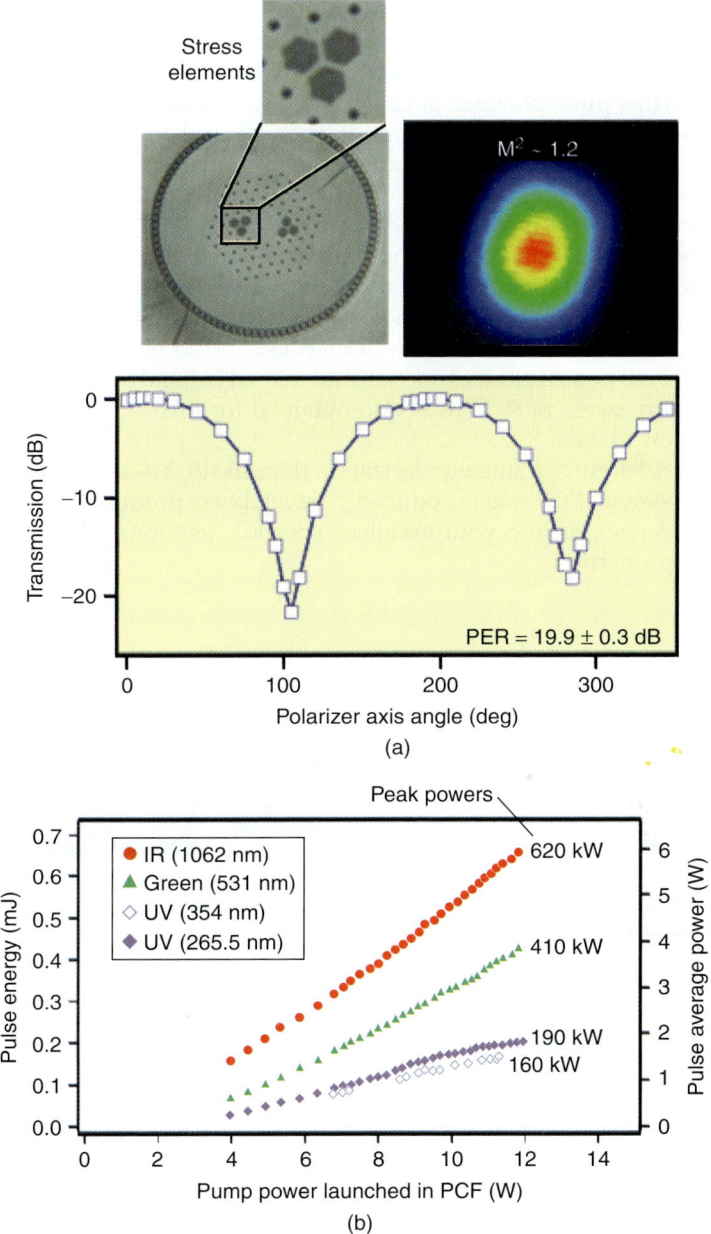

FIGURE 16.8 (a) Cross-sectional photograph of polarizing 40 μm-core Yb-doped PCF and recorded near-field image of its output beam. The measured polarization extinction ratio (PER) for light launched in the core of this fiber approaches 20 dB. (b) Pulse energy and corresponding average power vs. pump power for fundamental output (1062 nm) and harmonics, generated by two-stage MOPA featuring the polarizing 40-μm-core Yb-doped PCF as the final amplifier.

(25 kW output peak power) that was dual seeded by 1064- and 1059-nm-wavelength, pulsed diodes to produce terahertz radiation via difference-frequency generation.[48]

Further pulse power scaling in a linearly polarized format has been enabled by the extension of polarization control to very-large-core rod-type PCF. Figure 16.9 documents an exemplary performance of a single-polarization, Yb-doped, 70/200-μm core-/cladding-diameter, ~85-cm-long rod-type PCF incorporated in a MOPA driven by an actively triggered and pulse programmable, single-frequency 1064-nm diode laser. This source produced peak power and pulse energy in excess of 2 MW and 2 mJ, respectively (~0.75-ns pulses at 15-kHz PRF), within a near-diffraction-limited ($M^2 < 1.2$) output of greater than 14-dB polarization extinction and ~20-GHz bandwidth. Average power in excess of 80 W was also obtained for PRF = 50 kHz (pulse duration ~1.2 ns).[49]

Very recently, a single-polarization (Fig. 16.10), Yb-doped 100-μm-core rod-type PCF was introduced,[50] which bears promise for further pulse power scaling with excellent spectral, temporal, and spatial beam properties.

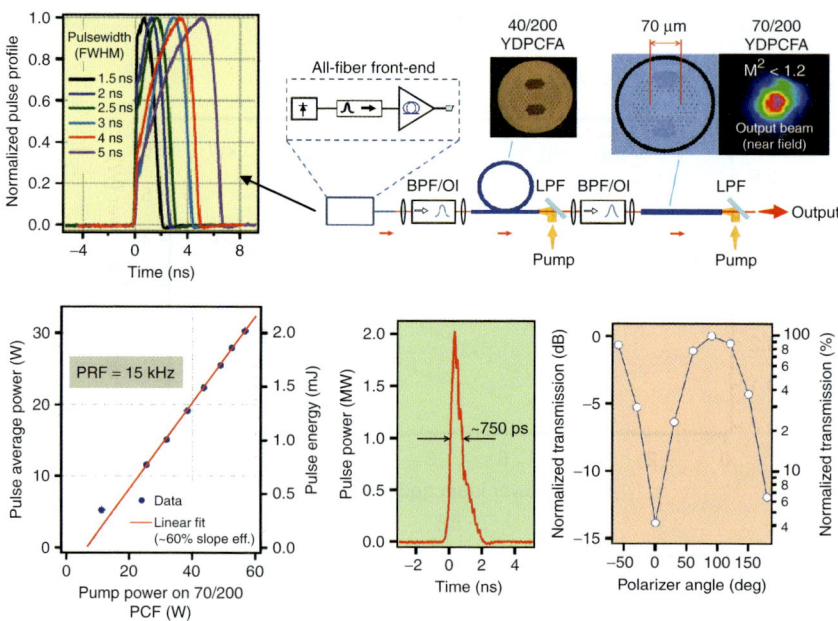

FIGURE 16.9 The illustrated performance amounts to the largest pulse energy and peak power produced by a single-polarization or polarization-maintaining fiber to date. FWHM: full width, half maximum; YDPCFA: Yb-doped PCF amplifier.

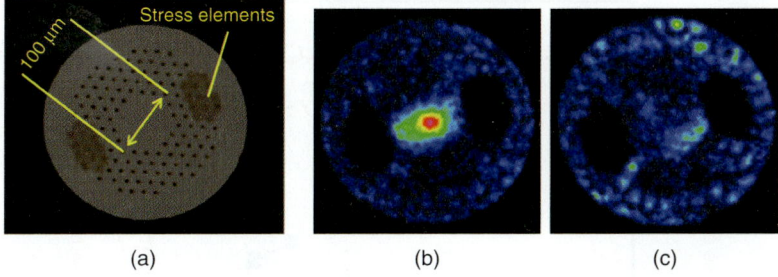

FIGURE 16.10 (a) Cross sectional photograph of 100 μm-core Yb-doped rod-type PCF. (b) Near-field image of PCF output when light is launched in the core with its polarization direction matching the PCF slow axis (intersecting the core and stress elements). (c) Corresponding near-field image for launched light polarized along the fast axis.

16.4.4 Eye-Safe, Pulsed Fiber Laser Sources

As discussed in Sec. 16.3.1, high-power pulse fiber laser source development at eye-safe wavelengths has been hampered by the relatively poor availability of effective and diverse fiber solutions, as compared with the 1-μm region. In the 1.5- to 1.6-μm region, the highest in-fiber peak power (1.2 MW) and pulse energy (1.4 mJ) has been produced by a 65 um/0.16 core diameter/NA, 9.5-m-long Er-doped fiber amplifier driven by an actively pulse-amplified, spectrally sliced Er-fiber ASE source, generating a 1567-nm adjustable-pulse seed (see Fig. 16.11). However, the amplifier output (1.1-ns pulses at PRF = 7.5 kHz) was emitted in a highly multimode beam of $M^2 = \sim 8.5$.[51] Another very-large-core (100-μm diameter), highly multimode Er-doped fiber has been implemented as the gain medium in an acousto-optically Q-switched, 980-nm diode-pumped, ring-cavity fiber laser, which generated pulse energy of ~11 mJ (the highest reported from a fiber-based source in the 1.5–1.6 μm wavelength range) when operated at PRF = 1 kHz (pulse duration ~140 ns), within an output beam of $M^2 = \sim 6$.[52]

Lower peak powers and pulse energies have been reported in Er-based fiber sources that retain high spectral brightness and stable output polarization, thus being amenable to efficient wavelength conversion. For example, a MOPA architecture similar to that in Ref. 51, but featuring a 15-μm-core PM, Er-Yb-codoped fiber power amplifier, was demonstrated to generate pulse energy greater than 100 μJ (100 kHz/8 ns PRF/pulse width) in a high spectral brightness (3-dB spectral line width less than 0.8 nm), linearly polarized (22-dB polarization extinction) near-diffraction-limited beam ($M^2 = \sim 1.5$). Such output was used to pump a periodically poled lithium niobate optical parametric oscillator (OPO), resulting in pulse average power in excess of 1 W in the 3.8–4.0-μm midinfrared wavelength region.[53]

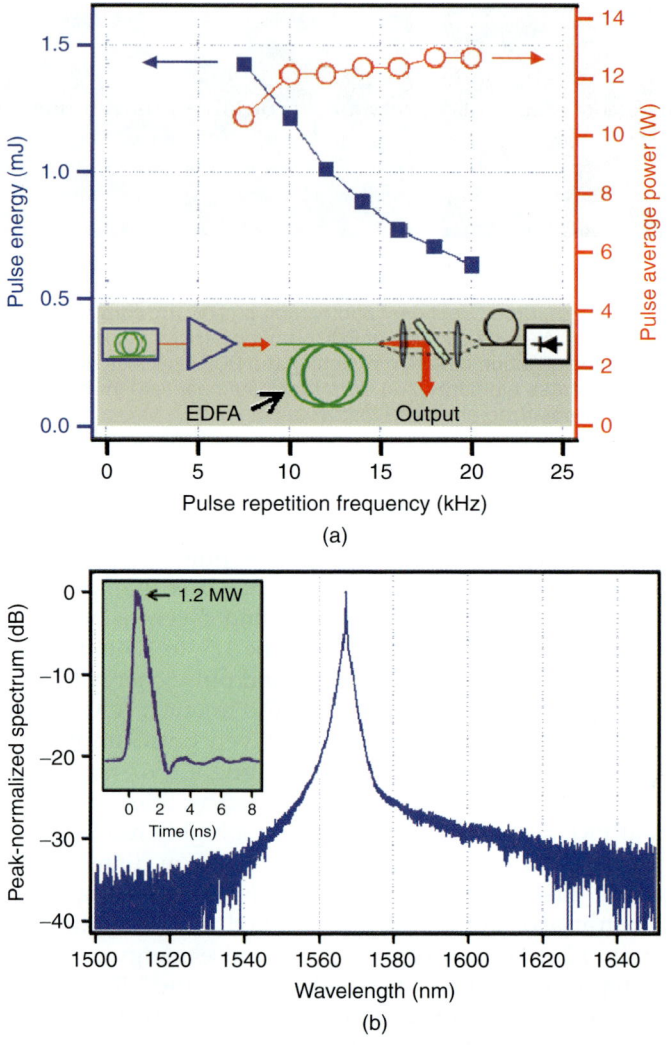

Figure 16.11 (a) Pulse energy and corresponding average power vs. pulse repetition frequency (PRF) for Er-doped fiber MOPA (Inset: Schematic architecture. EDFA: erbium-doped fiber amplifier); (b) Optical spectrum of the output beam, recorded at ~1.4 mJ and PRF = 7.5 kHz (Inset: Corresponding pulse temporal profile). EDFA: erbium-doped fiber amplifier.

Recently, advances attained in the development of efficient LMA Tm-doped fibers for high-power operation have led to significant progress in the generation of high pulse power in the 1.9–2.1 μm region. Very high pulse energies (up to almost 15 mJ) have been obtained via gain switching of Tm-doped fiber lasers, obtained by pulsed pumping using Ti:sapphire and Nd:YAG lasers.[54,55] Such schemes, however, implement inefficient high-brightness pump sources, which are undesirable for practical applications and which produce poorly controlled, very long (typically ~1-μs duration) pulses that exhibit an irregular, spiky profile due to relaxation oscillations. A more practical implementation of gain switching has been reported that consists of core pumping a short-cavity (< 30 cm) Tm-doped fiber with a pulsed Er-doped-fiber-based MOPA operating at 1550-nm wavelength. In such cases, smooth pulses as short as 10 ns could be obtained.[56] Because the short Tm fiber prevents significant energy storage and extraction, this type of laser has been used effectively as a seeder source for a 795-nm diode-pumped, LMA Tm-doped fiber-based MOPA, which generated 1995-nm-wavelength pulses of energy in excess of 140 μJ (30 ns/30 kHz pulse duration/PRF). These lasers were used to pump a $ZnGeP_2$ OPO for watt-power-level midinfrared generation in the 3.4–3.9 and 4.1–4.7 μm bands.[57] Q-switching is another method to produce energetic, relatively short pulses (tens of nanoseconds), which is compatible with CW diode-laser pumping in the Tm-glass high-optical-efficiency ~0.8-μm region. This approach was first reported by El-Sherif and King,[58] who obtained pulse energy and peak power of ~2.3 mJ and 3.3 kW, respectively, though at a very low PRF of 70 Hz; more recently, Eichhorn and Jackson[59] obtained pulse energy and peak power of ~270 μJ and 5.5 kW, respectively, at PRF of 10 kHz from a bidirectionally end-pumped 20-μm-core Tm-doped fiber. McComb et al.[60] reached ~350 μJ in ~115 ns pulses at a PRF of 20 kHz from a 25-μm-core Tm-doped fiber.

16.5 Conclusions and Outlook

This chapter has presented a review of the challenges, design solutions, and recent results pertaining to the power scaling of pulsed fiber laser (PFL) sources. The main focus of the review was on the near-nanosecond duration, multikilohertz PRF pulse format, obtained in fiber-based oscillators or MOPA, which concurrently maintain significant control on their spectral and spatial brightnesses. Over the past few years, the pace of progress in this area has been comparable to that of CW high-power fiber lasers and has resulted in the evolution of PFL concepts from low-power operation regimes appropriate for optical telecommunications to multi-millijoule pulse energies and multimegawatt peak powers comparable to bulk solid-state lasers. A key ingredient in such progress has been the research and development of new rare-earth-doped fibers, which have provided an effective tool for overcoming nonlinear optical parasitics, which had been the primary obstacle to pulse power scaling in the fiber medium.

The near future bears promise for further pulse energy and peak power scaling, though performance limits might already be in sight.

The most fundamental of such limits appears to be self-focusing, which is expected to ultimately clamp the peak power of individual fiber lasers and amplifiers at values only slightly above those obtained in recent years. Another difficult challenge is the management of thermal loads in fibers designed to be short, large-core, and very weakly guiding to minimize nonlinearity without beam quality penalty. For the many fiber concepts that converge to such common design features, the support for high-average-power operation is ultimately hampered by the thermo-optic distortion of their internal refractive index profile, which may disrupt wave guidance and eventually lead to a power-dependent beam quality reminiscent of bulk lasers. Finally, energy storage and extraction from a single fiber that can sustain high spatial and high spectral brightness appears very difficult to scale, in practice, to the tens of millijoule levels and, in any case, is unlikely to ever break the 100-mJ barrier.

Although future research and development may prove that overcoming some of these limiting factors is somehow possible, a promising way around them is represented by beam-combination schemes, which afford power scaling and concomitant operation of each component fiber within its limits (Chap. 19). As with CW operation, spectral multiplexing and coherent phasing are especially interesting in that they preserve spatial brightness. Although such concepts are by no means a novelty and can, in principle, be applied to any laser, the architecturally flexible, rugged, and compact nature of fibers represents an ideal framework for a truly practical implementation of them.

An example of spectral combination of nanosecond-pulse fiber-based sources was recently reported by Schmidt et al.,[61] who multiplexed with an external diffraction-grating four-pulsed diode-laser-driven, 5-nm-apart, temporally synchronized ~2-ns-pulse MOPAs, each terminated by a 80-μm-core Yb-doped, fundamental-mode rod-type PCF to achieve a combined beam with somewhat degraded spatial quality ($M^2 =$ ~2), 3.7-mJ pulse energy (~1.8 MW peak power) at a PRF of 50 kHz. Further development of this concept to attain combined peak powers in excess of the self-focusing critical power for a single fundamental-mode fiber is anticipated.

Coherently phased beam combining is, in principle, even more attractive in view of its unique ability to preserve the spectral brightness of individual fiber channels. Although an actual realization of coherently phased nanosecond-pulse fiber sources has yet to be reported, the recently demonstrated robust active phase locking of a 180-μJ-energy, 1-ns-pulse fiber amplifier to a master oscillator[62] suggests that this concept is viable and should be regarded as a promising power scaling scheme for the near future.

Acknowledgments

I wish to acknowledge former and present coworkers who have made important contributions to the developments and results reviewed in the forth section of this Chapter, including Christopher D. Brooks (Voxis Inc.), Sebastien Desmoulins (SPI Lasers), and Michael K. Hemmat and Eric C. Cheung (Northrop Grumman Aerospace Systems).

References

1. Dilley, C. E., Stephen, M. A., and Savage-Leuchs, M. P. "High SBS-Threshold, Narrowband, Erbium Codoped with Ytterbium Fiber Amplifier Pulses Frequency-Doubled to 770 nm," *Opt. Express*, 15: 14389–14395, 2007.
2. Leigh, M., Shi, W., Zong, J., Yao, Z., Jiang, S., and Peyghambarian, N., "High Peak Power Single Frequency Pulses Using a Short Polarization-Maintaining Phosphate Glass Fiber with a Large Core," *Appl. Phys. Lett.*, 92: 181108, 2008.
3. Agrawal, G. P., *Nonlinear Fiber Optics*, 4th ed. Burlington, MA: Academic Press, 2007. (See, in particular, Chap. 9.4.5 for a discussion of dynamic aspects of SBS and Chap. 7 for cross-phase modulation.)
4. Zenteno, L., Wang, J., Walton, D., Ruffin, B., Li, M., Gray, S., Crowley, A., and Chen, X., "Suppression of Raman Gain in Single-Transverse-Mode Dual-Hole-Assisted Fiber," *Opt. Express*, 13: 8921–8926, 2005.
5. Kim, J., Dupriez, P., Codemard, C., Nilsson, J., and Sahu, J. K., "Suppression of Stimulated Raman Scattering in a High Power Yb-Doped Fiber Amplifier Using a W-Type Core with Fundamental Mode Cut-Off," *Opt. Express*, 14: 5103–5113, 2006.
6. Alkeskjold, T. T., "Large-Mode-Area Ytterbium-Doped Fiber Amplifier with Distributed Narrow Spectral Filtering and Reduced Bend Sensitivity," *Opt. Express*, 17: 16394–16405, 2009.
7. Wang, Y., and Po, H., "Dynamic Characteristics of Double-Clad Fiber Amplifiers for High-Power Pulse Amplification," *J. Lightwave Technol.*, 21: 2262, 2003.
8. Schimpf, D. N., Ruchert, C., Nodop, D., Limpert, J., Tünnermann, A., and Salin, F., "Compensation of Pulse-Distortion in Saturated Laser Amplifiers," *Opt. Express*, 16: 17637–17646, 2008.
9. Limpert, J., Deguil-Robin, N., Manek-Hönninger, I., Salin, F., Schreiber, T., Liem, A., Röser, R., et al., "High-Power Picosecond Fiber Amplifier Based on Nonlinear Spectral Compression," *Opt. Lett.*, 30: 714–716, 2005.
10. Brooks, C. D., and Di Teodoro, F., "1-mJ Energy, 1-MW Peak-Power, 10-W Average-Power, Spectrally Narrow, Diffraction-Limited Pulses from a Photonic-Crystal Fiber Amplifier," *Opt. Express*, 13: 8999–9002, 2005.
11. Feve, J. P., Schrader, P. E., Farrow, R. L., and Kliner, D. A. V., "Four-Wave Mixing in Nanosecond Pulsed Fiber Amplifiers," *Opt. Express*, 15: 4647–4662, 2007.
12. Sintov, Y., Katz, O., Glick, Y., Acco, S., Nafcha, Y., Englander, A., and Lavi, R., "Extractable Energy from Ytterbium-Doped High-Energy Pulsed Fiber Amplifiers and Lasers," *J. Opt. Soc. Am. B*, 23: 218–230, 2006.
13. Smith, A. V., Do, B. T., Hadley, G. R., and Farrow, R. L., "Optical Damage Limits to Pulse Energy from Fibers," *IEEE J. Select. Topics Quantum Electron.*, 15: 153–158, 2009.
14. Fibich, G., and Gaeta, G. L., "Critical Power for Self-Focusing in Bulk Media and in Hollow Waveguides," *Opt. Lett.*, 25: 335, 2000.
15. Farrow, R. L., Hadley, G. R., Smith, A. V., and Kliner, D. A. V., "Numerical Modeling of Self-Focusing Beams in Fiber Amplifiers," *SPIE Proceedings Conf. on Fiber Lasers IV: Technology, Systems, and Applications*, 6453: 645309-1-9, 2007 (SPIE Photonics West, San Jose, CA).

16. Major, A., Yoshino, F., Nikolakakos, I., Aitchison, J. S., and Smith, P. W. E., "Dispersion of the Nonlinear Refractive Index in Sapphire," *Opt. Lett.*, 29: 602–604, 2004.
17. Koechner, W., *Solid-State Laser Engineering*, 6th ed. New York: Springer, 2006. (See Chap. 4.6.).
18. Schmidt-Uhlig, T., Karlitschek, P., Marowsky, G., and Sano, Y., "New Simplified Coupling Scheme for the Delivery of 20 MW Nd:YAG Laser Pulses by Large Core Optical Fibers," *Appl. Phys. B*, 72: 183–186, 2001.
19. Sun, L., and Marciante, J. R., "Filamentation Analysis in Large-Mode-Area Fiber Lasers," *J. Opt. Soc. Am. B*, 24: 2321–2326, 2007.
20. Fini, J. M., "Bend-Resistant Design of Conventional and Microstructure Fibers with Very Large Mode Area," *Opt. Express*, 14: 69–81, 2006.
21. Wielandy, S., "Implications of Higher-Order Mode Content in Large Mode Area Fibers with Good Beam Quality," *Opt. Express*, 15: 15402–15409, 2007.
22. Dong, L., Wu, T.-W., McKay, H. A., Fu, L., Li, J., and Winful, H. G., "All-Glass Large-Core Leakage Channel Fibers," *IEEE J. Sel. Topics Quantum Electron.*, 15: 47–53, 2009.
23. Liu, C., Huang, S., Zhu, C., and Galvanauskas, A., "High Energy and High Power Pulsed Chirally-Coupled Core Fiber Laser System" (paper MD2), *Advanced Solid-State Photonics*, OSA Technical Digest Series (CD), Optical Society of America, 2009.
24. Ramachandran, S., Nicholson, J. W., Ghalmi, S., Yan, M. F., Wisk, P., Monberg, E., and Dimarcello, F. V., "Light Propagation with Ultralarge Modal Areas in Optical Fibers," *Opt. Lett.*, 31: 1797–1799, 2006.
25. Fini, J. M., and Ramachandran, S., "Natural Bend-Distortion Immunity of Higher-Order-Mode Large-Mode-Area Fibers," *Opt. Lett.*, 32: 748–750, 2007.
26. Yoda, H., Polynkin, P., and Mansuripur, M., "Beam Quality Factor of Higher Order Modes in a Step-Index Fiber," *J. Lightwave Technol.*, 24: 1350–1355, 2006.
27. Limpert, J., Röser, F., Schimpf, D. N., Seise, E., Eidam, T., Hädrich, S., Rothhardt, J., et al., "High Repetition Rate Gigawatt Fiber Laser Systems: Challenges, Design, and Experiment," *IEEE J. Select Top. Quantum Elec.*, 15: 159–169, 2009.
28. Hädrich, S., Schreiber, T., Pertsch, T., Limpert, J., Peschel, T., Eberhardt, R., and Tünnermann, A., "Thermo-Optical Behavior of Rare-Earth-Doped Low-NA Fibers in High Power Operation," *Opt. Express*, 14: 6091–6097, 2006.
29. Goto, R., Takenaga, K., Okada, K., Kashiwagi, M., Kitabayashi, T., Tanigawa, S., Shima, K., et al., "Cladding-Pumped Yb-Doped Solid Photonic Bandgap Fiber for ASE Suppression in Shorter Wavelength Region" (paper OTuJ5), *Optical Fiber Communication Conference and Exposition and the National Fiber Optic Engineers Conference*, OSA Technical Digest (CD), Optical Society of America, 2008.
30. Siegman, A., *Lasers*, Sausalito, CA: University Science Books, 1986. (See Chap. 26.)
31. Di Teodoro, F., Koplow, J. P., Kliner, D. A. V., and Moore, S. W., "Diffraction-Limited, 300-kW Peak-Power Pulses from a Coiled Multimode Fiber Amplifier," *Opt. Lett.*, 27: 518–520, 2002.
32. Di Teodoro, F., and Brooks, C. D., "1.1 MW Peak-Power, 7 W Average-Power, High-Spectral-Brightness, Diffraction-Limited Pulses from a Photonic Crystal Fiber Amplifier," *Opt. Lett.*, 30: 2694–2696, 2005.
33. Farrow, R. L., Kliner, D. A. V., Schrader, P. E., Hoops, A. A., Moore, S. W., Hadley, G. R., and Schmitt, R. L., "High-Peak-Power (>1.2 MW) Pulsed Fiber Amplifier," *Proc. SPIE*, 6102: 61020K, 2006.
34. Di Teodoro, F., and Brooks, C. D., "Multistage Yb-Doped Fiber Amplifier Generating Megawatt Peak-Power, Sub-Nanosecond Pulses," *Opt. Lett.*, 30: 3299–3301, 2005.
35. Torruellas, W., Chen, Y., Macintosh, B., Farroni, J., Tankala, K., Webster, S., Hagan, D., et al., "High Peak Power Ytterbium-Doped Fiber Amplifiers," *Proc. SPIE*, 6102: 61020N, 2006.
36. Dawson, J. W., Beach, R. J., Jovanovic, I., Wattellier, B., Liao, Z. M., Payne, S. A., and Barty, C. P., "Large Flattened Mode Optical Fiber for High Output Energy

Pulsed Fiber Lasers" (paper CWD5), *Conference on Lasers and Electro-Optics/ Quantum Electronics and Laser Science Conference*, Technical Digest, Optical Society of America, 2003.
37. Di Teodoro, F., and Brooks, C. D., "Fiber Sources Reach Multimegawatt Peak Powers in ns Pulses," *Laser Focus World,* 42(11): 94–98, 2006.
38. Schmidt, O., Nodop, D., Limpert, J., and Tünnermann, A., "105 kHz, 85 ps, 3 MW Peak Power Microchip Laser Fiber Amplifier System" (paper WB23), *Advanced Solid-State Photonics, OSA Technical Digest Series* (CD), Optical Society of America, 2008.
39. Brooks, C. D., and Di Teodoro, F., "Multimegawatt Peak Power, Single-Transverse-Mode Operation of a 100 μm Core Diameter, Yb-Doped Rodlike Photonic Crystal Fiber Amplifier," *Appl. Phys. Lett.,* **89:** 111119, 2006.
40. Cheng, M.-Y., Chang, Y.-C., Galvanauskas, A., Mamidipudi, P., Changkakoti, R., and Gatchell, P., "High-Energy and High-Peak-Power Nanosecond Pulse Generation with Beam Quality Control in 200-μM Core Highly Multimode Yb-Doped Fiber Amplifiers," *Opt. Lett.,* 30: 358–360, 2005.
41. Di Teodoro, F., "Multi-MW Peak-Power, Multi-mJ Pulse-Energy, High-Spectral-Brightness Fiber Amplifiers," presented at the IEEE Lasers and Electro-Optics Society (LEOS) Summer Topicals, Quebec City, Canada, July 17–19, 2006.
42. Galvanauskas, A., Cheng, M.-Y., Hou, K.-C., and Liao, K.-H., "High Peak Power Pulse Amplification in Large-Core Yb-Doped Fiber Amplifiers," *IEEE J. Sel. Top. Quantum Electron.,* 13: 559–566, 2007.
43. Schreiber, T., Schultz, H., Schmidt, O., Röser, F., Limpert, J., and Tünnermann, A., "Stress-Induced Birefringence in Large-Mode-Area Micro-Structured Optical Fibers," *Opt. Express,* 13: 3637–3646, 2005.
44. Schreiber, T., Röser, F., Schmidt, O., Limpert, J., Iliew, R., Lederer, F., Petersson, A., et al., "Stress-Induced Single-Polarization Single-Transverse Mode Photonic Crystal Fiber with Low Nonlinearity," *Opt. Express,* 13: 7621–7630, 2005.
45. Di Teodoro, F., and Brooks, C. D., "Harmonic Generation of an Yb-Doped Photonic-Crystal Fiber Amplifier to Obtain 1 ns Pulses of 410, 160, and 190 kW Peak-Power at 531, 354, and 265 nm Wavelength" (paper ME3), *Advanced Solid-State Photonics*, Technical Digest, Optical Society of America, 2006.
46. Brooks, C. D., and Di Teodoro, F., "High Peak Power Operation and Harmonic Generation of a Single-Polarization, Yb-Doped Photonic Crystal Fiber Amplifier," *Opt. Commun.,* 280: 424–430, 2007.
47. Schrader, P. E., Farrow, R. L., Kliner, D. A. V., Fève, J.-P., and Landru, N., "High-Power Fiber Amplifier with Widely Tunable Repetition Rate, Fixed Pulse Duration, and Multiple Output Wavelengths," *Opt. Express,* 14: 11528–11538, 2006.
48. Creeden, D., McCarthy, J. C., Ketteridge, P. A., Schunemann, P. G., Southward, T., Komiak, J. J., and Chicklis, E. P., "Compact, High Average Power, Fiber-Pumped Terahertz Source for Active Real-Time Imaging of Concealed Objects," *Opt. Express,* 15: 6478–6483, 2007.
49. Di Teodoro, F., Potter, A. B., Hemmat, M. K., Cheung, E., Palese, S., Weber, M., and Moyer, R., "Actively Triggered, 2 mJ Energy, 80 W Average Power, Single-Mode, Linearly Polarized ns-Pulse Fiber Source," presented at Fiber Lasers VII: Technology, Systems, and Applications Conference (SPIE Photonics West 2009), San Jose, CA, January 24–29, 2009.
50. Di Teodoro, F., Hemmat, M. K., Morais, J., and Cheung, E. C., "100 micron Core, Yb-Doped, Single-Transverse-Mode and Single-Polarization Rod-Type Photonic Crystal Fiber Amplifier," *Proc. SPIE,* 7580: 758006, 2010.
51. Desmoulins, S., and Di Teodoro, F., "High-Gain Er-Doped Fiber Amplifier Generating Eye-Safe MW Peak-Power, mJ-Energy Pulses," *Opt. Express,* 16: 2431–2437, 2008.
52. Lallier, E., and Papillon-Ruggeri, D., "High Energy Q-Switched Er-Doped Fiber Laser," *European Conference on Lasers and Electro-Optics 2009 and the European Quantum Electronics Conference* (CLEO Europe—EQEC), 1-1, 2009.
53. Desmoulins, S., and Di Teodoro, F., "Watt-Level, High-Repetition-Rate, Mid-Infrared Pulses Generated by Wavelength Conversion of an Eye-Safe Fiber Source," *Opt. Lett.,* 32: 56–58, 2007.

54. Dickinson, B. C., Jackson, S. D., and King, T. A., "10 mJ Total Output from a Gain-Switched Tm-Doped Fiber Laser," *Opt. Commun.*, 182: 199–203, 2000.
55. Zhang, Y., Yao, B.-Q., Ju, Y.-L., and Wang, Y.-Z., "Gain-Switched Tm^{3+}-Doped Double-Clad Silica Fiber Laser," *Opt. Express*, 13: 1085–1089, 2005.
56. Jiang, M., and Tayebati, P., "Stable 10 ns, Kilowatt Peak-Power Pulse Generation from a Gain-Switched Tm-Doped Fiber Laser," *Opt. Lett.*, 32: 1797–1799, 2007.
57. Creeden, D., Ketteridge, P. A., Budni, P. A., Setzler, S. D., Young, Y. E., McCarthy, J. C., Zawilski, K., et al., "Mid-infrared $ZnGeP_2$ Parametric Oscillator Directly Pumped by a Pulsed 2 µm Tm-Doped Fiber Laser," *Opt. Lett.*, 33: 315–317, 2008.
58. El-Sherif, A. F., and King, T. A., "High-Energy, High-Brightness Q-Switched Tm-Doped Fiber Laser Using an Electro-optic Modulator," *Opt. Commun.*, 218: 337–344, 2003.
59. Eichhorn, M., and Jackson, S. D., "High-Pulse-Energy Actively Q-Switched Tm^{3+}-Doped Silica 2 µm Fiber Laser Pumped at 792 nm," *Opt. Lett.*, 32: 2780–2782, 2007.
60. McComb, T. S., Shah, L., Willis, C. C., Sims, R. A., Kadwani, P. K., Sudesh, V., and Richardson, M., "Thulium Fiber Lasers Stabilized by a Volume Bragg Grating in High Power, Tunable and Q-Switched Configurations" (paper AMB2), *Advanced Solid-State Photonics*, OSA Technical Digest Series (CD), Optical Society of America, 2010.
61. Schmidt, O., Andersen, T. V., Limpert, J., and Tünnermann, A., "187 W, 3.7 mJ from Spectrally Combined Pulsed 2 ns Fiber Amplifiers," *Opt. Lett.*, 34: 226–228, 2009.
62. Cheung, E. C., Weber, M., and Rice, R. R., "Phase Locking of a Pulsed Fiber Amplifier" (paper WA2), in *Advanced Solid-State Photonics*, OSA Technical Digest Series (CD), Optical Society of America, 2008.

CHAPTER 17
High-Power Ultrafast Fiber Laser Systems

Jens Limpert

Institute of Applied Physics, Friedrich Schiller University Jena, and Fraunhofer Institute for Applied Optics and Precision Engineering, Jena, Germany

Andreas Tünnermann

Institute of Applied Physics, Friedrich Schiller University Jena, and Fraunhofer Institute for Applied Optics and Precision Engineering, Jena, Germany

17.1 Introduction and Motivation

Sources of ultrashort and high-peak-power optical pulses have become extremely important for numerous applications, such as spectroscopy, remote sensing, or high-field physics. In fact, considerable progress has been made over the past decade in obtaining high-peak-power femtosecond (fs) sources based on Ti:sapphire lasers and amplifiers, as described in Chap. 12.[1] Such systems are the workhorses of ultrafast science and have proven to be a reliable option at low repetition rates. Ultrashort and intense laser sources are used in high-field physics to probe processes that often have very small probabilities of occurrence. As a consequence, the detection of induced processes requires sophisticated, sensitive apparatus. An increase of a few orders of magnitude in the repetition rate will provide a tool that would allow breaking the actual limits and that would open the door to in-depth investigation of such phenomena. The same holds true for other applications of ultrashort laser pulses, such as laser micromachining. Ti:sapphire systems provide the required parameters for creating very small, sophisticated high-quality structures without the need for any postprocessing.[2] However, the processing time suffers from the low repetition rate, which ultimately prevents the use of ultrafast lasers at industrial scale. Hence, the motivation for developing high-repetition-rate, high-peak-power laser systems is to give such interesting applications the chance to evolve from proof-of-principle experiments to real-world applications.

One approach for scaling the average power of conventional solid-state lasers is to improve the thermo-optical properties of the active medium by cryogenic cooling.[3] This solution, however, comes at the cost of further complexity and dependence on a laboratory environment. Consequently, in the past few decades, novel gain medium architectures have been developed to overcome thermo-optical issues. An outstanding geometry is the thin disk laser introduced in Chap. 10, whose application to ultrafast pulse generation is discussed in Chap. 13.[4] However, due to the active medium's short length, the gain per pass cannot be high. Therefore, regenerative amplification is required to extract a reasonable output, which leads to rather bulky and alignment-sensitive laser systems.

The opposite approach, in terms of gain medium dimensions, is the fiber concept. Having a gain medium that is both long and thin leads to outstanding thermo-optical properties. Fiber-based laser systems have the reputation of being immune to thermo-optical problems due to their special geometry. Their excellent heat dissipation results from the large ratio of surface to active volume. The beam quality of the guided mode is determined by the fiber core design and is, therefore, largely power independent. In continuous wave operation, powers as high as 10 kW in a diffraction-limited beam quality have already been demonstrated.[5] Due to the confinement of both the laser and the pump radiation, the intensity is maintained over the entire fiber length, resulting in very efficient operation of fiber laser systems that leads to large single-pass gain and low pump threshold values. In addition, complete integration of the laser process in a waveguide potentially confers inherent compactness and long-term stability to fiber lasers. In particular, Ytterbium-doped glass fibers, which have a quantum defect of less than 10 percent, can provide optical-to-optical efficiencies well above 80 percent and, therefore, with low thermal load. Ytterbium-doped fibers are especially interesting for high-power ultrashort-pulse generation and amplification because of several unique properties.[6] The most relevant of those properties are the fibers' broad amplification bandwidth of up to several tens of nanometers, depending on the center wavelength, and the fact that their absorption spectrum covers a wavelength range in which powerful diode lasers are available, In addition, the long fluorescence lifetime (~1 ms) results in a high-energy storage capability.

17.2 Nonlinear Effects as Basic Limitations of Ultrashort Pulse Amplification in Rare-Earth-Doped Fibers

The fiber geometry itself is responsible for most of the outstanding properties of rare-earth-doped fibers that make them attractive gain media. However, this geometry also promotes nonlinear effects by making the light propagate under tight confinement over considerably long lengths. In fact, in the context of ultrashort pulse amplification, nonlinearity is mostly harmful and imposes performance limitations in fiber laser systems.

Nonlinear effects in fibers can be manifold. The lowest-order nonlinear effects in standard optical fibers originate from the third-order susceptibility $\chi^{(3)}$. These effects can be divided into those related to an intensity-dependent refractive index and those resulting from stimulated inelastic scattering.[7] Self-phase modulation (SPM), four-wave-mixing (FWM), and self-focusing all fall into the first category, whereas stimulated Raman scattering (SRS) and stimulated Brillouin scattering (SBS) are effects of the second category.

In general, the nonlinearity coefficients in silica glass fibers are intrinsically small. Both the nonlinear index coefficient n_2 and the gain coefficients of SRS and SBS are at least two orders of magnitude smaller than in other common nonlinear media.[8] Nevertheless, due to the large product of intensity and interaction length inside the fiber core, nonlinear effects can be observed at very low peak power levels and can basically limit the performance of pulsed rare-earth-doped fiber systems.

All the aforementioned nonlinear effects scale both with the light intensity in the fiber core and with the interaction length between the optical radiation and the nonlinear medium—that is, the fiber. Hence, to reduce the impact of nonlinearity, temporal and spatial scaling are needed. Temporal scaling can be achieved by the well-known technique of chirped pulse amplification (CPA; see Chap. 12).[9] In this technique, ultrashort optical pulses from a mode-locked oscillator are stretched in time by a certain factor, which can be as large as 10000, by passing them through a dispersive delay line. Therefore, during amplification, the peak power of the pulses is considerably reduced, as are the nonlinear effects. After amplification, the stretched amplified pulses travel through a second dispersive delay line with the opposite sign of the stretcher dispersion, resulting in a recompression back to ultrashort pulse duration.

On the other hand, spatial scaling requires advanced fiber designs that present a large mode area of the actively doped core and an absorption length that is as short as possible, thus reducing the nonlinear interaction length. Consequently, there has been a pursuit of novel fiber designs with increased core dimensions that are still able to emit a stable fundamental mode. Conventional active step-index fibers with reduced numerical aperture, assisted by coiled or tapered sections to achieve single-mode operation, allow for core diameters of up to 40 μm.[10] Rare-earth-doped photonic crystal fibers (PCFs) allow for significantly larger core sizes of up to 100 μm[11–13] due to the significantly better control of the index step between a nanostructured core and the holey photonic crystal cladding. Due to the PCF cladding's design freedom, an additional functionality, such as polarizing or polarization maintaining properties, can be added.[14] As an alternative to this weakly confined large-mode PCF, other approaches have been investigated, including designs based on tailored propagation losses for higher-order modes, as demonstrated in the chirally coupled core (CCC) fibers[15] or in the leakage channel fibers (LCFs).[16]

A fundamental challenge of mode area scaling is the increased bend sensitivity, which scales rapidly with increasing mode size and is manifested by deformation and shrinkage of the guided mode. In a first-order approximation, this effect may not depend on the core's numerical aperture—that is, it does not rely on the underlying fiber design. The crucial parameter is the straight fiber's mode area. However, it should be mentioned that confinement losses in a bent fiber strongly depend on the fiber core's numerical aperture. Figure 17.1 shows the calculated mode area deformation of three different large-mode-area (LMA) fibers. As can be seen, a fiber with an initial mode

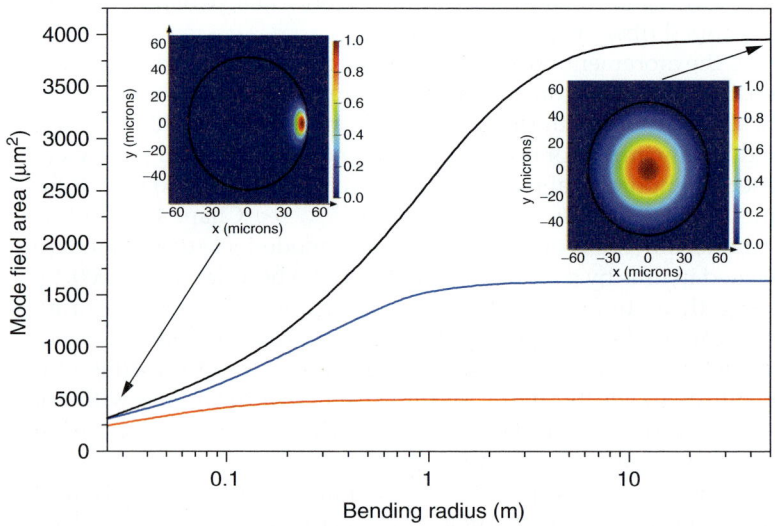

FIGURE 17.1 Simulated mode squeezing in bent fibers with different initial mode areas. Insets show the mode profile of a straight and a tightly coiled large-mode-area fiber.

area of about 400 µm² can be bent to radii of less than 10 cm without suffering from a significant reduction in mode area. In contrast, a fiber with a straight mode area of 1600 µm² experiences comparable mode shrinking at a bending radius of about 1 m, whereas in a fiber with 4000 µm², this already happens at a bending radius of about 5 m. As a consequence, to fully benefit from a large mode area over the entire fiber length, the active fiber should be kept straight. Note that higher-order modes (HOMs) in highly multimode fibers[17] do not suffer from extensive mode shrinking; however, the usefulness of such modes for active fibers is limited due to spatial inversion hole burning.

The path of core size increase has been followed with the development of a Yb-doped microstructured fiber known as a rod-type photonic crystal fiber. A cross section of the ultralarge mode area is shown in Fig. 17.2. The 200-µm inner cladding (pump waveguide) is surrounded by an air clad that consists of ninety 400-nm-thick and ~10-µm-long silica bridges. This structure provides a numerical aperture of ~0.6 at 976 nm, allowing for an efficient coupling of multimode pump radiation into the fiber. Nineteen missing holes in the center of the fiber form the Yb/Al codoped active core region, which has a corner-to-corner distance of 88 µm. Three rings of small, carefully dimensioned airholes (pitch Λ = ~14.9 µm; relative hole size d/Λ = ~0.1) around this core provide the confinement of the radiation in the doped core. The mode area of the fundamental mode is as large as 4000 µm². The small ratio of pump core area to active core area results in an enhanced pump light absorption of about 30 dB/m at 976 nm. The entire inner structure is surrounded by a stiff 1.5-mm diameter fused silica outer cladding. This large outer cladding is primarily introduced to keep the fiber straight, thus preventing bend-induced losses or distortions of the weakly guided fundamental mode. Thus, in such a straight structure, the large mode area is accessible over the entire fiber length, which is not the case in conventional LMA fibers. In addition, the outer cladding makes the

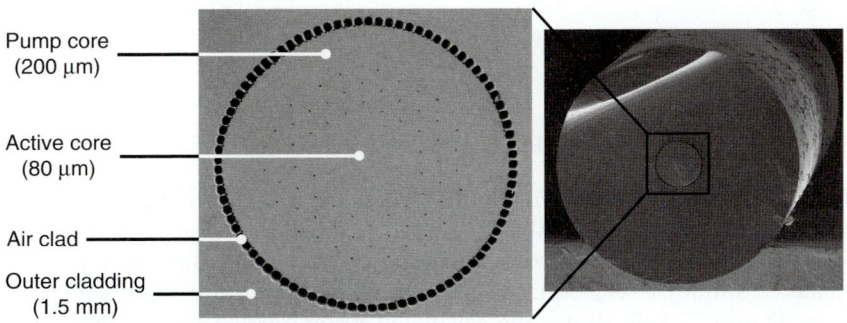

FIGURE 17.2 Cross section of the large-mode-area rod-type photonic crystal fiber (PCF), with a zoom into the embedded microstructured part.

fiber mechanically robust on its own, so that no extra coating material is required; this allows for straightforward high-power extraction. By applying this kind of fiber, novel performance levels have been achieved in various operation regimes, ranging from femtosecond to nanosecond pulses.[18–20]

Despite large temporal (CPA) and spatial scaling (LMA fiber designs), nonlinearity will still begin to have an impact at a certain peak power level in a fiber-based amplifier. Self-phase modulation is the nonlinear effect that arises first. As a consequence of the temporal Kerr effect, the pulse imposes on itself an extra phase term in the time domain given by the following (neglecting the effects of dispersion and amplification):

$$d\phi_{SPM}(z, T) = \gamma |A(z, T)|^2 dz \qquad (17.1)$$

Here, $A(z, T)$ represents the pulse field amplitude, z the propagation distance, T the time in the frame of reference moving with the pulse, and γ the nonlinear parameter defined by

$$\gamma = \frac{n_2 \omega_0}{c A_{eff}} \qquad (17.2)$$

with A_{eff} being the effective mode field area, n_2 the nonlinear refractive index coefficient, c the speed of light, and ω_0 the center angular frequency of the optical field. Under certain circumstances, SPM can be used to increase the peak power of ultrashort laser pulses, for example, by nonlinear pulse compression.[21] However, in the context of CPA, SPM is mostly harmful. The phase term imposed by SPM may lead to a nonlinear chirp, which cannot be compensated by standard dispersive elements. An important quantifier of the accumulated nonlinear phase of a pulse propagating through a fiber, and hence of the impact of SPM, is the so-called B integral, defined by

$$B = \frac{\omega_0}{c} \cdot \int_0^L n_2 \cdot I(z) \cdot dz \qquad (17.3)$$

where $I(z)$ is the pulse peak intensity varying over the fiber length L.[22]

A B integral smaller than π rad is considered to be linear propagation. Conversely, pulse quality degradation must be expected above this value. To illustrate the consequences of SPM, a CPA system is simulated assuming a sech-squared (sech^2) pulse shape during amplification. Figure 17.3 illustrates the calculated recompressed pulses after accumulating an extra nonlinear phase shift, in comparison with the transform-limited pulse sech^2 pulse). As revealed, SPM leads to pulse

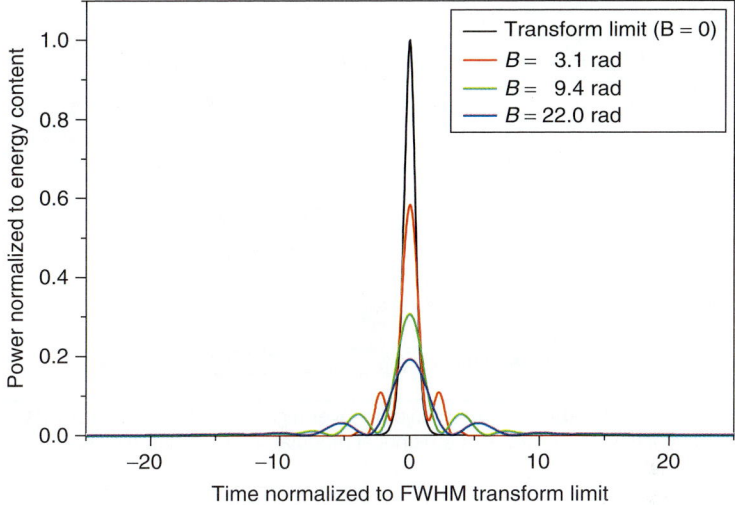

FIGURE 17.3 Impact of self-phase modulation (SPM) on the pulse quality of the recompressed pulses in a chirped pulse amplification (CPA) system, assuming a sech-squared pulse shape during amplification. FWHM: full width, half maximum.

broadening and to the development of wing structures that contain a considerable amount of pulse energy. Hence, even moderate SPM can significantly distort a recompressed pulse after amplification, with the result of reduced peak power and degraded pulse contrast.[23]

17.3 High-Repetition-Rate Gigawatt Peak Power Fiber Laser System

Based on the considerations of the impact of nonlinearity in fiber CPA systems, an ultrafast high-peak-power setup has been developed. The schematic setup of the high-energy high-average-power fiber CPA system is shown in Fig. 17.4.[24] It consists of a passively mode-locked ytterbium-doped potassium gadolinium tungstate (Yb:KGW) oscillator; a dielectric grating stretcher-compressor unit; an acousto-optical modulator used as a pulse selector; and two Yb-doped photonic crystal fibers, both used in single-pass configuration as amplification stages, providing an overall gain factor of approximately 25000.

The long-cavity Yb:KGW oscillator delivers transform-limited 400-fs pulses at a repetition rate of 9.7 MHz and with an average power of 1.6 W at 1030 nm center wavelength. The stretcher-compressor unit employs two 1740 line/mm dielectric diffraction gratings and stretches the 3.3-nm bandwidth pulses to 2 ns. A quartz-based acousto-optical modulator (AOM) is used to reduce the pulse repetition rate. The preamplifier comprises a 1.2-m-long 40-μm-core

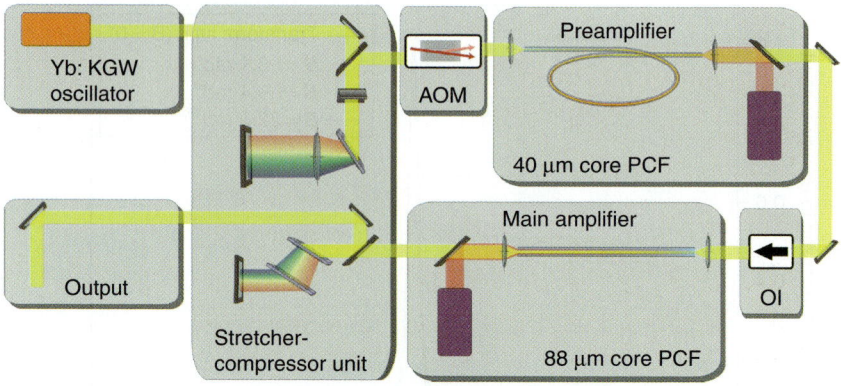

FIGURE 17.4 Schematic setup of the millijoule-level high-repetition-rate fiber CPA system. OI: optical isolator, AOM: acousto-optical modulator, PCF: photonic crystal fiber.

single-polarization air-clad PCF. The main amplifier is built using a 1.2-m-long piece of the low-nonlinearity air-cladding photonic crystal rod-type fiber, described earlier. The core of this fiber supports a reduced set of transverse modes; however, stable excitation of the fundamental mode alone is possible by seed mode matching. The result is a power-independent beam quality characterized by an M^2-value of less than 1.2.

Figure 17.5 shows the characteristics of the output after compression at a 200-kHz repetition rate. The main amplifier was seeded with 0.5 W average power, corresponding to 2.5 µJ pulse energy. At a

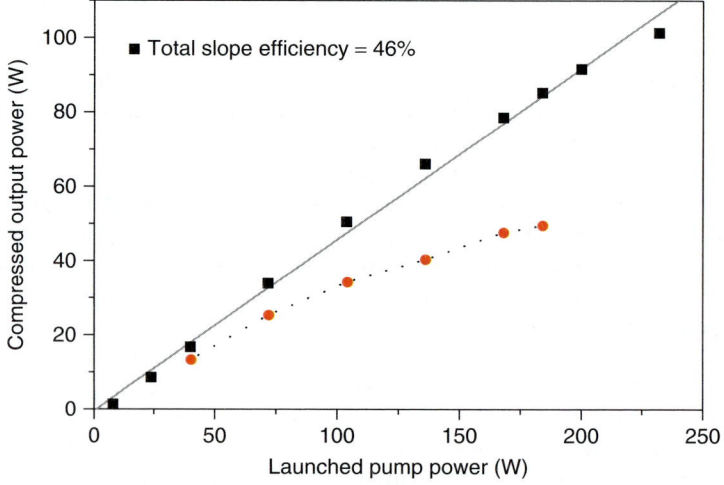

FIGURE 17.5 Average output power of the fiber CPA system after pulse compression at 200 kHz (black squares) and at 50 kHz (red circles).

launched pump power of 230 W, an average output power of 145 W is obtained with a slope efficiency as high as 66 percent. Because this fiber has no coating material and a stable fiber mount is used, no thermo-optical or thermomechanical issues are observed up to this average power level. The degree of polarization of the fiber amplifier output is 98 percent, allowing for an efficient recompression of the pulses. The compressor has a throughput of 70 percent, leading to a total slope efficiency of 46 percent and a compressed average power of 100 W, which implies pulse energies of 500 µJ.

The grating distance in the compressor is always adjusted for minimum autocorrelation width. At 500 µJ, the width is measured as 1.2 ps, as shown in Fig. 17.6 (dashed line), corresponding to a pulse duration of 780 fs, assuming a $sech^2$ pulse shape. For comparison, the autocorrelation trace at very low pulse energy is also shown (Fig. 17.6, dotted line). A wing structure growing with pulse energy appears, which can be attributed to the imposed nonlinear phase, as discussed in Sec. 17.2. The total B integral in this case is calculated to be 4.7 rad.

At a 50-kHz repetition rate and 70-mW seed power, we achieve 71 W of average power with a pump power of 180 W, corresponding to 1.45 mJ energy (red circles in Fig. 17.5). In this case, the B integral is as high as 7 rad. A further increase in pump power was avoided due to an increased risk of end-facet damage and pump light absorption saturation.[25] The compressed pulses exhibit an autocorrelation width of 1.23 ps (equivalent to 800-fs pulse duration), as is shown in

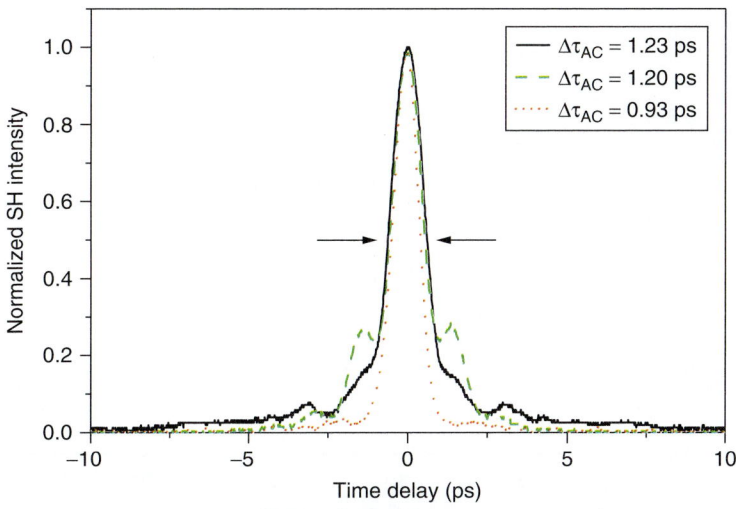

Figure 17.6 Measured autocorrelation traces of the compressed pulses. Dotted: at low pulse energy; Dashed: 200 kHz and 500 µJ; Solid: 50 kHz and 1 mJ.

Fig. 17.6, and a pulse energy as high as 1 mJ. The stronger nonlinearity is indicated by the wider spread in time of the pedestal. The corresponding pulse peak power is approximately 1 GW.

17.4 Peak Power and Pulse Energy Scaling Considerations

One figure of merit of the pulse quality degradation with increasing nonlinearity—the so-called effective peak power—can be defined as the compressed pulse peak power normalized to the peak power of a rectangular pulse that has the same energy and transform-limited pulse width (FWHM). Figure 17.7 shows the result of that figure of merit as a function of the B integral, assuming a significantly larger stretching phase than the SPM-induced phase term. In Fig. 17.7, the stretcher's and compressor's higher-order dispersion terms have been neglected. As revealed by Eq. (17.1), the SPM phase in a CPA system is proportional to the stretched pulse shape; therefore, the pulse shape determines the impact of SPM on the pulse quality. As a consequence, Fig. 17.7 contains curves for sech^2, gaussian and parabolic stretched-pulse shapes; it also reveals the reduction of the figure of merit with increasing B integral.

Indeed, the parabolic pulse appears to be immune to SPM-phase accumulation, because it possesses an effective peak power that is independent of the B integral. This observation can be understood because the SPM-induced nonlinear phase in this kind of pulse is also

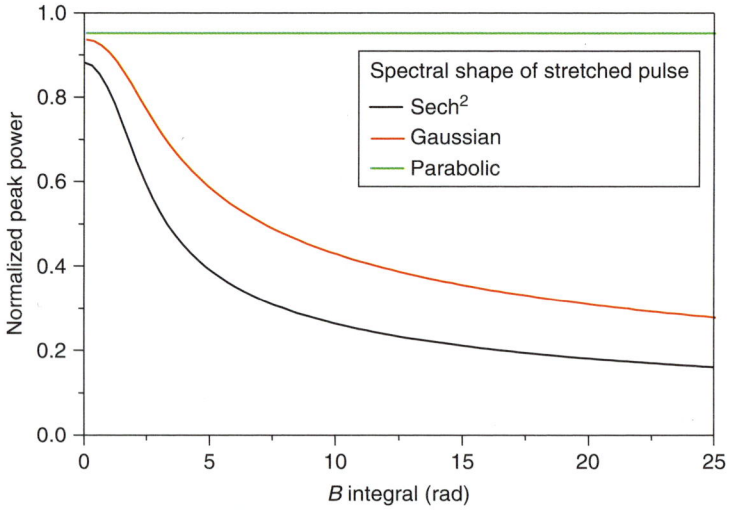

FIGURE 17.7 Degradation of the effective peak power in a nonlinear CPA system for different stretched pulse shapes.

parabolic [according to Eq. (17.1)], which corresponds to a perfect linear chirp that can be easily compensated by adaptation of the compressor (e.g., the grating separation). This effect has been experimentally observed in a fiber-based CPA system by amplifying actively shaped parabolic pulses to B integrals as high as 16 rad.[25] As an alternative, the amplification of asymmetric spectral amplitude profiles (generated by static filters) in a fiber-stretcher/grating-compressor CPA configuration, showing a third-order dispersion mismatch, has been used for the compensation of the phase term due to SPM.[26–29] However, in such a system, the total phase due to both second-order and third-order dispersion of the nonlinear CPA system must be compensated for by carefully choosing the right length of stretcher fiber and the corresponding separation in the grating compressor, as well as the peak power of the pulse during amplification. Therefore, there is only one ideal point of output power operation (corresponding to the appropriate B integral to balance the dispersion) for a chosen stretcher-compressor configuration. In that optimal configuration, the effective peak-power enhancement, as compared with perfectly matched stretcher-compressor configurations (red and black lines in Fig. 17.7), is approximately 30 percent.

As an alternative to amplitude shaping, spectral phase shaping is a promising technique for controlling the impact of nonlinear phase in a fiber CPA system.[30,31] Theoretically, a peak power corresponding to the transform-limited case is always achievable (intersection with ordinate in Fig. 17.7).

The impact of self-phase modulation on chirped pulses can be described by a spectral phase, which is given by the product of the B integral and the normalized spectrum $s(\Omega)$.[32] Thus, the application of an extra phase term of the form $\varphi = -Bs(\Omega)$ may result in transform-limited pulses at the output of the nonlinear CPA system. Experimentally, this is accomplished with a pulse shaper in a nonlinear CPA system. The setup is similar to the one shown in Fig. 17.4. However, the spectral bandwidth is smaller (2.6 nm root mean square), and the pulses are stretched to 1.3 ns. The pulse shaper is a spatial-light modulator (SLM) consisting of two layers of liquid crystals. The device permits phase-only shaping with a maximum phase shift of about 3π in transmission. The SLM is placed in the system before the preamplifier (compare with Fig. 17.4). The shape of the input spectrum is required for the calculation of the required compensation phase and can be measured with an optical spectrum analyzer. Then, the spectral phase for the compensation is produced with the SLM. The compensating phase term for this example is shown in the inset of Fig. 17.8. Using this technique, it is possible to generate nearly transform-limited pulses at the output of the nonlinear CPA system. The B integral of the system is 8 rad. The output pulse energy is about 840 µJ (after compression) at a repetition rate of 50 kHz. The autocorrelation FWHM of the output pulse is the same as for the linear CPA system,

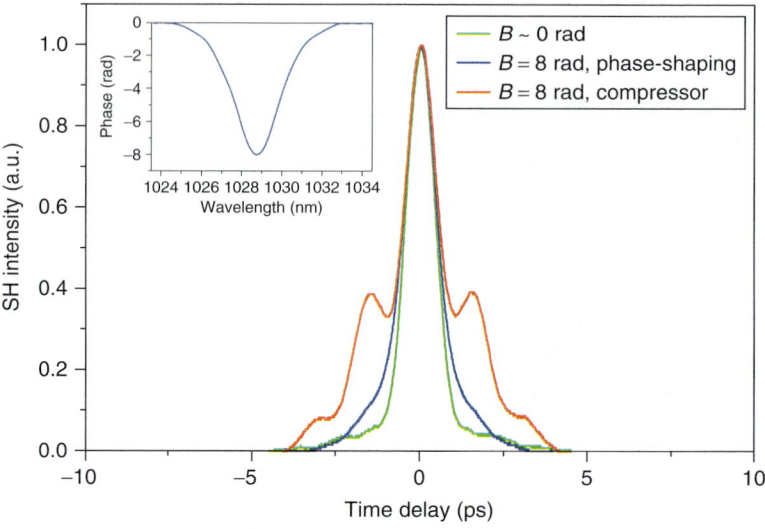

FIGURE 17.8 Improvement of the pulse quality by phase-only pulse shaping. Autocorrelation of the output pulses from a nonlinear CPA at a B integral of 8 rad. The inset shows the compensating phase, which is applied by the spatial-light modulator (SLM).

amounting to 1.05 ps. If the phase shaping is turned off and only the compressor is used for best compression, then the autocorrelation shows a strong wing structure, and its FWHM increases to about 1.4 ps. From Fig. 17.7, the increase in peak power by means of phase shaping is estimated to be higher than a factor of 2.

In addition to the detrimental impact of nonlinear phase accumulation, resulting from the envelope of the propagating pulse, weak initial spectral amplitude and phase modulations also lead to pulse contrast degradation with increasing B integrals by shifting energy to satellite pulses. Because such weak modulations can never be avoided in a real laser system, the accumulated nonlinear phase in the laser system should be as low as possible to generate high-quality and high-contrast pulses.[33,34]

For further peak power and energy scaling, stimulated Raman scattering must be considered as well. SRS is initiated at a certain threshold value. Above this threshold, energy is rapidly transferred to a frequency-downshifted Stokes wave. The threshold for the onset of SRS in a fiber amplifier is, to a rough approximation, given by Eq. (17.4), where g_R is the peak Raman gain and L_{eff} is the effective interaction length.[7] An accurate expression for SRS threshold in active fibers can be found in Ref. 35.

$$P_{\text{threshold}}^{\text{SRS}} \approx \frac{16 \cdot A_{\text{eff}}}{g_R \cdot L_{\text{eff}}} \qquad (17.4)$$

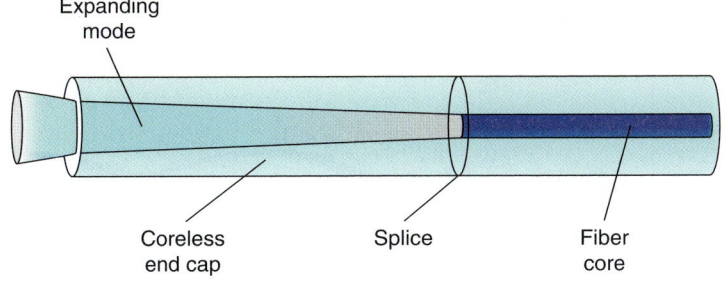

FIGURE 17.9 Coreless fiber end cap to avoid facet damage.

In addition to these basic nonlinear effects, fiber damage must also be considered as a scaling limitation of high-energy fiber laser systems. The surface damage fluence threshold of fused silica, which is significantly lower than the bulk damage fluence threshold, at a wavelength of ~1 µm can be estimated as

$$H_{\text{damage}} = 22 \cdot (\Delta \tau)^{0.4} \ \text{J/cm}^2 \quad (17.5)$$

where $\Delta \tau$ is the pulse duration in nanoseconds.[36] Therefore, in general, the damage threshold is significantly lower than the extractable energy. This problem can be solved by special treatment of the fiber end. One solution is to splice a coreless end cap on the output side of the fiber amplifier, as shown in Fig. 17.9. The expansion of the beam reduces the fluence and avoids fiber facet damage.

The maximum achievable pulse energy in fiber CPA systems is mainly limited by nonlinear effects and fiber damage of the amplifier, even though it also depends on the stretching ratio. A stretched pulse duration on the order of 2 ns is feasible. (This stretched pulse duration is assumed in the following discussion.) To identify the most crucial point, the different limiting effects have been summarized in Fig. 17.10 as a function of the mode field diameter, using Eqs. (17.3), (17.4), and (17.5). As pointed out, surface damage (dotted line) can be avoided by preparing the fiber with an end cap. Interestingly, a fiber CPA system in the linear regime ($B < \pi$ rad) with millijoule-level pulse energy is feasible; however, an actively doped fiber possessing a mode field diameter (MFD) larger than 100 µm is needed. Even higher pulse energies require a compensation (active or passive) of the accumulated nonlinear phase, for example, making an energy of about 3 mJ accessible with B integrals below 10 rad. Furthermore, an optimization of amplification bandwidth will allow for a possible decrease of compressed pulse duration below 300 fs. In turn, these pulse energy levels and durations should enable peak powers exceeding 10 GW out of a fiber CPA system. In combination with the high repetition

Fiber Lasers

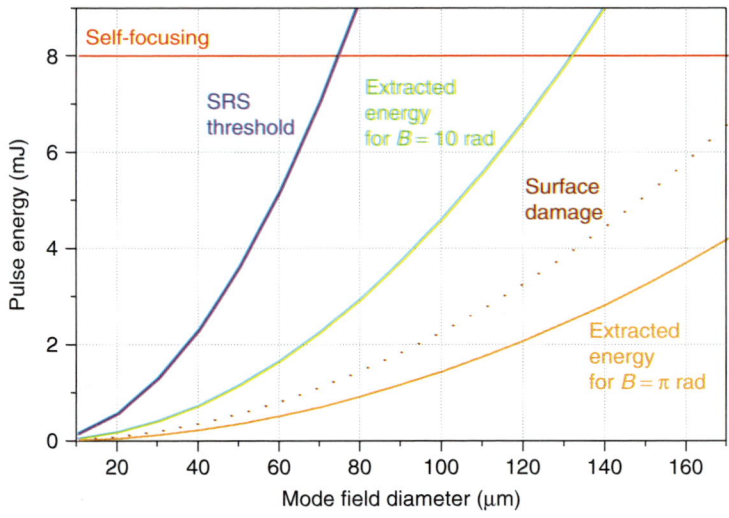

FIGURE 17.10 Overview of pulse energy limitations in a fiber CPA system as a function of mode field diameter, assuming a stretched pulse duration of 2 ns and a 1.2-m-long rod-type fiber with a gain of 30 dB.

rate a unique laser performance will be achieved. At the assumed pulse duration, a fundamental barrier around 8-mJ pulse energy is given by the self-focusing limit, which is independent of the mode area. (In fused silica, the critical peak power for linearly polarized light is about 4 MW at 1-μm wavelength.[37])

17.5 Average Power Scaling of Ultrashort Pulse Fiber Systems

In addition to the discussed potential in peak power increase, fiber amplifiers offer the possibility for producing pulses with femtosecond duration at immense average power levels.[38,39] To demonstrate this average power capability, a modified version of the CPA system depicted in Fig. 17.4 is used. In this modified version, the front end-oscillator produces 200-fs pulses at 1042-nm signal wavelength with 78 MHz pulse repetition frequency and 150 mW average power. The AOM is removed, and the pulses are stretched to 800-ps duration. The gratings in the stretcher-compressor unit have been replaced to increase the overall efficiency. Afterward, the remaining signal (120 mW) is amplified to 50 W, using two amplifier stages that comprise 1.2- and 1.5-m-long double-clad PCFs, both with 40-μm core and 170-μm pump cladding diameters. All fibers are pumped at 976-nm wavelength. The main amplifier fiber is a water-cooled 8-m-long double-clad fiber with 27-μm mode field diameter and 500-μm

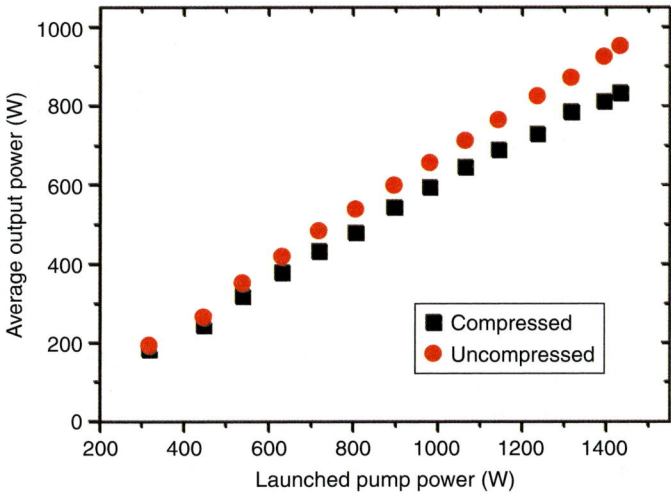

FIGURE 17.11 Uncompressed and compressed signal average powers, depending on the launched pump power.

air clad. Although this step-index fiber has no PCF structure, the core is nanostructured and consists of Yb- and F-doped glass rods with subwavelength diameters arranged in such a way that they allow tuning of the refractive index while also reducing the numerical aperture (NA). This LMA fiber was especially designed for high-average-power operation—that is, it possesses a large air clad and a signal that guarantees single-mode operation even at these average power levels. The resulting amplifying slope is depicted in Fig. 17.11.

At a launched pump power of 1450 W, the signal output power is 950 W, corresponding to 12.2-µJ pulse energy. The main amplifier's beam quality was measured to be $M^2 = 1.3$ at this power level. The resulting maximum compressed signal power is 830 W with 10.6-µJ pulse energy. The compression efficiency drops slightly from 95 to 88 percent at the maximum signal power, because of depolarization taking place inside the fiber. Owing to the acquired nonlinear phase (mainly in the main amplifier) with a calculated B integral of 11 rad, the autocorrelation width increases from 750 to 880 fs, and the corresponding peak power is approximately 12 MW.

17.6 Conclusion and Outlook

This chapter provides an overview of the main challenges of high peak and average power extraction in fiber-based laser systems. Among these challenges, overcoming the detrimental effects of self-phase modulation is currently the most urgent. The amount of accumulated nonlinear phase is determined, to a large extent, by the fiber of the main amplification stage. Thus, the use of low-nonlinearity rare-earth-doped double-clad fibers, such as the described 88/200 rod-type PCF, is a prerequisite for the extraction of high-energy pulses from fiber-based systems. The unique performance of millijoule pulses with femtosecond-pulse duration from a high-repetition-rate fiber CPA system has been experimentally demonstrated.

It should be pointed out that the gain bandwidth of the amplification stages (approximately 20 nm) theoretically supports significantly shorter pulses than those described in this chapter. We are convinced that femtosecond fiber CPA systems have the potential for several millijoules of pulse energy at megahertz repetition rates—that is, greater than kilowatt average power, which is the subject of ongoing developments. Such a laser source will open up new possibilities in applications ranging from fundamental science to industrial production.

References

1. Yamakawa, K., and Barty, C. P. J., "Ultrafast, Ultrahigh-Peak, and High-Average Power Ti:Sapphire Laser System and Its Applications," *IEEE J. Select. Top. Quant. Electron.*, 6: 658–675, 2000.
2. Nolte, S., Momma, C., Jacobs, H., Tünnermann, A., Chichkov, B. N., Wellegehausen, B., and Welling, H., "Ablation of metals by ultrashort laser pulses," *J. Opt. Soc. Am. B*, 14: 2716–2722, 1997.
3. Müller, D., Backus, S., Read, K., Murnane, M., and Kapteyn, H., "Cryogenic cooling multiplies output of Ti:sapphire lasers," *Laser Focus World*, 41: 65–68, 2005.
4. Giesen, A., Hugel, H., Voss, A., Wittig, K., Braucli, U., and Opower, H., "Scalable Concept for Diode-Pumped High Power Solid-State Lasers," *Appl. Phys. B*, 58: 365–372, 1994.
5. Fomin, V., Abramov, M., Ferin, A., Abramov, A., Mochalov1, D., Platonov, N., and Gapontsev, V., "10kW Single Mode Fiber Laser," SyTu-1.3, Symposium on High-Power Fiber Lasers, 14th International Conference, *Laser Optics*, 2010.
6. Paschotta, R., Nilsson, J., Tropper, A. C., and Hanna, D. C., "Ytterbium-Doped Fiber Amplifiers," *IEEE J. Quantum Electron.*, 33: 1049–1056, 1997.
7. Agrawal, G. P., *Nonlinear Fiber Optics*, San Diego, CA: Academic Press, 1995.
8. Alfano, R. R., *The Supercontinuum Laser Source*. New York: Springer-Verlag, 1989.
9. Strickland, D., and Mourou, G., "Compression of Amplified Chirped Optical Pulses," *Opt. Commun.*, 55: 447–449, 1985.
10. Jeong, Y., Sahu, J. K., Payne, D. N., and Nilsson, J., "Ytterbium-Doped Large-Core Fiber Laser with 1.36 kW Continuous-Wave Output Power," *Opt. Express*, 12: 6088–6092, 2004.
11. Limpert, J., Liem, A., Reich, M., Schreiber, T., Nolte, S., Zellmer, H., Tünnermann, A., et al., "Low-Nonlinearity Single-Transverse-Mode Ytterbium-Doped Photonic Crystal Fiber Amplifier," *Opt. Express*, 12: 1313–1319, 2004.
12. Limpert, J., Schmidt, O., Rothhardt, J., Röser, F., Schreiber, T., Tünnermann, A., Ermeneux, S., et al., "Extended Single-Mode Photonic Crystal Fiber Lasers," *Opt. Express*, 14: 2715–2720, 2006.
13. Brooks, C. D., and Di Teodoro, F., "Multimegawatt Peak-Power, Single-Transverse-Mode Operation of a 100 μm Core Diameter, Yb-Doped Rodlike Photonic Crystal Fiber Amplifier," *Appl. Phys. Lett.*, 89: 111119, 2006.
14. Schmidt, O., Rothhardt, J., Eidam, T., Röser, F., Limpert, J., Tünnermann, A., Hansen, K. P., et al., "Single-Polarization Ultra-Large-Mode-Area Yb-Doped Photonic Crystal Fiber," *Opt. Express*, 16: 3918–3923, 2008.

15. Liu, C.-H., Chang, G., Litchinitser, N., Galvanauskas, A., Guertin, D., Jacobson, N., and Tankala, K., "Effectively Single-Mode Chirally-Coupled Core Fiber" (paper ME2), *Advanced Solid-State Photonics,* OSA Technical Digest Series (CD), Optical Society of America, 2007.
16. Wong, W. S., Peng, X., McLaughlin, J. M., and Dong, L., "Breaking the Limit of Maximum Effective Area for Robust Single-Mode Propagation in Optical Fibers," *Opt. Lett.,* 30: 2855–2857, 2005.
17. Ramachandran, S., Nicholson, J. W., Ghalmi, S., Yan, M. F., Wisk, P., Monberg, E., and Dimarcello, F. V., "Light Propagation with Ultralarge Modal Areas in Optical Fibers," *Opt. Lett.,* 31: 1797–1799, 2006.
18. Ortaç, B., Schmidt, O., Schreiber, T., Limpert, J., Tünnermann, A., and Hideur, A., "High-Energy Femtosecond Yb-Doped Dispersion Compensation Free Fiber Laser," *Opt. Express,* 15: 10725–10732, 2007.
19. Limpert, J., Deguil-Robin, N., Manek-Hönninger, I., Salin, F., Schreiber, T., Liem, A., Röser, F., et al., "High-Power Picosecond Fiber Amplifier Based on Nonlinear Spectral Compression," *Opt. Lett.,* 30: 714–716, 2005.
20. Schmidt, O., Rothhardt, J., Röser, F., Linke, S., Schreiber, T., Rademaker, K., Limpert, J., et al., "Millijoule Pulse Energy Q-Switched Short-Length Fiber Laser," *Opt. Lett.,* 32: 1551–1553, 2007.
21. Eidam, T., Röser, F., Schmidt, O., Limpert, J., and Tünnermann, A., "57 W, 27 fs Pulses from a Fiber Laser System Using Nonlinear Compression," *Appl. Phys. B,* 92: 1, 9–12, 2008.
22. Perry, M., Ditmire, T., and Stuart, B., "Self-Phase Modulation in Chirped-Pulse Amplification," *Opt. Lett.,* 19: 2149–2151, 1994.
23. Schreiber, T., Schimpf, D., Müller, D., Röser, F., Limpert, J., and Tünnermann, A., "Influence of Pulse Shape in Self-Phase-Modulation-Limited Chirped Pulse Fiber Amplifier Systems," *J. Opt. Soc. Am. B,* 24: 1809–1814, 2007.
24. Röser, F., Eidam, T., Rothhardt, J., Schmidt, O., Schimpf, D. N., Limpert, J., and Tünnermann, A., "Millijoule Pulse Energy High Repetition Rate Femtosecond Fiber Chirped-Pulse Amplification System," *Opt. Lett.,* 32: 3495–3497, 2007.
25. Limpert, J., Roser, F., Schimpf, D. N., Seise, E., Eidam, T., Hadrich, S., Rothhardt, J., et al., "High Repetition Rate Gigawatt Peak Power Fiber Laser Systems: Challenges, Design, and Experiment," *IEEE J. Select. Top. Quantum Electron.,* 15: 159–169, 2009.
26. Schimpf, D. N., Limpert, J., and Tünnermann, A., "Controlling the Influence of SPM in Fiber-Based Chirped-Pulse Amplification Systems by Using an Actively Shaped Parabolic Spectrum," *Opt. Express,* 15: 16945–16953, 2007.
27. Kuznetsova, L., and Wise, F. W., "Scaling of Femtosecond Yb-Doped Fiber Amplifiers to Tens of Microjoule Pulse Energy via Nonlinear Chirped Pulse Amplification," *Opt. Lett.,* 32: 2671–2673, 2007.
28. Shah, L., Liu, Z., Hartl, I., Imeshev, G., Cho, G. C., and Fermann, M. E., "High Energy Femtosecond Yb Cubicon Fiber Amplifier," *Opt. Express,* 13: 4717–4722, 2005.
29. Zaouter, Y., Boullet, J., Mottay, E., and Cormier, E., "Transform-Limited 100 µJ, 340 MW Pulses from a Nonlinear-Fiber Chirped-Pulse Amplifier Using a Mismatched Grating Stretcher–Compressor," *Opt. Lett.,* 33: 1527–1529, 2008.
30. Yilmaz, T., Vaissie, L., Akbulut, M., Booth, T., Jasapara, J., Andrejco, M. J., Yablon, A. D., et al., "Large-Mode-Area Er-Doped Fiber Chirped Pulse Amplification System for High-Energy Sub-Picosecond Pulses at 1.55 µm," *Proc. SPIE,* 6873: 687354, 2008.
31. He, F., Hung, H. S. S., Price, J. H. V., Daga, N. K., Naz, N., Prawiharjo, J., Hanna, D. C., et al., "High Energy Femtosecond Fiber Chirped Pulse Amplification System with Adaptive Phase Control," *Opt. Express,* 16: 5813–5821, 2008.
32. Schimpf, D. N., Seise, E., Limpert, J., and Tünnermann, A., "Self-Phase Modulation Compensated by Positive Dispersion in Chirped-Pulse Systems," *Opt. Express,* 17: 4997–5007, 2009.
33. Schimpf, D. N., Seise, E., Limpert, J., and Tünnermann, A., "The Impact of Spectral Modulations on the Contrast of Pulses of Nonlinear Chirped-Pulse Amplification Systems," *Opt. Express,* 16: 10664–10674, 2008.

34. Schimpf, D., Seise, E., Limpert, J., and Tünnermann, A., "Decrease of Pulse-Contrast in Nonlinear Chirped-Pulse Amplification Systems Due to High-Frequency Spectral Phase Ripples," *Opt. Express,* 16: 8876–8886, 2008.
35. Jauregui, C., Limpert, J., and Tünnermann, A., "Derivation of Raman Threshold Formulas for CW Double-Clad Fiber Amplifiers," *Opt. Express,* 10: 8476–8490, 2009.
36. Köchner, W., *Solid-State Laser Engineering*. Berlin: Springer, 1999.
37. Brodeur, A., and Chin, S. L., "Ultrafast White-Light Continuum Generation and Self-Focusing in Transparent Condensed Media," *J. Opt. Soc. Am. B,* 16: 637–650, 1999.
38. Eidam, T., Hädrich, S., Röser, F., Seise, E., Gottschall, T., Rothhardt, J., Schreiber, T., et al., "A 325-W-Average-Power Fiber CPA System Delivering Sub-400 fs Pulses *IEEE J. Sel. Top. Quantum Electron.,* 15: 187, 2009.
39. Eidam, T., Hanf, S., Seise, E., Andersen, T., Gabler, T., Wirth, C., Schreiber, T., et al., "Femtosecond Fiber CPA System Emitting 830 W Average Output Power," *Opt. Lett.,* 35: 94–96, 2010.

CHAPTER 18
High-Power Fiber Lasers for Industry and Defense

Michael O'Connor
Director, Advanced Applications, IPG Photonics Corporation, Oxford, Massachusetts

Bill Shiner
Vice President, Worldwide Sales, IPG Photonics Corporation, Oxford, Massachusetts

18.1 Introduction

In the years since the collapse of the telecommunications market expansion of the 1990s, fiber lasers have dramatically changed. When demand for telecommunications devices declined, companies were in a position of having remarkable new technologies and capabilities but with only a few markets. This situation, coupled with a well-timed injection of funded development programs by the Defense Advanced Research Projects Agency (DARPA), a U.S. government technology development agency, resulted in an unprecedented run-up in fiber laser output power. In 2002, fiber lasers were available as commercial products at average power levels in the range of a few watts to a few tens of watts. By 2010, multikilowatt fiber lasers had become commercially available, and power levels up to 10 kW single mode and 50 kW multimode had been achieved. As of this writing, high-power fiber lasers are penetrating the industrial and defense laser markets at a rapid rate.

The relatively swift acceptance of fiber lasers in industrial and defense applications since 2002 has been for good reasons. Relative to other types of high-power lasers, fiber lasers offer numerous advantages:

- High reliability and ruggedness
- High efficiency
- Low maintenance requirements
- Small footprint and volume
- Capability for high continuous wave (CW) power and excellent beam quality
- Fast turn-on to full power without warm-up
- Fast modulation rate, up to about 10 kHz
- Flexible fiber delivery
- Low cost

18.2 Fiber Laser Engineering

When we speak of fiber lasers for high-power industrial and defense applications, we refer almost exclusively to ytterbium (Yb)-doped fiber lasers. Due to the small quantum defect between the ~1060 nm emission wavelength and the pump wavelength in the 915 to 975 nm range, Yb fiber lasers offer excellent optical-to-optical efficiency, which results in reduced heat generation in the gain medium. In addition, the long, thin optical fiber (typically on the order of 10 m long and 125 µm in diameter) has a very high surface area-to-volume ratio, which allows the fiber to expel heat rapidly. If one would like to scale a laser to high average power, this is a wonderful combination, because the Yb-doped fiber laser generates relatively little heat, and the heat it does generate is quickly shed. All of this has allowed for unprecedented CW scalability. The rate of power scaling for Yb fiber lasers versus carbon dioxide (CO_2) and neodymium-doped yttrium aluminum garnet (Nd:YAG) lasers is shown in Fig. 18.1.

Fiber lasers have benefited greatly from telecommunications industry investment. Many of the components used in present-day high-power fiber lasers were first developed during the telecommunications industry expansion. Perhaps most critical to the success of high-power fiber lasers has been the development of single-emitter pump diodes (Fig. 18.2). These diode packages are fiber coupled, are hermetically sealed, and have been developed at 9XX-nm wavelengths to very high levels of electrical-to-optical efficiency. The lifetime of these diodes is extraordinary, often predicted to be 50,000 hours or even greater than 100,000 hours. In comparison, the bar diodes traditionally used to pump solid-state lasers, though greatly improved in recent years, do not boast these levels of reliability or longevity. Periodic replacement of bar diodes and flash-lamp pumps increase maintenance time and cost. Single-emitter diodes, conversely, are generally maintenance free and do not require replacement

High-Power Fiber Lasers for Industry and Defense

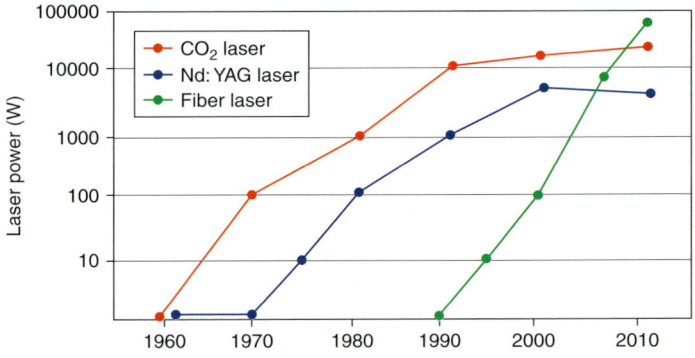

FIGURE 18.1 Evolution of output power for carbon dioxide (CO_2), Nd:YAG, and Yb fiber lasers. (European Automotive Laser Application, *Automatic Feed Co.*, Advanced Laser Applications for Welding 2009.)

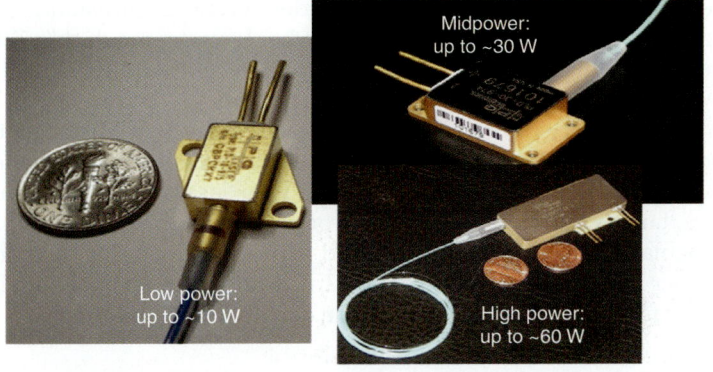

FIGURE 18.2 Telecommunications-style fiber-coupled, single-emitter pump diodes. Each single emitter typically provides ~10 W, so that higher-power packages contain multiple emitters coupled into a single fiber using micro-optics. (*Courtesy of IPG Photonics*)

during the life of the laser. Another advantage of single-emitter diodes is that they are conductively cooled and, thus, do not come into direct contact with cooling water or other cooling fluid. Finally, the fiber-coupled packages are a natural match for pumping fiber lasers. Typically, several or even many tens of these packages are combined via all-fiber combiners. The fiber-coupled direct current (DC) electrical-to-optical efficiency of these diodes is typically in the range of 50 to 55 percent. This high electrical-to-optical efficiency, coupled with the high optical-to-optical efficiency of Yb fiber lasers (in the range of 75 to 85 percent), results in the unprecedented overall electrical-to-optical efficiency of Yb fiber lasers in the range of 30 to 40 percent, including all losses.

In addition to high efficiency and reliability, Yb fiber lasers that use single-emitter diodes offer other practical advantages. For example, they do not require simmer or warm-up time, because fiber delivery enables single-emitter diodes to thermally decouple from one another and from the active fiber. Remarkably, they can go from a cold start to full power in less than 100 μs, and they can be modulated at rates up to 10 kHz. These features are particularly useful in industrial applications.

An area of concern regarding the reliability of Yb-doped fiber lasers is photodarkening, which involves the formation of "color center" defects in the fiber core and which is usually observed as a gradual degradation in fiber laser output power with operating time. These defects are thought to be created when the energies of multiple inverted ions combine to create emission in the visible or ultraviolet (UV) spectral range, rather than in the near infrared (IR). These energetic photons are thought to cause core defects that are highly absorptive in the UV and visible wavelengths. The absorption curve frequently extends into the near IR, resulting in reduced pump power available for inversion, as well as in increased loss at the output emission wavelength. Usually, this phenomenon is gradual, occurring over hundreds of hours; however, under certain circumstances, the degradation can be dramatic, with the fiber laser "going dark" within hours or even minutes in extreme cases. Recent studies indicate a power-law dependency of the rate of photodegradation versus the density of inverted ions.[1] Thus, gain stages with high Yb concentrations or with clustered Yb ions are far more susceptible to this effect. Manufacturers of Yb-doped fibers and fiber lasers have taken measures, largely proprietary, to inhibit or minimize photodarkening. In particular, certain codopants are used to prevent clustering of Yb ions in order to inhibit energy exchange. In addition, the level of gain in a stage can be reduced to minimize the density of inverted ions.

18.3 Power Scaling of Broadband Multimode Fiber Lasers

Broadband multimode fiber lasers have been scaled to output power of 50 kW, and it appears that up to several hundred kilowatts of output power is feasible. Unlike high-power, broadband, single-mode fiber lasers, in which power is amplified in gain stages arranged in series, high-power multimode fiber lasers are generally produced via the combining of multiple single-mode fiber lasers in parallel. A schematic of such a fiber laser is provided in Fig. 18.3. This configuration involves using an all-fiber combiner to merge the outputs of a multitude of single-mode fiber lasers. Power scaling of these lasers is currently limited by the all-fiber combiners employed in this laser configuration.

The all-fiber combiner employed in these high-power multimode fiber lasers is a form of what has become known in the industry as a tapered fiber bundle (TFB), which is another of the many fiber components developed during the telecommunications industry expansion of the 1990s. The TFB was originally developed to combine pump

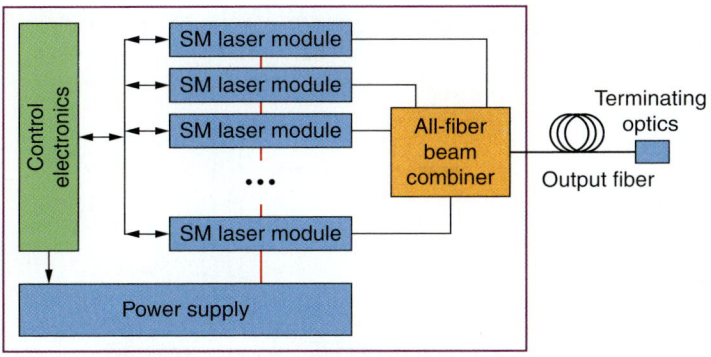

FIGURE 18.3 High-power multimode laser configuration. SM: single-mode.

and signal inputs into a double-clad gain fiber for higher power erbium-doped fiber amplifiers (EDFAs). The variant of the all-fiber combiner used for multimode fiber lasers involves the combination of several single-mode input fibers into a single multimode output fiber. Because the multiple inputs are combined incoherently, brightness is not increased; rather, it is diminished somewhat, owing to imperfections in the TFB and to slight underfilling of the TFB aperture and the numerical aperture (NA). The combined power, however, is largely conserved, except for small losses in the combiner. Typical losses are about 2 to 3 percent or less, resulting in about 97 percent or greater beam-combination efficiency. Table 18.1 shows several examples of demonstrated high-power multimode laser performance.

There are several advantages to this incoherent combination architecture:

1. The modular approach is easily scalable.
2. It is naturally redundant. In fact, additional reserve single-mode fiber modules may be incorporated and controlled so that they would turn on if one or more modules were to fail.
3. Each single-mode fiber laser can be derated to operate at a power well below its failure limit, resulting in a reliable, stable system.

MM Output Fiber Diameter (um)	Output Fiber NA	BPP (mm-mrad)	M^2	State-of-the-Art Output Power (kW)
50	0.09	2.3	7	7
100	0.09	5	15	20
200	0.09	10	30	50

TABLE 18.1 TFB Output Fiber Parameters, Beam Quality, and State-of-the-Art Output Power

FIGURE 18.4 50-kW multimode fiber laser. (*Courtesy of IPG Photonics*)

This scheme has proved highly reliable and scalable, with hundreds of multikilowatt fiber lasers sold into industrial and defense applications. A picture of a 50-kW fiber laser package is shown in Fig. 18.4, in which the vertically packaged 1-kW single-mode fiber laser modules are clearly visible below the power supplies.

Broadband multimode fiber lasers of this design lack the complexity of the broadband high-power single-mode lasers (described below). Due to the parallel configuration and large-core output fibers, power is not usually limited by stimulated Raman scattering (SRS) or available pump brightness. The primary limitation for these parallel multimode lasers is actually thermal issues within the fiber combiner. Because the combiner must take the full power load of all combined single-mode fiber lasers, even small percentage losses can result in significant power loss, resulting in high heat generation in a relatively small volume. Minimizing combiner loss and managing the heat load are critical to the reliability of a high-power multimode fiber laser of this design.

A secondary limitation for power scaling of multimode fiber lasers comes into play when attempting to scale power while simultaneously achieving high brightness. As shown in Table 18.1, beam quality for these lasers can be improved by using output fibers with smaller diameter. If the output power is sufficiently high, the power density in the output fiber core may reach a level such that SRS may limit the output fiber's length.

18.4 Power Scaling of Broadband Single-Mode Fiber Lasers

Four primary factors limit the power scaling of broadband single-mode fiber lasers:

- pump brightness
- nonlinear effects in the fiber, particularly SRS
- excessive heat generation in the final gain stage
- loss of fundamental mode power to higher-order modes

Each factor limits power scaling in the power amplifier stage (the final gain stage). Note that stimulated Brillouin scattering (SBS) is a limiting factor for spectrally narrow fiber lasers, but not for broadband fiber lasers, that is, those with a spectral bandwidth of several nanometers. SRS also limits the length of the output fiber (note that the SRS threshold is reduced as the fiber length increases and as the power density in the fiber core increases). Although each of these obstacles limits power scaling independently, they are also interrelated in that methods to reduce or avoid one may increase the effects of another. For example, one might develop or purchase brighter pump diodes to overcome the pump brightness issue, only to find that doing so increases the thermal problems in pump coupling or in the gain fiber. To reduce the heat generated per unit length, the gain fiber might be lengthened, but then the SRS threshold will be reduced. Due to the interdependent nature of these limiting factors, one must maintain a holistic approach to power-scaling strategies, taking into account all four factors in determining how to proceed.

Consider the power amplifier shown in Fig. 18.5. The gain fiber has a 20-μm core diameter and a 400-μm cladding diameter. Using

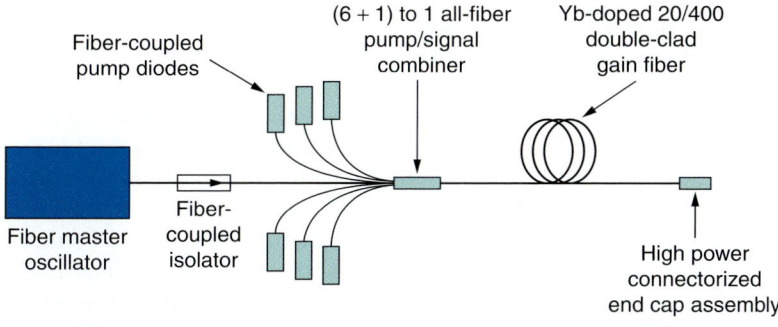

FIGURE 18.5 A typical high-power Yb fiber power amplifier configuration.

industry-standard pump combiners, which have six pump legs and a central signal fiber, there are six 200-μm, 0.22-NA pump ports. If one were to pump from both ends of the gain fiber, 12 such pump ports would be available. If fiber-coupled pumps were limited to 350 W for this fiber size, a total of 12×350 W, or 4.2 kW, of pump power would be available. If the input power to this power amplifier stage were 500 W, and the optical-to-optical efficiency were 75 percent, including all losses, the pump-limited output power would be 0.75×4.2 kW + 0.5 kW = 3.65 kW. To increase pump power to the gain fiber, one could argue that side couplers might be used to distribute more pump power along the gain fiber. However, at these power levels and fiber core size, the required lengthening of the gain fiber would result in a reduction of the SRS threshold. To increase the SRS threshold, the core size might be increased, resulting in reduced core power density and increased core-to-cladding ratio. This would, in turn, improve pump absorption, thus reducing fiber length. However, a reduced fiber length results in increased heat per unit length, and, as we have seen from Chap. 15, it is very difficult to obtain single-mode output with a fiber core larger than about 25 μm. Thus, the SRS limitation generally disallows distributed pumping and limits us to pumping only from both ends of the gain fiber.

Suppose now that our pump-brightness-limited amplifier is scaled by the availability of new, higher-brightness pumps with twice the power from the same size fibers. In theory, we could now scale the amplifier shown in Fig. 18.5 to 6.8 kW, or nearly double the output power. However, at this power level, we may exceed the SRS threshold, requiring a further shortening of the gain fiber. In order to absorb all this pump power in a very short gain fiber, the gain fiber's Yb concentration might be increased. However, at some point, doing so will exceed the thermal threshold due to excessive heat generated per fiber unit length, resulting in burning or degradation of the polymer fiber coatings. These examples show how the interrelated factors of pump brightness, SRS, heat load, and fundamental-mode guiding all limit the output power of broadband fiber lasers. Note that in addition to overcoming these primary factors, certain other fiber-based capabilities are prerequisite, including low-loss fibers and splices, fibers and gain stages designed to inhibit photodarkening, and low-index polymer fiber coatings capable of handling high pump power levels.

A novel approach to overcoming the four primary obstacles to power scaling broadband single-mode fiber lasers involves using fiber lasers, rather than pump diodes, to resonantly pump the final high-power amplifier. The pump brightness limitation was recently overcome by the development of single-mode Yb fiber lasers at 1018 nm with power output of up to 270 W. The schematic for a 10-kW single-mode fiber laser using these fiber pump sources is shown in Fig. 18.6.

The initial fiber master oscillator is diode laser pumped at 975 nm at the peak of the Yb absorption. The power amplifier is

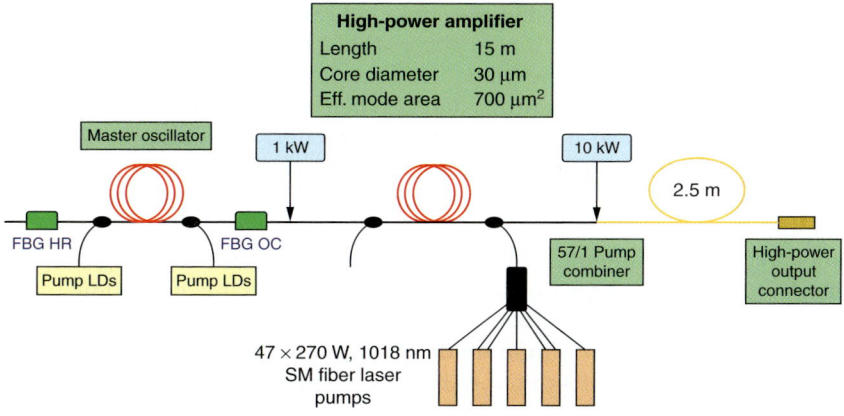

FIGURE 18.6 10-kW single-mode fiber laser configuration. FBG: fiber Bragg grating; LD: laser diode; HR: high reflector; OC: output coupler. (*Courtesy of IPG Photonics, Inc*)

pumped at 1018 nm on the red shoulder of the 975-nm absorption peak (Fig. 18.7). Although this pump wavelength results in a lower-absorption cross section, the brightness of the pumps is increased by more than two orders of magnitude, from about 30-W multi-mode diodes in 105-μm core (which was the limit for single-emitter fiber-coupled diode packages at the time of development) to 270-W single-mode fiber lasers. This increase allows for reduced

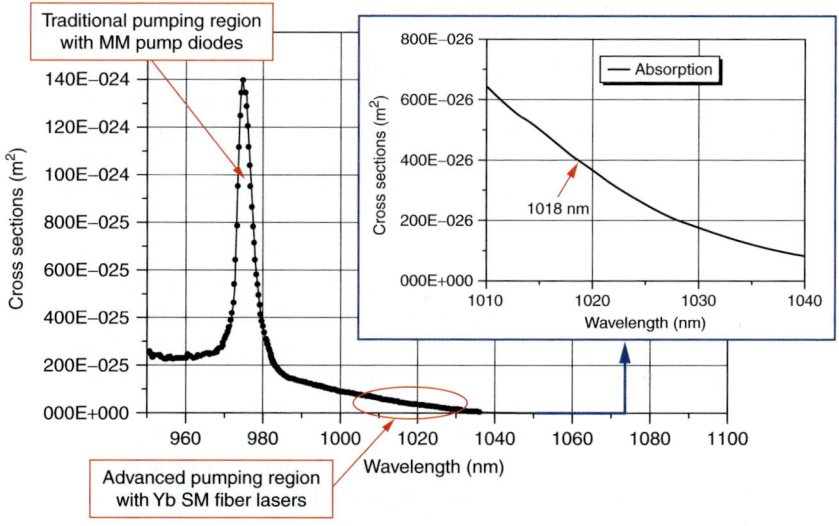

FIGURE 18.7 Absorption cross section for Yb^{3+} ions in silica glass with 1018 nm pumping versus 975 nm pumping. MM: multimode; SM: single mode.

cladding size, thus increasing the core-clad ratio and thereby compensating for the lower Yb core absorption cross section. In addition, the scheme is synergistic in that it addresses the heat generation issue. The quantum defect when pumping at 1018 nm and emitting at 1070 nm is less than 5 percent, versus approximately 9 percent for 975-nm pumping. Therefore, about half as much heat overall is generated in the gain fiber. The resonant pumping scheme has also been reported to improve mode guidance for the fundamental mode versus high-order modes.[2] Single-mode output power greater than 10 kW has been achieved to date using this resonant fiber laser pumping scheme.[3] It is speculated that even 20 to 30 kW single mode may be feasible, with Raman and thermal issues considered to be the ultimate limitations.[4]

18.5 High-Power Fiber Lasers in Industrial Applications

The benefits listed in previous sections, among others, have allowed for swift acceptance of high-power fiber lasers in the industrial laser markets. This rapid integration has been primarily driven by the replacement of existing lasers, such as relatively inefficient lamp-pumped and diode-pumped Nd:YAG lasers. A secondary mode of market infiltration has been due to the development of materials processing applications that take advantage of the combination of power and beam quality offered only by fiber lasers.

Fiber lasers are displacing other industrial lasers for several reasons. As a replacement technology, fiber lasers have proved able to retire older lasers and to provide improved throughput, while greatly reducing operating costs. Cost factors include reduction of downtime, reduction in service spares, and high electrical efficiency, often greater than 30 percent wall plug. Rising energy costs have only helped the case for switching to Yb-doped fiber lasers. When replacing welding CO_2 lasers with Yb fiber lasers, the elimination of increasingly expensive helium gas has been a major cost consideration. For new installations, integrators have taken advantage of the ease of beam delivery via flexible fiber lasers, using robots to reduce the capital investment for material processing cells. Companies have also been quick to adopt these lasers for multiuse applications. By utilizing the rapid switching of output fibers made possible by fiber lasers (via a bulk optics–based beam switch), customers can drive multiple work stations from a single laser source. Combination processing, including cutting, welding, and cladding, can also be arranged, dependent only upon the output fiber core diameter and the appropriate delivery optics. Fiber delivery also enables separation between work cells of 100 m or greater.

Competition is fierce within the materials processing markets. As a result, the total cost of ownership, including initial purchase price, setup costs, and long-term operating costs, is of great concern. Fiber lasers excel in all three areas. With ever-increasing demand, volume manufacturing has driven down fiber laser manufacturing costs. Recently, fiber laser purchase prices beat those of competing lasers. Setup costs with fiber lasers are low, due to several factors. A small footprint reduces the floor space requirement for the manufacturer. Fiber delivery of the output power makes integration with robots simple. Perhaps most important is reduced long-term operating costs. The unprecedented efficiency of fiber lasers not only reduces the cost over time of electricity needed to operate the laser, but also allows for smaller chillers to be used, which, in turn, reduces energy costs further. Long-term reliability and low maintenance also reduce long-term costs.

Continuous wave fiber lasers of moderate power level (tens of watts to a few hundred watts) have also achieved wide market acceptance. Like the pulsed fiber lasers, CW fiber lasers are usually air cooled, efficient, and compact. With a multimode step-index fiber, the mode at focus is converted into a top-hat spot rather than a Gaussian distribution. This change can make the application more consistent in many welding applications by making the focus and power adjustment less sensitive to change. For example, in spot welding of razor blades, the Gaussian distribution has made the threshold for welding versus cutting too close in both power sensitivity and focus adjustment. The top-hat distribution has proved to be much more robust in production. Single-mode fiber lasers with a 10-µm core are ideally suited for cutting thin materials and for microwelding applications. Modulation at rates up to 50 kHz enables such applications as microdrilling and microwelding with controlled heat management on thermally sensitive materials.

Pulsed fiber lasers with peak power up to 6 kW and average power of 400 W can operate in both the pulsed and the CW modes of operation. These units are very price competitive and are ideal for welding in the medical device and computer industries, which are currently dominated by high-peak-power, low-average-power lamp-pumped YAG lasers. A sampling of applications for these fiber lasers includes materials cutting (medical stents, surgical blades, solder masks, silicon solar panels); welding, sintering, and soldering; and marking and engraving.

High-power (>1 kW) single-mode and low-order multimode fiber lasers of good beam quality have become particularly disruptive in the industrial marketplace, as described earlier. The kilowatt and greater fiber lasers have gained wide acceptance because of their high reliability, huge reduction in maintenance cost, and low cost of operation. More important, however, they perform better and more consistently on all materials processing applications. They can cut up to five times faster than CO_2 lasers in materials up to 6 mm thick and slightly faster at 20 mm thickness. Fiber lasers are dominating the remote-weld area, where the use of high-speed scanners and robots take advantage of the long-focal-length focusing lenses to maximize

- Cutting of hydroformed automotive frames
- Blank welding for automotive industry
- Titanium welding of aircraft skins
- Laser cladding for aerospace and oil industries
- Battery welding for medical device industry
- Pacemaker welding for medical device industry
- Transmission welding for automotive industry
- Sheet metal cutting

TABLE 18.2   Industrial Applications for High-Average-Power Fiber Lasers

cutting speed. In 2009, the traditional original equipment manufacturers (OEMs) who offer large bed-cutting machines began offering fiber lasers as an alternative to CO_2 lasers, with the benefit of higher-cutting speed, much lower operating cost, much less optical complexity of the beam path, and the added benefit that aluminum, copper, and brass could be cut reliably with high quality. Fiber lasers can produce deep-penetration welds comparable to welds only produced with electron beam welding machines. For example, a 20-kW fiber laser can produce an electron beam quality weld in stainless steel 25 mm deep at a rate of 2 m/min.

A final growth area for fiber lasers is in the retrofit market. Companies that have already made large investments in their machine tools are finding that by replacing their old CO_2 or Nd:YAG lasers, they can cost-justify the new laser while increasing their manufacturing production. A sampling of kilowatt-class industrial applications are listed in Table 18.2.

18.6   Defense Applications of High-Power Fiber Lasers

Defense-directed energy applications have adopted Yb fiber lasers for many of the same reasons as for industrial applications. In particular, defense applications often require highly efficient, rugged, compact lasers. Fiber delivery of the laser light to a beam director is also a benefit. Fiber lasers have been used in a variety of directed-energy demonstrations, including counter–improvised explosive devices (IED), counter–rockets, artillery, mortars (RAM), and counter–unmanned aerial vehicles (UAV).

18.6.1   Fiber Lasers for Strategic versus Tactical Directed-Energy Applications

As a preface to a discussion of the requirements for lasers in various directed-energy applications, it is worthwhile beginning with a brief discussion of the effect of laser power and beam quality on irradiance and spot size on a distant target. Irradiance (another term for power density, usually given in watts per unit area) at a distant target depends not only on laser *power*, but also, more accurately, on laser *brightness*. It is important to note that brightness is proportional to laser power but inversely proportional to beam quality *squared*. Hence, as distance to the target increases, a laser of poor beam quality will become quickly inadequate, even if it has sufficient power. Conversely, for short-range applications, a laser of lesser beam quality may prove satisfactory.

Strategic directed-energy applications include such long-range applications as counter—intercontinental ballistic missiles (ICBM). These applications typically require lasers with output power in the hundreds of kilowatts or greater, with near-diffraction-limited beam quality. Single-mode fiber lasers are not expected to exceed about 25 kW, which is insufficient for such applications. Nevertheless, given the many compelling advantages of Yb fiber lasers, recent research has focused on combining multiple fiber lasers to meet the power and beam quality requirements. To combine the fiber lasers while maintaining beam quality, coherent or spectral methods are being investigated (see Chap. 19). These methods of beam combination (in contrast with incoherent beam combination) require relatively spectrally narrow lasers. As discussed in the fiber laser introduction, narrow-line fiber lasers are limited by SBS. The power limitation depends on line width. As of 2010, monolithic (all-fiber) Yb fiber lasers with line widths in the kilohertz range are limited to less than 200 W output power. Ytterbium fiber lasers in the 10-GHz range have been demonstrated up to about 1 kW of output power. A current DARPA program has a goal of 2 to 3 kW of output power for Yb fiber lasers with less than 10-GHz line width. It is believed that perhaps 100 such narrow-line lasers may be spectrally or coherently combined, resulting in a multi-hundred kilowatt laser with near-diffraction-limited beam quality.

While development of narrow-line fiber lasers for strategic applications continues, the power scaling of broadband Yb fiber lasers for industrial applications has been rapid. This growth has garnered the attention of directed-energy system developers for tactical applications with ranges less than about 10 km. To be effective, applications beyond about 3 to 5 km generally require single-mode beam quality. Multimode fiber lasers are sometimes adequate for applications of less than about the 3 to 5 km range. Using standard industrial Yb fiber lasers, successful demonstrations have been accomplished in applications such as counter-RAM, counter-UAV, and counter-IED. These successes have fueled an interest in power scaling of single-mode and multimode broadband fiber lasers and in improving beam quality for high-power multimode fiber lasers.

References

1. Joona Koponen, Mikko Söderlund, Hanna J. Hoffman, Dahv A. V. Kliner, Jeffrey P. Koplow, and Mircea Hotoleanu, "Photodarkening rate in Yb-doped silica fibers," *Appl. Opt.* 47: 1247-1256 2008.
2. Codemard, C. A., Nilsson, J., and Sahu, J. K., "Tandem Pumping of Large-Core Double-Clad Ytterbium-Doped Fiber for Control of Excess Gain" (paper AWA3), presented at the Advanced Solid State Photonics Conference, 2010.
3. O'Connor, M., Gapontsev, V., Fomin, V., Abramov, M., and Ferin, A., "Power Scaling of SM Fiber Lasers Toward 10kW" (paper CThA3), presented at the Conference on Lasers and Electro-optics, 2009.
4. Dawson, J. W., Messerly, J. M., Beach, R. J., Shverdin, M. Y., Stappaerts, E. A., Sridharan, A. K., Pax, P. H., et al., "Analysis of the Scalability of Diffraction-Limited Fiber Lasers and Amplifiers to High Average Power," *Opt. Express,* 16: 13240, 2009.

PART 5

Beam Combining

CHAPTER 19
Beam Combining

CHAPTER 19
Beam Combining

Charles X. Yu

Technical Staff, MIT Lincoln Laboratory, Lexington, Massachusetts

Tso Yee Fan

Associate Group Leader, MIT Lincoln Laboratory, Lexington, Massachusetts

19.1 Introduction

Recently, significant progress has been made in beam combining laser arrays, with total output power of 100 kW already achieved. Scaling analysis indicates that further scaling of average output power with good beam quality is feasible by using existing state-of-the-art lasers. Thus, knowledge of beam combining can be critical in high-power laser design. This chapter is divided into four sections. Section 19.1 provides a brief review of beam-combining metrics, while Sec. 19.2 describes several beam-combining techniques, including incoherent beam combining, coherent beam combining (CBC), wavelength beam combining (WBC), and hybrid beam combining (HBC). Section 19.3 presents a review of recent work with beam combining in fiber, semiconductor, and bulk solid-state laser systems, while Sec. 19.4 provides a brief chapter summary.

19.1.1 Motivation for Beam Combining

Lasers cannot be scaled to arbitrarily high power while maintaining their beam quality simply by supplying them with more pump power. In the case of solid-state lasers, thermal distortions, such as thermal lensing, leads to beam quality degradation at increased output power. In the case of fiber lasers, the damage threshold of bulk fused silica can be reached around 10–20 kW because of the inherent small mode-field diameter of single-mode fibers. Similar mode-field area and size constraints limit semiconductor laser powers to a few watts. Thus, multiple lasers are needed for applications that require power levels above that achievable from a single laser.

The motivation behind beam combining is simply to achieve higher power and brightness than that obtainable from a single laser emitter. For successful beam combining, the laser emitters require spectral control and/or mutual coherence preservation, depending on the beam-combining technique. These properties are usually not present in single laser emitters outputting at highest power. The basic laser emitter elements that are suitable for beam combining can often produce significantly less power than their non-beam-combinable counterparts. Beam-combining techniques thus must overcome this initial shortcoming to be of relevance.

19.1.2 Beam-Combining Performance Metrics

The goal of beam-combining systems is to produce essentially ideal laser beams from arrays of lasers. It is useful to have simple metrics to quantify the degree of ideality achieved. Two metrics commonly used to measure laser beam quality are M^2 and Strehl ratio S. A value of M^2 of 1 ($M^2 \geq 1$) corresponds to an ideal gaussian beam, while $M^2 > 1$ indicates poorer beam quality. A Strehl ratio of 1 ($S \leq 1$) corresponds to a uniform-phase, uniform-amplitude beam emitted by a circular aperture. Values of $S < 1$ indicate beam quality degradation.

An $M^2 = 1$ gaussian beam has a gaussian intensity pattern at all planes of propagation.[1] M^2 gives a measure of the second moment of a beam and is defined by

$$W^2(z) = W_o^2 + (M^2)^2 \left(\frac{\lambda}{\pi W_o}\right)^2 z^2 \qquad (19.1)$$

where z is the beam position, $W(z)$ is the second-moment beam width at position z, $2W_o$ is the beam waist, and λ is the wavelength.

The Strehl ratio S characterizes the on-axis far-field intensity of a beam propagated from a near-field hard aperture. S is defined as the ratio of this on-axis intensity to that of a uniform (in phase and amplitude), equal-power, top-hat beam that fills the same hard aperture.[2] A beam with a Strehl ratio of 1 (uniform amplitude and phase top hat) gives the classic airy pattern in the far field and is significant because such a beam has the highest on-axis intensity of any from a hard aperture. The Strehl ratio for small-wavefront variance and top-hat intensity can be expressed using the Marechal approximation as

$$S = \exp\left[-(2\pi/\lambda)^2 (\Delta\Phi)^2\right] \qquad (19.2)$$

where λ is the wavelength and $(\Delta\Phi)^2$ is the wavefront variance in length units.

One difficulty with these measures of laser beam quality is that an M^2 measure generally cannot be mathematically transformed into a Strehl ratio measure and vice versa. For example, an ideal gaussian beam ($M^2 = 1$) is infinite in extent, whereas a Strehl ratio is defined only for a finite aperture. To transform a gaussian beam into S requires a choice of a beam radius to clip the gaussian beam. In addition, an $S = 1$ beam has an undefined M^2. Although there is no exact transformation between M^2 and S, for good beam quality in either metric, the wavefront distortion must be relatively small—that is, for both $M^2 = 1$ and $S = 1$, the wavefront must be planar at the beam waist. Notionally, it can be thought that

$$S \sim 1/(M^2)^2 \tag{19.3}$$

The fundamental difficulty in transforming between M^2 and S, as well as in transforming among other beam-quality metrics, is the reduction in information inherent in using a single number to describe a laser beam with spatially dependent amplitude and phase. In the simplification, the information thrown away depends on the beam-quality metric definition.

Consequently, there is no single best definition of beam quality, and metrics other than M^2 and S are often used. The most appropriate metric for beam quality depends on the application. Another common definition of beam quality is power in the bucket (PIB).[2] This type of definition relies on hard apertures in the near and far fields. In the near field, a hard aperture of some size is defined, and the far-field hard aperture is expressed in terms of an angular divergence. The most commonly used beam-quality metrics are M^2 and S, which we will use in this chapter where possible.

The power and beam quality can be used to express the radiance (or brightness), defined as power per unit area per unit solid angle. The radiance B is

$$B = \frac{CP}{\lambda^2 (M^2)^2} \tag{19.4}$$

where P is the power, λ is the wavelength, M^2 is the beam quality (assuming a circularly symmetric beam), and C is a constant that depends on the definition of beam size and divergence angles. If a gaussian beam definition is chosen, then $C = 1$. According to the radiance theorem,[3] radiance of an individual laser is conserved as its output propagates through any lossless optical system. Because of this conservation and the fact that radiance accounts for both beam quality and power, it is useful to think of beam combining as combining radiance from individual laser sources.

19.2 Beam-Combining Techniques and Theories

There are many approaches to beam combining.[4] Beam-combining techniques on laser arrays can be characterized according to three broad classes: incoherent beam combining, CBC, and WBC. These are notionally illustrated in Fig. 19.1. In incoherent (side-by-side) beam combining, the radiance cannot be increased beyond that of a single element, whereas in the other two (CBC and WBC), the radiance can be increased relative to a single array element proportionally to the number of array elements N. Although technically, one can hybridize beam combining by using two or more of the above techniques, in this chapter, HBC refers only to a combination of CBC and WBC.

In incoherent beam combining, the array elements may (or may not) operate at the same wavelength; however, no attempt is made to control the relative spectra or phases of the elements. Although power increases as more elements are added, the overall beam quality decreases as the power increases, because adding elements increases the size of the aperture without a corresponding decrease in the beam's divergence angle. Thus, the combined beam's radiance is not increased.

Both CBC and WBC can increase the overall power while maintaining beam quality. In CBC, or phased arrays, all of the array elements operate with the same spectrum, and the relative phases of the elements are controlled such that there is constructive interference. Historically, most of the effort in laser beam combining has attempted to use this class of techniques to obtain good beam quality. CBC is the analogue of phased-array transmitters in the radio frequency and microwave portions of the electromagnetic spectrum; however, in the optical domain, CBC has proven difficult because of the shortness of an optical wavelength. The phases of the array elements must be controlled to a small fraction of a wavelength (2π phase); for the optical portion of the spectrum, the wavelength is on the order of 1 µm. Although CBC has been demonstrated for small arrays, it has been more challenging to identify robust, simple, phased-array approaches for combining large arrays (10's to 100's of elements) with nearly ideal beam quality has been more challenging, though progress is

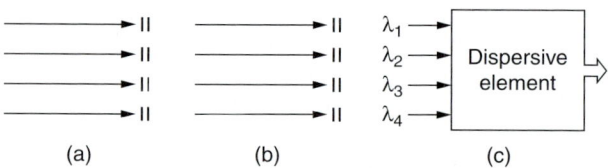

FIGURE 19.1 Three classes of beam combining: (*a*) incoherent beam combining, (*b*) coherent beam combining (CBC), and (*c*) wavelength beam combining (WBC). (Reprinted with permission from T. Y. Fan.[4] © 2005 IEEE)

being made. For example, 48 passive fibers have been coherently combined with a residual phase error of less than $\lambda/20$.[5]

In WBC, also known as spectral beam combining, the array elements operate at different wavelengths, and a dispersive optical system is used to overlap the beams from the elements in the near and far fields. WBC has also been used historically, though not nearly as often as CBC. Wavelength-division multiplexing in optical communications, in which multiple wavelengths are coupled into a single-mode optical fiber, falls into the WBC class. Again, the difficulty has been in identifying approaches to robustly combine large laser arrays in a simple manner.

Recent progress has demonstrated near-ideal combining ($M^2 = 1.3$) on a 100-element laser array, using an external-cavity, grating-stabilized beam-combining implementation.[6] Finally, although polarization multiplexing is a form of beam combining, it is not of interest for large arrays, because the improvement is limited to two. Therefore, we will not discuss polarization multiplexing in this chapter.

19.2.1 Incoherent Beam Combining

Incoherent beam combining refers to any configuration in which multiple lasers are placed side by side, without any attempt to control their spectra or phases in order to increase their net radiance. Conventional diode laser arrays (linear bars and two-dimensional arrays) fall into this class. For some industrial applications such as welding, in which the distance between the laser and the target is in centimeters, multiple diffraction-limited fiber lasers are coupled into a multimode fiber to increase the delivered energy. Although total power can easily be increased, the radiance of these types of sources cannot be any greater than the radiance of a single array element. For applications that rely on the far-field on-axis intensity, incoherent beam combining provides no benefit. Therefore, we will not go into details on incoherent beam combining in this chapter. Reviews on the topic can be found in Diehl[7] and Sprangle et al.[8]

19.2.2 Coherent Beam Combining

This section begins with a discussion of the fundamental requirements for ideal CBC.[3,9–19] This is followed by a review of some of the implementations[20–40] and sources of performance degradations.

In CBC, all elements of the laser array are forced to operate with the same frequency spectrum and with the same phase, so that when the beams are overlapped, the electric fields add to provide constructive interference. It is useful to understand the simplest two-element implementations of CBC, shown in Fig. 19.2a and 19.2b, which can then be generalized to large numbers of elements. As characterized by output formatting, the two subsets of CBC implementations are tiled-aperture and filled-aperture. In filled-aperture implementations,

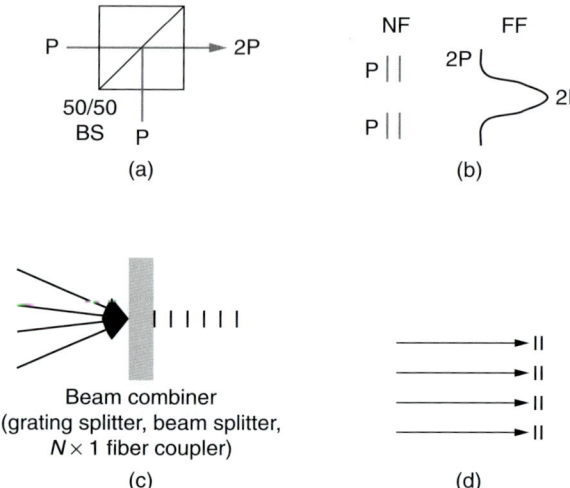

FIGURE 19.2 Two-element CBC: (a) filled aperture and (b) tiled aperture. N-element CBC: (c) filled aperture and (d) tiled aperture. NF: near field; FF: far field; BS: beam splitter. (Reprinted with permission from T. Y. Fan.[4] © 2005 IEEE)

the interference occurs in the near field. CBC of two elements at a 50/50 beamsplitter, shown in Fig. 19.2a, is an example of a filled-aperture implementation. The beam combiner in a filled-aperture system can be thought of as the inverse of a beamsplitter. In order to transfer all of the input power to the output port, several conditions must be met. The polarizations must be the same, the input powers must match, the optical phases must be controlled in order to have constructive interference, and the input beams must overlap in the near and far fields. In tiled-aperture implementations, as shown in Fig. 19.2b, individual elements have outputs that are adjacent to each other, and there is interference only in the far field. This type of implementation can be thought of as a synthesized plane wave. To minimize the power in side lobes and to obtain the maximum far-field intensity, the fill factor must approach unity (i.e., the spaces between tiles must be minimized). For both implementations, the combining elements must be mutually coherent to ensure constructive interference. Thus, the spectra from the combining elements must match each other, and the combining elements must be within their coherence length.

The tiled-aperture implementation differs from the filled-aperture implementation in the location of the interference. In Fig. 19.2a, beams are interfered in the near field, whereas in Fig. 19.2b, the beams are interfered in the far field. The conditions for ideal beam combining are essentially the same. For a tiled aperture, the beams must point in the same direction, and they must be tiled so that there is no gap between the beams, which essentially levies the same requirements on beam alignment as the beam-overlap condition in a filled aperture. Ideally, the radiance doubles by beam combining two elements—in other words, the beam-combining efficiency is unity. In the case of near-field interference, the near- and far-field patterns are unchanged compared with that of a single element, and the power is twice that of a single element relative to the beam combiner being not in place. In the tiled implementation, the power and the aperture size both double, while the divergence angle halves. Thus, the radiance doubles. This scaling can be generalized to the case of N laser elements, as shown in Fig. 19.2c and 19.2d. In Fig. 19.2c, N laser elements are combined in a filled configuration. The combiner can be, for example, a grating splitter, a series of free-space beam splitters, or an $N \times 1$ fiber coupler. In Fig. 19.2d, N laser elements are tiled side by side. In both cases, radiance scales linearly with the number of laser elements N. For a filled aperture, the near- and far-field patterns remain the same while the number of elements increases, and the power increases linearly, leading to the linear scaling of radiance. For an ideal tiled aperture, the Strehl ratio remains fixed as the number of elements increases, and the power increases linearly, leading to the linear scaling of radiance.

In filled-aperture implementations, however, in order to achieve ideal scaling, the beam-combining optics must be designed specifically for each N. For example, clearly the beam-combining element used to combine two elements in Fig. 19.2a (a 50/50 beam splitter) cannot be used to combine a larger number of elements. Conversely, a beam combiner designed for N elements but with inputs from only FN elements ($F < 1$) has a beam-combining efficiency of F and power into the combined beam of F^2.[13] Power that does not go into the combined beam ends up in side lobes, which can be seen from the two-element combining example of Fig. 19.2a. If only a single element operates, the power that goes into the combined beam is only one-quarter of that for both elements operating. Half of the power from a single element goes in the desired direction, and the other half goes to the undesired port. By analogy,[13] in a tiled-aperture system, the Strehl ratio scales as F, and the on-axis power scales as F^2, where F is the area fill factor of the aperture and where the size of the transmitting aperture is fixed (i.e., it is not allowed to change with N). Although this square-law scaling has, at times, been touted as an advantage, it is not. For example, the on-axis intensity in a tiled-aperture system goes as N^2 if the aperture size is allowed to scale with N; but this type of scaling can be achieved in any laser system if the aperture size increases proportionally to the power. Fundamentally, the radiance scales only with N, at best.

CBC has been applied to arrays of semiconductor, solid-state, gas, and fiber gain elements. Many of the CBC efforts throughout the

1990s can be described as marginally successful at best. Early demonstrations were not particularly robust against perturbations and had unclear scalability to large arrays and higher powers. Much of the work from the mid-1980s through the 1990s focused on efforts to coherently beam combine diode laser arrays, driven by the potential inherent simplicity of semiconductor systems. (References 3, 9, 15, and 18–20 provide excellent reviews on these efforts.) With advances in understanding of the requirements and in better analysis of specific implementations came a better appreciation of the difficulties in scaling to large arrays. This appreciation has, in turn, led to demonstrations of coherent combining, with the promise of scalability in both array size and power. Many CBC implementations have been reported, and these often fall into one or a combination of the following approaches: common resonator, evanescent- or leaky-wave coupling, self-organization, active feedback, and nonlinear optical effects, all of which are notionally illustrated in Fig. 19.3.

In common-resonator approaches (Fig. 19.3a), the array elements are placed inside an optical resonator, and feedback from the resonator is used to couple the elements together.[22–29] This approach can be viewed as being a spatially sampled version of a bulk resonator.

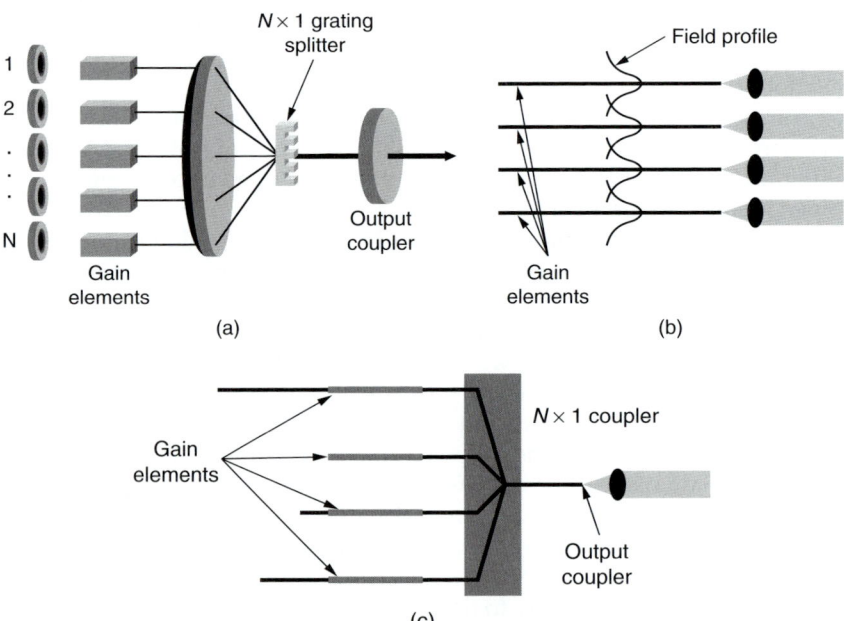

FIGURE 19.3 Various CBC implementations: (a) common resonator, (b) evanescent-wave/leaky-wave coupling, (c) self-organization, (d) active feedback, and (e) nonlinear optical effects. (Reprinted with permission from T. Y. Fan.[4] © 2005 IEEE)

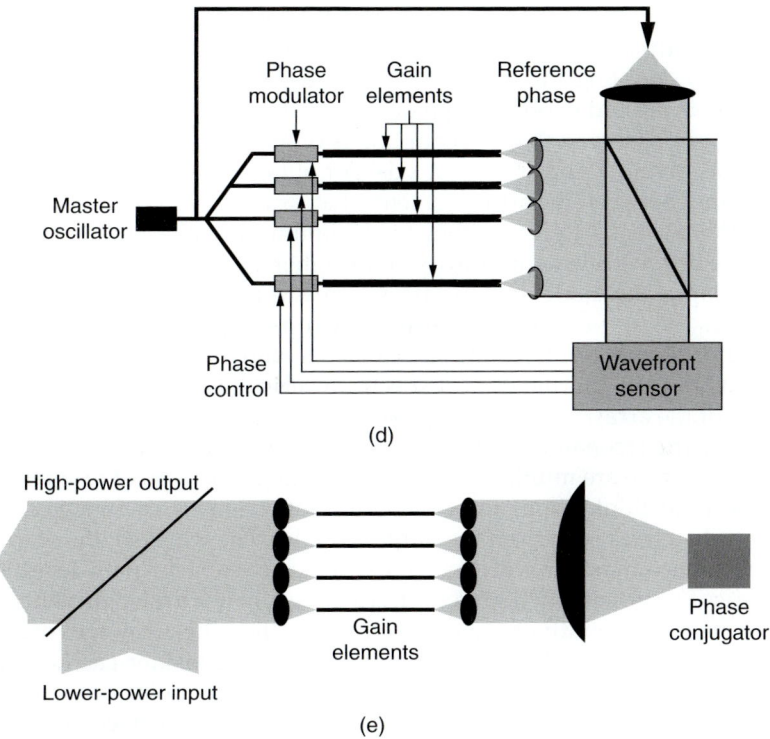

FIGURE 19.3 (*Continued*)

Consequently, as with a bulk resonator, the challenge for the resonator is to force lowest-order transverse-mode operation. In a bulk resonator, this might be done by using an intracavity spatial filter. In CBC using common resonators, mode selection has been done using intracavity spatial filters and the Talbot effect. Although these common-resonator approaches have been successful at low average power, as the power increases, there has been difficulty obtaining low-order transverse mode operation. One issue is differences in the optical path length, known as *piston error*, among the array elements, particularly at higher powers; this can be viewed as being the equivalent of wavefront distortion in a bulk optical element. Piston error makes it difficult to attain lowest-order transverse mode operation, similar to the distorted optical media in bulk lasers. This common resonator approach has been more successful with CO_2 lasers[26–28] than with diode or solid-state lasers, because of the much longer 10-µm wavelength of the CO_2 laser. This lower piston error (in number of waves) has enabled an 85-element CO_2 laser array to be phase locked.[28]

Evanescent-wave or leaky-wave coupling approaches[30–33] have been used extensively, particularly for CBC semiconductor laser arrays (Fig. 19.3*b*). In this approach, the array elements are placed sufficiently close together so that their field distributions overlap, thereby coupling the elements. In-phase coupling of the array elements is desired to obtain high on-axis far-field intensity; however, it has been observed that the coupling is often predominately π out of phase, giving a power null on axis. For out-of-phase coupling, there is a null between the array elements that, as compared with in-phase coupling, tends to lead either to minimum loss, particularly if the space between elements is lossy, or to higher gain, because the spatial overlap of the mode with the array elements is better.

In the self-organizing, also known as supermode, approach (Fig. 19.3*c*), the array comprises elements with very different optical path lengths, and the optical spectrum self-adjusts to minimize the loss of the array.[38,40–52] This approach is essentially a Michelson interferometric resonator.[38,40] generalized to arrays of more than two elements. There are multiple ways to understand this type of resonator. One is to think about the resonator's reflectivity as a function of wavelength as seen from the output coupler. The wavelengths of the reflectivity maxima will change as the array-element path lengths vary; if a sufficiently high reflectivity occurs at a wavelength within the gain bandwidth of the array elements, then the array will oscillate. Another way to view this approach is to consider each array element as a separate optical resonator (from the point of view of axial-mode frequencies). The array elements mutually injection lock at an optical frequency that is within the injection-locking range for every array element. This mutual injection locking forces a common optical frequency spectrum for the entire ensemble of array elements consistent with the lowest intracavity loss but cannot enforce an ideal phase relationship needed for efficient and high-brightness combining. Demonstrations have been performed using this technique with up to ten elements using fiber lasers. However, the beam-combining efficiency falls off as the number of elements increases. The prospects for scaling these self-organizing implementations to large arrays are poor[42,52] due to the lack of a mechanism that forces an ideal phase relation among elements; having the optical frequency as the only adjustable parameter leads to an overconstrained system.

Nonlinear optical approaches to beam combining have included phase conjugation and Raman beam combining.[53–65] Many of the CBC efforts using phase conjugation have relied on stimulated Brillouin scattering (SBS) in bulk media, which has a relatively high threshold that requires high-peak-power lasers. More recently, lower thresholds have been obtained by using guided-wave configurations.[65] Key issues with nonlinear optical beam combining include scaling to large numbers of elements, having low threshold, and handling the bandwidth and dynamic range of the required phase corrections.

The approaches listed above can all be considered as passive CBC. The phases themselves are not actively controlled; instead, a single device or mechanism is employed to enable the lowest-loss cavity mode when the laser elements are mutually phase locked. However, approaches that do not use active control of phase have not yet demonstrated scalability to large numbers. In active CBC implementations (Fig. 19.3d), laser elements are individually phase controlled, and feedback is used to equalize the optical path lengths modulo 2π.[5,66–69] This approach can be thought of as being equivalent to using a deformable mirror to actively correct the wavefront distortion in a bulk gain element. The implementation has been used mostly in master oscillator power amplifier (MOPA) architectures. Some of the key issues include defining the method of detection of optical-path length differences, understanding the dynamics of optical-path length variations, and designing a servo system that includes actuators with sufficient bandwidth and dynamic range to correct for these variations. Two types of servo loops—phase sensing[68,69] and multidither[5,67]—are commonly employed. In the former, the laser phases are detected and corrected via dedicated phased-locked loops (PLLs). In the latter, a single detector is used for all the lasers. Instead of detecting the phase, the laser elements are dithered to maximize the beam quality. For the servo loops to be effective, the closed-loop bandwidth must exceed the bandwidth of the laser elements' phase noise.[69] Because servo-loop design is very robust, active CBC can scale to very large numbers of laser elements. The drawback of active control is that an extra phase controller is needed for each laser.

One example of a phase-sensing control loop is the heterodyne wavefront sensor (Fig. 19.4a), in which a single master oscillator (MO) is split into multiple legs. One leg of the MO is frequency shifted and employed as the reference for all phase-detection systems. The remaining legs go through phase shifters and are amplified in multiple amplifiers. The amplifier outputs are placed side by side in a tiling geometry, and the reference beam interferes with all the amplifier outputs. The frequency shift between the reference and amplifier outputs generates temporal beat patterns that allow the electronic detection of optical phases relative to a common reference, provided the selected frequency shift itself is low enough to be detected by a photodetector. These phases are used as feedbacks for the control electronics to send appropriate signals to phase shifters in order to lock the beat frequency relative to the reference. The control electronics can be as simple as an XOR (exclusive or) gate followed by an integrator (Fig. 19.4b). This technique parallelizes CBC and is inherently scalable to many lasers. The drawback is that phasing more lasers requires more detection and electronic hardware.

19.2.3 CBC Performance Degradations

This last part of the section examines CBC performance degradations due to nonuniformities among the elements in the laser array and array fill factor. The nonuniformities include power variation (ΔA),

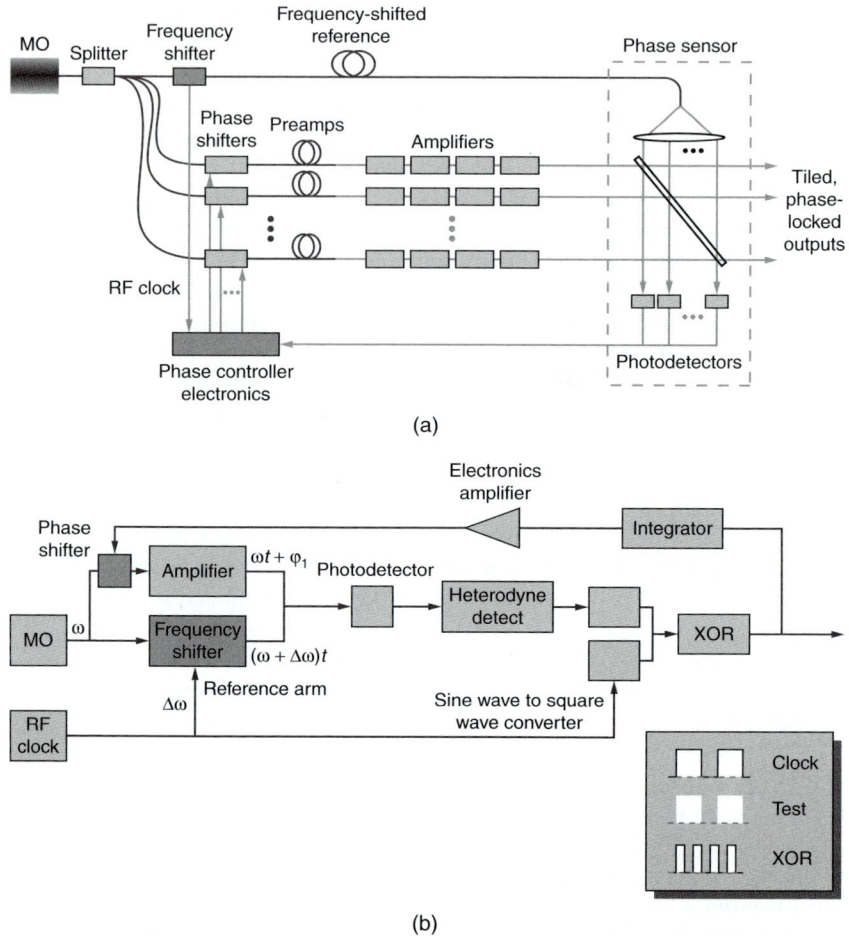

FIGURE 19.4 CBC using heterodyne wavefront sensor: (a) experimental schematic and (b) phase-control electronics for a single amplifier chain. MO: master oscillator; XOR: exclusive or; RF: radio frequency.

laser phase jitter ($\Delta\Phi$), laser-to-laser spacing error (Δx), and array-element pointing nonuniformity (Δp). The degradations are quantified in terms of the Strehl ratio for a tiled-aperture implementation, but this is entirely equivalent to the efficiency of beam combining for a filled-aperture implementation.[13]

Power variation among the array elements is quite common because different lasers can output slightly different power, even if they are designed to be identical. Thus, it is important to determine the degradation caused by power variation (in the absence of phase errors) in order to understand whether active equalization is needed to achieve good beam-combining efficiency. The analysis is presented by Fan[13]; for a large array and small ΔA, the expression is

$$n_{\Delta A} = 1 - \frac{\sigma_{\Delta A}^2}{4} \qquad (19.5)$$

where $n_{\Delta A}$ is the degraded Strehl and $\sigma_{\Delta A}$ is the standard deviation of ΔA.

As mentioned previously, active or passive phase-control techniques are employed to ensure that the elements in the laser array are in phase with each other. Thus, degradation due to residual phase jitter determines the phase control requirement. This analysis has been presented by multiple authors,[3,10] and for uncorrelated small $\Delta \Phi$ and a large array, the expression is

$$n_{\Delta \varphi} = 1 - \sigma_{\Delta \varphi}^2 \qquad (19.6)$$

where $\sigma_{\Delta \Phi}$ is the standard deviation of $\Delta \Phi$ and $n_{\Delta \phi}$ is the degraded Strehl, in the absence of power errors. Equation (19.6) is a linearization of the Marechal approximation presented in Eq. (19.2).

It is worth noting that even though Eqs. (19.5) and (19.6) look similar, $\Delta \Phi$ can vary over a much larger range than ΔA. Degradation from ΔA can be 10 times less severe than degradation from $\Delta \Phi$. Typical laser power uniformity is within 10 percent for nominally identical lasers, leading to a reduction in Strehl ratio of less than 1 percent; so their active equalization is not required. Controlling element phase is critical. Even a phase error as small as $\lambda/20$ results in Strehl degradation of 10 percent.

To examine the degradations caused by Δx and Δp, it is helpful to use the fiber array as an example. Fiber arrays are often mated with a matching microlens array to ensure collimated output beamlets. If the fiber-spacing errors lead to a mismatch with the microlens-array-element spacing, the collimated beamlets will not point in the same direction. This aggregate pointing error results in a degraded wavefront quality, which reduces S. The expression for Strehl reduction can be computed using Eq. (19.2). Thus, the fiber-spacing uniformity must be better than a fraction of the mode-field diameter (MFD). Using the same example, one can easily see that Δp degrades the Strehl by reducing the array's fill factor and by clipping individual beams at the edges of the microlenses. The fiber tilt must be controlled to within a small fraction of the fiber's numerical aperture (NA). Figures 19.5a and 19.5b plot the Strehl degradation as a function of Δx and Δp for standard single-mode fiber at 800 nm, which has MFD of 5.5 μm and NA of 0.12. Figure 19.5 shows that micrometer-size tolerance, or 20 percent of the fiber MFD, is needed for Δx; and the tolerance on Δp can be 20 mrad, or 20 percent of the fiber's NA. Of these, Δx is the more challenging to control.

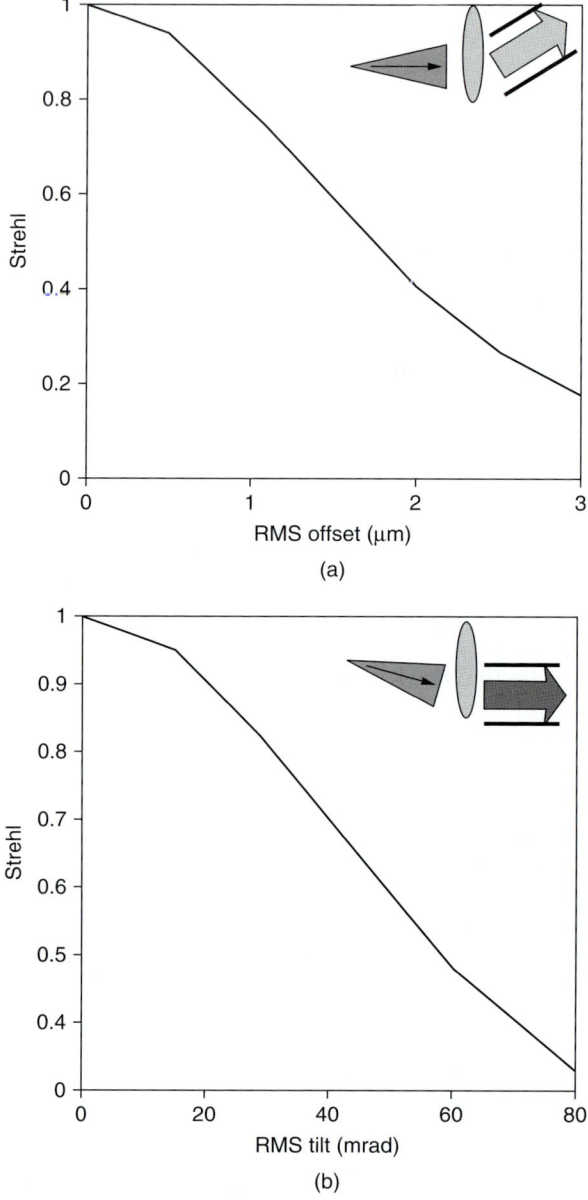

FIGURE 19.5 Strehl degradation for a single-mode fiber at 800 nm versus (a) RMS offset and (b) RMS tilt. All calculations were made using NA = 0.12 and pitch = 250 μm; ideal array has Strehl = 1.

Large-mode-area fibers used for fiber lasers can have MFD of 25 μm and NA of 0.03. Such fibers relax the Δx requirement at the expense of the Δp requirement.

Another source of Strehl degradation for the tiled configuration is the array fill factor. If no gap exists between the laser elements, the array fill factor will be unity, and the resulting combined beam will reproduce a single aperture. For laser arrays with fill factor less than unity, $S = c^2/b^2$, where b is the laser element spacing and c is the laser element size, assuming the tiles are ideal square patches with uniform intensity and phase. Thus, the Strehl ratio degradation is equal to the fill factor.[3]

In analyzing the effect of missing array elements for a multiport beam combiner used for filled-aperture configurations, a direct analogy can be made between the Strehl ratio for a tiled aperture and the beam-combining efficiency for a filled aperture.[13] The beam combiner in filled-aperture implementations can have unity efficiency only if all ports are populated and operational. For beam combiners that require all ports to have equal power for unity combining efficiency, the beam-combining efficiency when only a fraction F of the ports are operational is simply equal to F, which is in direct analogy to the Strehl ratio being equal to the fill factor. This leads to a beam-combined output power from a beam-combined system of F^2 relative to a fully populated and operational array.

19.2.4 Wavelength Beam Combining

The radiance of WBC systems scales linearly with the number of laser elements by scaling the spectral composition. The two subsets of its implementations, characterized by the beam combiner, are serial and parallel (see Fig. 19.6). An early implementation of wavelength combining that was proposed and demonstrated used dichroic interference filters to serially combine six diode lasers operating at different wavelengths.[70] In this case, each diode laser, or channel, operated at a different wavelength. The output of an individual diode was transmitted through an interference filter that passed its wavelength while reflecting all other wavelengths. One drawback of the serial approach

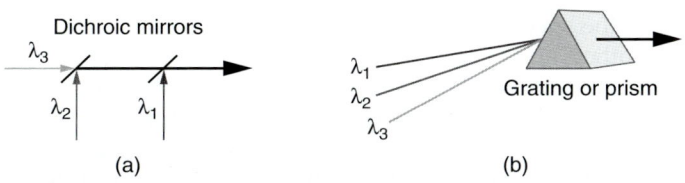

FIGURE 19.6 WBC schematic: (a) serial implementation and (b) parallel implementation. (Reprinted with permission from T. Y. Fan.[4] © 2005 IEEE)

is the number of beam combiners needed. Combining N beams requires $N-1$ beam combiners, each of which is spectrally different. Each laser must be individually aligned to an accuracy such that all N beams after $N-1$ stages still overlap. The serial approach has proven to be very effective for small N; however, scaling to large N appears difficult because of the required number of different beam combiners.

Concepts for parallel implementations of WBC for power and radiance scaling using gratings have also been developed. For example, a pair of gratings, called a *grating rhomb*, was used to wavelength combine diode lasers operating at different wavelengths.[71] The difficulty with this implementation was that it used conventional Fabry-Pérot diode lasers, and their output spectra were not stable, even with temperature control. The lack of sufficient wavelength stabilization degraded the output beam quality.

Wavelength-combining techniques have also been developed in the area of wavelength-division multiplexing (WDM) transmitters for optical communications.[70,72,73] The earlier WDM combiners were for a small number of wavelengths, such as four or eight. These combiners were mostly serial types, such as interference filters or fiber Bragg gratings. Because the serial approach was not deemed scalable, however, parallel approaches using free-space gratings[72] or arrayed-waveguide gratings[73] were then developed to accommodate a larger number of wavelength channels. A single-stage WDM combiner with as many as 512 wavelength channels has since been demonstrated.[73] However, in WDM transmitters, the focus is getting multiple wavelength channels into a single-mode fiber; consequently, power, brightness, and efficiency are not particularly important drivers. These WDM combiners are often lossy, which limits the combining efficiency. Furthermore, the power handling of these combiners is limited to less than 1 W.

Engineers at the Massachusetts Institute of Technology's Lincoln Laboratory have invented a low-loss free-space WBC implementation that simultaneously provides wavelength control and nearly ideal beam combination for large (hundreds of elements) laser arrays.[6,74–81] A schematic of this implementation with a linear diode laser array is shown in Fig. 19.7. Note that an array of fiber gain elements could be substituted for the diode laser array. By using optical feedback, the spectrum of each element is controlled to be different from the others and to be right for ideal beam combination. Each laser gain element is inside a laser resonator, in which one resonator mirror is on one end of the gain element and the output end of the laser resonator is the partially reflective output coupler. At the interface between the laser gain elements and the free space is an antireflection coating or an angled facet that prevents feedback from this interface. The transform lens, grating, and output coupler are common optical

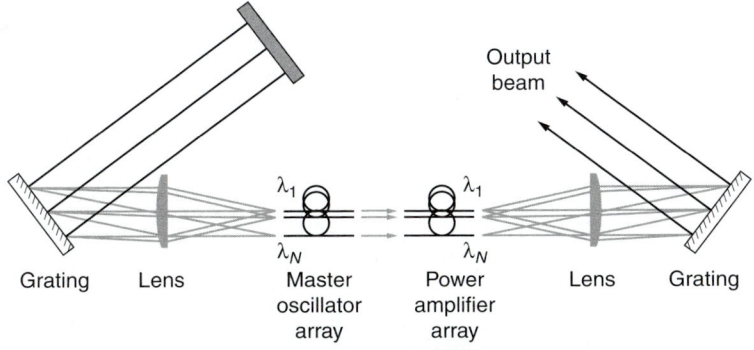

FIGURE 19.7 Schematic of master oscillator power amplifier (MOPA) architecture for a WBC with master oscillator (MO) wavelength control. (Reprinted with permission from T. Y. Fan.[4] © 2005 IEEE)

elements of the external resonator shared by each laser array element. The transform lens acts to transform the position of an array element into angle of incidence on the grating, provided that the lens is located one focal length from the array. Spatial overlap of the beams from each element is ensured by placing the grating one focal length away from the transform lens. Codirectional propagation of the individual beams is forced by the flat output coupler, since the directions of propagation of the output beams are all normal to this mirror. Because the incidence angles on the grating for the beams from each array element are different, the external resonator selects different wavelengths for each array element as needed to force coaxial propagation.

Another way to view the operating principle of this external-cavity laser is to consider a single array element. A single array element can be tuned in this resonator by translating the array element in the plane of the page and perpendicular to the optical axis of the lens. When this array element is translated, the output's propagation direction does not change, because the output coupler forces propagation normal to its surface. In addition, the position of the beam footprint on the grating does not change, because the grating is located one focal length away from the transform lens. Consequently, if we instead put additional array elements along this path, then each array element will operate at a different wavelength with beams that are coaxial with each other.

Yet another way of viewing the operation of this architecture is via analogy to a grating spectrometer. In a grating spectrometer, typically broadband radiation is incident on the grating (propagating in a direction opposite the combined laser output beam). The grating disperses wavelength into a diffraction angle off the grating; then a transform lens or mirror converts the propagation angle into position at the focal plane so that different wavelengths fall onto different locations. Essentially, the spectrally combined array can be viewed as a grating spectrometer run in reverse. This implementation works with one-dimensional arrays. This external-cavity implementation has been used to wavelength combine diode[6,76,77] and fiber laser arrays[75, 78–80] In principle, spectral combining can be extended to two-dimensional arrays by using crossed gratings, as is done in spectrometer systems that use charge-coupled device (CCD) imagers as detectors.[82]

We have performed some estimates of the number of array elements that could be combined using this architecture. It appears that hundreds to thousands of elements can be combined under reasonable assumptions. It can be shown that the dimensional extent of the gain element array d is related to the focal length of the transform lens f, the total wavelength spread of the optical output $\Delta\lambda$, and the dispersion of the grating $d\beta/d\lambda$ by the expression

$$d \approx f\left(\frac{d\beta}{d\lambda}\right)\Delta\lambda \qquad (19.7)$$

The dispersion of the grating relates the change in diffraction angle to the change in optical wavelength. This dispersion, in turn, is related to the grating groove spacing a and the diffraction angle β by

$$\frac{d\beta}{d\lambda} = \frac{1}{a\,\cos\beta} \qquad (19.8)$$

A typical value for dispersion for a 2000-line/mm grating at 1-μm wavelength is approximately 4 rad/μm. For f = 20 cm and a total wavelength spread of 25 nm across the array, which is achievable in fiber and semiconductor gain media near 1-μm wavelength, d is about 2 cm, assuming 4 rad/μm grating dispersion. For array elements spaced on 250-μm centers, such a design accommodates around 80 gain elements. Tighter element spacing or larger focal length would enable even larger arrays to be combined.

For WBC at high powers, the grating needs thermal management to avoid significant beam-quality degradation. Gold-coated gratings can have absorption as high as 1 percent. For a 100-kW beam, this translates into a large 1-kW heat load. However, the absorption of the recently available multilayer dielectric gratings is ~10^{-4}. The heat load for a 100-kW beam becomes a manageable 10 W if multidielectric gratings are employed.

19.2.5 WBC Performance Degradations

Serial and parallel WBC implementations pose challenges in spectrum control and element alignment for scaling to large arrays.

In serial approaches, the spectrum-control problem is both for array elements and for filters. As N increases and the wavelength spacing between elements decreases, manufacturing efficient filters becomes increasingly difficult. In addition, the series arrangement requires tight tolerances in the angular positioning of the interference filters, because the laser at the end of the array accumulates a large number of bounces. Errors in angular positioning lead to smearing of the output in the far field and degradation of the on-axis intensity. Clearly, the near and far fields must overlap to a small fraction of a diffraction-limited beam in order to achieve a combined beam with near-diffraction-limited output. However, this basic approach to serial combination is used in WDM of transmitters for fiber optic communication, which was enabled by developments in distributed-feedback lasers, fiber Bragg gratings, and single-mode optical fibers. Efficiency in WDM transmitter applications is less important than it is in power and radiance scaling applications, so losses are more tolerable. The errors in angular and near-field positioning are eliminated by the use of single-mode optical components, at the expense of optical loss if fiber couplers or splices are less than ideal.

In parallel implementations, such as those in Fig. 19.7, the individual laser element must have spectral width sufficiently narrow that the beam spreading by the grating is small relative to the diffraction-limited beam divergence. This constraint, due to a finite laser line width, has been quantified[74,83] as

$$\sin^{-1}\left[\left(\lambda + \frac{\Delta}{2}\right)g - \sin(\theta_i)\right] - \sin^{-1}\left[\left(\lambda - \frac{\Delta}{2}\right)g - \sin(\theta_i)\right] \ll \frac{4\lambda}{\pi d} \quad (19.9)$$

where d is beam width, Δ is laser line width, θ_i is incident angle of light on the grating, and g is the grating frequency, or the inverse of the grating groove spacing a in Eq. (19.8). Thus, the dispersion arising from the finite line width on the left side of Eq. (19.9) should be less than the natural diffraction of the beam on the right.

Furthermore, there is a requirement on the element spectrum that is coupled to the near-field placement. The placement of an array element in the near field must be controlled to be correct given the element's wavelength, or conversely, the element's wavelength must be controlled to be correct given the element's near-field placement. The use of optical feedback automatically controls array elements to operate at a wavelength and spectral extent set by the near-field placement in the array plane. This effectively lifts the requirement on near-field placement in this plane; equivalently, optical feedback control is used to adjust the wavelengths to match the near-field position. Placement in the orthogonal direction (out of the plane of the array) is important; "smile" effects in a linear array will lead to degradation in the beam quality in the noncombining plane. The need for diffraction-limited alignment implies that the far-field pointing of the elements and the optical system must be arranged so that the beams have good spatial overlap on the grating. If the transform optic is exactly one focal length from the array, then the far-field pointing of the array elements must be the same to within a small fraction of the far-field beam divergence of a single element.

Bochove[81] analyzed all such effects and found the expressions for degradation in beam quality resulting from fiber spectral bandwidth, tilt error, and lateral offset:

$$M_x^2 = \sqrt{\nu\langle 1+\omega_x^2\rangle + 6\left\langle\frac{\Delta\theta_x^2}{\theta_0^2}\right\rangle + 4\left\langle\frac{\Delta(\omega_x\theta_x)^2}{\theta_0^2}\right\rangle} \qquad (19.10)$$

$$M_y^2 = \sqrt{\langle 1+\omega_y^2\rangle + 4\left\langle\frac{\Delta\theta_x^2}{\theta_0^2}\right\rangle + 4\left\langle\frac{\Delta(\omega_y\theta_y)^2}{\theta_0^2}\right\rangle + 4\left\langle\frac{\Delta y^2}{\omega_0^2}\right\rangle} \qquad (19.11)$$

where

$$\langle\omega_x^2\rangle \approx \frac{\varepsilon^2}{Z_R^2} + \frac{\alpha_0\varepsilon W^2}{6fZ_R^2} + \frac{\alpha_0^2 W^4}{80f^2Z_R^2}$$

$$\langle\omega_y^2\rangle \approx \frac{\varepsilon^2}{Z_R^2} + \frac{\beta_0\varepsilon W^2}{6fZ_R^2} + \frac{\beta_0^2 W^4}{80f^2Z_R^2}$$

$$\nu = 1 + \frac{1}{2}\frac{\Delta\lambda_{Bn}^2}{\Delta\lambda_{Cn}^2}$$

and where W is the width of the array, α_0 is the incidence angle, β_0 is the diffraction angle, ε is the defocusing parameter or deviation of the array element from the focal plane of the transform lens ($\varepsilon = 0$ is the element at the front focal plane of the transform lens), f is the focal length of the lens, $Z_R = \pi w_0^2/\lambda$ is the Rayleigh length of the modal field, w_0 is the gaussian mode radius, θ_0 is the divergence angle of the emitter, θ_x and θ_y are the tilt angles along the x- and y-dimensions, $\Delta\lambda_{Bn}$ is the spectral bandwidth of the fiber element, $\Delta\lambda_{Cn}$ is the spectral bandwidth of the cavity, and Δy is offset along the vertical or noncombining dimension (better known as "smile" in semiconductor lasers). The brackets denote array averages:

$$\langle\chi^2\rangle = \sum_n P_n\chi_n^2$$

$$\langle\Delta\chi^2\rangle = \sum_n P_n\chi_n^2 - \left[\sum_n P_n\chi_n\right]^2 \qquad (19.12)$$

$$\sum_n P_n = 1$$

where P_n is the normalized output of the nth fiber element and χ_n is the quantity that is being averaged. It can easily be shown that the above formulas for beam quality apply to the configuration in which microlens array is included if the source is taken to be at the focal plane of the microlens array.

Finally, it is worthwhile noting that WBC has more graceful degradation against misalignments and laser element failure than does CBC because the addition is in power, rather than in electric fields. One consequence is that WBC efficiency in parallel implementation is independent of the number of elements in a fixed optical configuration, in contrast to CBC.

19.2.6 Hybrid Beam Combining

The beam-combining techniques listed in Secs. 19.2.1 through 19.2.3 can be combined to hybridize beam combining. The motivation behind the hybrid approach is to scale the number of beam-combining elements beyond the limits imposed by only one technique. Such limits can be imposed by laser physics, such as the laser medium's gain bandwidth, or by the current state of technology, such as the achievable control bandwidth of the phase-locked loop.

The simplest hybridization of beam combining is incoherent beam combining of multiple beam-combined laser systems. The power from a laser system is first increased via CBC or WBC. When the power limit is reached, more laser systems are fabricated and placed side by side. However, in this chapter, HBC refers to complementary use of WBC and CBC to scale both power and radiance.[84,85] This can be implemented as follows: First, CBC is employed to control the phases of an array of N laser elements to be aligned. The resulting laser array can then be treated as a single laser with only one wavelength. Next, multiple laser arrays M, each with a different wavelength, can be fabricated. These laser arrays are then combined via WBC. The total number of laser elements combined is $N \times M$. A graphical representation for a compact implementation is shown is Fig. 19.8, in which N master oscillators, each with a different wavelength, are used, and in which each master oscillator drives an array of M amplifiers. All the laser outputs pass through a transform lens and impinge onto a diffraction grating. The grating superimposes all the colors onto each other. Optical taps and dichroic filters are employed to extract the feedback signals necessary to close the multidither control loop. The control requirement of a hybrid system, in this case, will need to satisfy its coherent and wavelength components, respectively.

19.3 Beam Combining of Specific Laser Systems

Over the past few decades, much work has been done on beam combining of various laser systems. This section summarizes only the details from recent beam-combining advances related to high-average-power

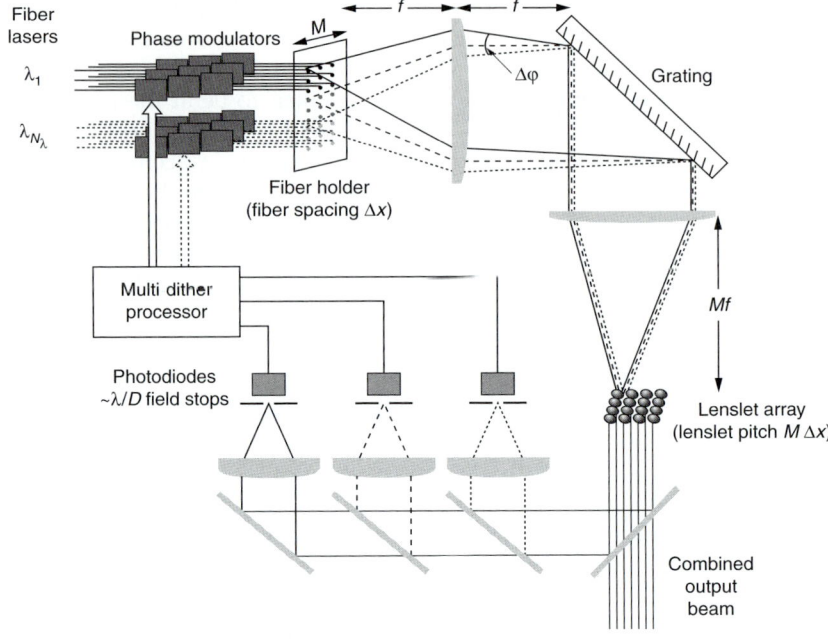

FIGURE 19.8 Hybrid beam combining (HBC) schematic in which N (WBC) × M (CBC) lasers are combined.

laser systems, while also putting those recent works in a historic context. We choose work in three specific areas—fiber laser, semiconductor laser, and solid-state laser systems—because they have seen the most significant advances.

19.3.1 Fiber Laser Beam Combining

Although rare-earth-doped glass was shown to be a good laser material as early as the 1960s, such a laser in single-transverse-mode fiber form did not appear until the 1980s.[86,87] The application that motivated this advance in a fiber laser is its use as an amplifier for optical communication transmission. Because the transmission window is at 1550 nm, most early work on fiber lasers concentrated on erbium (Er)-doped fibers. The pump light and the signal light both propagate in the fiber core; therefore, pump diodes with very high brightness are needed to achieve amplification. Because high-brightness pump diodes were limited in power, the output power from an Er-doped fiber amplifier (EDFA) was limited to hundreds of milliwatts.

Cladding pumping of a fiber amplifier was demonstrated in the early 1990s, opening the way for low-brightness pumping of such amplifiers.[88] The availability of high-power, low-brightness pump diodes led to an increase in the output power from fiber amplifier.

For example, ytterbium-doped fiber amplifiers (YDFAs) are now commercially available with multikilowatt diffraction-limited output.[89] A high-average-power laser system can be achieved by combining many lasers based on such amplifiers. Because of its guided nature, fiber laser output can be inherently diffraction limited. Thus, fiber lasers are excellent candidates for beam combining. To build a high-power laser system, tens to hundreds of fiber lasers must be combined with good beam quality. Although this is a challenging task, the analysis in the previous section indicates that the existing beam-combining techniques can scale to such numbers.

All forms of beam combining have been demonstrated using fiber lasers. Four fiber lasers, each at the kilowatt level, were independently steered and focused onto the same spot on a target, thus achieving incoherent beam combining.[8] The resulting beam quality is understandably poor. Nevertheless, such combining can find applications for targets very near the laser systems.

Both passive and active forms of coherent combining of fiber lasers have been demonstrated by multiple parties. In Hendow et al.,[90] multiple fiber amplifiers (0.5-W output each) were placed in a cavity. The outputs from all the amplifiers were focused through a pinhole to provide the feedback signal, which was fed back into the cavity (Fig. 19.9). The minimum cavity loss occurred when all the fiber amplifiers were phase locked so that this laser system passively combined all the fiber amplifiers in its cavity. Up to eight fiber amplifiers were coherently combined, though beam quality degradation of ~40 percent was observed for the eight-fiber case. An in-depth analysis[52] showed that overall cavity loss increases with the number of laser elements. Thus, this approach is not scalable to very large fiber-count combining. Alternatively, a self-Fourier cavity can be used to self-select the supermode in which all fiber amplifiers are phased locked.[51] Seven fibers at low

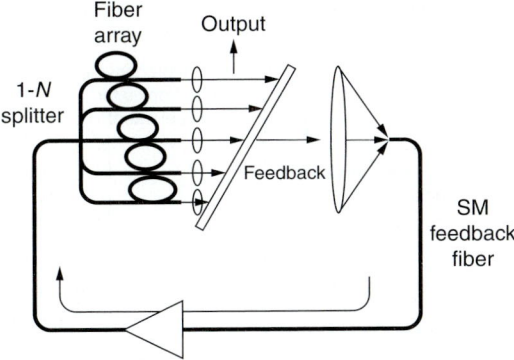

FIGURE 19.9 Passive fiber beam combining using a single cavity. (Reprinted with permission from S. T. Hendow et al.[90])

powers were coherently combined with a fringe contrast of 0.87 for a total power of 0.4 W. In-line fiber couplers were also used in Refs. 41–44 for passive CBC. In Minden et al.,[44] up to five fiber lasers, each of 7 W, were locked with five times Strehl improvement as compared with the unlocked case. However, further scaling attempts at 19 fiber elements did not successfully produce an in-phase supermode for all 19 elements. Analysis indicates that this approach is limited to ~10 laser elements.[42] A nonlinear optics approach using an SBS phase-conjugate mirror to achieve passive CBC was also demonstrated for two fiber amplifiers at watt level.[61]

Active CBC using fiber lasers has been demonstrated using MOPA configurations. A single MO is split into multiple legs and amplified in multiple power amplifiers (PAs). Each leg has its own phase controller, which accepts a feedback signal. In Anderegg et al.,[68] four narrow line width YDFAs were coherently combined using a standard PLL technique. Each amplifier required its own phase detector and PLL. The total output power was 470 W, and less than $\lambda/30$ peak-to-peak phase noise was observed. In Shay et al.,[66] five narrow line width amplifiers, 145 W each, were coherently combined. Each fiber amplifier was dithered with its own frequency. A technique similar to subcarrier demodulation was employed to obtain feedback signals for all the fibers from a single detector (see Fig. 19.10). This

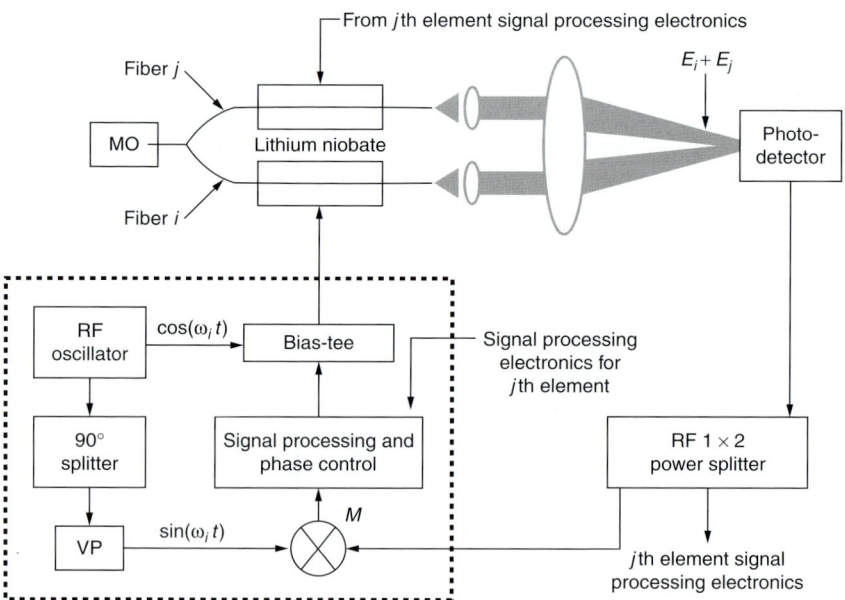

FIGURE 19.10 Schematic of locking of optical coherence by single-detector electronic-frequency tagging (LOCSET) phase-locking technique. (Reprinted with permission from T. M. Shay et al.[66])

technique is termed locking of optical coherence by single-detector electronic-frequency tagging (LOCSET). Again less than λ/30 phase noise was observed when the control loop was closed. This experiment represented the highest combined power to date. Efforts are under way to combine 16 amplifiers to reach 2-kW power.

In an experiment that involved the largest number of fibers,[5,67] 48 single-frequency fibers were combined for a total of 1-W power (the experimental schematic is shown in Fig. 19.11). A single-frequency Ti:sapphire laser was split into a reference fiber and 48 signal fibers. The 48 signal fibers were followed by 48 piezo fiber stretchers and aggregated with a two-dimensional fiber array. The two-dimensional array output was a rectangular fiber grid with four corners cut off. A portion of this output was tapped as the feedback signal. Both phase-sensing-based and multidither-based techniques were demonstrated. A single CCD was used to measure the phases of all the fibers, and that information was used to lock all the fibers in phase. Although using a single CCD was simpler to implement than using individual detectors, the relatively slow CCD limited the achievable bandwidth to less than 1 kHz. For the multidither approach, far-field on-axis intensity was selected as the metric, which was consistent with the case of mutual phase locking. Dithers were then applied to the laser elements to maximize the selected metric via a "hill climbing" type of

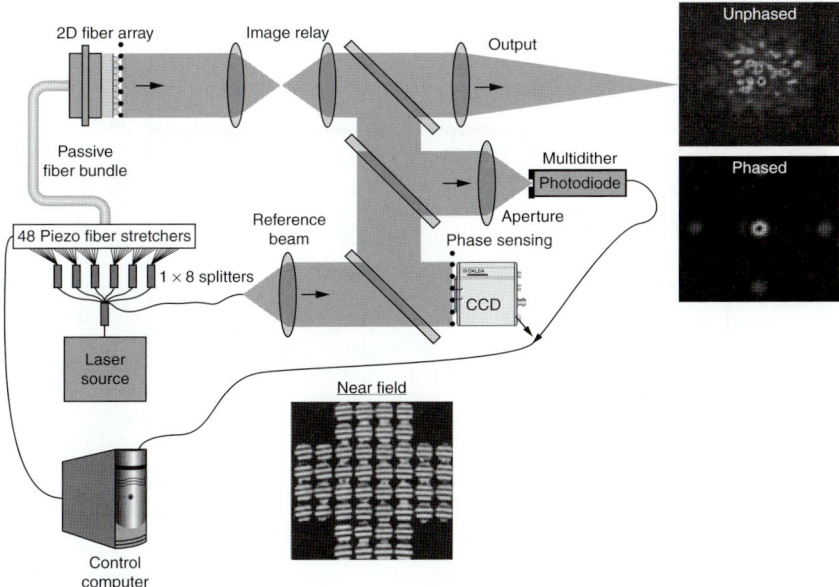

FIGURE 19.11 Schematic of 48-fiber CBC experiment using phase-sensing and multidither techniques. CCD: charge-coupled device. (Reprinted with permission from J. E. Kansky et al.[67])

algorithm. In both cases, the resulting Strehl degradation was around 1 dB and was dominated by imperfection in the optics, such as fiber alignment and focal length nonuniformities of the microlenses, rather than from the phase control loop.

It is important to note that although convenient, narrow line width amplifiers are not necessary for CBC. Because power handling in the fiber is limited by SBS and because increasing line width can increase SBS threshold, using broad line width fiber amplifiers enables higher power per fiber. In Augst et al.,[83] two 10-W fiber amplifiers were coherently combined using a PLL approach. Because the amplifier's line width was 25 GHz, the path lengths from the two amplifiers had to be matched to within a few millimeters. Fringe contrast greater than 90 percent was demonstrated once the path-length matching was successful. This broad line width approach allows CBC of fiber amplifiers whose output power is significantly higher than what has been demonstrated to date. In fact, it was recently shown that a 1.26-kW fiber amplifier with 21-GHz line width is compatible with CBC.[91]

It is also useful to understand the phase-control requirements for fiber MOPAs. The path-length variation (phase noise) of a commercial 30-W YDFA has been reported[83] (the results are shown in Fig. 19.12). At fiber amplifier turn-on, the path length goes through

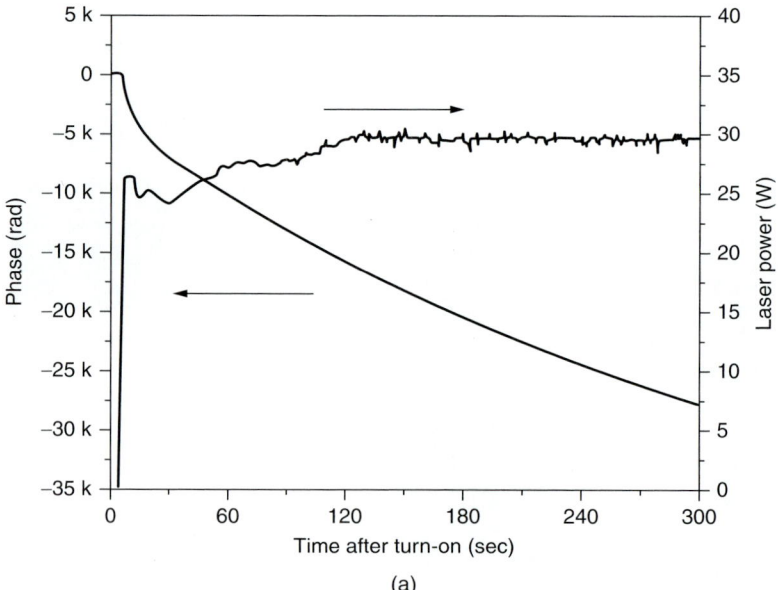

FIGURE 19.12 Phase noise of a 30-W fiber amplifier: (*a*) transient response and (*b*) steady-state power spectral density (solid curve) and integrated phase noise (dashed curve). (Reprinted with permission from S. J. Augst et al.[83])

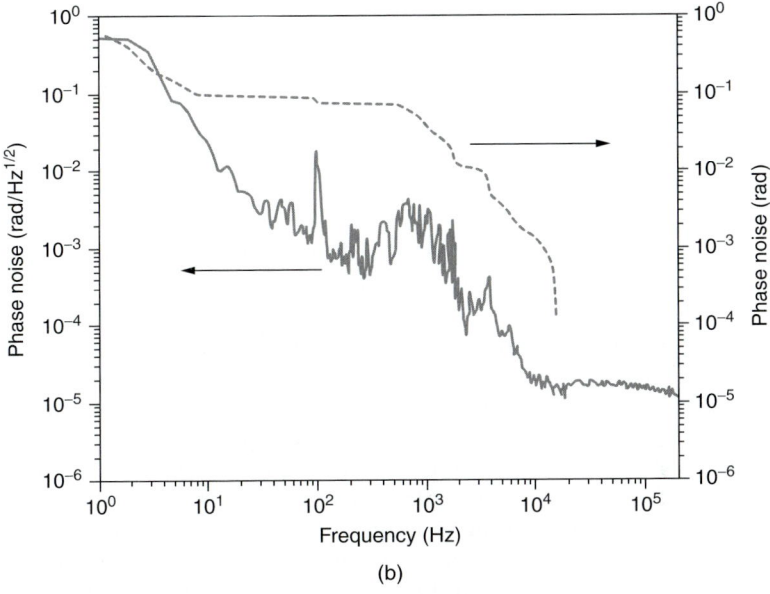

FIGURE 19.12   (Continued)

thousands of waves, primarily driven by heating the fiber, as shown in Fig. 19.12a. In thermal steady state (Fig. 19.12b), the path length in microsecond time scales varies a few tenths of a wave in a quiet laboratory environment, though this variation can be much larger in acoustically noisy environments. Clearly, these path-length changes are large enough that they must be compensated in order to successfully perform CBC. Any CBC implementation must be able to accommodate these types of fluctuations, in terms of both their bandwidth and the dynamic range. Fortunately, the phase noise is dominated by thermal and environmental causes, rather than by spontaneous emission. Thus, such noise is typically in the kilohertz range or lower, as shown in Fig. 19.12. This phase noise origin was confirmed in Jones, Stacey, and Scott,[92] in which efforts were made to minimize environmental noise on a 260-W fiber amplifier in a laboratory environment. The measured laser phase noise rolls off at around a few hertz.

For WBC applications, multiple high-power fiber lasers, each at a different wavelength, have their outputs superimposed either in parallel or in series. Unlike CBC, WBC does not need to control the phase of the individual lasers. Thus, phase noise has no effect on WBC performance. For the parallel configuration, a diffraction grating is the preferred choice because of its high efficiency and high-power-handling capabilities. WBC of fiber lasers has been demonstrated in both oscillator[75,79,80] and MOPA architectures.[78] MOPA architectures separate temporal and spectral waveform control from power generation, as it is often observed that fiber laser oscillators pulse and have undesirable spectral broadening effects. In Cook and Fan,[75] an external cavity was used to ensure that the wavelengths from individual fiber lasers corresponded to their respective spatial positions. Because of the advances in fiber laser power and the availability of low-loss multidielectric gratings, the output power from fiber laser WBC has dramatically increased recently. In 2003, an array of five 2-W YDFAs was combined with $M^2 = 1.14$.[78] A few years later, three YDFAs were combined at 522-W total power with $M^2 = 1.18$,[74] and four YDFAs were combined at 2065-W total power with $M^2 = 2.0$.[93]

Wavelength combining does not require the individual fiber lasers to be single frequency. Demonstrations in Refs. 74, 78, and 93 all used fiber amplifiers with gigahertz line width. Many used a diffraction grating as the dispersive element. For those cases, the beam-combining optics were designed so that dispersive beam broadening from the broad laser line width did not degrade the final output beam's M^2.[78] Volume Bragg gratings (VBGs) have also been used to implement the serial configuration (the experimental schematic is shown in Fig. 19.13). Five amplifiers were combined using four VBGs. The reported WBC output of 770 W had 91.7 percent power transmission efficiency, but the output beam quality was not reported.[94]

A few WBC examples used oscillator architecture. An array of four thulium-doped fiber lasers with 11-W output power and an unspecified beam quality was demonstrated[79]; and an array of three Yb-doped fiber lasers with 104 W and $M^2 = 2.7$ was demonstrated using a fused-silica transmission grating for the dispersive element.[80]

Fiber laser beam combining also provided the first demonstration for HBC, which was invented in 2004.[84] In this proof-of-principle demonstration,[85] two pairs of fiber lasers, each at 40 mW, were combined. Each pair had the same wavelength, and the two fibers in the same pair were first coherently combined. The coherently combined

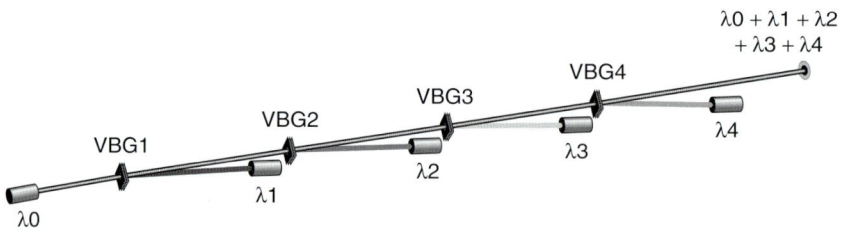

FIGURE 19.13 Five-element WBC experiment schematic using volume Bragg gratings (VBGs). (Reprinted with permission from O. Andrusyak et al.[94] © 2009 IEEE)

pairs were then wavelength combined. The output had an excellent beam quality of $M^2 = 1.15$.

Although the total fiber laser beam combined power has not exceeded those coming out of a single fiber laser, scalability analysis indicates that this goal should be reachable via CBC or WBC alone. In fact, the combined power should exceed a single fiber laser power by at least an order of magnitude. Currently, efforts are under way to demonstrate this beam-combining scalability using fiber lasers.[95]

19.3.2 Semiconductor Laser Beam Combining

Although semiconductor lasers have small size and high efficiency, they can only produce low to modest power (~1 W) with good beam quality ($M^2 < 1.2$). Because many high-power commercial applications do not require good beam quality, they can often be dominated by incoherent combining of semiconductor lasers. For such applications, multiple semiconductor lasers are fabricated side by side on the same substrate in a one-dimensional array configuration. Such a laser "bar" can contain hundreds of laser elements and output hundreds of watts. The beam quality is very poor—M^2 can be as large as 1000 along the dimension containing multiple lasers.[7]

To build a high-average-power laser system with good beam quality using semiconductor lasers, one must start with lasers that have good beam quality. Thus, such a system presents many challenges, in part because of the sheer number of lasers involved. At 1 W per laser, a high-power system requires successful combination of tens of thousands to hundreds of thousands of lasers, all while maintaining good beam quality. However, because of the attractiveness of a semiconductor-based system, many groups in the early to late 1990s worked toward such a goal.[3,9,18,19,31–37]

CBC of semiconductor lasers was demonstrated as early as 1975.[22] A significant amount of work followed.[15–24,31–37] Partly because a large number of such semiconductor lasers must be coherently combined to achieve the power goal, much of the early work[18,19] concentrated on fabricating monolithic phase-locked semiconductor laser arrays. Multiple devices, either gain guided or index guided, were fabricated on the same substrate. Instead of controlling the phase of each laser individually, various mode-selection mechanisms were inserted into the laser cavity to enable passive coherent combining. Such mechanisms included leaky-wave/evanescent-wave coupling between the laser strips and external feedback between laser ends, such as Talbot-effect coupling. Such efforts led to output power of 2 W,[18,19] of which 1.6 W was coherent and 1.15 W was in the central lobe. This power level is comparable to a single laser emitter.

Alternatively, active CBC was pursued in MOPA or injection-locking configuration.[20,34–36] A single master oscillator was split and injected into an integrated power-amplifier–phase-modulator array.

Feedback control was applied to the phase modulators to actively lock the amplifiers in phase. In No et al.,[34] a one-dimensional, 100-emitter amplifier array achieved a fringe contrast of 97 percent. However, partly because the collimating optics following the laser diodes could not capture the off-axis laser light, it only produced 1.63 W in the far-field central lobe, out of a total of 7.9 W measured in the near field. These one-dimensional arrays were then used as building blocks to produce a two-dimensional, 900-emitter (9×100) coherently combined laser array. The total output power from this two-dimensional array was 15 W, though its beam quality was not reported.[35] In Bartelt-Berger et al.,[37] an array of 19 fiber-pigtailed semiconductor lasers was injection locked to force the array elements to operate with the same spectrum. The fiber pigtails were brought together to a tiled aperture, and the fiber pigtail lengths were actively controlled to produce constructive interference in the far field. The on-axis intensity under servo control was measured to be 13 times that for no phase control (compared with an ideal of 19 times). Based on the far-field pattern, the fill factor in this demonstration was low.

In Saunders et al.,[33] passive and active CBC were both used to enable a coherently combined two-dimensional, 144-element, surface-emitting laser array. In this case, a laser array was phased locked via a Talbot cavity and a liquid crystal array to achieve a 1.4-W diffraction-limited output.

Recent beam-combining efforts have involved both CBC and WBC.[6,21,39] An optical phase-locked loop (OPLL) technique was used to lock two watt-level "slave" lasers to a common master laser in both filled and tiled configurations.[21] In the filled configuration, a 2×1 fiber combiner was used to obtain a single output at watt level. In the tiled configuration, the collimated outputs from two 60-mW distributed-feedback lasers were placed adjacent to each other, and the combined intensity increased by ~75 percent. In Huang et al.,[39] slab-coupled optical waveguide lasers (SCOWLs) were used in beam-combining experiments to increase the power per laser element to nearly 1 W. A 10-element, one-dimensional SCOWL array was placed in a Talbot laser cavity to enable successful passive CBC. Nearly diffraction-limited output was observed at 7.2 W of output power. Alternatively, the SCOWLs were then fabricated in an amplifier array. A single master oscillator seeded the 10-element SCOWL array, and active coherent combining was demonstrated for a total power of 4.9 W. The phasing algorithm was multidither, and the drive currents to the individual SCOWL array elements were adjusted to control the phase of each SCOWL amplifier. The SCOWL phase noise power density is shown in Fig. 19.14. The required control bandwidth for SCOWL is only a few hertz. The successful CBC of one-dimensional SCOWL arrays enables them to be used as building blocks for two-dimensional laser array combining.

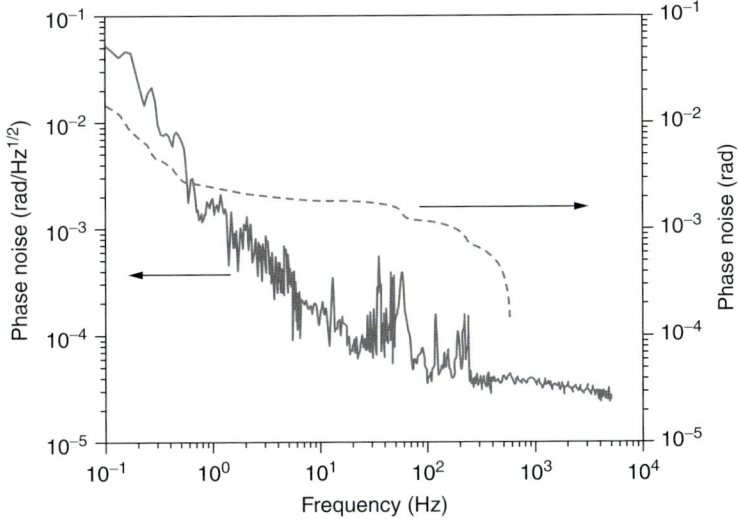

FIGURE 19.14 Phase noise of the slab-coupled optical waveguide laser (SCOWL) diode: power spectral density (solid curve) and integrated phase noise (dashed curve).

WBC using SCOWLs has also been demonstrated in Refs. 6 and 96. A 100-element, 100-μm-pitch SCOWL array was inserted into a beam-combining cavity (Fig. 19.15). The result showed 50 W of nearly diffraction limited output ($M^2 < 1.35$ in both dimensions). Figure 19.15a shows the spectrum as a function of position along the array, whereas

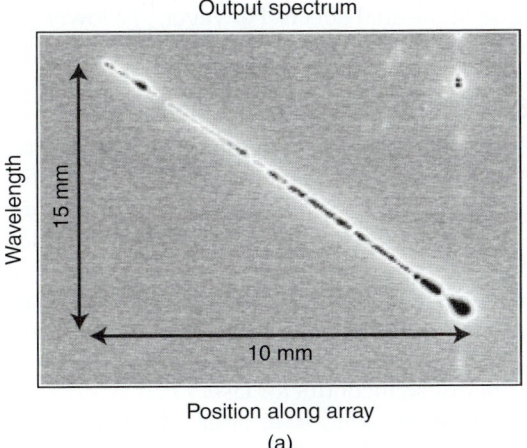

FIGURE 19.15 WBC experimental results with 100 laser diodes: (a) output spectrum and (b) output power and beam quality. (Reprinted from B. Chann et al.[6])

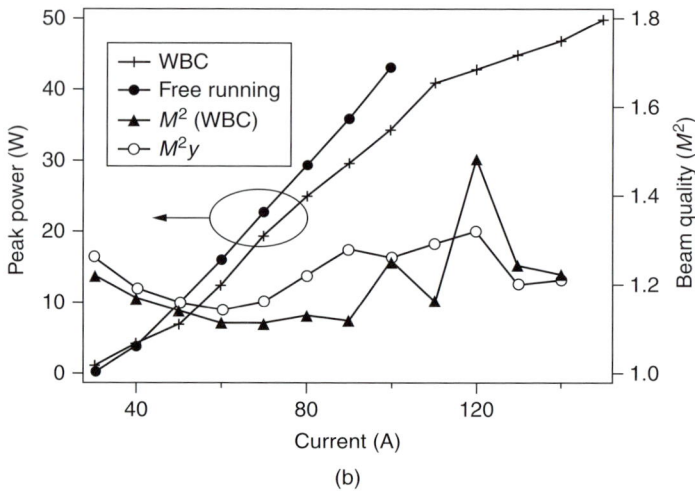

FIGURE 19.15 WBC experimental results with 100 laser diodes: (a) output spectrum and (b) output power and beam quality. (Reprinted from B. Chann et al.[6])

Fig. 19.15b shows the output power and beam quality as a function of current to the array. This is the highest diffraction-limited output from a semiconductor laser–based system, far exceeding the diffraction-limited power achievable from a single emitter. Another demonstration using the same basic approach, though involving 1400 semiconductor lasers, was presented by Hamilton et al.,[97] in which seven 200-emitter diode bars were combined in a Schmidt configuration for a total power of 26 W. Beam quality was not measured at full power because of thermal effects. However, at a lower 8.6 W, the combined beam had $M_x^2 = 4.4$ and $M_y^2 = 3.0$.

19.3.3 Solid-State Laser Beam Combining

Compared with fiber lasers and semiconductor lasers, bulk solid-state lasers are presently capable of producing significantly higher power. The drawback with solid-state lasers is that at high power, the laser can (and usually does) experience large thermal distortions, such as thermal lensing. By incorporating adaptive optics correction, good beam quality (~1.5 times diffraction limited) can be obtained at 10 kW of output power.[98] Therefore, a high-power laser system based on the solid-state laser requires combining fewer such lasers as compared with fiber or semiconductor lasers.

Passive CBC using solid-state lasers has been demonstrated by multiple groups.[47–50] Multiple lasers were constructed inside a single laser cavity by pumping different regions of a single laser rod. A supermode selection mechanism, such as a diffractive and/or interferometric element, was inserted inside the cavity to ensure that all the lasers in

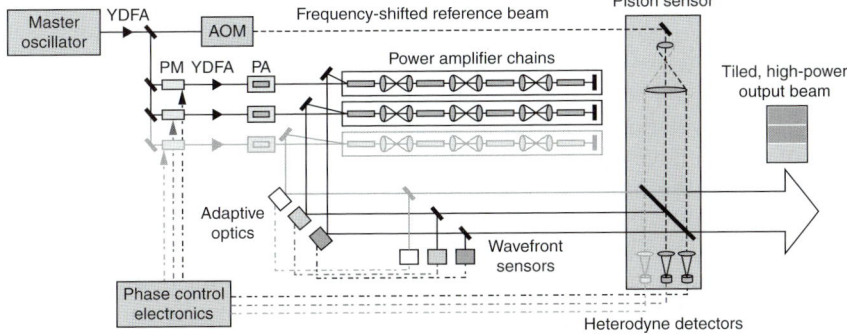

FIGURE 19.16 CBC schematic of high-power Nd:YAG amplifiers. AOM: acousto-optic modulator; PM: phase modulator; PA: pre-amplifier; YDFA: ytterbium-doped fiber amplifier. (Reprinted from G. D. Goodno et al.[100])

the cavity were phase locked. This implementation can be thought of as an approach to scaling the mode area of a single-transverse-mode resonator for a bulk laser. Coherent combining of up to 16 regions of the laser rod was successfully demonstrated in pulsed mode at low repetition rate. A beam-combining efficiency of 88 percent and M^2 of 1.3 were obtained at tens of milliwatts. For higher powers, thermo-optic effects, such as lensing and higher-order distortions, cause the beam-combining efficiency to degrade, much as the beam quality degrades at higher power in a conventional laser resonator. Lenses can be inserted inside the laser cavity to compensate for thermal lensing at a fixed operating point to improve the output power,[49] which is similar to conventional resonators for bulk lasers.

Although active phase locking of two lasers is often performed for many applications, such as metrology,[99] application of active CBC to achieve high-power solid-state laser systems was not demonstrated until very recently. In Goodno et al.,[100] a common master oscillator fed two Nd:YAG zigzag slab amplifiers (Fig. 19.16). The output power from each amplifier was ~10 kW. The two amplifiers were coherently combined to produce an output power of 19 kW with a beam quality of 1.73 times diffraction limited. This represents the brightest CW solid-state laser to date, exceeding that achievable with a single solid-state laser. Further power scaling has been achieved by coherent combining of seven such amplifiers at 15 kW per amplifier, for a total of 105 kW and a beam quality of 2.9.[101] These two lasers were discussed in more detail earlier in Chap. 8.

19.4 Summary

This chapter reviews laser beam-combining techniques such as WBC and CBC, the latter of which includes both passive and active beam combining. Although the CBC requirements for the passive approach still need further research, active combining, which requires additional hardware beyond what is already inside the laser cavity, is well understood. Active combining can be used to scale to a large number of laser elements at the expense of increased hardware components. WBC is also well understood at this point; its implementation requires use of a dispersive optical element, such as a diffraction grating. To further scale the number of lasers beyond coherent or wavelength combining alone, hybrid approaches have been invented to leverage both techniques. The chapter also surveys the state-of-the-art beam-combining work in three laser systems—fiber lasers, semiconductor lasers, and solid-state lasers. The beam-combining results for the three systems have all progressed significantly in recent years, currently achieving beam-combining output either exceeding or close to that from a single laser emitter of the same type.

Acknowledgments

The authors thank Steve Augst for reading this manuscript and providing Fig. 19.12 and Fig. 19.14, Hagop Injeyan for providing Fig. 19.4, and Bien Chann, Jan Kansky, Daniel Murphy, and Antonio Sanchez-Rubio for their contributions to beam combining.

This work was in part sponsored by the Department of the Air Force under Air Force Contract FA8721-05-C-0002. Opinions, interpretations, conclusions, and other recommendations are those of the authors and are not necessarily endorsed by the United States Government.

References

1. Siegman, A. E., *Lasers*, Mill Valley, CA: University Science Books, 1986.
2. Sasiela, R. J., *Electromagnetic Wave Propagation in Turbulence: Evaluation and Application of Mellin Transforms*, Bellingham, WA: SPIE Press, 2007.
3. Leger, J. R., "External Methods of Phase Locking and Coherent Beam Addition of Diode Lasers," *Surface Emitting Semiconductor Lasers and Arrays*, eds. G. A. Evans and J. M. Hammer, Boston: Academic Press, 1993, 379–433.
4. Fan, T. Y., "Laser Beam Combining for High Power, High Radiance Sources," *IEEE J. Quantum Electron.*, 11: 567–577, May 2005.
5. Yu, C. X., Kansky, J. E., Shaw, S. E. J., Murphy, D. V., and Higgs, C., "Coherent Beam Combining of Large Number of PM Fibers in a 2-D Fiber Array," *Electron. Lett.*, 42: 1024–1025, 2006.
6. Chann, B., Huang, R. K., Missaggia, L. J., Harris, C. T., Liau, Z. L., Goyal, A. K., Donnelly, J. P., et al., "High-Power, Near-Diffraction-Limited Diode Laser Arrays by Wavelength Beam Combining," *Opt. Lett.*, 30: 1253–1255, 2005.
7. Diehl, R., ed., *High Power Diode Lasers: Fundamentals, Technology, Applications*, Berlin: Springer, 2000.
8. Sprangle, P., Ting, A., Penano, J., Fischer, R., and Hafizi, B., "Incoherent Combining and Atmospheric Propagation of High Power Fiber Lasers for Directed Energy Applications," *IEEE J. Quantum Electron.*, 45: 138–148, 2009.
9. Chinn, S. R., "Review of Edge-Emitting Coherent Laser Arrays," *Surface Emitting Semiconductor Lasers and Arrays*, eds., G. A. Evans and J. M. Hammer, Boston: Academic Press, 1993, 9–70.

10. Nabors, C. D., "Effect of Phase Errors on Coherent Emitter Arrays," *Appl. Opt.*, 33: 2284–2289, 1994.
11. Cheston, T. C., and Frank, J., "Array Antennas," in *Radar Handbook*, ed., M. I. Skolnik, New York: McGraw-Hill, 1970, Chap. 11.
12. Leger, J. R., Holz, M., Swanson, G. J., and Veldkamp, W., "Coherent Beam Addition: An Application of Binary Optics," *Lincoln Lab. J.*, 1: 225–245, 1988.
13. Fan, T. Y., "The Effect of Amplitude (Power) Variations on Beam-Combining Efficiency for Phased Arrays," *IEEE J. Select. Top. Quantum Electron.*, 15: 291–293, 2009.
14. Khajavikhan, M., and Leger, J. L., "Modal Analysis of Path Length Sensitivity in Superposition Architectures for Coherent Laser Beam Combining," *IEEE J. Select. Top. Quantum Electron.*, 15: 281–290, 2009.
15. Glova, A. F., "Phase Locking of Optically Coupled Lasers," *Quantum Electron.*, 33: 283–306, 2003.
16. Leger, J. R., Swanson, G. J., and Veldkamp, W. B., "Coherent Laser Addition Using Binary Phase Gratings," *Appl. Opt.*, 26: 4391–4399, 1987.
17. Mecherle, G. S., "Laser Diode Combining for Free Space Optical Communication," *Proc. SPIE*, 616: 281–291, 1986.
18. Botez, D., "Monolithic Phase-Locked Semiconductor Laser Arrays," *Diode Laser Arrays*, eds., D. Botez and D. R. Scifres, Cambridge, UK: Cambridge University Press, 1994, 1–67.
19. Welch, D. F., and Mehuys, D. G., "High-Power Coherent, Semiconductor Laser, Master Oscillator Power Amplifiers and Amplifier Arrays," *Diode Laser Arrays*, eds., D. Botez and D. R. Scifres, Cambridge, UK: Cambridge University Press, 1994, 72–122.
20. Goldobin, I. S., Evtikhiev, N. N., Plyavenek, A. G., and Yakubovich, S. D., "Phase-Locked Integrated Arrays of Injection Lasers," *Sov. J. Quantum Electron.*, 19: 1261–1284, 1989.
21. Satyan, N., Liang, W., Kewitsch, A., Rakuljic, G., and Yariv, A., "Coherent Power Combination of Semiconductor Lasers Using Optical Phase-Lock Loops," *IEEE J. Select. Top. Quantum Electron.*, 15: 240–247, 2009.
22. Philipp-Rutz, E. M., "Spatially Coherent Radiation from an Array of GaAs Lasers," *Appl. Phys. Lett.*, 26: 475–477, 1975.
23. Leger, J. R., Scott, M. L., and Veldkamp, W. B., "Coherent Addition of AlGaAs Lasers Using Microlenses and Diffractive Coupling," *Appl. Phys. Lett.*, 52: 1771–1773, 1988.
24. Corcoran, C. J., and Rediker, R. H., "Operation of Five Individual Diode Lasers as a Coherent Ensemble by Fiber Coupling into an External Cavity," *Appl. Phys. Lett.*, 59: 759–761, 1991.
25. Glova, A. G., Dreizin, Yu. A., Kachurin, O. R., Lebedev, F. V., and Pis'mennyi, V. D., "Phase Locking of a Two-Dimensional Array of CO_2 Waveguide Lasers," *Sov. Tech. Phys. Lett. (USA)*, 11: 102–103, 1985.
26. Newman, L. A., Hart, R. A., Kennedy, J. T., Cantor, A. J., DeMaria, A. J., and Bridges, W. B., "High Power Coupled CO_2 Waveguide Laser Array," *Appl. Phys. Lett.*, 48: 1701–1703, 1986.
27. Kachurin, O. R., Lebedev, F. V., and Napartovich, A. P., "Properties of an Array of Phase-Locked CO_2 Lasers," *Sov. J. Quantum Electron. (USA)*, 15: 1128–1131, 1988.
28. Vasil'tsov, V. V., Golbev, V. S., Zelenov, Ye. V., Kurushin, Ye. A., and Filimonov, D. Yu., "Using Diffraction Optics for Formation of Single-Lobe Far-Field Beam Intensity Distribution in Waveguide CO_2-Lasers Synchronized Arrays," *Proc. SPIE*, 2109: 122–128, 1993.
29. Kono, Y., Takeoka, M., Uto, K., Uchida, A., and Kannari, F., "A Coherent All-solid-state Laser Array Using the Talbot Effect in a Three-mirror Cavity," *IEEE J. Quantum Electron.*, 36: 607–614, May 2000.
30. Youmans, D. G., "Phase Locking of Adjacent Channel Leaky Waveguide CO_2 Lasers," *Appl. Phys. Lett.*, 44: 365–367, 1984.
31. Scifres, D. R., Burnham, R. D., and Streifer, W., "Phase-Locked Semiconductor Laser Array," *Appl. Phys. Lett.*, 33: 1015–1017, 1978.

32. Mawst, L. J., Botez, D., Zmudzinski, C., Jansen, M., Tu, C., Roth, T. J., and Yun, J., "Resonant Self-Aligned-Stripe Antiguided Diode Laser Array," *Appl. Phys. Lett.*, 60: 668–670, 1992.
33. Saunders, S., Waarts, R., Nam, D., Welch, D., Scifres, D., Ehlert, J. C., Cassarly, W. J., et al., "High Power Coherent Two-Dimensional Semiconductor Laser Array," *Appl. Phys. Lett.*, 64: 1478–1480, 1994.
34. No, K. H., Herrick, R. W., Leung, C., Reinhart, R., and Levy, J. L., "One Dimensional Scaling of 100 Ridge Waveguide Amplifiers," *IEEE Photon. Technol. Lett.*, 6: 1062–1066, 1994.
35. Levy, J. L., and No, K. H., "Coherent Two Dimensional 1000-Emitter Phase Array," presented at Seventh Annual Diode Laser Technology Conference, Fort Walton Beach, FL, 1994.
36. Osinski, J. S., Mehuys, D., Welch, D. F., Waarts, R. G., Major, J. S., Dzurko, K. M., and Lang, R. J., "Phased Array of High-Power, Coherent, Monolithic Flared Amplifier Master Oscillator Power Amplifiers," *Appl. Phys. Lett.*, 66: 556–558, 1995.
37. Bartelt-Berger, L., Brauch, U., Giesen, A., and Opower, H., "Power-Scalable System of Phase-Locked Single-Mode Diode Lasers," *Appl. Opt.*, 38: 5752–5760, 1999.
38. DiDomenico, Jr., M., "Characteristics of a Single-Frequency Michelson-Type He-Ne Gas Laser," *IEEE J. Quantum Electron.*, QE-2: 311–322, 1966.
39. Huang, R., Chann, B., Missaggia, L., Augst, S., Connors, M., Turner, G., Sanchez, A., and Donnelly, J., "Coherent Combination of Slab-Coupled Optical Waveguide Lasers," *Proc. SPIE*, 7230: 7230-1G, 2009.
40. Smith, P. W., "Mode Selection in Lasers," *Proc. IEEE*, 60: 422–440, 1972.
41. Shirakawa, A., Saitou, T., Sekiguchi, T., and Ueda, K., "Coherent Addition of Fiber Lasers by Use of a Fiber Coupler," *Opt. Express*, 10: 1167–1172, 2002.
42. Kouznetsov, D., Bisson, J., Shirakawa, A., and Ueda, K., "Limits of Coherent Addition of Lasers: Simple Estimate," *Opt. Rev.*, 12: 445–447, 2005.
43. Sabourdy, D., Kermene, V., Desfarges-Berthelemot, A., Lefort, L., Barthélémy, A., Mahodaux, C., and Pureru, D., "Power Scaling of Fibre Lasers with All-Fibre Interferometric Cavity," *Electron. Lett.*, 38: 692–693, 2002.
44. Minden, M. L., Bruesselbach, H., Rogers, J. L., Mangir, M. S., Jones, D. C., Dunning, G. J., Hammon, D. L., et al., "Self-Organized Coherence in Fiber Laser Arrays," *Proc. SPIE*, 5335: 89–97, 2004.
45. Michaille, L., Bennett, C., Taylor, D., and Sheperd, T., "Multicore Photonics Crystal Fiber Lasers for High Power/Energy Applications," *IEEE J. Select. Top. Quantum Electron.*, 15: 320–327, 2009.
46. Liu, L., Zhou, Y., Kong, F., and Chen, Y. C., "Phase Locking in a Fiber Laser Array with Varying Path Lengths," *Appl. Phys. Lett.*, 85: 4837–4839, 2004.
47. Oka, M., Masuda, H., Kaneda, Y., and Kubota, S., "Laser-Diode-Pumped Phase-Locked Nd:YAG Laser Arrays," *IEEE J. Quantum Electron.*, 28: 1142–1147, 1992.
48. Menard, S., Vampouille, M., Desfarges-Berthelemot, A., Kermene, V., Colombeau, B., and Froehly, C., "Highly Efficient Phase Locking of Four Diode Pumped Nd:YAG Laser Beams," *Opt. Commun.*, 160: 344–353, 1999.
49. Ishaaya, A., Davidson, N., and Friesem, A., "Passive Laser Beam Combining with Intracavity Interferometric Combiners," *IEEE J. Select. Top. Quantum Electron.*, 15: 301–310, 2009.
50. Ishaaya, A. A., Davidson, N., Shimshi, L., and Friesem, A. A., "Intracavity Coherent Addition of Gaussian Beam Distributions Using a Planar Interferometric Coupler," *Appl. Phys. Lett.*, 85: 2187–2189, 2004.
51. Corcoran, C. J., and Durville, F., "Passive Phasing in a Coherent Laser Array," *IEEE J. Select. Top. Quantum Electron.*, 15: 294–301, 2009.
52. Bochove, E., and Shakir, S. A., "Analysis of a Spatial-Filtering Passive Fiber Laser Beam Combining System," *IEEE J. Select. Top. Quantum Electron.*, 15: 320–327, 2009.
53. Basov, N. G., Efmkov, V. F., Zubarev, I. G., Kotov, A. V., Mironov, A. B., Mikhailov, S. I., and Smimov, M. J., "Influence of Certain Radiation Parameters

on Wavefront Reversal of a Pump Wave in a Brillouin Mirror," *Sov. J. Quantum Electron. (USA)*, 9: 455–458, 1979.
54. Carroll, D. L., Johnson, R., Pfeifer, S. J., and Moyer, R. H., "Experimental Investigations of Stimulated Brillouin-Scattering Beam Combination," *J. Opt. Soc. Am. B*, 9: 2214–2224, 1992.
55. Kong, H. J., Lee, J. Y., Shin, Y. S., Byun, J. O., Park, H. S., and Kim, H., "Beam Recombination Characteristics in Array Laser Amplification Using Stimulated Brillouin Scattering Phase Conjugation," *Opt. Rev.*, 4: 277–283, 1997.
56. Rodgers, B. C., Russell, T. H., and Roh, W. B., "Laser Beam Combining and Cleanup by Stimulated Brillouin Scattering in a Multimode Optical Fiber," *Opt. Lett.*, 16: 1124–1126, 1999.
57. Vasil'ev, A. F., Mak, A. A., Mit'kin, V., Serebryakov, V. A., and Yashin, V. E., "Correction of Thermally Induced Optical Aberrations and Coherent Phasing of Beams During Stimulated Brillouin Scattering," *Sov. Phys.-Tech. Phys. (USA)*, 31: 191–193, 1986.
58. Rockwell, D. A., and Giuliano, C. R., "Coherent Coupling of Laser Gain Media Using Phase Conjugation," *Opt. Lett.*, 11: 147–149, 1986.
59. Gratsianov, K. V., Komev, A. F., Lyubimov, V. V., Mak, A. A., Pankov, V. G., and Stepanov, A. I., "Investigation of an Amplifier with a Composite Active Element and a Stimulated Brillouin Scattering Mirror," *Sov. J. Quantum Electron. (USA)*, 16: 1544–1546, 1986.
60. Gratsianov, K. V., Komev, A. F., Lyubimov, V. V., and Pankov, V. G., "Laser Beam Phasing with Phase Conjugation in Brillouin Scattering," *Opt. Spectrosc. (USA)*, 68: 360–361, 1990.
61. Grime, B., Roh, W. B., and Alley, T., "Beam Phasing Multiple Fiber Amplifiers Using a Fiber Phase Conjugate Mirror," *Proc. SPIE*, 6102: 6102-1C, 2006.
62. Sumida, D. S., Jones, D. C., and Rockwell, D. A., "An 8.2 J Phase-Conjugate Solid-State Laser Coherently Combining Eight Parallel Amplifiers," *IEEE J. Quantum Electron.*, 30: 2617–2627, 1994.
63. Moyer, R. H., Valley, M., and Cimolino, M., "Beam Combination Through Stimulated Brillouin Scattering," *J. Opt. Soc. Amer. B*, 5: 2473–2489, 1988.
64. Smith, M. H., Trainor, D. W., and Duzy, C., "Shallow Angle Beam Combining Using a Broad-Band XeF Laser," *IEEE J. Quantum Electron.*, 26: 942–949, 1990.
65. Russell, T. H., Willis, S. M., Crookston, M. B., and Roh, W. B., "Stimulated Raman Scattering in Multimode Fibers and Its Application to Beam Cleanup and Combining," *J. Nonlinear Opt. Phys. Mater.*, 11: 303–316, 2002.
66. Shay, T. M., Baker, J., Sanchez, A. D., Robin, C., Vergien, C., Zerinque, C., Gallant, D., et al., "High Power Phase Locking of a Fiber Amplifier Array," *Proc. SPIE.*, 7195: 71951M-1-8, 2009.
67. Kansky, J., Yu, C. X., Murphy, D., Shaw, S., Lawrence, R., and Higgs, C., "Beam Control of a 2D Polarization Maintaining Fiber Optic Phased Array with High-Fiber Count," *Proc. SPIE*, 6306: 63060G, 2006.
68. Anderegg, J., Brosnan, S., Cheung, E., Epp, P., Hammons, D., Komine, H., Weber, M., and Wickham, M., "Coherently Coupled High Power Fiber Arrays," *Proc. SPIE*, 6102: 61020U1-5, 2006.
69. Augst, S. J., Fan, T. Y., and Sanchez, A., "Coherent Beam Combining and Phase Noise Measurements of Ytterbium Fiber Amplifiers," *Opt. Lett.*, 29: 474–476, 2004.
70. Nosu, K., Ishio, H., and Hashimoto, K., "Multireflection Optical Multi/Demultiplexer Using Interference Filters," *Electron. Lett.*, 15: 414–415, 1979.
71. Minott, P. O., and Abshire, J. B., "Grating Rhomb Diode Laser Power Combiner," *Proc. SPIE*, 756: 38–48, 1987.
72. Yu, C. X., and Neilson, D. T., "Free Space Diffraction Grating Based Demux Using Image Plane Transformations," *J. Select. Top. Quantum Electron.*, 8: 1194–1201, 2002.
73. Takada, K., Abe, M., Shibata, M., Ishii, M., and Okamoto, K., "Low-Crosstalk 10GHz-Spaced 512-Channel Array-Waveguide Grating Multi/Demultiplexer Fabricated on a 4-in Wafer," *IEEE Photon. Technol. Lett.*, 13: 1182–1184, 2001.

74. Loftus, T., Thomas, A., Hoffman, P., Norsen, M., Royse, R., Liu, A., and Honea, E., "Spectral Beam Combined Fiber Lasers for High Average Power Application," *IEEE J. Select. Top. Quantum Electron.*, 13: 487–497, 2007.
75. Cook, C. C., and Fan, T. Y., "Spectral Beam Combining of Yb-Doped Fiber Lasers in an External Cavity," in *OSA Trends in Optics and Photonics, vol. 26, Advanced Solid-State Lasers*, eds., M. M. Fejer, H. Injeyan, and U. Keller, Washington, DC: Optical Society of America, 1999, 163–166.
76. Daneu, V,. Sanchez, A., Fan, T. Y., Choi, H. K., Turner, G. W., and Cook, C. C., "Spectral Beam Combining of a Broad-Stripe Diode Laser Array in an External Cavity," *Opt. Lett.*, 25: 405–407, 2000.
77. Hamilton, C., Tidwell, S., Meekhof, D., Seamans, J., Gitkind, N., and Lowenthal, D., "High Power Laser Source with Spectrally Beam Combined Diode Laser Bars," *Proc. SPIE*, 5336: 1–10, 2004.
78. Augst, S. J., Goyal, A. K., Aggarwal, R. L., Fan, T. Y., and Sanchez, A., "Wavelength Beam Combining of Ytterbium Fiber Lasers," *Opt. Lett.*, 28: 331–333, 2003.
79. Clarkson, W. A., Matera, V., Kendall, T. M. J., Hanna, D. C., Nilsson, J., and Turner, P. W., "High-Power Wavelength–Combined Cladding-Pumped Tm-Doped Silica Fibre Lasers," *OSA Trends in Optics and Photonics (TOPS)*, vol. 56, Conference on Lasers and Electro-optics (CLEO 2001), Technical Digest, Postconference Edition, Washington, DC: Optical Society of America, 2001, 363–364.
80. Reich, M., Limpert, J., Liem, A., Clausnitzer, T., Zellmer, H., Kley, E. B., and Tünnermann, A., "Spectral Beam Combining of Ytterbium-Doped Fiber Lasers with a Total Output Power of 100 W," *Europhysics Conference Abstracts*, 28C: Fib-10137, 2004.
81. Bochove, E. J., "Theory of Spectral Beam Combining of Fiber Lasers," *IEEE J. Quantum Electron.*, 38: 432–445, 2002.
82. Vogt, S. S., and Penrod, G. D., "HIRES: A High Resolution Echelle Spectrometer for the Keck 10-m Telescope," *Instrumentation for Ground-Based Optical Astronomy: Present and Future*, ed. L. B. Robinson, Berlin: Springer-Verlag, 1988, 68–103.
83. Augst, S. J., Ranka, J., Fan, T. Y., and Sanchez, A., "Beam Combining of Ytterbium Fiber Amplifiers," *J. Opt. Soc. Am. B*, 24: 1707–1715, 2007.
84. Fan, T. Y., Goyal, A., and Sanchez, A., "High Power, Spectrally Combined Laser Systems and Related Methods," US Patent 6,697,192, Feb. 24, 2004.
85. Fridman, M., Eckhouse, V., Davidson, N., and Friesem, A., "Simultaneous Coherent and Spectral Addition of Fiber Lasers," *Opt. Lett.*, 33: 648–650, 2008.
86. Desurvire, E., *Erbium-Doped Fiber Amplifiers, Principles and Applications*, New York: Wiley, 1994.
87. Digonnet, M. J. F., ed., *Rare-Earth-Doped Fiber Lasers and Amplifiers*, New York: CRC Press, 2001.
88. Po, H., Cao, J. D., Laliberte, B., Minns, R., Robinson, R., Rockney, B., Tricca, R., and Zhang, Y., "High Power Neodymium-Doped Single Transverse Mode Fiber Laser," *Electron. Lett.*, 29: 1500–1501, 1993.
89. Yeong, Y., Sahu, J. K., Payne, D., and Nilsson, J., "Ytterbium-Doped Large-Core Fiber Laser with 1.36 kW cw Output Power," *Opt. Exp.*, 12: 6088–6092, 2004.
90. Hendow, S. T., Shakir, S. A., Culver, B., and Nelson, B., "Passive Phasing of Fiber Lasers," presented at Solid State and Diode Laser Technology Review, Los Angeles, CA 2007.
91. McNaught, S. J., Rothenberg, J. E., Thielen, P. A., Wickham, M. G., Weber, M. E., and Goodno, G. D., "Coherent Combining of a 1.26-kW Fiber Amplifier" (paper AMA2), *OSA Topical Meetings on Advanced Solid-State Photonics*, 2010.
92. Jones, D. C., Stacey, C., and Scott, A. M., "Phase Stabilization of a Large-Mode-Area Ytterbium-Doped Fiber Amplifier," *Opt. Lett.*, 32: 466–468, 2007.
93. Wirth, C., Schmidt, O., Tsybin, I., Schreiber, T., Peschel, T., Bruckner, F., Clausnitzer, T., et al., "2 kW Incoherent Beam Combining of Four Narrow-Linewidth Photonics Crystal Fiber Amplifiers," *Opt. Exp.*, 17: 1178–1183, 2009.

94. Andrusyak, O., Smirnov, V., Venus, G., Rotar, V., and Glebov, L., "Spectral Combining and Coherent Coupling of Lasers by Volume Bragg Gratings," *IEEE J. Select. Top. Quantum Electron.*, 15: 344–353, 2009.
95. Seeley, D., "High Energy Laser Joint Technology Office Electric Laser Initiatives," presented at The Optical Fiber Communication Conference and Exposition (OFC) and the National Fiber Optic Engineers Conference (NFOEC), San Diego, CA (2009). http://www.ofcnfoec.org/conference_program/2009/images/09-1a-seeley.pdf.
96. Huang, R. K., Chann, B., Missaggia, L. J.,. Donnelly, J. P., Harris, C. T., Turner, G. W., Goyal, A. K., et al., "High-Brightness Wavelength Beam Combined Semiconductor Laser Arrays," *IEEE Photon. Technol. Lett.*, 19: 209–211, 2007.
97. Hamilton, C., Tidwell, S., Meekhof, D., Seamans, J., Gitkind, N., and Lowenthal, D., "High Power Laser Source with Spectrally Beam Combined Diode Laser Bars," *Proc. SPIE*, 5336: 1–10, 2004.
98. Goodno, G. D., Asman, C. P., Anderegg, J., Brosnan, S., Cheung, E. C., Hammons, D., Injeyan, H., et al., "Brightness-Scaling Potential of Actively Phase-Locked Solid-State Laser Arrays," *IEEE J. Quantum Electron.*, 13: 460–472, 2007.
99. Ye, J., and Hall, J. L., "Optical Phase Locking in the Microradian Domain: Potential Application to NASA Spaceborne Optical Measurements," *Opt. Lett.*, 24: 1838–1840, 1999.
100. Goodno, G. D., Komine, H., McNaught, S. J., Weiss, S. B., Redmond, S., Long, W., Simpson, R., et al., "Coherent Combination of High Power, Zigzag Slab Lasers," *Opt. Lett.*, 31: 1247–1249, 2006.
101. McNaught, S. J., Komine, H., Weiss, S. B., Simpson, R., Johnson, A. M. F., Machan, J., Asman, C. P., et al., "100 kW Coherently Combined Slab MOPAs" (paper CThA1), *Conference on Lasers and Electro-Optics*, 2009.

Index

A

ABL. *See* Airborne Laser
Absorption bleaching, 339
AC Stark effect, 349
AC Stark SESAM, 347
Accelerators. *See also* Continuous Electron Beam Accelerator; Stanford Linear Accelerator Center; Stanford Superconducting Accelerator; Thomas Jefferson National Accelerator Facility
   for FEL, 87–89
   laser-based particle sources and, 323
Acousto-optical modulator (AOM)
   in high-repetition-rate fiber laser, 505, 506, 512
   in JHPSSL, 203
   in q-switched fiber laser, 312
   in q-switched TDL, 255
   in regenerative amplifier, 311
   in solid-state CBC, 565
   in ultrashort pulse fiber lasers, 512
Active mode locking, 454–456
Active-matrix liquid crystal displays (AMLCD), 38
Active-matrix organic light-emitting diode displays (AMOLED), 38–39
Adaptive optics (AO), 182–184, 268
Adaptive optics associates (AOA), 215, 216
Advanced Research Projects Agency (ARPA), 62. *See also* Defense Advanced Research Projects Agency
AFRL. *See* Air Force Research Labs
AFWL. *See* Air Force Weapons Lab
Air Force Research Labs (AFRL), 72
Air Force Weapons Lab (AFWL), 65
Airborne Laser (ABL), 72
Alkali-laser pumping, 143

Alpha laser, 63–64
America Optical Company, 413
AMLCD. *See* Active-matrix liquid crystal displays
AMOLED. *See* Active-matrix organic light-emitting diode displays
Amplified spontaneous emission (ASE), 170
   energy extraction and, 471
   in fiber lasers, 424, 434, 442, 453
   HCL effects from, 278, 280–282
   HCL suppression mechanism for, 270
   LMA fibers and, 471
   Monte Carlo ray tracing for calculating, 242
   MOPAs and, 477
   Nd:YAG and, 169, 170, 184
   PFLs and, 464, 471–472, 477–478
   as power scaling limit, 248–250
   SSLs and, 165, 169
   in TDLs, 233, 242–250
   ThinZag laser and, 218, 219
   TZ-3 laser and, 218–219
   VECSELs and, 335, 340
Angular multiplexing, 204
Antiguiding optical fiber, 443
AO. *See* Adaptive optics
AOA. *See* Adaptive optics associates
AOM. *See* Acousto-optical modulator
Argus laser, 359
Army Space and Missile Defense Technical Center (U.S.), 222
ARPA. *See* Advanced Research Projects Agency
Arrayed-waveguide gratings, 548
ASE. *See* Amplified spontaneous emission
Atmospheric composition measurement, 321
Autocorrelation, 319–320

B

Backscatter physics, 396
Baikowski Japan Company, 270, 271
Bar diodes, 519
Barrier-pumped VECSELs, 332
Baseline Demonstration Laser (BDL), 62
Basoc, Nikolai, 18
Bathtub curve, 116, 117
BBO. *See* Beta barium borate
BDL. *See* Baseline Demonstration Laser
Beam combining. *See also specific types*
   classes of, 536
   of fiber lasers, 554–561
   motivation for, 533–534
   performance metrics for, 534–535
   of semiconductor lasers, 561–564
   for SSLs, 564–565
   strategic directed-energy applications and, 529
   techniques/theories for, 537–553
Beam conditioning
   in DF laser, 48
   frequency conversion and, 393–394
   methods of, 386
   in NIF, 360, 368, 387–395
   with phase plates, 387, 390–393
   spatial, 387, 390–393
   temporal, 393–395
Beam formation
   MOPAs and, 179
   in SSLs, 176–179
   stable resonators for, 177
   unstable resonators for, 177–179
Beam quality
   directed-energy applications and, 531
   HAP SSLs and, 164
   in high-power diode laser arrays, 141–142, 145–146
   in high-power fiber lasers, 445–447
   of high-power TDLs, 251
   HOM and, 478
   of lamp-pumped SSLs, 151
   of TDLs, 227, 251
   thermal effects and, 331
   of TZ-2 laser module, 216
Beam shaping
   cylindrical lens arrays for, 149, 150
   in NIF, 376
   polarization coupling for, 149, 150
   refractive optics for, 149
   Southampton beam shaper, 146–148
   step mirror FhG-ILT, 148–149
Beam steering
   for HCL, 290, 292
   in semiconductor laser diodes, 114
Beamlet, 365

Bell Laboratories, 413
Beta barium borate (BBO), 319
Boron nitride, 121
Brauch, Uwe, 226
Brewster-angle amplifier slabs, 361
Brewster-plate amplifiers, 187
Broadband multimode fiber lasers, power scaling of, 520–522
Broadband single-mode fiber lasers
   limiting factors for power scaling of, 523
   power scaling of, 523–526
Budker Institute of Nuclear Physics, 85, 93
Bulk resonators, 542–543
Burn-in
   of HCL pump tiles, 269
   of semiconductor laser diodes, 116–118
Burrus, C. A., 413

C

Capsule implosion experiments, 398
Carbon dioxide laser (CO_2 laser)
   applications for, 13–15
   basics for, 4–9
   cost factors disfavoring, 526
   diffusion-cooled, 9, 10–12
   energy transfer in, 4–6
   fast-flow, 9, 12–13
   gas discharge of, 6
   gas mixtures for, 7–8
   general characteristics of, 4
   heat removal in, 7
   high-power fiber lasers *vs.*, 527–528
   lifespan of, 8
   operating temperature of, 9
   RF discharge in, 6–7
   sealed-off, 9, 10
   small signal gain for, 6
   types of, 9–13
   uses of, 3
   vibration energy levels in, 5
   vibration modes in, 5
   water in, 7–8
Carbon monoxide lasers, 74
Catastrophic optical damage (COD), 124–125
Catastrophic optical mirror damage (COMD), 145
Cavity leaking, 256
Cavity-dumped operation of TDLs, 256–257
CBC. *See* Coherent beam combining
CCC. *See* Chirally coupled core fibers
CCEPS. *See* Conduction-cooled, end-pumped slab lasers
CEA. *See* Commissariat à l'Énergie Atomique

CEBA. *See* Continuous Electron Beam Accelerator
Center for Radiative Shock Hydrodynamics, 406
Chemical lasers. *See also specific types*
 classifications and background of, 44–45
 other concepts for, 73–74
Chemical oxygen iodine laser (COIL), 43, 45
 block diagram of, 66
 deactivation process, 69
 energy levels, 65–66
 energy pumping reactions for, 67–68
 iodine dissociation, 69
 performance, 71–72
 singlet oxygen generator, 65, 69–71
 small signal gain, 66–67
Chemical vapor deposition diamond, 121, 231
Chemically excited species generation, 53–55
Chernock, Joe, 187
Chinese Academy of Sciences, 267
Chip-on-submount (COS), 120, 122, 124
Chirally coupled core fibers (CCC), 442, 475, 502
Chirped fiber Bragg gratings, 448
Chirped pulse amplification (CPA), 501
 downchirped, 313–316
 in fiber laser system, 506–508, 510–512, 514
 introduction of, 306
 nonlinearity and, 504, 505, 508–510
 optical parametric, 317–319
 SPM and, 429, 504–505, 508–509
 TDLs and, 259, 261
 ultrafast pulse amplification and, 458
 ultrafast SSLs and, 307–309
Chromatic aberration, 310
Cladding pumping, 414, 554
CO_2 laser. *See* Carbon dioxide laser
COD. *See* Catastrophic optical damage
Co:GGG, 279, 282
Coherent beam combining (CBC), 535, 536
 AOM in solid-state, 565
 common-resonator, 540–541
 evanescent-wave/leaky-wave, 541–542
 fiber arrays and, 545
 of fiber lasers, 555–558
 filled-aperture *vs.* tiled-aperture, 537–539
 in HBC, 553

Coherent beam combining (CBC) (*Cont.*):
 MOPA and, 545, 558, 560–561, 563–564
 passive *vs.* active approaches, 544–545
 performance degradations in, 545–549
 self-organizing (supermode), 544
 of semiconductor lasers, 563–564
 of SSLs, 566–567
Coherent Inc., 29, 33
COIL. *See* Chemical oxygen iodine laser
Cold ablation, 33
Colgate, Stirling, 360
Combination processing, 528
COMD. *See* Catastrophic optical mirror damage
Commissariat à l'Énergie Atomique (CEA), 364
Common-resonator CBC, 542–543
Computer-controlled fabrication, 181
COMSOL Multiphysics, 235
Conduction-cooled, end-pumped slab lasers (CCEPS), 198–202
Constructive interference, 539
Continuous Electron Beam Accelerator (CEBA), 88
Continuous phase plates (CPP), 357, 388
Continuous wave fiber lasers
 background for, 445–450
 market acceptance of, 527
 power scaling of, 451–452
 pump coupling for, 450–451
 SBS in, 465
 scalability of, 518
 single-frequency, 452–453
Continuous wave lasers (CW), 43, 414. *See also specific types*
 FELs and, 85
 high-power diode laser arrays as, 141, 145
 pressure recovery for, 62
Continuous wave TDLs
 fundamental mode of, 251–254
 high average power, 250–251
 power output of, 338
Copper-tungsten metal matrix (CuW), 227, 231, 235
COS. *See* Chip-on-submount
Counter-intercontinental ballistic missile applications, 529
CPA. *See* Chirped pulse amplification
CPP. *See* Continuous phase plates
Cross-phase modulation (XPM), 468
Cryogenic cooling, 170, 316, 502

Index

Cutting processes, 3
  CO_2 lasers, 13–15
  in combination processing, 526
  in dental surgery, 153
  fusion, 13
  HCL in, 297
  high-power fiber lasers in, 526–528
  high-power TDLs for, 251
  laser plasma, 15
  oxidation, 13–15
  SSLs for, 153, 158
  sublimation, 13
  ultrafast SSLs in, 304
CuW. *See* Copper-tungsten metal matrix
CW. *See* Continuous wave lasers
Cyclops laser, 359
Cylindrical lens arrays, 149, 150

D

DAPKL. *See* Diode-Array Pumped Kilowatt Laser
DARPA. *See* Defense Advanced Research Projects Agency
DBR. *See* Distributed Bragg reflector
Dead zones, 188
Defense Advanced Research Projects Agency (DARPA), 195, 196, 519, 531
Deformable mirrors (DM)
  in AO, 183–184, 268, 287, 543
  modeling, 378
  OPD correction with, 204
  in ThinZag laser, 216–217, 220
  wavefront control with, 290, 293–294, 367
Dental surgery, 154, 155, 158
Department of Defense, U.S. (DOD), 222
Depolarization
  in HCL, 282, 285, 288–290
  MOPAs limited by, 179
  in NIF, 390
  optical fibers and, 419, 443, 513
  thermal effects and, 166, 331
Detuning length, 90
Deuterated KDP (dKDP), 379, 387
Deuterium fluoride laser (DF), 43, 45
  chemically excited species generation for, 53–55
  energy levels in, 47, 49–51
  fluid mechanics/nozzle design for, 57–60, 62
  kinetic processes/deactivation/energy transfer, 55–57
  layout of, 46, 48
  overtone, 62
  performance of, 63–65
  pressure recovery for, 62

Deuterium fluoride laser (DF) (*Cont.*):
  pulsed, 63
  resonator in, 47, 48
  small signal gain of, 51–53
Deuterium-tritium nuclear fusion, 357
DF. *See* Deuterium fluoride laser
DFB. *See* Distributed feedback
DF-CO_2 transfer devices, 73
DFG. *See* Difference frequency generation
Diamond-like carbon (DLC), 39
Die bonding, 120–121
Difference frequency generation (DFG), 319
Diffraction
  grating spectrometer and, 549–550
  Kerr nonlinearity and, 430
  optical phase distortion and, 236
  in zigzag slab lasers, 192
Diffusion-cooled CO_2 lasers, 9
  high-power, 10–12
Diode lasers. *See also* High-power diode laser arrays; Semiconductor laser diodes
  bar assembly for, 134–136
  efficiency improvements in, 133, 134
  fiber coupling of, 146–149
  heat sinks on, 104, 110, 113, 118, 119, 123
Diode pumping, 184, 211, 415–416
  HAP SSLs and, 173
  single-emitter diodes for, 521–522
  SSLs and, 173–174, 225, 328
Diode-Array Pumped Kilowatt Laser (DAPKL), 192, 196–198
Diode-pumped solid-state lasers (DPSS), 225, 328, 463, 474
Direct patterning, 37
Directed-energy applications, 528–529
Direct-energy weapon systems, 417
Discharge circuits, in excimer lasers, 20–23
Distributed Bragg reflector (DBR), 334, 452–453
Distributed feedback (DFB), 454–453
dKDP. *See* Deuterated KDP
DLC. *See* Diamond-like carbon
DM. *See* Deformable mirrors
DOD. *See* Department of Defense, U.S.
Downchirped pulse amplification (DPA), 313–316
DP25 laser, 195, 196
DPA. *See* Downchirped pulse amplification
DPSS. *See* Diode-pumped solid-state lasers
Dynamic gain saturation, 337, 341

Index

E

École Polytechnique, 360, 361
EDFAs. *See* Erbium-doped fiber amplifiers
Edge bars, 193
Effective mode index, 417–418
Electro-optic modulator (EOM), 311–312
End-pumped slab lasers
 architecture/technical issues in, 199–200
 performance of, 200–201
 power scaling of, 201–204
Energy extraction
 ASE and, 473
 in FEL, 78, 81–83
 HOM fibers and, 475
 MPA and, 313
 power scaling and, 433–435
 in TDLs, 246–247
Energy recovery, in FELs, 91–92
EOM. *See* Electro-optic modulator
Equation of motion
 fundamental, 240, 243
 for TDLs, 240–241
Erbium-doped fiber amplifiers (EDFAs), 114, 415–416, 521, 554
ETH Zurich, 254
EUV. *See* Extreme ultraviolet light
EUV lithography, 324
Evanescent-wave coupling, 541–542, 563
Excimer lasers
 applications of, 33–40
 application-specific designs, 28–33
 beam profile of, 26–27
 discharge circuit, 20–23
 gas circulation/cooling/replenishment, 23–25
 gas media/emission wavelengths for, 18
 high-power, 28–31, 33–40
 high-resolution micromachining, 35–38
 LASIK, 32–33
 microlithography, 31–32
 preionization in, 20–21
 principal design/technology of, 19–20
 principle of operation, 17, 19
 resonator in, 25–28
 scaling laser volume in, 21
 window lifetimes in, 25
ExciStar XS, 33
Expansion-matched microchannel heat sinks, 138–139
Extreme ultraviolet light (EUV), 324–325
Eye-safe pulsed-fiber laser sources, 493–495

F

Failure mechanisms, 124
Fan, T. Y., 226
Faraday isolators
 in DAPKL, 198
 in MOPAs, 179
 regenerative amplifiers and, 312
 YDFAs and, 202
Fast-flow CO_2 lasers, 9, 12–13
FBG. *See* Fiber Bragg gratings
FEA. *See* Finite element analysis
FEL. *See* Free-electron lasers
Femtosecond sources, 499
Fiber amplifiers
 beam combining and, 556
 counterpumped, 429, 435
 effective length for, 17, 428
 energy extraction and, 435–437
 Erbium-doped, 114, 415–416, 523, 556
 gain-staged MOPAs, 485–490
 high pulse energy/peak power, 480–493
 in NIF, 371, 372
 optical, 415–416
 polarization-maintaining MOPAs, 490–492
 pump-brightness-limited, 526
 simulation of, 426–427
 single-stage, 482–485
 wavelength conversion and, 490–492
 Yb-doped, 202, 557, 558, 562
Fiber arrays
 CBC performance degradation and, 547
 fiber laser beam combining and, 557, 559
Fiber Bragg gratings (FBG), 448–451, 550
Fiber combiners, 524
Fiber coupling
 lamp-pumped SSLs and, 151, 152
 of laser diode bars, 146–149
 semiconductor laser diode package design for, 122–125
 Southampton beam shaper for, 146–148
 step mirror FhG-ILT for, 148–149
Fiber CPA system, 506–508, 510–514, 516
Fiber damage, 513
Fiber lasers. *See also* Continuous wave fiber lasers; Pulsed fiber lasers; *specific types*
 advantages of, 417–418, 519–520
 ASE and, 426, 436, 444, 455
 beam combining for, 556–563
 broadband multimode, 522–524
 broadband single-mode, 525–528
 CBC for, 557–561

Index

Fiber lasers (*Cont.*):
  directed-energy applications of, 530–531
  engineering of, 520–522
  HBC for, 562
  high-power, 437–447
  history of, 415–417
  nonlinear limits of, 428–432
  power scaling of, 428–437, 519, 520, 522–524, 525–528
  pumping with, 526
  pumping/beam quality of, 445–447
  q-switched, 312, 455–456, 481
  as replacement technology, 528
  retrofit market and, 530
  single-frequency, 454–455
  SPM and peak power of, 436–437, 451
  telecom and, 520–521
  WBC for, 561–562
Fiber master oscillator, 525, 526, 527
Fiber-coupled high-power diode laser devices, 151–152
Fiber-coupled package design/processes, 122–125
Field enhancement, 348
Filamentation, 321, 362
Fill factor, 142, 150, 362
Final optics assembly (FOA), 360, 367, 368, 385, 389
Finite element analysis (FEA), 235, 236, 238
Flash pumping, 172–173
Fluence beam contrast, 372
Fluorescence
  spectral distribution of, 244
  TDLs and, 232–234
FOA. *See* Final optics assembly
FOPAI. *See* Fractional power above intensity
Forward-rotational Raman scattering, 362
Foster, John, 362
Four-wave-mixing (FWM), 470–472, 503
Fractional power above intensity (FOPAI), 391–393
Fraunhofer Institute for Laser Technology, 150
Free-electron lasers (FEL). *See also* IR Demo FEL; IR/UV Upgrade FEL; Linac Coherent Light Source; Regenerative-Amplifier FEL
  accelerators for, 87–89
  continuous wave, 85
  development of, 77
  energy extraction in, 78, 81–83
  energy recovery, 91–92
  fourth-generation, 324
  gain/bandwidth, 79, 81–82

Free-electron lasers (FEL) (*Cont.*):
  hardware implementation, 83–92
  injectors, 84–87, 323
  laser-plasma-based electron sources for, 323
  optical cavity, 82, 89–90
  physical mechanism, 77–78
  practical considerations, 82–83
  small signal gain in, 79, 81
  status of, 93–95
  wavelength, 78–79
  wigglers, 77, 89
Free-space gratings, 550
Frequency combs, 459
Frequency resolved optical gating (FROG), 320
FROG. *See* Frequency resolved optical gating
Fs-oscillators, 254
Fundamental space-filling mode, 440
Fusion cutting, 15
FWM. *See* Four-wave-mixing
FWM coherence length, 471
FWM pump beam, 470

G

Gadolinium gallium garnet (GGG), 268, 284–285
Gain-staged MOPAs, 485–490
Gas dynamic lasers (GDL), 9, 44
Gas replenishment, 23–25
Gate turn-offs (GTOs), 22
GDD. *See* Group delay dispersion
GDL. *See* Gas dynamic lasers
Gekko XII laser, 363
Geometrical multipass amplifiers, 260
German Aerospace Center, 226
GGG. *See* Gadolinium gallium garnet
Giesen, Adolf, 226
Gires-Tournois interferometer mirrors (GTI), 258, 259
Graded index separate confinement heterostructure (GRINSCH), 104
Graded reflectivity mirror (GRM), 215
Grating rhomb, 550
Grating spectrometer, 551–552
GRENOUILLE, 320
GRINSCH. *See* Graded index separate confinement heterostructure
Grism stretcher, 315
GRM. *See* Graded reflectivity mirror
Group delay dispersion (GDD), 340, 350
GTI. *See* Gires-Tournois interferometer mirrors
GTOs. *See* Gate turn-offs

Index

H

Hair removal, 155, 158
Half-wave plate (HWP), 375
Hall, J. L., 327
Handbook of Laser Materials Processing (Laser Institute of America), 13
Hänsch, T. W., 327
HAP SSL. *See* High-average-power solid-state lasers
Harmonic conversion, 362
HBC. *See* Hybrid beam combining
HCL. *See* Heat capacity lasers
Heat capacity lasers (HCL), 176
 antiaircraft applications of, 297–298
 applications/related experimental results of, 296–300
 ASE effects in, 278, 280–282
 ASE suppression in, 270
 beam steering for, 290, 292
 configuration of, 293
 current state of art of, 290, 293–295
 cutting with, 297
 depolarization in, 282, 285, 288–290
 gain module of, 268, 269–271
 heat buildup in, 274–277, 279
 instantaneous pump spectral intensity for, 274
 on military vehicle, 299–300
 mine clearing with, 298–299
 operation of, 272–273
 performance modeling for, 273–290
 power extraction from, 290
 pump absorption/gain/extraction in, 273–277
 pump arrays for, 267–269, 272
 pump tile burn-in, 269
 QCW operation of, 269, 271
 rapid material removal with, 297
 scaling approaches for, 296
 system architecture for, 267–273
 temperature/stress calculations for, 284–285, 286
 wavefront calculations for, 285–288
 wavefront control for, 290, 293–295
 wavefront distortion in, 282, 284–288
Heat removal
 in CO_2 lasers, 7
 diode laser array product platforms and, 139–141
 from fiber lasers, 433–434
 for high-power diode laser arrays, 136–139
 in SSLs, 174–176
 ultrafast TDL power scaling and, 333
 water guidelines for, 137–138

Heat sinks
 on diode lasers, 104, 110, 113, 118, 119, 123
 on high-power diode laser arrays, 135–141, 144–145, 150
 in high-power TDLs, 238
 materials for, 231
 microchannel, 137, 138–139
 minichannel, 136–138, 140–141
Heat treating, 157
Heterodyne wavefront sensor, 545, 546
HF. *See* Hydrogen fluoride laser
HHG. *See* High harmonic generation
High Energy Laser Joint Technology Office, 222
High harmonic generation (HHG), 323–326, 350
High-average-power solid-state lasers (HAP SSL), 163–164
 beam quality of, 164
 diode pumping and, 173
 gain materials for, 167–168
 heat removal from, 175
 optical damage resistance and, 184
 PCMs in, 181–182
 SPPs in, 181
High-field science, 350
Highly multimode fibers, 505
High-order mode fibers (HOM), 477–478
High-power broad-area laser diodes, 110–112
High-power diode laser arrays
 applications of, 153–159
 bar assembly for, 134–136
 beam delivery for, 146
 beam quality/brightness, 141–142, 145–146
 continuous wave, 141, 145
 defense applications of, 154–155, 158–159
 device performance for, 141–145
 emitter counts on, 142
 fiber coupling of, 146–149
 fiber-coupled devices, 151–152
 fill factor for, 142, 150
 HCL pump arrays, 267–269, 272
 heat removal from, 136–139
 heat sinks on, 135–141, 144–145, 150
 heat treating with, 157
 industrial applications of, 153, 155–157
 laser brazing with, 157, 158
 lifetime/reliability of, 144–145
 medical applications of, 154–155, 158
 multimode, 142
 operating modes of, 141
 power scaling in, 150–151
 power/efficiency of, 141

High-power diode laser arrays (*Cont.*):
  product performance for, 145–152
  product platforms for, 139–141
  spectral stability of, 143, 144
  time-dependent center wavelength in, 273–274
  wavelength locking in, 142–144
  wavelength of, 141
  welding with, 155–157
High-power excimer laser
  applications of, 33–40
  design of, 28–31
High-power fiber lasers
  CO_2 lasers *vs.*, 529–530
  defense applications of, 530–531
  industrial applications of, 528–530
  LCFs for, 441–443
  LMA fibers for, 437–439
  other large-core fibers for, 443–445
  PCFs for, 439–441
  pumping/beam quality of, 445–447
High-power laser diodes, 101
  attributes of, 102–104
  bars of, 112–114, 134–136
  broad-area, 110–112
  single-mode, 114–116
High-power single-mode laser diodes, 114–116
High-power TDLs
  beam quality of, 251
  continuous-wave operation of, 250–254
  cutting processes with, 251
  design study for, 238–239
  heat sinks in, 238
  pulse durations in, 348
High-power ultrafast fiber laser systems
  average power scaling of ultrashort pulsed, 514–515
  fiber damage as scaling limitation, 513
  high-repetition rate gigawatt peak power, 507–510, 514
  nonlinearity in, 503–507
  peak power/pulse energy scaling in, 510–514
  thermo-optical problems avoided by, 502
High-repetition-rate gigawatt peak power fiber laser system, 507–510, 514
High-temperature superconductor (HTS), 39
High-voltage dc guns, 84–85
Hohlraums, 361, 365, 366, 398, 399, 402, 407
HOM. *See* High-order mode fibers

HTS. *See* High-temperature superconductor
Hughes Research Lab, 172, 360
HWP. *See* Half-wave plate
Hybrid beam combining (HBC), 535, 538, 555, 556, 562
Hybrid resonators, 11, 201
Hydrogen fluoride laser (HF), 43, 45
  chemically excited species generation for, 53–55
  energy levels in, 47, 49–51
  fluid mechanics/nozzle design for, 57–60, 62
  kinetic processes/deactivation/ energy transfer, 55–57
  overtone, 62
  performance of, 63–65
  pressure recovery for, 62
  pulsed, 63
  small signal gain of, 51–53
Hydrogen halide devices, 73

I

IBM, 32
ICF. *See* Inertial confinement fusion
IGBTs. *See* Insulated gate bipolar transistors
Incoherent beam combining, 535, 538, 539, 555, 557
Incoherent combination architecture, 523
Indirect-fusion studies, 365
Indium tin oxide (ITO), 35
Inertial confinement fusion (ICF), 361, 363, 401, 402
Injectors, for FEL, 84–87, 323
Ink-jet nozzle drilling, 36
INO. *See* National Optics Institute
Institute of Laser Engineering, 363
Insulated gate bipolar transistors (IGBTs), 22
In-well pumped VECSELs, 332, 337
IR Demo FEL, 89, 92, 93
IR/UV Upgrade FEL, 93–95
ITO. *See* Indium tin oxide

J

JAEA. *See* Japan Atomic Energy Agency
Janus laser, 361
Japan Atomic Energy Agency (JAEA), 93
JDS Uniphase, 117, 416
Jefferson Lab. *See* Thomas Jefferson National Accelerator Facility
Jenoptik, 226, 261
JHPSSL. *See* Joint High Power Solid-State Laser

Index

Joint High Power Solid-State Laser (JHPSSL), 202, 203, 209, 222

K

Kane, Dan, 320
KDP. *See* Potassium dihydrogen phosphate
Keller, Ursula, 254
Kerr effect, 304–305, 314, 432, 467, 469, 506
Kerr lens mode locking, 339
Kerr nonlinearity, 432
Kidder, Ray, 360, 361
Kinks, 114–116
Kinoform phase plates (KPP), 390
KMS Fusion, 362
Koester, C., 415
Konoshima Chemical Company, 270, 271
KPP. *See* Kinoform phase plates
Kramers-Kronig relation, 445

L

Laboratoire pour l'Utilisation des Lasers Intenses (LULI), 363
Laboratory for Laser Energetics, 363
Labsphere integrating sphere power meter, 212
Lambda Physik, 30
Lambertian scattering surfaces, 211
Lamp-pumped SSLs, 151, 153, 172–173
LANL. *See* Los Alamos National Laboratory
Lanthanide contraction, 347
Large-mode-area fibers (LMA)
 ASE and, 473
 high average power-specific, 515
 for high-power fiber lasers, 437–439
 mode shrinking due to bending of, 504–505
 spatial scaling and, 503, 506
 ultra-, 505
Laser boring/ablation, 297
Laser brazing, 157, 158
Laser cutting, 3
Laser diode bars, 112–114
 assembly of, 134–136
 beam delivery for, 146
 emitter counts on, 142
 fiber coupling of, 146–149
 heat removal for, 136–139
 lifetime/reliability of, 144–145
 power scaling and, 150–151
Laser fusion, 361
Laser gain materials
 cross section/lifetime of, 164–166
 for HAP SSL, 167–168

Laser gain materials (*Cont.*):
 for high-pulse-energy/peak power SSLs, 170–171
 host materials, 166–167
 Nd:YAG, 168–169
 for SSLs, 164–171, 207–208
 Yb:YAG, 169–170
Laser Institute of America, 13
Laser Mégajoule (LMJ), 364
Laser performance operations model (LPOM), 368
 calibration results for, 369
 frequency conversion performance in, 380–386
 main laser performance verification for, 378–380
 MPA model in, 377
 power *vs.* energy operating envelopes, 370–372
 pulse shaping from, 373
Laser plasma cutting, 15
Laser welding, 155–157
Laser-assisted in situ keratomileusis (LASIK), 32–33
Laser-based photon/particle sources, 323
Laser-induced breakdown spectroscopy (LIBS), 321
Laser-plasma coupling, 397
Laser-plasma interactions, 398
Laser-plasma-based electron sources, 323
Laser-pumped solid state lasers, 152, 153
Lasers and Electro-Optics Society (LEOS), 226
LASIK. *See* Laser-assisted in situ keratomileusis
Lasnex, 361
Lawrence Livermore National Laboratory (LLNL), 193, 267, 268, 270, 290, 297, 298, 360, 361
Lawrence Radiation Laboratory. *See* Lawrence Livermore National Laboratory
LCF. *See* Leakage channel fibers
LCLC. *See* Linac Coherent Light Source
LDC. *See* Low duty cycle
Leakage channel fibers (LCF), 441–443, 477, 504
Leaky modes, 420
Leaky-wave coupling, 543–544, 563
Lebedew Institute of Physics, 18
LEOS. *See* Lasers and Electro-Optics Society
LIBS. *See* Laser-induced breakdown spectroscopy
Lidar. *See* Light detection and ranging
Light detection and ranging (Lidar), 321

Light-induced polymerization, 322–323
Ligne d'Integration Laser (LIL), 364
LIL. *See* Ligne d'Integration Laser
Linac Coherent Light Source (LCLS), 83, 84
Lincoln Laboratory, 550
LLNL. *See* Lawrence Livermore National Laboratory
LMA. *See* Large-mode-area fibers
LMJ. *See* Laser Mégajoule
Locking of optical coherence by single-detector electronic-frequency tagging (LOCSET), 558, 559
LOCSET. *See* Locking of optical coherence by single-detector electronic-frequency tagging
Los Alamos National Laboratory (LANL), 83, 84
Low duty cycle (LDC), 219–220
Low-temperature polycrystalline silicon (LTPS), 38–39
LPOM. *See* Laser performance operations model
LTPS. *See* Low-temperature polycrystalline silicon
LULI. *See* Laboratoire pour l'Utilisation des Lasers Intenses
LuxxMaster, 143

M

Mach-Zehnder interferometer, 319, 320
Magnetorheological finishing (MRF), 181, 390
Maiman, Ted, 172, 360
Martin, Bill, 187
Massachusetts Institute of Technology (MIT), 226, 550
Master oscillator (MO/MOR)
  in DAPKL, 197
  fiber, 525, 526, 527
  in JHPSSL, 202
  in NIF, 367, 373–374
Master oscillator power amplifier (MOPA)
  ASE management with, 479
  CBC and, 545, 558, 560–561, 563–564
  in DAPKL, 196
  gain-staged, 485–490
  in JHPSSL, 202, 203
  in microlithography system, 32
  OPCPA and, 319
  PCMs and, 181–182
  phase-control requirements for fiber, 560–561
  polarization-maintaining, 490–492
  power oscillators *vs.*, 480–482
  SSL beam formation and, 179
  WBC and, 561

MBE. *See* Molecular beam epitaxy
MEMS. *See* Microelectromechanical systems
Metal inert gas welding (MIG), 156
Metal organic chemical vapor deposition (MOCVD), 102, 104
Michelson interferometric resonator, 544
Microbending, 441
Microchannel heat sinks, 137, 138–139
Microelectromechanical systems (MEMS), 35
Microfabrication, 322–323
Microlithography, 31–32
Micromachining
  high repetition rates and, 435
  high-resolution, 35–38
  short UV for, 34
  ultrafast SSLs for, 321–323
  ultrafast TDLs for, 329, 344
Microstructured gain fibers, 479
Mid-Infrared Advanced Chemical Laser (MIRACL), 60, 61, 63
MIG. *See* Metal inert gas welding
Mine clearing applications, 298–299
Mine-resistant, ambush-protected vehicle (MRAP), 299–300
Minichannel heat sinks, 136–138, 140–141
MIRACL. *See* Mid-Infrared Advanced Chemical Laser
MIT. *See* Massachusetts Institute of Technology
MIXSEL. *See* Mode-locked integrated external-cavity surface emitting lasers
MO. *See* Master oscillator
MOCVD. *See* Metal organic chemical vapor deposition
Mode locking
  active, 456–458
  Kerr lens, 339
  passive, 305, 458–459
  Q-switched, 343, 345
  SESAM, 338–350
  Soliton, 340
  SPM and, 340, 344
  in TDLs, 340
  in ultrafast TDLs, 342
  in VECSELs, 341, 344–345
Mode-locked fiber lasers, 456–460
  active mode locking for, 456–458
  frequency combs for, 459
  passive mode locking for, 458–459
  ultrafast pulse amplification in, 459
Mode-locked integrated external-cavity surface emitting lasers (MIXSEL), 329, 351
Mode-locked lasers, 416

Index

Molecular beam epitaxy (MBE), 102, 104
Monte Carlo ray tracing, 242, 278
MOPA. *See* Master oscillator power amplifier
MOR. *See* Master oscillator
MPA. *See* Multipass amplification
MRAP. *See* Mine-resistant, ambush-protected vehicle
MRF. *See* Magnetorheological finishing
Multidither servo loops, 545, 559
Multimode fiber lasers, 531
Multimode fibers, 444, 476
Multimode lasers, 142
Multipass amplification (MPA)
 energy extraction and, 313
 geometrical, 260
 LPOM model for, 377
 in NIF, 376, 377
 thermal mitigation and, 316–317
 ultrafast SSLs and, 312–313
Multiplexing
 angular, 204
 polarization, 179
 power scaling and, 151, 152
 spatial, 126–127
 wavelength, 152, 447
 wavelength-division, 416, 422, 447, 448, 458, 539, 550
MZA Associates, 220

N

NACL. *See* Navy-ARPA Chemical Laser
National Ignition Campaign (NIC), 364, 398, 406
National Ignition Facility (NIF), 170, 172
 beam energetics of, 369, 370
 beam shaping in, 376
 beamlines of, 364–365, 367
 bundle qualification for, 368
 depolarization measurements for, 390
 fiber amplifiers in, 373, 374
 final optics assembly of, 360, 367, 368, 385, 389
 fluence beam contrast for, 372
 focal spot beam conditioning in, 360, 368, 387–395
 frequency conversion of conditioned pulses, 395–396
 frequency conversion performance of, 380–386
 historical background for, 360–364
 layout of, 367–368
 main laser performance verification, 378–380
 master oscillator for, 373–374, 376

National Ignition Facility (NIF) (*Cont.*):
 MPA in, 376, 377
 optics of, 365, 366
 overview of, 359–360, 364–368
 PAM in, 373, 374, 376–377
 PDS in, 368
 performance qualification shots, 373–386
 phase plates in, 387, 389
 power balance in, 402–403, 404
 power *vs.* energy operating envelopes for, 370–372
 project completion criteria for, 398, 399, 400
 pulse shaping for ignition experiments in, 387, 388, 395–397
 pulse shaping system of, 373–374, 375, 406–407
 pulse-shape reproducibility in, 402–404
 regenerative amplifier in, 374–376
 spatial beam conditioning in, 390–393
 status of/experiments at, 397–407
 temporal beam conditioning in, 393–395
 temporal pulse shaping, 396–397
 third harmonic generation in, 380–384
 wavefront errors in, 378
National Institute of Standards and Technology (NIST), 34, 369
National Optics Institute (INO), 215
Naval Postgraduate School, 89
Navy-ARPA Chemical Laser (NACL), 62
Nd:glass, 170–171
Nd:YAG, 168–169, 170, 184, 191, 196, 200, 207, 254, 270, 279, 528
Nd:YVO, 254
NIC. *See* National Ignition Campaign
NIF. *See* National Ignition Facility
NIST. *See* National Institute of Standards and Technology
NLEs. *See* Nonlinear effects
Nonlinear effects (NLEs)
 as basic limitations of ultrashort pulse amplification, 503–507
 Kerr nonlinearity, 432
 as limits of fiber lasers, 428–432
 PFL power scaling and, 466–472
 thresholds for, 430
Nonlinear phase modulation, 469–470
Nonlinear self-focusing, 432, 433
Normalized frequency, 419
Nova laser, 362–363
Novette laser, 362
Novosibirsk Recuperator, 93, 94
Nuckolls, John, 360
Nuvonyx, 211

Index

O

OLED. *See* Organic light-emitting diodes
Omega EP laser, 363
Omega laser, 363
1D SSD. *See* Smoothing by spectral dispersion in one dimension
OPA. *See* Optical parametric amplifier
OPCPA. *See* Optical parametric chirped pulse amplification
OPD. *See* Optical path difference; Optical phase distortion
Operational and performance qualification (OQ-PQ), 368
Ophir power meter, 212
OPLL. *See* Optical phase-locked loop
OPO. *See* Optical parametric oscillator
OPSL. *See* Optically pumped semiconductor lasers
Optical cavity
　for FELs, 82, 89–90
　passive mode locking and, 305
　in regenerative amplifier, 311
　in ThinZag laser, 210, 212–216, 219–220
Optical damage
　HAP SSLs and, 184
　PFL power scaling and, 474–476
　power scaling and, 434–435
　at seal contact areas, 193–194
　in zigzag slab lasers, 193–194
Optical fiber amplifiers, 415–416
Optical fibers
　basics of, 418–422
　depolarization in, 421, 445, 515
Optical parametric amplifier (OPA), 317, 319
Optical parametric chirped pulse amplification (OPCPA), 317–319
Optical parametric oscillator (OPO), 493
Optical path difference (OPD)
　AOs for, 184
　DMs for correcting, 204
　eliminating, 180
　non-zigzag axis temperature nonuniformity and, 193
　thermal expansion and, 175, 178, 189
　unstable resonators and, 178
Optical phase distortion (OPD)
　diffraction and, 236
　in TDLs, 236, 237
Optical phase-locked loop (OPLL), 564
Optical scrambler, 211
Optically pumped semiconductor lasers (OPSL), 328
OQ-PQ. *See* Operational and performance qualification
Organic light-emitting diodes (OLED), 38–39
Osaka University, 363
Osram diodes, 219
Oxidation cutting, 13–15

P

Package-induced failure (PIF), 125
PAM. *See* Preamplifier module
Parasitic losses, 218
Passive mode locking, 305, 458–459
PCF. *See* Photonic crystal fibers
PCM. *See* Phase conjugate mirrors
PDS. *See* Precision diagnostic system
PEEK. *See* Polyether ether ketone
PET. *See* Positron emission tomography
PFL. *See* Pulsed fiber lasers
Phase conjugate mirrors (PCM), 181–182
Phase distortion, 312
Phase matching condition, 471
Phase modulation. *See also* Self-phase modulation
　nonlinear, 469–470
Phase plates
　beam conditioning with, 387, 390–393
　continuous, 359, 390
　kinoform, 390
　in NIF, 387, 389
　random, 390
　spatial, 180–181
　thermal focusing correction with, 181
Phase sensing servo loops, 545, 559
Phased-locked loops (PLL), 545, 558, 560
Phase-sensing control loop, 545
Phebus laser, 363
Photodarkening, 522, 526
Photodynamic therapy, 154, 158
Photoinjector electron sources, 323
Photon flux density, calculating, 243–244
Photonic crystal fibers (PCF)
　for high-power fiber lasers, 439–441
　in high-repetition-rate system, 508
　in PFLs, 477
　polarization-maintaining MOPAs and, 490–492
　rare-earth-doped, 504
Photorefractive eye surgery, 32. *See also* Laser-assisted in situ keratomileusis
PIF. *See* Package-induced failure
Plasma generation, 323
Plasma-coupling experiments, 364

Index

Plasma-electrode Pockels cell (PEPC), 367
PLL. *See* Phased-locked loops
PLM. *See* Precision laser machining
PM fiber. *See* Polarization maintaining fiber
Pockels cell, 256, 257, 258, 375
Polarization coupling, 149, 150
Polarization maintaining fiber (PM fiber), 421, 439
  in PFLs, 477
Polarization multiplexing, 179
Polarization smoothing (PS), 359, 387, 389
Polarization-maintaining MOPAs, 490–492
Polyether ether ketone (PEEK), 35
Positive chirp, 470
Positron emission tomography (PET), 35
Potassium dihydrogen phosphate (KDP), 319, 380–381. *See also* Deuterated KDP
Power oscillators
  MOPA *vs.*, 480–482
  series, 219
  single-aperture, 208–209, 217, 220
Power scaling
  ASE as limit of, 248–250
  of broadband multimode fiber lasers, 522–524
  of broadband single-mode fiber lasers, 525–528
  CCEPS and, 198
  in continuous-wave fiber lasers, 453–454
  of end-pumped slab lasers, 201–204
  energy extraction and, 435–437
  fiber damage limiting, 513
  of fiber lasers, 428–437, 519, 520, 522–528
  of HCL, 296
  in high-power diode laser arrays, 150–151
  of high-power ultrafast fiber lasers, 510–515
  laser diode bars and, 150–151
  multiplexing and, 151, 152
  nonlinear limits and, 428–432, 466–472
  optical damage and, 434–435, 474–476
  PFLs and, 466–476
  pump coupling and, 525
  of side-pumped slab lasers, 194
  of TDLs, 230–250, 346–348
  thermal effects and, 333, 433–434
  of ultrafast TDLs, 333, 350
  ultrashort pulse amplification and, 514–515
  of zigzag slab lasers, 189–192, 194

Preamplifier module (PAM), 373, 374, 376–377
Precision diagnostic system (PDS), 368, 397, 402
Precision laser machining (PLM), 195
Preionization, 20–21
Probe laser, 213
PS. *See* Polarization smoothing
Pulse compression, 508, 509
Pulse measurement
  short-, 217, 218
  in ultrafast SSLs, 319–320
Pulse shaping, 396–397
  LPOM and, 373
  for NIF ignition experiments, 387, 388, 395–397
  NIF system for, 373–375, 406–407
  reproducibility of, in NIF, 402–404
  for SESAM mode locking, 339
  temporal, 396–397
Pulsed fiber lasers (PFL), 466
  AOM in ultrashort, 514
  ASE and, 466, 473–474, 479–480
  eye-safe, 493–495
  fiber types for high-power, 476–479
  gain-staged MOPAs, 485–490
  market acceptance of, 529
  NLEs and, 466–472
  optical damage and, 474–476
  PCFs in, 477
  PM fiber in, 477
  polarization-maintaining MOPAs, 490–492
  power scaling challenges for, 466–476
  self-focusing and, 474–475
  SPM in, 469–470
Pulsed laser deposition, 39–40
Pump absorption
  in HCL, 273–277
  in TDLs, 239–240
  in zigzag slab lasers, 191
Pump combiners, 525–526
Pump coupling
  in continuous-wave fiber lasers, 452–453
  power scaling and, 525
Pump diodes, 184, 211, 415–416
  single-emitter, 521–522
  SSLs and, 173–174, 225, 328
Pump-induced thermal gradients, 294

Q

QCW. *See* Quasi-continuous wave laser arrays
QD-SESAM. *See* Quantum dot SESAMs
QML. *See* Q-switched mode locking

Index

Q-switched lasers, 416
  AOM in, 255, 312
  fiber, 312, 455–456, 481
  TDL, 255–256
Q-switched mode locking (QML), 343, 345
Q-switching instabilities, 343–344
Quantum dot SESAMs (QD-SESAM), 344–345
Quantum well SESAMS (QW-SESAM), 344
Quantum wells (QW), 104
Quarter-wave plates (QWP), 376
Quasi-continuous wave laser arrays (QCW), 140, 141, 145, 269, 271
Quasi-phase-matched cross talk, 472
Quenching, 333
QW. *See* Quantum wells
QWP. *See* Quarter-wave plates
QW-SESAM. *See* Quantum well SESAMS

R

Radiation modes, 420
Radioisotope production, 323
RAFEL. *See* Regenerative-Amplifier FEL
Random phase plates (RPP), 390
Rapid material removal, 297
Rare-earth-doped fibers
  basics of, 422–423
  history of, 415–417
  limitations of ultrashort pulse amplification in, 503–507
  optical fiber basics, 418–422
  PCFs as, 504
  properties of, 422–427
  self-focusing in, 474–475
  spectral properties, 423–426
Rayleigh scattering, 436, 474
Refractive optics, beam shaping with, 149
Regenerative amplification, 260, 502
  advantages of, 311–312
  AOM/EOM in, 311
  Faraday isolators in, 312
  geometrical MPAs and, 260
  in NIF, 374–376
  optical cavity in, 311
  phase distortion in, 312
  TDLs and, 257–258
  thermal mitigation and, 316–317
  ultrafast pulses and, 310
  ultrafast TDLs and, 311–312
Regenerative-Amplifier FEL (RAFEL), 83
Relative refractive index, 419
Relay imaging, 362

Repetition rate, 435, 502
Resonators
  bulk, 542–543
  CBC with common, 542–543
  in DF laser, 47, 48
  in excimer lasers, 25–28
  hybrid, 11, 201
  Michelson interferometric, 544
  nonlinearity in, 346–347
  in Q-switched TDLs, 255–256
  ring, 90
  SESAM mode-locking and, 342
  stable, 13, 177, 202
  in TDLs, 226, 239, 251–252, 254
  in ultrafast TDLs, 329–331
  unstable, 90, 177–179, 195–196, 201, 277, 283
  in VECSELs, 346–347, 348
  WBC and, 550–551
RF discharges, 6–7
Ring resonators, 90
Rofin-Sinar, 226, 261
RPP. *See* Random phase plates

S

SBS. *See* Stimulated Brillouin scattering
Scaling laws, 189–192
Scan FROG, 320
SCH. *See* Separate confinement heterostructure
SCOWLs. *See* Slab-coupled optical waveguide lasers
SCP. *See* Surface corona preionization
Scrapers, 90
SDI. *See* Strategic Defense Initiative
SDL. *See* Semiconductor disk lasers
Sealed-off CO_2 lasers, 9, 10
Second harmonic generation
  in cavity-dumped TDLs, 256–257
  in CW TDLs, 251–254
Self-focusing, 401
  nonlinear, 432, 433
  PFLs and, 474–475
Self-organizing (supermode) CBC, 544
Self-phase modulation (SPM), 503
  in CPA system, 431, 506–507, 510–511
  fiber laser peak power and, 436–437, 451
  as NLE in fiber lasers, 430–432, 503, 506–507
  in PFLs, 469–470
  SESAM mode locking and, 340, 344
  spectral phase description of, 511
  TDL pulse-energy scaling and, 346, 348
Semiconductor disk lasers (SDL), 328

Semiconductor laser diodes, 101
  attributes of high-power, 102–104
  beam steering for, 114
  burn-in and reliability of, 116–120
  efficiency of, 108–110
  fiber-coupled package design/
    processes, 122–125
  geometry/fabrication process of,
    104–106
  heat sinks on, 104, 110, 113, 118, 119,
    123
  high-power, single-mode, 114–116
  high-power bars, 112–114
  high-power broad-area, 110–112
  historical growth of power of, 102
  performance attributes, 125–126
  qualification and reliability of,
    127–129
  spatially multiplexed high-
    brightness pumps, 126–127
  submount design/assembly,
    120–122
  vertical/lateral confinement
    structures, 106–108
Semiconductor lasers. See also Diode
    lasers
  bars of, 563
  beam combining for, 563–566
  CBC for, 563–564
  WBC for, 564–566
Semiconductor saturable absorber
    mirrors (SESAM), 254, 304, 305,
    328
Separate confinement heterostructure
    (SCH), 104
Series aperture power oscillator, 219
Servo loops, 545
SESAM. See Semiconductor saturable
    absorber mirrors
SESAM mode locking, 336–348
  average power in, 341–343
  operation regimes, 340–350
  outlook for, 350–351
  pulse duration in, 347–350
  pulse energy in, 346–347
  pulse formation mechanisms,
    338–340
  pulse shaping/stabilization for,
    339
  recovery time in, 338–339
  repetition rate in, 343–346
  resonators for, 342
  SPM in, 340, 344
Shack-Hartmann sensor, 183, 204, 215,
    220
Shiva laser, 359–360
Short-pulse measurement, 217,
    218
Sibbett, Wilson, 304

Side-pumped slab lasers
  architecture/technical issues,
    192–194
  edge bars in, 193
  non-zigzag axis temperature
    nonuniformity, 193
  OPD and, 193
  optical damage at seal contact areas,
    193–194
  performance of, 195–198
  scaling limitations of, 194
Siegman, A. E., 457
Silica glass fibers, nonlinearity
  coefficients in, 503
Simmtec, 268, 269, 270, 271
Single array element, 551
Single-aperture power oscillator,
  208–209, 217, 220
Single-emitter pump diodes, 521–522
Single-frequency lasers, 416
  fiber, 454–455
Single-mode optical fiber lasers, 416
Single-stage fiber amplifiers, 482–485
Singlet oxygen generator, 65, 69–71
SIS. See Synchronous image scan
Slab fabrication, 192
Slab lasers. See End-pumped slab
    lasers; Side-pumped slab lasers;
    Zigzag slab lasers
Slab sizes, 192
Slab-coupled optical waveguide lasers
    (SCOWLs), 564–565
SLAC. See Stanford Linear Accelerator
    Center
SLM. See Spatial-light modulator
Small signal gain
  in CO_2 laser, 6
  in COIL, 66–67
  in FEL, 79, 81
  general expression for, 51
  in HF/DF lasers, 51–53
Smoothing by spectral dispersion
    (SSD), 359, 373, 389, 391
Smoothing by spectral dispersion in
    one dimension (1D SSD), 387
SMPS. See Switched-mode power
    supply
Sm:YAG, 270
Snitzer, E., 415
Soft x-ray holography, 324
Soft x-ray microscopes, 326
Soft x-rays, 324–326
Solid-state lasers (SSL). See also specific
    types
  AO integrated in, 184
  ASE limiting pulse energy in, 165,
    169
  beam combining for, 564–565
  beam formation in, 176–179

588   Index

Solid-state lasers (SSL) (*Cont.*):
  beam quality of lamp-pumped, 151
  CBC for, 564–565
  cutting processes and, 153, 158
  diode-pumped, 173–174, 225, 328
  fiber coupling of, 151, 152
  gain materials for, 164–171, 207–208
  geometries for, 175–176
  heat removal in, 174–176
  high-average-power, 163–164
  lamp-pumped, 151, 153, 172–173
  laser extraction for, 174–176
  laser-pumped, 152, 153
  materials for high pulse-energy/
    peak-power, 170–171
  MOPAs in, 179
  PCMs in, 181–182
  pump sources for, 163, 172–174
  pumping/cooling/thermal effects,
    171–176
  quenching in, 333
  SPPs in, 181
  stable resonators, 177
  unstable resonators, 177–179
  wavefront correction for, 180–184
Soliton mode locking, 338
Sorenson SFA-150 power supply, 220
Southampton beam shaper, 146–148
Spatial beam conditioning, 385,
  388–391
Spatial chirp, 310
Spatial filtering, 359
Spatial light modulators, 362
Spatial multiplexing, 126–127
Spatial phase plates (SPP), 180–181
Spatial-light modulator (SLM), 511,
  510
Spatially multiplexed high-brightness
  pumps, 126–127
Spectra Diode Labs, 102
Spectra of Diatomic Molecules, 50
Spectra Physics, 150
Spectral combining, 550
Spectral filtering, 477
Spectral Phase Interferometry for Direct
  Electric-field Reconstruction
  (SPIDER), 320
Spectral phase shaping, 511
Spectral stability, of high-power diode
  laser arrays, 143, 144
Spherical aberration, 310
SPIDER. *See* Spectral Phase
  Interferometry for Direct Electric-
  field Reconstruction
SPM. *See* Self-phase modulation
SPP. *See* Spatial phase plates
Springing, 192
SRF. *See* Superconducting RF
SRS. *See* Stimulated Raman scattering

SSD. *See* Smoothing by spectral
  dispersion
SSL. *See* Solid-state lasers
Stable resonators, 13, 177, 202
Stanford Linear Accelerator Center
  (SLAC), 83, 84
Stanford Superconducting Accelerator,
  88
Static x-ray imagers (SXI), 404, 406
Step mirror FhG-ILT, 148–149
STI Optronics, 89
Stimulated Brillouin scattering (SBS),
  181, 196, 371, 426–428, 449,
  465–466, 501, 523, 558
Stimulated Raman scattering (SRS),
  359, 428, 449, 466–467, 501, 510,
  522–524
Stokes beams, 465
Stokes red shift, 466
Stone, J., 413
Strategic Defense Initiative (SDI), 64
Strategic directed-energy applications,
  529
Streaked x-ray detectors (SXD), 405,
  404
Streh degradation, 545–547
Stress-induced birefringence, 227
Sublimation cutting, 13
Superconducting RF (SRF), 86, 88
Surface corona preionization (SCP),
  20, 21
Swamp Optics, 320
Switched-mode power supply (SMPS),
  23
SXD. *See* Streaked x-ray detectors
SXI. *See* Static x-ray imagers
Synchronous image scan (SIS), 36–37
Synchrotron light sources, 324
System shot, 368

━━━ T ━━━

Tactical High-Energy Laser (THEL), 64
Talbot-effect coupling, 561
Tapered fiber bundle (TFB), 520–521
Tattoo removal, 155, 158
TCO. *See* Transmissive conductor
  oxide
TDL. *See* Thin-disk lasers
TEA lasers. *See* Transversely excited
  atmospheric pressure lasers
Telecordia GR-468-CORE, 126
Temporal beam conditioning, 391–393
Teramobile, 321
Textron Defense Systems, 208–209
TFB. *See* Tapered fiber bundle
TFT. *See* Thin film transistor
THEL. *See* Tactical High-Energy Laser
Thermal choking case, 58

Index

Thermal distortion, 310, 316–317
Thermal effects
  depolarization and, 166, 331
  laser performance/beam quality and, 333
  OPD and, 175, 178, 189
  power scaling and, 331, 431–432
  wavefront distortion and, 166
Thermal focusing, 175, 181
Thermal lensing, 564
  MOPAs limited by, 179
  in TDLs, 227, 235–237
  ultrafast SSLs and, 316
  zigzag slab lasers and, 190
Thermal mitigation, 316–317
Thermal resistance, 231, 233
Thermal shock parameter, 232
Thermally induced stress
  in TDLs, 234–235
  YAG fracture from, 167
Thermo-optical problems, fiber-based laser systems and, 500
THG. *See* Third harmonic generation
Thin film polarizer, 312
Thin film transistor (TFT), 38
Thin-disk lasers (TDL), 187, 194, 500.
    See also High-power TDLs;
    Ultrafast TDLs
  amplification of short pulses in, 257–259
  ASE and, 233, 242–250
  average temperature in, 230–232
  beam quality of, 227, 251
  cavity-dumped operation of, 256–257
  continuous-wave, 250–254, 340
  coupled quasi-static numerical model, 241–242
  CPA and, 259, 261
  deformation/stress in, 235–237
  design of, 227
  design study for high-power, 238–239
  energy extraction in, 246–247
  equation of motion for, 240–241
  fluorescence and, 232–234
  gain saturation in, 343
  high average power, 250–251
  high pulse energy, 259–261
  history of, 225–226
  industrial applications of, 261
  influence of fluorescence in, 232–234
  interaction of ASE and excitation in, 243–245
  materials for, 229–230
  mode locking in, 340
  modeling gain/excitation in, 239–240
  mounting designs for, 238

Thin-disk lasers (TDL) (*Cont.*):
  numerical modeling/scaling of, 230–250
  OPD in, 236, 237
  power scaling of, 230–250
  principles of, 226–229
  pulsed operation of, 254–261
  pulse-energy scaling of, 344–346
  pump absorption in, 239–240
  pumping of, 228–229
  q-switched operation of, 255–256
  q-switching instabilities in, 341–342
  regenerative amplification and, 257–258
  resonators in, 226, 239, 251–252, 254
  scaling limit from ASE in, 248–250
  SHG in, 251–254, 256–257
  SPM and pulse-energy scaling of, 346, 348
  temperature gradients in, 226, 227
  thermal lensing in, 227, 235–237
  thermally induced stress in, 234–235
  time-resolved numerical model, 245–248
  VECSEL *vs.*, 332, 333
ThinZag laser, 192
  ASE and, 218–219
  configuration of, 208–209
  coupling TZ-3 modules, 217–222
  DMs in, 216–217, 220
  optical cavity in, 210, 212–216, 219–220
  TZ-1 module development, 209–210
  TZ-2 module development, 211–216
  TZ-3 module development, 216–217
Third harmonic generation (THG), 380–384
Thomas Jefferson National Accelerator Facility (Jefferson Lab), 85, 88, 93
Three-dimensional patterning, 36–37
Thyristors, 22
TIG. *See* Tungsten inert gas welding
Ti:sapphire, 171, 303–304, 312, 319, 499, 559
Townsend discharge, 6
Transmissive conductor oxide (TCO), 34, 35
Transparent ceramics, 269–271
Transverse Brillouin scattering, 361
Transversely excited atmospheric pressure (TEA) lasers, 9
Trebino, Rick, 320
Trumpf Laser, 226, 250, 261
Tungsten inert gas welding (TIG), 156
TZ-1 laser, 209–210
TZ-2 laser, 211–216
TZ-3 laser
  coupling 3 modules of, 217–222
  development of, 216–217

U

Ultrafast oscillators, 305–306
Ultrafast pulse amplification, 459–460
Ultrafast SSLs
  aberrations and, 310
  advances in, 303–304
  amplification techniques for, 306–316
  amplifier schemes for, 310–311
  applications of, 321–326
  CPA for, 307–309
  cutting processes with, 304
  downchirped pulse amplification in, 313–316
  filaments and, 321
  high harmonic generation with, 323–326
  Kerr effect and, 304–305
  micromachining with, 321–323
  MPAs in, 312–313
  oscillators for, 305–306
  as photon/particle sources, 323
  precision machining with, 321–323
  pulse measurement in, 319–320
  radiological applications of, 323
  regenerative amplification in, 311–312
  thermal distortions in, 316–317
  thermal lensing in, 316
  thermal mitigation in, 316–319
Ultrafast TDLs, 329
  average power of, 339–341
  heat removal and, 329
  high harmonic generation with, 348
  micromachining with, 329, 342
  mode-locked, 340
  outlook for, 348–349
  power output of, 339
  power scaling of, 331, 348
  pulse duration in, 345–350
  pump geometry for, 329–330
  resonators in, 329–330
  SESAM mode locking, 335–348
  thermal management, 331–336
Ultralarge mode area fibers, 503
Ultrashort pulse amplification
  average power scaling and, 512–513
  nonlinear effects as basic limitations of, 501–505
  in rare-earth-doped fibers, 501–505
Ultrasonic welding, 156
University of Michigan, 406
University of Rochester, 360
University of St. Andrews, 304
University of Stuttgart, 226
Unstable resonators, 177–179, 195–196, 201, 277, 283
  ring, 90
Upconversion lasers, 414

V

VBG. *See* Volume Bragg gratings
VECSEL. *See* Vertical external cavity surface-emitting lasers
Vertical external cavity surface-emitting lasers (VECSEL), 328–329
  ASE and, 335, 340
  average power of, 339–341
  barrier-pumped, 330
  gain in, 330
  in-well pumped, 330, 335
  mode-locked, 339, 342–343
  pulse duration in, 345–348
  pulse energy of, 344–345
  repetition rate scaling in, 343
  resonators in, 344–345, 346
  SESAM mode-locked, 339
  TDL disk *vs*., 332, 333
  thermal management of, 334–335
VHG. *See* Volume holographic gratings
Vibration welding, 156
Volume Bragg gratings (VBG), 143, 144, 560
Volume holographic gratings (VHG), 143
Voss, Andreas, 226

W

Water
  in CO_2 lasers, 7–8
  heat removal with, 137–138
Water window region, 325–326
Wavefront correction/control
  AOs for, 182–184
  DMs for, 290, 293–294, 365
  in end-pumped slab lasers, 204
  in HCL, 290, 293–295
  PCMs for, 181–182
  SPPs for, 180–181
  in SSLs, 180–184
Wavefront distortion, 166, 282, 284–288
Wavefront errors, 375
Waveguide lasers, 9, 11
Wavelength beam combining (WBC), 533, 536–537
  of fiber lasers, 559–560
  in HBC, 553
  MOPA and, 559
  parallel, 547, 548, 551–552
  performance degradations in, 550–553
  resonators for, 548–549
  of semiconductor lasers, 562–566
  serial, 547–548, 550–551
Wavelength locking, 142–144
Wavelength multiplexing, 152, 445

Wavelength-division multiplexing
 (WDM), 414, 420, 445, 446, 456,
 539, 550
WaveScope, 216
WBC. *See* Wavelength beam
 combining
WDM. *See* Wavelength-division
 multiplexing
Welding
 conventional methods of, 156
 with lasers, 155–157
 of plastics, 156–157
 ultrasonic, 156
 vibration, 156
Wigglers, 77, 89
Wittig, Klaus, 226

Xinetics, Inc., 183, 215, 216
XPM. *See* Cross-phase modulation

YAG
 cryogenic cooling of, 170
 fracture limit of, 190
 gain coefficient modeling for, 283
 Nd-doped, 168–169, 170, 184, 191,
  196, 200, 207, 254, 270, 279, 528
 temperature calculations for,
  284–285
 thermal stress fracture in, 167
 Yb-doped, 169–170, 174, 194, 200,
  226, 229, 254
Yb fiber lasers
 cost factors favoring, 526–527
 heat and, 518
 optical-to-optical efficiency of,
  518–519
 reliability of, 520
 single-emitter diodes in, 519–520
Yb-doped fiber amplifiers (YDFA),
 202, 555, 556, 560

Yb:KGW oscillator. *See* Ytterbium-
 doped potassium gadolinium
 tungstate oscillator
Yb:YAG, 169–170, 174, 194, 200, 226,
 229, 254
YDFA. *See* Yb-doped fiber amplifiers
Ytterbium-doped glass fibers,
 500, 503
Ytterbium-doped potassium
 gadolinium tungstate oscillator
 (Yb:KGW oscillator), 505

Z

Zeolite, 62
Zewail, A., 327
Zigzag slab lasers
 advantages over rod geometries of,
  190–191
 aspect-ratio scaling of, 191
 dead zones in, 188
 diffraction in, 192
 edge bars in, 193
 end-pumped, 198–204
 fracture limits of, 190
 geometry of, 187–188
 non-zigzag axis temperature
  nonuniformity, 193
 OPD and, 193
 optical damage at seal contact areas,
  193–194
 performance of side-pumped,
  195–198
 principles/advantages of, 187–192
 pump absorption, 191
 scaling laws for, 189–192
 scaling limitations of side-pumped,
  194
 slab fabrication for, 192
 slab size in, 192
 thermal lensing resistance in, 190
 traditional side-pumped slabs,
  192–198